The Routledge Handbook of the Polar Regions

The Routledge Handbook of the Polar Regions is an authoritative guide to the Arctic and the Antarctic through an exploration of key areas of research in the physical and natural sciences and the social sciences and humanities. It presents 38 new and original contributions from leading figures and voices in polar research, policy and practice, as well as work from emerging scholars.

This handbook aims to approach and understand the Polar Regions as places that are at the forefront of global conversations about some of the most pressing contemporary issues and research questions of our age. The volume provides a discussion of the similarities and differences between the two regions to help deepen understanding and knowledge. Major themes and issues are integrated in the comprehensive introduction chapter by the editors, who are top researchers in their respective fields. The contributions show how polar researchers engage with contemporary debates and use interdisciplinary and multidisciplinary approaches to address new developments as well as map out exciting trajectories for future work in the Arctic and the Antarctic.

The handbook provides an easy access to key items of scholarly literature and material otherwise inaccessible or scattered throughout a variety of specialist journals and books. A unique one-stop research resource for researchers and policymakers with an interest in the Arctic and Antarctic, it is also a comprehensive reference work for graduate and advanced undergraduate students.

Mark Nuttall is Professor and Henry Marshall Tory Chair of Anthropology at the University of Alberta, Canada.

Torben R. Christensen is Professor of Arctic Biogeochemistry at Aarhus University, Denmark.

Martin J. Siegert is Professor of Geosciences at Imperial College London, UK.

The Routledge Handbook of the Polar Regions

Edited by
Mark Nuttall, Torben R. Christensen
and Martin J. Siegert

Routledge
Taylor & Francis Group

LONDON AND NEW YORK

First published 2018 by Routledge

2 Park Square, Milton Park, Abingdon, Oxon, OX14 4RN
605 Third Avenue, New York, NY 10017

Routledge is an imprint of the Taylor & Francis Group, an informa business

First issued in paperback 2020

British Library Cataloguing-in-Publication Data
A catalogue record for this book is available from the British Library

Library of Congress Cataloging-in-Publication Data
A catalog record has been requested for this book

ISBN: 978-1-138-84399-8 (hbk)
ISBN: 978-0-367-73387-2 (pbk)

Typeset in Bembo
by Swales & Willis Ltd, Exeter, Devon, UK

Contents

Contents

PART III
Polar politics and resource futures

263

Contents

Figures

Figures

Tables

Contributors

Dag Avango is a researcher at KTH-Royal Institute of Technology in Stockholm, Sweden. He holds a PhD in the History of Technology and specializes in industrial heritage research. His research has primarily dealt with the relationship between resource extraction, science and geopolitics in the Polar Regions, and the effects of such interactions on environments and societies, from a long-term historical perspective. A related field of research is on industrial heritage and how it can contribute to sustainable development in post-industrial communities. His research is situated at the interface between archaeology and history, based on the theoretical assumption that material objects and environments play an active role in society and therefore should be considered in explanations of historical change, and following this the methodological approach of combining archival studies with archaeological fieldwork. He is part of the leadership of the Nordic Centre of Excellence REXSAC (Resource Extraction and Sustainable Arctic Communities).

Affiliation: KTH-Royal Institute of Technology, Stockholm, Sweden.

Elizabeth A. Bagshaw is a glaciologist and Lecturer in the School of Earth and Ocean Sciences at the University of Cardiff, UK. Her research interests lie in glaciology, biogeochemistry, geomicrobiology, sensors development and environmental monitoring. She has particular interests in biogeochemical processes in the cryosphere, and in the development and testing of new technologies to monitor them. She has conducted over ten seasons of fieldwork in Antarctica and Greenland, monitoring the impact of physical processes on microbial communities through geochemical changes in meltwater. She has undertaken fieldwork in numerous Arctic and Antarctic locations, including the Dry Valleys in Antarctica, in order to understand how microbial communities survive in seemingly extreme and hostile environments.

Affiliation: School of Earth and Ocean Sciences, University of Cardiff, UK.

Rory Bingham is Lecturer in Physical Geography at the University of Bristol, UK. He is a member of the Bristol Glaciology Centre, which is a formal research group of the School of Geographical Sciences. His research is in the area of physical oceanography, including the measurement of ocean elevation using satellite remote sensing, and the flow of ocean water. He is particularly interested in processes at the interface between ice sheets and the ocean as they are key to how sea levels may rise in the coming century. Before moving to Bristol in 2012 he worked as a research scientist at the University of Newcastle, UK.

Affiliation: School of Geographical Sciences, University of Bristol, UK.

Terry V. Callaghan has worked in every Arctic country and in almost each year since 1967. His research includes ecosystem science and environmental change. He has developed several scientific fields and led many initiatives, contributing to major Arctic and global activities. He developed networking skills within the International Biological Programme. He was a founder of the UK NERC Arctic Station and coordinated its first science programme. For 14 years, he led the Abisko Research Station and in 2001 developed a network of research stations which became INTERACT, with 83 stations in 14 countries. With over 430 scientific publications, Terry is a world-wide "Most Cited and Influential Researcher" (Web of Science). He has Honorary PhDs from Sweden, Finland and Russia, medals from HM the King of Sweden, HM Queen of the United Kingdom, and the international Arctic community, and was included in the joint award of the Nobel Peace Prize to IPCC in 2007. He currently holds Professorships in the UK and Russia.

Affiliation: Department of Animal and Plant Sciences, University of Sheffield, UK, and Department of Botany, Tomsk State University, Russia.

Sanjay Chaturvedi is Professor of International Relations, South Asian University, New Delhi, India. He specializes in theories and practices of geopolitics and IR, with special reference to Polar Regions and the Indian Ocean region. He was awarded the Nehru Centenary British Fellowship to pursue post-doctoral research at the Scott Polar Research Institute (SPRI), University of Cambridge, UK (December 1991 to January 1993), which was followed by the Leverhulme Research Grant (June 1993 to June 1995). He is a regional editor of *The Polar Journal* (Routledge), and a Member of the International Executive Committee (ex officio) SCAR Antarctic Humanities and Social Sciences Expert Group (Geopolitics). He is author and co-author of several books including *The Polar Regions: A Political Geography* (Wiley 1996) and *Climate Terror: A Critical Geopolitics of Climate Change* (Palgrave Macmillan 2015, and co-authored with Timothy Doyle).

Affiliation: Department of International Relations, South Asian University, New Delhi, India.

Torben R. Christensen is Professor of Arctic Biogeochemistry at Aarhus University, Denmark and is also affiliated with the Department of Physical Geography and Ecosystem Science at Lund University, Sweden. His research focuses on Arctic ecosystem ecology and trace gas biogeochemistry with special focus on carbon dioxide and methane exchange in northern terrestrial environments. He holds a PhD from the University of Cambridge, UK, and has carried out extensive research in Greenland, Alaska, Svalbard, Siberia and northern Sweden. He was coordinator of the Nordic Center of Excellence, DEFROST, and participates in a wide range of international projects. He is author and co-author of more than 120 peer-reviewed scientific publications, including papers in both *Science* and *Nature* as well as several book chapter contributions. He is currently Associate Editor of the Springer journal *Biogeochemistry* as well as co-editor of a special issue between the EGU Copernicus journals *The Cryosphere* and *Biogeosciences*. He has contributed to a range of different international assessments related to climate-ecosystem interactions including the Arctic Climate Impact Assessment, the Millennium Ecosystem Assessment, the AMAP SWIPA assessments and the Nobel Prize-winning Fourth Assessment of the Intergovernmental Panel on Climate Change.

Affiliation: Department of Bioscience, Aarhus University, Denmark.

Contributors

Anita Dey Nuttall is Associate Director of UAlberta North (and is on the faculty of the Department of Earth and Atmospheric Sciences) at the University of Alberta, Canada, having previously been Associate Director (and also Acting Director) of the Canadian Circumpolar Institute. She also served as the previous Chair of the Canadian Committee for Antarctic Research. She studied History at Delhi University and International Relations at Jawaharlal Nehru University, India, and holds a PhD in Polar Ecology and Management from the Scott Polar Research Institute, University of Cambridge, UK. She has also been a visiting researcher at the Thule Institute, University of Oulu in Finland. Her research interests include the history and management of national Antarctic programmes and the politics-science interface in the polar regions, Canada's geopolitical interests in the Arctic and Canada's strategy for polar science. Among her publications, she is co-editor of *International Security and the Arctic: Understanding Policy and Governance* (Cambria Press 2014).

Affiliation: UAlberta North and Department of Earth and Atmospheric Sciences, University of Alberta, Canada.

Klaus Dodds is Professor of Geopolitics at Royal Holloway, University of London and Fellow of the Academy of Social Sciences. He is the author of a number of books including the co-authored *The Scramble for the Poles* (2016 with Mark Nuttall), *Ice: Nature and Culture* (Reaktion 2018) and a co-edited *Handbook on the Politics of Antarctica* (Edward Elgar 2017). He is currently a Major Research Fellow (2017–20) funded by the Leverhulme Trust and served as a specialist adviser to the House of Lords Select Committee on the Arctic (2014–15).

Affiliation: Department of Geography, Royal Holloway University of London, UK.

Anne Merrild Hansen is a Professor with the Arctic Oil and Gas Research Centre at Ilisimatusarfik, University of Greenland and Associate Professor with the Danish Centre for Environmental Assessment at Aalborg University, Denmark. Her research is focused on social impact assessment, especially in the North, and the planning and management of energy projects. Currently her research is concerned with public participation processes, inclusion of traditional knowledge in impact assessments and conflicts related to development of extractive industries. She has been involved in the practice of social impact assessments in the oil and gas sector and collaborated with various stakeholders such as NGOs, indigenous communities, industry proponents and authorities.

Affiliation: AAU Arctic, Department of Planning, Aalborg University, Denmark.

Ilisimatusarfik, University of Greenland, Greenland.

Heidi Hansson is Professor of English Literature at Umeå University, Sweden. Over the last few years, her research has been concerned with the representation of the North in travel writing, fiction and popular culture from the late eighteenth century onwards. She was the leader of the interdisciplinary research programme Foreign North: Outside Perspectives on the Nordic North where her own work concerned gendered visions and accounts of the North and has contributed to several research projects on aspects of the North and the Arctic in literature and culture. She is currently co-editing a transnational collection of articles about the Arctic in literature for children and young adults. She is a member of the board of Arcum, the Arctic Research Centre at Umeå University.

Affiliation: Department of Language Studies, Umeå University, Sweden and Arcum, Umeå University, Sweden.

Andrew Hodgkins completed a PhD in Educational Policy Studies at the University of Alberta, Canada, in 2013. His research focused on indigenous vocational education and training programmes in the Northwest Territories and Alberta. He is a project lead investigator with Resources and Sustainable Development in the Arctic (ReSDA), examining Inuit training and employment programmes with a mining company in Nunavut. He is currently employed as a high school science teacher with Edmonton Public Schools.

Affiliation: Edmonton Public Schools, Edmonton, Alberta, Canada.

Richard Hodgkins is a Senior Lecturer in Physical Geography at Loughborough University, UK, and Senior Fellow of the UK Higher Education Academy. He has been researching polar, glacial meteorology and hydrology since 1991, and has published over 40 peer-reviewed papers on aspects of both Arctic and Antarctic environments. By using water quality, water balance and time-series analyses, including the first, comprehensive water balance for an Arctic proglacial zone, his work has helped elucidate the drainage systems of high-latitude glaciers and their implications for hydrological fluxes, glacier dynamic behaviour and environmental change response. In parallel to research, and in addition to undergraduate and postgraduate teaching of polar-related topics since 1998, he has pursued outreach activities for audiences ranging from primary school children to senior citizens, with a conviction that there is one world, and we all have to find the best way to live on it together.

Affiliation: Department of Geography, Loughborough University, UK.

Alf Håkon Hoel is Professor, College of Fisheries Science, UiT – Norway's Arctic University, in Tromsø, Norway. He was recently on leave, working as counsellor for fisheries and oceans at the Royal Norwegian Embassy in Washington, DC. A political scientist by training, his main research interest is international co-operation in the management of oceans and their resources. He has published widely on these issues; recent publications include a study on how the 1995 UN Fish Stocks Agreement has impacted the development of marine science and a study of the process to develop a fisheries regime for the central Arctic Ocean.

Affiliation: Institute of Marine Research, Tromsø, Norway.

Margareta Johansson is a researcher in the Department of Physical Geography and Ecosystems Science at Lund University, Sweden, as well as the Coordinator of INTERACT – a consortium of circum-Arctic field stations. She studies permafrost in subarctic Sweden and manipulates snow conditions to determine how the land will respond to changes in climate. She has broad experience in Arctic research, ranging from glaciology/climatology to Arctic ecology, and for more than a decade has been focusing on permafrost in a changing climate in northern Sweden. Her research experience includes helping to coordinate major environmental assessments on terrestrial ecosystems, and international networks such as SCANNET, a circum-Arctic network of terrestrial field bases.

Affiliation: Department of Physical Geography and Ecosystems Science, Lund University, Sweden.

Tom A. Jordan joined the aerogeophysics team at the British Antarctic Survey as a research scientist after completing his doctorate at the University of Oxford, UK. He has taken part in numerous aerogeophysical campaigns across East Antarctica, West Antarctica and the Antarctic Peninsula. His main interest lies in the investigation of the geological and tectonic evolution of Antarctica, using geophysical techniques including airborne radar, gravity and magnetics. This has led to publications in a broad range of fields including crustal architecture and rifting, magma emplacement, plate kinematics, and the geomorphological evolution of the hidden sub-ice topography of Antarctica.

Affiliation: British Antarctic Survey, Cambridge, UK.

Sheila Kirkwood is a professor of atmospheric physics based in Kiruna, at 68°N in Arctic Sweden, where she has lived and worked for more than 30 years. She works primarily with radar to study the polar atmosphere in both the Arctic and in Antarctica. Author of more than 100 papers in the scientific literature, she has studied the electrodynamics of the northern lights using the European Incoherent-Scatter (EISCAT) radars in northern Scandinavia, and the effects of high-energy particle precipitation on the polar upper atmosphere using EISCAT and two smaller VHF radars – ESRAD at Esrange, Kiruna and the Moveable Atmospheric Radar for Antarctica (MARA) in Queen Maud Land, Antarctica. She has also used ESRAD and MARA to study mountain waves, turbulence and stratosphere-troposphere exchange in the polar lower atmosphere.

Affiliation: Polar Atmospheric Research, Swedish Institute of Space Physics, Kiruna, Sweden.

Timo Koivurova is a Research Professor and Director of the Arctic Centre at the University of Lapland in northern Finland. He specializes in various aspects of international law applicable in the Arctic and Antarctic regions. Among his numerous publications is *Environmental Impact Assessment in the Arctic: A Study of International Legal Norms* (Ashgate 2002). Increasingly, his research work (and on which he has published extensively) addresses the interplay between different levels of environmental law, the legal status of indigenous peoples, the law of the sea as applicable to Arctic waters, integrated maritime policy in the EU, the role of law in mitigating/ adapting to climate change, the function and role of the Arctic Council in view of its future challenges, and the possibilities for an Arctic treaty. He has been involved as an expert in several international processes globally and in the Arctic region.

Affiliation: Arctic Centre, University of Lapland, Rovaniemi, Finland.

Sanne Vammen Larsen is an Associate Professor at the Danish Centre for Environmental Assessment at Aalborg University and is also affiliated with the cross-faculty centre AAU Arctic. She has researched and taught extensively within the field of impact assessment both in a Danish and an Arctic context during her ten years at Aalborg University. Her research is focused particularly on risk, uncertainty and use of knowledge in planning and impact assessment as well as on the inclusion of social impacts. She has worked with sectors such as climate change planning, water management, infrastructure and renewable energy.

Affiliation: Department of Planning, Aalborg University, Aalborg, Denmark.

Elizabeth Leane is an Associate Professor of English at the University of Tasmania, Australia, where she holds a research fellowship split between the School of Humanities and the Institute for Marine and Antarctic Studies. With degrees in both science and literature, she is interested in bringing the insights of the humanities to the study of the Antarctic region. Her other research areas include the relationship between literature and science, with a particular focus on popular science writing, and human-animal studies. She is the author of *South Pole: Nature and Culture* (Reaktion 2016), *Antarctica in Fiction* (Cambridge 2012) and *Reading Popular Physics* (Ashgate 2007), and the co-editor of *Considering Animals* (Ashgate 2011) and *Imagining Antarctica* (Quintus 2011). A former Australian Antarctic Arts Fellow (2003–04), she is Arts and Literature editor of

The Polar Journal and co-chair of the Humanities and Social Science Expert Group of the Scientific Committee on Antarctic Research.

Affiliation: School of Humanities/Institute for Marine and Antarctic Studies, University of Tasmania, Hobart, Australia.

Daniela Liggett is a social scientist with a background in environmental management, Antarctic politics and tourism. She is currently involved in collaborative research on the topics of Antarctic futures, Antarctic gateway cities, the use and provision of polar environmental forecasts and Antarctic science-policy interactions. She has been actively involved in the Association of Polar Early Career Scientists (APECS) – as a President for one term, as a member of the Executive Committee for two terms, and as a member of the Advisory Committee since 2012. She has also contributed to the Scientific Committee on Antarctic Research's First Antarctic and Southern Ocean Horizon Scan, is a co-chair of the Scientific Committee of Antarctic Research's (SCAR) Humanities and Social Sciences Expert Group and a member of both SCAR's Standing Committee on the Antarctic Treaty System and its Capacity Building, Education and Training Committee. She is also one of the co-chairs of the Societal and Economic Research and Applications (SERA) subcommittee of the World Meteorological Organization's (WMO) Polar Prediction Project (PPP) and also serves on the PPP Steering Group.

Affiliation: Gateway Antarctica, University of Canterbury, New Zealand.

Keith Makinson is an oceanographer and drilling engineer with the Polar Oceans Programme at the British Antarctic Survey and has been involved in numerous field campaigns in Antarctica, Greenland and the Alps since 1988. He has a BEng in Mining Engineering from the University of Nottingham, UK, and was awarded his PhD in Oceanography from The Open University, UK, in 2002. His research interests are within the fields of oceanography and glaciology. These include the measurement and modelling of sub-ice shelf ocean circulation, numerical modelling of ice-ocean interactions and sub-ice shelf ocean mixing, in particular the influence of ocean tides in these environments. He designed, developed and operated a number of bespoke hot-water drills to provide deep subglacial access holes, mostly to acquire oceanographic data and marine sediments from beneath thick floating ice shelves and enable deployment of long-term instrumentation.

Affiliation: British Antarctic Survey, Cambridge, UK.

Gareth Marshall is a research scientist in the Atmosphere Ice and Climate team at the British Antarctic Survey in Cambridge, UK, who specializes in understanding climate change and variability in Antarctica, with specific relevance to the Antarctic Peninsula, and also the Arctic. His work utilizes a combination of numerical modelling of atmospheric processes, satellite observations of climate change and field measurements.

Affiliation: British Antarctic Survey, Cambridge, UK.

Arthur Mason is an anthropologist and is currently Associate Professor in the Department of Social Anthropology at Norwegian University of Science and Technology, Trondheim, and Adjunct Associate Professor in the Department of Cultural Anthropology at Rice University, USA. His research addresses the exchange of petro-industry information in elite premium networking spaces and the broader context of energy knowledge provisioning for visualizing the future. His previous work examines Alaska Native history, including moments of identification and political recognition via forms of heritage work, citizenship and academic expertise. His forthcoming research focuses on aesthetics in hydrocarbon imagery and interconnections between the Arctic's changing environmental and cultural systems and other regions of the world, and explores the reliance of government and industry on energy consultant expertise during a period of interest in Arctic natural gas development.

Affiliation: Department of Social Anthropology, Norwegian University of Science and Technology, Norway.

Nadya Matveyeva works in the Komarov Botanical Institute in St. Petersburg, Russia. Following her first field trip to the European North in 1964, her attention focused on flora and vegetation of the Taymyr Peninsula for the next 25 seasons. Gradually the other Arctic regions (Alaska, Canadian Arctic, North Land, Spitsbergen, Yakuta, and again the European North) fell into the sphere of her interests. She has published more than 160 scientific journal articles and book chapters. Her main research interest is with the structure of plant cover (from intra-community mosaics up to zonal divisions of the Arctic) as well as with the problems of vegetation classification; she has turned her attention also to dynamic processes of Arctic landscapes after re-visiting the IPB sites where she began her career as a young "recruit" to polar research.

Affiliation: Komarov Botanical Institute of Russian Academy of Sciences, St. Petersburg, Russia.

John McCannon is associate professor of history at Southern New Hampshire University, USA. He teaches Russian and European history and has a longstanding interest in Arctic studies. He is the author of *Red Arctic: Polar Exploration and the Myth of the North in the Soviet Union, 1932–1939* (Oxford University Press 1998) and *A History of the Arctic: Nature, Exploration and Exploitation* (Reaktion Books and University of Chicago Press 2012).

Affiliation: Department of History, Southern New Hampshire University, USA.

Robert McKay is a glacial geologist at the Antarctic Research Center of Victoria University of Wellington, New Zealand. His research involves acquiring and studying geological archives of the Antarctic continent and Southern Ocean during past warmer climates to provide important clues as to how this region may respond to future anthropogenic warming scenarios. He has been a member of the multinational ANDRILL programme for several years. He is also a member of a drilling expedition in the Wilkes Land continental margin (a previously un-drilled sector off East Antarctica) through the Integrated Ocean Drilling Program (IODP). His research is presently involved in reconstructing an ocean-climate history from sediment cores collected from offshore eastern NZ, to investigate Antarctic water masses entering the Pacific Ocean basin, and how these changes may influence global climate.

Affiliation: Antarctic Research Center, Victoria University of Wellington, New Zealand.

Mal McMillan is an academic research fellow at the University of Leeds, UK, and a member of the UK Centre for Polar Observation and Modelling (CPOM). His research uses satellite observation to study Earth's polar regions, with a focus on the Greenland and Antarctic ice sheets. He specializes in using the techniques of radar altimetry and interferometry to monitor current ice sheet evolution, and to investigate the mechanisms which drive ice sheet change.

Affiliation: School of Earth and Environment, University of Leeds, UK.

Hans Meltofte worked on weather stations in Northeast Greenland for several years. This was followed by more than 35 expeditions and trips to the Arctic and 5 trips to the Antarctic. Since 1973, he has worked as a freelance ornithologist. He earned a DSc degree from the University of Copenhagen in 1994 on a thesis analysing Western Palearctic/African shorebird migration strategies. He was a driving force behind the establishment of the Zackenberg Research Station in Northeast Greenland in 1995, where he worked for 11 seasons. Currently, he is Senior Scientist Emeritus at the Department of Bioscience at Aarhus University, Denmark, primarily engaged with research in Arctic ecology and waterbird biology. He has authored or co-authored about 600 scientific and popular articles/reports and produced 14 books.

Affiliation: Department of Bioscience, Aarhus University, Denmark.

Helle Møller was educated and trained as a nurse and anthropologist in Denmark and Canada. She has worked as a nurse, a consultant for the tuberculosis programme and an educator of medical interpreters in the Canadian Arctic and taught nursing students in Greenland. Currently she is an Associate Professor in the Department of Health Sciences and the Associate Director of the Centre for Rural and Northern Health Research at Lakehead University, Canada, where she teaches in the Master of Public Health Program. Her research broadly centres on Indigenous, rural, remote, northern and Arctic health, health-care and health education framed within a social determinants and social justice point of view.

Affiliations: Department of Health Sciences and Centre of Rural and Northern Health Research, Lakehead University, Canada.

Bram Noble is Professor at the University of Saskatchewan, Canada. His research is focused on environmental assessment, especially in the North, and planning for and managing the cumulative effects of resource development. Much of his current work is also focused on strategic environmental assessment and the development and application of frameworks and decision support tools to facilitate energy transition. He is also actively involved in the practice of environmental assessment and has been engaged in regulatory reviews, policy and training for a variety of government agencies, industry proponents and Indigenous communities.

Affiliation: Department of Geography and Planning, University of Saskatchewan, Canada.

Mark Nuttall is Professor and Henry Marshall Tory Chair in the Department of Anthropology at the University of Alberta, Canada. With a focus on climate change, resource development and human-environment relations, he has carried out extensive research in Greenland, Canada, Alaska, Finland and Scotland. He also holds a visiting professorship at the University of Greenland and Greenland Institute of Natural Resources. He is author and editor of several books, including *Climate, Society and Subsurface Politics in Greenland: Under the Great Ice* (Routledge 2017), *The Scramble for the Poles: The Geopolitics of the Arctic and Antarctic* (Polity 2016; co-authored with Klaus Dodds), and *Anthropology and Climate Change: From Actions to Transformations* (Routledge 2016; co-edited with Susan Crate). He is Arctic Regional Editor of *The Polar Journal*, Fellow of the Royal Society of Canada, a member of the Norwegian Scientific Academy for Polar Research, and received the University of Alberta's J. Gordin Kaplan Award for Excellence in Research in 2016.

Affiliations: Department of Anthropology, University of Alberta, Canada.

Andreas Østhagen is a Research Fellow at the Fridtjof Nansen Institute in Oslo, Norway. He is also a PhD candidate at the University of British Columbia in Vancouver, Canada, and a Senior Fellow and Leadership Group member at The Arctic Institute in Washington DC. Previously, he worked for the Norwegian Institute for Defence Studies in Oslo, and at the North Norway European Office in Brussels. From Bodø, North Norway, he has been concerned with Arctic-related issues for a decade. Currently, his work focuses on maritime boundaries and resource management, under the larger framework of international relations. In addition, he has worked on questions concerning security and coast guard co-operation across the Arctic states and regions. Andreas holds an MSc from the London School of Economics in European and International Affairs.

Affiliation: Fridtjof Nansen Institute, Oslo, Norway.

Frans-Jan W. Parmentier is a research scientist whose scientific interest focuses on the Arctic carbon cycle and how it is affected by declines in the cryosphere, such as sea ice loss, snow cover reductions and permafrost thaw. While these changes in the arctic cryosphere can be quite prominent, subsequent impacts on arctic vegetation growth and the exchange of greenhouse gases remain uncertain – not least because of complex interactions that connect the various processes together across temporal and spatial scales. In his research, he aims to clarify these issues with computer models and to improve these models with what we learn from observations.

Affiliations: Department of Physical Geography and Ecosystem Science, Lund University, Sweden, and Department of Geosciences, University of Oslo, Norway.

Birger Poppel is Project Head of the Survey of Living Conditions in the Arctic, SLiCA at Ilisimatusarfik, the University of Greenland, where he now holds an emeritus position. He holds an MA in Economics. He served as Statistics Greenland's Chief Statistician from 1989–2004 and has since 2004 been affiliated with Ilisimatusarfik and is engaged in a number of national and international research projects including SLiCA, ArcticChallenge and Economies of the North, ECONOR. His research interests are living conditions, well-being and quality of life of Inuit, Sámi and other indigenous peoples of the Arctic; the economic, social, cultural and political development in the Arctic region; and the demographic changes, sustainable development and local, regional and circumpolar impacts of climate change and resource development. He is the author/co-author of more than 40 peer-reviewed articles, book chapters and encyclopaedia entries. He has edited/co-edited three books, most recently *SLiCA: Arctic Living Conditions. Living Conditions and Quality of Life Among Inuit, Sami and Indigenous Peoples of Chukotka and the Kola Peninsula* (Nordic Council of Ministers 2015).

Affiliation: Department of Social Sciences, Ilisimatusarfik, the University of Greenland, Greenland.

Ursula Rack is a researcher in polar history at Gateway Antarctica, University of Canterbury, New Zealand. She conducted her doctoral research on the social history of polar expeditions at the Alfred-Wegener-Institute for Polar and Marine Research in Bremerhaven, Germany, and the University of Vienna, Austria, and was awarded a PhD in 2009. Her research interests are social and environmental history based on personal accounts such as diaries and correspondence. She is the recipient of a COMNAP fellowship for the research project "Reconstructing historic Antarctic climate data from logbooks and diaries of the Heroic era" in 2012–13. Following her expertise, she was involved in the Deep South National Science Challenge: ACRE Antarctica

Data Rescue until September 2017. She has been awarded the New Zealand Winston Churchill Memorial Fellowship 2018 for her research project: "Frozen history: researching, collecting and communicating Antarctic history."

Affiliation: Gateway Antarctica, University of Canterbury, New Zealand.

Donald R. Rothwell is Professor of International Law at the ANU College of Law, Australian National University where he has taught since July 2006. His research has a specific focus on law of the sea, law of the polar regions and implementation of international law within Australia as reflected in 22 books and over 200 articles, book chapters and notes in international and Australian publications. His recent authored, co-authored or edited books include *International Law in Australia* 3rd edition (Thomson Reuters 2017) edited with Emily Crawford; *The International Law of the Sea* 2nd edition (Bloomsbury 2016) co-authored with Tim Stephens; and *The Oxford Handbook of the Law of the Sea* (OUP 2015) edited with Alex Oude Elferink, Karen Scott and Tim Stephens. He is co-editor of the *Australian Year Book of International Law* and Editor-in-Chief of the *Brill Research Perspectives in Law of the Sea.*

Affiliation: ANU College of Law, Australian National University, Australia.

Jessica M. Shadian is the Director of Arctic 360 and a Distinguished Senior Fellow at the Bill Graham Centre for Contemporary History, Trinity College, Canada and Munk School of Global Affairs, University of Toronto, Canada. Her research and publications concentrate on the intersection between Arctic and indigenous governance and law with a focus on resource and infrastructure development and SAR. Her last book, *The Politics of Arctic Sovereignty: Oil, Ice, and Inuit Governance* (Routledge 2014), is the first in-depth history of the Inuit Circumpolar Council (ICC) and Inuit sovereignty in global politics reaching back to pre-European discovery.

Affiliation: Munk School of Global Affairs, University of Toronto, Canada.

Martin J. Siegert is Professor of Geosciences and co-Director of the Grantham Institute, Imperial College London, UK. He was formerly Head of the School of GeoSciences and Assistant Principal for Energy and Climate Change at the University of Edinburgh, UK, and director of the Bristol Glaciology Centre at the University of Bristol, UK. He was educated at Reading University, UK, where he gained his degree in Geological Geophysics, and at Cambridge University, UK, where he was awarded his PhD in the numerical modelling of large ice sheets, at the Scott Polar Research Institute. His research interests are in the field of glaciology. He uses geophysical techniques to quantify the flow and form of ice sheets both now and in the past. Using airborne radar, he has identified and located ~400 subglacial lakes, has discovered ancient pre-glacial surfaces hidden beneath the existing ice and has demonstrated how sub-ice water is generated and interacts with the flow of ice above. He leads the UK NERC Lake Ellsworth Consortium, which aims to directly measure and explore an ancient subglacial lake in West Antarctica, to search for life in its water and comprehend records of climate held in sediments. In 2007 he was elected a Fellow of the Royal Society of Edinburgh. He was awarded the 2013 Martha T. Muse Prize in Antarctic Science and Policy.

Affiliation: Grantham Institute and Department of Earth Science and Engineering, Imperial College London, UK.

Emma J. Stewart is a human geographer and an Associate Professor in Parks and Tourism at Lincoln University, New Zealand, as well as a Research Associate with the Arctic Institute of

North America, Canada. Her research interests include polar tourism, cruise tourism, climate change, parks and protected areas and participatory research. With over 20 years of polar experience, she is one of a few polar tourism researchers with research experience at both Poles. She has visited Scott Base, Antarctica on six occasions working on a variety of social science projects, and she has considerable experience with community-based research projects in Arctic Canada. She has published widely on the topic of polar tourism including over 25 peer-reviewed journal articles, 2 co-edited special issues and 2 co-edited books. She is a member of the Societal and Economic Research and Applications (SERA) subcommittee of the World Meteorological Organization's (WMO) Polar Prediction Project (PPP) and a long-standing committee member of the International Polar Tourism Research Network (IPTRN).

Affiliation: Faculty of Environment, Society and Design, Lincoln University, New Zealand.

Julienne Stroeve is a senior research scientist at the US National Snow and Ice Data Center in Boulder, Colorado, USA, and Professor of Polar Observation and Modelling at University College London (UCL), UK. Her research interests lie in the remote sensing of snow and ice in the visible, infrared and microwave wavelengths. She specializes in measuring and modelling Arctic and Antarctic sea ice thickness and extent, with particular emphasis on how these measurements have changed in the past, and will change in future, under global warming.

Affiliation: National Snow and Ice Data Center, Cooperative Institute for Research in Environmental Sciences, University of Colorado, USA, and Department of Earth Sciences, University College London, UK.

Pippa Whitehouse is a Research Fellow in the Department of Geography at Durham University, UK. She studies the interactions and feedbacks between ice sheet dynamics, sea-level change and the deformation of the solid Earth. Her expertise lies in building computer models of these processes, which are then tuned to fit geological and geomorphological evidence for past ice sheet and sea-level change around the world. She works closely with glaciologists, geodesists and seismologists to understand the physical processes that control ice flow and Earth deformation, and this cross-disciplinary approach has led to the recent development of a new suite of computer models that incorporate more realistic Earth structures into glacial rebound calculations. Although her background lies in mathematics and computer modelling, she is equally at home in the field, and she enjoys carrying out fieldwork in Antarctica, making measurements of solid Earth rebound that allow us to infer contemporary rates of ice sheet change.

Affiliation: Department of Geography, Durham University, UK.

Jeremy Wilkinson is a sea ice physicist with the British Antarctic Survey. He is an expert on sea ice dynamics, thermodynamics and mechanics, ocean wave propagation through sea ice, and deep-convection and water mass modification. He represents the UK on the Arctic Ocean Science Board (AOSB) / International Arctic Science Committee's (ISAC) Marine Science Working Group and is a member of the Programme Advisory Board for Arctic Science for the UK funding agency NERC. His expertise extends to a broad range of techniques, from the remote sensing and in-situ monitoring of sea ice, through to its visualization through the use of upward-looking sonars, and the modelling of sea ice and the flow of oil spilled under sea ice. His scientific fieldwork, both in the Arctic and Antarctic, has been performed from many different platforms, including autonomous underwater vehicles, ice-breakers, helicopters, aeroplanes and ice camps. In addition, his experience extends to the planning, organization and participation of polar field campaigns and logistics and plays a crucial role in guiding the technological development of autonomous instrumentation, including automatic weather stations, drifting buoys, tilt-meters and sea ice mass balance buoys.

Affiliation: British Antarctic Survey, Cambridge, UK.

Kaitlin Young is a PhD student at the University of Alberta in Edmonton, Canada. Her graduate work has focused on a review of the regulatory processes in Greenland surrounding the social impacts of potential mining projects, along with the role of non-governmental organizations as counter-hegemonic groups providing places for resistance outside the political sphere. She continues her work in Greenland and is now focusing on the visual semantics of ownership through mapping, tourism and media in the country. Outside of her graduate work, she volunteers at numerous community outreach programmes and has recently started a participatory mapping workshop with a drop-in centre that services marginalized populations in Edmonton, Alberta.

Affiliation: Department of Anthropology, University of Alberta, Canada.

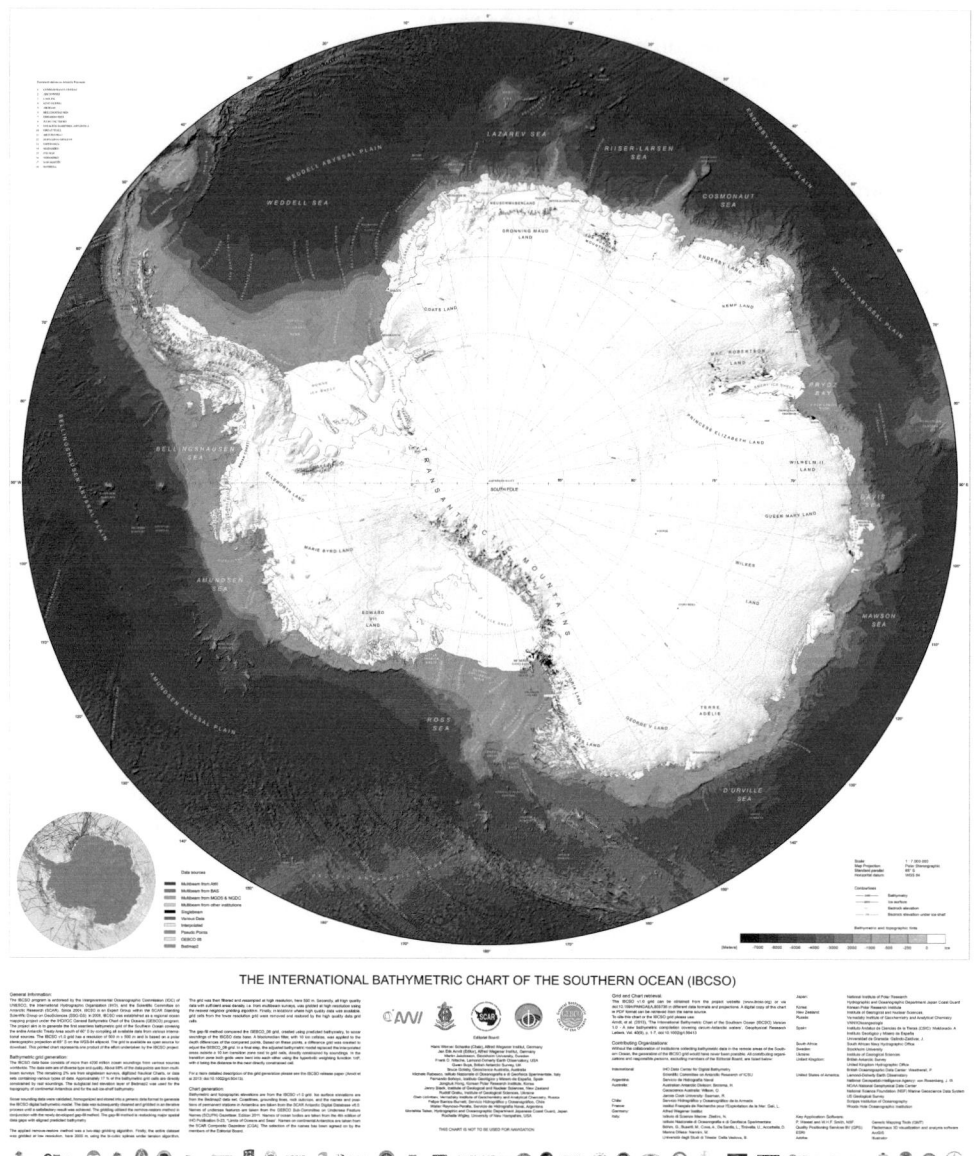

THE INTERNATIONAL BATHYMETRIC CHART OF THE SOUTHERN OCEAN (IBCSO)

Figure 0.1   Antarctica and the Southern Ocean

Source: GEBCO – General Bathymetric Chart of the Oceans (www.gebco.et)

Citation: Arndt, J.E., H.W. Schenke, M. Jakobsson, F. Nitsche, G. Buys, B. Goleby, M. Rebesco, F. Bohoyo, J.K. Hong, J. Black, R. Greku, G. Udintsev, F. Barrios, W. Reynoso-Peralta, T. Morishita and R. Wigley. 'The International Bathymetric Chart of the Southern Ocean (IBCSO) Version 1.0: a new bathymetric compilation covering circum-Antarctic waters', 2013, *Geophysical Research Letters* 40: 3111–3117, doi: 10.1002/grl.50413.

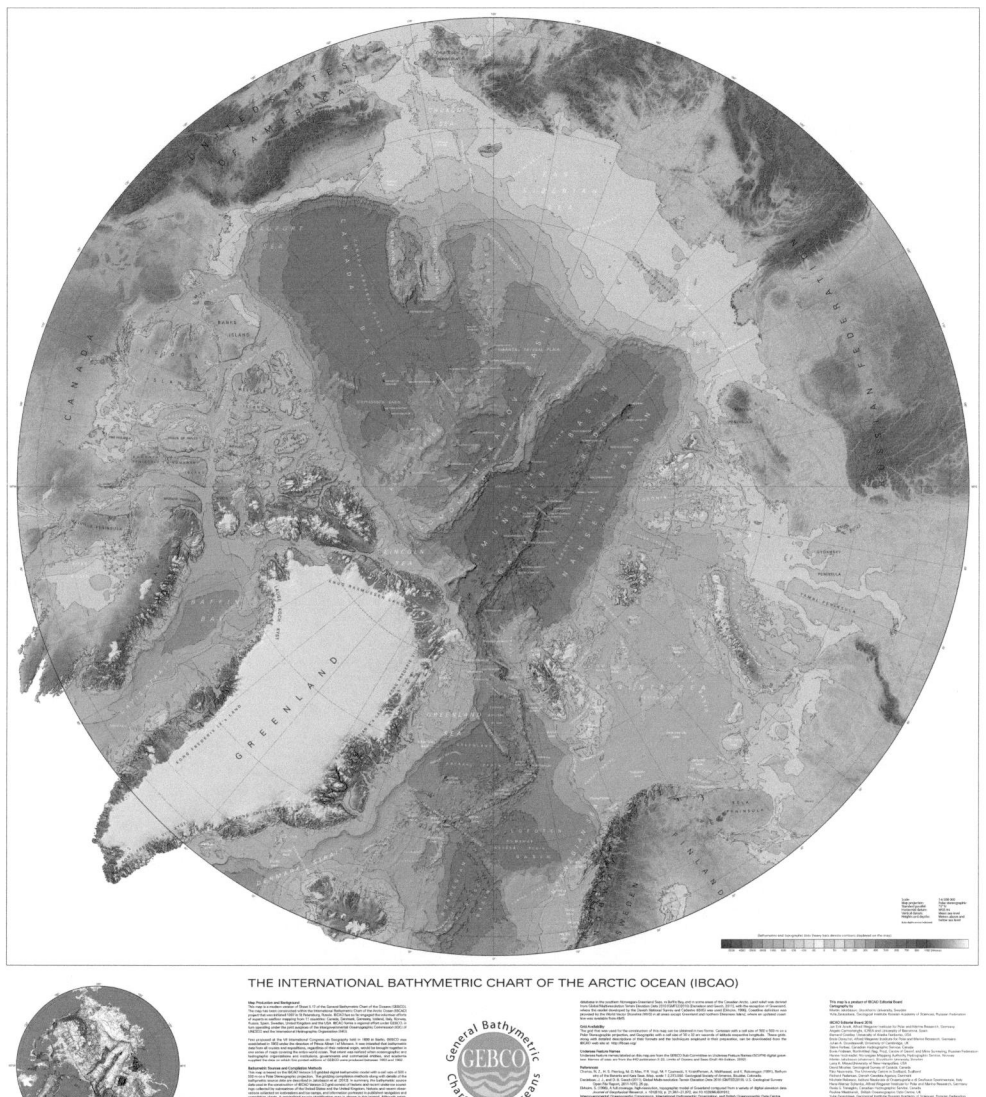

THE INTERNATIONAL BATHYMETRIC CHART OF THE ARCTIC OCEAN (IBCAO)

Figure 0.2 The Arctic

Source: GEBCO – General Bathymetric Chart of the Oceans (www.gebco.net).

Citation: Jakobsson, M., L. Mayer, B. Coakley, J.A. Dowdeswell, S. Forbes, B. Fridman, H. Hodnesdal, R. Noormets, R. Pedersen, M. Rebesco, H.W. Schenke, Y. Zarayskaya, D. Accettella, A. Armstrong, R.M. Anderson, P. Bienhoff, A. Camerlenghi, I. Church, M. Edwards, J.V. Gardner, J.K. Hall, B. Hell, O. Hestvik, Y. Kristoffersen, C. Marcussen, R. Mohammad, D. Mosher, S.V. Nghiem, M.T. Pedrosa, P.G. Travaglini and P. Weatherall. 'The International Bathymetric Chart of the Arctic Ocean (IBCAO) Version 3.0', 2012, *Geophysical Research Letters* 39, L12609, doi:10.1029/2012GL052219.

Introduction

Locating the Polar Regions

Mark Nuttall, Torben R. Christensen and Martin J. Siegert

In November 2017, scientists reported on transformations in marine life on the seafloor beneath the Ross Ice Shelf in Antarctica. Exploring an area of darkness where the presence of marine creatures was previously cited as being sparse, divers found a richer variety of species, including sea cucumbers, deep-sea sponges and starfish, than they had done on their last visit some eight years earlier. This, the researchers said, suggested that changes had been unexpectedly rapid in an environment where extremely low temperatures mean ecological processes move at a slow pace. This greater species diversity now characteristic of the area is thought by the scientists working there to result from climate change – shifts in sea ice over the last few years have meant that the deep waters around parts of Antarctica are now receiving more sunlight which, in turn, means higher ecosystem productivity.[1] That same month, media also reported on how researchers from the British Antarctic Survey had published a map in the journal *Geophysical Research Letters* that illustrated the extent of geothermal heat flux, or the warmth originating and rising up from the rocks beneath the Antarctic ice sheet. Understanding the heat coming from the interior of the Earth is critical for an understanding of ice sheet dynamics and ice flow, allowing scientists to model the possible effects of climate change in Antarctica in relation to how the ice behaves (Martos et al. 2017). Again in November 2017, an article in *The New York Times* described how the thinning and loss of sea ice along the coast of Labrador in eastern Canada was having significant social and economic impacts on Inuit livelihoods – it highlighted how, as has been reported in many other places across the North American Arctic, the residents of Labrador coastal communities such as Rigolet are facing challenges in accessing hunting and fishing areas or travelling to other communities on ice that is increasingly unpredictable and treacherous.[2] A few days before this piece appeared, the BBC news website ran a story on the latest Russian activities in the Arctic Ocean in the form of the Iceberg Project, an initiative to use new underwater technology for resource exploration and development in extreme environments; the *Belgorod*, an unarmed nuclear submarine – and, at 600 feet long, the world's largest – is at the centre of the programme and will be used to deploy technologies (many of which have yet to be devised) to map and probe beneath the seabed and exploit hydrocarbons and minerals.[3] And on the last day of November, against a backdrop of growing concern over the exploitation of Arctic waters as sea ice disappears, the five Arctic coastal states (the United States, Canada, Norway, Denmark and Russia), along with Japan, the Republic of Korea, China, Iceland and the European Union reached an agreement to prohibit fishing in the central Arctic Ocean for at least sixteen years once that agreement comes into force.[4]

These reports appeared as we were putting the final touches to this handbook and we can certainly point to many others in the subsequent months, but they are only a few examples of numerous other stories published in conventional and online media not just at the time, but increasingly so over the past few years. Indeed, public interest in the Polar Regions has grown considerably during the last decade, but so has research activity in the Arctic and Antarctic, and it is often the results from such research that news stories cover. It sometimes appears difficult to keep up with all the news emerging from the Polar Regions – each week seems to bring reports of extreme weather events, revised estimates of when the Arctic Ocean will be ice-free during summer, of polar bears struggling to survive in warming Arctic waters, the rapid release of methane from tundra soils and thawing permafrost, or ice shelves breaking off from Antarctic ice sheets, calls for stronger conservation measures and species protection, or of countries such as China investing in Arctic resource projects or new Antarctic research bases.[5] But there are also stories of remarkable scientific discoveries to marvel at – of scientists probing beneath Antarctic ice sheets and mapping subglacial landscapes, or gazing into the skies from observatories at polar latitudes and advancing understanding of the atmosphere, the upper edge of the mesosphere, and auroral activity, of encounters with new species in the darkest depths of polar seas, and of archaeological and genetic research shedding new light on the lives of the earliest indigenous peoples who moved across Arctic tundra, water and ice several thousand years ago.

There is now greater awareness of the regional and global impacts of climate change witnessed in the world's circumpolar latitudes – such as sea level rise from mass loss in the polar ice sheets – but there is speculation about the resource potential of the Arctic and Antarctic, as well as international concern about the threats posed by environmental change and resource development to the ecosystems of both regions. Debates concerning rights over access to some Arctic lands and waters, as well as who has ownership of resources and rights to their exploitation, continue to dominate some national and international discussions on the future of the region. Land claims and self-government have given indigenous peoples ownership of traditional lands, but also of subsurface resources in many parts of the North American Arctic. But just as Arctic peoples assert their rights, and articulate their own views of sovereignty, sustainable livelihoods and northern futures, many other interests are at play and often rub up against each other as non-Arctic states become more engaged in the high latitudes of the world. We are challenged more than ever to understand the Polar Regions, the precarious nature of a changing cryosphere, the needs of northern societies, and how to balance resource use and economic development with environmental conservation. But there are big scientific questions too that drive research in the Arctic and Antarctic – whether related to ice cores, whale behaviour, migratory birds, glacial surges and flows, human health and medical science, engineering in cold regions, telecommunications, or human-environment interactions.

The Polar Regions and global change

Often, such contemporary attention on the Polar Regions crystallizes around a narrative of a rush to map continental shelves, exploit polar resources and claim territory, seemingly made possible as global warming melts the frozen seas of today and transforms them into the open waters of tomorrow (see Dodds and Nuttall 2016 for a discussion of how this nourishes discussion of scrambles and scrambling for resources and territory in the Arctic and Antarctic). This is particularly evident in the Arctic, which is increasingly represented as an emerging global region of major resource extraction and development projects, criss-crossed by new oceanic routes and ice-free navigable passages, especially in the northern parts of Canada and Russia (and possibly in the Arctic Ocean in the coming decades). Arctic seas have become environments of international scientific focus and concern as the uptake of anthropogenic CO_2 increases the acidification of the Arctic Ocean (Di et al. 2017) and as global warming continues to push the region towards a possible ice-free future. At the same time, many parts of the Arctic are attracting increased interest as sites for extractive industries – oil, gas and mining projects – and cruise ship tourism is expanding and reaching into remoter areas. As it transitions from being characterized by less frozen and more fluid states, the Arctic is increasingly spoken about as being vulnerable and at risk. This continues to inform the geopolitics of and about the Arctic. For many indigenous residents and local communities across the Arctic, life at the ice-edge, on eroding coasts, or on the tundra seems increasingly uncertain and precarious, but the scale and extent of the impacts of Arctic climate change are worldwide, as melting ice sheets raise sea levels and global precipitation patterns are affected.

Although the Northwest Passage has not yet emerged as an important route for international shipping, increasing numbers of cruise ships do venture there (as they do in northern seas around Greenland, Svalbard and Russian Arctic islands), while the occasional commercial vessel makes a summer voyage through these northern Canadian waters. Russia's Arctic waters present more likely possibilities for shipping, however. Russia actively advertises and promotes its Northern Sea Route as a shorter sea passage for transporting cargo between Asia and northern Europe than through the Suez Canal. Each year sees more ships travelling along Russia's Arctic coasts, from the Kara Sea to the Bering Strait, encouraged by rapidly changing sea ice and assisted by nuclear-powered icebreakers in enabling transit. Ironically, while climate change is contributing to the environmental conditions that allow more shipping to travel in Arctic waters, the shorter routes may lead to more efficiency in fuel consumption, thereby reducing maritime transport emissions. In August 2017, a Russian tanker, *Christophe de Margerie*, became the first ship to transit the complete length of the Northern Sea Route without the assistance of icebreaker vessels. Carrying liquefied natural gas, it made the journey from Hammerfest, Norway to Boryeung in South Korea in nineteen days (the Northern Sea Route section of the voyage took six and a half days).[6] Various media outlets reported that the voyage highlighted how climate change is opening up the High Arctic, how the tanker had been built to take advantage of diminishing sea ice, and emphasized the risks to a pristine environment.[7] Although the commercial conditions and prospects remain uncertain in a region undergoing a transformation enabling new mobilities for goods, objects, technologies and people, Chinese, Japanese and South Korean companies are interested in shipping cargo containers regularly through Russia's Arctic waters. Open polar seas and new Arctic shipping corridors – and the processes that entangle northern waters in the globalization of the oceans – have the potential to change the geographies of production and distribution, global supply chains, and the networks of ports and flows of cargo that have already been affected dramatically by containerization over the past fifty or sixty years (e.g. Broeze 2002; Notteboom and Rodrigue 2008; Birtchnell, Savitzsky and Urry 2015). In January 2018, China released a white paper outlining its Arctic strategy, setting out the country's position as a "near-Arctic state" and outlining its interests in the Arctic, the challenges the region faces, and the contributions China feels it can make to governance and sustainable development. Significantly, the white paper emphasizes that as:

> [a]n important member of the international community, China has played a constructive role in the formulation of Arctic-related international rules and the development of its governance system. The Silk Road Economic Belt and the 21st-century Maritime Silk

Road (Belt and Road Initiative), an important cooperation initiative of China, will bring opportunities for parties concerned to jointly build a "Polar Silk Road", and facilitate connectivity and sustainable economic and social development of the Arctic.[8]

At the same time, these northern seas and passages are the focus of national and international efforts and campaigns by conservationists and environmental groups, as well as indigenous peoples' organizations, for the designation of Arctic marine ecosystems as protected areas (for example in Arctic Canada and northern Greenland), for bans on oil drilling near marginal ice zones (for example in Norway's Arctic waters and in the Barents Sea), or for new technologies that bring less invasive bottom-trawling in Arctic fisheries, while, in response to current maritime traffic and in anticipation of increasing shipping, the International Maritime Organization (IMO) adopted the International Code for Ships Operating in Polar Waters (otherwise known as the Polar Code) as an international framework to regulate shipping in both the Arctic and Antarctic. The Polar Code came into force in January 2017 and has set out mandatory standards covering aspects of ship design, construction and equipment, as well as operational, training and safety, and environmental protection matters. The IMO has also agreed to move towards the phasing out of heavy fuel oil in Arctic shipping.

With each ship that transects the Northern Sea Route or Northwest Passage, with each crack and fissure in an Antarctic ice shelf, with each satellite image of glacial melt on the surface and edges of the Greenland inland ice, with each submission by an Arctic country of geological data to the Commission on the Limits of the Continental Shelf (CLCS), with each environmentalist intervention and protest over Arctic oil exploration, with each report of the loss of polar bear habitats or increased levels of contaminants found in seals or Arctic freshwater fish, or with each designation of a new protected area in Antarctica, the Polar Regions are increasingly located in global geopolitics and the global imagination. A word of caution is necessary, however. While affected by climate change and airborne and seaborne pollutants, while wildlife and ecosystems are threatened by human activities and resource development, and while much needs to be done in terms of conservation and environmental protection, the Polar Regions are not quite the empty, unexplored wilderness areas or unregulated frontiers they are often imagined to be.

Protecting the Polar Regions

Reports in the media, politics and in academic research often dramatize and sensationalize the effects of climate change on the Arctic as opening up the region to a "cold rush" for hydrocarbons and mineral wealth as the ice melts, while an increasing number of countries are viewed with suspicion as to the real reasons (commercial, strategic?) for asserting their interests in the Arctic and also in Antarctica. China, in particular, has been written about as "flexing its muscles in Antarctica" (McCallum 2017), and whose emergence as a polar power provokes concerns over military interests and economic influence in the Polar Regions (Brady 2017). The narrative, as it unfolds in the circumpolar North – and which often sees the Arctic as synonymous with only the Arctic Ocean, and so being empty of human presence – also verges dangerously close to one that predicts conflict over circumpolar lands, waters and resources. This is not to downplay those concerns over possible conflict in the Arctic, or between Arctic states in other parts of the world which could have a bearing on Arctic political relations, but to point to the importance of understanding current arrangements and agreements that make up an extensive circumpolar-wide system of international co-operation. Politics, sovereign rights, indigenous land claims and existing ownership and entitlements (including indigenous harvesting rights and ownership of subsurface resources in parts of the North American Arctic) are also often ignored in such discursive coverage of the Polar Regions under threat. So too, as several chapters in this book make clear, are international conventions, agreements and legal regimes. The Arctic states have often emphasized that they have agreed to co-operate on Arctic matters with reference to the framework of many of these agreements. And while China's white paper on the country's Arctic policy sets out its intentions to build a "Polar Silk Road" by developing Arctic shipping routes, it also states that it abides by the Polar Code and that Arctic shipping should operate in accordance with international law. While we must remain concerned over the future of the Polar Regions and the global impact of such change, we should also take note of national and international efforts in conservation, environmental regulation and sustainable development, and how polar diplomacy through the work of Arctic states and organizations such as the Arctic Council, or the Antarctic Treaty System and its various instruments, can work effectively in co-operation with other international bodies. However hesitant or precarious, there are initiatives which should make us feel optimistic about the possibility of safeguarding the Polar Regions and the future of the planet, even if the challenges of ensuring international sustainability goals, and meeting greenhouse gas emission commitments under the Paris Agreement on climate change, make it hard to be so.

The new agreement safeguarding the central Arctic Ocean from commercial fisheries, for instance, builds on the United Nations Convention on the Law of the Sea (UNCLOS) and the UN Fish Stocks Agreement and represents an additional building block in the development of a global framework for fisheries management. It also obligates the ten signatories to making significant commitments in fisheries and ecosystem research and monitoring. While newspaper headlines describe how coastal states are rushing to claim the Arctic Ocean, the truth is rather more nuanced (Dodds and Nuttall 2016; and as various chapters in this book make clear). UNCLOS grants a coastal state sovereign rights over the resources of the legal continental shelf, which can in most cases be equated with the continental margin (which is not the geophysical continental shelf) 200 nautical miles from coastal baselines. Claims can be made for features extending 350 nautical miles and submission must be made to the CLCS within ten years of becoming a party to UNCLOS. And this is precisely what Arctic coastal states are doing (save for the US, which has not ratified UNCLOS – however, all five Arctic coastal states affirmed the importance of working within the rules and procedures of UNCLOS with the Ilulissat Declaration in 2008). Some states are also setting aside large areas of Arctic territory for protection – in August 2017, for example, the Canadian government and the Qikiqtani Inuit Association in eastern Nunavut agreed on the boundaries for a future marine conservation area in Tallurutiup Imanga/Lancaster Sound in Nunavut. Indigenous peoples too are increasingly asserting rights so as to have a say in decision-making processes concerning development in the Arctic and are working to set agendas for community-based monitoring and environmental management. In the marine conservation area in Tallurutiup Imanga/Lancaster Sound, for example, Inuit from Canada's eastern Arctic expect to have a major role to play in the defining of conservation principles and in the overall management of the area.

Antarctica is subject to the rules, regulations and procedures of the Antarctic Treaty and its associated instruments, such as the Convention for the Conservation of Antarctic Marine Living Resources (CCAMLR, which was established in 1982), and the Protocol on Environmental Protection (signed in 1991 and entered into force in 1998). On 1 December 2017, the Ross Sea Region Marine Protected Area (MPA) entered into force, one year after it had been created by CCAMLR. This is the world's largest marine protected area and includes the 500,000 square kilometer (193,000 square mile) Ross Ice Shelf. It provides protection for the foraging ranges of large populations of Adélie and emperor penguins, Weddell and crabeater seals, killer whales, Antarctic petrels and Antarctic toothfish. While supposedly safeguarding Antarctica as a continent for science, ensuring conservation measures are in place, and marking it off as out of bounds to mining, instruments such as CCAMLR and the Protocol on Environmental Protection also have an uncertainty about them. CCAMLR sets out to conserve Antarctic marine life (and the need for CCAMLR in the first place was precipitated by a concern with commercial interests in Antarctica's marine environment, particularly over fishing for Antarctic krill and the historic over-exploitation of other marine resources). Yet, it does not exclude the harvesting of resources from Antarctica's marine ecosystems; instead it seeks to conserve those resources through an ecosystem-based approach to conservation. Any fishing, for instance, has to be practised in a sustainable manner and take the effects on other parts of the ecosystem into account. And while the environmental protocol prohibits mineral exploitation, it will be open for review in 2048, fifty years after it came into force.

Imagining, approaching and placing the Polar Regions in contemporary world affairs

Throughout the long history of human encounters with the Polar Regions, these vast parts of the planet have been subject to different, often conflicting and contested representations, images and ideas, which have influenced and shaped the ways we think about circumpolar worlds, how we imagine them (and how we make them imaginary places and populate them with extraordinary things), how we have ventured into polar spaces, how we have sought to exploit or protect them, and how we have chosen to act towards animals and, in the case of northern lands, the indigenous peoples who have lived there for millennia. These representations, images and ideas have also been profoundly influential in shaping, producing and sustaining different histories, stories and mythologies about the Arctic and Antarctic. They furnish cultural narratives and accounts of them as exceptional and extraordinary places to travel to, in which to live, to move around and to get to know. They have influenced ideals and ideologies about exploration, discovery, science, settlement and development; in the Arctic, they have influenced attitudes towards, and relations with, Northern societies (the rich and extensive historical and contemporary movement of indigenous peoples around the North has often been ignored); and they continue to inform debates about the future of circumpolar lands and peoples (McGhee 2004; Dodds and Nuttall 2016).

Today, the Polar Regions have been firmly located within research agendas concerned with planetary boundaries and understanding the current epoch when humans have introduced changes to the Earth system, which we increasingly refer to as the Anthropocene. There is much for us to learn from research in the Arctic and Antarctic about living on an Earth undergoing constant change, a hazardous planet at risk (and a planet of risk) increasingly revealed as bearing the scars of the activities of humans as agents of environmental disruption and geological rupture. The global gaze has fallen on what were once thought of as remote, inaccessible places that only a few scientists and adventurers visited (and in the case of the Arctic, in which small populations of indigenous peoples lived). Yet popular ideas and stereotypes persist, and they obscure the histories and contemporary realities of the Polar Regions. The Arctic and Antarctica may have long been imagined, represented and described as many different things, most notably as distant and often inaccessible places at the very ends of the Earth, but they are not peripheral to world events, despite popular images that suggest otherwise. The Polar Regions were firmly drawn in to support growth in the industrialized-global economy at least two centuries ago (although parts of the Arctic were entangled in North Atlantic, North Pacific and Eurasian trade activities even earlier) and have been subject to the effects of increasing globalization and other global processes since then. This is apparent in the case of the fur trades and other historical trade networks in northern North America and northern Eurasia, or fisheries and whaling in both the Arctic and Antarctic.

In Europe, geographical societies played a significant role in their active promotion of polar exploration and their support of expeditions during the nineteenth century – whether to seek Arctic sea passages to Asia or discover the North Pole and South Pole. In the southern circumpolar regions, many of the early voyages that sought to map and chart portions of the Antarctic continent and its surrounding seas were led and supported by people connected to whaling interests, and it was sealers who were largely active in the early discoveries, surveys and mappings of sub-Antarctic islands and the Antarctic Peninsula, while they were seeking new waters in which to hunt. Indeed, it was the crew of *Cecilia*, an American sealing vessel under the command of Captain John Davis, which made the first known landing on the Antarctic continent at Hughes Bay on the west coast of the Antarctic Peninsula on 7 February 1821 (the first documented landing was at Cape Adare in 1895, however). The bay, though, was later named after Edward Hughes, who was master of the sealing vessel *Sprightly*, owned by the London-based whaling company operated by the Enderby family, and which explored the area in 1824–25. British and American sealing ships, however, were already active in the South Shetlands at the time *Cecilia*'s crew were making their first forays on Antarctic land and their intensive pursuit of fur seals meant the population had seriously declined in southern seas by 1830. By then, the Antarctic had become a major region for exploiting marine resources and was caught up in the international rivalry and competition, especially between the British and Americans, that characterized commercial sealing and whaling enterprises elsewhere in the world. As if to emphasize the importance of whalers and sealers for discovery and exploration, claiming territory and sovereignty, many place names in the Antarctic and sub-Antarctic islands commemorate the captains of ships (e.g. Hughes Bay), the ships themselves (e.g. Sprightly Island in Hughes Bay), and the whaling and sealing companies and owners who combined hunting with exploration (e.g. Enderby Land). There were some exploratory voyages to Antarctica concerned with geographical discovery and with locating specific polar spaces and places on charts and maps, such as that of James Clark Ross, who discovered the Ross Sea and established that the South Magnetic Pole was inland in 1841, but throughout much of the nineteenth century Antarctica, sub-Antarctic islands and the surrounding seas were in many ways the sole preserve of sealers and whalers. By the end of the nineteenth century, whaling firms had improved the commercial potential of the Antarctic whaling industry although exploitation of Antarctic waters was more akin to a non-renewable extractive industry. By the early twentieth century, whaling had become the chief economic reason to venture south and explore Antarctica and it was the activities of the whaling industry that led to the first claims to territorial sovereignty, although this was overshadowed somewhat in the first years of the century by various discovery expeditions aiming to be the first to reach the South Pole and which defined an era of heroic exploration (Dodds and Nuttall 2016).

Until the middle decades of the twentieth century, an increasing number of nations became interested in Antarctica and they undertook expeditions of geographical exploration and scientific discovery. Governments began to lay claim to parts of Antarctica – Britain was the first to do so in 1908, followed later by Norway, France, New Zealand, Australia, Argentina and Chile (Britain, Argentina and Chile each asserted sovereignty over the Antarctic Peninsula and these overlapping claims remain to this day). Scientific activity intensified – notably following the end of World War II and especially after the International Geophysical Year (IGY) of 1957–58 – with the establishment and maintenance of permanent national Antarctic science programmes by countries such as the United Kingdom, the United States, Norway, France, Australia and New Zealand, all with the aim of giving some legitimation to territorial claims.

Signed in 1959 by the twelve nations active in the IGY (Argentina, Australia, Belgium, Chile, France, Japan, New Zealand, Norway, South Africa, the Soviet Union the United

Kingdom and the United States) and having previously formed the Scientific Committee on Antarctic Research (SCAR) in 1958, the Antarctic Treaty entered into force in 1961. It set aside Antarctica for peaceful purposes and international collaboration in science. Since then, forty-one other nations have joined the Antarctic Treaty, full membership of which is limited to states that "demonstrate substantial scientific interest" in the continent (there are twenty-nine Consultative Parties, which comprise the full membership). The nature of national scientific activities continues to evolve, but countries are expected to maintain a permanent scientific programme. The growing number of countries involved in Antarctic politics and the corresponding increase in the scientific activities conducted on the continent has led to concern over the management of human activities, and how best to avoid impacts on Antarctica's terrestrial and marine environments while at the same time allowing for its use for the global good. As mentioned, extractive industries in the Antarctic are prohibited for fifty years by the Protocol on Environmental Protection (more commonly known as the Environmental Protocol and previously also known as the Madrid Protocol, named after the city in which it was signed in 1991; it came into force seven years later), but this does not preclude efforts to assess the Antarctic continent's potential for having significant mineral-bearing geologies (although current assessments suggest that sizeable deposits known to exist are rare, relatively inaccessible and not economically viable).

The various and diverse circumpolar lands and waters that make up the Arctic are tightly tied politically, economically and socially to the national mainstream of eight northern nations (Russia, Finland, Sweden, Norway, Iceland, Denmark/Greenland, Canada and the United States) and are connected in complex ways to the global economy. In recent decades, the Arctic has also transitioned from a Cold War space divided between a Soviet Arctic and a Western Arctic to one characterized by co-operation between Arctic states on issues relating to the environment and sustainable development. Indigenous peoples had already been living in Alaska, Canada, Greenland and the Eurasian Arctic for several thousand years when Europeans first "discovered", "explored" and lay claim to these regions. From the late fifteenth century, much Arctic exploration originally had as its goal the discovery of northern routes to Asia and the Far East – mariners sought to find a northwest passage across North America and a northeast passage north of Russia. Mapping of Arctic coasts, rivers and lands later followed as a result of fur trading and whaling activities and as industrial activities and large-scale resource exploitation expanded into the far north (e.g. McGhee 2004). Indigenous people's knowledge of the lands and waters they hunted and fished and travelled around was seldom taken into account, and Arctic places were seen as empty spaces to name, claim and exploit. From the sixteenth century, but especially during the nineteenth and early twentieth centuries, the colonization of the Arctic and the commercial exploitation of its resources, such as whales, seals, fish and fur-bearing land mammals, led to frequent and extended contact between indigenous peoples, outsiders and, increasingly, global mercantile institutions and influences.

Over the past 70–100 years in particular, indigenous cultures and societies across the circumpolar North have been affected and transformed by dramatic and far-reaching social, economic and political changes, while challenges persist from the legacies of the resettlement of indigenous peoples by the state, residential schools and the health effects of changes in lifestyle and society, including increasing urbanization and industrial development. Yet despite this, many Arctic indigenous communities continue to depend on the harvesting of terrestrial and marine resources and maintain a strong connection to the environment through these activities. The politicization of indigenous peoples since the 1970s, the assertion of rights to self-determination and, in some cases, the settlement of land claims and introduction of forms of self-government, have re-shaped relations between indigenous peoples and the state, but have also reshaped the maps of some circumpolar countries. In Canada and Greenland (and more quietly so in Norway and Sweden), discussions about reconciliation are seeking to confront and redress historic mistreatment of indigenous peoples, while some northern regions such as Greenland are moving towards greater autonomy and aspire to independence (Nuttall 2017). However, the policies and actions of the state have a powerful presence in the lives of indigenous peoples, global influences continue to penetrate deep into Arctic societies and economies, and rapid social, economic and demographic change, resource development, trade barriers, conservation and environmental management, and animal-rights campaigns continue to have their impacts on hunting, herding, fishing and gathering activities. Furthermore, the Arctic's climate is being affected by a warming trend unknown in its recent history that is affecting the functioning of its ecosystems and its resources, while marine resources have been affected by contaminants in several parts of the region. For indigenous peoples, these changes and environmental issues challenge traditional knowledge and understanding of the environment, affect harvesting, food security and community economies, hinder mobility across ice, water and land, and make prediction, travel safety and resource access more difficult.

To obtain a comprehensive scientific understanding of the impacts of a changing climate on the ecosystems of the Arctic, long-term monitoring of key climate and ecosystem parameters are necessary. Many such monitoring initiatives have developed under the auspices of the Arctic Council working group the Arctic Monitoring and Assessment Programme (AMAP). While cross-disciplinary long-term monitoring programmes that document climate, hydrology, phenology and population dynamics of birds and mammals are rare, the Greenland Ecosystem Monitoring programme (GEM, www.g-e-m.dk) that started at the Zackenberg research station has recorded ecosystem dynamics for more than twenty years in Northeast Greenland, with comparative monitoring efforts that started in West Greenland in the late 2000s. Such cross-disciplinary monitoring is important for the evaluation, assessment and understanding of the impacts of climate change as well as for the interactions between the different parameters and ultimately possible feedback effects they jointly may cause in the global climate system. Greenland is particularly well positioned for such cross-disciplinary and co-ordinated ecosystem monitoring as major components such as the ice sheet, glaciers, rivers and downstream lakes, terrestrial ecosystems and, ultimately, the near-coastal marine environment, are all located and reachable within a relatively short distance from each other. Other work in the Arctic includes atmospheric greenhouse gas monitoring, which is now extending to the furthest north through continuous measurements in Svalbard and Station Nord in Greenland.

Recent discussion about the future of the circumpolar North has also focused on it becoming a more militarized and securitized region, yet this is contested by those who consider such concerns to be overstated and who point instead to the significant achievements made in recent decades to nurture international co-operation on Arctic matters. The militarization of the Arctic began during World War II and increased during the Cold War, which effectively divided the region into two sectors – the Soviet Arctic and the Western Arctic. For example, a Soviet military base was established on Franz Josef Land, the largest detonation of a nuclear bomb took place on Novaya Zemlya in July 1961 (the Tsar bomb, with a yield of 50 mega tonnes of TNT), and the US established a permanent air force presence in Greenland (as well as embarking on secret military projects on the inland ice, such as Camp Century, with the intent of stationing missiles that could be launched at the Soviet Union). National perspectives on the Arctic were dominated by questions of security as the region emerged as a zone of potential hostile confrontation. A significant turning point in Arctic international relations came in 1987, when President Mikhail Gorbachev gave a speech in Murmansk in which he called for the Arctic to be recognized as a zone for peace, and also proposed greater international co-operation among

Arctic countries. Finland took the lead in developing an initiative for discussing the nature of such co-operation – what form it could take and how it should proceed – on a formal level. The outcome of this was the Arctic Environmental Protection Strategy (AEPS), adopted by the eight Arctic states by declaration at a ministerial conference in Rovaniemi, Finland in 1991. With the AEPS, the USA, Canada, Denmark/Greenland, Iceland, Norway, Sweden, Finland and Russia agreed to focus on issues of environmental protection, most notably on the conservation of flora and fauna and the assessment of pollution and contaminants in the Arctic. Some countries argued, though, that more formal, government-endorsed and supported arrangements were needed to facilitate and strengthen Arctic international co-operation and promote sustainable development, and that such arrangements would require governments to make a far stronger commitment to taking action in a region increasingly understood to be undergoing rapid change. The eight Arctic countries created the Arctic Council at a ministerial conference in Ottawa, Canada, in September 1996. The Arctic Council functions as a high-level forum to promote co-operation, co-ordination and interaction among the Arctic states, with the involvement of Arctic indigenous communities and other Arctic inhabitants on common Arctic issues, in particular issues of sustainable development and environmental protection in the Arctic. The Arctic Council has recognized the importance of indigenous issues in the Arctic, related to the environment, conservation and sustainable development, and six indigenous peoples' organizations (the Inuit Circumpolar Council, Saami Council, Russian Association of Indigenous Peoples of the North, Arctic Athabaskan Council, Gwich'in Council International and the Aleut International Association) have status as Permanent Participants. The chairmanship of the Arctic Council rotates for terms of two years. In addition, a growing number of non-Arctic states (such as the UK, France, Germany, Italy, Japan, China, Singapore and India), intergovernmental and inter-parliamentary organizations (such as the Nordic Council of Ministers, International Council for the Exploration of the Sea and the United Nations Development Programme), and non-governmental organizations (such as the International Union for Circumpolar Health, the International Work Group for Indigenous Affairs and the World Wide Fund for Nature) have Observer status at the Arctic Council.[9]

The Arctic and Antarctic, then, have long been international political regions, but today they are the focus of contemporary geopolitical and environmental concerns over the effects and impacts of climate change, threats to the cryosphere, biodiversity and human security, and debates about resource futures. They are subject to a reconfiguration of ideas (many of them conflicting) of space, territory and place. Research in the Polar Regions is thus of global importance, and global processes and global interconnections are being highlighted by scientific research and by indigenous and local communities. The need to put in place measures for conservation and environmental protection is also emphasized through the work and diplomacy of the various instruments of the Antarctic Treaty or the activities of the Arctic Council. Such measures, though, often need to apply beyond the Polar Regions as part of global conservation and management initiatives. Protecting Arctic wetlands or coastlines, for instance, as crucial habitats for Arctic migratory birds, is not enough if the critical habitats along their global migratory pathways are also under threat. The Arctic Council's Conservation of Arctic Flora and Fauna (CAFF) working group has developed an Arctic Migratory Birds Initiative (AMBI) that sets out to address that kind of Arctic-global issue. Identifying flyways between the Arctic regions and more southerly places, CAFF is working to ensure that habitat for birds such as spoon-billed sandpipers and bar-tailed godwits are not just protected and managed in Arctic Russia, Alaska and northern Canada, but that coastal ecosystems in China are protected or management plans in Singapore are implemented. This illustrates the significance of the Arctic Council as a forum in which dialogue between Arctic states and non-Arctic states can proceed on matters of international conservation. It is also a counterbalance to views that are often expressed about non-Arctic states becoming increasingly influential in Arctic affairs. If anything, by allowing non-Arctic states such as China and South Korea into the Arctic Council as Observers, the Arctic Council has stronger possibilities to engage with them on circumpolar issues that are global priorities. And while it is in the interests of Arctic states that, for example, the tidal flats of the Jiangsu coast ecosystem are protected, it is probably just as much in China's interest that intertidal areas of the West Kamchatka coast or the Yukon-Kuskokwim Delta are also protected. Conservation and environmental management processes and initiatives are never entirely free of a political dimension, and while conservation in the Polar Regions must also be examined within the context of geopolitical and strategic interests, it requires us to frame those interests in a broader context of polar geopolitics.

In appreciation of the global significance of the Polar Regions, in recent years there have been several attempts to understand future research needs in terms of scientific drivers, and logistical and engineering capabilities, as well as the changing nature of how research is done with indigenous communities (and what research priorities indigenous communities identify). A joint fourth International Polar Year (IPY) in 2007–08 (further to the IGY mentioned earlier, preceded by International Polar Years in 1882–83 and 1932–33), organized under the auspices of the International Council for Science (ICSU) and the World Meteorological Organization (WMO), focused on a lack in our understanding of the Polar Regions in relation to climate change, and important new research was conducted as a result. SCAR, as a further example, undertook a formal twenty-year international Horizon Scan, in which the top eighty scientific questions that need answers by 2035 were compiled (Kennicutt et al. 2015), followed by an assessment of how the challenges of the science required can be met by national logistics providers (Kennicutt et al. 2016). In the Arctic too, there have been several national and international plans for co-operation in future research like the IPY mentioned above and, most notably, under the auspices of the International Arctic Science Committee (IASC), while member states of the Arctic Council have undertaken to improve co-operation on Arctic science through the Agreement on Enhancing International Arctic Scientific Cooperation, a legally-binding agreement signed on 11 May 2017 at the Arctic Council Ministerial meeting in Fairbanks, Alaska. The agreement aims to facilitate access for researchers, whether to Arctic areas for fieldwork, to infrastructure, facilities and equipment, or to data. These types of aims are shared by several international processes to improve the accessibility and mobility of Arctic researchers between different geographical settings in the circumpolar North. One example of this is the European Union funded INTERACT programme, which helps facilitate wider use and exchange of scientists between currently eighty-three arctic research bases and infrastructures (www.eu-interact.org). At the same time, the needs of Arctic communities are beginning to influence and shape research agendas and practices.

Locating polar research

This handbook is an authoritative guide to some of the most significant contemporary issues pertinent to the Polar Regions, as well their global significance, influence and reach, through an exploration of key areas of research in both the natural/physical sciences and the social sciences/humanities. The aim is not to provide a subject-by-subject overview of specific disciplinary topics, however. Nor do we divide the Arctic and Antarctic into separate sections. Our intention is to approach and understand the Polar Regions as places that are at the forefront of global conversations about some of the most pressing issues of our age, including a changing cryosphere, shifting weather patterns, threats to biodiversity, and human–environment interactions in the

Anthropocene. This handbook sets out to provide comprehensive and critical reviews and appraisals of the current state of the art in polar research. In doing so, a range of chapters written by polar scholars from a number of different disciplines in the social sciences, humanities, and physical and natural sciences cover a large range of topics. We aim to encourage and foster dialogue across and beyond disciplinary boundaries, and see our book as providing a baseline for the development of conceptual and theoretical approaches to understanding the Polar Regions. The call for interdisciplinary research to address linked social and environmental issues is not new, of course. But at a time when rapid changes are affecting the Arctic and Antarctic, the need to be thinking and working across the boundaries of the respective academic disciplines in which we are trained is an urgent responsibility. Such collaboration is not without theoretical, methodological and practical challenges, but interdisciplinary efforts and interfaces enhance understanding of the complex interactions between human societies and the environment, as well as the actions and agency of the non-human and more than human things that comprise, constitute and fill polar environments, such as marine mammals, fish, rocks, ice, snow, wind, atmospheric phenomena and so on, and the relations between humans and the non-human. The value of this kind of co-operation is also underscored by its relevance for policy making. Collaboration between the natural sciences and social sciences (and, in the Arctic, with indigenous peoples and local communities) is an essential step for improved communication and collaboration between researchers, communities, stakeholders and policy-makers.

The various chapters of this handbook contribute to our increased understanding of where and how we locate the Polar Regions in relation to other parts of the globe, but also in broader research endeavours concerned with global environmental change and policy processes, as well as in national and international political debates. There are a number of different ways, often according to scientific discipline and environmental and climatic criteria, by which both the Arctic and Antarctic can be defined, and we allow this diversity to become apparent throughout the book. Although they are often assumed to be regions worthy of comparative study, the Arctic and Antarctic are polar opposites in several important ways. They are fundamentally different geographically, of course, in that a large portion of the Arctic consists of the ice-covered Arctic Ocean (although the extent of that ice is diminishing), which is surrounded by many islands and archipelagos, and the northern parts of the mainland areas of the North American and Eurasian continents, whereas the Antarctic is an ice-covered landmass surrounded by an ocean. They also have different environmental patterns, climatic systems and wildlife habitats – for instance, there is much lower terrestrial biodiversity in the Antarctic than in the Arctic – and there are no indigenous peoples living in Antarctica. Antarctica has also been subject to an international framework for environmental management and conservation under the Antarctic Treaty System for almost sixty years, whereas no such regime exists in the northern circumpolar regions. The Arctic regions comprise the northern parts of eight nation states (and it is probably more accurate to say that the Arctic is not one region, but several places), whereas, while the international community does not recognize sovereignty by states over any portion of Antarctica, seven countries have made territorial claims to parts of the continent. In some ways, the very ideas that constitute our understanding of what and where the Polar Regions are have been relatively recent constructs, as we have classified ecosystems, places and peoples according to particular criteria – often seeing them as unique and exceptional. The foundations of the Antarctic Treaty System and recent Arctic co-operation in the form of the Arctic Council are cemented in ideas of circumpolar connections and flows, but also in regional and polar distinctions from the rest of the world.

Nonetheless, and despite the differences between the Polar Regions, the connections between places within circumpolar systems, and the emphasis on the particular, exceptional and distinctive, their influence over the function of the Earth system is emphasized in the scientific literature. Research in both Polar Regions is increasingly valued for the wider contribution it makes to how we understand global processes and global issues. For instance, climatologists have long understood the importance of studying the Polar Regions because of the ways they influence the Earth's weather systems – for example, the sea ice in both the Arctic and Antarctic is a major element in the global climate system, driving the global thermohaline flow of the ocean in which heat is transferred across the globe, while the Southern Ocean plays a significant role in processes of biogeochemical cycling and exchange of gases between the ocean and the atmosphere. The Arctic and Antarctic are both sites of international political (and geopolitical) interest and scientific activity, as well as ecologically-sensitive regions that conservationists work to protect. An increasing number of non-Arctic states are expressing interest in both the resource potential and the governance of the Arctic Ocean, with some states and non-state actors such as conservation organizations calling for the creation of an Arctic treaty system, or a series of internationally-governed protected spaces, to oversee the management of the region and its resource development. Indigenous peoples are increasingly assertive in demanding they are involved in decision-making in the Arctic. This is well-illustrated by the Inuit Circumpolar Council's (ICC) Circumpolar Inuit Declaration on Sovereignty in the Arctic, released in 2009, which states that issues of sovereignty and sovereign rights in the Arctic have become inextricably linked to issues of self-determination in the region, that the rights, roles and responsibilities of Inuit must be fully recognized and accommodated in discussions on matters linked to Arctic sovereignty, including climate change and resource development, and that Inuit and Arctic states need to work together closely and constructively to chart the future of the region.

The legal basis for all territorial claims in Antarctica has been grounded in a combination of exploration and geographical discovery, the dispatch of large-scale expeditions, symbolic performances supporting national endeavours, administrative acts, occupation and science. While increasing and enhancing scientific knowledge has provided an argument for continuous activity in Antarctica, along with the presence of permanent bases and scientific stations, growing awareness of the impacts of climate change and suspicions that states are interested in the resource potential of the continent mean that scientific activities and political motives for being involved in Antarctic research are increasingly coming under global scrutiny. The Antarctic and the Arctic are also attracting large numbers of tourists – most of whom travel through polar waters on cruise ships – eager to experience what are represented as some of the world's last wilderness areas. Ironically, while diminishing sea ice hinders indigenous mobilities in the North or affects vital habitat for Antarctic wildlife, it enables the movement of tourists on polar expedition-styled adventure packages. Tourism operators routinely advertise trips to the Polar Regions as offering the chance to experience the Arctic and Antarctic before traditional northern cultures, sea ice and polar wildlife disappear. An encounter with Greenland's Ilulissat Icefjord, for instance, is not one which just allows people to marvel at the Arctic environment, but one that tour companies promise to be an encounter with tipping points and earthly forces in a changing cryosphere. But, the presence of large cruise ships also provokes anxiety amongst environmentalists over the impacts increasing numbers of visitors have on the environment and wildlife habitat, while indigenous and local communities in the Arctic are not always welcoming of their arrival.

We have organized this handbook in four Parts. Locating the circumpolar worlds of the Arctic and Antarctic, Part I is concerned with polar mappings and imaginaries and begins with chapters on the history of exploration of the Arctic and Antarctic. It then moves on to a consideration of the place of the Polar Regions in literature and the popular imagination, before examining a number of issues of concern to indigenous peoples in the Arctic – self-determination and land claims, human-environment relations, environmental change, health and well-being, and education – before concluding with a chapter on sites of archaeology and heritage in the Arctic and Antarctic. Part II discusses the natural environments of the Polar Regions. Through introductions on biodiversity, geology, oceanography, sea ice, glaciology and permafrost, it describes how these elements integrate with wider global systems. In doing so, it reveals how the Polar Regions influence, and are influenced by, climate and environmental change processes with examples from the near and recent past, and in predictions of the future. The chapters in Part III consider polar politics and resource futures and range from an exploration and assessment of polar governance and international co-operation – for instance the Antarctic Treaty, the Arctic Council, the law of the sea, environmental governance and conservation – to discussions of indigeneity and sovereignty, geopolitics and security, sustainable development, national Antarctic science programmes, tourism, energy and social and environmental impact assessments, Arctic fisheries, and bioprospecting and the future of Antarctica. Part IV deals with polar scientific frontiers and how we can advance scientific discoveries in the Polar Regions in the coming years. It examines the role of technology in building knowledge in both historical terms and in view of future needs. The Polar Regions contain some of the most important records of climate change and extracting them at greater resolution and covering deeper time is a major scientific theme. Ice cores from the Antarctic and Greenland ice sheets, for example, are vital archives revealing evidence of past climate but are also important for modelling climate futures – as such they are not only key for our understanding of environmental history but have become enrolled in environmental representations and earthly geopolitics (Antonello and Carey 2017). The Polar Regions also contain some of the most extreme and unexplored parts of the planet, in particular within and beneath the large ice sheets in Greenland and Antarctica, and in the polar seas and oceans, as well as in the skies and atmosphere above the poles. And there are many habitats unknown to science in Antarctic waters. For example, while the greater species diversity discovered recently in deep parts of the Ross Sea, and which was noted at the beginning of this introduction to the handbook, may be due to climate change, new species of deep-sea starfish have been found in hydrothermal vents in the East Scotia Ridge in the Southern Ocean.[10] Another area of future work revolves around the use of the Polar Regions as vantage points from where purposely built and positioned apparatus can lead to major scientific discoveries. Telescopes and air sampling tall towers located on polar plateaus benefit from the high elevations and clean atmosphere. This may be in the form of the neutrino array at the South Pole that measures sub-atomic particles emanating from deep space, which will continue to help us understand profound issues, such as the creation of, and our place within, the Universe. The IceCube Collaboration, which involves forty-nine institutions from twelve countries has built the IceCube Neutrino Observatory at the Amundsen-Scott South Pole station. The observatory searches for neutrinos in a quest to probe and understand the cosmos – exploding stars, gamma-ray bursts, black holes, the nature of dark matter, as well as the nature and the properties of neutrinos themselves.[11] But as well as seeking answers to big questions in physics about the atmosphere and space, the IceCube Observatory is looking within the Earth, too, and is using a process called neutrino tomography to image the Earth's interior. And, as the recent discovery of a 280-million-year-old fossil forest in the Transantarctic Mountains reveals, geologists working in the Polar Regions are contributing to our understanding of deep time and the age of the Earth.

We thank all the contributors to this volume for their fine work and are enormously grateful for the support at Routledge we have received from Andrew Mould, Egle Zigaite and Faye Leerink. We also appreciate the patience of our authors and publishers while the handbook has developed. We wish to thank our respective families for their support, patience and understanding as this book has taken shape and taken up our time. We hope this handbook will prove useful both for polar specialists and for people with a general interest in the Polar Regions. As such it may help create awareness about the important messages the Polar Regions can bring when assessing the current status of our planet and inform discussion concerned with how we think about its future.

Notes

1 www.independent.co.uk/environment/sea-life-changes-beneath-below-antarctic-ross-ice-shelf-global-warming-climate-change-new-harbour-a8076901.html.
2 www.nytimes.com/interactive/2017/11/25/climate/arctic-climate-change.html.
3 www.bbc.com/future/story/20171121-why-russia-is-sending-robotic-submarines-to-the-arctic.
4 www.nytimes.com/2017/11/30/world/europe/russia-arctic-ocean-fishing-thaw.html.
5 In addition to reports appearing in conventional media, online sources and social media outlets also provide extensive coverage on polar topics. In the Arctic, for example, Arctic Now is a partnership with a number of media organizations around the circumpolar North (such as *High News North* [based in Bodø, Norway], *The Arctic Journal* [based in Nuuk, Greenland], and *Nunatsiaq News* [based in Iqaluit, Nunavut], among others) that provide news coverage of a changing region. See: www.arcticnow.com/.
6 *Christophe de Margerie* was built by South Korea's Daewoo Shipbuilding and Marine Engineering and is the first in a series of fifteen icebreaking LNG vessels designed to serve northern Russia's Yamal LNG project from the port of Sabetta.
7 For example, see *The Guardian*'s coverage on 24 August 2017: www.theguardian.com/environment/2017/aug/24/russian-tanker-sails-arctic-without-icebreaker-first-time.
8 The full text of "China's Arctic Policy" white paper was published in English on the China Daily Asia website on 26 January 2018: www.chinadailyasia.com/articles/188/159/234/1516941033919.html. The Belt and Road Initiative is a Chinese government development strategy to connect China and Eurasian countries and regions to a Chinese-centred network of trade and markets. The initiative partly revives ancient trading routes and will build infrastructure for land and maritime transport; the Polar Silk Road can be seen as a key Arctic element to the initiative and Chinese investment may likely be seen in transportation hubs along the Northern Sea Route and the Nordic Arctic.
9 See the Arctic Council website for more information about members, permanent participants and observers (www.arctic-council.org/index.php/en/), as well as Chapter 22 by Timo Koivurova in this handbook.
10 www.bas.ac.uk/media-post/first-new-family-of-starfish-discovered-in-hydrothermal-vents/.
11 http://icecube.wisc.edu/.

References

Antonello, A. and M. Carey 2017. 'Ice cores and the temporalities of the global environment' *Environmental Humanities* 9(2): 181–203.
Birtchnell, T., S. Savitsky and J. Urry (eds.) 2015. *Cargomobilities: Moving Materials in a Global Age*. London and New York: Routledge.
Brady, A-M. 2017. *China as a Polar Great Power*. Cambridge, UK: Cambridge University Press.
Broeze, F. 2002. *The Globalisation of the Oceans: Containerisation from the 1950s to the Present*. St. John's, NL: International Maritime Economic History Association.
Di, Q., L. Chen, B. Chen, Z. Gao, W. Zhong, R.A. Feely, L.G. Anderson, H. Sun, J. Chen, M. Chen, L. Zhan, Y. Zhang and W-J. Cai 2017. 'Increase in acidifying water in the western Arctic Ocean' *Nature Climate Change* 7: 195–9.
Dodds, K. and M. Nuttall 2016. *The Scramble for the Poles: The Geopolitics of the Arctic and Antarctic*. Cambridge, UK: Polity.
Kennicutt, M.C., and 75 others 2015. 'A roadmap for Antarctic and Southern Ocean science for the next two decades and beyond' *Antarctic Science* 27: 3–18.
Kennicutt, M.C., and 57 others 2016. 'Enabling 21st century Antarctic and Southern Ocean science' *Antarctic Science* 28: 407–423.
Martos, Y.M., M. Catalan, T.A. Jordan, A. Golynsky, D. Golynsky, G. Eagles and D.A. Vaughan 2017. 'Heat flux distribution of Antarctica unveiled' *Geophysical Research Letters* 6 November 2017, DOI: 10.1002/2017GL075609.

McCallum, A. 2017. 'As China flexes its muscles in Antarctica, science is the best diplomatic tool on the frozen continent' *The Conversation*. 13 November, http://theconversation.com/as-china-flexes-its-muscles-in-antarctica-science-is-the-best-diplomatic-tool-on-the-frozen-continent-86059.

McGhee, R. 2004. *The Last Imaginary Place: A Human History of the Arctic World*. Toronto, ON: Key Porter Books.

Notteboom, T. and J-P. Rodrigue 2008. 'Containerisation, box logistics and global supply chains: the integration of ports and liner shipping networks' *Maritime Economics and Logistics* 10(1–2): 152–74.

Nuttall, M. 2017. *Climate, Society and Subsurface Politics in Greenland: Under the Great Ice*. London and New York: Routledge.

Part I
Circumpolar worlds

1

Exploring and mapping the Arctic
Histories of discovery and knowledge

John McCannon

To most minds, the phrase "Arctic exploration" is likely to evoke sepia-toned impressions of grizzled adventurers – storied figures such as Nansen and Peary, or roisterous prospectors straight from the pages of Jack London and Robert Service – racing on ships, on skis, with dogs, and soon enough on airships and airplanes, toward the North Pole, or in search of gold and other riches on a legendary scale. As large as it looms in the public imagination, though, this "heroic" age, generally considered to have coincided with the late 1800s and early 1900s, constitutes only a narrow sliver of the long history of Arctic exploration. Whether we take "discovery" – a loaded term in the field for more than a generation now – to mean all forms, indigenous or external, of understanding a geographical space, or a more systematic mapping and study of a place unfamiliar to oneself (but not to its original inhabitants), processes of discovery have unfolded in the circumpolar North for centuries at a minimum, and arguably for millennia, although older and non-Western examples are harder to know with certainty. The breadth of this timespan is matched by that of the range of motivations which drove Arctic exploration: migration and colonial settlement; the quest for wealth, whether it was to be gained from minerals, animals, or creatures of the sea; the imperatives of war, national security, and state-building; the proselytizing crusades of Christian churches; sheer scientific and navigational curiosity; and more. Rather than attempting a comprehensive survey of the topic, a difficult challenge for even a book-length study, this chapter will provide a chronological overview, highlighting key themes and trends for each period, illustrating these with selected examples, both well-known and less familiar. Due to considerations of space, and intending no slight to newer and more expansive readings of the terms "discovery" and "exploration", it will concentrate principally on navigational, cartographic, and scientific endeavours that conform more closely to traditional understandings of those concepts.

Toward hyperborea: early "discoveries" in the Arctic

Human beings are thought to have first reached the Arctic Circle as early as 28,000 years ago, and while the origin of those peoples who presently inhabit the far north forms a complex subject of its own, broad consensus has it that most of these groups, or at least their ancestors, had moved into their homelands, or were migrating toward them, by around the years between 3000 BCE and 1000 CE.

Without a doubt, these peoples – be they the Eskimo and Inuit of North America and Greenland; the Sámi and Komi of Fennoscandia and northwest Russia; or the dozens of native peoples in northern Siberia, including the Koryaks and the Chukchi – explored, and did so extremely well. Their survival depended on a precise and thorough understanding of the physical layout and natural workings of the ecosystems they inhabited. There is little point in quibbling ungenerously over the semantics of whether such knowledge should be seen as "lore" or, instead, "science" of the sort Westerners typically associate with exploration. The main problem here is that the near-total absence of written documents, combined with the difficulty of reconstructing from oral tradition the worldview of societies from the distant past, leaves us poorly equipped to trace in detail how and when indigenous northerners in the pre-contact era accumulated their geographical knowledge, and in what forms – other than oral transmission and the sharing of lived experience – they may have preserved and transmitted it. When it comes to visualizing and recording geographical data, the best-known instance among Arctic indigenous people is the creation by Eskimo and Inuit peoples of so-called driftwood maps, in which the contours of a given coastline are carved into the edge of the wood. Surviving examples, however, are of comparatively recent origin, and how far back in time this practice might extend is a matter of speculation. The *fact* that indigenous peoples explored the north is beyond dispute, but recovery of the details may belong more properly to the sphere of anthropology than to that of history. A final note to make here concerns the incalculable debt that Euro-American explorers would later owe to the Arctic's indigenous inhabitants: native understanding of the northern wilderness proved indispensable to outsiders who came to travel there. It can be said with certainty that without native expertise, little of the Arctic "discovery" carried out by the West would have been possible.

Non-native awareness of a distinctly northern region, or "hyperborea", is considered to have begun during the Greco-Roman era, as Herodotus wrote in his *Histories* of a perpetually wintry realm beyond "civilized" bounds, and as the Greek voyager Pytheas of Massalia, around 330 BCE, claimed to have rounded the mass of mainland Europe and entered the frozen seascape he dubbed "Thule". By Europe's medieval period, exploratory ventures to the Arctic were more numerous, but in many ways still shrouded in uncertainty. Chronicles from the middle ages have marked limitations as sources, and many of those who travelled to the Arctic had every incentive *not* to report their "discoveries" in a way that can be captured by the historical record: why, for example, would a whaler reveal the location of a favourite hunting ground, or a merchant a new trade route?

Prior to Columbus, European exploration of the Arctic was by necessity confined to the North Atlantic islands, the Fennoscandian far north, and the high latitudes of northwest Russia. Viking excursions to the west brought the Faeroes (settled by the Norse around 800) and Iceland (discovered in 870) permanently into the Scandinavian sphere. Erik the Red pushed out even farther in the 980s, reaching Greenland, where Norse colonists farmed and fished on its southern and southwestern shores – trading and warring with the local Inuit (or "skraelings", as the newcomers called them) – then withered during the late 1300s and early 1400s, dying out for reasons still not completely understood. Vikings also sailed north and east to crest the Arctic rim of Scandinavia, continuing on from there into the region they called "Bjarmaland", which consisted of the Kola Peninsula and Russia's White Sea coast. Among these early freebooters was the Norse hunter Ottar, who, around 890, reached what was probably the mouth of the North Dvina, traded with the local Sámi and Komi, and brought back a bounty of walrus tusks, seal pelts, and furs that gave him bragging rights when he visited the court of Alfred the Great. Others came in Ottar's wake, whether to hunt, to fish rich grounds like the Lofoten Islands (off Norway's coast near 70°N), or to Christianize the natives.

Venturing farther to the east were the Russians, especially from the city-state of Novgorod, a successful expansionist power until its absorption by Muscovy in the 1400s. In search of amber, ivory, and furs, Novgorodians and other Russians followed northwestern rivers like the Dvina through taiga and tundra, arriving at the White Sea coast in the 900s and 1000s. There they built ports and salt works, and then continued eastward along the coast. They charted the southern tip of Novaya Zemlya during the eleventh century and attained over the coming decades a decent geographical understanding of the Arctic shore between the Kola Peninsula and the mouth of the Ob River. Churchmen arrived alongside the trappers, sea-hunters, and settlers, both to preach to the natives – Stephen of Perm began his mission to the Komi in the 1380s – and to seclude themselves as much as possible from worldly distractions, a goal easily achieved in such remote spiritual outposts as the famed Solovetsky Monastery, established on the islands of the White Sea in 1429 (and destined in centuries to come to serve as one of the USSR's most feared prison camps).

After Columbus: the sixteenth and seventeenth centuries

Although Christopher Columbus never voyaged near the high northern latitudes, his 1492 encounter with the Americas marks a watershed in the history of Arctic exploration, in that it opened up a new set of frontiers for Europeans to "discover" and provided new incentives to spur those discoveries on. Arctic exploration during the 1500s and 1600s was impelled by the same economic and religious motivations that had driven it in earlier centuries, and now increasingly by state-building and colonization. It now proceeded, however, on a larger scale, with greater scientific precision, and in multiple directions. The newest variable was the desire of nations like France, England, and others to undercut the monopoly held by Portugal and Spain on the New World's most desirable territories and on sea routes to Asia. Were the higher latitudes of North America, neglected by the Spanish, worth exploiting? And could European ships reach China and the East Indies by sailing through Arctic waters, opening a Northeast Passage along the coast of Russia or a Northwest Passage through the islands and waterways of the country we now know as Canada?

In the west, John Cabot led the way with his 1497 survey of Newfoundland, and more such expeditions, such as Jacques Cartier's mapping of the St. Lawrence River in the 1530s and 1540s, boosted European familiarity with the terrain and waterways that led upward to the North American Arctic. In the 1570s and 1580s, the Dano-Norwegian kingdom re-established its old colonial claim to Greenland (including its Inuit inhabitants) and more thoroughly explored its southern coastlines. Knocking even harder on the door of the Northwest Passage were Martin Frobisher and John Davis, each of whom, in a series of pivotal voyages during the 1570s and 1580s, entered the chilly waters between western Greenland and the Canadian archipelago, with Davis sailing as far as 72°N. Both men stimulated further exploration of the region by reporting on the vast quantities of bowhead whales and cod they had spotted, and regarding the viability of a Northwest Passage, Davis opined that it was "most probable, the execution easie".[1]

Also in the 1500s, fishing, whaling, and seal- and walrus-hunting, along with the prospect of opening a Northeast Passage to Asia, brought larger numbers of European ships to the Norwegian, White, and Barents seas in the 1500s. A costly but consequential voyage set out from England in 1553, under the command of Hugh Willoughby; although two of Willoughby's ships were lost, Richard Chancellor landed on the White Sea coast, made his way to the court of Ivan the Terrible, and negotiated terms of trade that led to the formation of the Muscovy Company. (This in turn gave birth, in 1589, to one of the circumpolar world's most important ports: Arkhangelsk.) In 1556, Stephen Burrough became the first West European to sail past Novaya Zemlya and into the Kara Sea, and the western sectors of Russia's Arctic coast were charted more thoroughly over the next several decades. On land, parallel efforts extended Russia's political reach – and geographical knowledge – into western Siberia. Following river valleys northward from the sub-Arctic, as well as inching along the shoreline, Russian hunters and travellers mapped most of the northern littoral from the Kola Peninsula to the Yamal Peninsula and the gulf formed by the Ob, and even to the mouth of the Yenisei by the mid-1500s. All this was helped by the fact that Ivan the Terrible, with his conquests of the Tatar khanates of Kazan and Astrakhan in the 1550s, had opened the Siberian subcontinent to a wave of conquest that eventually took the Russians all the way to the Pacific. Full-scale invasion began with Yermak's assault across the Urals in the 1580s; although the Cossack warlord fell in battle in 1585, his campaign led to the establishment of Russia's first Siberian cities, Tiumen and Tobolsk, and the Russian network of townships and settlements soon included numerous outposts along the northern reaches of the Ob and Yenisei rivers, including the Arctic port of Obdorsk, founded in 1595.

Closing out the century along the Northeast Passage was the Dutch mariner Willem Barents. Between 1594 and 1597, Barents led three expeditions to the sea that now bears his name and recorded an astonishing "farthest north" of 79°49′N. He failed in his goal, which was to locate an ice-free path through the Kara Sea, and he and many of his men lost their lives to these voyages: the majority perished while stranded on Novaya Zemlya during the winter of 1596–1597. Still, the survivors returned with a wealth of information about both seas, and it was Barents who formally charted Spitsbergen, the chief island of the Svalbard archipelago. The state of knowledge here had greatly improved since the 1530s, when Sweden's Olaus Magnus depicted a somewhat distorted version of the North Atlantic and "Bjarmia" on his famed *Carta marina*.

The search for both passages intensified in the 1600s, as did European appetites for fishing and hunting in the high north. Essaying the northeast and northwest routes alike was Henry Hudson of England, who sailed variously for his own country and for the Dutch. Convinced like Gerhard Mercator and Abraham Ortelius, the premier mapmakers of the day, that beyond the icy seas of the Arctic lay a body of warm water surrounding a pillar-like North Pole, Hudson attempted to pierce this boreal barrier, first for the Muscovy Company in 1607, travelling as far as Spitsbergen, then in 1608 and 1609. These last two voyages took him only as far as Novaya Zemlya, and during the 1609 venture, which he undertook for the Dutch East India Company in the *Half Moon*, frustration led him to turn to the west, all the way to Newfoundland, and then to Cape Cod and the river that bears his name – a "discovery" that prompted the Dutch to claim the surrounding territory and establish New Amsterdam, now New York. In 1610, Hudson returned to the Americas, this time for England in the *Discovery*, and charted Hudson Bay: a watershed moment in the search for the Northwest Passage and the colonization of the Canadian Arctic. Unfortunately, the expedition ended in mutiny and the death of Hudson and those loyal to him. Meanwhile, Samuel de Champlain of France founded the cities of Port Royal (1605) and Quebec (1608), solidifying the foundation from which French explorers would move upward into the Canadian north.

In the northeast, the Muscovite state discouraged foreign voyages east of Arkhangelsk, as shown by its 1619 closure of the Arctic port of Mangazeya, on the Ob Gulf. None of this kept English, French, Dutch, and Scandinavian ships, drawn northward by seals, walrus, and whales, out of the waters surrounding Novaya Zemlya, Spitsbergen, and other nearby islands. East of the Ob, however, exploration and exploitation fell almost exclusively to the Russians, who forged a path to the Pacific, both overland and along the Northeast Passage, which they came to call the Northern Sea Route. From Siberian centers like Yakutsk and Irkutsk, and from smaller outposts, or *ostrogi*, numerous surveys mapped the course of the Lena, Yana, Indigirka, and Kolyma rivers to where they opened into the Arctic Ocean. In 1638–1639, the explorer Ivan Moskvitin reached the Sea of Okhotsk, and farther to the north, Semyon Dezhnev sailed around the Chukchi Peninsula in 1648–1649, entering what we now know as the Bering Strait, but most likely without consciously recognizing that he had shown Eurasia and North America to be separate continents. The expansionist will of Muscovy's tsars drove this process in part, but the true engine behind it was the hunt for fur-bearing creatures of all kinds, most of all the majestic sable. In many places throughout northern Siberia, Russian rule consisted mainly of forcibly Christianizing native groups and subjecting them to the dreaded *yasak*, the tribute system by which they harvested furs for the state under severe compulsion. By the end of the century, Russian interlopers took their initial steps into Arctic regions that would long resist centralizing authority, such as the Taimyr, Siberia's northernmost extremity; Chukotka in the far northeast, where Chukchi and Koryak natives would fight sustained wars of resistance during the 1700s; and the volcano-studded but resource-rich and logistically important peninsula of Kamchatka, which Vladimir Atlasov entered in 1697–1699.

Developments in the Americas' high north proceeded along similar lines, if somewhat more slowly. By the mid-1600s, European fishermen and sea-mammal hunters, having drastically depleted the waters of the Fennoscandian and northwest Russian Arctic, were swarming westward across the North Atlantic, to Jan Mayen Island (discovered around 1614), to the Greenland Sea, and then to western Greenland and eastern Canada, well into Davis Strait and beyond, seeking fresh sources of prey. Their rapacity was matched by French and English hunters on land, who quested in search of beaver pelts and other furs, enlisting – especially in the case of the French *coureurs* and *voyageurs du bois* – Native American and First Nations groups to guide and assist them. The search for the Northwest Passage continued, and the European presence in and around Hudson Bay steadily expanded. In the 1650s and 1660s, the staggering quantities of fine furs brought back to Europe by trappers like Pierre Radisson (who worked both for his native France and the English) prompted the 1670 formation of the Hudson's Bay Company (HBC), which played a two-century role in exploring and settling much of northern and Arctic Canada.

During the mid-1600s, in a comment both typical of its time and predictive of the attitudes that would prevail in the years to come, the chancellor of Sweden, Axel Oxenstierna, referred exultantly to his country's Arctic possessions as "a new India".[2] More and more, statesmen and colonizers saw in the high north opportunities to strengthen their realms politically and militarily, to enlarge their populations by absorbing native subjects, and to amass new resources – not just furs, fish, and ivory, but metals, timber, and other assets increasingly coming to light. Exploration of the Arctic would soon be suffused with a more modern spirit of imperial acquisitiveness.

The Arctic and the Age of Reason

The eighteenth century brought new levels of European intrusion to the Arctic. As this happened, a new priority – scientific understanding – took its place beside older ones such as governmental self-aggrandizement and the accumulation of wealth. As Enlightenment enthusiasms swept all aspects of Europe's intellectual life, they transformed the practice of overseas exploration. Cartography became a more exact science. Surveys of the high north's flora, fauna, and climate, and the languages and folkways of its native peoples, took on a more systematic and professional character. This was hardly a case of scholarship for its own sake; practical applicability remained paramount. But if commercial or political purposes had to be served, knowledge in and of itself started to count for more as the century wore on.

This new bundling of imperatives became evident early on. In Greenland, the Danish pastor Hans Egede, determined to tighten his country's colonial hold on the island and to preach the gospel to the Inuit there, landed on the western shore in 1721 and founded the settlement of Godthåb (now Nuuk). He explored the coastline northward, almost to the Arctic Circle, and, an accomplished naturalist, studied the region's plants and animals. During the 1730s, Carl Linnaeus of Sweden, creator of the modern system of taxonomy, conducted field research in Norrland and the Scandinavian Mountains, as far north as the 67th parallel, and his countryman, the astronomer Anders Celsius – best known for the system of temperature measurement named after him – travelled to the polar latitudes in 1736 to measure the arc of a degree of latitude. This calculation was compared with similar measurements taken farther to the south by way of confirming that the earth flattens slightly at the poles.

Such ventures became commonplace throughout the high north – to the point that, in the 1730s and 1740s, the Arctic became the staging ground for one of the most ambitious scientific endeavours ever undertaken. This was the two-stage expedition organized by Vitus Bering, a Danish captain in Russian service, to prove that Eurasia and North America were divided by open water. During the First Kamchatka Expedition of 1725–1730, Bering and his lieutenants, including Alexei Chirikov, a notable explorer in his own right, concentrated narrowly on this goal by sailing north through what is now called the Bering Strait and turning westward along Russia's northern coast – but did not proceed far enough to satisfy skeptics. It was to answer these critics that Bering staged the larger and more comprehensive Second Kamchatka Expedition, better known as the Great Northern Expedition. This lasted from 1733 to 1743 and sent forth more than 900 personnel, not just to make Bering's case about the separateness of the continents, but to survey and chart Russian's northern coast, which for this purpose was subdivided into five sectors between Arkhangelsk and Siberia's northeastern-most tip. The Aleutians and the Sea of Okhotsk were mapped more minutely as well. Bering himself died in 1741, but he and Chirikov demonstrated beyond doubt that Eurasia and North America were distinct landmasses. The coastal surveys returned a gigantic mass of useful geographical, botanical, and zoological knowledge, and invaluable research was carried out by the German naturalist Georg Wilhelm Steller.

By mid-century, the pace of exploration was accelerating in the Arctic, and in most cases, especially marine expeditions, it had become standard routine to bring along at least one scientist. In Canada, agents of the HBC and the British crown had been mapping the eastern waters and the territory surrounding Hudson Bay, while the French explored westward into the sub-Arctic Canadian interior. After 1763, when victory in the French and Indian War placed Canada fully in their hands, the British redoubled their efforts. The inimitable James Cook surveyed the Newfoundland coast in the 1760s. Seaborne explorers pushed further into the mazelike waters of the Canadian archipelago, seeking the Northwest Passage, while trappers and surveyors approached from landward.

Between 1769 and 1772, Samuel Hearne, on behalf of the Hudson's Bay Company, trekked up the Coppermine River, reaching Coronation Gulf on the Arctic Ocean, becoming the first European to reach Canada's north coast from the interior. Probing farther to the west was Alexander Mackenzie, hired by the North West Company, the HBC's chief rival for fur-trade dominance. Tasked in 1788 with opening a route to connect the Canadian interior with the Pacific, Mackenzie eventually succeeded, approaching via British Columbia in 1792–1793. One of his false starts, however, took him into the Arctic in 1789, from Great Slave Lake along the river known to the Athabaskan-speaking natives as the Deh Cho. Instead of bending westward, the Deh Cho flowed north into the Arctic Ocean. A frustrated Mackenzie called the river "Disappointment", but it turned out to be Canada's longest river and one of the high north's most vital waterways, and soon came to bear the Scotsman's name. Before the end of the century, the Canadian and Alaskan Pacific were charted with the greatest diligence yet. After visiting the Hawaiian islands in 1778, James Cook sailed eastward in the *Discovery* and *Resolution* to Vancouver Island, and from there embarked on a huge coastal survey that included southern Alaska, the Aleutians, and Alaska's western shore, all the way through the Bering Strait – which was thus named by Cook in the older explorer's honour. Cook then attempted to find the western end of the Northwest Passage, but could not get past Alaska's Icy Cape, at the 70th parallel. He turned back to winter in Hawaii, where he would be killed in a quarrel with natives the following February.

The European Arctic and Siberia's upper reaches likewise fell more squarely under the scientist's gaze during the late 1700s. Attempts to reach the North Pole or crack the Passage's mysteries led to impressive high-latitude voyages in the neighborhood of Novaya Zemlya and the Svalbard archipelago, as in the cases of Vasily Chichagov of the Russian navy, who reached 80°N in 1765 and 1766, and England's Lord Mulgrave, whose naturalist performed the first scholarly autopsy of the polar bear in 1773. (Among those on this adventure was a fourteen-year-old Horatio Nelson, and as an intriguing bookend to that fact, a young Napoleon Bonaparte applied in 1785 to sail with Jean-François de la Perouse on his attempted circumnavigation of the world – a voyage that conducted useful surveys in northern waters near Alaska, Sakhalin, and Kamchatka – but was rejected, thus sparing him when the expedition came to fatal grief in the South Pacific.)

Farther to the east, during the 1770s, Ivan Lyakhov started mapping the remote New Siberian Islands, first visited by outsiders in the 1710s, but not fully charted until the twentieth century. Between 1785 and 1795, Joseph Billings, an English officer in the pay of Russia's empress, Catherine the Great, brought a number of scholars to conduct a multidisciplinary study of Chukotka, the Aleutians, Kodiak Island, and western Alaska. This expedition did not, as had been hoped, complete the Northeast Passage, but it returned with the finest maps made to date of these regions. Billings also reported on Russian ill-treatment of native Siberians and Aleuts, and warned as well about the terrible ravages that Russian hunters had visited upon the North Pacific's sea otter population, which had been heavily targeted since the 1740s to make up for the dwindling number of sables in mainland Siberia. In the meantime, other mariners had been active in these parts: aside from James Cook's aforementioned travels here, Spanish voyagers mapped Canadian and Alaskan waters between the 1770s and 1790s. By century's end, a quadrilateral competition over the northernmost Pacific was shaping up between Great Britain, the new United States, Spain, and Russia. The boldest thrust was made by Russia, several of whose fur-hunting companies ventured into Kamchatka, Chukotka, and the islands of Beringia, coursing after sables, sea otters, and – following the 1786 voyage of Gerasim Pribiloff to the Aleutians and the isles that bear his name – fur seals. Entrepreneur-colonizers such as Alexander Baranov and Grigorii Shelikhov worked during the 1780s and 1790s to anchor a military and commercial presence in the Aleutians, Kodiak, and Alaska proper, leading to the formation of the Russian-American Company and the empire's seven-decade effort to maintain a North American colony.

The Heroic Age? Arctic exploration in the 1800s and early 1900s

In many respects, nineteenth-century efforts to explore the Arctic followed the same pattern as during the 1700s. Stalled for more than two decades by the Revolutionary and Napoleonic wars, the search for the northern sea passages continued, and while most of the northern circumpolar map's basic contours had been sketched out by now, many fine-grained details had yet to be resolved, especially at the highest latitudes, where many blank spots remained, and where the possible existence of an open sea or some large, unknown landmass was still an open question. Among the things that changed over these years were the scale and frequency of expeditions to the high north, as well as their growing sophistication and reliance on industrial-era technology and scientific equipment. Also different was the emergence of mineral wealth – the goldfields of the Klondike, coal on Spitsbergen, the Kiruna iron mines in northern Sweden – as a motivation behind them. The scholarly disciplines of ethnography and anthropology came into their own and also spurred many of the era's signature expeditions. Most of all, it was during this century – widely remembered as the "heroic" age of polar exploration – that Arctic and Antarctic adventuring became a public craze, stirred by nationalist obsessions and driven by an increasingly modern mass-media complex.

Space permits only the briefest catalogue of the dozens of Euro-Americans seeking to uncover what they hoped were the high north's final mysteries. In the Anglo-American historiography, the person credited with reviving Arctic efforts after Napoleon's downfall is John Barrow of the Admiralty, who declared it a matter of urgency that Great Britain be first to open the two passages. In 1818, Barrow dispatched two ships, one under David Buchan and John Franklin, who were to enter the Northeast Passage via Svalbard, and the other under John Ross and William Parry, who were ordered to enter Baffin Bay, complete the Northwest Passage, and rendezvous with Buchan and Franklin in the Pacific. The overall mission failed, but Ross and Parry sailed up Davis Strait far enough to chart new waterways like Smith Sound and Melville Bay, and Parry's optimism persuaded the British to keep probing in this direction. Parry returned four times between 1819 and 1827, bulling his way through more than half the westward distance and reaching a record 82°45′N. Ross and his nephew, James Clark Ross (renowned for his future role in Antarctic exploration), took the steam-driven vessel *Viceroy* to the Arctic in 1829–1833, and while they had to abandon ship, they calculated the location of the North Magnetic Pole.

Concurrently, Russian surveyors, many of them sponsored by Count Nikolai Rumyantsev – Barrow's Slavic analogue – mapped Novaya Zemlya; the White Sea and Barents shorelines; the Laptev Sea and the New Siberian Islands; and the waters and islands of the Bering and Chukchi seas. Among those active here were Mathias von Hederström, Yakov Sannikov, Otto von Kotzebue, and Fyodor Litke. The Danes investigated Iceland and Greenland more closely. Another venture from this period was the La Recherche voyage of 1838–1840, led by Joseph Paul Gaimard of France, but with substantial assistance from Norway and Sweden. Gaimard surveyed a wide stretch of northern waters, from the Faroes to Spitsbergen.

The mid-1800s witnessed the transformation of this risky but routine (and generally little-known) work into a global sensation, and also the burgeoning of a debate about the best methods for exploring polar climes. The figure whose tragic end in the 1840s did most to trigger these changes was John Franklin of England. Veteran of Buchan's 1818 voyage to the Northwest Passage, Franklin gained further notoriety in 1819–1822 as "the man who ate his boots", thanks to a disastrous overland expedition up the Coppermine and through Canada's Barren Grounds, where almost half his men died and where the survivors subsisted on rotting provisions and leather scraps. Franklin had better luck during an 1825–1827 expedition up the Mackenzie, mapping portions of Canada's and Alaska's northern coast, but he exemplified an older approach: large, ponderous expeditions, often organized along military lines and inflexible in their importation of travel and survival methods better suited to other conditions, rather than learning how to adapt to the polar environment. Parry, Ross, and many Scandinavians and Russians were departing from this model – and were beginning to adopt techniques learned from indigenous northerners – but Franklin did not, much to his grief and that of those following him as he sailed forth in 1845, in the *Erebus* and *Terror*, hoping to complete the Northwest

Passage. Geographically, this seemed an easier task than before, with only a gap between Lancaster Sound and Boothia Peninsula remaining unexplored, but in 1846–1847, the ships became trapped in the ice off King William Island. All members of the expedition perished: some, including Franklin, stayed with the ships, others struck out on a doomed attempt to reach the nearest HBC outpost. Most of the mysteries surrounding Franklin's disappearance have been resolved in the 1990s and 2000s, and the seabed locations of the wrecked *Erebus* and *Terror* have been discovered (in 2014 and 2016, respectively), but in 1848, all that a horrified public knew was that he had vanished – and was either dead or in dire peril.

Dozens of expeditions, mainly from Britain and the United States, set out in the late 1840s and the 1850s to find Franklin and his men. Taken together, these ventures uncovered the Northwest Passage in 1850–1854 (credit going to Robert McClure, who lost the *Investigator* in the process), mapped most of the Canadian archipelago, and caused many explorers – including Francis McClintock and John Rae, both of whom, between 1852 and 1859, found enough traces of Franklin's failure that the expedition was finally declared lost – to organize on a smaller scale and take lessons from northern natives about clothing, hunting, and transport.

With Franklin presumed dead, Arctic expeditions began prioritizing other goals. One was to clarify whether an open polar sea existed high to the north, a view surviving from the days of Mercator and Ortelius, and still espoused by scholars as respected as German cartographer August Petermann (who held in addition that a peninsula called "Transpolarland" extended across the top of the world from northern Greenland to the recently-discovered Wrangel Island, off Siberia's northeast coast) and US Navy oceanographer Matthew Fontaine Maury. Acolytes of the open polar sea theory included England's Edward Inglefield, who claimed Ellesmere Island as British territory, and Elisha Kent Kane of the Second Grinnell Expedition (1853–1855), whose contingent included the Greenlandic Inuit guide Suersaq, or Hans Hendrik, and which relied on native methods to run dogsled teams above 80°N. Kane's conflict-ridden enterprise mapped much of the gap between Ellesmere Island and northwest Greenland, but its claims to have sighted signs of an open sea to the north later proved unfounded. The same is true of the reports made by Kane's fellow American and associate turned rival, Isaac Israel Hayes, who searched for open water between Greenland and Canada in 1860–1861 and claimed – erroneously – to have sledded as far north as 82°N.

More precise charting was also the order of the day, whether for commercial purposes or nation-states' perceived need to assert sovereignty over "their" Arctic possessions. (Scandinavia underwent key political changes during the 1800s, requiring many territorial renegotiations in the north, and Canada, hemmed in by Danish Greenland to the east and American-owned Alaska to the west, built up its police presence on land and sponsored voyages such as the Dominion Government Expedition of 1903–1904 and the maritime patrols of Joseph Elzéar Bernier between 1906 and 1911.) Scholarship and adventure played their part as well, as did the growing desire to reach the North Pole for its own sake. A battery of expeditions continued thrusting up between Ellesmere and Greenland. These included Charles Francis Hall's failed 1871–1873 voyage on the *Polaris* – which led to his own death and marooned most of his fellows on the ice for over half a year, kept alive only by the survival skills of the Inuit guide Hans Hendrik – and George Nares's more fruitful British Arctic Expedition of 1875–1876, which set a new "farthest north" by crossing the 83rd parallel. Adolphus Greely's attempt in 1881–1884 to break Nares's record went hideously awry, with only six men returning and rumors of cannibalism dogging the survivors.

East of Greenland, islands such as Novaya Zemlya (first circumnavigated in 1870) and Spitsbergen (where hints of rich coal deposits beckoned) received more visitors, and the Northeast Passage came under closer scrutiny. German voyagers, boosted by August Petermann, sailed this way in 1868 and 1869–1870, and it was an Austro-Hungarian expedition, led by Julius Payer and Karl Weyprecht in 1872–1874, that located Franz Josef Land. George Washington DeLong of the US Navy, sailing the *Jeannette* westward along the Siberian coast between 1879 and 1881, established that Wrangel Island was indeed an island, and not the tip of Petermann's theoretical "Transpolarland", but his ship was caught fast in the ice, and he himself perished. The true giant here was the geologist-adventurer Nils Adolf Erik Nordenskiöld, a Finnish-born Swede most famous for having organized the first successful voyage through the Northeast Passage – in the *Vega* from 1878 to 1879. He also conducted valuable surveys of Spitsbergen, the Kara and Barents seas, and Greenland between the 1850s and 1880s. Eduard Toll of the Russian Academy of Sciences focused on the New Siberian Islands and the Taimyr Peninsula during the 1880s and 1890s, until his disappearance in 1902.

Greenland became a front of its own in the struggle to tame the Arctic. Henrik Rink of Denmark laid important groundwork in the 1840s and 1850s with his surveys of its northern districts, and Nordenskiöld's discovery of airborne pollutants scattered on the island's icecaps provided Western scientists with one of their first clues about industrialization's impact on the Arctic ecosystem. Still, as late as the 1880s, Greenland's forbidding interior had been penetrated only to a distance of 150 miles from the coast. It was the quest to push farther inward that brought to center stage two of Arctic exploration's most titanic figures: Fridtjof Nansen, the robust and cerebral scholar-humanitarian from Norway, and the implacable Robert Peary, consumed with winning polar fame for the United States and, most of all, for himself. Peary attempted to traverse the island in 1886 and would return many times as he developed the skills that catapulted him toward the North Pole in a string of increasingly urgent treks during the 1890s and 1900s. Nansen stunned the world – and launched his own distinguished polar career – by crossing Greenland in 1888. By now, Greenland was seen as hiding the answers to significant scientific mysteries, mainly pertaining to the fluctuation of weather patterns and the production of icebergs, and numerous expeditions worked there as the century turned. The German geophysicist Alfred Wegener, best remembered today for originating the theory of continental drift, took part in the Mylius-Erichsen Expedition of 1905–1908 and, with Johan Peter Koch in 1912–1913, became one of the first individuals to overwinter on the Greenland icecap.

By this stage, efforts in the Arctic had undergone a noticeable bifurcation. Although the line dividing them was never absolute, one camp viewed polar exploration as a means to larger scientific and scholarly ends, while another grew obsessed with breaking records and racing for the North Pole as if it were a boreal Holy Grail. These viewpoints were never mutually exclusive – Nansen and Nordenskiöld, for instance, would have been overjoyed to reach the Pole, however strong their commitment to science – but the difference was palpable enough to influence work in the high north during the late 1800s and early 1900s.

The pro-science attitude was epitomized by Karl Weyprecht of Austria-Hungary, who dismissed the fixation with setting records and dashing to the Pole as a wasteful "international steeplechase", and he was echoed by Clements Markham of the Royal Geographical Society, who declared in 1894 that "merely to reach the North Pole, or to attain a higher latitude than someone else", were "objects unworthy of support".[3] It was this that led Weyprecht, in conjunction with Georg von Neumayer of the German Marine Observatory, to coordinate the first International Polar Year (1882–1883), which established twelve main stations throughout the Arctic – and synchronized many smaller efforts as well – particularly to benefit the fields of meteorology and geophysics. It was in a similar spirit that Nansen took the *Fram*, a schooner purpose-built for Arctic navigation, on a high-latitude drift that lasted from 1893 to 1896 and amassed a trove of new information about pack ice and oceanic currents in the Arctic.

(These scholarly aims did not keep Nansen from leaving the ship in the hands of its captain, Otto Sverdrup, and attempting his own dogsled dash to the Pole. He failed to reach his target but shattered the "farthest north" record by going above the 86th parallel.) Roald Amundsen of Norway did much the same, but in Canadian waters, piloting the *Gjoa* through the Northwest Passage in 1902–1906 and becoming the first to sail through the Passage in its entirety. The work of those like Wegener and Nordenskiöld should be mentioned in this context as well.

So should the ethnographers who now ventured to the high north. A pivotal figure here was the German-American Franz Boas, who played an enormous role in developing anthropology as a modern academic discipline. In 1883–1884, Boas lived amidst the Inuit of southern Baffin Island, but even more important was his leadership of the Jesup North Pacific Expedition (1897–1902), which coordinated numerous ethnographers in their study of native peoples on both sides of the Bering Strait. Some of Russia's most acclaimed anthropologists, such as Vladimir Bogoraz-Tan, took part. (Ironically, many were in a position to help because they had been exiled to Siberia for their political opposition to the tsarist regime.) Knud Rasmussen began his work among the Inuit of western Greenland in 1902–1904, and Vilhjalmur Stefansson of Canada lived among the Mackenzie Inuit in 1906–1907. In 1908–1912, with Rudolph Anderson, he covered the wide arc between Alaska's Point Barrow and the Coppermine River.

Attracting most attention by far, however, was the media-driven, nationalism-fuelled "race for the poles". Assaults against the North Pole were made by sportsmen and scientists alike, but the one who outstripped them all in ferocity of spirit and psychological compulsion was Robert Peary. In a succession of North Pole attempts during the 1890s and 1900s, Peary inched ever closer to his target, powering through to the 87th parallel by 1906 and perfecting the "Peary system": an elaborate method using base camps and advance parties to lay down supply caches in preparation for the final dash northward. Readily adapting to native modes of survival and enlisting Inuit assistance on a scale never seen before, Peary also displayed a racialist condescension that stood out even by the standards of his time.

Dogged by fears that his financial support would dry up or that one of his rivals would overtake him, Peary set forth on his final Arctic venture in 1908–1909, asserting upon his return that he had reached the Pole on 6 April 1909. He found to his horror that a conflicting claim had already been submitted by a fellow American, Frederick Cook, who, if he spoke true, had won the Arctic crown in April 1908. The resulting scandal consumed the public, ruined Cook, and ended with formal declarations by the US Congress, the Explorers' Club, and the National Geographic Society, awarding victory to Peary. A near-universal consensus now holds that Cook falsified his claim, but the process of debunking it has also brought to light inconsistencies in Peary's story, especially regarding the absence of anyone on his last push qualified to verify his navigational readings and the astounding rates of speed he purported to have achieved on his return journey. Today, the commodore's victorious reputation seems suspended, like Schrödinger's cat, in a state of indeterminacy: none of the major bodies that credited Peary with having reached the Pole has reversed its decision, and yet a steadily growing proportion of historians has come to believe that neither Cook nor Peary achieved the triumph they sought.

In many ways, Arctic expeditions prior to (and during) World War I foreshadowed trends that the twentieth century would bring to greater prominence. Stefansson continued his far-ranging ways during the Canadian Arctic Expedition of 1913–1916, simultaneously sparking controversy by setting off on his own when his flagship, the *Karluk* – subsequently lost with many lives – became trapped in the ice. The American Donald MacMillan, a significant but underappreciated figure from Arctic history, gained early experience during the Crocker Land Expedition of 1913–1917. In the Eurasian Arctic, the one sizable landmass still unknown to science, Severnaya Zemlya, was charted by tsarist Russia's last polar undertaking, the Arctic Ocean Hydrographical

Expedition (1910–1915), led by Boris Vilkitsky and charged with surveying the Northern Sea Route from the Bering Strait to the Yenisei River. New technological possibilities made their mark as well. Not only had the transition from sail to power-driven vessels been underway in the Arctic since mid-century, special designs were coming into vogue. Ships like Nansen's *Fram* and Amundsen's *Gjoa* were constructed particularly for polar voyaging, and Russian mariners commissioned some of the first modern icebreakers, including the *Yermak*, built at the request of Admiral Stepan Makarov in 1897–1899. Russia similarly fitted other ships out for Arctic research, such as the *Taimyr* and *Vaigach*, which spearheaded Vilkitsky's hydrographical expedition.

Even the Arctic skies were opening up. In 1897, Salomon Andrée of Sweden made the first attempt to reach the North Pole by air, in the balloon *Eagle*, a quixotic flight that led to a fatal forced landing near the 83rd parallel. That tragedy notwithstanding, Ferdinand von Zeppelin, the airship king, came to Spitsbergen in 1910 to test out the dirigible's potential as a vehicle for high-latitude aviation. In 1914, Russia's Jan Nagurski, flying the *Pechora* to 76°N, became the first person to pilot an airplane above the Arctic Circle. Sooner than most at the time imagined, both airships and airplanes would assume outsized roles in Arctic exploration and development.

Air, undersea, and space: mapping the Arctic in the 1900s and beyond

During the interwar period and afterward, Arctic exploration grew steadily less centered on straightforward geographical discovery, and more heavily involved with supporting useful occupation of the region. In 1930, the journal *Science* proclaimed that polar exploration had "ceased to be a blind and adventurous wandering into the unknown". "Scientific laws", the article continued, were now "the really big game of the polar hunt".[4] So too were the imposition of governmental authority, the extraction of resources, and the creation and expansion of infrastructural networks, both for transportational and military purposes. Although work in the Arctic still posed sizable risks and entailed sizable efforts, it was, throughout the century, becoming routinized, mechanized, and grander in scope.

During the 1920s and 1930s, anxieties over Arctic sovereignty heightened, as most of the circumpolar map now lay uncovered, and more of the islands recently charted seemed both accessible and strategically or economically desirable. The presence of coal, diamonds, gold, an array of industrially useful metals, and, coming to prospectors' notice between the 1920s and the 1940s, oil and gas – all coupled with the traditional importance of fishing grounds – sharply raised political stakes in the north. The scientist Lord Kelvin had once predicted that the Arctic Ocean might someday be transformed into a "polar Mediterranean", and now, with varying degrees of intensity and success, nations possessing Arctic territory embarked on campaigns of full-scale northern development.[5] Vilhjalmur Stefansson, a proponent of polar industrialization, adopted Kelvin's motto and added his own – the "friendly Arctic" – as he advocated for the modernization of the Canadian north, and while his government may not have shared all his enthusiasms, national institutions like the navy and the Royal Canadian Mounted Police (RCMP) pushed higher and further into the Arctic to make it more Canada's own. While other countries followed suit, none surpassed the new Soviet state in Russia when it came to fusing exploratory efforts with gigantic developmental initiatives, particularly with the formation during the 1930s of the Main Administration of the Northern Sea Route. This, under Otto Schmidt, the so-called "commissar of ice", controlled not just polar research and exploration, but all non-military and non-secret police functions along Russia's Arctic coast and in all territory east of the Ural mountains and north of the 62nd parallel – leading one foreign journalist to compare it to a socialist version of the British East India Company.[6]

Several technologies immensely extended the operational range and mapping capacity of Arctic explorers during the interwar years. One was radio, which allowed mutual communication among explorers and maintained contact between expeditions and the mainland. Radio's feasibility was demonstrated by Donald MacMillan, who included wireless operator Don Mix on an expedition of 1921, and its usefulness became immediately and universally evident. In the 1930s, Soviet radioman Ernst Krenkel turned himself into a national celebrity, serving as the voice of the famed "Cheliuskinites" – castaways dramatically airlifted from the pack ice in 1934 after their vessel sank off the Siberian coast – and the SP-1 expedition, which established a four-man research outpost near the North Pole in 1937.

Even more revolutionary was the advent of aerial exploration, which despite obvious risks, eased the act of travel itself and aided explorers in their task of surveying, photographing, and mapping vast amounts of territory with greater accuracy than ever. While naysayers from the "golden" age of polar adventuring disparaged Arctic flying as insufficiently heroic ("it was a damn sight harder when we did it!" complained Matthew Henson, who trekked northward with Robert Peary),[7] most leaped at the opportunity to exploit this new asset. For some time, the airship and the airplane appeared to compete over which would emerge as the premier technology for Arctic work. Nansen favored the airship, founding an international organization in 1924 to sponsor its use – the famed Aeroarctic – and for now, airships had the advantage of range and cargo capacity, while also making excellent platforms for aerial photography and the collection of meteorological and aerological readings. Roald Amundsen, Lincoln Ellsworth, and Umberto Nobile overflew the North Pole in the dirigible *Norge* in May 1926, and the German pilot Hugo Eckener, working with Russian scientist Rudolf Samoilovich, masterfully demonstrated the *Graf Zeppelin*'s potential as a tool for research during a meticulously-executed flight to the Soviet Arctic in 1931.

On the other hand, airplanes were more nimble than airships and far more versatile for the purposes of developmental and infrastructural work. It was Richard Byrd's airplane, the *Josephine Ford*, that narrowly beat the airship *Norge* to the Pole in May 1926 (although some dispute Byrd's claim to have reached his target), and the catastrophic wreck of Nobile's *Italia* in 1928 darkened the airship's prospects in the Arctic – a debacle compounded by the death of Amundsen, who took part in the effort to save the *Italia*'s downed crew. Dirigibles in general fell out of favour after the *Hindenburg* disaster of 1937, and in the Arctic, the future belonged to the airplane. Pilots of many nations scouted new routes, wrestled with the peculiarities of cold-weather flying and navigation so near to the magnetic North Pole, and aimed to break records whenever possible – from Byrd and Amundsen to bush pilots like Carl Ben Eielson of Alaska and the cadre of Soviet aviators who lent wings to Stalin's subjugation of the Arctic. Valery Chkalov, hailed as the "Russian Lindbergh", and Mikhail Gromov each broke the world distance-flying record in 1937 by flying over the North Pole from Moscow to the west coast of the United States. Also in 1937, a Soviet air expedition led by Mikhail Vodopianov landed four scientists within twenty miles of the North Pole to establish the SP-1 drifting research station.

At the same time, the Soviets took another technological lead, deploying the world's most powerful icebreaker fleet, including the 10,000-ton *Stalin*, which entered service in 1939. In addition to icebreakers, specially-fitted vessels, such as the Soviet ice-forcer *Sadko*, followed the example of Nansen's *Fram* and Amundsen's *Maud*, embarking on high-latitude drifts that significantly benefitted sciences such as oceanology, oceanography, geophysics, and meteorology.

These fields, along with glaciology, received particular emphasis during the Second International Polar Year (1932–1933). They were pursued by scientists working at the dozens of polar stations erected in the 1920s and 1930s throughout the region. As before, ethnographers played their own part in strengthening Euro-American familiarity with Arctic geography. The standout during these years was Knud Rasmussen, whose wide-ranging Thule expeditions helped to solidify the now-accepted wisdom that the Eskimo-Inuit peoples of North America and Greenland arrived from Asia and migrated eastward across the Arctic coasts. Danger still loomed, even as Arctic work became more regularized: high-profile deaths included that of Alfred Wegener on the Greenland ice in 1930 and Rasmussen's during the Seventh Thule Expedition in 1933.

The militarization of the circumpolar north during World War II and the Cold War brought the Arctic more than ever before within the intellectual compass of Euro-American outsiders. The infrastructural spike caused by the world war – the expansion of Russia's Northern Sea Route, the construction of the Alaska-Canadian Highway, the establishment of new harbours and airfields from Greenland to the Aleutians – was followed by an even deeper commitment by the Cold War superpowers to a permanent, high-tech military presence in the Arctic. Installations such as the Distant Early Warning (DEW) Line in North America and the sprawling naval bases that housed the Soviet Union's Northern Fleet on the Kola Peninsula vividly illustrate this inclination. Also motivating the acquisition of new Arctic knowledge was awareness of the large reserves of oil and natural gas hidden underground and undersea throughout the high north, be it in Alaska, Siberia, or off the coast of Norway.

New research facilities appeared throughout the region – the Naval Arctic Research Laboratory in Barrow, for example, or the "SP" series of drifting stations resumed by the USSR in 1950 – and the International Geophysical Year (1957–1958) further intensified polar scientific efforts. As this postwar wave of research surged on, one key trend involved a series of new "firsts", mainly having to do with travel to the North Pole in one form or another. The first airplane flight universally considered to have landed at 90°N was carried out in 1952 by William Benedict and Joseph Fletcher of the US Air Force (although many Russians consider Alexander Kuznetsov to have flown there successfully in 1948). In 1958, the US Navy's submarine *Nautilus* became the first vessel to reach the Pole. The Soviets sent the first surface vessel to the Pole, the atomic icebreaker *Arktika*, in 1977. In 1968, a snowmobile expedition led by America's Ralph Plaisted and Canada's Jean-Luc Bombardier completed the second overland journey to the Pole – or the first, depending on what one believes about Peary's trek in 1908–1909.

More profoundly, however, postwar Arctic science unlocked new frontiers as it probed the region's remaining physical secrets. One new instrument, adding literally another dimension to the field of polar exploration, is the submarine, along with other submersible vehicles and devices, both manned and remote-controlled. Submarines offered physical access to the Arctic depths immediately after World War II, and on a smaller scale, submersibles have enabled the gathering of oceanographical and oceanological data in ways and in quantities undreamed of before, except in the pages of Jules Verne's *Twenty Thousand Leagues under the Sea*: photographs, soundings, temperature and salinity readings, seabed maps, and more.

Mapping backward in time – into a fourth dimension, so to speak – has been made feasible by advances in sciences such as dendrochronology, paleobotany, and, most germane to the polar realms, deep ice-core sampling – used to especially good effect in the Greenland ice sheet. Such pictures from the distant past shed light on the Arctic's present and provide hints as to its future. Arguably the most transformative aspect of Arctic exploration in recent decades has involved the growing capacity to survey and monitor the region from above, by means of high-altitude and space-based telemetry. Satellite imagery of the Arctic became available as early as 1959, but true breakthroughs emerged with more sophisticated tools like Landsat, run by NASA upon its development in 1972 and now used by the National Ocean and Atmospheric Administration

(NOAA). Such macroscopic perspectives have allowed Arctic research to proceed on a truly global scale – among other things, helping to trace the rapid progress of climate change. To the extent the future of Arctic exploration can be predicted, the most important developments are likely to grow out of innovations high up in space and deep below the ocean's surface.

Notes

1 Cited in Fred Bruemmer, *The Arctic World* (New York, 1985), 94.
2 Cited in John McCannon, *A History of the Arctic* (London, 2012), 94.
3 Karl Weyprecht, "Scientific Work of the Second Austro–Hungarian Polar Expedition", *Journal of the Royal Geographical Society of London* XLV (1875): 32–33; Clements Markham, "The Promotion of Further Discovery in the Arctic and in the Antarctic Regions", *The Geographical Journal* IV/1 (1894): 7.
4 Isaiah Bowman, "Polar Exploration", *Science* LXXII/1870 (1930): 441.
5 W. T. Kelvin, *Popular Lectures and Addresses* (Cambridge, 2011), 282.
6 H. P. Smolka, *Forty Thousand Against the Arctic* (New York, 1939), vii–x.
7 *Polar Times* 5 (October 1937): 2.

References

Armstrong, T. E. 1958. *The Russians in the Arctic: Aspects of Soviet Exploration and Exploitation of the Far North, 1937–57*. London: Methuen.
Armstrong, T. E., G. W. Rogers and G. Rowley 1978. *The Circumpolar North: A Political and Economic Geography of the Arctic and Sub-Arctic*. London: Methuen.
Baird, P. D. 1965. *The Polar World*. New York: Longmans Green and Co.
Berton, P. 1988. *The Arctic Grail: The Quest for the North West Passage and the North Pole, 1818–1909*. Toronto, ON: Anchor Canada
Bruemmer, F. 1985. *The Arctic World*. New York: Firefly Books.
Glines, C. V. 1964. *Polar Aviation*. New York: F. Watts.
Hall, S. 1987. *The Fourth World: The Heritage of the Arctic and its Destruction*. New York: Alfred Knopf Inc.
Kirwan, L. P. 1959. *A History of Polar Exploration*. New York: W.W. Norton and Co.
Lincoln, W. B. 1993. *The Conquest of a Continent: Siberia and the Russians*. New York: Random House.
Lopez, B. 1986. *Arctic Dreams: Imagination and Desire in a Northern Landscape*. London: Macmillan.
McCannon, J. 1998. *Red Arctic: Polar Exploration and the Myth of the North in the Soviet Union, 1932–1939*. Oxford, UK: Oxford University Press.
McCannon, J. 2012. *A History of the Arctic*. London: Reaktion Books.
McDougall, W. A. 1993. *Let the Sea Make a Noise: A History of the North Pacific from Magellan to Macarthur*. New York: Basic Books.
McGhee, R. 2004. *The Last Imaginary Place: A Human History of the Arctic World*. Toronto, ON: Key Porter Books.
Nuttall, M. (ed.) 2005. *Encyclopedia of the Arctic*. 3 vols. London and New York: Routledge.
Officer, C. and J. Page. 2001. *A Fabulous Kingdom: The Exploration of the Arctic*. Oxford, UK: Oxford University Press.
Robinson, M. F. 2006. *The Coldest Crucible: Arctic Exploration and American Culture*. Chicago, IL: University of Chicago Press.
Vaughan, R. 1994. *The Arctic: A History*. Phoenix Mill, UK: A. Sutton.

2

Exploring and mapping the Antarctic

Histories of discovery and knowledge

Ursula Rack

Introduction

Discovery and knowledge of the Antarctic has been acquired through activities inspired by a range of different reasons and motives. Antarctica is and has always been a place of and for the imagination, a place of personal challenges, and a site for national, territorial and economic ambitions; however, research and science have become more evident and entrenched over time as the principal activities for nation states with interests in the continent. Modern science developed from the Enlightenment onwards and inspired many researchers to pursue an evolving understanding of our planet. Antarctica became a focal point where exploration and science, especially over the past century, have been pursued for purposes of discovery and knowledge as opposed to activities driven by economic reasons. As such, Antarctica has been marked out as a continent for science. In the nineteenth century and early twentieth century, expeditions to Antarctica, often supported by learned societies, brought back knowledge which greatly enhanced understanding of the continent, resulting in new maps and charts and raising yet more questions about its geography and ecology which, in turn, inspired further exploration. This chapter discusses moments of discovery and their motives, the acquisition of knowledge and understanding of one of our world's most intricate and dramatic natural systems, and the ways in which various geographic and scientific approaches to the continent have developed over time. In particular, the history of discovery and knowledge of Antarctica is explored through a consideration of mapping and expeditions from the earliest times, as well as the people behind these activities. Our view of Antarctica today is partly a product and reflection that has been inherited from the earlier days of exploration. Images from the early Antarctic photographers such as Herbert Ponting (1870–1935) and Frank Hurley (1885–1962), for instance, still form a basis for many people's ideas of Antarctica as a pristine and fascinating place.

The earliest concepts of and about Antarctica were Greek attempts to "balance the world" before there was any knowledge of its scale and its dimensions. This did not involve cartography as much as imagination and speculation. When early explorers reached the waters around Antarctica, this idea was replaced by a need for maps that attempted to be true and accurate geographical representations to enable navigation in the mercantile search for riches. In other words, economics became a key driving force for mapping. The Age of Enlightenment, from about the early eighteenth century, raised the first scientific questions as we know them today. From the early nineteenth century, learned societies became more powerful and supported the specification and practice of scientific disciplines. The need for and production of accurate maps took place by means of conquering and colonizing new lands, with economic and imperialistic motives as driving factors. Alongside nationalistic motives, France, the United States and Britain dispatched scientific expeditions from the 1840s to the far corners of the world, to claim land and obtain control of maritime traffic routes, but also to acquire knowledge of, for example, magnetism, meteorology and geography. At the same time, geographical societies became more powerful and not only oversaw map making and the collection of observations and curation of objects, but also supported the development of the specification of science disciplines.

In the late nineteenth century, the need for coordinated research approaches in Antarctica came to be recognized as both necessary and as a priority and early research initiatives were instigated largely by national geographical societies. Antarctic expeditions were sent South for varying motives, such as attaining and conquering the South Pole and also to undertake a wide range of research activities. This variability of approach continued through the great depression and two world wars until the International Geophysical Year (IGY) of 1957–1958 cemented the modern approach to research in the Antarctic. This chapter will show the development up to this point and introduce some of the events and individuals that are closely linked to our understanding of the Antarctic today.

Early mapping of Antarctica: from imaginings to the first observations

Early maps of Antarctica began with the ancient Greeks. These maps were clearly not based on observations at high southern latitudes but represented imaginings of what might be in and at such a location. One theory was that there should be a continent in the far South in order to, in some way, balance or counterbalance the presence of the huge Eurasian landmass in the northern hemisphere. Indeed, this imagined southern continent, which some thought was abundant in life and wealth, and inhabited by strange creatures, persisted in a number of European maps of the world through the medieval and early modern period.

It was not until the circumnavigation of Antarctica by Captain James Cook, during his second voyage of 1772–1775, and his reports of a frigid world where ice and high winds dominated the climate of a southern ocean, that the dream of a rich continent was banished. Cook did not reach or even see the Antarctic coast (Figure 2.1). Although his most southerly point was 60°2′S, he did not see a reason for going further. He wrote: "I will not say it was impossible anywhere to get in among this Ice . . . but the bare attempting it would have been a dangerous . . . enterprise",[1] and he proceeds: "whoever has resolution and perseverance to clear up to this point by proceeding farther than I have done, I shall not envy him the honour of the discovery but I will be bold to say that the world will not be benefited by it".[2]

European and American sealers and whalers, however, saw economic potential in the waters around the Antarctic from about the 1790s and during the early decades of the nineteenth century, as their traditional hunting grounds in the Arctic became progressively less profitable due to over-exploitation. Mapping also went along with increasing marine mammal hunting activity in Antarctic waters. Much of this early geographical knowledge was, however, kept secret as it conveyed economic advantage to the sealers and whalers.

In contrast to this atmosphere of secrecy was the initial growth of scientific interest in Antarctica. One example is the German whaler and sealer Eduard Dallmann (1830–1896). Dallmann undertook a whaling expedition on behalf of the Hamburgian businessmen Albert

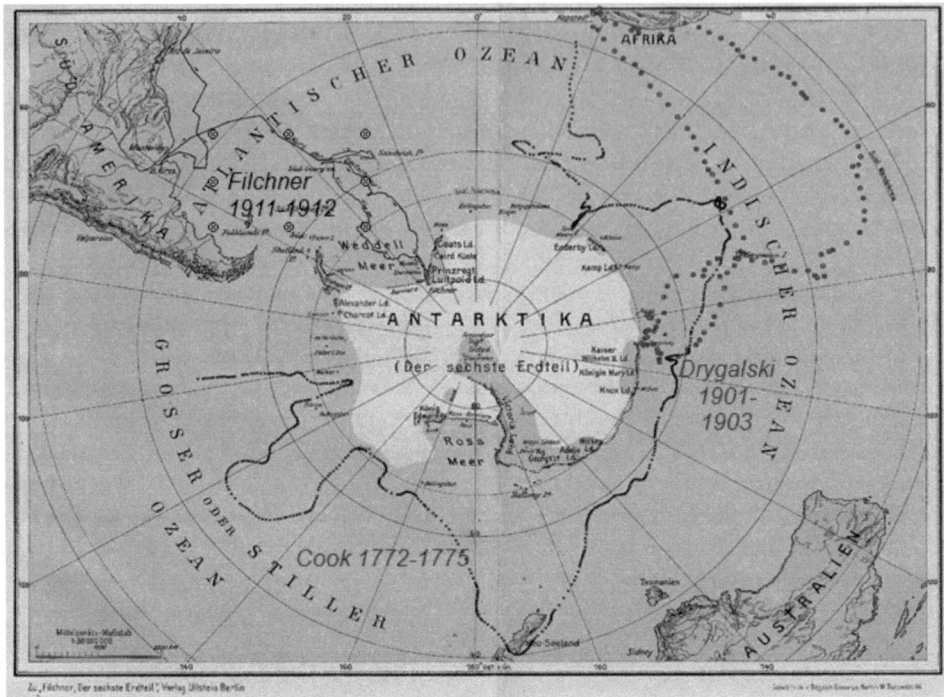

Figure 2.1 Map from Filchner's narrative, 1923, with modified coordinates (by Dr. Wolfgang Rack) for Cook's second voyage around the Antarctic, the Drygalski expedition, and the Filchner expedition.

Rosenthal (1828–1882), to take the German whaling business into the Antarctic in 1873–1874. The expedition only caught one whale but became profitable through sealing; the idea of a German whaling industry in the Antarctic was abandoned, however. Nevertheless, the expedition was successful in geographical terms, because Dallmann discovered and charted many islands (e.g. Deception Island, Liege and Kaiser-Wilhelm Islands) offshore of what is now known as the Antarctic Peninsula and also mapped parts of the peninsula's coastline.[3] Dallmann's discoveries were added to the "Südpolar-Karte" (South Polar Map) and published and made accessible in Petermanns Geographische Mitteilungen 1875. In this way, Dallmann's whaling expedition supported the advancement of knowledge.

Nineteenth-century expeditions: from piecemeal science to co-operation and the influence of geographical societies

The need for accurate maps fitted well with national aims of exploring new land and becoming increasingly interested in scientific knowledge. Beginning in the 1820s, national expeditions were operating in Antarctic waters to discover and consequently claim land on that continent for their respective countries, and to search for iconic but scientifically significant locations such as the South Magnetic Pole. In the 1840s Britain, France and the United States all sent expeditions to the South. James Clark Ross (1800–1862) for Britain, Jules Dumont d'Urville (1790–1842) for France, and Charles Wilkes (1798–1877) for the US, each led expeditions which claimed islands in the sub-Antarctic and fulfilled a range of scientific work. These expeditions contributed to improving and extending maps and to the understanding of the Antarctic. In the course of these activities some new research disciplines emerged. One of these was geography.

Geography became an instrumental link between science, economy and territorial claims in support of colonialism and imperialism.[4] In Germany and Britain, it became an independent discipline in the university systems in the 1860s. At the same time geographical societies became more powerful in official decision-making from the mid-nineteenth century onward, linked to the exploitation of the wealth of the colonies of European nations and to the strengthening of territorial claims. However, over time the geographical societies also became interested in different science disciplines. The connection between land claims, military action, economy and politics is evident for this time; however, global scientific interests, mainly in meteorology, magnetism, oceanography, geology and biology appeared to bridge these disparate interests to a certain extent. Increasing technical developments in measuring instruments as well as more effective transport were important for increasing momentum, as the first expeditions with a detailed scientific programme headed South by the end of the nineteenth and beginning of the twentieth century, supported by geographical and other learned societies.

The origin of international collaboration to explore the Polar Regions lies in the First International Polar Year (IPY) of 1882–1883. Karl Weyprecht (1838–1881), the leader of the Austro-Hungarian Arctic expedition in 1882–1874, recognized that useful observations of temperature, wind, pressure, radiation, evaporation, clouds and humidity could only be achieved from scientific stations, rather than from moving ships and through seeking the geographical poles. In the first IPY two stations were established in the southern hemisphere: a German station on South Georgia and a French station at Tierra del Fuego. Georg von Neumayer (1826–1909), a German geophysicist and meteorologist, was not satisfied that only two stations undertook observations in the sub-Antarctic and southernmost South America. His goal was to extend scientific observations onto the Antarctic continent. At the Sixth International Geographical Congress in London in 1895, Neumayer gave a passionate and detailed presentation on the importance of Antarctic research. An intense meeting followed, which Sir Clements Markham (1830–1916), president of the Royal Geographical Society, opened with these words:

> The completion of such a work as that outlined in Dr. Neumayer's admirable paper will result in one of the grandest discoveries of the nineteenth century, and our warm and most hearty thanks are due to him for the able manner in which he has brought together all the scientific results which will accrue therefrom.[5]

At the congress a rough plan was outlined and three major scientific problems were identified:

> In the first place, the key to the future knowledge of terrestrial magnetism lies in the determination of the exact position of the south magnetic pole; [. . .] The second of these great problems is the meteorology of the Antarctic area, of which we know the barest outlines only. The third problem is the geology of the region in question.[6]

After a plea that this "Congress will be the means of inducing an international co-operation in polar discovery"[7] a resolution was agreed to investigate the Antarctic region. Neumayer and Markham and the responsible geographical societies in Germany and Britain promoted the collaboration of two expeditions which were planned to begin in 1901.

However, before the expeditions were finally undertaken, extensive preparations were required. There were perceptions and reports on how life in the ice would be, derived from previous experiences, mainly in the Arctic. The idea of the "type of man", who should undertake such endeavours in the South, should be considered in the following context. A number of practical problems needed to be taken into account when planning the expeditions. The *Antarctic Manual*[8] was written for Captain Scott's British expedition in 1901–1904 based on the existing *Arctic Manual*. As pointed out in its preface, these instructions included "much suggestive information".[9] Further advice in the manual came from extracts of expedition narratives such as those of Jules Dumont d'Urville, John Biscoe (1794–1843), and Charles Wilkes. For sledges travelling over ice and snow, the accounts of Sir Leopold McClintock (1819–1907) based on his Canadian Arctic experiences were also incorporated into the *Antarctic Manual*. These are examples to demonstrate the vision the organizers of these Antarctic expeditions had and how they compared the South with knowledge they already had from the North. The Arctic was also chosen as a training ground in preparation for Antarctic expeditions. For example, Wilhelm Filchner (1877–1957) travelled in Spitsbergen in 1910 before he undertook the Second German Antarctic Expedition of 1911–1912. Many members of Antarctic expeditions, however, had little or no experience of what could be challenging in Antarctica.[10] It was a general expectation of the time that men would strive and conquer the elements and circumstances with or without training.

The Polar Regions were seen as a place of manhood, camaraderie and vigour. The aspect of camaraderie was a strong focus for perceived Edwardian virtues of forbearance and sacrifice that became pervasive during the time leading up to and including the Great War. The deaths of Robert F. Scott (1868–1912) and his four companions, and particularly of Captain Lawrence Oates (1880–1912), were regarded as noble acts of sacrifice for comrades, King and Country. Another factor was that the Antarctic had no indigenous people to fight against and subdue in contrast to most colonies; the environment, however, was harsh enough for men to "prove" themselves. Another idea was that the Antarctic could also be a place for returning to "a second childhood"[11] in the sense that the daily constraints from home were loosened and the "adventurer" could appear. This debatable romantic perception was carried in media articles and novels. The explorers themselves strengthened this vision in their published narratives, as stated in Douglas Mawson's (1882–1958) account of his expedition of 1911–1914. The foreword of his narrative reads: "Science and exploration have never been at variance; rather, the desire for the pure elements of natural revelation lay at the source of that unquenchable power the 'love of adventure.'"[12] A part of Mawson's story is that of his survival alone after he became separated from his two companions. With little food left, and no dogs, he cut his sledge in half to be able to pull it and, ill and exhausted, he reached the hut where his companions welcomed him. They nursed him back to health, and he was then able to undertake some of the expedition's scientific observations as planned.

Despite these complex social and psychological perspectives, science was the focus according to the *Antarctic Manual*. The manual gave a clear indication of what work was expected to be undertaken in the Antarctic to fulfil scientific missions. Instructions for taking measurements in many research areas were shown in the table of contents: astronomy, tidal observations, pendulum observations, terrestrial magnetism, Antarctic climate (e.g. Drygalski's meteorological station, Figure 2.2), wave observations, Aurora, atmospheric electricity, chemical and physical notes on the instruments, geology, volcanoes, ice observations, zoology, botany, and instructions for collecting rocks and minerals. It was described how to use certain instruments and to note measurements systematically. The same instructions were to be followed by expeditions which took place simultaneously in order to provide comparability.

The German expedition led by Erich von Drygalski (1865–1949) and the British expedition under Robert F. Scott were finally underway in 1901–1904.[13] A Swedish expedition,

Figure 2.2 Drygalski's ship, *Gauß*, with weather station in the foreground. ("Gauß" von
Süden [*Gauß* from the south] 12.09., Leibniz Institute for Regional Geography,
Archive for Geography, DSE Collection, Dry 1768).

led by Otto Nordenskjöld (1869–1928), and a Scottish expedition led by William S. Bruce
(1867–1921) were also involved in the wider scientific programme. Finally, a French expedition, led by Jean-Baptiste Charcot (1867–1936), contributed to the international collaboration.
One of the most comprehensive scientific outcomes was delivered by the German expedition
and was published in twenty volumes covering the results from many scientific disciplines and
incorporating observations from the other four expeditions for comparison.

The Heroic Age (1901–22): national aspirations, personal ambitions and science

Governments became increasingly interested in reaching the South Pole and opening up new
territory for imperialistic reasons and tended only to fund scientific work when this aim might
be achieved. The so-called "Heroic Age" of Antarctic exploration was, on one hand, marked
by these national interests, represented by individuals such as Scott and Roald Amundsen
(1872–1928), and, on the other, by the desire of scientists for a better understanding of the
Antarctic. Several national expeditions visited the Antarctic in the first two decades of the twentieth century, but international collaboration beyond these expeditions was not achieved as in
the five expeditions mentioned above. Nonetheless, a number of scientists were in co-operation
internationally and played important intellectual roles in the expeditions of other nations.

The first decade of the 1900s was marked by what has been termed "the race to the Pole".
There was, however, scientific reason to reach the South Pole and to traverse the entire Antarctic

continent, in terms of gaining knowledge of the geography of Antarctica, much of which was still completely unexplored. Different ideas of Antarctica's makeup were current. Filchner's theory was that Antarctica was an enormous rigid plate covered by a thick ice sheet. Sir John Murray (1841–1914) and Ernest Shackleton (1874–1922) assumed that the Antarctic was a single landmass, covered by ice. Fridtjof Nansen (1861–1930) argued for an accumulation of islands, building a sort of atoll. After the Swedish expedition in 1901–1904, Otto Nordenskjöld and Gunnar Andersson (1874–1960) hypothesized that Antarctica was connected to the South American Cordillera. Markham, Nordenskjöld, and Albrecht Penck (1858–1945) also proposed that the Antarctic was divided into West- and East-Antarctica by an arm of sea.[14]

Driven by this discussion, Filchner organized the "Second German Antarctic Expedition" in 1911–1912. Filchner was a Bavarian army officer and aspired to bring Germany to the fore through novel Antarctic exploration and research and engaged a broad spectrum of researchers. Unfortunately, some of his expedition members had no previous polar experience and no training in the cold regions and consequently struggled with the harsh Antarctic environment. At some time during the long austral night, the experienced Norwegian ice pilot Paul Bjørvik wrote in his diary:

> I have heard that the mood during overwintering should become not good. I, for myself, did not observe that in all the times I overwintered. But those on board living astern, are possibly affected by the Austral night, although it has just begun. Because, when I meet them on the ice, they look at each other like cattle with the only difference that cattle roar and these here are silent.[15]

To make matters worse, when Filchner's ship *Deutschland* was trapped in the ice, the relationship between the expedition members deteriorated to one of controversy and mistrust. After ship's captain Richard Vahsel (1868–1912) died, the men divided into two groups and the rivalry made life on board almost unbearable. Some feared for their lives and slept with loaded guns in their cabins. When they finally broke free of the ice and reached Grytviken in South Georgia, mutiny broke out and the expedition came officially to an end.[16] Somewhat surprisingly, given the difficult circumstances, each expedition scientist fulfilled the scientific programme they were involved in and produced remarkable results in meteorology, geography and oceanography which were published in the relevant scientific communities. Discoveries including the Filchner-Ronne Ice Shelf and Prinzregent Luitpold Land brought new knowledge to the topography of the Antarctic. Deep-sea oceanographic measurements were a major research area, and the Filchner Trench was discovered.

Another important expedition was undertaken by the Australian Douglas Mawson. To obtain sufficient funding for this 1912–1914 expedition, Mawson emphasized the economic benefits of exploring for minerals. More broadly, however, he designed an ambitious scientific programme focused on magnetism, biology, geology and meteorology. A geologist himself, he aimed to finish what he had begun on Shackleton's *Nimrod* expedition of 1907–1909. Mawson's expedition was not only epic in terms of his personal survival but was also one of the most successful scientific expeditions of its time.

The "Heroic Age" was also the time of "firsts", and Mawson's expedition could claim the first, although limited, use of a radio in the Antarctic. The expedition was able to receive news on a regular basis via a relay station on Macquarie Island which provided a link to the outside world. Mawson engaged Frank Hurley, a talented photographer who later joined Shackleton's *Endurance* expedition of 1914–1917. Hurley's photographs, together with those taken by Herbert Ponting, who had gone South with Scott, provided previously unimagined images of

Antarctica, both of its unique landscape and of the men who were attempting to 'conquer' it. Such images were vital in visualizing the stories of Scott's, Shackleton's, and Mawson's expeditions in an engaging way – today, these early photographs of Antarctica and Antarctic exploration remain iconic.[17]

Shackleton's death on South Georgia during his *Quest* expedition in 1922 marked the end of the Heroic Age expeditions. Governments and economists were more focused on rebuilding and re-establishing the world order after the chaos caused by the Great War. For many, Antarctica now seemed too far away to be a place of importance.

A time of higher priorities: the Interwar period and World War II

The Pole had been achieved in 1911, and in a post-war time where funds were in very short supply, both governments and individuals, whose wealth was being rapidly eroded, were uninclined to support further southern ventures. Nevertheless, there was a desire to keep science alive. In Germany, the "Notgemeinschaft der Deutschen Wissenschaft" (Emergency Association of German Science) was established in 1920 in order to, as stated in their founding document, "prevent the danger of total collapse of German scientific research due to the economic crises". With very limited resources, the "Notgemeinschaft" supported the *Meteor* expedition of 1925–1927 which systematically examined large parts of the Southern Atlantic Ocean's floor by depth sounding. One result was a bathymetric map that gave new insights into the morphology of the ocean bed.

Some better-supported expeditions also took place, utilizing new technologies to widen the scope of research. In particular, rapid developments in aviation led to the first aerial sorties into Antarctica and the beginning of mapping using aerial photographs. Scientists also became more dependent on external logistical support.[18] The American Richard E. Byrd (1888–1957) organized two Antarctic expeditions in 1928–1930 and 1933–1935 which reflected the new opportunities provided by aeroplanes, aerial cameras, snowmobiles, and radio-communication equipment. The interior of the 13 million square kilometre continent could be explored and mapped in a much more efficient way with the new aerial survey techniques, first developed for military purposes in the Great War. As well as work with aircraft, scientists on Byrd's expedition undertook the first seismic measurements in the Antarctic for gathering data on ice thickness, a technique that had already been used successfully on the German Greenland Expedition in 1930–1931. In addition to these scientific efforts, Byrd's expedition also had a political significance, with the US signalling its first significant interest in the continent since the 1840s.

Britain was concerned about the growing American interest in Antarctica, and British claims in parts of the Antarctic were reinforced through the British Graham Land Expedition (BGLE; 1934–1937). Funding was meagre, but the scientists were enthusiasts to the point that they even worked without payment for the duration. The programme of science and exploration of the BGLE was ambitious. An aeroplane was used for reconnaissance and surveying, although dog sledging remained the main method of transport. Initial aerial surveys suggested that a group of islands existed in the area called Graham Land. Detailed land-based surveys with travel using dogs, however, demonstrated that Graham Land was a peninsula – the Antarctic Peninsula. Along with this major geographical discovery, many geological and biological samples were collected by the BGLE, building an important addition to the scientific knowledge of the continent.[19]

Economic motives and scientific approaches were often closely linked during the interwar period, in some cases causing political concerns. An example is provided by the whaling industry, which was an ongoing issue for Norway and Britain. The demand in Germany for whale oil was immense, among other requirements, for supplying the margarine market. The British government put pressure on Norway to stop delivering whale oil to Germany.[20] The German government's attempts to establish their own whaling industry in the far South, however, interfered with British and Norwegian interests. The German *Schwabenland* expedition went South in 1938 to secure whaling grounds and fulfil a scientific programme. Securing whaling grounds was largely unsuccessful, but the expedition members managed a prodigious amount of geographical exploration of a previously unmapped sector of Antarctica in only three weeks through the use of airborne reconnaissance and photographic flights over what is now known as Dronning Maud Land. Unfortunately, their findings were not recognized due to the outbreak of World War II[21] and because the results were published in German, which meant that they only became recognized by the wider scientific community decades later.

Post-war exploration and the Antarctic Treaty: Cold War interests and emerging international collaboration

Byrd continued exploring the Antarctic during and after World War II and led several expeditions until his death in 1957. However, after World War II, the Cold War affected a divided world. Two powerful blocs, driven by national ambitions and ideology combined with economic motives, each worked to establish a bridgehead on the Antarctic continent. Scientists from the Western and Eastern Blocs were, however, a significant driving force behind the international collaboration of the International Geophysical Year (IGY) of 1957–1959. The exceptional international collaboration and research activities that marked the IGY, involving Western and Soviet scientists, can be considered as the beginning of the way in which Antarctic research is conducted today under the auspices of the Antarctic Treaty System (ATS) and the Scientific Committee on Antarctic Research (SCAR). Today, permanently occupied, summer-only and unmanned automatic stations over much of the Antarctic continent are delivering continuously data on a wide range of research topics. New technologies such as satellite-based monitoring allow novel ways of imaging the Antarctic. These methods contribute to the acquisition of knowledge on the continent and, in turn, its influence on global climate. New discoveries are made regularly by undertaking research in this unique place which helps us to obtain a better understanding of our world – the discovery of the ozone hole is one clear example.

Paradoxically, the atmosphere of negativity and confrontation during the Cold War coincided with a positive outcome for Antarctica in terms of empowering research in an international context during the IPY that was sustained with the establishment of the Antarctic Treaty System in 1959 (see Chapter 23 by Dey Nuttall and Chapter 20 by Dodds in this volume).

Conclusion

Perceptions of Antarctica have changed dramatically over the last two millennia. Although our understanding of this exceptional place still presents many challenges, it remains influenced by the early explorers and scientists who discovered and mapped the continent and engaged in the first scientific pursuits there. Their observational records are still valuable for our contemporary research questions, especially in providing a baseline against which to measure present and future environmental change. The meteorological measurements of early Antarctic explorations are a clear example, providing the first points on a graph of continuing climate change and variability.

In the nineteenth and early twentieth centuries, big exploration and science questions about the earth system were important, although they were far from being the only stimuli for sending ships and men South. Those men had their own perceptions of the space that was and remains Antarctica, and presented the continent in a way that still affects us today. The history of discovery and knowledge in the Antarctic has just begun and exploring and mapping remains an ongoing process – we still know little of the processes taking place beneath the kilometres-thick Antarctic ice sheet or the dimensions of the huge water-filled cavities under the floating ice shelves (see Chapter 33 by Siegert in this volume).

With this in mind, Antarctica remains a place where political power and economic motives are a reality. Fisheries and the potential for minerals exploitation are issues that are presently managed by regulation under the Antarctic Treaty System and its environmental protocols. Some motives, however, have varied little since the beginning of Antarctic exploration, and international collaboration remains vital in the investigation the Antarctic environment and its global linkages today.

Notes

1 Beaglehole, J. C. 1961. *The Voyages of the* Resolution *and* Adventure *1772–1775. The Journals of James Cook on his Voyages of Discovery*, vol. 2. Cambridge, UK: Cambridge University Press, p. 323.
2 Beaglehole, p. 646.
3 See: Krause, R. and U. Rack (eds.). 2006. Schiffstagebuch der Steam-Bark Groenland geführt auf einer Fangreise in die Antarktis im Jahre 1873/1874 unter der Leitung von Capitain Ed. Dallmann (Logbook of the German Steam Bark Groenland written during a sealing and whaling campaign in Antarctica in 1873/1874 under the command of Captain Ed. Dallmann). In: Berichte zur Polar- und Meeresforschung (Reports on Polar and Marine Research) 530/2006. http://epic.awi.de/26705/1/BerPolarforsch2006530.pdf.
4 See: Hudson, B. 1972. 'The new geography and the new imperialism 1870–1918'. *Antipode* 9(2): 140–53.
5 Royal Geographical Society 1896. *Report of the Sixth International Geographical Congress*. London, https://ia601702.us.archive.org/1/items/jstor-196860/196860.pdf, p. 163.
6 ibid.
7 ibid.
8 Murray, G, (ed.) 1901. *The Antarctic Manual for the use of the Expedition of 1901*; London, https://archive.org/details/antarcticmanual00britgoog.
9 ibid. p. viii.
10 Rack, U. 2010. pp. 192–205.
11 Griffiths, T. 2007. *Slicing the Silence. Voyaging to Antarctica*. Cambridge, MA and London: Harvard University Press. p. 252.
12 Mawson, D. 2009. *The Home of the Blizzard: Being the Story of the Australian Antarctic Expedition, 1911–1914*. Ebook #6137, www.gutenberg.org/files/6137/6137-h/6137-h.htm#link2H_INTR.
13 See: Lüdecke, C. 2003. 'Scientific collaboration in Antarctic (1901–1903): a challenge in times of political rivalry'. *Polar Record* 39(1): 35–48.
14 See: Lüdecke, C. 1995. *Die deutsche Polarforschung seit der Jahrhundertwende und der Einfluß Erich von Drygalskis. (German polar research since the turn of the century and the influence of Erich von Drygalski)*. In: Berichte zur Polarforschung (Reports on Polar Research) 158/1995, Bremerhaven, Germany p. 59. http://epic.awi.de/26336/1/BerPolarforsch1995158.pdf.
15 Rack, U. 2010, p. 204 (translation by the author).
16 See: Rack, U. 2010, pp. 192–205.
17 See: Mundy, R. 2014. 'Pioneering Antarctic photography: Herbert Ponting and Frank Hurley'. *The Polar Journal* 4(2): 37–41.
18 See: Fogg, G. E. 1992. *A History of Antarctic Science*. Cambridge, UK: Cambridge Press. pp. 130–54.
19 See: Lintott, B. 2011. *The British Graham Land Expedition, 1934–1937*. Cambridge, UK: SPRI.
20 See: Rack, U. 2010. pp. 31–5.
21 See: Lüdecke, C. and Summerhayes, C. 2012. *The Third Reich in Antarctica. The German Antarctic Expedition 1938–39*. Norwich, UK: Erskin-Press.

References

Beaglehole, J. C. 1961. *The Voyages of the* Resolution *and* Adventure *1772–1775. The Journals of James Cook on his Voyages of Discovery*, vol. 2. Cambridge, UK: Cambridge University Press.

Filchner, W. 1923. *Zum Sechsten Erdteil. Die zweite deutsche Südpolarexpedition.* Berlin: Ullstein.

Fogg, G. E. 1992. *A History of Antarctic Science.* Cambridge, UK: Cambridge University Press.

Griffiths, T. 2007. *Slicing the Silence. Voyaging to Antarctica.* Cambridge, MA and London: Harvard University Press.

Hudson, B. 1972. 'The new geography and the new imperialism 1870–1918' *Antipode* 9(2): 140–53.

Krause, R. and U. Rack (eds.) 2006. *Schiffstagebuch der Steam-Bark Groenland geführt auf einer Fangreise in die Antarktis im Jahre 1873/1874 unter der Leitung von Capitain Ed. Dallmann (Logbook of the German Steam Bark* Groenland *written during a sealing and whaling campaign in Antarctica in 1873/1874 under the command of Captain Ed. Dallmann).* In: Berichte zur Polar- und Meeresforschung (Reports on Polar and Marine Research) 530/2006. http://epic.awi.de/26705/1/BerPolarforsch2006530.pdf.

Lintott, B. 2011. *The British Graham Land Expedition, 1934–1937.* Cambridge, UK: SPRI.

Lüdecke, C. 1995. *Die deutsche Polarforschung seit der Jahrhundertwende und der Einfluß Erich von Drygalskis (German polar research since the turn of the century and the influence of Erich von Drygalski).* In: Berichte zur Polarforschung (Reports on Polar Research) 158/1995, Bremerhaven, Germany.

Lüdecke, C. 2003. 'Scientific collaboration in Antarctic (1901–1903): a challenge in times of political rivalry' *Polar Record* 39(1): 35–48.

Lüdecke, C. and C. Summerhayes 2012. *The Third Reich in Antarctica. The German Antarctic Expedition 1938–39.* Norwich, UK: Erskin-Press.

Mawson, D. 2009. *The Home of the Blizzard: Being the Story of the Australian Antarctic Expedition, 1911–1914.* Ebook #6137, www.gutenberg.org/files/6137/6137-h/6137-h.htm#link2H_INTR.

Mundy, R. 2014. 'Pioneering Antarctic photography: Herbert Ponting and Frank Hurley' *The Polar Journal* 4(2): 37–41.

Murray, G. (ed.) 1901. *The Antarctic Manual for the Use of the Expedition of 1901.* London, https://archive.org/details/antarcticmanual00britgoog.

Rack, U. 2010. *Sozialhistorische Studie zur Polarforschung anhand von deutschen und österreich-ungarischen Polarexpeditionen zwischen 1868–1939 (Social-historic study of polar exploration based on German and Austrian-Hungarian polar expeditions between 1868–1939).* In: Berichte zur Polar- und Meeresforschung (Reports on Polar and Marine Research), vol. 618. Bremerhaven, Germany: Alfred-Wegener-Institute. http://hdl.handle.net/10013/epic.35941.

Royal Geographical Society 1896. *Report of the Sixth International Geographical Congress. London.* https://ia601702.us.archive.org/1/items/jstor-196860/196860.pdf.

3

The Arctic in literature and the popular imagination

Heidi Hansson

The Walt Disney animated short film *Polar Trappers* from 1938 offers one of the best misrepresentations of the Polar Regions in popular culture. It is notable for being the first cartoon in which Disney's characters Donald Duck and Goofy appear without Mickey Mouse, and the two are shown in a frozen world living in an igloo and running the Donald and Goofy Trapping Co. But is this the Arctic or the Antarctic? One cannot be entirely sure. The assumption though is that it is the Arctic (the snow house – or igloo – is perhaps the clue), but objects, images and animals are mixed and jumbled. Along with the igloo – considered the quintessential Arctic dwelling in which Inuit supposedly live – there are references to walrus and a cast of penguins appear. The opening frame shows an icebound shipwreck, hinting at the fate of many expeditions of polar exploration in general and the Franklin expedition in particular. The next segment shows Goofy setting up a trap for walrus, suggesting how the Arctic has been exploited for its natural resources as well as emphasising its status as a natural, as opposed to a civilised world. At the same time, Goofy's song "We don't know why we catch them but we bring them back alive" draws attention to the arbitrary causes for Arctic exploitation. The cartoon then moves to Donald Duck, heating up beans in the igloo that serves as their home. The segment succinctly captures the absent presence of the indigenous population in many popular representations of the Arctic by referencing iconic markers, but completely leaving out – in this case – the Inuit. Unhappy with the constant diet of beans, Donald looks out the window and sees a penguin that transforms in his mind to a grilled chicken (this possibly references the scene in the 1925 Charlie Chaplin film *The Gold Rush*, when a hungry Big Jim imagines the tramp as a chicken) and sets out to catch the bird.

Polar signposting in general is both stereotypical and uncertain, with the Arctic and the Antarctic conflated and understood in terms of a generic geography and fauna. The mistaken metonymy of the penguin is not unusual in popular representations.[1] In Disney's cartoon, the penguin is depicted as a young girl, and Donald first courts her to get close enough to her for the kill. The story then unfolds with Donald as the pied piper attempting to catch the entire penguin colony, but in typical Donald Duck plot resolution he ends up trapped, together with Goofy, in their own cage. The short cartoon contains almost every image of the Arctic in popular culture. Although partly undercut by the constant failure of Donald Duck's pursuits, the region is imagined as the last frontier, a space of alternative livelihoods and opportunities where people can remake themselves. It is the site for heroic quest and exploitation, with the iced-over ship a reminder of the price for heroism. It is gendered in a complex manner as a feminine space available for conquest but a landscape that fights back, emasculating its conqueror. Finally, it is an ultimately futile goal.

The Arctic as site for heroic endeavour

At the beginning of September 1909, the sensational world news was that the American doctor Frederick Albert Cook and his two Greenland Inuit companions Ahwela and Etukishook had reached the geographical North Pole a little more than a year earlier, on 21 April 1908.[2] Only a few days later, the news communication was that Commander Robert Edwin Peary and

AN UNDISPUTED CLAIM.

American Eagle. "MY POLE, ANYWAY!"

Figure 3.1 'An Undisputed Claim'. A satirical cartoon published in *Punch*, obliquely commenting on the controversy between Frederick Albert Cook and his Inuit companions Ahwela and Etukishook on the one hand and Commander Robert Edwin Peary and Matthew Henson on the other as to which party was the first to discover the North Pole in 1909. Reproduced with permission of Punch Ltd., www.punch.co.uk.

Matthew Henson had discovered the Pole on 6 April 1909 and that there had been no signs of previous human presence at the location (Figure 3.1). Cook's claims were dismissed fairly soon, but towards the end of the twentieth century there were also doubts over whether Peary and Henson were indeed the true discoverers. In the last few decades there have been reservations as to whether any of the early explorers actually arrived at 90°N and growing consensus that the first confirmed trip to the North Pole on foot, and without mechanised transport, was undertaken by Sir Walter William Herbert as late as 1969.

Regardless of who really was first to reach it, the North Pole has been depicted as a desirable goal throughout Western history and members of the numerous expeditions attempting that goal have been portrayed as heroes. As the last frontier, the Arctic is where boys become men, pioneers are separated from imitators, and leaders distinguished from the rank and file. In Anglophone popular culture, the scientific gains take second place to the elements of competition and national honour.[3] The very same day that Peary's claims were published, the English writer Frank Hubert Shaw, or Captain Frank H. Shaw as his name usually appears, received a request from his publishers for an adventure story to dramatise the event. He wrote *First at the Pole* in a week, and the book was available in bookstores on 1 November, less than two months after Peary's announcement (Richards 1992: 83). The American writer Edward Stratemeyer, or perhaps the Stratemeyer syndicate that was founded in 1905, likewise packaged the discovery for juvenile consumption in *First at the North Pole; or Two Boys in the Arctic Circle*, with a foreword dated 15 November 1909. As stated in the preface, Stratemeyer's purpose was "to show what pure grit and determination can do under the most trying of circumstances", to give "an insight into Esquimaux life" and to "relate what great explorers like Franklin, Kane, Hall, DeLong, Nansen, Cook, and Peary have done to open up this weird and mysterious portion of our globe" (Stratemeyer 1909: vi), in a sum of the components of the heroic discourse.

Both Stratemeyer's and Shaw's books were aimed at boys and are typical examples of their time and genre. The message is that the final triumph can only be reached through determination under duress and unselfish sacrifice:

> They crept forward, footstep by footstep, their feet aching, blood frozen in their shoes. They suffered agonising tortures, for the slightest wound became a ghastly horror in a few hours; and the pain weakened them. Added to this was the lack of sufficient food; and yet not a single dissentient voice was raised. The Pole must be reached, no matter what happened afterwards.
>
> *(Shaw 1909: 272)*

In Shaw's story, the expedition is brought about because of the terms of a will. The first man to reach the Pole is to inherit a great deal of money, but the main character is motivated primarily by the opportunity to do something for his country: "I'm going mainly because I want an Englishman to be first to reach the Pole. That's why – our countrymen have tried hard for the prize – and – we deserve to win" (ibid.: 9). Books chronicling Arctic adventures were considered particularly suitable as school and Sunday school prizes at the end of the nineteenth century and beginning of the twentieth century, because of their emphasis on endurance, patriotism and moral, upright manliness (David 2000: 197). Praiseworthy characteristics were to be able to think and act fast in situations of crisis, place honour and the ultimate goal of discovery above personal safety, to be able to issue or occasionally follow orders, and to be a good loser as well as a generous winner. The protagonists normally came from the well-educated upper or upper middle classes, while indigenous Arctic people, whalers and sea captains filled the roles of helpers, or provided negative contrasts or comic relief. An Arctic explorer was an officer and a gentleman and, at least in the case of Frank H. Shaw, an important purpose was to foster a new generation of soldiers and empire builders with the help of thrilling and inspiring adventure stories (Richards 1992: 81, 106). Cultural expressions like dandyism, decadence, aestheticism, political movements such as the struggle for female suffrage and a general increase in the standard of living, led to a widespread fear of degeneration and emasculation around the turn of the twentieth century. The Arctic stories being published at the time served to counter such fears, and their glorification of manly strength and courage had a morally rousing function not least in connection with the First World War when the stoic, self-sacrificing hero was needed as an ideal. The polar hero was evidence that the danger of degeneration was considerably overstated.

A common image of polar exploration in Anglophone popular culture is thus masculinist and imperialist in nature, although the imperial narrative develops unevenly. Analysing the anonymous 1825 children's book *Northern Regions*, Erika Behrisch Elce argues that the story advocates a cultural relativity rarely expressed in the exploration narratives it is based on (Elce 2015: 326). Alongside the idea of the Arctic as a testing place for men, there are representations where the symbolic meaning of the North Pole is changed, and its status as a desirable goal is undermined. The Cook-Peary controversy in 1909 gave rise to a great deal of humorous and satirical commentary in genres from cartoons in comic periodicals to music hall and silent film, but an anti-heroic discourse in dialogue with the dominant paradigm can be traced back at least to the early nineteenth century and probably much further back. The main terms of this alternative Arctic discourse were the fundamental meaninglessness of polar discovery, the description of the eventual arrival at the North Pole as only a matter of happenstance and a general debunking of the Arctic hero-myth.

The Arctic as futile goal

An early example of Arctic satire is the anonymous *Munchausen at the Pole* (1819) which parodies John Ross's 1818 expedition in pursuit of the Northwest Passage.[4] In some measure it is also a general response to what Janice Cavell (2008: 24) has termed the "onslaught" of information about Arctic exploits in the early nineteenth century:

> I am given to understand that some other discoverers have returned (previous to my arrival in England) unsuccessful, from an attempt in which my exertions have been crowned with such glorious success, and that they are preparing for publication a ponderous account of their disappointments. I beg the public to take notice that this work has no concern with theirs, which I have never seen, or wish to see, and know not anything of whatever.
>
> (Munchausen *1819: 6*)

Unlike the protagonist of Frank H. Shaw's boys' story, Munchausen is entirely motivated by money: "The spark of ambition which was nearly extinguished, revived afresh, and at the mention of £20,000 reward, burst into an unquenchable flame" (ibid.: 5). Science is reduced to a random collection of items for a curiosities cabinet, including a "*cable* – made from the hairs of an *East* Greenlander's beard" and four bottles of different-coloured snow (n. p.), and the North Pole itself turns out to be a sprout from "the Axletree of the World" (ibid.: 91). Going beyond the Pole, Munchausen finds a passage to India, where his extravagant romp through history and myth continues. While he raises the British flag on the top of the Pole and proclaims George III "Monarch of the Polar Regions" (ibid.: 94), this act is followed closely by his comment on arrival in Ceylon that "the governor of this island was occupied in the interior, suppressing a rebellion, that is, in plain English, an attempt of a native prince to recover his hereditary possessions" (ibid.: 99). Alongside the enumeration of useless discoveries and artefacts, such colonialist critique is a common component of exploration satire and comedy.

Most Arctic expeditions, at least from the nineteenth century onwards, were thus transformed into media events, sometimes from the very moment of setting out. In comic-satirical and indeed most popular discourse, clichés stand in for real knowledge, giving undue attention to polar bears and icebergs but ignoring the existence of an Arctic summer. Linguistic and domestic frames of reference also influence what is selected as Arctic metonymies. In German-language reception, the seal looms large, as Johan Schimanski and Ulrike Spring explain, since the German word "Seehund", sea-dog, provides opportunities for satirical drawings where seals are depicted with dogs' heads, on leashes and domesticated (Schimanski and Spring 2010: 28). In English-language culture, wordplay on geographical Poles, political polls and literal poles is standard fare, sometimes adopted by the explorers themselves as in the first telegram from Robert Peary to his wife, Josephine Diebitsch-Peary, which ran "I have the D. O. P.", that is "the darned old pole", according to the *New York Times* ("I have the D. O. P.", 3). The light-hearted tone gives the impression of a modesty otherwise rarely discernible in either Peary or his country when it comes to the Arctic enterprise. In his book about the expedition, Peary copies one of the notes he deposited at the site after the discovery:

> I have to-day hoisted the national ensign of the United States of America at this place, which my observations indicate to be the North Polar axis of the earth, and have formally taken possession of the entire region, and adjacent, for and in the name of the President of the United States of America.
>
> *(Peary 1910: 297)*

The discovery is framed as a national triumph. In the English comic periodical *Punch, or the London Charivari*, such imperialist overtones are frequently the butt of the jokes, and Peary's flag-raising is held up as the opposite of national humility:

> Judge Woodward, of New York, holds the opinion that, while American people many years ago were probably over boastful, the pendulum has now swung the other way, and the average American is too modest in asserting the glories of his native land. But this was said before the Stars and Stripes had been run up at the North Pole.
>
> *("Charivaria" 8 September 1909: 163)*

A cartoon by Bernard Partridge in the following week's issue of the periodical develops the theme, depicting a cigar-smoking American eagle proudly poised at the top of the Pole-pole with the flag between its claws (Partridge 1909: 191). But the scene is a desolate snow-desert, inviting the question of the value of an uninhabitable wasteland far from civilisation.

The Cook-Peary controversy concerning who had been first to reach the Pole was obviously irresistible material for comedy and satire. Both expeditions were downgraded to tourist trips in *Punch*: "The tourist season opened early this year, our first visitor arriving on April 6. It will be remembered that last year the rush for the Pole commenced on April 21" ("Arctic Items", p. 185). The *New York Times* had contributed to financing Peary's expedition and obviously supported the Pearyites, but the notice "The News from Peary" grudgingly accepted the possible validity of Cook's claim in the concluding remark that whatever the outcome, it was at least certain that the Pole had been discovered by an American ("The News from Peary" p. 8). Popular culture was quick to catch on, as in the popular song *A Yankee Always Gets There First* (1909) where Scottish Sandy, English John and Irish Pat make a bet as to who will be the first to reach the North Pole, only to be met on their arrival by an American sliding down the Pole and taking the jackpot (Bateman and Hyde 1909). In a similar, although slightly more serious vein, *Punch* published some verses where the then editor Owen Seaman expresses his relief that no Englishman had taken part in the race since the result would then have been Arctic war:

> Meanwhile at home we well may thank
> Our stars that it did not occur
> To one of you to be a Yank
> And one by birth a Britisher;
> U. S. would now be arming for
> A long and bloody Polar war.
> *(Seaman 1909: 182)*

The corresponding occurrence in Frank H. Shaw's novel is represented as a moral example of honourable behaviour under adversity, with the losing English and the winning US expedition leaders shaking hands across the Pole (ibid.: 293). There is nothing gentlemanly about either Cook or Peary – or Spook and Query as they are named in the music hall number *How I Climbed the Pole* (Wells 1909) – as they are depicted in popular culture. They are even mentioned as examples of fraud and deceit in school plays, as in the children's operetta *Arcticania* (Willard and Eldridge 1916) where Prince Polar and Queen Aurora reject both their claims:

> Reporter.
> Are you sure, that in your book
> You haven't somewhere, the name of Cook?
> Prince.
> Cook! Cook! for goodness sake!
> We thought you knew he was only a fake.
> Reporter.
> But there is yet another claim,
> Peary, I believe is the fellow's name –
> Prince.
> I've heard it said that he was here,
> But his report is somewhat queer –
> No one knows of his advent,
> When he came, nor when he went.
>
> *(p. 11)*

The punchline in many of the verses, songs and commentaries is that neither Cook nor Peary can prove their success because the Pole-pole has already been removed, by a lower-class character as in A. A. Milne's "An Unconvincing Narrative" (1909: 188), by a rogue as in the silent film *How I Cook-ed Peary's Record* (Booth 1909) where Baron Munchhausen collaborates with a polar bear to take down the pole and bring it to civilisation, or completely by coincidence.[5] The dominant paradigm throughout the nineteenth and most of the twentieth century is that the Arctic is open for conquest by national, polar heroes, but alongside this there is an alternative tradition where the North Pole as the ultimate Arctic symbol is represented as an anti–climax and the explorer as an anti–hero. In some measure, these comic and satirical representations can be understood as reactions to societies in flux, when it comes to the struggle for national supremacy as well as in gender terms.

The Arctic as gendered space

As is typical of the nineteenth-century geographical imagination, the Arctic is frequently represented as a woman, silently awaiting her lover-conqueror. A characteristic example are the verses published in *Punch* on the departure of the British Arctic Expedition of 1875–1876, led by George Strong Nares and Henry Stephenson, where the goal is described as an expectant bride:

> At her feet the Frozen Ocean, round her head the Auroral lights,
> Through cycles, chill and changeless, of six month-days and nights,
> In her bride-veil, fringed with icicles, and of the snowdrift spun,
> Sits the White Ladye of the Pole, still waiting to be won.
>
> *("Waiting to be Won" p. 248).*

The British captains are conceived as active seducers and the Arctic as a passive virgin, in line with the prevalent, although increasingly unstable gender ideology. A more ambiguous example of feminisation occurs in a later *Punch* cartoon, "The Sleeping Beauty of the North" from 1896, which depicts Fridtjof Nansen in some kind of Sámi costume peeping in at a scantily dressed woman who reclines on a bed of ice with a polar bear at her back (Figure 3.2). Here, the North Pole has become a boudoir, in a move that domesticates the Arctic but also feminises Nansen. Unlike the British explorers whose masculine superiority is not in doubt, Nansen's achievement is doubly questioned, first because the idea of the North as a courtesan reduces his achievement, and second, because of the association with the indigenous Sámi who, like other distant or colonised people, were generally understood in feminine terms at the time.

The virginal Ice Maiden, however, had her counterpart in the idea of a formidable Ice Queen with the power to defeat the explorers daring to enter her realm. In *Munchausen at the Pole* she appears as "*the spirit of the Pole, who presides over and protects the sacred magnet*", able to expel the intruder "by terror from her dominions" (p. 21, italics in original).[6] She reappears as the evil *Snow Queen* in H.C. Andersen's tale of that title, and is resurrected as the White Witch in C. S. Lewis's *The Lion, the Witch and the Wardrobe* (1950) and the evil mother Mrs Coulter in Philip Pullman's *The Golden Compass* (1995), among others. An iconic image is the large woodcut "A Cold Reception", published in *Punch* in November 1876 after the failure and return of the British Arctic Expedition. The regal Ice Queen still appears in diaphanous veils, but towers over the explorers, weapon-like spire in her hand, as the ultimate cause of their suffering in the treacherous snow-fields. Unlike in the Nansen cartoon, this idea of the feminine Arctic does not feminise the explorers. Instead, her hostile nature emphasises their fortitude and reinforces a model of heroic masculinity. In some measure, the Ice Queen hints at the possibility of a reversed gender order, but as a figure of femininity, she is exceptional, and like the powerless Ice Maiden she ultimately supports a conventional gender ideology. Imaginative experiments with alternative social organisations are instead conducted in utopian fictions set in worlds accessed through the Arctic.

The Arctic as otherworld

Many seventeenth-century European maps show the water of the oceans emptied out of the world at the North Pole which creates a vortex where ships are sucked down. The subgenre of hollow-earth utopias connects to this tradition, and Munchausen's passage via the Pole to the Indian Ocean (p. 98) has its counterparts in a flurry of lost or secret world romances set in a lush otherworld beyond the Arctic. In this tradition, the Arctic is a transitional space, crucially important as the place of disorientation that readies the protagonist for an alien

THE SLEEPING BEAUTY OF THE NORTH.

The Arrival.

"All precious things discover'd late | And draws the veil from hidden 'North.'
To those that seek them issue forth, | *   *   *   *
For 'pluck' in sequel works with fate, | "The many fail: the one succeeds!"—*Tennyson.*

["Dr. Nansen has reached North Pole, found land, and is returning."]

Figure 3.2 'The Sleeping Beauty of the North'. The North Pole imagined as a courtesan in a *Punch* satire of the Norwegian explorer Fridtjof Nansen's success in 1896 in reaching further north than any other recorded polar expedition. Reproduced with permission of Punch Ltd., www.punch.co.uk.

experience. The extreme natural conditions expose the visitor to an environment where familiar ways of interacting with the surrounding world are challenged both physically and mentally. Society is left behind when the power relations between nature and culture are reversed, which leads to a loss of self. Finally, as one of the last unknown places on the earth the Arctic strips away the visitor's cultural constraints and certainties in preparation for cognitive reorientation.[7]

A common feature in utopian fiction is that the alternative world is found through an accident, frequently caused by the loss or malfunction of equipment. The geographical North Pole is north of the magnetic North Pole which means that the northern point of the compass will point south at the geographical Pole and the compass will be unpredictable at the magnetic Pole. Foreshadowing the process of reorientation that is to take place, the malfunctioning object is therefore often the compass, as the most obvious metaphor for our ability to orient ourselves in the world. Thus, the fictional manuscript of *The Third World: A Tale of Love and Strange Adventure* (1895) by Henry Clay Fairman is found in the Arctic grave of the last survivor of John Franklin's tragic fourth expedition, together with a "broken mariner's compass inscribed 'The Terror'" (p. 6). Shipwreck leads to finding the new world in possibly the first example of the genre, a prose work with the title *The Description of a New World Called the Blazing-World* by Margaret Cavendish, Duchess of Newcastle, published in 1666. Vera Zarovitch who is the protagonist of Mary E. Bradley's *Mizora: A Prophecy* (1890) actually has two shipwrecks, whereas one suffices in William George Emerson's, *The Smoky God* (1890) and William R. Bradshaw's, *The Goddess of Atvatabar* (1892). The crash of a hot-air balloon occasions the discovery in William Shaw Jenkins's *Under the Auroras* (1888) and Robert Ames Bennet's *Thyra, A Romance of the Polar Pit* (1901). "Had I started out with a resolve to discover the North Pole, I should never have succeeded" is a typical comment (Lane 1890: 8). The accidental and ultimately meaningless North Pole arrival parallels the bumbling polar discoveries of the comical tradition.

The Blazing-World is an odd mixture of adventure story, romance, philosophical tract, scientific study and autobiography. It is too much to call it a proto-feminist work, but it could be said to begin a tradition of imagining the Arctic as the entrance to a place of female rule. As with the incarnations of the Ice Queen, the radical potential of these fantasies varies considerably. The gender order is frequently dissolved or reversed, but class structures, essentialised gender definitions and nineteenth-century bourgeois ideals often remain intact. Thus, in the Christmas story *The Finding of the Ice Queen* (1879) by Frank Barrett, feminist separatism is rejected as an aberration. The feminist and socialist utopia in Lane's *Mizora* is undeniably racist and the goddess cult and strong feminine role models in *The Goddess of Atvatabar* (1892) are off-set by the conventional marriage plot. The genre is occasionally used parodically, as in Mrs. J. Wood's [probably William Mill Butler], *Pantaletta: A Romance of Sheheland* (1882), a cross-dressing fantasy where the women, called shehes, rule and the men, or the heshes, are slaves. The story ends with General Gullible's escape from marriage to the president of Petticotia, and his heartfelt wish, at a time when feminism is beginning to have influence in American society, that conventional gender relationships might prevail:

> O, my native land [. . .] May thy sons and daughters still be sons and daughters, and thy men and women, men and women. And may the day never dawn when amateur world-builders, or vainglorious demagogues, shall, out of thy matchless civilization, shape abortions like the shehes and heshes of Sheheland!
>
> *(p. 239)*

Using the Arctic as a setting for utopia relies on the same thought patterns as when the region is thought of as available for exploitation: it is a blank space, open and ready for inscription. Therefore it meets the necessary conditions for imagining, critiquing and tolerating something that is entirely new, whether it is social, spiritual or actual.

A recent development is so-called cli-fi, or climate fiction, usually dystopian stories about the effects of and attempts to halt global warming. Instalments range from realistic fiction like Ian McEwan's *Solar* (2008), referencing modern history and science and partly set in Svalbard, to the Siberian western of Marcel Theroux's *Far North* (2009) and the mass market eco-thriller *Arctic Drift* (2008) by Clive Cussler and Dirk Cussler, where finding the lost Franklin expedition is the clue to saving the world from ecological disaster. Unlike the utopian novels of the late nineteenth and early twentieth century, cli-fi employs the Arctic as a setting, not only the transition to an alternative world. With few exceptions, however, the underlying preservation ideal perpetuates the representation of the Arctic as pristine nature under threat. It is only rarely that the region is depicted in social terms, as a site of modernity and future possibility.

New Arctic, new fiction

Writing about the farthest North is almost exclusively a matter of writing for the outsider which requires much more of an attachment to agreed-upon narrative paradigms. The conventional format for Arctic fiction is "human-against-nature", followed by the plot pattern "human-against-human" in frontier stories. So far, plots based on "human-against-society" or even "human-in-society" have been largely absent, since the Arctic has rarely been acknowledged as a social environment. Crime fiction is however a genre that makes no sense without the various legal and moral systems that govern a society, which means that the emergence of Arctic detective stories is gradually changing both the genre of crime writing and the character of Arctic literature. The Arctic setting introduces nature as a major factor, but now the Arctic is populated with characters who are not only representatives of different indigenous or settler communities, but complex individuals driven by personal motives.

Crime fiction is a genre of the present, with plots and characters that reflect current fears and anxieties and investigation techniques that respond to the most recent social, scientific and technological developments. At the same time, the history of the detective novel is not linear or unidirectional, but multi-layered, which means that older varieties of the genre co-exist with new forms. One of the early examples of Arctic crime fiction is Dana Stabenow's mystery series about the Aleut investigator Kate Shugak, set in an Alaskan national park. The books borrow from the cosy detective genre, with characters reappearing from book to book, producing the sense that the stories describe a community populated with familiar figures rather than a wilderness devoid of human presence. But the criminal threat is neither from within, as in Golden Age crime writing of the 1930s and 1940s, nor from without, as in the sexually-motivated serial crimes of present-day noir, but often the result of outside exploitation and greed. Environmental issues are addressed and environmental crimes are condemned, and the fate and circumstances of indigenous populations is a common sub-theme. At the same time, Shugak's detective abilities are near over-shadowed by her outdoor survival skills, and the final defeat of the criminal is frequently the result of his or her failure to adapt to Alaskan conditions.[8]

A regional setting creates market value and the most marketable asset is exoticism. Book titles and cover art of Arctic crime fiction suggests that at least the publisher wishes primarily to highlight the ethnic content and the exigencies of the climate. On the whole, this also appears to be what readers are looking for. As the reviewer Beth Jones notes about Melanie McGrath's crime novel *White Heat* (2011), "the most addictive character – both hero and villain of the piece – is the Arctic itself" (Jones 2011). These expectations and marketing strategies create concerns regarding tokenism and raises fears of clichéd exoticisation, but the other side of the coin is continued narrative absence for Arctic people. McGrath's stories about the investigator Edie Kiglatuk set in fictional Craig Island close to Ellesmere Island contain details that both sustain and question the exotic discourse. *White Heat* is liberally peppered with Inuktitut words and phrases and Edie is shown to be part of the land, performing an "Inuit search" for evidence by walking barefoot in circles over the crime scene, trusting her body over her eyes (pp. 140–142). On the other hand, it is the outsider Bill Fairfax who is wearing "traditional caribou mukluk boots, a sure sign of a man in the grip of Arctic nostalgia" (p. 69). Like the main character of Peter Høeg's *Miss Smilla's Feeling for Snow* (1992), Edie is a threshold figure, part Inuk and part *qallunaaq*, or white, and a central theme in the novel is how traditional skills and attitudes are disappearing as a result of alcoholism, suicide, the availability of junk food and an education inappropriate for Arctic life. McGrath is well aware of the tradition she is breaking, describing a series of Arctic prints "presenting the sanitized, picture-perfect, people-free Arctic fantasy beloved of southern photographers and artists" (p.339). Her Arctic, in contrast, is insistently populated, troubled, between tradition and modernity and in a losing battle against both the present and the past. The novel expands its genre by problematising the question of the victim as not only the people murdered, but the entire community where the crimes take place.

The history of the Arctic in the popular imagination runs an uneven course, from rampant hero worship in a *terra nullius* icescape, through satirical scepticism regarding the value of reaching a symbolic position on the map to imaginary social arrangements and present-day ecological concerns. Iconic images include Arctic polar bears as well as Antarctic penguins, underscoring both the continual circulation of the Arctic as a natural world and the lack of actual knowledge of its properties. The various representations form mediascapes shaped by their audiences, genres, geographical place of production and historical context. In the end, this means that they turn back on themselves and say as much about their cultures of origin as the Arctic.

Notes

1 On the IMDB website, the cartoon is described as set in Antarctica. This would be supported by the penguins' presence but contradicted by the reference to caribou and walrus and the presence of the igloo. www.imdb.com/title/tt0030595/.
2 Some of the material in this article has previously appeared in Swedish, in "Nordpolen enligt Puh. Alternativa arktiska diskurser i brittiska populära framställningar 1890–1930", *Reiser og ekspedisjoner i det litterære Arktis*, ed, Johan Schimanski, Cathrine Theodorsen and Henning Howlid Wærp (Trondheim, Norway: Tapir Akademisk Forlag, 2011), pp. 239–261.
3 My discussion is concerned with outside representations of the Arctic, primarily in Anglophone culture. Indigenous representations are certainly different, not least because they proceed from a position where the Arctic is "home".
4 The first-person narrator has taken the name Captain Munchausen because of his similarity to the fictional Baron Hieronymus Karl Friedrich von Münchhausen, and is not intended to be the same character (*Munchausen at the Pole*: 6). Note the variant spelling of "Munchausen".
5 Note the variant spelling of "Munchhausen". The main character of the film is supposed to be the fictional Baron von Münchhausen.
6 It should be noted, though, that later in the text, the "Spirit of the Pole" is imagined in masculine terms (*Munchausen at the Pole*: 94–95).
7 Some of the material in this section has previously appeared in different form in "Arctopias: The Arctic as No Place and New Place in Fiction", *The New Arctic*, ed. Birgitta Evengård, Joan Nymand Larsen and Øyvind Paasche, Heidelberg, Germany: Springer, 2015, 69–78.
8 For a more extensive discussion of Stabenow's detective novels, see Heidi Hansson, "Arctic Crime Discourse: Dana Stabenow's Kate Shugak Series", *Arctic Discourses*, ed. Anka Ryall, Johan Schimanski, and Henning Howlid Wærp, Newcastle, UK: Cambridge Scholars, 2010, 218–239.

References

Anonymous. 1824. *Munchausen at the Pole*; or *The Surprising and Wonderful Adventures of a Voyage of Discovery etc.* 1819. London: Printed for J. Johnson.
"Arctic items", *Punch, or the London Charivari* 15 Sep. 1909: 185.
Bateman, Edgar and Will Hyde. 1909. *A Yankee Always Gets There First*. London: Francis, Day and Hunter.
Booth, Walter R. (dir.) 1909. *How I Cook-ed Peary's Record* [alt. title *Capturing the North Pole*]. London: Charles Urban Trading Company.
Cavell, Janice. 2008. *Tracing the Connected Narrative: Arctic Exploration in British Print Culture, 1818–1860*. Toronto, ON: University of Toronto Press.
"Charivaria", *Punch, or the London Charivari* 8 Sep. 1909: 163.
"A cold reception (Arctic Regions 1875)", *Punch, or the London Charivari* 11 Nov. 1876: 203–204.
David, R.G. 2000. *The Arctic in the British Imagination 1818–1914*. Manchester, UK: Manchester University Press.
Elce, E.B. 2015. '*Northern Regions* (1825): A new template for imperial children' *The Lion and the Unicorn* 39(3): 311–330.
Fairman, H.C. 1895. *The Third World: A Tale of Love and Strange Adventure*. Atlanta, GA: The Third World Publishing Co.
Hansson, H. 2010. 'Arctic crime discourse: Dana Stabenow's Kate Shugak series' in A. Ryall, J. Schimanski and H.H. Wærp (eds) *Arctic Discourses*. Newcastle, UK: Cambridge Scholars, pp. 218–239.

"'I have the D. O. P.', Peary cables wife", *New York Times* 7 Sep. 1909: 3. http://query.nytimes.com/mem/archive-free/pdf?res=9C0CE7D91F31E733A25754C0A96F9C946897D6CF.

Jones, B. 2011. '*White Heat* by MJ McGrath: Review' *Telegraph* 13 Mar. 2011, www.telegraph.co.uk/culture/books/8371383/White-Heat-by-MJ-McGrath-review.html.

Lane, M.E. Bradley 1890. *Mizora: A Prophecy*. New York: G.W. Dillingham.

McGrath, M. 2011. *White Heat*. London: Pan Macmillan.

Milne, Alan Alexander. 1909. 'An unconvincing narrative' *Punch, or the London Charivari* 15 Sep. 1909: 188.

"The news from Peary", *New York Times* 7 Sep. (1909): 8. http://query.nytimes.com/mem/archive-free/pdf?_r=1&res=9D06E6D91F31E733A25754C0A96F9C946897D6CF.

Partridge, Bernard. 1909. 'An undisputed claim' *Punch, or the London Charivari* 15 Sep. 1909: 191.

Peary, R.E. 1910. *The North Pole: Its Discovery in 1909 Under the Auspices of the Peary Arctic Club*. New York: Frederick A. Stokes, 1910.

Polar Trappers [dir. Ben Sharpsteen]. Walt Disney Productions, 1938.

Richards, J. 1992. 'Popular imperialism and the image of the army in juvenile literature', in J. MacKenzie (ed.) *Popular Imperialism and the Military, 1850–1950*. Manchester, UK: Manchester University Press, pp. 80–108.

Schimanski, J. and U. Spring. 2010. 'A black rectangle labelled "polar night": imagining the Arctic after the Austro-Hungarian expedition of 1872–1874' in A. Ryall, J. Schimanski and H.H. Wærp (eds.) *Arctic Discourses*, Newcastle, UK: Cambridge Scholars, pp. 19–42.

Seaman, Owen. 1909. 'The battle of the Pole', *Punch* 15 Sep. 1909: 182.

Shaw, F.H. 1909. *First at the Pole: A Romance of Arctic Adventure*. London: Cassell.

"The Sleeping Beauty of the north" ["Dr. Nansen discovering the North Pole"], *Punch, or the London Charivari* 22 Feb. 1896: 86.

Stratemeyer, E. 1909. *First at the North Pole, or Two Boys in the Arctic Circle*. Boston, MA: Lothrop, Lee & Shepard.

"Waiting to be won", *Punch, or the London Charivari* 5 Jun. 1875: 248.

Wells, G. 1909. *How I Climbed the Pole*. London: Francis, Day & Hunter.

Willard, J. Bassett (libretto) and H. C. Eldridge (music), *Arcticania, or Columbia's Trip to the North Pole: An Operetta in Two Acts*. Franklin, OH: Eldridge Entertainment House, 1916.

Wood, J. [William Mill Butler] 1882. *Pantaletta: A Romance of Sheheland*. New York: American News.

4

The Antarctic in literature and the popular imagination

Elizabeth Leane

The narrator of Julian Barnes' novel *Flaubert's Parrot*, a retired doctor named Geoffrey Braithwaite, at one point makes a list of literary genres that he would like to ban. High among his dislikes are predictable, overused settings: his strictures include a twenty-year hiatus on "novels set in Oxford and Cambridge" and a quota system on "fiction set in South America" with its standard magic-realist motifs: "the opera house now overgrown by jungle". Conversely, "Novels set in the Arctic and the Antarctic will receive a development grant" (Barnes 1985: 99). For this literary joke to work, author, narrator and reader must share the assumption that the Polar Regions are highly incongruous settings for literary texts. Braithwaite ironically promotes polar icescapes because of their very resistance to the clichés of literary fiction. They harbour neither opera houses – high culture – nor jungles – growth, fertility, natural abundance. None of the usual fodder for fiction, it seems, can be found in the polar deserts. This is a view with which environmental historian Stephen Pyne, writing around the same time as Barnes, roughly concurs: in his seminal interdisciplinary history of Antarctica, *The Ice*, he draws a parallel between the bareness of the southern icescape and the paucity of creative responses to it, arguing that the region has "been largely a wasteland for imaginative literature" (Pyne 1986: 153).

Behind such characterizations of Antarctic (and Arctic[1]) literature as a sparsely populated terrain lies an assumption about what 'counts' as literature. By the mid-1980s, Antarctica had in fact featured as a setting for literally hundreds of novels, as well as a good number of short stories, poems and plays, and a handful of narrative films. Some very popular authors set their novels in the far south: in *At the South Pole* (1870), boys' adventure writer W.H.G. Kingston sends a whaling crew into high-southern latitudes, where they endure a volcanic eruption as well as attacks from bears and giant walruses; Hammond Innes, whose adventure thrillers sold in the tens of millions, maroons the company of a whaling factory ship (including a psychopathic killer) on unstable ice floes in *The White South* (1949), which was rapidly adapted for film as *Hell Below Zero* (1954), starring Alan Ladd. Innes started a trend: the following few decades saw the publication of dozens of thrillers set in the Antarctic. These examples, however, fall into the category of popular or 'genre' fiction. They are not the kinds of novels that Braithwaite is concerned with, nor those that, generally speaking, receive development grants.

Since *Flaubert's Parrot* and *The Ice* were published, there have been significant changes in access to Antarctica and hence to the nature of literary responses. Ironically, given Barnes' joke, numerous novelists and other creative writers have indeed received development grants in the form of residencies in Antarctica sponsored by various national programmes – something that had been happening on an ad hoc basis since the 1960s but began to be formalized towards the end of the twentieth century. Partly as a response to this, and partly due to new kinds of approaches to older texts, the last quarter of a century has produced growing recognition among critics of an important body of imaginative texts engaged with Antarctica. These texts have now been the subject of several anthologies (see e.g. Manhire 2004; Spufford 2007); three extensive online bibliographies;[2] a number of large-scale academic studies in several languages (e.g. Glasberg 1995; Lenz 1995; Wijkmark 2009; Essigmann 2010; Guijarro Ceballos 2010; Leane 2012); as well as many shorter ones.

Most of this recognition continues to concentrate on high cultural texts – those judged largely on aesthetic rather than commercial criteria. This is a continent that has seen visits not only from explorers, scientists, dignitaries and tourists, but also Biggles, Dr Who, Mulder and Scully, Scooby-Doo, the Hardy Boys, G.I. Joe, Caspar the Friendly Ghost, Tarzan, and just about every superhero. However, the diverse popular cultural responses to Antarctica – these include not only genre fiction but also illustrated children's books, comics, off-Broadway musicals, Hollywood films, board games, online games, advertisements and many other forms – have so far received little critical attention.[3]

This neglect is surprising, for cultural texts written in response to Antarctica – whether 'high' or 'low' – can tell us a lot about this place; they both reflect and affect perceptions and experiences of the continent. Literary critics and cultural geographers alike now accept that texts are not merely passive mirrors of human interaction with particular locations, but rather an active part of that relationship – a key component of the hybrid of materiality, social practice, and meaning that comprises the concept of 'place' (Cresswell 2014: 9). With Antarctica being unique among the continents in its lack of indigenous inhabitants and a permanent population, texts become even more central to human awareness of the region. Excepting the handful of people born on the continent, everyone's experience of Antarctica is filtered first – and most often only – through textual encounters: still and moving images, aural and written accounts. And while this body of texts is highly heterogeneous, similar tropes, themes and narrative arcs recur, creating an Antarctic imaginary that shapes human experiences with this place.

Pyne argues that the continent's 'great literature' was penned not by creative writers but rather explorers (ibid.: 168). In particular, narratives that came out of Robert Falcon Scott's last expedition – the leader's posthumously published *Journals* (1912) and Apsley Cherry-Garrard's *The Worst Journey in the World* (1922) – are recognized as classic literary texts in their own right. Authorship did not necessarily come naturally to expedition leaders, though, and some – such as Ernest Shackleton and Douglas Mawson – leant heavily on journalists or fellow expedition members to bring their books to fruition and eventual publication. Nonetheless, expedition accounts are receiving increasing attention from literary critics, interested in the way they contribute to Antarctica's place-identity, or their relationship to other literary works, or what they reveal of cultural practices of their time (e.g. Teorey 2004; Brazzelli 2011; Freer 2011). With the development of cruise-ship tourism (available since the 1960s but expanding massively since the 1990s) and official writers' residencies, acclaimed contemporary meditations on far southern journeys, such as Jenny Diski's *Skating to Antarctica* (1997) and Sara Wheeler's *Terra Incognita* (1996), began to appear. Such texts have generated their own body of criticism (e.g. Rosner 1999; Price 2015) and have reinforced the continuing importance of the non-fiction travel narrative as a mode of literary engagement with Antarctica.

These works of literary non-fiction should not, however, displace imaginative responses in our attempts to understand humanity's relationship with the continent, but rather should be read and analysed alongside them. Fiction and non-fiction are entangled in the Antarctic imaginary.

Travel texts such as those described above are not only full of allusions to literary texts, they also adopt narrative structures established in fiction (see e.g. Leane 2009); conversely, creative writers have looked to non-fiction accounts to ground their work in realistic detail and have also retold, recontextualized and reinvented the famous stories of exploration. Some prominent examples of the latter are Kåre Holt's *The Race* (1976), originally published in Norwegian in 1974; Ursula Le Guin's 'Sur' (1982) and *The Left Hand of Darkness* (1969); Beryl Bainbridge's *The Birthday Boys* (1991); and Rebecca Hunt's *Everland* (2014).

There are many ways of organizing an examination of imaginative textual engagements with the Antarctic, each of which captures some aspects of this relationship and neglects others. A historical approach would show that creative responses to the Antarctic have tended to come in waves, responding to particular historical events: expeditions, discoveries, adventures, tragedies, controversies. A thematic approach would look at the way that literary texts cluster around particular preoccupations – extremity, temporality, purity, whiteness, transformation, fragility (to name a few). My approach here, which sometimes overlaps with these others, focuses on the resonances between the place-identity of Antarctica and particular textual forms and genres. I begin with fiction, the largest body of imaginative works dealing with Antarctica, with an eye to giving equal weight to both literary and popular texts, before discussing more briefly poetry, drama, and film. I do not, of course, promise a comprehensive picture of Antarctica's significance in the contemporary imagination (I focus, for one thing, mainly on English-language sources, and leave aside the visual arts, music and many other creative forms of response). My aims are to give an indication of the diversity of creative texts dealing with Antarctica; to show how these texts together form something like a whole – a group of long-standing, intertwined traditions of representing the far south; and to argue for the importance of the imagination in understanding humanity's ongoing relationship with this region.

Fiction

Because Antarctica is the only continent where humans have never lived permanently, the archetypal Antarctic narrative – both fiction and non-fiction – takes the form of a journey. Antarctica (or at least, the geographical region where the continent was eventually found to be) first appears in English literature in the fantastic journeys of the sixteenth and seventeenth centuries, inspired by the cartographic idea of a *Terra Australis Incognita*. Perhaps the best known of these is Robert Paltock's *The Life and Adventure of Peter Wilkins* (1750), which deals with a shipwrecked sailor's encounter with a race of flying people inhabiting the far south. Inherent in these narratives is the notion of the south polar regions as not only remote from but also oppositional to those of the north: the potential home of societies which are or might be very different from – even inversions of – the author's own. This tradition developed, in the nineteenth century, into the utopia and the 'lost race romance'. Both genres found in Antarctica an ideally blank space in an increasingly mapped world. Utopian satires such as James Fenimore Cooper's *The Monikins* (1835) and James De Mille's *Manuscript Found in a Copper Cylinder* (1888) were joined by more cornucopian visions, such as George McIver's *Neuroomia* (1894), which speculated on what a far southern continent might have to offer by way of land, resources, technology and social ideas. The lost race genre flourished in the late nineteenth and early twentieth centuries, often drawing on anxieties about race and evolution: Antarctic examples include Eugene Bisbee's *Treasure of the Ice* (1898), Charles Stilson's *Polaris of the Snows* (1915), and Edison Marshall's *Dian of the Lost Land* (1935). Increasing geographical knowledge of the continent produced techno-optimistic visions of mining for resources or terraforming the icescape for inhabitation: pulp science fiction magazines of the 1930s published stories such as Peter van Dresser's 'South Polar Beryllium, Limited' (1930) and I.R. Nathanson's 'The Antarctic Transformation' (1931) (both appeared in *Amazing Stories*).

Utopian visions of the far south were by this time premised on building, rather than finding, a new and improved (at least in the text's own terms) civilization. Narratives appeared in which a journey to Antarctica did not necessarily entail a journey home. Beall Cunningham's *The Wide White Page* (1936), David Poyer's *The White Continent* (1980) and Kim Stanley Robinson's *Antarctica* (1997) are examples of the diverse ways in which utopian attempts at permanent inhabitation of the continent have been imagined. Countering these visions are dystopian narratives of environmental devastation and/or political chaos, shifting into (as the twentieth century continued) post-apocalyptic scenarios in which the far south becomes a last hope: Valery Bryusov's 'The Republic of the Southern Cross' (1918; originally published in Russian in 1905); David Graham's *Down to a Sunless Sea* (1979); John Calvin Batchelor's *The Birth of the People's Republic of Antarctica* (1983); and Kevin Brockmeier's *A Brief History of the Dead* (2006). Of all Antarctic genres, the utopia (including its satirical and negative forms) is the longest standing and perhaps the most significant; the influence of utopian thinking in regard to the continent is evident in a wide range of contexts (see Hemmings 2013).

Most of these texts deal with the possibilities Antarctica holds for human inhabitation or exploitation on the scale of a whole society. Another parallel literary tradition focuses on the impact of a southern journey on the individual psyche. In these stories, the journey south becomes a process of transformation, insight or self-discovery. The "pattern of the southern journey" writes Victoria Nelson, is an "archetypal sea trip from a bustling port (consciousness) to Terra Australis Incognita (unconsciousness), where a transcendental encounter takes place that initiates either the integration of the self or the possibility of psychic (and physical) annihilation" (Nelson 2001: 148). In journeys of the first kind, protagonists are refreshed, regenerated and positively transformed by their Antarctic experiences. James Fenimore Cooper's *The Sea Lions* (1849) inaugurates this tradition, which informs later twentieth- and twenty-first-century novels such as Graham Billing's *Forbush and the Penguins* (1965), Nikki Gemmell's *Shiver* (1997) and Maria Semple's bestselling comic novel *Where'd You Go, Bernadette?* (2012), as well as nonfiction travel narratives. In the second kind, which draws on the convention of the gothic, the Antarctic traveller is disturbed, haunted and sometimes destroyed by the sublime insights the far south offers. Beginning with Samuel Taylor Coleridge's poem 'The Rime of the Ancient Mariner' (1798), this pattern reappears in later texts such as Edgar Allan Poe's 'MS. Found in a Bottle' (1833) and *The Narrative of Arthur Gordon Pym of Nantucket* (1838), and H.P. Lovecraft's *At the Mountains of Madness* (1936). In the twentieth century, it shades into the test of moral (and male) character, as in Joseph Conrad's short story 'Falk' (1903), John Presland's *Albatross* (1931), and Charles Neider's *The Grotto Berg* (2001).

In the relatively few cases when narrative fiction in Antarctic is not premised on some kind of journey, it usually deals with its opposite: claustrophobia and entrapment. As cultural geographer Yi-Fu Tuan has argued, in polar icescapes there is no buffer between the safety of the home and the 'otherness' of the ice: "homeplace is the hut and immediately beyond is alien space" (Tuan 1993: 154). There are no front or back yards, no gardens, no parks, no cities or towns, no streetscapes as they are normally understood. "Ordinary" human existence happens indoors, in a series of very limited interior spaces. John W. Campbell's science fiction story "Who Goes There?" (1938) and its various film adaptations deal with the paranoia engendered by constant enforced cohabitation, literalizing the 'alien space' of the Antarctic environment as a space alien: dug up from the ice, the ancient extraterrestrial revives and invades not only the home place of the base (itself built into the ice) but also the men's bodies and minds.

Usually classed as science fiction and/or horror, Campbell's story also, as Pyne notes, draws on conventions of the 'locked room' murder mystery (ibid.: 192). Combining the potential for suspense of a very isolated community with the opportunities for physical adventure and danger offered by an extreme environment, the thriller has become perhaps the most dominant Antarctic genre, in terms of numbers, of the last half-century. This genre can trace its roots back to imperial adventure novels such as Kingston's *At the South Pole*, through adventure thrillers like Innes's *The White South*, to early environmentally focused thrillers such as David Burke's *Monday at McMurdo* (1967), and thence the eco-thriller: Robinson's *Antarctica*, Louis Charbonneau's *The Ice* (1991), Bob Reiss's *Purgatory Road* (1996), and L.A. Larkin's *Thirst* (2012), to name a few. The Antarctic thriller, more than any other popular genre fiction set in the far south, deals directly with the contemporary geopolitics of the region: plots circulate around secret mining projects, espionage, sabotage, national claims, treaty violations, nuclear weapons and global warming. Their representation of the international intrigue in the Antarctic – the nationality of heroes and villains, the issues over which they clash, the ways in which resolution is achieved – can both reveal and affect popular understandings of the region's politics. While thrillers are often dismissed as escapist 'airport novels', the sheer volume of these texts read, and their potential impact on readers, means they – and other popular novels set in the far south – need to be taken seriously.

Poetry

Central to the Antarctic imaginary is, ironically, the sense that the continent is unimaginable: that it exceeds language or renders it redundant; that humanity, having never permanently inhabited the region, lacks the requisite vocabulary to describe it; that writing somehow corrupts the purity of the ice. The point here is not, as in Julian Barnes' literary joke, that the region seems to offer too little for the literary imagination to work with, but rather too much. Writers attempting to describe the Antarctic icescape become, in the words of Helen Garner, "control freaks, spoiling things for everyone else, colonising, taming, matching their egos against the unshowable, the unsayable" (Garner 1998: 18). The continent is, for Australian poet John Rowland Russell, "only itself / The last unwritten page in our planet's book". The challenges of finding a language equal to 'the ice' produce both ironic reflections on the continent's resistance to poetry – as in David Wheatley's 'The Antarctic Poetry School' (2008) – and a fascination with existing attempts, whether by scientists or travellers, to find words to describe the environment – as in Katherine Coles's 'Use / / in a Sentence' (in *The Earth is Not Flat*, 2013) and Elizabeth Bradfield's 'Notes on Ice in *Bowditch*' (in *Approaching Ice*, 2009).

Early explorers were unaware of or unworried by the continent's unwritability: they produced 'sledging songs' as they marched along, and published serious and comic verse in their house newspapers (such as Scott's *South Polar Times*). Ernest Shackleton addressed the London Poetry Society in 1911; one of his biographies is entitled *Shackleton: A Life in Poetry*, its author arguing that poetry was "at the very core of who [Shackleton] was" (Mayer 2014: x). Conversely, T.S. Eliot drew on Shackleton's own (ghost-written) narrative *South* (1919) in what is arguably the most important twentieth-century poem in English, *The Waste Land* (1922). Early expeditions also inspired literary outpourings from dozens of occasional or amateur poets, published in newspapers or magazines, or simply shared privately: Scott's papers include numerous copies of poems – some printed, some handwritten, some by children – sent to his family by their authors after his death.[4]

Many of the published poetic responses to Antarctica in the last century – Douglas Stewart's verse play for radio *The Fire on the Snow* (1944), Donald Finkel's *Endurance* (1978), Melinda Mueller's *What the Ice Gets* (2000) – are long narrative works, inspired or influenced by

'Heroic-Era' expeditions. Even shorter lyric poems often engage – sometimes obliquely – with journeys of exploration: Derek Mahon's 'Antarctica' (in *Antarctica*, 1985), Anne Michael's 'Ice House' (in *Skin Divers*, 1999), Dorothy Porter's 'Wilson's Diary' and 'Oates's Diary' (in *Driving Too Fast*, 1989), for example, all draw on the Scott tragedy. Poets renew perceptions of events and figures in the continent's history by recontextualizing them: Bradfield's collection includes not only poems about famous expedition leaders, but also titles such as 'Polar Explorer Capt. John Cleves Symmes (1820)' and 'Polar Explorer: Lynne Cox' – thus opening up the ways in which, and the people by whom, the polar regions might be explored.

Not all of Antarctic poetry focuses on exploration, geographic or otherwise. Poems can be found dealing with all aspects of human experience of Antarctica: the aesthetics of its environment, the paradoxes of its tourist industry, the complexities of its politics, the insights of its science, the fragility of its icescapes. Acclaimed poets such as Les Murray ('Antarctica', in *Dog Fox Field*, 1990), Bill Manhire ('Antarctic Field Notes' in *Collected Poems*, 2001), Pablo Neruda ('Antarctic Stones' first published in Spanish in 1961; translated in Manhire's *The Wide White Page*), and Diane Ackerman ('The White Lantern: Poems of Antarctica' in *Jaguar of Sweet Laughter*, 1993) have all turned to the far south at some point. The compiler of an Antarctic poetry anthology would be spoiled for choice.

The most influential poem about Antarctica, however, remains the earliest one: Samuel Taylor Coleridge's 'The Rime of the Ancient Mariner', first published in 1798 (and revised several times between then and Coleridge's death in 1834). Coleridge's long gothic ballad tells of an old sailor whose ship, blown off course, becomes trapped in the Antarctic ice – a surreal, oppressive world in which the only living thing is an albatross, which the crew befriend. After the mariner unaccountably shoots the seabird, he suffers prolonged and spectacular punishment by a polar avenger, a "spirit who bideth by himself / In the land of mist and snow" (l. 402–3). The albatross, one critic suggests, acts as a "figure of primal innocence", closely associated with the Antarctic, itself imagined as "a landscape embodying originary purity" (Moss 2006: 213, 215). Yet the ice is figured not as a neutral or benign medium, but rather a threatening living creature: "It cracked and growled, and roared and howled" (l. 61). Unlike previous imaginative writers representing the far south – such as Paltock with his flying people – Coleridge evokes an Antarctic uninhabited by human or human-like beings. The natural environment, and particularly ice, play a correspondingly large part in the poem – even its title, with its archaic spelling of 'rime', evokes frost and snow. Between Paltock and Coleridge, James Cook's second expedition had circumnavigated Antarctica, pushing down past the seventieth parallel; as well as other polar sources (north and south), Coleridge had Cook's narrative to draw on, and his own schoolteacher William Wales had been navigator on the circumnavigation (see Smith 1992: 135–71). The Antarctic region, if not the continent, was by this time becoming known empirically as well as imaginatively, and the impact of this different kind of knowledge is evident in the 'Rime'. With its infusion of a compellingly evoked icescape with supernatural and gothic elements; its narrative of a life forever transformed and haunted by a journey south; and its prescient representation, in the figure of the albatross, of the Antarctic's vulnerability to thoughtless human action, Coleridge's poem more than any other imaginative work might make a claim to being the continent's foundational narrative.

Drama

Land-based exploration of Antarctica has long had a strong element of performance, due in no small part to the necessity of generating media interest to help fund expeditions. Leaders needed to be showmen, giving speeches before they left and lectures on their return.

Expedition members had to pander to cameramen, pre-enacting their achievements: four of the five men who eventually made up Scott's party went through the ritual of pitching tent, cooking and getting into their sleeping bags for the benefit of Herbert Ponting's cinematograph a few days before they set off, with the leader noting how enjoyable it would be to watch the scene in the cinema on their return (*90° South*). And to occupy time, polar explorers (in addition to writing poetry) put on plays and concerts: Scott's first expedition (1901–04) turned a storage hut into the 'Royal Terror Theatre' to stage a nineteenth-century farce about convicts; men of the Australasian Antarctic Expedition (1911–14) wrote and performed an opera entitled 'The Washerwoman's Secret', the 'programme' for which was carefully pasted into diaries and reproduced in the official expedition narrative, *The Home of the Blizzard* (1915).

Attempts to stage professional productions set in Antarctica were fewer – perhaps partly due to the difficulty of representing the polar ice in a theatre – and the first examples look, from a twenty-first century perspective, bizarre and surreal. A melodrama featuring (according to its playbill) a 'splendid ice landscape' and a penguin attack was staged in Hobart, Tasmania, in 1841: inspired by the visit to the city of the British Antarctic expedition led by James Clark Ross, it featured some expedition members as characters, and was watched by others. Another Australian production, a pantomime staged in Sydney at the turn of the twentieth century, centred on an airship journey to Antarctica, and included a bear attack, a comic ballet of polar animals, a dramatic whirlpool scene and the annexation of the South Pole by the northern colonizers. A 1937 twelve-tone opera performed in Germany under the Third Reich retold the story of "Scott of the Antarctic" using a chorus of singing, dancing penguins.[5]

Scott's tragedy and other famous Heroic Era journeys attracted playwrights as the twentieth century continued, although – with polar exploration now seen through feminist and postcolonial lenses – they often satirized these stories or put them to new purpose. These works include Howard Brenton's excoriating *Scott of the Antarctic* (1971),[6] which brought a pantomimic and banal flavour to the polar heroes by staging the story on an urban ice rink; Manfred Karge's Brechtian *Conquest of the South Pole* (1988, translated from the original German); and Patricia Cornelius's award-winning *Do Not Go Gentle* (2010). The latter two plays move the action away from the polar plateau to far more mundane locations – an attic in an industrial town in Germany, and an unspecified retirement home. Each group of characters acts out the exploration narratives of Amundsen and Scott respectively as a way of grappling with purpose and identity in less exotic but similarly challenging environments.

Most recently, increased access to the continent for creative artists has seen the production of a more diverse range of performances – some of them multi-media. Mojisola Adebayo's *Moj of the Antarctic* (2007), for example, combines images taken of the actor/playwright in character in Antarctica with live performance. The play is premised on two journeys: Adebayo's own on a tourist ship, and a nineteenth-century journey of a black lesbian slave who escapes, cross-dressed as a white man, from the American South to Britain, and thence (on a whaling vessel) to Antarctica. This scenario enables Adebayo to explore the politics of gender, sexual orientation and race in relation to the continent's literary and exploration history. Through the device of an African storyteller, the action is placed within a mythic frame, which underlines Africa and Antarctica's ancient history as one continent, and also predicts the melting of the ice and flooding of the land in the future due to global warming. In this and other recent plays, such as Lynda Chanwai-Earle's *Heat* (2004), the ice is no longer imagined as simply a backdrop – an impressive spectacle or a minimalist white surface to be marched across – but becomes an agent in its own right – fragile, unstable, unpredictable, threatened and threatening, intrinsically connected with the rest of the globe.

Film

While drama, poetry and prose are all forms that predate human encounters with Antarctica, film developed alongside it: "The classical era of polar exploration and the start of motion pictures took place at almost the same time" (McKernan 2000: 92). Many of the famous Heroic-Era expeditions, including those led by Carsten Borchgrevink, Jean-Baptiste Charcot, Nobu Shirase, Douglas Mawson, Scott, Shackleton and Amundsen included a kinematograph. Capturing the exotic icescape and its wildlife, as well as the expedition personnel and living conditions, was central to generating popular interest in the expedition, which was in turn important for raising funds to cover the inevitable debts. Footage would be shown in lectures on the expedition's return, and sometimes made into autonomous films, such as Ponting's *90° South* (1933) and Frank Hurley's *South* (1919). Many of the best-known filmic responses to Antarctic come out of this documentary tradition. Again, the journey narrative dominates, even in a wildlife documentary such as *March of the Penguins* (2005) – the emphasis on the birds' long trek to and from the ice-edge echoes tales of human sledging journeys. A marching bird also features prominently in Werner Herzog's *Encounters at the End of the World* (2007) – a "demented" penguin that inexplicably heads off "towards certain death" in the continent's interior provides an ironic (and typically Herzogian) visual comment on both *March of the Penguins* and Heroic-Era endeavours.

If you look hard enough, you can find dozens of full-length fiction films partly or wholly set in the far south, dating back to the 1920s: an early example is *Dirigible* (1931), directed by Frank Capra and starring (among others) Fay Wray.[7] Many are adaptations of novels or short stories: *Hell Below Zero* (1954), mentioned above; *Quick Before it Melts* (1964), a screwball comedy based on Philip Benjamin's novel of the same name and date; *Mr Forbush and the Penguins* (1971), an adaptation of Graham Billing's book starring John Hurt; *Fukkatsu No Hi* (1980), released to English-speaking audiences as *Virus*, a post-apocalyptic narrative based on the 1964 novel by Sakyo Komatsu (translated into English in 2012); and *Whiteout* (2009), a thriller adapted from the 1999 graphic novel of the same name by Greg Rucka and Steve Lieber. Most notable is John Carpenter's 1982 adaptation of Campbell's "Who Goes There?" as *The Thing*. With its ominous score by Ennio Morricone, and state-of-the-art (for its time) special effects, this science-fiction horror has become perhaps the most iconic of all Antarctic films. Antarctic base personnel are said to screen it when the last boat or plane has departed, leaving a small, isolated wintering crew like the one devastated in the film. An earlier adaptation, *The Thing from Another World* (1951), was set in the Arctic; a later one, *The Thing* (2011), is both a prequel and homage to Carpenter's. Also popular are biopics and docudramas based on historical expeditions or events: *Scott of the Antarctic* (1948), which features a score by Ralph Vaughan Williams; the television mini-series *Shackleton* (2002), starring Kenneth Branagh; the Japanese film *Nankyoku Monogatari* (1983), which depicts the remarkable survival of two abandoned dogs in Antarctica for nearly a year, and the Hollywood remake for children, *Eight Below* (2006). Intriguing in this context is the Dutch film *Forbidden Quest* (1993), a fictional documentary in which footage of real polar expeditions from the early twentieth century is spliced into a narrative of the ill-fated (and imaginary) 1905 *Hollandia* expedition.

Few of these films feature live-action footage shot in Antarctica, although establishing shots of icescapes and animals may be incorporated. John Hurt's eight-week stay, with a 14-strong crew, on the Antarctic Peninsula, living in a hut near a colony of 160,000 penguins ('Actor Shares Screen' 1972), was highly unusual. In most cases, Canada, Alaska or other icy landscapes are substituted for the Antarctic. More recently, computer generated imagery (CGI) has enabled the digital generation of the Antarctic ice, most notably in the blockbuster children's penguin films, *Happy Feet* (2006) and *Happy Feet 2* (2011). As with fiction, the constraints that the Antarctic environment makes on human activity and inhabitation render certain genres of film more tenable than others: speculative narratives and remakes of heroic journeys predominate; comedies and romances are few. In 1984, one Australian television channel bravely aired a sit-com set in an Antarctic station: entitled *Brass Monkeys*, and filmed entirely in the studio, it lasted only one season. The difficulty of working in the Antarctic environment and the comparatively limited range of stories that can be realistically filmed there, along with the expense of sending actors and crew south, means that outside of docudrama and genres such as science fiction, horror and disaster in which CGI is already heavily incorporated, imaginative narratives of Antarctica on screen remain relatively limited.

Conclusion

Creative textual responses to Antarctica – whether literary or popular – are not decorative trimming, marginal to the real concerns of science, law and politics. In *Geopolitics: A Very Short Introduction*, Klaus Dodds examines the "interconnection between popular culture and geopolitics", emphasizing the two-way relationship between these realms. Critics, he suggests, need to go beyond the simplistic view of imaginative texts as mere representations of the 'real world' and ask whether "artistic interventions help to constitute public understandings of key actors and places, and are they all the more significant when watched and engaged with by audiences that are not likely to have any experience of the places cited?" (Dodds 2014: 121). While Dodds is not focusing on Antarctica here, this and other questions he raises are immediately relevant to a region that so few people ever directly experience. This chapter has provided a glimpse of, and offered some ways into, a body of popular and literary texts – far larger and richer than it is possible to show here – through which such questions might be investigated.

Notes

1 The relationship between the Arctic and Antarctic imaginaries – too large a topic to be dealt with here – is touched on in my book *Antarctica in Fiction* (2012: 12–16).
2 See the late Fauno Cordes' 'Tekeli-li or Hollow Earth Lives: A Bibliography of Antarctic Fiction', available via (www.antarctic-circle.org/fauno.htm) with updates at (www.antarctic-circle.org/fiction.htm); my own 'Representations of Antarctica' (www.utas.edu.au/representations-of-antarctica), which includes drama, poetry and other categories as well as fiction; and Laura Kay's website 'LK's Polar Collections' (which has a separate category for genre fiction) available until recently at www.phys.barnard.edu/~kay/polar/.
3 This is slowly changing – see, for example, Nielsen 2017; Kay 2011; Campbell 2012.
4 See Scott Polar Research Institute, MS 1453/31.
5 The relevant texts are *The South Polar Expedition* (1841, anon.); *Australis, or the City of Zero* (1900, by J.C. Williamson and Bernard Epinasse); and *Das Opfer* (1937, by Reinhold Goering and Winfried Zillig). See Leane (2013) for more details.
6 Dates provided for plays refer to the year of first production. With the exception of *The South Polar Expedition*, all of the plays mentioned here exist as published scripts.
7 See Valmar Kurol's bibliography "Dramas/Fictional Movies about Antarctica" accessible at (www.antarctic-circle.org/movies.htm).

References

'Actor shares screen with 160,000 penguins'. 1972. *Sarasota Journal* 11 July: 11.
Barnes, J. 1985. (orig. pub. 1984). *Flaubert's Parrot*. London: Pan Books.
Brazzelli, N. 2011. 'A symbolic geography of the ice: Apsley Cherry-Garrard, the worst journey in the world and modernity' in Massimo Bacigalupo and Luisa Villa (eds) *The Politics and Poetics of Displacement*. Pasian di Prato, Italy: Campanotto. 45–57.

Campbell, C. 2012. 'Between the ice floes: imagining gender, fear and safety in Antarctic literature for young adults' *International Research in Children's Literature* 5(2): 151–166.

Coleridge, Samuel Taylor. 1987. (first pub. 1798). 'The Rime of the Ancient Mariner' in J. Leonard (ed.) *Seven Centuries of Poetry in English*. Melbourne: Oxford University Press. 229–243.

Cresswell, T. 2014. 'Place'. *The SAGE Handbook of Human Geography*. Vol. 1. Roger Lee et al. (eds). London: SAGE. 3–21.

Dodds, K. 2014. *Geopolitics: A Very Short Introduction*. 2nd ed. Oxford, UK: Oxford University Press.

Essigmann, J.M. 2010. 'Ein kleiner, schwarzer Punkt am weisslichen Himmel: Antarctica & Ice in German Expressionism'. Master's thesis. University of Tennessee, Knoxville, USA.

Freer, S. 2011. 'The lives and modernist death of Captain Scott' *Life Writing* 8(3): 301–315.

Garner, H. 1998. 'Adrift in the floating world' *The Age*. 30 May: 'Good weekend' 12+.

Glasberg, E. 1995. 'Antarctica of the imagination: American authors explore the last continent 1818–1982'. PhD thesis. Indiana University, USA.

Guijarro Ceballos, J. 2010. *Melancholía del hielo: Textos e imágenes sobre la Antártida*. Mérida, Spain: Editora Regional de Extremadura.

Hemmings, A. 2013. 'Utopian framings of the Polar Regions' *The Polar Journal* 3(2): 273–276.

Kay, L. 2011. 'It was a very long dark and stormy night: "bad" Antarctic fiction from the pulps to the self-published' in Ralph Crane, Elizabeth Leane and Mark Williams (eds) *Imagining Antarctica*. Hobart, Australia: Quintus. 89–103.

Leane, E. 2009. 'Eggs, emperors and empire: Apsley Cherry-Garrard's "worst journey" as imperial quest narrative' *Kunapipi* 31(2): 15–31.

Leane, E. 2012. *Antarctica in Fiction: Imaginative Narratives of the Far South*. Cambridge, UK: Cambridge University Press.

Leane, E. 2013. 'Icescape theatre: staging the Antarctic' *Performance Research: A Journal of Performing Arts* 18(6): 18–28.

Lenz, W. 1995. *The Poetics of the Antarctic: A Study in Nineteenth-Century American Cultural Perspective*. New York: Garland.

Manhire, B. (ed.) 2004. *The Wide White Page: Writers Imagine Antarctica*. Wellington, NZ: Victoria University Press.

Mayer, J. 2014. *Shackleton: A Life in Poetry*. Oxford, UK: Signal.

McKernan, L. 2000. 'The great white silence: Antarctic exploration and film' in *South: The Race to the Pole*. London: National Maritime Museum.

Moss, S. 2006. *Scott's Last Biscuit: The Literature of Polar Travel*. Oxford, UK: Signal.

Nelson, V. 2001. *The Secret Life of Puppets*. Cambridge, MA: Harvard University Press.

Nielsen, H. 2017. 'Selling the south: commercialization and marketing of Antarctica' in K. Dodds, A. Hemmings and P. Roberts (eds) *Handbook on the Politics of the Antarctic*. Cheltenham, UK: Edward Elgar. 183–198.

Playbill for Hobart Performance. 1841. *The South Polar Expedition*. National Library of Scotland, GB/C.235.

Pyne, S.J. 1986. *The Ice: A Journey to Antarctica*. Iowa City, IA: University of Iowa Press.

Rowland, J.R. 1994. 'Antarctica' in *Granite Country: Poems*. Deakin, Australia: Brindabella. 49.

Smith, B. 1992. *In the Wake of the Cook Voyages*. Carlton, Australia: Melbourne University Press.

Spufford, F. (ed.) 2007. *The Ends of the Earth: An Anthology of the Finest Writing on the Arctic and Antarctic*. Vol. 2. New York: Bloomsbury.

Teorey, M. 2004. 'Sir Ernest Shackleton's miraculous escape from Antarctica as captivity narrative' *English Literature in Transition* 47(3): 273–291.

Tuan, Y.F. 1993. 'Desert and ice: ambivalent aesthetics' in Salim Kemal and Ivan Gaskell (eds) *Landscape, Natural Beauty and the Arts*. Cambridge, UK: Cambridge University Press.

Wijkmark, J. 2009. '"One of the most intensely exciting secrets": the Antarctic in American literature, 1820–1849'. PhD thesis, Karlstad University, Sweden.

5

Self-determination and indigenous governance in the Arctic

Mark Nuttall

Across the Arctic today, indigenous peoples live as citizens within the political and administrative systems and legal regimes of seven of the circumpolar North's eight nation-states. Iceland had no indigenous population when the first Norse settlers arrived there from Scandinavia in the second half of the ninth century, but the northern reaches of the other Arctic countries have constituted indigenous homelands for several thousand years. Diverse indigenous societies and cultures inhabit the circumpolar North, with livelihoods often finely-tuned to the rhythms, movements, and seasonal flows of different worlds of ice, water, coast, tundra, forest and mountain environments, but they share similar experiences with the histories and long-lasting effects of colonization and social and cultural change. For some, indigenous movements for land claims and self-government over the last fifty years have led to political processes which have brought greater indigenous autonomy, especially in the North American Arctic, while in other parts of the circumpolar North, such as in Russia, indigenous peoples continue to struggle to have their rights recognized and, in some cases, remain at risk of being dispossessed of traditional lands and livelihoods.

From the sixteenth century (in the case of Greenland and parts of the North American Arctic, and earlier in the Eurasian Arctic), but especially during the nineteenth and early twentieth centuries, exploration and the exploitation of northern lands and waters meant frequent and prolonged contact between indigenous peoples and outsiders, often in the form of seasonal visitors such as crews on voyages of geographical discovery, whalers, and the establishment of longer-term trade relations initially, but later in colonization processes and the presence of the state in everyday life in the North. Explorers, whalers, traders and missionaries brought new economic, cultural and religious influences, institutions and systems with them, but they also often carried, unwittingly perhaps, infectious diseases to northern regions to which indigenous peoples had little or no immunity and resistance. Over the past 70–100 years in particular, indigenous societies and cultures across the Arctic have been transformed by significant and often disruptive social, economic and political changes, including relocation by the state into more permanent settlements, the curtailment of nomadic practices, and limits placed on traditional resource use. Despite this, a remarkable continuity and social and cultural resilience is evident in how indigenous peoples today rely on the procurement of terrestrial and marine resources (such as by hunting and fishing), and how they maintain strong relationships to the environment through these activities.

Today, the Arctic's indigenous peoples include the Iñupiat, Yup'iit, Alutiit, Aleuts and Athabaskans of Alaska; the Inuit, Inuvialuit, Athabaskans and Dene of northern Canada; the Kalaallit and Inughuit of Greenland; the Sámi of northern Fennoscandia and northwest Russia's Kola peninsula; and the Chukchi, Even, Evenk, Nenets, Nivkhi, Yukaghir and many other groups of the Russian Far North and Siberia. For many people in contemporary indigenous communities (including, in some cases, larger urban centres in Greenland, Alaska or Russia), the living resources of land and sea, or freshwater environments, provide the basis for livelihoods as hunters, fishers and reindeer herders, as they have done for centuries. For Inuit in Canada and Greenland, for example, the most commonly harvested species are marine mammals such as seals, walrus, narwhals, beluga, fin and minke whales, and polar bears and land mammals such as caribou, reindeer and musk–ox; and fish such as cod, salmon and Arctic char. While, for some communities, they are primarily subsistence resources essential for family livelihoods, food security and household consumption, living marine and land resources also figure prominently in the cash–economy of households and communities, or form the basis for wider regional economies and participation in international markets, such as those centred on commercial fisheries in Greenland or reindeer herding in Fennoscandia and Russia.

Many Northern communities, though, struggle with the legacies and contemporary realities of rapid social, cultural and economic change, of resettlement from small hunting, fishing and reindeer herding camps and villages into newly created towns, of increasing urbanization, the effects of globalization, the erosion of tradition and a struggle for cultural survival and political autonomy and economic independence. Some of these legacies and effects are evident in the health and well-being of indigenous populations (see the chapter by Helle Møller in this handbook). And as a number of chapters in this handbook describe, social, economic and environmental changes affect Arctic environments, wildlife, and communities in profound ways. In many parts of Alaska, northern Canada, northern Fennoscandia and the Russian North, tundra and boreal environments have been disturbed by extensive industrial development, such as oil extraction facilities, pipelines and trails from seismic surveys, mining projects (abandoned mine sites have their own toxic legacies – see, e.g., Keeling and Sandlos [2015]), or commercial forestry and clear-cut logging. Decommissioned airstrips, the Distant Early Warning (DEW) line, and other military installations, also bear witness to the strategic importance of the circumpolar North during the Cold War. Arctic communities have also experienced, and are still experiencing, stress from a number of other different processes that threaten to restrict harvesting activities and sever these relationships to animals and the environment. Melting and disappearing ice, for instance, hinders indigenous mobilities and resource use patterns in northern Greenland, Canada, and Alaska, while multiple use conflicts, such as forestry, mining, hydropower development, and tourism, have implications for reindeer herding in Norway, Sweden and Finland. Wildlife management, conservation policies, environmentalist interventions, and increasing global interest in the Arctic, evident in resource development, for instance, also bring their own quite considerable challenges.

From the late 1960s onwards, and in response to such tremendous and often disruptive change, influences and restrictions, indigenous political and cultural organizations across the Arctic have emerged and have focused on the right to self-determination based on the historical and cultural rights to the ownership and use of lands and resources, as well as seeking redress for historic abuses by the state and colonial agents. Land claims campaigns and struggles for self-government are often at the heart of these movements. Today, new challenges arise from the transformative effects of environmental and climate change and other global processes. The Arctic regions are tightly tied politically, economically and socially to the national mainstream of circumpolar states and are linked to and embedded within global economic systems, institutions and practices, which bring opportunities but also further disrupt traditional subsistence economies. Social, economic and demographic change has been rapid and sweeping in some places, and urbanization is an increasing trend while resource development and extractive industries, trade barriers and animal–rights campaigns (which influence international attitudes and markets for furs and sealskins) have all had their impacts on hunting, trapping, herding, fishing and gathering activities in recent decades. In pushing for effective forms of autonomy, indigenous peoples are also confronted with the need for governance institutions to respond to and deal with the effects of social, economic and environmental change, but to meet the opportunities such transformations can also bring.

Land claims and self-government

Notwithstanding the effects of government policies of assimilation, processes of marginalization and forced resettlement in some cases, one defining feature of contemporary Arctic indigenous politics is an assertion that indigenous peoples continue to exist as distinct political and cultural communities. Through indigenous rights movements and indigenous peoples' organizations they seek to regain and exert control over their lives and lands, and aspire to determine their futures guided by indigenous ideas of governance and sovereignty (see the chapter by Jessica Shadian in this volume).

Some Arctic states have recognized the need to settle some of the claims indigenous peoples advance for land and self-government, or have sought to pass legislation that recognizes the need for dialogue on land claims and resource rights. In Canada, for example, the federal government established a policy framework in 1973 for the negotiation and settlement of land claims by Aboriginal people. Canada's 1982 Constitution Act recognizes and affirms the Aboriginal and treaty rights of the country's indigenous peoples, i.e. the Inuit, First Nations, and Métis, and the Inherent Right Policy of 1995 set out how self-government negotiations and arrangements can be considered as part of comprehensive land claims. Despite this, the process of dealing with land claims has been extremely slow, involving lengthy and bureaucratic negotiation processes that go on for decades in some cases. Most agreements made to date have been with Inuit groups in the Arctic and with First Nations in the Northwest Territories and Yukon.

Since the 1970s, a number of significant settlements have been negotiated between the governments of nation–states and indigenous populations that have resolved (either wholly or in part) land claims issues, or which have granted indigenous communities varying degrees of self-government. Notable among these are the Alaska Native Claims Settlement Act (ANCSA) of 1971, Greenland Home Rule in 1979 and Self Rule in 2009, and in Canada the James Bay and Northern Quebec Agreement (1975), the Inuvialuit Agreement (1984), comprehensive land claims agreements with the Gwich'in and Sahtu Dene in the early 1990s, and the creation of the new territory of Nunavut in 1999. For the Sámi of northern Fennoscandia, Sámi Parliaments have been established in Norway, Sweden and Finland. They promote political initiatives and manage directives and laws delegated to them by national governments and authorities, rather than acting as autonomous decision-making bodies. In Finland, for example, Sámi are recognized as an indigenous people in the Finnish Constitution. Effectively, they have had constitutional self-government concerning their language and culture in their homelands since 1996. They have a right to maintain and develop Sámi language and culture as well as traditional livelihoods such as reindeer herding. Some Sámi leaders argue, however, that this allows Finnish governments to consider Sámi more as a linguistic minority than an indigenous people. Moreover, land use issues are key to relations between the Sámi and Finnish government – Sámi have no secure land rights because 90% of what Finnish Sámi consider and claim to be traditional land belongs to the state. In Norway and Sweden, negotiations are ongoing between national governments and their Sámi populations over land, water, and resource use rights, and the effects of mining and other major development projects remain issues of significant concern for Sámi communities and organizations. There is no official census, but some 100,000 Sámi are estimated live in the Nordic Arctic and Northwest Russia (c. 50,000 in Norway, 40,000 in Sweden, 9,000 in Finland, and 2,000 in Russia).

The situation regarding land claims, land and resource rights, and indigenous autonomy in Russia differs greatly from the North American Arctic. Although indigenous minorities of the Russian North and Siberia were given certain rights and privileges under the Soviets, these have not always been recognized and implemented and many indigenous groups continue to work on efforts to achieve regional autonomy. In Russia, indigenous peoples are not defined or recognized as such by national legislation, but 40 peoples are legally recognized as "indigenous, small-numbered peoples of the North, Siberia and the Far East". This status is tied to the specific conditions that a group or people has no more than 50,000 members (this numerical qualification is what "small-numbered" refers to), maintains a traditional way of life, inhabits certain remote regions of Russia and identifies itself as a distinct ethnic community. A definition of "indigenous" without the numerical qualification does not exist in Russian legislation. Based on these criteria, for example, Evenks, Sámi, Yup'ik, Nenets, Chukchi and Yukaghir are among the peoples recognized as indigenous, whereas other peoples of northern Russia and Siberia such as Sakha, Buryat, Komi, and Khakass do not hold this status because of their larger populations, although this numerical definition is not consistent with how people self-identify as indigenous. The criteria of tradition and remoteness also act to essentialize the idea of who is, and what it means to be, indigenous, and categorize people as indigenous only if they carry out subsistence practices based on living resources in forests, tundra and along the coast. The small-numbered indigenous peoples comprise approximately 260,000 individuals in total – less than 0.2% of Russia's population (ethnic Russians, for instance, account for 78%). The small-numbered indigenous peoples are protected by Article 69 of the Russian Constitution and three federal framework laws that establish cultural, territorial and political rights. However, the implementation of regulations contained in these laws is complicated by subsequent changes to natural resource legislation and government decisions on natural resource use in the North. Indigenous peoples in Russia and Siberia increasingly feel that indigenous rights are gradually being eroded. In recent years, this has been experienced through the ways regional and national indigenous associations are subject to increasing surveillance and control by the state, the fact that groups concerned with indigenous rights risk being declared as 'foreign agents', and the distribution and allocation of land to settlers from other parts of the Russian Federation in the Russian Far East.

No land claims or self-government agreement has yet been so extensive as to give an indigenous group full political independence – and, with the exception of Greenland (which has a population of 57,000 – some 89% of which is Greenlandic Inuit), perhaps, this is not what indigenous peoples necessarily aspire to in much of the Arctic – but the decentralization of specific administrative functions and responsibilities to local and regional governments has given some indigenous peoples responsibility for overseeing and managing self-governance arrangements and institutions. The arrangements where indigenous peoples have the greatest autonomy are to be found in the models of tribal sovereignty/self-government and land claims in Alaska, in comprehensive land claims agreements in Canada, and in the system of extensive self-government in Greenland. In Canada, for example, land claims and self-government negotiations are often a matter of devolution whereby some matters of political control and decision-making authority are transferred from the federal government to First Nations or indigenous regional governments, but they also put in place the possibilities for the emergence of distinctive forms of indigenous governance. The extent to how far governments – and hence what social, cultural and political rights people have had to be in positions of power, to assert that power, and to govern – have existed in indigenous societies historically has been a subject of considerable debate. For example, in the Arctic the emergence of the idea of a common Greenlandic society and nation, and the eventual formation of political parties and a parliamentary government were themselves products of colonialism in a land where formal leadership had been practically absent in Inuit society (Dahl 2005).

It is important, though, to make a distinction between government and governance. Government is usually understood to refer to structures and arrangements of specific kinds of public and state institutions that are vested with authority by the state to make decisions on behalf of an entire community, country, or nation, whereas governance, while including institutions and instruments of government, also encompasses other social forms, practices, institutions and non-governmental organizations that play a role in making decisions. In a sense, governance encompasses both formal and informal aspects of decision-making, and involves aspects of civil society. Writing about indigenous forms of governance, Gail Fondahl and Stephanie Irlbacher-Fox describe governance:

> [a]s the exercise of legitimate authority within a group to make decisions regarding the allocation of resources and the coordination and management of communal and, to some extent, individual activities. The term refers to the principles, institutions and practices that a collective employs to regulate relations among its members, and between its members and the external world. Governance stipulates how resources are shared and managed; it guides social relations. It is informed by endogenous norms and practices mediating such relations.
> *(Fondahl and Irlbacher-Fox 2009: 2)*

The ability of indigenous peoples to govern, they argue, and the ability to have the power and autonomy to make authoritative decisions about the use and management of the environment and its resources, and of social practices, "depends on the knowledge of one's environment, and the demonstration of this knowledge through skilled practices" (ibid: 6).

The political changes accompanying land claims and self-government agreements often include structural, bureaucratic and institutional changes in the ways that living and non-living resources are managed, utilized, and traded and distributed, such as the development of co-management models which allow indigenous peoples and communities active involvement in the management and conservation of lands and resources (such as ecosystems, water, caribou, and whales, for instance). A far greater degree of local participation and involvement in resource use management decisions has been introduced, including in some cases the actual transfer of decision-making authority and environmental monitoring to the local community, local institutions, or regional level organizations, thus enabling the use of indigenous knowledge in matters of governance. As self-government is about being able to practice and enact autonomy, the devolution of decision-making authority to community and regional institutions and organizations, and the introduction of forms of co-management for wildlife and the environment, allow indigenous peoples opportunities to improve how management and the regulation of resource use considers and incorporates indigenous views and traditional resource use systems (Huntington 1992). Co-management projects – at least in terms of how they are negotiated and how they are supposed to operate in practice – involve greater recognition of indigenous rights to resource use and present opportunities for collaboration between indigenous peoples, scientists, and policy-makers concerned with the sustainable use and management of living resources. For example, governance mechanisms introduced in northwestern Canada through the Inuvialuit Final Agreement of 1984 are helping Inuvialuit to negotiate and manage the impacts of environmental change. For instance, the five co-management bodies established by the agreement provide a way for Inuvialuit communities to communicate with regional, territorial, and federal governments and, indeed, to wider circumpolar bodies concerned with conservation and sustainable development such as the Arctic Council.

Land claims movements in Arctic North America emerged at a time when a number of major oil, gas and mining projects were being planned in Alaska, northern Canada and Greenland, and indigenous peoples were anxious about the threats they posed to the environment, living marine and terrestrial resources, and to indigenous rights, interests and entitlements. In the 1970s, the Greenlandic Home Rule movement was concerned partly with the possible social and environmental impacts of mining and oil and gas exploration, while the Berger Inquiry in Canada assessed the prospects of a Mackenzie Valley pipeline project – the main recommendation to the federal government being that a moratorium should be placed on pipeline development and oil and gas activities until indigenous people had their land claims sorted out, as such development would not likely benefit them. The discovery of oil at Prudhoe Bay on Alaska's North Slope, for example, together with indigenous concerns over other large-scale industrial development projects, led to the establishment of the Alaska Federation of Natives (AFN) in 1967. Given the prospect of large-scale oil development and the construction of the trans-Alaska oil pipeline system, AFN lobbied the United States Congress for the appropriate settlement of land claims for Alaska Natives and, in 1971, Congress passed the Alaska Native Land Claims Settlement Act (ANCSA). ANCSA did not recognize an indigenous claim to the whole of the state of Alaska, but it did establish twelve regional Native corporations, giving them effective control over one-ninth of the state. ANCSA extinguished Native claims to the rest of Alaska and $962.5 million was given in compensation. In effect, ANCSA made Alaska's Native people shareholders in corporate-owned land and required them to establish business models for community and regional development (there are around 120,000 indigenous Alaskans in the state).

Self-determination and self-government: some examples from Canada and Greenland

Inuit self-government in Canada

In response to disruptive social change precipitated by the state and agents of colonial activity, cultural rupture and the prospects of large-scale economic development, Canadian Inuit political leaders have worked to advance efforts for the recognition of self-determination and to achieve degrees of self-government in recent decades. In 1969, faced with oil and gas exploration and development in the Mackenzie Delta and Beaufort Sea, the Inuvialuit of the northwestern Arctic formed the Committee of Original People's Entitlement (COPE), and in 1971 the Inuit Tapirisat of Canada (ITC, and now called Inuit Tapiriit Kanatami, or ITK) was founded in Ottawa as a national voice for Inuit throughout Canada's North. Key to the work of ITC and other Inuit community and regional organizations was the initiation of land use and occupancy studies. The Inuit Land use and Occupancy Project, for instance, was carried out in the Northwest Territories in 1974–75, while another land use and occupancy project was carried out in Labrador in 1975–76. These extensive studies mapped and gathered knowledge of historic and contemporary Inuit use of land, waters and ice, emphasized social and cultural relationships with the environment and with animals, and formed the basis for subsequent land claims negotiations and the delineation of Inuit homelands.

In 1984 the Inuvialuit Final Agreement accorded title to over 90,000 square kilometres of the Northwest Territories to the Inuvialuit, together with financial compensation and other rights (including gas, petroleum and mineral rights for almost 13,000 square kilometres of subsurface) as a final settlement of territorial claims. In 1975 the Inuit of northern Quebec signed a land claims agreement against the backdrop of controversy surrounding hydroelectric development in James Bay. In 1992 the Tunngavik Federation of Nunavut (now Nunavut Tunngavik Incorporated) and the Government of Canada signed an agreement on Inuit land claims and harvest rights committing the federal government to establishing Nunavut ("our land") in the Canadian Eastern Arctic. The Nunavut Land Claims Agreement (NLCA) was signed on 25 May 1993 and the territory of Nunavut, comprising 2.093m km^2 of northern Canada, was inaugurated on 1 April 1999. Under the NLCA, Inuit have title to 350,000 km^2 of Nunavut (and mineral rights to 35,257 km^2). Nunavut Tunngavik Incorporated administers the land claim on behalf of Inuit beneficiaries.

Canadian territories and provinces have extensive public self-government within the limits defined by the Canadian constitution and it is important to point out that the Nunavut settlement did not create a new ethnically-defined Inuit region, but public government. The population of Nunavut is around 80% Inuit, however, and the government is effectively Inuit-led. At the time, the Nunavut settlement was significant in that it gave the Inuit of the eastern Arctic a greater degree of autonomy and self-government than enjoyed by any other aboriginal group in Canada. The Inuit homeland of Nunavik in northern Quebec is currently subject to negotiations for status as a self-governing region within the province. An agreement was signed in December 2007 between the Inuit of Quebec (through their legal representative body, the Makivik Corporation), the provincial government of Quebec and the federal government of Canada, to create a new regional government in Nunavik, although this will be public government, not a form of Inuit self-government, representing all citizens of Nunavik. In northern Labrador, following ratification of the Labrador Inuit Land Claims Agreement, the government of Nunatsiavut was established in 2005 to represent the rights of Labrador Inuit. Nunatsiavut is distinctive in that the Labrador Inuit Land Claims Agreement included provisions for self-government within the land claim, making Nunatsiavut the first of Canada's Inuit regions to achieve self-government (Nunavut, by contrast, is a territory and the government is a form of public government, while the Nunavut land claim and regional Inuit organizations mean Inuit in the territory have specific rights, such as access to lands and resources, that other residents of Nunavut do not). While Nunatsiavut remains part of the province of Newfoundland and Labrador, the Inuit regional government has authority over many areas of governance including culture and language, education, health, justice, and community matters. The four Inuit regions of Canada – the Inuvialuit region, Nunavut, Nunavik and Nunatsiavut – constitute Inuit Nunangat, a term Inuit feel is more encompassing in how it describes not just land, but water, ice, and air (the total Inuit population in Canada is around 50,000). In February 2017, the Government of Canada signed the Inuit Nunangat Declaration on the Inuit-Crown Partnership with Inuit Tapiriit Kanatami, Inuvialuit Regional Corporation, Makivik Corporation, Nunatsiavut Government, and Nunavut Tunngavik Incorporated. The declaration aims to renew the Inuit-Crown relationship based on the recognition of rights, respect, co-operation, and partnership as part of a broader goal of achieving reconciliation between the federal government and the indigenous peoples of Canada.

Self-government and First Nations in Canada's Yukon Territory

In Canada's Yukon Territory, approximately 26% of the 33,000 people living there are Aboriginal. There are fourteen First Nations groups. Yukon Territory has four levels of government: federal, territorial, First Nation, and municipal. The federal government has ownership and control of the territory's public land, water and resources. First Nation land claims settlement negotiations have been ongoing since 1973 and the Yukon territorial government has been negotiating the transfer of federal programmes to local and regional control. An example relevant to resource development and environmental protection is the devolution to the Yukon territorial government and First Nations of responsibility for environmental assessment under the Yukon Environmental and Socio-Economic and Assessment Act (YESAA). Under YESAA development assessment legislation is a process required of every Yukon First Nation Final Agreement and ensures that the concerns and aspirations of Aboriginal people in the assessment of development projects are recognized. Specifically, First Nations can use YESAA in their discussions and negotiations with industry and government to make them aware of Aboriginal concerns about social, cultural and health issues. The Umbrella Final Agreement provided a comprehensive framework for Yukon First Nations final and self-government agreements which have subsequently been realized for eleven First Nation governments which now operate as self-governing jurisdictions under the federal Yukon First Nation Self-Government Act (1995). They have responsibility for the administration of land claims rights and benefits (Leas 2005; Roddick 2006). Another three Yukon First Nations, still negotiating their land claim settlements, operate as band councils under the federal Indian Act. Most Yukon First Nation governments also participate in one or more regional tribal organization. The largest regional body, the Council of Yukon First Nations (CYFN), represents nine self-governing Yukon First Nations. The Kaska Dena Council represents five-member governments in south eastern Yukon and British Columbia, and the Gwich'in Tribal Council represents four communities in northern Yukon and the Mackenzie River Delta area of the Northwest Territories.

Greenland: colonization, the emergence of an Inuit nation and self-government

In 1721, Hans Egede, a Norwegian–Danish Lutheran priest, arrived on Greenland's west coast. Establishing a trade and mission station near present-day Nuuk, Greenland's capital, his activities marked the beginning of over 230 years of Danish colonial rule over the indigenous Inuit inhabitants. The Danish authorities assumed responsibility for trade in 1726, hoping to establish a viable and lucrative trade network based on marine mammal products (mainly seal and whale oil, sealskins, fish and fox furs) and the transfer of trading rights in Greenland to independent companies. This proved unsuccessful in the long term, so the Danish government formed the Kongelige Grønlandske Handelskompagni, or KGH (Royal Greenland Trade Company) in 1774, thus establishing a Danish trade monopoly in Greenland that was to last until after the end of World War II (Nuttall 1994).

By 1814, the majority of Greenland's indigenous Inuit were involved in a trading economy controlled and monopolized by Denmark. Yet, some Inuit populations were still relatively isolated, having little or no contact with Europeans. The Inughuit (or Polar Inuit) were visited in northwest Greenland by John Ross in 1818, Douglas Clavering met and traded with groups of Inuit in northeast Greenland in 1823 (indeed this is the only recorded contact between Northeast Greenlanders and Europeans in what is now an uninhabited part of Greenland), and in 1884 Gustav Holm wintered with the people of Ammassalik on the east coast. Despite the changes to Inuit society and culture that inevitably occurred a result of Danish involvement in Greenland, the majority of the Inuit population continued a hunting and fishing lifestyle – indeed this was encouraged and promoted by the KGH to sustain its activities. Seals and other marine mammals provided the mainstay of the local economy, with skins and blubber underpinning the trade economy (Nuttall 1992).

The idea of Greenland as a country and Greenlanders as a people with a common identity inhabiting an emerging nation began to take shape during the nineteenth century. The establishment of a printing house in Nuuk during the 1850s was an important development for Greenlandic as a written language. From it the newspaper *Atuagagdliutit* was first published in 1861. Originally, *Atuagagdliutit* came out monthly and its significance lay not only in being a source of news in Greenlandic about Greenland and the outside world, but in providing a medium for cultural expression and playing a role in political and intellectual development. It paved the way for the beginning of a Greenlandic literary tradition and a nascent nationalist movement that began to argue that Greenlanders should be involved in the government of their own land.

Following the end of World War II, Denmark ended its isolationist policy toward Greenland and began a process of modernization. Colonial status was superseded in 1953 when Greenland became an integral part of the Kingdom of Denmark, thus giving Greenlanders equal status to Danes. The ending of colonial rule marked the beginning of another era characterized by profound and extensive social, economic and political changes in Greenlandic society. Improved health care meant the population began to increase, and economic development was now based almost entirely on a commercial fishing industry. During the 1960s, the Danes implemented controversial policies of centralization and urbanization: many Inuit were moved from small, remote settlements and relocated in the growing west coast towns.

By the late 1960s and early 1970s, Greenlandic society had been transformed from one based primarily on small-scale hunting and fishing to a modern export-oriented economy based on commercial fishing. The majority of the Inuit population was now living in urban centres and this demographic transition brought its own social and economic problems. These upheavals and profound changes led to the politicization of Inuit culture and identity and the formation of political parties and the beginnings of a movement for Home Rule. The Danes were well aware of the extent of dissatisfaction felt by Greenlanders and recognized that a change in the relationship between Denmark and Greenland was both necessary and desirable. A Home Rule Commission was set up in 1975, followed by the passing of the Home Rule Act three years later. Greenland Home Rule was established by referendum in January 1979, and the first Greenlandic government was elected in April of that year. Legislative and administrative powers in a large number of areas and public institutions were quickly transferred to the Home Rule authorities. Greenland left the EC in January 1985 (it had joined with Denmark in 1973), but negotiated Overseas Countries and Territories Association status, which allows favourable access to European markets.

On 25 November 2008, 75.5% of those who went out to vote in a referendum on self-governance did so in favour of greater autonomy. Greenlanders thus gave their political leaders a mandate for significant political and economic change. Thirty years after Home Rule came into being, it ended on 21 June 2009 when the new political arrangement of Self-Rule was instituted. Areas that could come under Greenlandic control include the justice system, police system, prison affairs, and the coastguard. Greenland will also be able to represent itself on the stage of international affairs increasingly in the future, although foreign policy remains a matter of Danish government jurisdiction. Greenlandic (*Kalaallisut*) has now become the country's official language. The Self-Rule agreement also recognizes that Greenlanders are a nation with an inherent right to political independence if they choose it. Despite the challenges ahead, and irrespective of whether it means eventual independence, Self-Rule is a statement expressing growing cultural and political confidence in a country of only 57,000 people, over 80% of whom are Inuit (Nuttall 2008).

Greenland has often been considered a model for indigenous self-government, but it has been a process of nation-building rather than an ethno-political movement (Nuttall 1992, 2017). Its relevance goes beyond that of self-determination for indigenous peoples and says much about the aspirations for autonomy in small political jurisdictions and stateless nations (Nuttall 2008, 2017). Independence from Denmark dominated the campaign leading up to the general election in April 2018, but questions dominate Greenlandic politics about how the country will pay for the other responsibilities it has started to take over from the Danish state and how it will lay the foundations for a sustainable economy. The Greenlandic economy remains dependent for 60% of its budget revenue on a 3.5 billion DKK annual block (around 470 million EUR) grant it receives from Denmark, with the balance coming from local taxes. The main challenge to securing greater self-government and economic independence is to replace the block grant with revenues generated from within Greenland. This requires the development of new economic initiatives and industries and, for the last few years, extractive industries have often dominated political discussion about economic development. The Danish-Greenlandic Self-Rule Commission was established in 2004 to negotiate the terms of greater self-government. It considered Greenland's claim to mineral rights, its ownership of subsoil resources, and right to the revenues from non-renewable resource development. The commission concluded that minerals in Greenland's subsoil belong to Greenland and that the country has a right to their extraction. Under the Self-Rule agreement, the income generated by subsurface resource development would be administered by Greenland, with the level of the block grant being reduced by Denmark by an amount corresponding to 50% of the earnings from minerals and energy extraction once they exceed 75 million DKK. Future revenues from oil and mineral resources will then be divided between Greenland and Denmark while the annual block grant is reduced further and eventually phased out. Political and public debates continue to focus on whether non-renewable resource development could provide the economic basis for greater autonomy and possible independence. A number of mining projects are in the planning stages (a ruby mine opened in southwest Greenland in May 2017) and oil exploration has been taking place in coastal waters. Extractive industries and public participation in decision-making processes concerning the development of subsurface resources remains a contested political, economic, social, and cultural issue (Nuttall 2017).

Environmental change and indigenous governance

Barry Scott Zellen argues that land claims in the circumpolar North represent "the first concrete step in the process of decolonizing the North by devolving decision-making authority from what many northerners long perceived to be far away colonial centers of administration and decision-making to local communities" (Zellen 2008: 8). But, he argues, "by letting go, central authorities were in fact strengthening their hand, gaining greater political legitimacy through their new collaboration, co-management, and devolutionary policies" (ibid.). In both Canada and Alaska, Zellen suggests, traditional ideas of sovereignty have been entwined with ideas of national sovereignty and while indigenous peoples have gained legal title to their homelands – and in the process have become integrated more into mainstream Canadian and Alaskan society – settling land claims has an advantage for the nation state and sovereign claims to the North.

Zellen has a point, and there is certainly no shortage of critics of the way land claims in Alaska and Canada have been negotiated, the political reasons for negotiating them in the first place, and of the corporate, profit-making nature of the institutions some land claims processes ended up creating, especially ANCSA (and which often rub up against indigenous notions of human-environment and human-animal relations). However, traditional lands, as Anderson, Schneider and Kayseas (2008), argue, cannot be separated from indigenous people and indigenous identities and cultures. Traditional resource use practices and indigenous perceptions of the environment and attitudes towards animals remain vital for expressing and maintaining social relationships and cultural identity in indigenous societies. Hunting, herding, fishing, and gathering activities in today's Arctic, for example, are based on continuing social relationships between people, animals, and the environment. MacKay states that

> For indigenous peoples, secure and effective collective property rights are fundamental to their economic and social development, to their physical and cultural integrity, and to their livelihoods and sustenance. Secure land and resource rights are also essential for the maintenance of their worldviews and spirituality and, in short, to their very survival as viable territorial and distinct cultural collectivities.
>
> *(MacKay 2004: 49)*

This underscores the significance of indigenous understandings of land and human–environment and human–animal relations as not merely something that can be possessed and utilized in a material and productive way, but as encompassing cultural meanings and spiritual elements (MacKay ibid.), and how indigenous peoples think of their surroundings as being composed of human and non-human entities and elements. As Ortiga (2004: vi) points out, "land is not only a physical asset with some economic and financial value, but an intrinsic dimension and part of people's lives and belief systems". For Inuit communities, for example, living marine resources not only sustain livelihoods across the North American Arctic in an economic and nutritional sense, the relationships people have with animals and the non-human elements of Arctic places underpin and inform social identity and are essential to cultural survival (Nuttall 2017). Land, waters, resources and traditional knowledge become central to movements for autonomy and sovereignty as they are foundations upon which indigenous peoples argue that communities can secure rights to continue customary practices and ensure sustainable livelihoods, so that they can improve their socio-economic circumstances and participate in regional and global processes.

Indigenous homelands, people's livelihoods and traditional resource use in the Arctic are also being challenged today, however, by environmental change such as witnessed in the effects of climate variability and a rapidly warming Arctic. The Arctic is warming at a rate that alarms the scientists who study and monitor climatic processes, and changing weather patterns are having considerable effects on northern environments, on indigenous and local livelihoods, and on wider northern economies (AMAP 2017; Henshaw 2009; Krupnik and Jolly 2002; Marino 2015). These effects are increasingly apparent through diminishing glacial ice, the retreat and thinning of sea ice during winter as well as summer, thawing permafrost, coastal erosion, an intensification of stormy weather, and changes to the migration routes and population sizes of a number of animal and fish species (ACIA 2005; AMAP 2012, 2017).

Increasingly, the impacts of climate change on Arctic environments and wildlife populations will also have implications for renewable resource harvesting activities. Because many northern species of terrestrial and marine mammals as well as freshwater and ocean fish are a cornerstone of local community and regional economies, climate change poses significant threats and risks to local community and wider regional food security in the circumpolar North because it influences and effects the distribution and health of animal populations, the human ability to access wildlife, and the safety and quality of wildlife for consumption (Meakin and Kurvits 2009). For example, in northern Canada, residents of First Nations communities in both Yukon Territory and the Northwest Territories have been witness to changes in climate that affect the presence of wildlife in local community areas and are experiencing difficulties in hunting and fishing. There has been a corresponding decline in the nutrient intake from traditional foods (Guyot et al. 2006). In Alaska, northern Canada and Greenland climate change has increased the cost and risk of hunting and fishing activities. On the coast of northern Alaska, for example, where the ice pack has retreated a significantly greater distance from land, North Slope hunters have to cross an ever greater expanse of open water to reach hunting grounds. The increased time and distance added on to a hunting trip adds to the cost and risk of accessing marine mammal resources. Fuel and maintenance costs are greater because of the longer distance to travel, which also decreases the use and expectancy of the technology used (boats, engines, rifles). For safety reasons, boats with larger engines are required, adding strain to limited budgets (Nuttall et al. 2005).

In Northwest Greenland, where the effects of climate change are evident in the daily lives of people, the entire land fast ice regime has undergone significant changes over the past few decades. Because the marine ecosystem supports the livelihoods of communities in the region, changes in sea ice have far-reaching effects for hunting and fishing activities, for mobility, and for local economies. Projections of future sea ice conditions point towards continued declining drift ice in Baffin Bay and the decline and continued thinning of land fast ice. Climate change is rapidly altering the physical environment, leading to constant gradual changes in seasonal hunting activities and transportation patterns, which again has social, cultural, economic and ultimately political consequences (Hastrup 2009; Nuttall 2017). Hunters throughout Northwest Greenland report that the period of travel by dog sledge on good, solid sea ice is now only around three months during winter and spring (a significant decrease), with decent, but somewhat fluctuating conditions for another month or so. Near Melville Bay communities, hunters say that sea ice is now only best in March and April, reducing the amount of time for hunting and fishing by dog sledge significantly. Hunters also report that sea ice is of a different texture and consistency. The ice edge near Qaanaaq, for example, is a place of constant shifts and movement. This instability makes ice-edge hunting more difficult and dangerous. Changes in snow cover present difficulties in accessing hunting and fishing areas by dog sledge or snowmobile, making local adjustments in winter travel, and in hunting and fishing strategies, necessary. An increasing level of risk in travelling far out on the ice is highlighted by community members. The longer open water period in the region, however, opens up new possibilities in terms of hunting from boats and kayaks. In some seasons the open water period near the fjord mouth may be almost doubled in length for hunters from Qaanaaq. Around Kullorsuaq in the Upernavik district, hunters have, in recent years, been hunting by boat during some periods when there is open water during winter and spring. Local adaptive strategies include exploring new fishing grounds, seeking alternative sources of income, and greater reliance on boats during the increasingly ice-free water periods in winter (Nuttall 2017). Even if responding and adapting to an ever-changing sea ice environment or anticipating the possibilities of successful engagement with surroundings being transformed by climate change, is inherent in Inuit culture and local socio-economic practices, the implications are many.

Conclusions

As the global climate warms, dramatic changes are occurring at high latitudes regularly and with great speed. For many northern residents, climate change is a lived experience that affects local societies, cultures and economies. Accompanying concern over a melting region, however, the Arctic is no longer being viewed as an icy, remote space at the top of the world, but a dynamic region that is 'open for business'. The resilience of Arctic communities is not only challenged by climate change, but by governance systems and institutions that often inhibit and constrain locally-specific, long-term resource availability around communities and the entitlement of individuals and rights of communities to access those resources.

Local, national, circumpolar and global agendas and interests interact, and often conflict, but they are influencing and shaping trends and processes that will have a significant bearing on the future of the Arctic. The future of indigenous livelihoods and economies too, and which are often based on the living marine and terrestrial resources of the Arctic, will not only depend on ecosystem diversity, but on how far the institutional rules which manage wildlife and the environment, and which govern social and economic systems, enable conditions for sustainability and allow for indigenous participation in new business initiatives. The question of who has rights over access to Arctic lands and waters, as well as who has ownership of resources and rights to their exploitation, has shaped historical and contemporary relations between indigenous peoples and Arctic states. Land claims and self-government have given indigenous peoples ownership of traditional lands and waters, but also of subsurface resources in some parts of the Arctic. Sovereignty also means different things from an indigenous perspective. But just as Arctic peoples assert their rights, many other interests are at play in the high latitudes of the world as non-Arctic states and different organizations become more engaged in the region. Conservation and ideas concerning the protection of Arctic regions and Arctic species, for example, often differ from indigenous understandings of animals and the environment (Nuttall, Chapter 32, this volume). Ice, waters and lands are being re-imagined as special ecosystems under threat. Controversies over future conservation and management of Arctic landscapes and marine areas, as well as Arctic wildlife such as polar bears, whales and caribou, often pit the cultural interests and rights of Inuit and other indigenous peoples against those of scientists, environmentalists and nation states.

What does this mean for indigenous livelihoods? How can indigenous knowledge and scientific knowledge be integrated? Can indigenous governance institutions create additional opportunities to increase resilience, flexibility and the ability to deal with change and with the increasing global interest in the Arctic? How can, for example, new environmental governance mechanisms which incorporate local knowledge and allow local participation in decision-making help people negotiate and manage the impacts of climate change, or ensure biodiversity conservation and effective wildlife management; or how are indigenous economies going to benefit from growing business and commerce activities in the circumpolar North? The answer to these questions will depend on a range of factors, including the importance of understanding the nature of the relationships between people, communities and institutions if effective policy responses are to be developed. Thinking about the future of the Arctic also requires an exploration and understanding of different ways of seeing and experiencing Arctic lands, waters, icescapes, and animals. In a rapidly changing Arctic, it is more crucial than ever that indigenous peoples participate in decisions and policy-making about the range of possible futures of the region, its ecosystems, wildlife, and diverse societies, and scenarios for economic development, and in doing so determine the course of their own lives and safeguard indigenous homelands for successive generations.

References

ACIA 2005. *Arctic Climate Impact Assessment: Scientific Report*. Cambridge, UK: Cambridge University Press.

AMAP 2012. *Arctic Climate Issues 2011: Changes in Arctic Snow, Water, Ice and Permafrost*. Oslo: Arctic Monitoring and Assessment Programme.

AMAP 2017. *Snow, Water, Ice and Permafrost in the Arctic: Summary for Policy-Makers*. Oslo: Arctic Monitoring and Assessment Programme.

Anderson, R.B., B. Schneider and B. Kayseas 2008. *Indigenous Peoples' Land and Resource Rights*. Vancouver, BC: National Centre for First Nations Governance.

Dahl, J. 2005. 'The Greenlandic Version of Self-Government' in K. Wessendorf (ed.) *An Indigenous Parliament? Realities and Perspectives in Russia and the Circumpolar North*. Copenhagen: IWGIA.

Fondahl, G. and S. Irlbacher-Fox 2009. *Indigenous Governance in the Arctic: A Report for the Arctic Governance Project*, prepared for the Walter and Duncan Gordon Foundation, available at www.arcticgovernance. org/compendium.137742.en.html.

Guyot, M., C. Dickson, C. Paci, C. Furgal and M.M. Chan 2006. 'Local observations of climate change and impacts on traditional food security in two northern Aboriginal communities' *International Journal of Circumpolar Health* 65(5): 403–415.

Hastrup, K. 2009. 'Arctic hunters: climate variability and social flexibility' in K. Hastrup (ed.) *The Question of Resilience: Social Responses to Climate Change*. Copenhagen: Royal Danish Academy of Science and Letters.

Henshaw, A. 2009. 'Sea ice: the sociocultural dimensions of a melting environment in the Arctic' in S.A. Crate and M. Nuttall (eds) *Anthropology and Climate Change: From Encounters to Actions*. Walnut Creek, CA: Left Coast Press.

Huntington, H.P. 1992. *Wildlife Management and Subsistence Hunting in Alaska*. London: Belhaven Press.

Keeling, A. and J. Sandlos (eds.) 2015. *Mining and Communities in Canada: History, Politics, and Memory*. Calgary, AB: University of Calgary Press.

Krupnik, I. and D. Jolly (eds.) 2002. *The Earth is Faster Now: Indigenous Observations of Arctic Environmental Change*. Fairbanks, AK: ARCUS.

Leas, D. 2005. 'Self-government in the Yukon', in K. Wessendorf (ed.) *An Indigenous Parliament? Realities and Perspectives in Russia and the Circumpolar North*. Copenhagen: IWGIA.

MacKay, F. 2004. 'Indigenous peoples' right to free, prior and informed consent and the World Bank's extractive industries review' *Sustainable Development Law and Policy* 4(2): 43–65.

Marino, E. 2015. *Fierce Climate, Sacred Ground: An Ethnography of Climate Change in Shishmaref, Alaska*. Fairbanks, AK: University of Alaska Press.

Meakin, S. and T. Kurvits 2009. *Assessing the Impacts of Climate Change on Food Security in the Canadian Arctic*. Report prepared by GRID-Arendal for Indian and Northern Affairs Canada. Arendal, Norway: GRID-Arendal.

Nuttall, M. 1992. *Arctic Homeland: Kinship, Community and Development in Northwest Greenland*. Toronto, ON: University of Toronto Press.

Nuttall, M. 1994. 'Greenland: emergence of an Inuit homeland' in Minority Rights Group (ed.) *Polar Peoples: Self-Determination and Development*. London: Minority Rights Group.

Nuttall, M. 2008. 'Self-rule in Greenland: towards the world's first independent Inuit state?' *Indigenous Affairs* 1–2/8: 64–70.

Nuttall, M. 2017. *Climate, Society and Subsurface Politics in Greenland: Under the Great Ice*. London and New York: Routledge.

Nuttall, M., F. Berkes, B. Forbes, G. Kofinas, T. Vlassova and G. Wenzel 2005. 'Hunting, herding, fishing and gathering: indigenous peoples and renewable resource use in the Arctic' in ACIA *Arctic Climate Impact Assessment*. Cambridge, UK: Cambridge University Press.

Ortiga, R.R. 2004. *Models for Recognizing Indigenous Land Rights in Latin America*. Washington, DC: The World Bank Environment Department.

Roddick, D. 2006. 'Yukon First Nations and the Alaska Highway gas pipeline' *Indigenous Affairs* 2–3/6: 12–19.

Zellen, B.S. 2008. *Breaking the Ice: From Land Claims to Tribal Sovereignty in the Arctic*. Lanham, MD: Lexington Books.

6

Indigenous cartographies of Arctic places and spaces

Kaitlin Young

Introduction

There is a rich diversity in the histories, cultures, economics and forms of social organization of the Arctic's indigenous peoples. However, a special attachment to the environment and the animals they depend on for their livelihoods and household and community economies is a common experience for the various indigenous groups across the circumpolar North. This special relationship is reflected in traditional knowledge, ontologies, oral histories and human-environment relations (Nuttall 2000), but it is also articulated in indigenous movements and political declarations of self-determination and sovereignty. Indigenous identities, livelihoods and societies are often inextricably linked to traditional lands and resources as noted in the Kimberley Declaration, which emerged from the International Indigenous Peoples Summit on Sustainable Development in 2002:

> Our lands and territories are at the core of our existence – we are the land and the land is us; we have a distinct spiritual and material relationship with our lands and territories and they are inextricably linked to our survival and to the preservation and further development of our knowledge systems and cultures, conservation and sustainable use of biodiversity and ecosystem management . . . We are the original peoples tied to the land by our umbilical cords and the dust of our ancestors.

Indigenous expressions and articulations of attachment and belonging to landscape and place offer unique perspectives on locality and the connections between place, identity and culture in the Arctic (Nuttall 1992; Sejersen 2004). Politically, the concept of locality has been an impetus for developing tools of empowerment in discourses revolving around contested relations between center and periphery, especially in relation to land claims and wildlife management (see the chapters in this volume by Nuttall on land claims and environmental governance), with the terms "local knowledge" and "traditional ecological knowledge" (TEK) being strong examples of how indigenous narratives seek to advance indigenous rights, but also seek to incorporate indigenous perspectives in policy and practice. Over the last few decades, a sense of place and expressions of a strong attachment to locality and Arctic regions as indigenous homelands have become critical components of land claims movements as well as indigenous knowledge and sustainability discourses. A sense of place, though, is not necessarily intrinsic to a physical geographical setting itself but emerges from an intricacy of human experiences and long-term engagement with one's surroundings (Nuttall 1992; Stedman 2003). While the physical characteristics of a location (and what is at that location, what it commemorates, or what it promises to contain) may affect how a person feels, a sense of place refers to an individual's (as well as a community's) connection, memories and experience with a particular locality (Mueller 2011; Relph 1997; Vanclay 2008).

In northern circumpolar regions, counter-mapping efforts (including land use and occupancy studies) have become one way by which various indigenous groups have attempted to protect their lands and advance their rights to self-determination, as well as express their sense of attachment to Arctic places and spaces. This chapter provides an overview of some of these efforts and how, by drawing on indigenous expressions and experiences, different ways of mapmaking can capture the intricacies of movement and memory within and across circumpolar landscapes, icescapes and waters, and give local knowledge increasing resonance and importance within strategies for land claims, conservation and environmental protection, and social and environmental impact assessments.

Arctic and northern mapping

In her work on Inughuit communities in Northwest Greenland, Kirsten Hastrup draws attention to how the topographies, textures and contours of Arctic landscapes are often represented and seen as a mixture of emptiness and solidity, of sparse and isolated indigenous populations separated by immense distances, intersected by moments of intense sociality (Hastrup 2009). Arctic life is often precarious and inundated with surprise and extreme weather events, and nothing is ever to be taken for granted in a world that is constantly moving (Nuttall 2017). There is a sense, Hastrup argues, that people must constantly be aware of the environment and be acutely attentive and sensitive to the moments in which they live in these surroundings and how they move across ice, water and land, and that Arctic topography gives rise to a particular topophilia, defined by Tuan (1990: 4) as "the affective bond between people and place or setting". Such intimate attachment to the land – which is often expressed in place names and human-environment relations – emphasizes local bonds to place (Hastrup ibid.; Nuttall 1992, 2017). Elaborating further on the relationship between indigenous peoples and the environment, much recent anthropological research (as well as indigenous land claims movements) has emphasized the importance of focusing on indigenous place names and mapmaking, as well as the importance of understanding the Arctic as a homeland rather than an empty frontier or wilderness (Nuttall 1992; Sejersen 2002). Movement throughout Arctic landscapes involves more than physical movement across water, ice or tundra, though; it involves journeys through time and memory.

A map is perhaps the most quintessential geographic form of the representation of places and spaces. It allows us to simplify, visualize and understand the complexities of the world and to order, classify, define and categorize it (Hanson 1997). But it can also be used as a way of asserting power over territory and controlling the movement of people and animals. Ingold (2000) maintains that people and their lives are inscribed in the landscape and do not simply ascribe meanings to it. The integration of people with landscapes and places illuminates the physicality of place and the importance of orientation and location, and Ingold argues that it is impossible to know or sense a place except by being in it and dwelling in it. One can embody a place through mapping, however, and the intimate and close-up imagery of mapmaking techniques – and the products of maps and atlases – allows viewers to experience and consume places visually (Pocock 1976). In this way, however, maps can be mere representations that reduce and distill a sense of place, but they can also be sites of contestation – powerful metaphors, influencing and shaping our perceptions and understandings of the world – along with sites of identity construction and representation (Godlewska 1997; Urry and Larsen 2011). The meanings and uses of maps are, like all human activities and actions, set in a cultural context of values and beliefs that reinforce, and are reinforced by, the act of mapping itself (Rundstrom 1991). As many scholars argue, maps can often lack depth, however, and compress the complexities and intricacies of a landscape into something flat, empty and meaningless. This often excludes human experience as well; the map freezes the landscape so that it depopulates land and erases its history (Ryden 1993).

Most historical work on mapping has been focused on the development of the map in the Western scientific tradition, but research reveals important historical developments of mapping in other cultures and localities. Colonization in the Arctic has left a legacy of exploitation on indigenous cultures, and mapping practices were essential accomplices in colonizing processes, ignoring and eliminating native toponyms and an indigenous sense of place. As Julie Cruikshank puts it "The very language we use ('our knowledge, their values') is problematic. Knowledge implies certain absolutes; values suggest relativity" (Cruikshank 1984: 19). The validity of indigenous knowledge is less contested today, however, and indigenous cartography has emerged as a way for indigenous people to communicate their knowledge of their surroundings in a meaningful way (Collignon 2006).

Since the late 1990s, the historical study of maps has extended beyond understanding just geographical representations of the world. At least three approaches have been championed by scholars: the map as a cognitive system, the map as material culture, and the map as a social construction (Woodward and Lewis 1998). Indigenous peoples of the Arctic have long been recognized as being particularly able mapmakers. For the indigenous peoples of northern North America, for example, cultural information encoded within indigenous maps goes beyond that of merely representing landscapes or relating to the natural world and provides information about "individuals and groups of people; dwellings and settlements; routes and journeys; hunting, trapping, and fishing activities; clearings and fields; domesticated animals; battles, powwows, and councils; and very occasionally, boundaries" (Lewis 1998: 179). The utilization of maps and navigational markers to outline topographic features and travel routes is a striking feature of indigenous geographic knowledge. A complex and culturally distinctive set of place-based spatial conceptions is combined with technical practices that illustrate space in a way that resembles and goes beyond Western geographic and cartographic practice (Whitridge 2004). Indigenous mapping, though, is often elemental, taking place as a distinctive niche in the context of indigenous culture, and indigenous maps provide information about travel, detailed descriptions of sea ice conditions, landmarks, navigational routes, and the ever-important place names that fill localities (Whitridge ibid.).

Among early Arctic scientists who lived and worked among indigenous people, Franz Boas (who researched with Inuit groups in southern Baffin Island in the 1880s) was one of the first to recognize and emphasize the importance of indigenous place names and toponymic knowledge, and that they should be recorded on conventional maps (Collignon 2006). In recent years, interest in the place names used by local people has led to a greater understanding of human-environment relations in the Arctic. Place names crystallize the history and identity of the people who name their surroundings, as well as the landscape itself (Sejersen 2004). Tilley (1994: 18) has written about the process of naming places and things as one where they become "captured in social discourses and act as mnemonics for the historical actions of individuals and groups". Place names often reflect existing knowledge of the presence of animals or hunting practices or provide insight into local practices and use and people's experiences, as well as coding information about weather and climate. In his discussion of locality as memoryscape, Mark Nuttall (1992) has argued that place names are essential for an understanding of a sense of belonging, locality, community and cultural continuity in Greenland. Drawing on extensive work in the Upernavik and Avanersuaq areas of Northwest Greenland, Nuttall has described how there are many multilayered and textured layers of meaning within landscapes and the place names given to features of the environment. He shows how place names are multidimensional and have much more meaning than just that of the name itself. They are narratives about people's surroundings and about how people's lives are embedded within those surroundings and how they live and move spatially and temporally (Nuttall ibid.).

Towards critical cartography

Anthropological approaches to understanding landscape often have a specific focus on the practical, everyday and socio-political aspects of people's relationships to the natural and built environments. Place has often been theorized about in relation to ideas about space; the local and the socially and culturally meaningful are frequently opposed to the universal and the objective (Whitridge ibid.). This practice creates a space/place dichotomy that problematizes a nature/culture divide. The inadequacy of the space/place divide, however, is illustrated through the richness and complexity of indigenous knowledge and human–environment relations. Although these relationships embody numerous features of socially-embedded practices and place-based relations to the land, they also support instrumental navigation concepts and a variety of techniques for mapping that physically inscribe meaning to the environment (Whitridge ibid.). When cartographers make maps, they are engaged in an imaginative project which sets about representing the world through particular ways of knowing and perceiving and through particular technologies. When maps of the Arctic, though, draw on a specific set of knowledge claims and create blank white spaces from which indigenous people and their lives are omitted (Eades 2015), Harris (2002) refers to this practice as "cartographic erasure".

Simultaneously, indigenous peoples emphasize their attachment to the land, and the cultural meanings of such attachment, while science works to become increasingly detached from the local (Sejersen 2004). As Ingold, among others, has written about, the global ontology of land detachment takes priority over local ontologies of engagement in scientific or conventional mapmaking (Ingold 1993: 41). In the pursuit of scientific agendas or maps that fulfill a certain set of parameters, indigenous peoples find themselves marginalized where they see their interests and perspectives continually overruled, ignored and undervalued (Sejersen 2004). In applying post-structuralist approaches to understanding and analyzing power relations and cartography, many theorists including J.B. Harley (1989), have argued that an instrumental use of mapping serves to reinforce authoritarian hegemonic practices concerning space. "This atlas is *an* atlas and not *the* atlas"; instead it is one of many potential atlases created by a variety of artists, architects and others employed in cartographic discourses (Mogel and Bhagat 2010: 6). This discursive exercise mimics and reinforces the practice of referring to and approaching the Arctic as a "frontier". The conflicts between scientists and indigenous peoples in mapmaking are closely related to a colonial situation whereby indigenous narratives and experiences of the landscape cannot be understood without this political dimension. This underscores the importance of presenting and representing Arctic places as *homelands*, rather than as spaces of colonial expansion and *frontier* activities (Berger 1977).

As noted by Anna Tsing "A frontier is an edge of space and time: a zone of not yet – not yet mapped, not yet regulated". She states that a frontier is "a zone of unmapping: even in its planning, a frontier is imagined as unplanned" (Tsing 2005: 28). In northern contexts, such imaginative "unmapping" has sustained ideas of the Arctic as vast, empty and open to development.

The landscape is continually emptied of its inhabitants through such representations of northern environments (Nuttall 2017). Similarly, conventional cartography – especially that which draws upon satellite images of sea ice loss – depicts the Arctic as a wilderness and as one that does not truly represent the deeply embedded cultural aspects of place imbued through its use by indigenous peoples. Critical cartography, however, has democratized mapping discourses and practices to reflect the voices and experiences of marginalized and forgotten groups (Crampton 2010). Critical cartography (also referred to as counter-mapping) is the "one-two punch of new mapping practices and theoretical critique" and has become necessary to take the power held within maps and return it back to the hands of those directed by them (Crampton and Krygier 2006: 56). Nuttall's (1992) additional layers of meaning and usage are now being included on the two-dimensional surfaces of conventional cartography.

Indigenous counter-mapping

Most maps are often accepted as unproblematic depictions and representations of the world. However, there are instances of maps being rejected as early as the sixteenth century, in the mental mapping movement of the 1960s, and in indigenous and bioregional mapping initiatives. The desire to regain political, cultural and linguistic influence has resulted in counter-mapping and in new indigenous place-making efforts. As Hugh Brody put it in *Maps and Dreams*, his eloquent account from fieldwork in northeastern British Columbia, "The Indians say, with their maps, that they continue to use or need all of their territory" (Brody 1981: 174). Mapping projects have been integral components of indigenous political strategies to improve land claim negotiations where sense of place is turned into localism (Sejersen 2004). A new image of place can result in the actual building of an alternative possibility. Mapping can become a graphic tool that allows for the complexities of indigenous knowledge to become integrated into various geographic representations of place (Aberley 1993).

Indigenous peoples argue that they need to have the ability to protect themselves from the commodification and misinterpretation of indigenous knowledge. As Absolon and Willett argue, false representations perpetuate a false consciousness through conventional mapmaking that create "artificial contexts" that act to further disconnect indigenous people from their surroundings (Absolon and Willett 2004: 9–10). The project of indigenous counter-mapping is part of a wider trend in critical cartography that is based on promoting new ways of engaging with maps and mapmaking. The focus, though, is not solely based on the deconstruction of maps and the processes and politics behind mapmaking, but rather, reconstruction – that is, reconstructing ways of knowing and mapping. The numerous meanings that are negotiated within and ascribed to the landscape are all a part of the process of claiming and legitimizing rights to the land. Critical cartography then allows a way for indigenous attachment to land and indigenous human-environment relations to be expressed and understood. A number of issues remain that need to be explored and confronted, for once one "unleashes mapping's good magic", there are often implications and consequences following in its wake (Chapin et al. 2005: 630). First, cartographic representations of indigenous cultural knowledge can suffer from issues of translation due to differing ontological and epistemological cartographic structures (Louis 2007), but other questions focus attention, for example, on how the ownership of information should be handled, what the risks of exposing mapping practices entail, and why women are underrepresented in mapping projects.

Beyond the deconstructionist and enlightenment tasks set out by theorists such as Harley, there are opportunities for cartographers to put theory into practice. One way is to serve as a facilitator for enfranchising the disenfranchised, providing understanding for those who wish to depart from the margins and inscribe re-presentations of place and space by subverting the cartographic texts of the enfranchised (Rundstrom 1991). The origins of the contemporary indigenous mapping movement can be traced to the early 1970s, specifically with the publication of the landmark three-volume *Inuit Land Use and Occupancy Project* (ILUOP) in 1976, which pioneered the use of individual map biographies (Freeman 1976). Other initiatives, such as Hugh Brody's (1981) seminal work on mapping the lands of indigenous peoples throughout the Canadian subarctic, have meant that participatory mapping (as a form of critical cartography) has played a pivotal role in the demarcation and negotiation of land claims (Crampton and Krygier 2006). The following discussion of indigenous counter-mapping from the Canadian Arctic provides a good example.

Inuit maps and land claims

Harley (1989) emphasizes the importance of toponyms in his call for the revision of cartography. Maps can often reflect an insensitivity on the part of mapmakers and early cartographers; toponyms perceived as offensive have inspired movements to change local usage and eradicate what are regarded as cartographic insults (Monmonier 1995). The story of the Inuit encounters and relations with outsiders, government agencies and settlers in Canada's Arctic highlights the value of interpreting North American mapping practices as part of the 500-year dialogue between original inhabitants and newcomers (Rundstrom 1991).

For more than 150 years, the Inuit of northern Canada have lived in a world dominated by place names given by explorers, traders and surveyors and which have been officially sanctioned by government authority. Meanwhile, Inuit names in their native language, Inuktitut, persisted in almost subversive fashion by their passing through generations (e.g., through drawing maps, indigenous narratives and knowledge and story-telling). Colonists utilized maps to enfranchise themselves and disenfranchise the Inuit without actually taking over the land. Inuit experiences in Canada epitomize colonial mapping as a form of appropriation and power-knowledge (Rundstrom 1991). The Inuit traditionally disliked using government maps as they depicted an unrecognizable and foreign world. Forced relocations in the 1950s and the 1960s from traditional camps to permanent settlements and replacement of dog teams with snowmobiles altered indigenous travel techniques and practices of navigation.

The Inuit residents of the Nunavik region of northern Quebec became the first in Canada to negotiate a land claim in the form of the James Bay and Northern Quebec Agreement of 1975. This was followed by the establishment of the Avataq Cultural Institute (ACI), which initiated a systematic survey of Inuit toponyms in the Nunavik Region in the mid-1980s (Rundstrom 1991). More than 2,200 Inuit names had been officially accepted and approved by the Commission de toponymie due Quebec by June 1990 (a prototype map of the 1: 50,000 Inuit Place Name Map Series of Nunavik was issued in January that year). Complex toponymic and cartographic policy ensures Inuit/*Qallunaat* (whites) and Anglophonic/Francophonic mapping dialogues continue to take place. In essence, the Inuit *re*-appropriated a small portion of Arctic land and text from government and reimposed expressions of their own sense of place and toponymy onto it. In doing so, they have resituated themselves as part of the cartographic establishment and engage the enfranchised culture on their own terms, employing its technology and politics for their own ends.

Inuit maps were also critical for an assertion of Aboriginal title to two million square kilometers of Canada and formed a basis for the Nunavut Land Claims Agreement (NCLA) of 1993. The Inuit Land Use and Occupancy Project had provided a baseline for the NCLA and contributed to discussions about the demarcation of the political boundaries of the territory of

Nunavut, which was carved from the central and eastern parts of the Northwest Territories and established in 1999. Its primary aim was to map, document and explain the ways in which Inuit used their lands within living memory in the Northwest Territories (Brody 1981; Freeman 1976). The project was a combination of local oral history, early tales of white explorers, and archaeological explorations contributed to a succession of Arctic material cultures. This primary information was then used as a benchmark for establishing individual map biographies. In these:

> [h]unters, trappers, fishermen, and berry pickers mapped out all the land they had ever used in their lifetimes, encircling hunting areas species by species, marking gathering locations and camping sites – everything their life on the land had entailed that could be marked on a map.
>
> *(Brody 1981: 147)*

The study of Inuit land use and occupancy of the Northwest Territories led to the initiation of numerous other similar projects on the lands of the Inuit, Settlers, and Naskapi-Montagnai Indians of Labrador, the Dene of the Mackenzie River basin, the Indians of the Yukon, the Inuit and Cree of northern Quebec, and two Ojibway communities in northwest Ontario (Brody ibid.). The ILUOP led to the creation of the Nunavut Atlas, which signaled a new phase in the struggle for Inuit self-determination (Kral and Idlout 2006). Indigenous counter-mapping is a tool by which indigenous groups can re-present the world in ways which destabilize dominant and prevailing representations.

Counter-mapping has been a decisive strategy for indigenous communities and organizations in their struggles for political, economic, and territorial rights. As illustrated, it has been an efficient tool to appropriate the state's techniques and modes of representation and reinforce the legitimacy of indigenous claims. The resistance and the struggle for emancipation have developed primarily within the epistemological framework of the decolonization of indigenous methodologies (e.g., see Tuhiwai Smith 2002). While these Canadian Inuit cases are examples of mapping projects in specific indigenous landscapes (see also the Pan Inuit Trails atlas, which provides a view of Inuit mobility and occupancy of land, water and ice –www.paninuittrails.org), further research endeavours into counter-mapping elsewhere in the Arctic and elsewhere in the world means researchers are beginning to understand in greater depth the ways in which indigenous cartographies articulate in varying ways efforts to challenge dominant power relations that traverse indigenous landscapes (Sletto 2009). More specifically, we can begin to unpack how the process of mapmaking can influence the struggle for indigenous justice throughout the world, and specifically in the Arctic (Sletto 2009).

Conclusions

Indigenous peoples are increasingly expressing their interest in preserving their knowledge, which has the capacity to contribute to solving a range of problems that they are now faced with confronting (Brooke 1993). Heightened concerns over rapid and profound social and cultural change, potential megaproject resource development and the impacts of globalization in the Arctic have induced indigenous people to establish unique strategies for self-determination and governance (Nuttall 1992; Nuttall, Chapter 5 this volume). At the same time, wildlife management and conservation strategies are also often framed within a context of scientific discourse that excludes indigenous knowledge (see Chapter 32 on conservation and environmental governance by Nuttall in this volume). Critical cartography and indigenous counter-mapping can be pursued as tools to advance these strategies and complement an understanding of indigenous ontologies of place. As local, regional and global concerns over industrial development, resource exploitation and climate change increase, indigenous people throughout the Arctic have been developing and implementing new participatory approaches to environmental strategies and policies (Nuttall 2000). Counter-mapping provides a culturally-appropriate method to aid indigenous people in furthering their strategies for environmental protection and sustainable development (see Chapter 24 by Poppel this volume) as well as improving social and environmental impact assessment processes (see Chapter 29 by Hansen, Larsen and Noble in this volume). With climate change leading to environmental transformation, it is important to turn to indigenous map-makers of the Arctic to resist a reconfiguration of ideas, space and territory for political means (Nuttall 2014). Critical cartography offers a chance to re-engage with a place-making, rights-affirming, and culturally-apposing technique that can allow indigenous perspectives on place to be acknowledged through the process of mapmaking. As increasing interest in the Arctic and its resources continues to bring with it a wealth of challenges, we need more tools such as counter-mapping techniques and practices to help promote the saliency of environmental knowledge possessed by the indigenous peoples of the circumpolar North.

References

Aberley, D. 1993. 'The lure of mapping: an introduction' in D. Aberley (ed.) *Boundaries of Home: Mapping for Local Empowerment*. Gabriola, BC: New Society Publishers. 1–7.

Absolon, K. and C. Willett 2004. 'Aboriginal research: berry picking and hunting in the 21st century' *First Peoples Child and Family Review* 1: 5–17.

Berger, T. 1977. *Northern Frontier, Northern Homeland: The Report of the Mackenzie Valley Pipeline Inquiry*. Ottawa: Supply and Services Canada.

Brody, H. 1981. *Maps and Dreams: Indians and the British Columbia Frontier*. Long Grove, IL: Waveland Press.

Brooke, L. 1993. *The Participation of Indigenous Peoples and the Application of Their Environmental and Ecological Knowledge in the Arctic Environmental Protection Strategy*. Ottawa: Inuit Circumpolar Conference.

Chapin, M.Z. Lamb and B. Threlkeld 2005. 'Mapping indigenous lands' *Annual Review of Anthropology* 34: 619–638.

Collignon, B. 2006 'Inuit place names and sense of place' in P. Stern and L. Stevenson (eds.) *Critical Inuit Studies: An Anthology of Contemporary Arctic Ethnography*. London and Lincoln, NE: University of Nebraska Press. 187–205.

Crampton, J.W. 2010. *Mapping: A Critical Introduction to Cartography and GIS*. Chichester, UK: John Wiley & Sons.

Crampton, J.W. and J. Krygier 2006. 'An introduction to critical cartography' *ACME: An International E-Journal for Critical Cartographies* 4(1): 11–33.

Cruikshank, J. 1984. 'Oral tradition and scientific research: approaches to knowledge in the North' *Social Science in the North, Communicating Northern Values* 9: 3–23.

Eades, G.L. 2015. *Maps and Memes: Redrawing Culture, Place and Identity in Indigenous Communities*. Montreal, QC and Kingston, ON: McGill-Queen's University Press.

Freeman, M. 1976. *Inuit Land Use and Occupancy Project*. Ottawa: Ministry of Supply and Services.

Godlewska, A. 1997. 'The idea of the map' in S. Hanson (ed.) *Ten Geographic Ideas that Changed the World*. New Brunswick, NJ: Rutgers University Press. 15–39.

Hanson, S. 1997. 'Introduction: ten geographic ideas that changed the world' in S. Hanson (ed.) *Ten Geographic Ideas that Changed the World*. New Brunswick, NJ: Rutgers University Press. 1–14.

Harley, J.B. 1989. 'Deconstructing the map' *Cartographica* 26(2): 1–20.

Harris, C. 2002. *Making Native Space: Colonialism, Resistance and Reserves in British Columbia*. Vancouver, BC: University of British Columbia Press.

Hastrup, K. 2009. 'The nomadic landscape: people in a changing Arctic environment' *Geografisk Tidsskrift – Danish Journal of Geography* 109(2): 181–189.

Ingold, T. 1993. 'Globes and spheres: the topology of environmentalism' in K. Milton (ed.) *Environmentalism: The View from Anthropology*. New York: Routledge. 13–42.

Ingold, T. 2000. *The Perception of the Environment: Essays on Livelihood, Dwelling and Skill*. London: Routledge.

Kimberley Declaration. 2002. *Declaration Formulated at the International Indigenous Peoples Summit on Sustainable Development*. 20–23 August 2002. Kimberley, South Africa.

Kral, M.J. and L. Idlout 2006. 'Participatory anthropology in Nunavut' in P. Stern and L. Stevenson (eds.) *Critical Inuit Studies: An Anthology of Contemporary Arctic Ethnography*. London: University of Nebraska Press. 54–70.

Lewis, M. 1998. 'Maps, mapmaking, and map use by Native North Americans' in D. Woodward and G.M. Lewis (eds.) *Cartography in the Traditional African, American, Arctic, Australian, and Pacific Societies*. Chicago, IL: Chicago University Press. 152–184.

Louis, R.P. 2007. 'Can you hear us now? Voices from the margin: using indigenous methodologies in geographic research' *Geographical Research* 45(2): 130–139.

Mogel, L. and A. Bhagat 2010. *An Atlas of Radical Cartography*. Los Angeles, CA: Journal of Aesthetics and Protest Press.

Monmonier, M. 1995. *Drawing the Line*. New York: Henry Holt and Company.

Mueller, K.B. 2011. *Implications of Sense of Place for Recovery of Atlantic Salmon and Other Imperiled Fish* (doctoral dissertation). www.linkedin.com/pub/katrina-mueller/12/76/639.

Nuttall, M. 1992. *Arctic Homeland: Kinship, Community and Development in Northwest Greenland*. Toronto, ON: University of Toronto Press.

Nuttall, M. 2000. 'Indigenous peoples, self-determination and the Arctic environment' in M. Nuttall and T. V. Callaghan (eds.) *The Arctic: Environment, People, Policy*. Amsterdam: Harwood Academic Publishers. 377–410.

Nuttall, M. 2014. 'Introduction: Arctic geopolitics and resource futures' in M. Nuttall and A. Dey Nuttall (eds.) *Arctic Geopolitics and Resource Futures*. Oulu, Finland: Thule Institute, and Edmonton, AB: Canadian Circumpolar Institute. 5–14.

Nuttall, M. 2017. *Climate, Society and Subsurface Politics in Greenland: Under the Great Ice*. London and New York: Routledge.

Pocock, D.C.D. 1976. 'Some characteristics of mental maps: an empirical study' *Transactions of the Institute of British Geographers* 1(4): 493–512.

Relph, E. 1997. 'Sense of place' in S. Hanson (ed.) *Ten Geographic Ideas that Changed the World*. New Brunswick, NJ: Rutgers University Press. 205–226.

Rundstrom, R.A. 1991. 'Mapping, postmodernism, indigenous people and the changing direction of North American cartography' *Cartographica* 28(2): 1–12.

Ryden, K. 1993. *Mapping the Invisible Landscape: Folklore, Writing, and Sense of Place*. Iowa City, IA: University of Iowa Press.

Sejersen, F. 2002. *Local Knowledge, Sustainability and Visionscapes in Greenland*. Copenhagen: University of Copenhagen, Department of Eskimology.

Sejersen, F. 2004. 'Horizons of sustainability in Greenland: Inuit landscapes of memory and vision' *Arctic Anthropology* 41(1): 71–89.

Sletto, B. 2009. 'Special issue: Indigenous cartographies' *Cultural Geographies* 16: 147–152.

Stedman, R.C. 2003. 'Is it really just a social construction? The contribution of the physical environment to sense of place' *Society and Resources* 16(8): 671–685.

Tilley, C. 1994. *A Phenomenology of Landscapes. Places, Paths and Monuments*. Oxford, UK: Berg Publishers.

Tsing, A. 2005. *Friction: An Ethnography of Global Connection*. Princeton, NJ and Oxford, UK: Princeton University Press.

Tuan, Y.F. 1990. *Topophilia: A Study of Environmental Perception, Attitudes, and Values*. New York: Columbia University Press.

Tuhiwai Smith, L. 2002. *Decolonizing Methodologies: Research and Indigenous People*. London: Zed Books.

Urry, J. and J. Larsen 2011. *The Tourist Gaze 3.0*. Thousand Oaks, CA: Sage.

Vanclay, F. 2008. 'Place matters' in F. Vanclay, M. Higgins and A. Blackshaw (eds.) *Making Sense of Place*. Canberra: National Museum of Australia. 3–12.

Whitridge, P. 2004. 'Landscapes, houses, bodies, things: "Place" and the archaeology of Inuit imaginaries' *Journal of Archaeological Method and Theory* 11(2): 213–250.

Woodward, D. and M.G. Lewis 1998. 'Introduction' in D. Woodward and G.M. Lewis (eds.) *Cartography in the Traditional African, American, Arctic, Australian, and Pacific Societies*. Chicago, IL: The University of Chicago Press. 1–12.

7

Circumpolar health and well-being

Helle Møller

Health and well-being in the circumpolar regions is, as elsewhere, closely tied to the Social Determinants of Health (SDoH). The SDoH are shaped by the "inequitable distribution of power, money and resources" globally, nationally and locally (WHO 2012: 48). They are largely responsible for health inequities, defined as "the unfair and avoidable differences in health status seen within and between countries" (WHO n.d.: para. 5). It is recognized that circumpolar nations and peoples are heterogeneous and that there are significant variations in health outcomes within and between them (Larsen et al. 2014). Some commonalities exist, however. One is that Indigenous peoples of the circumpolar North (except to some degree the Sámi of Fennoscandia and northwest Russia) experience unfavourable situations of health and well-being when compared to majority or settler populations (Young and Chatwood 2015). Another is that there are SDoH that affect Indigenous peoples in particular, such as colonization and colonialism (Kelm 1998; Mowbray 2007). Even though most Indigenous people surveyed through the Survey of Living Conditions in the Arctic (SLiCA) project:

> [r]ate their own health as good or excellent . . . and are satisfied with life in their communities, indigenous people also acknowledge widespread social problems: unemployment, alcohol abuse, suicide, drug abuse, family violence and sexual abuse are considered major social problems by more than six indigenous respondents out of ten.
>
> *(Poppel 2015: 55)*

SDoH can be proximal, intermediate or distal. Proximal determinants are those that are seen as directly affecting physical, mental and spiritual health including: health behaviour such as alcohol consumption, smoking and physical activity; physical environments such as availability, affordability and quality of housing and transportation; levels of education, employment and income or socioeconomic status (SES); food (in)security (Loppie Reading and Wien 2009); and "the impact of widespread and devastating land degradation and climate change" (Mowbray 2007: 41). The intermediate determinants include healthcare and educational systems; community infrastructure and capacities; environmental stewardship; and cultural and linguistic continuity (Inuit Tapiriit Kanatami (ITK) 2014; Loppie Reading and Wien, 2009). Last but by no means least are the distal determinants, "within which all other determinants are constructed" (Loppie Reading and Wien 2009: 20). For Indigenous peoples, distal determinants revolve around colonialism, including the "severance of ties of Indigenous Peoples to their land, and resources" (Mowbray 2007: 41), "racism, social exclusion [and] . . . repression of self-determination" (Loppie Reading and Wien 2009: 20).

With a focus on Indigenous populations, and taking its departure from the SDoH, the aim of this chapter is to review important threats to health and well-being in the northern circumpolar world, and to discuss the promising ways in which they have been, are and may be, addressed.

Proximal determinants

For all circumpolar Indigenous peoples, societal and social changes over the last several decades, including a move to more Western lifestyles with significant changes in diet, alcohol and tobacco consumption and levels of physical activity, have contributed to a considerable rise in what is often called lifestyle diseases or non-communicable diseases such as cancer, cardiovascular diseases (CVD) and diabetes. As noted earlier, the proximal determinants for Indigenous peoples are constructed within a frame of colonization, colonialism, racism and discrimination and must be understood as such. Viewed in this frame, what may be seen as unhealthy behaviours can be understood as ways to cope with an unhealthy environment and subjectively increase well-being (Allan and Smylie 2015; Møller, 2011). Conversely, when unfavourable determinants are improved, for example with increased self-determination and culturally safe education and healthcare, health and well-being also improve (Allan and Smylie 2015; Fraser Health 2009).

Non-communicable diseases and substance use

Except for the Sámi (Norway, Sweden, Finland and Russia), Indigenous peoples in the Canadian North (Inuit, Inuvialuit, Gwich'in and Dene), Greenland (Kalaallit), Alaska (Aleut, Yupi'k, Inuit [Iñupiat] and Athabaskan) and Russia (which include Nenets, Khanty, Evenk, Chukchi and Yakuts) have a higher prevalence of tobacco use than the non-Indigenous peoples living in the same regions. For example, in 2014, 62% of Inuit 12 years of age and older were daily smokers, four times the Canadian average (Statistics Canada 2016a). In addition to a high prevalence of lung cancer, current and historical prevalence of tobacco use by Inuit are reflected in statistics on lower respiratory tract infections (LRTI) in infants and tuberculosis (TB) in the population as a whole that are among the highest in the world (Banerji et al. 2013; Health Canada and the Public Health Agency of Canada [PHAC] 2014).

CVD – affiliated with smoking and also with diet – is reported to be significantly higher in First Nations People in Canada that among Canadians overall. For example, acute myocardial infarction has been reported to be 76% higher for residents from high–First Nations areas than for residents from low-Indigenous areas. In addition a third of residents from high–First Nations areas traveled more than 250 km to get care compared to only 8 percent of residents from low-Indigenous areas, and they were less likely to undergo specific cardiac procedures (Canadian Institute for Health Information 2013; Wallace 2014). The Conference board of Canada (2016), on the other hand, reported that between 2009–2011 Nunavut, with a population of approximately 85% Inuit, had the lowest rate of mortality due to stroke and heart disease among Canadian provinces and territories. A factor in this may be that Inuit on average live 10 years less than Canadians overall and the probability of death from heart disease significantly increases with age. Between 2011 and 2013 high blood pressure was reported to be three times more common among Inuit (12%) than among the Canadian population (4%) (Canadian

Institute for Health Information 2013; Wallace 2014). Changes in diet and physical activity have also, through complex and sometimes non-intuitive ways, contributed to changes in body composition and levels of diabetes among Inuit. In 2012, 28% of Inuit aged 18 and over were overweight and 24% were obese; 5% of Inuit reported living with type two diabetes (TTD), which is slightly above the prevalence in Canada as a whole at 4% (Wallace 2014). Historically, TTD was almost unknown in Inuit.

In regard to alcohol consumption, which also contributes to higher incidence of non-communicable diseases, fewer Inuit than Canadians overall drink on a regular basis and more Inuit are abstinent (Korhonen 2004). Inuit across Canada who do drink alcohol report heavy drinking more often than Canadians overall (26% vs 18%) (Canadian Institute for Health Information 2013; Wallace 2014); there are, however, significant differences between Inuit communities. In 2014 the percentage of Nunavummiut (i.e. people of Nunavut; sing. Nunavummiuq) who were heavy drinkers was 14.3% lower than the Canadian average of 17.9% (Taylor 2016). Most communities in Nunavut have chosen to be either dry, meaning that alcohol cannot be bought or sold, or restricted, meaning that alcohol can only be accessed through an alcohol committee. Some now believe that easier accessibility would lower binge drinking and Iqaluit, Nunavut's capital, recently decided to allow a store selling beer and wine (Hopper 2014).

Tobacco and alcohol are but two substances that may be used and abused – often as a way to self-medicate to cope with abusive, oppressive or otherwise difficult circumstances and histories (Kaweionnehta Human Resource Group 1993; Møller 2011). Others include hashish or cannabis (Møller 2010), volatile solvents such as gasoline, lighter-gas and glue (Dell 2006) and hard drugs such as crystal meth and cocaine. In 2007–2008 43% of Nunavummiut 18 years and older and 60% of 18- to 29-year-olds reported having used hashish or cannabis in the last 12 months – six times as many as among all Canadians 15 years of age and older. Some 3% of adult Nunavummiut reported having tried to get high using various solvents, and 5% that they had used cocaine or crack (Galloway and Saudney 2012). The 2004 Nunavik Inuit Health Survey found that 5.9% of the population 15 years of age and above and 13.9% in the age group 15–19 used solvents to get high in the past 12 months from the date of the survey (Muckle et al. 2007). Inuit recognize that many determinants of health including colonization and colonialism contribute to drug and substance use and abuse and see it as a serious threat to health and well-being, connected to social and cultural disruption, the breakdown of families, high rates of violence, accidental injuries and suicide (Møller 2010, 2011; Poppel, 2015). With some support from the Canadian federal government, local governments make considerable efforts to help Inuit who abuse substances to deal with the underlying causes through several programmes and services run by Indigenous peoples (Government of Canada 2015).

Inhabitants of Greenland 14 years of age and older consumed the equivalent of 10.4 litres of 100% alcohol in 2010 (Aage 2012), less than the 11.3 litres/person consumed in Denmark (Danmarks Statistik 2011). More people living in Greenland than Denmark are abstinent, but more, similar to Nunavut, are heavy drinkers (Aage 2012). Some 25% of 16-year-olds had tried sniffing solvents in 2010 and 29% of 17-year-olds had tried smoking hashish. When compared to Denmark, solvent and hashish use is high; the use of other illicit drugs such as cocaine, heroine and ecstasy is not, however (United Nations Office on Drugs and Crime 2011). Smoking rates in Greenland are even higher than in Nunavut. The Greenlandic population health survey from 2005–2008 reports that 66% of the population in Greenland 18 years and older smokes almost three times as much as in Denmark in 2007 (Bjerregaard and Eidt 2010; Sundhedsstyrelsen 2008). High rates of smoking are reflected in high rates of lung cancer. The risk of getting lung cancer in Greenland and Denmark before the age of 75 is 8.5% and 4.8% respectively, and the age standardized death rates for lung cancer are 68.5 and 39.1 per 100,000 respectively (Engholm et al. 2015). Thus, not only do a much higher percentage of Greenlanders get lung cancer, a much higher proportion of those who do, die from the disease. This can be understood with an eye to smoking statistics but importantly also by looking at preventative and curative health care measures, which are a lot less available in Greenland than Denmark, and remembering Greenland's colonial history, both of which are discussed further below.

Similar to Nunavut, Greenland has a very high prevalence of infant respiratory tract infections (Koch et al. 2002) and TB has been on the rise in both children and adults since the mid 1980s, climaxing in 2010 with an incidence rate of 205 per 100,000 people. In 2013 the Greenlandic average was 165 per 100,000 people, but there were significant differences across the country. In 2013, the incidence rate in East Greenland was 931 per 100,000 people (Chief Medical Officer of Health in Greenland 2013).

Researchers noted in 2003 that Greenlanders' cardiovascular health was comparable and, in some cases, worse than that of Danes (Bjerregaard et al. 2003), and a recent review reiterated this in relation to coronary artery disease when comparing Inuit (including Greenlandic Inuit) to non-Inuit populations (Fodor et al. 2014). Similar to Nunavut, societal changes and with them changes in diet and physical activity, with an increase in obesity and being overweight, have also contributed in Greenland to an increase in CVD and a noteworthy increase in TTD. In the 1950s TTD was reported to be almost non-existent; now the prevalence of TTD in Greenland is more than double that of Denmark (Jørgensen 2014).

As a response to these statistics, the Public Health Department in Greenland (Paarisa 2013) launched a public health program in 2013 with a focus predominantly on alcohol, hashish and tobacco use, but also on diet and physical activity (Paarisa 2013). In addition, a taskforce was struck to evaluate the cost of alcohol and hashish abuse in Greenland, the gain that would result from addressing the issue, and ultimately develop an action plan (Government of Greenland Department of Health 2015).

The smoking prevalence in Indigenous populations in Alaska is significantly lower than in Greenland and Nunavut. Still, smoking among the Indigenous populations in Alaska is more than double that of non-Indigenous Alaskans (42% vs 19%) (Peterson et al. 2015). In this connection it is significant that Non-Native adults aged 25 to 64 who are of "low SES are nearly three times as likely as those of higher SES to be smokers (38% versus 14%)" (ibid., p. 2). Despite a higher smoking prevalence than the non-Indigenous population, Indigenous Alaskans do not have higher lung cancer prevalence (Centre for Disease Control and Prevention 2015). Indigenous Alaskans who do experience lung cancer more often die from the disease than their non-Indigenous counterparts, particularly in more northern and remote areas due to less available and accessible preventative and curative healthcare measures (Plescia 2014).

With adult age-adjusted diagnosed percentages of coronary heart disease at 8.1% and 6.2% respectively, 'American Indians'/'Alaska Natives' are more likely to be diagnosed with heart disease than their White (non-Hispanic) counterparts. American Indians/Alaska Native adults are also more likely to be obese and have high blood pressure than White adults (Blackwell et al. 2014). This must, however, be viewed in light of other determinants of health such as decreased access to affordable healthy food sources, lower levels of physical activity, and higher levels of alcohol consumption and substance abuse (Blackwell et al. 2014), which can, as noted earlier, be connected with colonialism, racism and discrimination and coupled with lower access to health care services leading to significantly higher drug and alcohol induced mortality rates (Trust 2015).

There are, however, some very positive developments in Alaska. Since 1999 'Alaska Native people' have been in control of their own healthcare system as "customer-owners". Through the implementation of Southcentral Foundation's Nuka System of Care based in Anchorage, Alaska, great improvement in a number of core health care indicators have occurred significantly improving health and well-being among Alaska's Indigenous population; a few examples include: in "75% of the HEDIS measures (national standards), Southcentral Foundation is in the 75th percentile or better, and for many, like diabetes care, in the 95th percentile" (Gottlieb 2013: 5). Staff turnover is down, childhood immunization is up, and "customer satisfaction with respect for their cultures and traditions [is] at 94%". Health and well-being increase with increased self-determination.

It is extremely difficult to obtain reliable population statistics about the socioeconomic conditions, health and well-being for Russia generally and for Russia's Northern Indigenous populations in particular (Rohr 2014). This means that what is reported here and elsewhere must be read with caution.

Smoking levels in Russia are reported to be the highest in all OECD countries with 34% of adults and 60% of adult men smoking (OECD 2014). Studies from across the Russian North have found that populations start to smoke at young ages and many women smoke while pregnant (Douglas et al. 2012). For example, in the Koryak Autonomous Area almost one in two children in grade ten smoke (Skryagin 2005), in Yakutia 33% of Yakut and Evenk teenaged girls smoke, as do 50% of their mothers.

There was an estimated 30% decrease in alcohol consumption in Russia between 1990 and 2011, when it was about 11.5 litres/person, but according to the 2015 SLiCA report, *all* Indigenous people in Chukotka and the Kola Peninsula reported that alcohol abuse is a problem in their communities and more than 95% that drug abuse was. In the other Arctic regions, it is approximately 85% and 66% respectively (Poppel 2015). When we look at the level of neglect and discrimination that the Indigenous peoples of Northern Russia experience, discussed later, this may not be surprising (Rohr 2014). Larissa Abryutina of the Russian Association of the Indigenous Peoples of the North (RAIPON) noted that the life expectancy of the Indigenous peoples in Chukotka has fallen to between 40 and 45 years due to environmental pollution, alcoholism and poor health care (as cited in George 2010), and Rohr (2014: 32) writes that "Just over one-third of Indigenous men (37.8%) and less than two-thirds of indigenous women (62.2%) in Russia reach the age of 60. At national level, the figures are 54% for men and 83% for women".

The mortality rates from cardiovascular disease are significantly higher across Russia than in the other Arctic regions (Rautio et al. 2014: 305), and the Indigenous peoples of Yakutia have levels of "arterial hypertension, ischaemic heart disease, and cerebrovascular disorders" that are significantly higher than among the non-Indigenous groups (Burtseva et al. 2014). At the same time, the Northern Indigenous peoples of Russia have great difficulty accessing health care, which with the political reform of the early 1990s was centralized, depleting outlying and remote communities further. This could indicate that mortality from these diseases might be even higher as causes of death might not be registered (Rohr 2014).

While there is variation in tobacco use among Sámi dependent on age, where they live (Sweden, Norway, Finland, Russia), SES, and whether they are more or less acculturated or marginalized, it appears that Sámi use of tobacco is similar to or less than the majority populations where they live (Eliassen et al. 2013; Spein 2008), as is alcohol consumption. Levels of physical activity are generally reported to be higher and obesity and overweight lower (Sjølander 2011). It is therefore not surprising that morbidity and mortality among the Sámi in relation to CVD, diabetes and cancer are reported to be at lower or similar levels to those of the non-Indigenous populations living in the same areas (Sjølander 2011), a fact that may be connected with many Sámi having a strong cultural identity today (Hansen 2015), as discussed further later.

Housing

The access to appropriate and affordable housing, like most other determinants, varies immensely between and within circumpolar nations and their regions. The Indigenous peoples of the Russian North have by far the most unfavourable housing conditions when compared to the other circumpolar regions and to the non-Indigenous people of the Russian North. In Yakutia, for example, 32.2% have no indoor amenities at all (Burtseva et al. 2014). This is particularly prevalent in the Indigenous regions and among the Indigenous peoples living there, a fact reflected in the overall satisfaction with their living conditions. Only 15% of people in Yakutia reported their living conditions as good. The Anarbarsky region, however, where the highest number of people were reported to live with no amenities, has the highest percentage of people who are satisfied with their general living conditions (23%). This is less surprising when acknowledging that this region has the highest number of people reporting to live nomadic lifestyles as reindeer herders, hunters and tent workers (17%) (Burtseva et al. 2014). A tent worker (*chumrabotnitsa*) is an individual, generally a woman, who travels with the reindeer herders as a cook and spends most of her time in the tent preparing meals and cleaning up afterwards (Vitebsky 2005). While not as extreme as in Yakutia, Inuit and Greenlanders also experience serious housing issues and issues of overcrowding and at levels up to ten times more than their Danish and non-Indigenous counterparts. Of significance is that there are noteworthy inequalities between the Indigenous populations and the settlers in the two regions, with settlers being the more privileged (Møller 2011; Riva et al. 2014a, 2014b).

Overcrowding (more than one person per room) contributes considerably to respiratory (for example tuberculosis) and other infectious diseases, as well as non-infectious diseases such as CVD. Riva et al. (2014a, 2014b) add that it also contributes to the development of chronic stress and decreased mental health, binge drinking and thus pathophysiology. Little research has explored these issues in an Arctic context. Addressing this research gap is imperative to inform public policies targeting housing conditions as a key strategy to improving Indigenous health, particularly, in the circumpolar region (Riva et al. 2014a, 2014b).

Employment and SES

Average incomes are for all circumpolar Indigenous peoples lower than for their non-Indigenous counterparts. Low average annual earnings in turn are connected to a lack of employment opportunities, high unemployment rates and a gap in education between settlers and Inuit.

Due to the high cost of importing food to the Canadian North, the average cost of groceries for a household with children in the Inuit Nunangat in 2007–2008 was $19,760 per year, yet 49% of Inuit adults earned less than $20,000 (Expert Panel on the State of Knowledge of Food Security in Northern Canada [EP] 2014). Inuit Nunangat is the homeland of Inuit of Canada. It includes the communities located in the four Inuit regions: Nunatsiavut (Northern coastal Labrador), Nunavik (Northern Quebec), the territory of Nunavut and the Inuvialuit region of the Northwest Territories. These regions collectively encompass the area traditionally occupied by Inuit in Canada (Statistics Canada 2016b). The Sivuliqtiksat Program which provides on-the-job training for Inuit in Nunavut who want to work in management, is an initiative that improves these statistics. It has been in place since 2001. So far 50% of government jobs are held by Inuit, which means there is a considerable way to go as 85% of the population is Inuit (Eggertson 2015). The rate of unemployment among Inuit living in Inuit Nunangat is, at 20%, more than three times that of non-Indigenous Canadians (Indigenous and Northern Affairs Canada 2013).

Although healthy food is subsidized in Greenland and made available in remote areas at prices that resemble town levels (Rex et al. 2014), food prices are still higher than those in Denmark, while average income is lower. Unemployment is also higher at 9.4%, almost double that of Denmark (4.9%) (Danmarks Statistik 2015; Økonomisk Råd 2014).

It is difficult to find data on Indigenous Alaskans as Native Americans and Alaska Natives are generally grouped together. However, unemployment among Alaska Natives has been reported as being three times the Alaska average at more than 20%; about 22% of Alaska Natives and 13% of all Americans live below the poverty line and the Alaska Native average income is two-thirds that of all Americans (Martin and Hill 2009).

The incomes of the Indigenous peoples of Russia are two to three times lower than the Russian national average, and the country's Federal Accounts Chamber reports that "unemployment among indigenous peoples is 1.5 to 2 times the Russian average, and ranges from 24.5% among indigenous peoples of Yamal-Nenets Orug to 47.8% among the indigenous population of Amur oblast" (as cited in Rohr 2014: 34).

Although initiatives such as on-the-job training have been implemented, unemployment, underemployment and average incomes that are significantly below the average of the non-Indigenous peoples inhabiting the same regions as Indigenous peoples continue to contribute to Indigenous peoples having less access to adequate food and housing and comparatively unfavourable levels of physical and mental health and well-being. This situation, though, also contributes to a message that Indigenous peoples are less able and have less value than non-Indigenous peoples, a message that with enough repetition becomes internalized. Another word for this is structural violence (Farmer 2009). Climate change is another sort of structural violence. It affects Indigenous peoples more than other peoples, even though they have had least responsibility for its origins, causes and intensification, and may be among those with least power to control the actions necessary to curb it.

Climate change, disease and food insecurity

All Indigenous circumpolar peoples experience the effects of climate change and subsequent concerns about water and food security, infectious diseases (Brubaker et al. 2011; Brustad et al. 2014; Dudarev et al. 2013; Willox et al. 2015), healthcare and healthcare infrastructure (Rautio et al. 2014), and the mental health consequences that living with these experiences have in the form of addictions, violence and suicide (Willox et al. 2015). "Decreased access to safe food and water is associated with an increase in infections such as gastroenteritis, respiratory infections and vector borne diseases" (Rautio et al. 2014: 309). The warming climate may also introduce new host species and pathogens into new habitats that may lead to infections in humans and domestic and wild animals alike.

While food insecurity is strongly connected to low income and high food prices, it is also related to Indigenous peoples' higher reliance on traditional food sources. Traditional food is food obtained from the land or sea through gathering, hunting and fishing activities that are made increasingly difficult and dangerous with climate change and subsequently less predictable weather patterns and more frequent and more severe weather events (changes in temperature patterns, freeze/thaw cycles, snow/rain quality and amounts, wind strength and directions) (Nasmith and Sullivan 2010; Willox et al. 2015).

In addition to being an important food source, traditional food is for many Indigenous peoples central to cultural identity, health and well-being (Møller 2011; Nuttall 2000; Rohr 2014). When traditional food is less easy to obtain (and less safe due to contaminants), populations rely more heavily on imported food, the price of which is often exorbitant and beyond many people's means, and the variety and availability limited in many circumpolar communities. Obtaining fresh produce can be particularly challenging because of limitations to regional and community infrastructure, again more so in the Russian North than any other circumpolar region. The least expensive imported food is often processed, calorie dense and nutrient poor, leading to significant health issues (Jeppesen et al. 2011).

In Nunavut, 70% of the population has reported issues of food insecurity with 35.1% experiencing severe food insecurity and 35.1% experiencing moderate food insecurity (EP 2014). Niclasen et al. (2013), who examined food insecurity among Greenlandic children, report that although data are inconsistent and collected differently, Greenlandic children seem to experience food insecurity issues at levels similar to some Arctic regions (such as in Canada's Yukon Territory) while lower than others (for example, Nunavut and Alaska). The authors found that 17% of children in Greenland often or always went to bed hungry. Food insecurity across the Arctic is connected to a westernization of the diet, compounded by poverty and traditional food sources being contaminated through both local and global industry, resource extraction and agriculture (Anaya 2010; Niclasen et al. 2013). Particularly in Russia where Indigenous peoples have returned to eating a more traditional diet, often due to poverty, the lack of available commercial food sources and neglect from the government, the effects of pollutants and contamination, and continually decreasing rights and ability to use traditional lands for hunting, gathering and fishing, are extraordinarily serious issues (Rohr 2014). Even in Greenland and Nunavut, where Indigenous rights to hunt, fish and gather are 'secured', this right is still "subject to the control and regulation of local, regional and national authorities" and even global interests (Nuttall 2000: 9).

It is evident that the proximal determinants continue to significantly affect the health and well-being of circumpolar Indigenous peoples despite efforts and initiatives aimed at changing this fact. However, in many instances, the proximal determinants are but outcomes of, or strongly connected to, the intermediate and particularly the distal determinants. Thus, initiatives that target the proximal determinants may not be very effectual if the intermediate and distal are not targeted prior to or at least in concert with the proximal.

Intermediate determinants

Healthcare and educational systems

Access to formal education is an important contributor to population health, well-being and self-determination. For Indigenous peoples, educational success is connected to culturally and linguistically appropriate education. This is reflected in Article 14 of the United Nations Declaration on the Rights of Indigenous Peoples (UNDRIP) which states that "Indigenous peoples have the right to establish and control their educational systems and institutions providing education in their own language, in a manner appropriate to their cultural methods of teaching and learning" (United Nations General Assembly 2007: 7). In most circumpolar regions, providing formal education opportunities generally, and providing culturally and linguistically appropriate education specifically, remain a challenge (Hirshberg and Petrov 2015; Møller 2011; see also Hodgkins, Chapter 8 this volume, for a survey and discussion of education in the circumpolar North); nowhere is this as evident as it is among the Indigenous peoples of Russia's North.

In 2002, 48% of Indigenous people in Russia had only elementary education and 17% of the Indigenous peoples of the North were illiterate. For all of Russia, the numbers were 8% and 0.5% (Anaya 2010). With reduced funding, school consolidations and many school closures leaving some northern communities without a school (Anaya 2010; Rohr 2014), the situation looks bleak. Sámi access to education varies depending on the country in which they reside. Sámi living in Russia's Kola Peninsula experience conditions similar to other Indigenous peoples of Russia. Access to education and educational attainment of Sámi in Sweden is very different. While the level of education among reindeer herding Sámi men is significantly lower than among Swedes living in the same region, Sámi women affiliated with reindeer herding, as well as non-reindeer herding Sámi men and women, have levels of education comparable to other Swedes in the same geographic setting (Sjølander 2011). In addition, cultural, ethnic and linguistic identity levels remain high among Sámi living in most Sámi regions, which may be connected to the overall higher levels of health and well-being among Sámi people generally (Sjølander 2011).

Access to education and levels of educational attainment in Inuit Nunangat and in Greenland are significantly lower than in Canada and Denmark respectively. In 2011, of Inuit aged 25–64 living in Inuit Nunangat, 28.2% had a postsecondary qualification and of these, 5.1% had a university degree. In comparison, 64.7% of Canada's non-Indigenous population aged 25 to 64 had a postsecondary qualification and of these 26.5% had a university degree (Statistics Canada 2016a).

In Greenland educational statistics, while a little more in favour of the Greenlandic population, are very similar to those of Nunavut. The more favourable statistics must be seen in light of formal education systems having been operating for a lot longer and Greenlandic educators having been part of the educational system for a more extensive period.

Many factors contribute to low educational levels, including that schools were, and to a large degree continue to be, developed and controlled by colonizing institutions and governments, with the language of instruction and educational culture being that of the colonizer rather than using, drawing from or sensitive to Indigenous languages and cultures (Møller 2011, 2013; Rohr 2014). Resent research in Greenland and Nunavut shows that the majority of those who succeeded in the nursing education offered in Nuuk and Iqaluit were from an elite of Inuit/Greenlanders "who possess the linguistic and cultural capital required to excel in the Southern framed educational systems that have prevailed in the Arctic during their elementary, secondary and post secondary schooling" (Møller 2011: 289).

The training of healthcare providers and researchers with a focus on the North occurs both locally – for example, the nursing educations mentioned and the midwifery education in Rankin Inlet – and in several Arctic centres, which are linked together through the University of the Arctic. Despite these efforts, it remains a challenge to recruit and retain health professionals, particularly to the more remote communities (Møller 2011; Rautio et al. 2014). This impacts population health.

Cultural and linguistic continuity

On account of extensive school closures in the Indigenous territories in Russia, Indigenous parents have been forced to send "their children to distant boarding schools which, for the most part, are poorly attuned to indigenous children's cultural needs and threaten family bonds, intergenerational transfers of culture, knowledge, language and skills" (Rohr 2014: 33), rendering the traditional language and culture of the Indigenous children in Russia's North severely threatened, as is their health and well-being (Rohr 2014: 33). This is very reminiscent of Residential Schools in Canada and the United States, which also existed in Inuit Nunangat and Alaska (Brody 2000; La Belle et al. 2005) and led to the loss of culture and a trauma that persists, impacting physical and mental health today (Juutilainen et al. 2014).

Some steps have been taken to help secure the languages and cultures of the Inuit living in Inuit Nunangat. Among these are establishing a teacher education at both bachelor and master's levels, employing elders to share their linguistic and cultural knowledge and abilities through several book series and their presence in the public schools. Still much more can be done – for example, the creation of Inuktitut text-books. The percentage of young people who speak, read and write the Inuit languages is decreasing; the use of English, depending on the region of the Inuit Nunangat, is increasing in part because high-school teachers are predominantly individuals from the southern provinces in Canada who speak only English (Berger 2008), and in part due to increasing globalization and access to electronic and social media. Unfortunately, low educational attainment and unemployment are, along with other factors, often connected to suicide.

Suicide

All northern circumpolar regions face high suicide rates, particularly Chukotka, Greenland and Nunavut where the proportion of Indigenous peoples are highest at 30%, 90% and 85% respectively. For all regions, increased colonialism, westernization, industrialization and marginalization appear to happen in concert with an increase in suicide rates (see the *International Journal of Circumpolar Health* special issue on Suicide and Resilience in Circumpolar Populations, published in March 2015).

Suicide rates in Russia are generally high; however, the rate in Chukotka, particularly among women, exceeds all other regions. The only study on an Indigenous group in Arctic Russia, the Nenets in the Nenets Autonomous Area in northwestern Russia, also shows a higher mortality than non-Nenets in the region with 78 per 100,000, versus 49.2 among the non-Indigenous population (Sumarokov et al. 2014).

The suicide rate in Alaska averaged nearly twice that of the United States as a whole between 2005 and 2014. In 2013 the age adjusted suicide rate for Indigenous Alaskans (46.8 per 100,000) was almost double that of non-Indigenous Alaskans (23.5 per 100,000) (Trust 2015) There is, however, significant regional variation. Most suicides appear to be connected with marginalization in the form of unemployment, low educational attainment and substance abuse (Young et al. 2015). The highest rate was in the Iñupiat region of Norton Sound where the rate was four times that of non-Indigenous peoples.

The incidence of suicide in Nunavut is ten times that of Canada as a whole, a fact reflected by the position of Nunavut's chief coroner who, on 28 September 2015, called suicide a public health state of emergency in the territory, which if declared as such would support the territory financially as a state of emergency releases federal government funding, which is otherwise not available, to address the emergency (Eggertson 2015).

Coinciding with increased westernization, suicide rates for both men and women in Greenland increased from 1960 to 1980 "and have remained around 100 per 100,000 person-years since then" (Bjerregaard and Larsen 2015: 1). There are great regional differences with a trend connecting levels of marginalization with rates of suicide (Bjerregaard and Larsen 2015).

For most health indicators, disparities between Sámi and non-Sámi are very small or non-existent (Sjølander 2011); this does not, however, hold true in relation to suicide. In a Swedish cohort, Sámi men had a 17% higher rate of suicide than other Swedes, and in a Finnish cohort, the rate for Sámi men was 2.5 times higher while for Sámi women no cohorts showed any increased suicide risk (Young et al. 2015: 5). Similarly, although more than 50% of young Sámi experience maltreatment due to ethnicity and have an increased level of suicidal ideation compared to Swedish counterparts, an increase in attempted suicide has not resulted (Omma et al. 2013). The increased suicide rates among Sámi men appear connected to being reindeer herders and the uncertainty, anxiety, stress and alcohol consumption affiliated with this in the face of land degradation and disputes (Kaiser et al. 2011; Kaiser and Renberg 2012).

The inquest into an 11-year-old Nunavummiuq girl taking her own life exposed the risk factors which are likely contributing to suicide across circumpolar regions. These include the effects of "historical trauma and its symptoms" including "high rates of child sexual abuse, alcohol and drug use, poverty, high school dropout rates, and the cultural losses brought about by residential schools and forced relocations", all of which could be viewed as the effects of colonization (Eggertson 2015). Research documents the necessity to address suicide ideation and attempts with culturally safe interventions, which take their point of departure in knowledge originating from the Indigenous peoples themselves. While this knowledge is available, and in some regions has been collected, for example, Greenland, and Nunavut, in other areas this is not the case, for example, in Russia. To decrease suicide and suicide attempts in the circumpolar regions further research is required to determine what is required in some areas while funding and political will are required in others (Eggertson 2015).

Distal determinants

Colonization, racism, discrimination, social exclusion and repression of self-determination

The colonization of Indigenous peoples, and subsequent continued colonialism, discrimination and racism, continue to impact the health and well-being of circumpolar Indigenous peoples on a large scale (Allan and Smylie 2015). They affect the ways in which political, educational and healthcare systems are developed and how these and land and resources are managed.

There are many individuals, groups and organizations working on the documentation, evaluation, reduction and elimination of disparities and towards positive adaptation to change (Rautio et al. 2014). Still, much more can be done including the governments of the colonizing nations accepting their responsibility, signing international declarations and acting according to these declarations.

Most countries endorsed the UNDRIP (the Declaration) in 2007. The United States and Canada were initially among countries against the Declaration; Russia abstained and continues to do so. The United States endorsed the Declaration in 2010 as did Canada, albeit reluctantly and with reservations. These reservations were removed in May 2016 when Indigenous Affairs Minister Carolyn Bennet addressed the Permanent Forum on Indigenous Issues at the United Nations in New York with the words "We are now a full supporter of the declaration, without qualification" (Fontaine 2016: para 2). The nations that share the traditional lands of the Sámi (save Russia) all endorsed the Declaration.

The work towards Sámi rights started already in the early 1900s, and progress in Sámi rights has occurred up to the present day. Sámi Parliaments, political bodies for the Sámi people, have been established in Finland (1973), Norway (1989) and Sweden (1993) (Sapmi 2015; Strømgren 2011). In addition, the Norwegian king and Church as well as the Swedish and Finnish Church have accepted their responsibility for mistreatment of and guilt towards the Sámi people and have apologized, and a reconciliation process was commenced by the Swedish Church in 2012 (Haglund 2013).

Inuit have been working with the Canadian government towards increasing self-determination since the mid 1900s, and the Inuit regions of the Inuit Nunangat including territorial and regional governments were established between 1999 and 2005. In 2007, a Truth and Reconciliation process began between the Indigenous peoples living in Canada and the

Canadian Government. In 2008 the prime minister of Canada, Stephen Harper, apologized to the Indigenous peoples of Canada for the profound negative impact the Residential School system had had and continues to have on individuals', families' and communities' lives, health and well-being (Harper 2008). The Truth and Reconciliation Committee (TRC) ended their work in June 2015 and published the comprehensive final report with numerous recommendations and 94 calls to actions in December 2015. These calls to action include seven focused on health. However, all calls would if implemented positively impact the health and well-being of both Indigenous and non-Indigenous peoples in Canada (TRC 2015). In 2009 the United States, as a first step towards reconciliation, apologized to all the Indigenous peoples of the United States for the wrongdoings that the American government and people had brought upon the nation's Indigenous peoples through colonization and beyond (First Peoples Worldwide 2010). While a nationwide Truth and Reconciliation process is not in effect in the United States, one state (Maine) started its own process in 2013. A commission was established as a collaborative effort between Maine's governor and several Indigenous tribal groups – at the time of writing, it was still ongoing (Maine Wabanaki-State Child Welfare Truth & Reconciliation Commission 2016).

Greenlanders have, as Inuit in Canada, been working towards increasing self-determination and independence from Denmark since the mid 1900s. This resulted in Greenlandic Home Rule in 1979 and expanded Greenlandic self-government in the form of self rule instituted in 2009. Some in the Greenlandic government under Aleqa Hammond (premier from September 2013 to September 2014) were working towards a stated aim of secession from Denmark, which appears to be less urgent under the current premier, Kim Kielsen (Breum 2015).

Denmark has not, however, expressed interest in participating in any reconciliation process with Greenland. In early 2014, Greenland appointed a reconciliation commission with the aim of scrutinizing Denmark's colonization of Greenland. According to Sejersen (2014: 22), Aleqa Hammond, the Greenlandic premier at the time, argued that "colonialism has had an impact on people's self-perception and she claims that some of the self-destructive behaviour among parts of the Greenlandic population can be explained by the colonial experience". The response of the then-Danish Prime Minister, Helle Thorning-Schmidt, to the initiative was that "Denmark had no need to engage in a process of reconciliation with Greenland" (Sejersen 2014).

Russia also has no intention of offering an apology or starting a reconciliation process. The Russian constitution, its government and the majority of the Russian nation does not acknowledge Indigenous peoples or the notion that Russia should bear any "historical guilt towards Indigenous peoples who are therefore entitled to some form of rehabilitation" (Rohr 2014: 56–57).

Conclusion

The Indigenous peoples of the circumpolar North have lived through many years of colonization and continued colonialism and continue to experience the devastating effects of these (Watt-Cloutier 2000). This has contributed to situations of health and well-being that for most are unfavourable when compared to the non-Indigenous peoples living in the same regions. While much remains that must be changed, Indigenous peoples have continued to fight for their rights and have gained important terrain in terms of self-determination and influence on the political arenas nationally and internationally, especially in the areas of health, education and cultural revival. As is evident with the situation of the Indigenous peoples of the Russian North, Indigenous peoples progress towards gaining recognition of their circumstances and achieving rights is seriously halted if national governments do not recognize their own responsibilities for the situations of Indigenous peoples and act as partners in and advocates for Indigenous peoples' rights. In order that the health and well-being of Indigenous peoples can continue to improve, national and international governments and organizations must recognize that Indigenous peoples and their governments and organizations know best in matters relating to them, their lives, health and futures.

References

Aage, H. 2012. 'Alcohol in Greenland 1951–2010: consumption, mortality, prices' *Int J Circumpolar Health* 71: 18444. http://dx.doi.org/10.3402/ijch.v71i0.18444, 11 pgs.

Allan, B. and J. Smylie. 2015. *First Peoples, Second Class Treatment: The Role of Racism in the Health and Wellbeing of Indigenous Peoples in Canada*. Toronto, ON: The Wellesley Institute.

Anaya, J. 2010. *Report of the Special Rapporteur on the Situation of Human Rights and Fundamental Freedoms of Indigenous Peoples. Addendum: Situation of Indigenous Peoples in the Russian Federation*. Fifteenth Session of the Human Rights Committee (23 June 2010), UN Doc. A/HRC/15/37/Add.5. https://documents-dds-ny.un.org/doc/UNDOC/GEN/G10/147/79/PDF/G1014779.pdf?OpenElement.

Banerji, A., V. Panzov, J. Robinson, M. Young, K. Ng and M. Muhammad. 2013. 'The cost of lower respiratory tract infections: hospital admissions in the Canadian Arctic' *Int J Circumpolar Health* 72: 2195, http//dx.doi.org/103402/ijch.v72i0.21595.

Berger, P. 2008. 'Inuit visions for schooling in one Nunavut community'. Doctoral Dissertation. Lakehead University, Thunder Bay, Ontario, Canada.

Bjerregaard, P. and E.C. Eidt. 2010. *Levevilkår, Livsstil og Helbred: Befolknings-undersøgelsen i Grønland 2005–2009*. (*Living Conditions, Lifestyle and Health Status: The Greenlandic Population Study 2006–2009*). Nuuk: Statens Institut for Folkesundhed and Government of Greenland.

Bjerregaard, P. and C.V.L. Larsen. 2015. 'Time trend by region of suicides and suicidal thoughts among Greenland Inuit' *Int J Circumpolar Health* 74, 10.3402/ijch.v74.26053.

Bjerregaard, P., T.K. Young and R.A. Hegele. 2003. 'Low incidence of cardiovascular disease among the Inuit: what is the evidence?' *Atherosclerosis* 166: 351–357.

Blackwell, D.L., J.W. Lucas and T.C. Clarke. 2014. 'Summary health statistics for U.S. adults: National Health Interview Survey, 2012' National Center for Health Statistics. *Vital Health Stat* 10(260). www.cdc.gov/nchs/data/series/sr_10/sr10_260.pdf.

Breum, M. 2015. 'Grønland Udskyder Selvstændighed: Uafhængighed står ikke længere øverst på den Politiske Dagsorden i Nuuk' (Greenland delays independence: independence no longer a top political priority in Nuuk) *Information Online* 22 April 2015. www.information.dk/530849.

Brody, H. 2000. *The Other Side of Eden: Hunters, Farmers and the Shaping of the World*. Toronto, ON: Douglas & McIntyre.

Brubaker, M., J. Berner, R. Chavan and J. Warren. 2011. 'Climate change and health effects in northwest Alaska' *Global Health Action* 4, http://dx.doi.org/10.3402/gha.v4i0.8445.

Brustad, M., K.L. Hansen, A.R. Broderstad, S. Hansen and M. Melhus. 2014. 'A population-based study on health and living conditions in areas with mixed Sami and Norwegian settlements: the SAMINOR 2 questionnaire study' *Int J Circumpolar Health* 73: 23147, http://dx.doi.org/10.3402/ijch.v73.23147.

Burtseva, T.E., T.E. Uvarova, M. Tomsky and J. Odland. 2014. 'The health of populations living in the indigenous minority settlements of Northern Yakutia' *Int J Circumpolar Health* 73: 25758, http://dx.doi.org/10.3402/ijch.v73.25758.

Canadian Institute for Health Information. 2013. *Hospital Care for Heart Attacks among First Nations, Inuit and Métis*. Ottawa: Canadian Institute for Health Information.

Centre for Disease Control and Prevention (CDC). 2015. *Lung Cancer Rates by Race and Ethnicity*. Atlanta, GA: CDC. www.cdc.gov/cancer/lung/statistics/race.htm.

Chief Medical Officer of Health in Greenland. 2013. *2013 Annual Report from the Chief Medical Officer in Greenland*. Nuuk: Author.

Conference Board of Canada. 2016. *Mortality Due to Heart Disease and Stroke. How Canada Performs: Provincial and Territorial Ranking*. www.conferenceboard.ca/hcp/provincial/health/heart.aspx.

Danmarks Statistik. 2011. *Forbrug af Alkohol og Tobak 2010 (Alcohol and Tobacco Consumption 2010)*. Copenhagen: Author. www.dst.dk/nytudg/14774.

Danmarks Statistik. 2015. Stort set uændret Ledighed januar 2015. (Basically unchanged unemployment January 2015). www.dst.dk/da/Statistik/NytHtml?cid=18772.

Dell, C.A. 2006. *Youth Volatile Solvent Abuse FAQs*. Ottawa: Canadian Centre on Substance Abuse. www.ccsa.ca/Eng/Pages/SearchResults.aspx#k=Youth%20Volatile%20Solvent%20Abuse%20FAQs.

Douglas, N.I., T.U. Pavlova, T.E. Burtseva, Y.G. Rad, P.G. Petrova and J.O. Odland. 2014. 'Women's reproductive health in the Sakha Republic (Yakutia)' *Int J Circumpolar Health*, 73(1): 25872. DOI: 10.3402/ijch.v73.2587.

Dudarev, A., P.A. Alloyarov, V.S. Chupakhin, E.V. Dushkina, Y.N. Sladkova, V.M. Dorofeyev, T.A. Kolesnikova, K.B. Fridman, L.M. Nilsson and B. Evengård. 2013. 'Food and water security issues in Russia I: food security in the general population of the Russian Arctic, Siberia and the Far East, 2000–2011' *Int J Circumpolar Health* 72, 10.3402/ijch.v72i0.21848.

Eggertson, L. 2015. '*Nunavut should declare state of emergency over suicide crisis*' *CBC News North*. 27 September 2015. www.cbc.ca/news/canada/north/nunavut-suicide-1.3245844.

Eliassen, B-M., M. Melhus, K.L. Hansen and A. Ragnhild Broderstad. 2013. 'Marginalisation and cardio-vascular disease among rural Sami in northern Norway: a population-based cross-sectional study' *BMC Public Health*, 13: 522. www.biomedcentral.com/1471-2458/13/522.

Engholm, G., J. Ferlay, N. Christensen, A.M.T. Kejs, R. Hertzum-Larsen, T.B. Johannesen, S. Khan, M.K. Leinonen, E. Ólafsdóttir, T. Petersen, L.K.H. Schmidt, H. Trykker and H.H. Storm. 2015. *NORDCAN: Cancer Incidence, Mortality, Prevalence and Survival in the Nordic Countries*, Version 7.1 (09.07.2015). Association of the Nordic Cancer Registries. www.ancr.nu.

Expert Panel on the State of Knowledge of Food Security in Northern Canada. 2014. *Aboriginal Food Security in Northern Canada: An Assessment of the State of Knowledge*. Ottawa: Council of Canadian Academies.

Farmer, P. 2009. 'On suffering and structural violence: a view from below' *Race/Ethnicity: Multidisciplinary Global Contexts* 3(1): 11–28.

First Peoples Worldwide. 2010. *President Obama Acknowledges Need for Native American Apology*. First Peoples Worldwide. http://firstpeoples.org/wp/tag/the-apology-to-the-native-peoples-of-the-united-states/.

Fodor, J.G., E. Helis, N. Yazdekhasti and B. Vohnout. 2014. '"Fishing" for the origins of the "Eskimos and heart disease" story: facts or wishful thinking?' *Can J Cardiol* 30(8): 864–868.

Fontaine, J. 2016. '*Canada officially adopts UN Declaration on Rights of Indigenous Peoples*' *CBC News, Indigenous*, 10 May 2016. www.cbc.ca/news/indigenous/canada-adopting-implementing-un-rights-declaration-1.3575272.

Fraser Health. 2009. *Fraser Health Aboriginal Health. Improving Health for Aboriginal Peoples*. BC Cancer Agency: Author.

Galloway, T. and H. Saudney. 2012. *Inuit Health Survey 2007–2008. Nunavut Community and Personal Wellness*. Ste-Anne-de-Bellevue, QC: Centre for Indigenous Peoples' Nutrition and Environment McGill University.

George, J. 2010. 'Indigenous people of Russia battered by hardships' Nunatsiaq News. *Around the Arctic* 20 May 2010 – 4:44 pm. Online. Available: http://nunatsiaq.com/stories/article/98789_indigenous_people_of_russia_battered_by_hardships.

Gottlieb, K. 2013. 'The Nuka system of care: improving health through ownership and relationships' *International Journal of Circumpolar Health*. 72:10.3402/ijch.v72i0.21118. doi:10.3402/ijch.v72i0.21118.

Government of Canada. 2015. *The National Native Alcohol and Drug Abuse Program*. http://healthycanadians.gc.ca/anti-drug-antidrogue/funding-financement/hc-sc-nnadap-pnlaada-eng.php.

Government of Greenland Department of Health. 2015. *Omkostninger til Misbrug Grønland. (The cost of Addictions in Greenland)*. Nuuk: Naalakkersuisut Government of Greenland Department of Health.

Haglund, M. 2013. 'Samisk Kyrka? En Studie av Førsoningsprocessen mellan Svenska Kyrkan och Samerna med Jæmførende Utblicker mod Norge' (Sami Church? An examination of the reconciliation process between the Swedish Church and the Sami with Norway in comparative perspective). Master's thesis. The Faculty of Humanities and Theology. University of Lund, Sweden.

Hansen, K.L. 2015. 'Access to health services by indigenous peoples in the Arctic region' In United Nations. *The State of the World's Indigenous Peoples. Second Volume. Indigenous Peoples Access to Health Services*. Geneva: United Nations. 58–80.

Harper, S., The Right Honourable. 2008. *Statement of Apology to Former Students of Indian Residential Schools*. Ottawa: Indigenous and Northern Affairs Canada. www.aadnc-aandc.gc.ca/eng/1100100015644/1100100015649.

Health Canada & the Public Health Agency of Canada (PHAC). 2014. *Tuberculosis Prevention and Control in Canada: A Federal Framework for Action*. Ottawa: Her Majesty the Queen in Right of Canada.

Hirshberg, D. and A. Petrov. 2015. 'Education and Human Capita' In J.N. Larsen and G. Fondahl (eds) *Arctic Human Development Report: Regional Processes and Global Linkages*. Copenhagen: Nordic Council of Ministers. 347–396.

Hopper, T. 2014. 'Iqaluit hopes to curb alcoholism and binge-drinking by opening city's first beer store in 38 years' *National Post*. 21 September 2014. Online edition of newspaper. http://news.nationalpost. com/news/canada/iqaluit-hopes-to-curb-alcoholism-and-binge-drinking-by-opening-citys-first-beer-store-in-fifteen-years.

Indigenous and Northern Affairs of Canada. 2013. *Fact Sheet – 2011 National Household Survey Aboriginal Demographics, Educational Attainment and Labour Market Outcomes*. Ottawa: Author, Government of Canada. www.aadnc-aandc.gc.ca/eng/1376329205785/1376329233875.

Inuit Tapiriit Kanatami (ITK). 2014. *Social Determinants of Inuit Health in Canada*. Ottawa: ITK.

Jeppesen, C., P. Bjerregaard and K. Young. 2011. 'Food based dietary guidelines in circumpolar regions' *Int J Circumpolar Health, Circumpolar Health Supplements*. Suppl. 8.

Jørgensen, M.E. 2014. 'Diabetes i Grønland: fra Alfred Bertelsen til Molekylær-Diagnostik i 2014' (Diabetes in Greenland from Alfred Bertelsen to molecular-diagnostic in 2014). *Ugeskr Læger (Danish Medical Journal)* 176(22): 2066–2068.

Juutilainen, S., R. Miller, L. Heikkila and A. Rautio. 2014. 'Structural racism and indigenous health: what indigenous perspectives of residential school and boarding school tell us? A case study of Canada and Finland' *The International Indigenous Policy Journal* 5(3). http://ir.lib.uwo.ca/iipj/vol5/iss3/3 DOI:10.18584/iipj.2014.5.3.3.

Kaiser, N., A. Nordstrom, L. Jacobsson and E.S. Renberg. 2011. 'Hazardous drinking and drinking-patterns among the reindeer-herding Sami population in Sweden' *Subst Use Misuse* 46(10): 1318–1327.

Kaiser, N. and E.S. Renberg. 2012. 'Suicidal expressions among the Swedish reindeer-herding Sami population' *Suicidology Online* 3: 102–113.

Kaweionnehta Human Resource Group. 1993. *First Nations and Inuit Community Youth Solvent Abuse Survey and Study*. Ottawa: National Native Alcohol and Drug Abuse Program/Addictions and Community-Funded Programs.

Kelm, M.E. 1998. *Colonizing Bodies: Aboriginal Health and Healing in British Columbia, 1900–50*. Vancouver, BC: UBC Press.

Koch, A., P. Sørensen, P. Homøe, K. Mølbak, F.K. Pedersen, T. Mortensen, H. Elberling, A. Eriksen, O. Rosing Olsen and M. Melbye. 2002. 'Population-based study of acute respiratory infections in children, Greenland' *Emerg Infect Dis*. [serial on the Internet]. June. https://wwwnc.cdc.gov/eid/article/8/6/01-0321_article.

Korhonen, M. 2004. *Alcohol Problems and Approaches: Theories, Evidence and Northern Practices*. Ottawa: National Aboriginal Health Organization (NAHO).

La Belle, J., S.L. Smith, C. Easley and G.P.C. Kanaqlak. 2005. *Boarding School: Historical Trauma among Alaska's Native People*. Anchorage, AL: National Resource Center for American Indian, Alaska Native and Native Hawaiian Elders.

Larsen, J.N., P. Schweitzer, A. Pertov and G. Fondahl, G. 2014. 'Tracking change in human development in the Arctic' In J. Nyman Larsen, P. Schweitzer and A. Petrov (eds) *Arctic Social Indicators. ASI Implementation*. Copenhagen: Nordic Council of Ministers. 15–54.

Loppie Reading, C. and F. Wien. 2009. *Health Inequalities and Social Determinants of Aboriginal Peoples' Health*. Prince George, BC: National Collaborating Centre for Aboriginal Health.

Maine Wabanaki-State Child Welfare Truth & Reconciliation Commission. 2016. *About*. Rockport, ME: Author. www.mainewabanakitrc.org/about/.

Martin, S. and A. Hill. 2009. *The Changing Economic Status of Alaska Natives, 1970–2007*. Note no 5, July 2009. Anchorage, AL: Institute of Social and Economic Research University of Alaska Anchorage.

Mowbray, M., ed. 2007. *Social Determinants and Indigenous Health: The International Experience and Its Policy Implications*. Geneva, Switzerland: World Health Organization Commission on Social Determinants of Health.

Muckle, G., O. Boucher and D. Laflamme. 2007. *Alcohol, Drug Use and Gambling Among the Inuit of Nunavik: Epidemiological Profile*. Nunavik Inuit Health Survey 2004/Qanuippitaa? How are we? Quebec: Institut national de santé publique du Québec and Nunavik Regional Board of Health and Social Services.

Møller, H. 2010. 'Tuberculosis and colonization: current tales about tuberculosis in Nunavut' *Journal of Aboriginal Health* 5(1): 38–48.

Møller, H. 2011. '"You need to be double cultured to function here": toward an anthropology of Inuit nursing in Greenland and Nunavut'. PhD Dissertation. Department of Anthropology, University of Alberta, Canada.

Møller, H. 2013. 'Double culturedness: the "capital" of Inuit nurses' *Int J of Circumpolar Health* 72 (Suppl 1). DOI: 10.3402/ijch.v72i0.21266.

Nasmith, K. and M. Sullivan. 2010. *Climate Change Adaptation Plan – Hamlet of Arviat, Nunavut.* Ottawa: Canadian Institute of Planners.

Niclasen, B., M. Molcho, S. Arnfjord and C. Schnohr. 2013. 'Conceptualizing and contextualizing food insecurity among Greenlandic children' *Int J Circumpolar Health* 72, 19928. 10.3402/ijch.v72i0.19928.

Nuttall, M. 2000. 'Barriers to sustainability: the Arctic in the global economy' in M. Nuttall (ed.) *The Arctic is Changing.* www.thearctic.is/articles/overviews/changing/enska/kafli_0300.htm.

OECD. 2014. 'How does the Russian Federation compare?' *OECD Health Statistics 2014 – Country Notes.* www.oecd.org/russia/oecd-health-statistics-2014-country-notes.htm.

Økonomisk Råd (Greenland's Finance Board). 2014. *Grønlands Økonomi 2014. (The Greenlandic Economy 2014).* Nuuk: Author.

Omma, L., M. Sandlund and L. Jacobsson. 2013. 'Suicidal expressions in young Swedish Sami, a cross-sectional study' *Int J Circumpolar Health*, 72: 19862. http://dx.doi.org/10.3402/ijch.v72i0.19862.

Paarisa (Greenlandic Department of Public Health). 2013. *Inuuneritta II (The New Public Health Program).* Nuuk: Paarisa. http://old.paarisa.gl/home/inuuneritta.aspx?lang=da.

Peterson, E., K. Pickle and C. Bushmore. 2015. *Alaska Tobacco Facts. 2015 Update.* Anchorage, AL: Alaska Department of Health and Social Services.

Plescia, M. 2014. 'Lung cancer deaths among American Indians and Alaska Natives, 1990–2009' *American Journal of Public Health* 104(S3): s388–s395.

Poppel, B. 2015. *SLiCA: Arctic Living Conditions – Living Conditions and Quality of Life Among Inuit, Saami and Indigenous Peoples of Chukotka and the Kola Peninsula.* Copenhagen: Nordic Council of Ministers.

Rautio, A., B. Poppel and K. Young. 2014. 'Human health and well-being' in J.N. Larsen and G. Fondahl (eds) *Arctic Human Development Report Regional Processes and Global Linkages.* Copenhagen: Nordic Council of Ministers. 297–346.

Rex, K.F., N.H. Larsen, H. Rex, B. Niclasen and M.L. Pedersen. 2014. 'A national study on weight classes among children in Greenland at school entry' *Int J Circumpolar Health* 73, 25537, http://dx.doi.org/10.3402/ijch.v73.25537.

Riva, M., Larsen, C.V. and P. Bjerregaard. 2014a. 'Household crowding and psychosocial health among Inuit in Greenland' *Int J Public Health* 59: 739. doi:10.1007/s00038–014–0599-x.

Riva, M., P. Plusquellec, R-P. Juster, E.A. Laouan-Sidi, B. Abdous, M. Lucas, S. Déry and E. Dewailly. 2014b. 'Household crowding is associated with higher allostatic load among Inuit' *Journal of Epidemiology & Community Health* 68(4): 363–369.

Rohr, J. 2014. *IWGIA Report 18: Indigenous Peoples in the Russian Federation.* Copenhagen: IWGIA.

Sapmi. 2015. *The Sami Parliaments in Finland and Norway.* www.samer.se/4609.

Sejersen, F. 2014. 'Region and country reports. The Arctic: Greenland' in C. Mikkelsen. ed. *The Indigenous World 2014.* Copenhagen: International Working Group for Indigenous affairs.

Sjølander, P. 2011. 'What is known about the health and living conditions of the indigenous people of northern Scandinavia, the Sami?' *Global Health Action* 4, 8457, DOI: 10.3402/gha.v4i0.8457.

Skryagin, A. 2005. 'Children of the north: our pain' *Ansipra Bulletin, 13 and 13a, English Language Edition*: 9–11. http://ansipra.npolar.no/english/.

Spein, A.R. 2008. 'Substance use among young indigenous Sami: a summary of findings from the north Norwegian youth study' *Int J Circumpolar Health* 67(1): 122–134.

Statistics Canada. 2014. *Table 105-0503: Health Indicator Profile, Age-Standardized Rate, Annual Estimates, By Sex, Canada, Provinces and Territories, Occasional*, CANSIM (database). www5.statcan.gc.ca/cansim/a26?lang=eng&id=1050503.

Statistics Canada. 2016a. *Smokers, By Sex, Provinces and Territories.* Ottawa. Government of Canada. www.statcan.gc.ca/tables-tableaux/sum-som/l01/cst01/health74b-eng.htm.

Statistics Canada. 2016b. *Area of Residence: Inuit Nunangat.* Ottawa: SC www12.statcan.gc.ca/nhs-enm/2011/ref/dict/pop149-eng.cfm.

Statistics Canada. 2016c. 'The educational attainment of aboriginal peoples in Canada'. *Author.* www12.statcan.gc.ca/nhs-enm/2011/as-sa/99-012-x/99-012-x2011003_3-eng.cfm.

Strømgren, J. 2011. 'Sapmi' in K. Wessendorf (ed.) IWGIA, *The Indigenous World 2011.* Copenhagen: The International Work Group for Indigenous Affairs.

Sumarokov, Y.A., T. Brenn, A.V. Kudryavtsev and O. Nilssen. 2014. 'Suicides in the indigenous and non-indigenous populations in the Nenets autonomous Okrug, northwestern Russia, and associated socio-demographic characteristics' *Int J Circumpolar Health* 73: 24308. http://dx.doi.org/10.3402/ijch.v73.24308

Sundhedsstyrelsen. 2008. *Danskernes Rygevaner 2008. (The Danes' Smoking Habits 2008)*. https://sundhedsstyrelsen.dk/da/sundhed/tobak/tal-og-undersoegelser/danskernes-rygevaner/2007.

Taylor, G. 2016. *The Chief Public Health Officer's Report on the State of Public Health in Canada 2015. Alcohol Consumption in Canada*. Ottawa: Public Health Agency of Canada.

Trust (Alaska Mental Health Trust Authority). 2015. *Alaska Scorecard: Key Issues Impacting Alaska Mental Health Trust Beneficiaries*. Juneau, AL: State of Alaska Department of Health and Social Services.

Truth and Reconciliation Commission of Canada. 2015. *Honouring the Truth, Reconciling for the Future: Summary of the Final Report of the Truth and Reconciliation Commission of Canada*. Ottawa: Author.

United Nations General Assembly. 2007. *United Nations Declaration on the Rights of Indigenous Peoples* (UNGA Document No. A/RES/61/295). New York: United Nations.

United Nations Office on Drugs and Crime (UNODC). 2011. *World Drug Report 2011*. (United Nations Publication, Sales No. E.11.XI.10). Vienna: United Nations Office on Drugs and Crime.

Vitebsky, P. 2005. *The Reindeer People: Living with Animals and Spirits in Siberia*. New York: Houghton Mifflin Harcourt.

Wallace, S. 2014. *Inuit Health: Selected Findings from the 2012 Aboriginal Peoples Survey*. Ottawa: Ministry of Industry and Statistics Canada. www5.statcan.gc.ca/access_acces/alternative_alternatif.action?l=eng&loc=/pub/89–653-x/89–653-x2014003-eng.pdf.

Watt-Cloutier, S. 2000. 'Honouring our past, creating our future: education in northern and remote communities' in M.B. Castellano, L. Davis and L. Lahache (eds) *Aboriginal Education: Fulfilling the Promise*. Vancouver: UBC Press. 114–128.

Willox, A.C., E. Stephenson, J. Allen, F. Bourque, A. Drossos, S. Elgarøy, M.J. Kral, I. Mauro, J. Moses, T. Pearce, J.P. MacDonald and L. Wexler. 2015. 'Examining relationships between climate change and mental health in the circumpolar north' *Reg Environ Change* 15: 169–182.

World Health Organization (WHO). 2012. *World Conference on Social Determinants of Health: Meeting Report, Rio de Janeiro, Brazil, 19–21 October 2011*. Geneva. WHO. www.who.int/social_determinants/sdhconference/en/.

World Health Organization (WHO). n.d. *What are Social Determinants of Health? Geneva*: WHO. www.who.int/social_determinants/sdh_definition/en/.

Young, K. and S. Chatwood. 2015. 'Comparing the health of circumpolar populations: patterns, determinants, and systems' in B. Evengård, J.N. Larsen and Ø. Paasche (eds) *The New Arctic*. Cham, Switzerland: Springer. 203–211.

Young, K., B. Revic and L. Soininen. 2015. 'Suicide in circumpolar regions: an introduction and an overview' *Int J Circumpolar Health* 74: 27349, http://dx.doi.org/10.3402/ijch.v74.27349.

8

Education in the Arctic

Trends, challenges, and possibilities

Andrew Hodgkins

Introduction

Why study education in the Arctic? At first blush the shear breadth and diversity of both the region and subject area pose a daunting challenge for those attempting to study it. Not only are Arctic societies diverse, but so too is the study of education, considering it comprises both disciplinary and social knowledge. While the former is associated with institutional structures (schools), transmission (pedagogy), and content (curriculum), the latter occurs outside the school yard gate and is essential to cultural transmission. There are several important reasons for assessing the state of education. In addition to its intrinsic benefits, associated with developing an individual's knowledge, skills, and values, are collective benefits, including the adaptive capacities of societies to respond to rapid social change brought on by modernity's wage labour economy. Relatedly, differential access to, and acquisition of, education occurring between genders, regions, and ethnic groups – as measured by a variety of indicators including graduation rates, migration, labour market participation, and income – impact the circumpolar region as a whole. Consequently, an analysis of education provides important insights into life in the Arctic, as well as political trends and economic contexts.

Using both quantitative and qualitative sources of information, the purpose of this chapter is to present some key themes, trends, and challenges characterizing formal education in the various circumpolar regions. My contribution stems from, and is informed by, a lifelong association in the Arctic. My formative years were spent in Canada's Northwest Territories (NWT) and Nunavut, before returning there to begin my career as a teacher. My research, which stems from graduate studies, examines northern aboriginal vocational education and training partnerships. My theoretical orientation is influenced by *social realism* – a relatively new and emerging field in the sociology of education which emphasizes the universal (*emergent*) nature of disciplinary knowledge – something to which I shall return later in the chapter when assessing how education has been conceptualized in the literature.

I begin by presenting a review of related literature, which includes a summary of recent cross comparative statistical analysis, as well as the methodological and conceptual frameworks used to collect and make sense of the data. This review provides a general assessment of access to education and achievement levels across the different circumpolar regions. In the second part,

I put a human face on these statistics by sharing a story about a young Dene couple who left their remote community in Canada's Northwest Territories in order to gain the necessary education and training to become a journeyperson in the skilled trades. Their experience provides insights into the learning-to-work transitions accompanying northern indigenous peoples, as well as insights into the politics of training programmes that have formed in recent years in partnership with resource extractive industries. As such, the story serves to tie together common themes, challenges, and possibilities covered in this chapter, as well as link them to other topics covered in this book.

Literature review

For the most part, Arctic scholarship has traditionally focused on the region's indigenous peoples and their social knowledge. In contrast to local indigenous knowledge (LIK) which is considered vital to cultural continuity and social well-being, in many parts of the Arctic formal education is considered a Janus face of sorts, representing the hegemonic vehicle of colonial assimilation, but also the entrance way into both the labour market and wider world of ideas (Kruse et al. 2008). At the same time, formal education is also relatively new in many parts of the Arctic, especially in northern Canada, having only recently taken root in the post-World War II modernization era. Consequently, there has existed relatively few studies examining education per se, with even fewer endeavouring to compare and contrast the different and very diverse regions of the circumpolar North. However, since the late 2000s, significant efforts have been made to broaden the scope of social science approaches to northern education, in part owing to the International Polar Year (2007–2008) which represented an intense burst of research activity in all things Arctic. The *Arctic Human Development Report* (AHDR), which was first published in 2004 under the auspices of the Arctic Council, was the first report of its kind to undertake such a task – devoting a whole chapter to the subject of education in the circumpolar North. The follow-up chapter on education, published in the second AHDR in 2015, was more comprehensive, responding to some of the knowledge gaps identified in the first report, and providing a much more detailed analysis drawn from an extensive literature review.

Research supporting the AHDR reports stems from international collaborative efforts initiated through various projects developed out of northern research centers. Of particular note are the contributions made by the Arctic Social Indicators (ASI) and the Survey of Living Conditions in the Arctic (SLiCA) projects. Findings help inform decision-making, and also serve as a baseline of knowledge to gauge rapid environmental and social change. Considering that the Arctic social science community is rather small, many of the same names appear on all three sets of publications, indicating the high degree of crossover in resources being used to inform our current knowledge. For the purposes of this chapter, these reports provide a useful starting point to identify major themes, trends, and challenges in education. What follows is a summary of key findings, with some initial commentary describing the methodological and conceptual frameworks used.

Methodology

Both SLiCA and ASI draw extensively on statistical sources of information and interview survey data. They are not exclusively concerned with education, but education nonetheless represents a significant social indicator used to gauge human development and track changes. In developing a cross comparative analysis, researchers sought to standardize indicators – a significant feat in-and-of itself, especially considering the diverse and demographically sparse populations being measured, as well as the differences in the way regions record and report information. SLiCA, which initially emerged in 1997 out of a need to access comprehensive data in Greenland in response to it gaining self-government, developed a comparative study based upon a survey of about 200 questions examining both the material and non-material components of well-being. By 2015, almost 8,000 people had been interviewed. While most people surveyed identified as Inuit (6,900), the research also included Sámi in the Nordic Arctic regions, and indigenous peoples of Chukotka and the Kola Peninsula in Russia. ASI, on the other hand, was initiated by the Stefansson Arctic Institute in Akureyri, Iceland, in response to the first AHDR. Its focus is broader in scope, in the sense that it has undertaken to survey both indigenous and non-indigenous peoples. In identifying education as one of six domains of human development and well-being, ASI researchers focus on the ratio of students successfully completing post-secondary education. This indicator was chosen for its relative ease of access; it was also reasoned that completion of post-secondary schooling also indicates the quality of K–12 education received in a given region.

Two prominent theories undergirding much of this literature include human capital theory and social constructivism. As the second AHDR notes, human capital is the "stock of knowledge and skills embodied in a human population that has economic value" (Hirshberg and Petrov 2015: 364). According to the same report, "Arctic regions demonstrate substantial gaps in terms of development of human capital" (p. 365), the antidote to which is the investment in people through education. It is reasoned, by raising the standard of education (which also includes LIK), societies will increase their level of well-being and self-sufficiency by being more productive. Social constructivism, on the other hand, assumes that knowledge is localized or situated, and that people learn once knowledge is made relevant to them (Rasmussen, Barnhardt and Keskitalo 2010). Whether or not these theories adequately capture the purpose of education or the nature of knowledge warrants further attention, considering the (neo)classical economic assumptions undergirding human capital theory that posit a unidirectional causality between education and economic output (Livingstone 2009),[1] and the tendency of social constructivism to conflate "knowledge to the knower", and pedagogy to the curriculum which hold implications for what should (and should not) be taught in schools (Barrett and Rata 2014; Rata 2012).

Findings

As the second AHDR notes, people are more educated in the Arctic than they were a decade ago (Hirshberg and Petrov 2015: 379). Nevertheless, significant challenges remain that relate to access and achievement. Educational achievement, which is measured by graduation rates and highest levels of schooling, varies according to three demographic variables: population size (village, city), ethnicity (indigenous, non-indigenous), and gender. Notably, a pronounced achievement gap exists between indigenous and non-indigenous groups – a problem exacerbated by limited access to educational opportunities for smaller and remote communities where school closures reflect a trend towards outmigration, ageing demographics, and decreased funding (Hirshberg and Petrov 2015; Kruse et al. 2008). Whereas larger regional centers whose population is mostly non-indigenous enjoy higher levels of education and better services (including access to educational opportunities), the reverse holds true for smaller communities where resources are limited and the majority of the population is indigenous. Consequently, those with less education and financial resources must leave their home communities in order to access education and jobs – a situation that contributes to a brain drain of local talent and resources. Contributing to these challenges is a gender gap, whereby the vast majority of people pursuing post-secondary education are women.

As noted to occur in many parts of the circumpolar North, including the Faroe Islands, northern Fennoscandia, parts of Iceland, and the Russian far north, the absence or loss of a community school means that students must leave their home communities and attend boarding schools in regional centers. Greenland, as a case in point, has only four academic high schools, whereas the Faroe Islands have only nine (and no boarding schools) (Hirshberg and Petrov 2015). However, other regions are bucking this trend. For instance, until the late 1970s, school facilities on Alaska's North Slope were available only up to the ninth grade; since then new school programmes have been introduced (Broderstad, Eliassen and Melhus 2015: 140). Nevertheless, even if access improves, the *quality* of education in smaller communities remains tenuously linked to the ability of local educational authorities to recruit and retain qualified and experienced teachers (Hirshberg and Petrov 2015; Rasmussen et al. 2010).

At the post-secondary level, significant differences also exist. As of 2011, there were 136 institutions of higher education, including 25 major universities and 111 branch campuses across the Arctic (Hirshberg and Petrov 2015: 370). The Nordic countries and Russia boast the most developed tertiary system in terms of enrolments, course offerings, and research centers. At the other end of the spectrum is Canada, which holds the dubious distinction of being the only circumpolar country to not have a university north of the 60th parallel. Instead, its three territories maintain a college system, each containing regional campuses and community learning centers. This difference is more a symptom of demographics than neglect; unlike other circumpolar regions which contain urbanized centers such as the Russian city of Murmansk (pop. ~ 300,000) where its State Technical University has an enrolment of 8,000 students, it is more pragmatic to maintain a college system in sparsely populated regions. Opportunities to enter into degree programmes (e.g. teacher education, nursing) that partner with universities in larger centers further south, as well as participate in online courses through the University of the Arctic[2] provide additional opportunities for students in northern Canada.

Lennert (2015: 238) notes that the proportion of the Greenlandic population having a formal education has increased from 28% to 47% over the span of 30 years. Similarly, the AHDR notes that the number of graduates in Greenland increased by 37% between 2005 and 2008; however, the number of drop-outs has also increased by about the same amount (Hirshberg and Petrov 2015: 358). In Alaska, the native drop-out rate is 34.7%, and native students make up only 22.7% of students in grades 7–12; the 2013 graduation rate was 71.8% for non-natives and 57% for native students (Hirshberg and Petrov 2015: 358). The graduation rate for Nunavummiut (Inuit of Nunavut) ranged from 32% to 38% (over four years) (Hirshberg and Petrov 2015: 358), with half the individuals over the age of 25 not holding a graduation certificate (Morin, Edouard and Duhaime 2015: 209). In neighbouring NWT, 38.7% of the aboriginal population holds a high school diploma, compared with 79.1% of all Canadians in 2009; high school attainment ranges from 11.8% to 80.9% across the different communities, a vast spread which is reflected in differences between small aboriginal communities and cities like Yellowknife, where the majority of the population is non-aboriginal (Petrov, King and Cavin 2014: 116).

Gender disparity in post-secondary education participation, while a global phenomenon as reported in OECD countries, is also noted to be markedly pronounced in most minority groups. With the possible exception of some parts of northern Canada, women had become a majority in higher education participation in all regions of the Arctic by the late 1990s, accounting for more than 70% of the students in several universities (Rasmussen et al. 2010: 79). For instance, in 2003, the University of Alaska enrolled 157 female students for every 100 male students; of those who were Alaska Natives, 69% were female; amongst rural Native students this figure increased to 73% (Kleinfeld and Andrews 2006: 430). And while very little is known about the levels of Sámi education, owing to the way in which ethnicity is reported in their respective home countries (Beach, Lewis, Rasmussen and Roto 2015), in 2003 Sámi women outnumbered men by more than 5:1 at Norway's Sámi University (Kleinfeld and Andrews 2006: 429). Similar ratios have also been reported in Ilisimatusarfik, the University of Greenland in Nuuk (M. Poppel 2015).

Reasons cited for gender disparity include speculation that formal schooling does not reflect traditional male roles (Kleinfeld and Andrews 2006), and that males are not as adaptive to changes associated with a transition from primary to tertiary sector employment (Rasmussen et al. 2010: 74). Regardless of these reasons, as more females vacate their home communities to gain employment that fits their qualifications, a distorted female-to-male ratio has developed in a number of regions, including Alaska, Greenland, the Faroe Islands, Newfoundland, Iceland, and the Sakha Republic in Russia (Fondahl, Crate and Filippova 2014; Rasmussen et al. 2010).

Intra-ethnic analysis of indigenous peoples also indicates a marked difference in highest levels of schooling achieved. Notably, 44% of Inuit adults in the Canadian Arctic have no education beyond elementary school. This figure stands in stark contrast to Inuit in Greenland, Alaska, and Chukotka, where these comparative figures are respectively 10%, 13%, and 26% (B. Poppel 2015). Conversely, those reporting completion of vocational or college education were reported respectively as being 14% (Nunavut), 46% (Greenland), 42% (Chukotka), and 25% (Alaska). In other words, Inuit in Greenland are three times more likely to have completed college or vocational training certificates than their neighbours to the west in nearby Nunavut. One possible reason for this discrepancy may be due to the historical differences in the provision of formal education and training. Unlike Nunavut, where education is relatively new, Greenland had its first vocational education programme (kayak building) in 1784; a training school for midwives was created in 1800; and in 1847 a teachers' college, Ilinniarfissuaq, was established; by 1905, some students were sent to Denmark for further education (Hodgkins 2010).

Journeys and journeying in the Canadian Arctic

The link between education and jobs can be clearly seen in vocational education and training programmes developed for northern aboriginal people in response to labour markets. In many ways these programmes and partnerships that form between the various stakeholders capture the hopes, tensions, and contradictions characterizing the rapid transitions shaping northern societies. The Berger Inquiry into a proposed gas pipeline along the Mackenzie Valley of the Northwest Territories in the 1970s was one of the most significant events in garnering international attention and support for aboriginal rights – pitting the interests of oil companies eager to tap the region's natural resource wealth against those of the aboriginal peoples, whose traditional way of life was being threatened. While industry and government proponents raised the promise of local employment, aboriginal opponents wanted to ensure their lands would be protected first – a standoff which resulted in the landmark decision to put a moratorium on development until land claims were settled. Since then, the politics of protest character-izing the Inquiry have transformed into a politics of partnerships, which characterized the public hearings that began in 2006 for this mega-project when it was revived as the Mackenzie Gas Project. With land claims settled, newly formed aboriginal governments and their indus-try partners resurrected the hopes for the project. Public-private training-to-employment programmes with a federal price tag of CAN$12.7 million dollars were developed in 2003, with the intention that aboriginal northerners would benefit from the jobs the pipeline would bring.[3] What then became of these initiatives? Here, I recount one couple's story that helps capture the challenges in transitioning to what ostensibly is an ultimate end goal of these programmes – the attainment of journey person status in a skilled trade.

In the spring of 2010 while pursuing my doctoral research (Hodgkins 2013), I visited a Gwich'in community nestled in the picturesque and remote region of northwest Canada's Beaufort Delta. During my visit, I interviewed several people about their experiences with education and training, including a young woman who shared her experiences involving her husband's career path. For purposes of this discussion I shall use pseudonyms to convey her story, as told through the narrative of Alice, the wife of Jordan.

According to Alice, Jordan was the top student in the trades access programme he was registered in at the regional college in Inuvik. When it came time to being apprenticed, several employers wanted Jordan to work with their oil company. As Alice recalls, "They picked Jordan as the first person they wanted to apprentice. They went around the table like it was a draft and he picked [company] as he considered it more promising for him". However, this was only the first step in what proved to be a long and arduous career and life transition path for the couple. After making the momentous decision, the couple and their young children relocated to Grande Prairie, a small city in the northwestern part of the province of Alberta 1600 miles away, where they were to live for the next six years.

Far removed from the "comfort zone and support" of their community, with its tightly-knit network of extended family and friends, the young couple struggled to adapt to their new life. Alice felt that the company had put them into an environment where people "don't give a shit"; requiring them to adjust to a life most people take for granted, including having to live by schedules, take transportation (the couple had to learn how to take a bus rather than paying for expensive taxis), and placing their children in childcare where they did not know the caregivers. Jordan and Alice eventually moved to a neighbourhood where there were more aboriginal people and aboriginal students in the school. Alice also met aboriginal student help workers who visited her in their home, and with whom she still keeps in touch.

As for work, Jordan awoke every morning at 5 a.m. and drove to the company gas plant for an eight-hour shift, returning home by 4 p.m. each day. Meanwhile, Alice registered at a local college to train as a carpenter but was not successful owing to childcare responsibilities that eventually caused her to drop out of the programme. Adding to her decision was a perception that the trades were an uninviting prospect, as it is "difficult for a woman to get into a man's world" because women are "considered a distraction" on worksites.

The couple eventually separated. Alice returned to her home community and is now employed with the local hamlet office. When the gas plant laid workers off, Jordan also returned north, taking a government job as a journeyman in Inuvik. However, I was also told by someone else that Jordan experienced a "subtle racism" at the work-site, which is why he quit and returned home.

According to Alice, Jordan now comes home from Inuvik "to do the family thing" on weekends. While the transition was a significant life-changing event for them, the experience also instilled a sense of confidence and entrepreneurial spirit. As a certified journeyman, Jordan is now working on his Red Seal and plans on getting dual-ticketed, which will increase his employability by both allowing him to work anywhere in Canada and gaining a second certificate.[4] Owing to his past contacts, Alice thinks that he will be able to start up his own company if oil and gas development occurs in the region. Alice also plans on returning to college to study political science. Aside from career aspirations, the transition also changed perceptions of their local community and region. According to Alice, Jordan "has never taken anything from their [aboriginal government] regarding money or entitlements, or student financial assistance, and now they want to use him as a role model". In a region with few success stories and role models, efforts to profile him are understandable. However, there is a degree of cynicism reflected in Alice's comments as she feels that a culture of entitlement pervades the region, which partly stems from the home with parents not holding their children accountable to get out of bed and off to school. As she asserts matter-of-factly, "You want it – you work for it". In a similar vein, Jordan considers the apprentices he now mentors to be "kind of lazy". When he gets his Red Seal the couple plan to move out of the region to the city of Whitehorse in the Yukon, where they feel that there is better education for their children.

Cynicism concerning the quality of education offered at local schools was commonly voiced and cited as a reason why some people had relocated to larger centers on account of perceptions that regional schools offered better education. A key source of contention was the policy of "social pass" which was popularly scorned as setting students and communities up for failure. In essence, this policy allows students to be passed on to the next grade with their peers regardless of their achievement. However, once they move into high school, students often become frustrated and drop out as they come to realize their significant educational deficits. To reinforce these problems in her community, Alice points to a picture on her office wall of the most recent graduating class. Of the fifteen students in the photograph, twelve of them have returned home, with only three students enrolled in post-secondary courses.

While the testimony of Jordan and Alice captures some of the challenges associated with education and the learning-to-work transitions experienced by northern aboriginal people, the story also yields important insights into how partners relate to their investments in human capital. By chance, later that same day when I returned to Inuvik, I happened to converse with the same oil company that had apprenticed Jordan. The party of three was in the region visiting a local community during an annual spring celebration where they had sponsored some of the festivities. Through a mutual acquaintance the company bought us dinner. During our conversation enquiries were made about Jordan: What was he doing? Where was he living? What were his plans? One member of the group mused out loud that they should get in touch with him to see if he would be willing to set up a company if development in the region proceeded.

This chance encounter illustrates how resource extraction companies constantly mine regions for capital – be it mineral or human. While constituting a small but significant segment of Alice's story, the couple's experience provides insights into the arduous journey to *journey* for aboriginal northerners, as well as the political nature by which successful aboriginal trades people are situated as pawns in a larger power play over resource development. In this case, Jordan was a "poster boy" adorning walls of various regional government and industry office buildings; an emblem to hang both the collective pride and partnership angst on; to in effect legitimize the oil company's regional presence and proclaim: "We did it; we partnered – we created a successful aboriginal person!"

Conclusion

This chapter has sought to focus on key themes, trends, and challenges associated with formal education in the Arctic, as well as to offer different ways of conceptualizing education and knowledge. Having focused primarily on issues relating to access and achievement, further attention is needed to take up a third theme in the literature that relates to blending social knowledge with disciplinary knowledge as a way of "indigenizing" education. As we have seen, many of the same themes presented in the broad statistical survey on the state of education are reflected in the comments and perceptions shared in the story. The personal challenges many aboriginal people face in leaving home in order to gain an education often-times result in them permanently uprooting to regional centers. Hence, a knowledge paradox ensues: by leaving their home communities in order to gain access to education and jobs, people lose the social knowledge that serves to strengthen bonds and cultural continuity. However, other elements also emerge from the story that are absent in the literature – notably, the mismatch between education and jobs. As the story shows, people need more than educational opportunities to be successful; they must also have access to training opportunities tied to committed and stable employers. In this case the training was tied to speculative development; education did not translate into jobs, thus illustrating conceptual weaknesses with human capital theory, or the reliance on statistics, where the numbers being reported clearly did not add up.

Notes

1 See also Duhaime, Édouard, and Bernard (2015) for a discussion on the correlation between Inuit income and education, and the paradox between material living conditions and extra-economic indicators of subjective well-being.
2 The University of the Arctic is a virtual university, comprising numerous member institutes that offer online courses in circumpolar studies.
3 See Hodgkins (2015) for a critical analysis of the way in which training programme outcomes were publicly reported.
4 The term Red Seal refers to certification that is recognized across Canada, rather in just one provincial or territorial jurisdiction; dual-ticketed refers to gaining a second trade certification.

References

Barrett, B. and E. Rata (eds) 2014. *Knowledge and the Future of the Curriculum: International Studies in Social Realism*. London: Palgrave Macmillan.
Beach, H., D. Lewis, R. Rasmussen and J. Roto 2015. 'The Survey of Living Conditions in the Arctic (SLiCA) as deployed in Sweden: initial issues' in B. Poppel (ed.) *SLiCA: Arctic Living Conditions and Quality of Life among Inuit, Saami and Indigenous Peoples of Chukotka and the Kola Peninsula*. Copenhagen: Nordic Council of Ministers. 319–384.
Broderstad, A., B. Eliassen and M. Melhus 2015. 'Prevalence of self-reported suicidal thoughts in SLiCA' in B. Poppel (ed.) *SLiCA: Arctic Living Conditions and Quality of Life among Inuit, Saami and Indigenous Peoples of Chukotka and the Kola Peninsula*. Copenhagen: Nordic Council of Ministers. 131–143.
Duhaime, G., R. Édouard and B. Bernard B 2015. 'Economic stratification and living conditions in the Canadian Arctic' in B. Poppel (ed.) *SLiCA: Arctic Living Conditions and Quality of Life among Inuit, Saami and Indigenous Peoples of Chukotka and the Kola Peninsula*. Copenhagen: Nordic Council of Ministers. 169–193.
Fondahl, G., S. Crate and V. Filippova 2014. 'Sakha Republic (Yakutia), Russian Federation' in J. Larsen, P. Schweitzer and A. Petrov (eds) *Arctic Social Indicators II: Implementation*. Copenhagen: Nordic Council of Ministers. 57 – 92.
Hirshberg, D., A. Petrov . 2015. 'Education and human capital' in J. Larsen and G. Fondahl (eds) *Arctic Human Development Report II: Regional Processes and Global Linkages*. Akureyri, Iceland: Stefansson Arctic Institute, doi: http://norden.diva-portal.org/smash/get/diva2:788965/FULLTEXT03.pdf.
Hodgkins, A.P. 2010. 'Bilingual education in Nunavut: Trojan horse or paper tiger?' *Canadian Journal for New Scholars in Education* 3(1): 1–10.
Hodgkins A.P. 2013. 'Regulation of vocational education and training fields in northern Canada' Unpublished PhD thesis Department of Educational Policy Studies, University of Alberta. Retrieved from http://hdl.handle.net/10402/era.34072.
Hodgkins,, A.P. 2015. 'The problem with numbers: an examination of the Aboriginal Skills and Employment Partnership programme' *Journal of Vocational Education and Training* 67(3): 257–273.
Kleinfeld, J. and J. Andrews 2006. 'The gender gap in higher education in Alaska' *Arctic* 59(4): 428–434.
Kruse, J., B. Poppel, L. Abryutina, G. Duhaime, S. Martin, M. Poppel, M. Kruse, E. Ward, P. Cochran, and V. Hanna 2008. 'Survey of living conditions in the Arctic' in V. Moller, D. Huschka and A. Michala (eds) *Barometers of Quality of Life Around the Globe*. Dordrecht, Netherlands: Springer. 107–134.
Lennert, M. 2015. 'Education in Greenland 1973–2004/06: an analysis based on three living conditions surveys' in B. Poppel (ed.) *SLiCA: Arctic Living Conditions and Quality of Life among Inuit, Saami and Indigenous Peoples of Chukotka and the Kola Peninsula*. Copenhagen: Nordic Council of Ministers. 237–257.

Livingstone, D.W. (ed.) 2009. *Education and Jobs: Exploring the Gaps*. University of Toronto Press, Toronto.

Morin, A., R. Edouard and G. Duhaime 2015. 'Beyond the harsh, objective and subjective living conditions in Nunavut' in B, Poppel (ed.) *SLiCA: Arctic Living Conditions and Quality of Life among Inuit, Saami and Indigenous Peoples of Chukotka and the Kola Peninsula*. Copenhagen: Nordic Council of Ministers. 197–232.

Petrov, A., L. King and P. Cavin 2014. 'The Northwest Territories, Canada' in J. Larsen, P. Schweitzer and A. Petrov (eds) *Arctic Social Indicators II: Implementation*. Copenhagen: Nordic Council of Ministers. 93–138.

Poppel, B. 2015. 'The Inuit world: measuring living conditions & subjective wellbeing – monitoring human development using Survey of Living Conditions in the Arctic (SLiCA) to augment ASI for the Inuit World' in J. Larsen, P. Schweitzer P. and A. Petrov (eds) *Arctic Social Indicators II: Implementation*. Copenhagen: Nordic Council of Ministers. 225–272.

Poppel, M. 2015. 'Changes in gender roles in Greenland and perceived contributions to the household' in B. Poppel (ed.) *SLiCA: Arctic Living Conditions and Quality of Life among Inuit, Saami and Indigenous Peoples of Chukotka and the Kola Peninsula*. Copenhagen: Nordic Council of Ministers. 297–318.

Rasmussen, R., R. Barnhardt and J. Keskitalo 2010. 'Education' in J. Larsen, P. Schweitzer and G. Fondahl (eds.) *Arctic Social Indicators: A Follow Up to the Arctic Human Development Report*. Copenhagen: Nordic Council of Ministers. 67–90.

Rata, E. 2012. *The Politics of Knowledge in Education*. New York: Routledge.

9

Historical sites and heritage in the Polar Regions

Dag Avango

Introduction

Over the last decades the Polar Regions have moved into an increasingly central place in political and public debates about climate change, new shipping routes across the Arctic Ocean and increasing interest in the extraction of Arctic natural resources to supply growing markets with metals and fossil fuels. Debates on these changes have resulted in efforts to develop responsible policies for mitigating social and environmental impacts. However necessary, these debates have tended to reinforce old colonialist narratives about the Polar Regions as empty spaces – pristine environments, untouched by human hands and unaffected by human action. Such notions are far from the truth. The Polar Regions of today are full of the material remains from several thousands of years of human activities, from the waves of peoples settling there in the distant past to the more recent activities by explorers, scientists, extractive industries, sealers, whalers and militaries. For indigenous populations in the Arctic, struggling to make their voices heard and considered, the notion of a pristine empty space is problematic for cultural and political reasons. For others, such as tourist visitors in the Arctic and Antarctic who anticipate wilderness, the material traces of past human action there can become a disappointment. Finally, for policy makers trying to assess the benefits and problems embedded in current processes of change, the history of past human interactions with the Polar Regions – and the material remains of those interactions – can become a resource.

The objective of this chapter is to give an overview of the material historical remains of human activities in the Polar Regions and to explain under which circumstances such remains have been recognized as cultural heritage, and a resource for the present and future. In the chapter, I will make a distinction between material historical remains and cultural heritage sites. I define material historical remains as material residues of human activities in the past. I will also use the terms archaeological record and archaeological sites to describe the same thing. Cultural heritage sites on the other hand, I define as material historical remains, or places of memory, which actors for various reasons have defined as cultural heritage and have sought to protect and manage (Harrison 2013). These actors can be state organizations that protect and manage historical sites within the framework of national or international legislation, they can be local historical societies who protect and narrate the remains of an activity that they feel is vital to their identity and they can be sites which tourism companies brand as heritage in order to bring visitors there and turn them into resources for their businesses. Under which circumstances do historical remains in the Polar Regions become heritage and why? This is a key question for anyone trying to build sustainable futures in both the Arctic and Antarctic.

Archaeological sites in the Polar Regions

As has been pointed out elsewhere in this volume, although both places have compelling similarities, it is hard to speak about the Arctic and Antarctic as Polar Regions with common characteristics. Without falling into the trap of determinism, I believe it is fair to state that the difference in environmental and geographical circumstances at each end of the globe have influenced the character of human activities there and therefore also the material historical remains of those activities. Because of the fact that the Arctic is to a large extent made up by an ocean surrounded by continents, it has been more attractive (and feasible) for humans to inhabit it and utilize it. As geographer David Sugden emphasized back in 1982, the situation is different in the Antarctic (Sugden 1982). The fact that this is a continent surrounded by a sea, with vast distances and unforgiving weather, made it difficult for people to even get there before more recent times. For this reason, the archaeological record in parts of the Arctic contains the material historical remains of more than 20,000 years of human activities and habitation, while in the Antarctic (including the sub-Antarctic) the archaeological record consists of remains from the last 250 years or so. In the following I will present an overview of the different types of archaeological sites present in the Arctic and Antarctic. Due to the limitations of space, it is impossible to provide an all-encompassing description of site categories. Instead I will describe broad categories of sites, with presentations of a few cases.

Archaeological sites in the Arctic

Archaeological sites in the Arctic can be categorized according to the kind of societies that created them – a categorization that, even if blurred, also has a temporal component. The first settlers to arrive in the Arctic – the ancestors of the present-day indigenous peoples – had a lifestyle based on hunting and gathering. They moved north into the Eurasian Arctic, in several waves, and subsequently across the Bering Strait into northern North America and from there to Greenland. This movement took place over an extensive time period, starting around 30,000 years ago. From the medieval period, people in parts of the Arctic also introduced animal husbandry and farming as part of their subsistence. From the 1500s, colonizers from the south established the first settlements based on industrial-scale natural resource exploitation – first whaling and from the 1600s mining, and from the mid-nineteenth century, oil and gas extraction. In the following I have divided the archaeological sites they left behind into the categories pre-industrial and industrial.

Pre-industrial archaeological sites in the Arctic

The first humans who crossed and settled north of the Arctic Circle did so around 28000–32000 BCE, practicing a gatherer and hunter life style (catching mammoth, horse, bison and reindeer) in unglaciated lowlands in present-day Siberia (McCannon 2012: 27–76).[1] This took place during the last ice age, when much of the northern hemisphere had environmental conditions similar to those we associate with the High Arctic today. The material remains from these peoples are vague, consisting of residues and imprints from temporary camps where the most significant physical remains consist of stone tipped tools. A second phase of settlement in the far north occurred from 19000 BCE until 11000 BCE, in present-day central and eastern Siberia. Just as those who lived in these areas before them, these peoples lived from the gathering of plants and hunting and trapping of the mega-fauna available at the time, as well as smaller animals like fox and boar. These people left behind archaeological remains from settlements, including famous sites such as Dyuktai (in eastern Siberia) and Mal'ta (in central Siberia), with remains of houses made out of animal bones, antlers and skins, and the remains of an impressive stone technology consisting of micro-blades which they most likely used as tools and for arrows. The people associated with those sites were the first ones to settle in North America, by crossing the Bering Strait which, at the time, was a land bridge because of the relatively low sea water levels during the ice age. Around 11000 BCE, archaeologists have identified a third wave of human expansion into the Arctic, the so-called Sumnagin culture who ventured further north than anyone before them, living mainly from reindeer hunting (Gómez Coutouly 2016; Pitul'ko and Pavlova 2016). During the same time period groups of settlers established themselves along the ice-free coasts in the Arctic parts of present-day Fennoscandia, leaving material remains of settlements behind as well as rock art, often interpreted as the ancestors of the present-day Sámi indigenous population in this region (Burenhult 2014; Ojala 2009).

In North America, the settlers who had arrived from Eurasia, which archaeologists have labelled Paleo-Arctic, settled in present-day Alaska and the Yukon over the period 11000–5000 BCE. After them a second wave of settlers arrived and established themselves in the period 2500–1700 BCE, which archaeologists have labelled the Arctic Small Tool tradition. They successively spread further eastward in the North American Arctic and along the western coast of Greenland. Over time these inhabitants developed different lifestyles, expressed by differing material cultures, settlement patterns and economies ranging from caribou hunting and fishing to whaling and sealing. Archaeologists have given these cultures different names – Saqqaq, Alaskan and Dorset. The Dorset, whose culture dominated in the present-day Canadian Arctic and Greenland from around 700 BCE lived from fishing and the hunting of seals, whales, musk-ox and caribou and are particularly known for their advanced stone technology and spectacular art. A third large change took place 1,000 years ago, when the Thule culture spread rapidly from the western to the eastern part of North American Arctic and throughout Greenland. The Thule settlements were mostly in coastal locations and their economy was based on whaling from umiaks and kayaks. Greenlanders today commonly trace their ancestry from the Thule culture (Buonasera et al. 2015; McCannon 2012: 27–76; Pauketat 2012: 115–123). A fourth event which has left an imprint on the archaeological record in the Arctic was the Norse settlements in southwestern Greenland, established in the tenth century CE and declining in the fourteenth century. Much effort has been spent on determining why the Norse disappeared from Greenland. Equally interesting is the fact that the Norse represented a partly new subsistence economy in the Arctic, involving farming and animal husbandry (Kintisch 2016; McCannon 2012: 72–73; Sugden 1982: 198–203).

Archaeological sites from large-scale resource exploitation in the Arctic

A significant part of the archaeological sites in the Polar Regions consists of remains from large-scale natural resource exploitation. The oldest of these are remains from the activities of European whalers in the early modern period. The first whalers to venture north were Basques, who established whaling stations along the coast of present-day Labrador (at Red Bay in particular) in the 1530s, responding to reports of rich whale grounds there and demands for whale oil in Europe. The Basques hunted black (right) and bowhead whales using hand held harpoons from small open boats, flensing the whales along the side of their boats and thereafter extracting whale oil from the blubber at shore-based whaling stations. It is the remains of those stations that form the archaeological record from whaling. The sites are small in scale compared to whaling stations of the nineteenth and twentieth centuries, consisting of remains from blubber ovens, cask production and housing. During the 1560s and 1570s, the industry grew as Basque whalers operated at least twelve whaling stations and, according to Barkham, a fleet of more than twenty ships during peak years (Barkham 1984: 518). Thereafter the industry declined, likely as a consequence of several factors – overharvesting of whale populations, competition, political turmoil and the discovery of hunting grounds further north in the European Arctic (Barkham 1984, 1978; Grenier, Bernier, and Stevens 2007; Tuck and Grenier 1989).

One such area was Spitsbergen, discovered in 1596 by the Dutch explorer Willem Barents. Subsequent visitors reported back on the presence of large whale populations, and over the decades that followed, whalers from Western Europe more or less emptied the fjords of the archipelago of bowhead whales. Building on the knowledge of the Basque whalers, they designed their production system in a way similar to that in Labrador. The archaeological record consists of a large number of whaling stations. The historical archaeologist Louwrens Hacquebord has shown how the whalers produced whale oil there by stripping whale carcasses from blubber at primitive flensing decks made out of boats turned upside-down and boiling the blubber in increasingly sophisticated ovens. The whaling companies turned some of the stations into small industrial communities with big, advanced furnaces and log buildings for housing employees. Some of the stations were turned into regional centers, with state representatives present and fortifications. The stations also functioned as markers of territorial claims which the companies used for competing for control over resources and political influence. The largest whaling stations on Spitsbergen was Smeerenburg (Dutch) and Edge-point (British). During this period, Dutch whalers also operated whaling stations at Jan Mayen and along the west coast of Greenland. From about 1650, the whaling companies changed their strategies as a consequence of excessive hunting in the bays of Spitsbergen and of a changing climate, moving their whaling operations out into the sea and abandoning their whaling stations. In the first half of the nineteenth century, the whaling activities declined in the European High Arctic, first and foremost as a result of depleted whale stocks but also because of competition from new inventions such as gas lights, mineral oils and whale oil from American whaling (Avango, Hacquebord and Wråkberg 2014; Hacquebord 2001, 1997, 1984; Hacquebord and Avango 2009; Hacquebord, Steenhuisen and Waterbolk 2003).

In the 1860s, a new period of intensive whaling began in the European Arctic, utilizing a new technological system based on inventions such as the harpoon gun and exploding harpoon based on purpose built steam ships, and shore-based whaling stations with processing plants for utilizing the whale carcasses not only for cooking whale oil, but also for producing guano, cattle feed and glue. Norwegian sea captain and ship owner Svend Foyn, together with other actors in the whaling industry, played a crucial role in establishing this system and based his whaling industry in Finnmark in northern Norway (Gustafsson 2012; Jacobsen 2008). In 1904 the Norwegian government introduced a total ban on whaling in Norway. The Norwegian whaling companies responded by moving their whaling stations and catcher fleets to other regions – the east coast of North America (Newfoundland) and the Antarctic (South Shetland, South Orkney, South Georgia and the Kerguelen Islands). Some of the whaling companies that remained active in the Arctic moved their operations to Spitsbergen (Green Harbour, Bell Sound) and Bear Island (Tönnessen and Johnsen 1982). Today the remains of these stations form an important part of the archaeological record of the modern whaling industry in the Arctic (Avango et al. 2009; Gustafsson 2010). In the American Arctic many US whaling companies were active, but they used the older technologies of boiling blubber in home harbours or onboard their ships, thus leaving little trace of their activities onshore. One exception is the whaling station (and government outpost) at Herschel Island, off the coast of Canada's Yukon Territory (Davis and Gallman 1993; Davis, Gallman and Gleiter 1997).

A second category of archaeological remains of industry in the Arctic are mining sites. The circumpolar Arctic region contains minerals which different actors, most often from outside of the Arctic, have mined during different time periods, depending on fluctuating global demands for metals and fossil fuels. The indigenous populations in the Arctic region were the first to utilize minerals, although on a limited scale and for local needs (Cooper 2007; Sejersen 2014: 41). Large-scale mineral extraction started in the context of early modern colonialism in the Arctic. Mineral prospectors from the south, with state backing, explored geologies in the Arctic parts of Fennoscandia from the seventeenth century and a number of mines were opened – silver, copper and iron ore (Hansson 2015). From the 1880s, mining companies from the south, backed by state actors and encouraged by a growing notion of the Arctic being an inexhaustible source of wealth, opened up large-scale mining operations and built massive related infrastructures – from Norrbotten in Sweden to Sørvaranger in Norway, the White Sea area of northwest Russia and later also in northern Finland. In the opening decades of the twentieth century this mining boom continued its northward course to Bear Island and Spitsbergen (today Svalbard), where mining towns were established for the extraction of coal. The mining industry at Svalbard has, with heavy Norwegian and Russian state subsidies, been in operation up until recent years (Avango 2005; Avango, Hacquebord and Wråkberg 2014; Elenius et al. 2015: 235–246; Hansson 1998; Sörlin 2002, 1988).

In Greenland the first mining operations began in the 1850s when a Danish company started mining cryolite at Ivittuut in southwest Greenland, a mine that was kept in operation until 1987, providing for a long time the key resource for the manufacture of aluminum. Other Danish companies followed in the opening decades of the twentieth century, mining copper and coal. After World War II, mining on a larger scale for zinc, lead and other minerals was started, also by non-Danish companies (Sejersen 2014; Stenfoss and Taagholt 2012: 79–86; Vikström and Högselius 2017).

Mining in Arctic North America began with the discovery of gold in Alaska from 1880 and the placer mine gold rush in the Yukon in 1896, followed by a slow establishment of copper, silver, lead and coal mines. A larger expansion of the mining industry began after World War II, when companies started to extract lead, zinc and uranium, and later also nickel and gold mines (Piper 2009; Wynn 2007). In Russia, from the late nineteenth century, the state sent scientists to the western part of its Arctic region to map minerals. A central purpose was to create economic development, but another objective was to strengthen Russian sovereignty over this region. Several mines and mining communities were established. In the opening decade of the twentieth century, Russian mining interests also headed further north to Novaya Zemlya (an archipelago in the Arctic Ocean separated from the Russian Arctic mainland by the Kara Strait) in 1909 and Spitsbergen from 1912. The Soviet state established mines further east across Siberia from the 1930s as part of the Gulag prison camp system (Bruno 2011; Josephson 2014; Laijus 2013, 2004).

Some of the mines and mining towns mentioned here are still in operation, important examples being Kiruna and Malmberget in Arctic Sweden. Many of the mines have been closed and abandoned, however. I could make a long list but will mention just a few – Qullissat, Maamorilik and Ivittuut in Greenland; Pine Point, Polaris and Port Radium in North America; Stekenjokk, Laver, Nautanen and Sørvaranger in Scandinavia; and Pyramiden, Advent city and Grumant city on Svalbard. After leaving the scene, the mining companies not only left behind ghost towns but also vast technological systems for post-processing of ores, storage and transports as well as underground and open pit mines. Due to the massive scale of these imprints, they are one of the most visible parts of the material historical remains in the Arctic.

Another category of material historical remains that needs to be mentioned is the remains from hundreds of years of scientific research and exploration in the Arctic. The remains of the scientific undertaking of European and later North American scientists consists of everything from small-scale depots for expeditions, to shipwrecks from failed attempts to cross the Northwest Passage, to research stations from different time periods and various infrastructures for scientific research in the form of instruments, measuring points and temporary shelters. These remains are spread across the entire circumpolar Arctic, including well-known sites such as the Kapp Thordsen and Polhem research stations at Svalbard; the Northumberland House or the shipwreck of HMS *Terror* in Arctic Canada; Malye Karmakuly at Novaya Zemlya from the first International Polar Year and Tichaya Buchta geophysical station at Franz Josef Land in the Russian Arctic; Tarfala research station in Arctic Sweden; and the Arctic Station at Qeqertarsuaq (the two latter still in use) (Barr 1995: 96, 2012; Blanchette, Held and Jurgens 2008; Lajus 2013; Liljequist 1993; Stenfoss and Taagholt 2012: 67, 415).

A final type of historical remains in the Arctic includes those that are the result of military conflicts. Among the oldest are fortresses in the Arctic parts of the western Eurasian mainland – the sixteenth-century Pustozersk and Kola forts in the western Russian Arctic, or the Vardøhus fort in Arctic Norway from the same period, built to bolster state authority in the region and to tax the Sámi indigenous population. Another significant fortification in the same region is the Boden fortress in Arctic Sweden, built as part of an inland defence line in Sweden in general but to protect the iron ore mines at Kiruna and Malmberget and its infrastructures in particular (Elenius et al. 2015: 85–108, 240f).

Another significant type of military material remains is the leftovers from former weather stations which Nazi Germany established in the Arctic from Novaya Zemlya to Eastern Greenland during World War II (Barr 1995: 100–104; Jensen and Krause 2014; Lüdecke 2008; Selinger 2001). From the Cold War period the amount of historical remains is even larger, both in numbers and in scale. During this period the Arctic was one of the most important arenas for any potential Third World War, because of the relatively short distance between the main opponents in the world in this period – the Soviet Union and the United States. Therefore, along the coastlines and inlands of Arctic North America, Greenland, Jan Mayen, Siberia, Novaya Zemlya and Franz Josef Land, there are military bases (navy, air force and army), various early warning systems such as the Distant Early Warning (DEW) line and launch sites for intercontinental nuclear missiles (Coates et al. 2008: 82–109; Tamnes 1991). Among the more spectacular of these was the US Camp Century base, built deep into the Greenland inland ice as a launch site for nuclear missiles destined for the Soviet Union (Martin-Nielsen 2013; Nielsen and Nielsen 2016). After the end of the Cold War, many of these military bases entered the archaeological record as abandoned sites. With growing international tensions in the world at the time of writing this chapter, some of these bases are being brought back to life again for military purposes.

Archaeological sites in the Antarctic

The image of the Antarctic as a pristine natural environment, untouched by human presence, is even stronger than its mirror image in the Arctic. This is especially true in the brochures of the tourism industry operating in Antarctica but is also expressed in the 1991 Protocol on Environmental Protection to the Antarctic Treaty, which governs human action south of the 60th parallel. Just like in the Arctic, however, the Antarctic continent and the archipelagos that surround it bear the imprints of human activity across centuries – archaeological sites resulting from sealing, whaling and scientific research.

The first people to visit the Antarctic, as far as is known today, did so in the early 1800s. They were sealers searching for new hunting grounds and European explorers searching for the so-far undiscovered continent. While the earliest explorers left no physical imprints on the continent or its surrounding archipelagos, the sealers certainly did. The first sealers arrived in the southern hemisphere in the mid-1700s and were active along the coasts of Tierra del Fuego, Patagonia and the Falkland Islands/Malvinas. In 1775 British explorer James Cook visited South Georgia and reported having seen big seal populations, which encouraged sealers to move their operations into the peri-Antarctic. From 1786 sealers from Britain and the USA started seal hunting at South Georgia and kept returning, in periods, over the entire nineteenth century. From 1819, sealers also moved south of the 60th parallel, starting at the South Shetland Islands and from there successively moving their hunting grounds to other archipelagos around the Antarctic continent. Archaeologists and historians have explained this movement as a consequence of their completely unregulated overharvesting of seal populations (Basberg 2006: 290; Headland 1984: 32–36; Zarankin and Senatore 2005: 44–45).

The sealing industry left behind a large number of archaeological sites by the coastlines on the archipelagos where they were active, at locations with some possibilities for shelter. As Andres Zarankin and others have shown, the most substantial remains at these sites are large try-pots, occasionally graves, rock shelters or open-air shelters which had roofs made of seal furs and whale bones for furniture. Zarankin's archaeological excavations at Livingstone Island have revealed clothes, wine bottles and board games, which provide clues about social life at the sealing stations (Senatore and Zarankin 2013: 600; Zarankin and Senatore 2005: 47–49). The sealing industry also left an imprint on the surrounding ecosystem, since it almost eradicated the population of fur and elephant seals in the region, none of which have fully recovered.

The next wave of human activity, which also left a substantial archaeological record, was the whaling industry. Whaling began in the Antarctic in the late 1800s in the form of a series of exploratory expeditions, but the first whaling operations to leave behind an archaeological record started when the Argentinean whaling firm Compagnia Argentina de Pesca established a whaling station at Grytviken at South Georgia in 1904. Pesca's operations resulted in substantial production figures and economic profit, which encouraged other whaling entrepreneurs from Norway and the UK to build six more whaling stations at South Georgia. From 1905, other whaling companies built stations in the archipelagos surrounding the Antarctic Peninsula, at Deception Island in the South Shetlands and at Signy Island in the South Orkneys. The Antarctic whaling industry grew and peaked during World War I and its immediate aftermath, and thereafter declined as the global economy went through a series of crises, while overharvesting of the whale populations decreased the catches year by year. From the end of the 1920s to the great economic depression year of 1931, all whaling stations in the Antarctic closed. Some of them were reopened for shorter periods thereafter, the last one – Grytviken at South Georgia – finally closing in 1966. The Antarctic whaling industry lived on, however, but based its operations on so-called factory ships, which had all the production facilities of the former stations but the advantage of being able to move and therefore access whale populations off the more inaccessible coast lines of the Antarctic continent (Basberg 1993, 1998, 2006; Hart 2006, 2001; Tönnessen and Johnsen 1982; Wexelsen, Basberg and Ringstad 1993).

The whaling companies are unrivalled when it comes to the footprint they left in the Antarctic. The whaling stations at South Georgia and Deception Island have suffered from the destruction of a harsh environment, visits by vandalizing British military and other ship-based looters. They are still there, however, with everything from whale oil and guano production systems, service buildings, housing units and even soccer fields (Basberg 2004; Gustafsson et al. 2011). Despite its ship-based character, the pelagic whaling industry has also left an archaeological record, mostly from the early years of the twentieth century when less-developed factory ships used land-based resources such as fresh water for their operations. These consist of storage depots with barrels, water tanks and anchor points at sheltered bays (Avango et al. forthcoming). From the mid-1920s, however, the factory ships no longer needed any facilities on land and therefore left no archaeological record behind. Just like the activities of the seal hunters, the whaling industry also left a substantial footprint in the ecosystem, almost eradicating entire species of whales, some of which have yet to recover. Their absence is a part of the archaeological record, as well as a healthy reminder of the consequences resource booms may have on the environment in the Polar Regions.

A third category of human activity which has left material historical remains in the Antarctic is exploration and scientific research. The earliest explorers at South Georgia – Antoine de la Rouché (1675), Gregorio Jerez (1756) and James Cook (1775) – did not leave any material remains there (Headland 1984: 21–31). The same is true for the first explorers to visit the Antarctic continent from the 1820s onwards, such as Nathaniel Palmer, Thaddeus von Bellingshausen, James Clark Ross, Dumont d'Urville, Charles Wilkes and Henrik Bull – the first two who may have seen the continent and the last mentioned who made the first confirmed landing there. The first explorers and scientists who spent time on land in the Antarctic are associated with the so-called heroic era of Antarctic research, starting in the late 1890s, who left a number of archeological sites behind – shelters and stations (Roberts 2017). Over the twentieth century, in particular after the International Geophysical Year 1957–1958 and the subsequent establishment of the Antarctic Treaty, scientist and research organizations from around the world have established stations in Antarctica. These are not only located along the coasts, like the sealing and whaling sites, but also in the inland areas of the continent. Related to these research facilities are the remains and imprints of infrastructures that were built to support station-based research – harbour facilities, systems for supplying fresh water, radio stations, airstrips and even roads. Another related category of remains are shelters, refuges and depots established for mobile expeditions. Today many of these materialities of science are still in use, but there are also a significant number of science stations and infrastructures that are abandoned, some of which are subject to removal under the Protocol on Environmental Protection of the Antarctic Treaty. Together with the remains of whaling, they represent elements in what is nothing less than a cultural landscape in the Antarctic. Not a pristine wilderness.

Cultural heritage in the Polar Regions

Our world is littered with material objects which people constructed in the past, from traces of societies that existed thousands of years ago to remains of activities that took place in the recent past, some abandoned, some re-used and embedded in the material world people interact with on a daily basis. Despite the fact that many material remains from the past have been important to people, only a few of them have been defined as cultural heritage and the subject for legal protection, conservation or other forms of management. This is as true for material historical remains in the Polar Regions as anywhere else – some of them are constructed as heritage, some are not. The objective of the final section of this chapter is to discuss which archaeological sites have been interpreted as cultural heritage, by whom and why. When answering these questions, I will use a constructivist approach to cultural heritage, in which the concept will include not only sites which state organizations or expert bodies such as the International Council on Monuments and Sites (ICOMOS) and The International Committee for the Conservation of Industrial Heritage (TICCIH) have defined as heritage, but also sites which other actors in society define as cultural heritage for various reasons.

Archaeological sites as heritage in the Arctic

The process of designating and protecting cultural heritage in the Arctic is first and foremost determined by the fact that the region is divided among eight sovereign nation states, with different heritage legislations, different principles for evaluating what material historical remains should be defined as heritage, and different principles and practices for managing already designated heritage sites. A feature which occurs in the heritage legislation in some of the eight Arctic states, is an automatic legal protection of material remains of a certain age as cultural heritage. What that age is, however, is different in different countries. In Sweden archaeological remains older than 1850 are automatically protected by the law (SFS 1998: 950). In Norway the year is 1537 for archaeological remains, 1650 for standing buildings and 100 years for archaeological remains of Sámi origin – the indigenous population of northern Norway as well as the northern parts of Sweden, Finland and northwest Russia (LOV-1978-06-09-50). In Finland the heritage law also protects all archaeological remains from the ancient past, but without determining a year. However, single finds of ancient objects and shipwrecks are legally protected if they are older than 100 years (Muinaismuistolaki 295./1963). Greenland's heritage legislation protects remains of human activity that occurred before 1900. Greenland's national museum, which is the state agency in charge of protecting and managing cultural heritage, can also decide to designate more recent remains as protected heritage, such as buildings or larger complexes of remains (Inatsisartulov nr. 11 af 19. Maj 2010; Inatsisartulov nr. 8 af 3. Juni 2015). At Svalbard, governed by Norway in accordance with the Treaty Concerning Spitsbergen (often called the Svalbard Treaty), cultural heritage protection is part of a Norwegian environmental protection law for the archipelago. Through the cultural heritage act of 1992, all remains of human activity at Svalbard that are older than 1946 are automatically protected. In addition, the law includes the possibility to protect even younger traces of human activity as heritage (Marstrander 1999).

In Canada, the USA and Russia there is no automatic protection for historic sites older than a particular year, but age nevertheless plays a role in the evaluation process. In Canada, official heritage designation is regulated by the Historic Sites and Monuments Act. Cultural heritage, or a historic place as the Act calls it, is defined as a "site, building or other place of national historic interest or significance, and includes buildings or structures that are of national interest by reason of age or architectural design". The Parks Canada agency and a board made up of members of said agency and province representatives, makes decisions on what to protect and how (Historic Sites and Monuments Act, R.S.C., 1985, c. H-4). In Alaska, the federal government is responsible for heritage designation and management, under the provisions of the National Preservation Act of 1966 and its amendments. Sites defined as cultural heritage are included in a "National register of historic places", managed by the US National Parks Service. Anyone – from individuals to historic societies and government agencies – can nominate a site for inclusion in the register. In the evaluation process particular attention is paid to factors of age (older than fifty years being a common requirement), integrity and significance (www.nps.gov/nr/about.htm, accessed 2017–10–11).

In Russia, where there is a broad definition of heritage objects, material remains from the past must be at least forty years old to become defined as heritage. The overarching responsibilities for cultural heritage protection are on the federal level, and regulated by Federal Law 73 of 2002 (which was modified most recently in December 2017), where overarching policies and sites of national significance are dealt with (such as world heritage sites and ancient archaeological sites). Sites of regional or local significance are evaluated by state organizations on the regional level and legal protection issued by regional governments. Thus, it is both regional and federal authorities that define historical remains as heritage in the Russian Arctic (Federova and Kochelyaeva 2013).

Through the above-mentioned legislations, states select and designate historical remains in the Arctic as cultural heritage and manage these sites in some level of accordance with national priorities and evaluation practices. Given the focus on age and significance in the heritage legislation of Arctic states, most of the historical sites that are protected as cultural heritage there are remains of early inhabitants in the region, often ancestors of current indigenous populations. Another type of remains that falls into the category of sites protected because of age and perceived significance, was constructed during the mediaeval and early modern periods. In southern Greenland, the remains of Norse settlements are protected, narrated on site and in museums, and the subject of substantial research efforts. At Labrador, West Greenland, Jan Mayen and Svalbard, authorities are protecting the remains of the stations built by the early modern whaling industry (Fylkesmannen Nordland 2000; Marstrander 1999). On the initiative of the Canadian government, a large concentration of early modern whaling stations at Red Bay, Labrador, have been protected as national monuments since 1979. In 2013, UNESCO inscribed the area on the world heritage list, the outstanding universal value being the unique example of the Basque whaling industry and proto-industry of whale oil production (http://whc.unesco.org/en/list/1412, accessed on 2017–08–17). Across the North American Arctic, also a number of fur trapper and trading posts are preserved as cultural heritage (www.nps.gov/nr/research/index.htm, accessed 2017–10–11; www.pc.gc.ca/en/lhn-nhs, accessed 2017–10–11). In the Arctic parts of Sweden, heritage authorities protect remains from the early modern mining industry because of their age – silver, copper and iron ore mines, associated structures for transport and energy production, as well as smelters and steel works (Länsstyrelsen i Norrbottens Län 2000).

In some cases, national heritage authorities have also defined more recent, late modern remains of human activities as cultural heritage. At Svalbard, as a consequence of the above-mentioned law designating all sites older than 1946 as heritage, a large number of remains of scientific research stations, late modern whaling stations and systems for mining – prospecting camps, mines, infrastructure for transport and storage, and entire settlements – are defined and protected as cultural heritage (Barr 2011; Marstrander 1999). Another example is the Swedish Arctic, where heritage authorities have designated the iron ore mines in Kiruna and Gällivare municipalities, together with transport infrastructures, hydropower stations and defense works as a mega-system representing a monument over Swedish industrial history and a national interest for cultural heritage protection. Even though these measures do not automatically mean that such historical remains will be protected against destruction or alteration, it nevertheless reveals an interest to emphasize the historical role of resource extraction in this part of the Arctic (Avango and Roberts 2017b).

In other parts of the Arctic, authorities have shown far less interest to think of legacies of more recent activities, such as extractive industries and fisheries, as cultural heritage. In Greenland, the abandoned mining sites of Qullissat, Maamorilik/Black Angel and Ivittuut are not listed as cultural heritage (Avango and Roberts 2017b). In the Russian Arctic, the Russian Research Institute for Cultural and Natural Heritage has worked with former industrial sites as heritage since the early 2000s (http://whc.unesco.org/archive/websites/arctic2008/russia.html, accessed 2017–08–31). In North America, as Lisa Cook has shown, historic mining sites associated with the gold rush in the Yukon are widely interpreted as cultural heritage sites and protected as such (Cook 2013). However, as Arn Keeling and John Sandlos have shown, abandoned mining sites are often considered as an unwanted legacy of a problematic past, with environmental remediation being the desired future (Keeling and Sandlos 2009, 2017).

There is, however, a much wider range of people and organizations that define various historical sites as heritage in the Arctic. One important category are historical societies who engage both in historical research and efforts to preserve what they consider as cultural heritage; often remains from activities they once were part of in local settlements (Isacson 2013). Another category are actors from indigenous communities who define historical material remains of importance to them as cultural heritage, remains which national heritage authorities have yet to define as cultural heritage. Arn Keeling, John Sandlos and Anne-Mette Jørgensen have shown examples of this from former mining communities with an indigenous work force in Rankin inlet, Arctic Canada (Cater and Keeling 2013), and Qullissat in Greenland (Jørgensen 2017). A growing category of actors is to be found within the tourism industry, which have been able to brand not only officially inscribed heritage sites but also other historical remains, as visitors' sites (Roura 2009, 2011). By branding and narrating, the tourism industry re-economizes these previously abandoned places and simultaneously constructs them as cultural heritage. A final group of actors that should be mentioned in this context are states with particular interests to protect the Arctic (Avango and Roberts 2017a). Apart from states with sovereignty over territories in the Arctic, there are also non-Arctic states with an increasing interest in influencing the future of the region. As Eric Paglia has shown, these states do so by constructing themselves as stakeholders in the Arctic and they often use history and heritage as tools in their strategies (Paglia 2016; Roberts and Paglia 2016). This political use of history and heritage has clearly increased, relating to the growing interest in the future of the Arctic region.

Archaeological sites as heritage in the Antarctic

In Antarctica, states which have signed the Antarctic Treaty have listed sites they deem to be cultural heritage from 1972 and onwards. From 2005, the Antarctic Treaty Consultative Meeting (ATCM) included heritage protection in the Environmental Protocol of the Antarctic Treaty, annex five and six (www.ats.aq/e/ep.htm; www.ats.aq/documents/cep/handbook/Protocol_e.pdf; www.ats.aq/documents/recatt/Att004_e.pdf; www.ats.aq/documents/recatt/Att249_e.pdf; www.polarheritage.com/content/library/50.pdf, all accessed on 2017–08–14). The protocol itself was signed in 1991 and entered into force in 1998. Annex five, on area protection and management, came into effect in 2005 and regulates the procedure for establishing historic sites as cultural heritage in the region governed by the Antarctic Treaty – south of the 60th parallel. According to annex five, any party to the Antarctic Treaty can propose a historic site in Antarctica to be listed as a Historic Site and Monument (HSM). The proposals are then evaluated and decided upon by the Antarctic Treaty Consultative Parties (ATCP) at the re-occurring ATCM's (Roura 2008, 2011, 2017).

By 2017, the number of sites listed as HSM was 92. These include a wide variety of sites. Some 42% consist of buildings or remains of buildings originally built for scientific research and exploration, as well as artefacts (another 10%) originating from such activities. In addition, a smaller number of graves and a shipwreck, also related to science and exploration, have been designated as HSM. It is worth noting that only one complex of historical remains from whaling have been designated as HSM – the remains of a whaling station located at Whalers Bay, Deception Island in the South Shetland Islands – despite the rather large number of sites with historical remains of whaling that exist in the Antarctic Treaty area. The same can be said about historical remains of sealing. The second largest category of HSM is made up of various monuments – cairns, statues, crosses and stones – commemorating past events. Some of them were established at the time of an historic event (e.g. where an expedition landed) while some have been built later. The monuments make up no less than 42% of all HSM (based on the current list of HSM at IPHC, www.polarheritage.com/polarheritage/Sitelist01up/, accessed on 2017–07–30).

There are also a number of historical remains protected as cultural heritage in the part of the Antarctic situated north of the 60th parallel – often called the peri-Antarctic or sub-Antarctic. These sites are not governed by the Antarctic Treaty, but by the states that claim sovereignty over these islands. The islands are Crozet Islands and Kerguelen Islands (claimed by France), Heard Island, the McDonald Islands and Macquarie Island (claimed by Australia), Prince Edward Islands (claimed by South Africa), and South Georgia and the South Sandwich Islands (claimed both by Britain and Argentina, but governed by Britain). All of these contain historical remains from sealing, science and exploration and some also from whaling.

Most of the states which govern the islands protect some of these remains as cultural heritage. At South Georgia, the British authorities, in close co-operation with the South Georgia Heritage Trust (SGHT), protect and manage the numerous sealing and whaling sites at the islands and other historical remains pertaining to exploitation of living marine resources, and remains from science, exploration and British governance (not least remains and monuments pertaining to the activities of British explorer Ernest Shackleton). After the whaling companies closed their stations in the 1930s and 1960s, scientists occasionally used them as shelters, and during the Falklands War in 1982, the Argentine and British armed forces turned some of them into strongholds. At this point in time, no-one considered them as cultural heritage, however. The British authorities and actors in the community of British Antarctic research started to define them as heritage only after the Falklands War had ended. Today, as cultural heritage, the whaling stations are an important component of the British governance of South Georgia (Avango 2013, 2017).

In 1997 UNESCO inscribed Heard Island, the McDonald Islands and Macquarie Island on the world heritage list, based on outstanding universal values pertaining to their natural environment. As a consequence of this and their isolated location, the remains of nineteenth-century sealing activities and twentieth-century science and exploration there are relatively well protected (http://whc.unesco.org/en/list/577; http://whc.unesco.org/en/list/629, both accessed 2017–10–11). The French authorities managing the Kerguelen and Crozet islands as part of the Terres Australes et Antarctiques Françaises (TAAF) have, since the early 2000s, laid down laws protecting historic sites pertaining to science, exploration, sealing and whaling there. In recent years, the French authorities have invested significant efforts both in conservation and in research of these sites – not least the remains of the whaling station at Port Jeanne d'Arc (Le Mouël 2004; Moreigneaux 2012). The exception in terms of cultural heritage protection in the sub-Antarctic is the South African Prince Edward Islands which do contain remains of nineteenth-century sealing and twentieth-century science and exploration, none of which is designated or protected as cultural heritage today.

There are also other actors in the Antarctic that are actively branding historical remains as cultural heritage – the tourism industry. Over the last decades, tourism in the Antarctic has gone through a rapid expansion, and in their efforts to include a wider variety of visitor sites to market to their customers, the tourism companies are including an increasing number of historical sites on their itineraries (Basberg 2010; Fletcher 2012; Nuttall 2010; Roura 2011). Some of these are already state designated cultural heritage sites, but far from all are. By bringing visitors to these sites and narrating those in the form of guided tours, signboards, local museums, printed materials and websites, the tourism industry de facto involves these sites in heritage processes.

Conclusions

The material remains of past activities in the Polar Regions provide archaeologists and historians with important sources about the history of these regions and points of departure for thinking about how these past activities have influenced the present. The waves of people who settled across the Arctic before the emergence of extractive industries there left thousands of archaeo-logical sites. The regional and temporal differences between the settlements and the material objects they left behind are so big that there is no point in talking about them as representing the history of an Arctic region. Just as today, there were many Arctics and just as everywhere else on earth, these societies and cultures changed over time in different ways. If there is one common characteristic of the pre-industrial archaeological sites in the Arctic, it is their relatively small imprints on local environments. The vague character of the traces is of course related to time – there are less remains from activities that took place 20,000 years ago than 800 years ago. However, with no intention to fall into the colonialist narrative about nature peoples, it is beyond doubt that the indigenous peoples in the Arctic had lifestyles which had less immediately visible impacts on their local environment compared to the industrialists that arrived in their homelands later.

The archaeological sites remaining from the late modern era in the Arctic and Antarctic, such as industrial-scale resource extraction, scientific research and military activities, are on a much larger scale, often in the form of systems that span over vast expanses of land and who have left a substantial environmental footprint. They were built by societies with a somewhat different approach to the environments they encountered – it was a resource that had to be explored, mapped, defined and sometimes extracted, if market demand and institutional contexts worked in favour of doing so. For the indigenous populations in the Arctic, the environment was a foundation for subsistence economies and trading that were meant to last and a land they valued as their home.

The physical remains of human activity provide material evidence through which we can try to explain why people have opted to settle there, how their societies have changed and why they have changed. They also provide evidence on how resource-oriented industries have been able to establish themselves in the Polar Regions, by adapting technologies, settlement planning and strategies for gaining control over resources, to the challenging environmental, geographical and political conditions in the Polar Regions. They also give witness to the environmental consequences of large-scale natural resource exploitation on the environment. The archaeological remains from science and exploration, in turn, provide archaeologists with unique information about practical issues and how changing ideologies have influenced the way polar research has been carried out and why. To preserve as much as possible of this archaeological record is therefore of crucial importance for scholars who use them as a source for knowledge (see also Avango 2016; Hacquebord and Avango 2016).

Archaeologists and historians are not the only actors who speak in favour of preserving archeological sites, however. As I have shown in this chapter, there are many stakeholders in the Polar Regions who, for different and sometimes conflicting reasons, argue that the archaeological sites represent a cultural heritage. These stakeholders can be state authorities on national/federal, regional and local levels in the Arctic, who are trying to develop ways to diversify local economies in the north by using heritage as a resource for tourism or to enhance local identity and thereby quality of life. State authorities can also have geopolitical motives for designating heritage in parts of the Arctic that have been subject to sovereignty conflicts – for example Svalbard. Other stakeholders can be historical societies, or it can be indigenous communities fighting for recognition of historical rights to lands, resources and political influence. In the Antarctic there are fewer stakeholders and their motives are somewhat different. Some are state actors wishing to enhance historical ties with the great southern continent and the islands surrounding it. In the sub-Antarctic region, cultural heritage management can be considered as good governance, but is at the same time a way of exercising sovereignty in largely unpopulated places. On the Antarctic continent it can be a way of promoting political influence there. Many of the actors involved in Antarctic heritage designation have less calculating motives though – people who are fascinated and inspired by events in the past and who want to use heritage sites as anchor points for telling the outside world about why they feel those events mattered.

Note

1 If not otherwise stated, the following builds on John McCannon's summary of current research on the early history of the Arctic (McCannon 2012: 27–76).

References

Avango,, D. 2005. *Sveagruvan: Svensk gruvhantering mellan industri, diplomati och geovetenskap.* Stockholm: Jernkontoret.

Avango, D. 2013. 'Heritage in action: historical remains in polar conflicts' in S. Sörlin (ed.) *Science, Geopolitics, and Culture in the Polar Regions: Norden Beyond Borders.* Farnham, UK: Ashgate, 329–356.

Avango, D. 2016. 'Acting artefacts: on the meanings of material culture in Antarctica' in P. Roberts, A. Howkins and L-M van der Watt (eds) *Antarctica and the Humanities.* London: Palgrave Macmillan, 159–179.

Avango, D. 2017. 'Working geopolitics: sealing, whaling, and industrialized Antarctica' in K. Dodds, A.D. Hemmings and P. Roberts *Handbook on the Politics of Antarctica.* Cheltenham, UK and Northhampton, MA: Edward Elgar Publishing, 485–506.

Avango, D., S. DePasqual, U. Gustafsson, H. de Haas, L. Hacquebord, C. Hartnell, and F. Kruse 2009. *LASHIPA 5: Archaeological Expedition on Spitsbergen 27 July – 17 August.* Groningen, the Netherlands: Arctic Center, University of Groningen.

Avango, D., U. Gustafsson, L. Hacquebord and G. Rossnes, forthcoming. *LASHIPA 8: Archaeological Expedition to South Orkney, South Shetland and the Antarctic Peninsula 6 March – 2 April 2010.* Groningen, the Netherlands: Arctic Center, University of Groningen.

Avango, D., L. Hacquebord and U. Wråkberg 2014. 'Industrial extraction of Arctic natural resources since the sixteenth century: technoscience and geo-economics in the history of northern whaling and mining' *Journal of Historical Geography* http://dx.doi.org/10.1016/j.jhg.2014.01.001.

Avango, D. and P. Roberts 2017a. 'Heritage, conservation, and the geopolitics of Svalbard: writing the history of Arctic environments' in L.A. Körber, S. MacKenzie and A. Westerståhl Stenport (eds) *Arctic Environmental Modernities: From the Age of Polar Exploration to the Era of the Anthropocene.* Cham, Switzerland: Palgrave Macmillan, 125–143.

Avango, D. and P. Roberts 2017b. 'Industrial heritage and Arctic mining sites: material remains as resources for the present – and the future' in R.C. Thomsen and L. Rastad Bjørst (eds) *Heritage and Change in the Arctic: Resources for the Present, and the Future.* Aalborg, Denmark: Aalborg University Press, 127–158.

Barkham, S. 1978. 'The Basques: filling a gap in our history between Jacques Cartier and Champlain' *Canadian Geographical Journal* 96(1): 8–19.

Barkham, S. 1984. 'The Basque whaling establishments in Labrador 1536–1632: a summary' *Arctic: Journal of the Arctic Institute of America* 137(4): 515–519.

Barr, S. 1995. 'The history of western activity in Franz Josef Land' in S. Barr (ed.) *Franz Josef Land.* Oslo: Norsk Polarinstitutt, 59–106.

Barr, S. 2011. 'Arctic and Antarctic: different, but similar – challenges of heritage conservation in the high Arctic' in S. Barr and P. Chaplin (eds) *Polar Settlements: Location, Techniques and Conservation.* Oslo, IOCMOS IPHC, 14–23.

Barr, S. 2012. 'Polar heritage: neglected child becomes international talking point' in D. Munroe (ed.) *Managing Industrial & Cultural Heritage: South Georgia in Context.* Dundee, UK: The South Georgia Heritage Trust, 6–15.

Basberg, B. 1993. 'Survival against all odds? Shore station whaling in the pelagic era, 1925–1960' in E. Wexelsen, B. Basberg and J.E. Ringstad (eds) *Whaling and History: Perspectives on the Evolution of the Industry.* Sandefjord, Norway: Sandefjordmuseene.

Basberg, B. 1998. 'The floating factory: dominant designs and technological development of twentieth-century whaling factory ships' *Northern Mariner* 8(2): 21–37.

Basberg, B. 2004. *The Shore Whaling Stations at South Georgia: A Study in Antarctic Industrial Archaeology.* Oslo: Novus.

Basberg, B. 2006. 'Perspectives on the economic history of the Antarctic region' *International Journal of Maritime History* 18(2) 285–304.

Basberg, B. 2010. 'Antarctic tourism and maritime heritage' *International Journal of Maritime History* 22(2): 1–20.

Blanchette, R.A, B.W. Held and J.A. Jurgens 2008. 'Northumberland House, Fort Conger and the Peary huts in the Canadian high Arctic: current condition and assessment of wood deterioration taking place' in S. Barr and P. Chaplin (eds) *Historical Polar Bases: Preservation and Management*. Oslo, ICOMOS IPHC.

Bruno, A.R. 2011. 'Making nature modern: economic transformation and the environment in the Noviet North' PhD thesis, University of Illinois at Urbana-Champaign, USA.

Buonasera, T.Y., A. Tremayne, C.M. Darwent, J.W. Eerkens and O.K. Mason 2015. 'Lipid biomarkers and compound specific d13C analysis indicate early development of a dual-economic system for the Arctic small tool tradition in northern Alaska' *Journal of Archaeological Science* 61: 129–138.

Burenhult, G. 2014. *Arkeologi i Norden*. Stockholm, Natur & Kultur.

Cater, T. and A. Keeling 2013. '"That's where our future came from": mining, landscape, and memory in Rankin Inlet, Nunavut' *Etudes/Inuit/Studies* 37(2): 59–82.

Coates, K.S., W.P. Lackenbauer, W.R. Morrion and G. Poelzer 2008. *Arctic Front: Defending Canada in the Far North*. Toronto, ON: T. Allen Publishers

Cook, L. 2013. 'North takes place in Dawson, Yukon, Canada' in D. Jørgensen and S. Sörlin (eds) *Northscapes: History, Technology and the Making of Northern Environments*. Vancouver, BC: University of British Columbia Press, pp.223–246.

Cooper, H.K. 2007. 'The anthropology of native copper technology and social complexity in Alaska and the Yukon Territory: an analysis using archaeology, archaeometry, and ethnohistory' PhD thesis, University of Alberta, Canada.

Davis, L.E. and R.E. Gallman 1993. 'American whaling, 1820–1900: dominance and decline' in B. Basberg, J.E. Ringstad and W. Wexelsen (eds) *Whaling and History*. Sanderfjord, Norway: Sandefjordmuseene, 55–66.

Davis, L.E., R.E. Gallman and K. Gleiter 1997. 'In pursuit of Leviathan technology, institutions, productivity, and profits in American whaling, 1816–1906' in *NBER Series on Long-Term Factors in Economic Development*. Chicago, IL: University of Chicago Press.

Elenius, L., H. Tjelmeland, M. Lähteenmäki and A. Golubev 2015. *The Barents Region: A Transnational History of Subarctic Northern Europe*. Oslo: Pax Forlag.

Federova, T. and N. Kochelyaeva 2013. 'Country profile: Russian Federation' in *Compendium of Cultural Policies and Trends in Europe*. Council of Europe/ERICarts.

Fletcher, D. 2012. 'South Georgia's industrial heritage: a bonus for tourists' in D. Munroe (ed.) *Industrial and Cultural Heritage: South Georgia in Context*. Dundee, UK: South Georgia Heritage Trust, 52–55.

Fylkesmannen Nordland 2000. *Miljøhandlingsplan for Jan Mayen*. Bodø, Fylkesmannen Nordland, Miljøvernavdelingen.

Gómez, C.Y. 2016. 'Migrations and interactions in prehistoric Beringia: the evolution of Yakutian lithic technology' *Antiquity* 90(349): 9–31.

Grenier, R., M-A. Bernier and W. Stevens (eds) 2007. *The Underwater Archaeology of Red Bay: Basque Shipbuilding and Whaling in the 16th Century*. Ottawa: Parks Canada.

Gustafsson, U. 2010. 'Industrialising the Arctic: settlement design and technical adaptions of modern whaling stations in Spitsbergen and Bear island' in J.E. Ringstad (ed.) *Whaling and History III*. Sandefjord, Norway: Vestfoldsmuseene IKS, 47–58.

Gustafsson, U. 2012. 'A science and technology studies (STS) approach on the evolution of the modern whaling industry' in L. Hacquebord (ed.) *LASHIPA: History of Large Scale Resource Exploitation in Polar Areas*. Groningen, the Netherlands: Barkhuis Publishing, 113–126.

Gustafsson, U., D. Avango, B. Basberg and G. Rossnes 2011. *LASHIPA 6: Archaeological Expedition on South Georgia 3 March – 12 April*. Groningen, the Netherlands: Arctic Center, University of Groningen.

Hacquebord, L. 1984. *Smeerenburg: het verblijf van Nederlandse walvisvaarders op de westkust van Spitsbergen in de zeventiende eeuw (The sojourn of Dutch whalers on the west coast of Spitsbergen in the seventeenth century* [with a summary in English]). Amsterdam: University of Amsterdam.

Hacquebord, L. 1997. 'Whaling stations as bridgeheads for exploration of the Arctic regions in the sixteenth and seventeenth century' in J. Everaert and Parmentier, J. (eds.) *International Conference on Shipping, Factories and Colonization: (Brussels, 24-26 November 1994)*. Brussels: Academie Royale des Sciences d'Outre-Mer, pp.289–297.

Hacquebord, L. 2001. *Three Centuries of Whaling and Walrus Hunting in Svalbard and Its Impact on the Arctic Ecosystem*. Cambridge, UK: The White Horse Press.

Hacquebord, L. and D. Avango 2009. 'Settlements in an Arctic resource frontier region' *Arctic Anthropology* 46(1–2): 25–39.

Hacquebord, L. and D. Avango 2016. 'Industrial heritage sites in Polar regions: sources of historical information' *Polar Science* 10(3): 433–440.

Hacquebord, L., F. Steenhuisen and H.J. Waterbolk 2003. 'English and Dutch whaling trade and whaling stations in Spitsbergen (Svalbard) before 1660' *International Journal of Maritime History* 15(2): 117–134.

Hansson, S. 1998. 'Malm, räls och elektricitet' in P. Blomkvist P and A. Kaijser (eds) *Den konstruerade världen: tekniska system i historiskt perspektiv*. Eslöv, Sweden: Symposium, 45–76.

Hansson, S. 2015. *Malmens land: gruvnäringen i Norrbotten under 400 år Luleå*. Tornedalica.

Harrison, R. 2013. *Heritage: Critical Approaches, Heritage Studies*. Abingdon, UK and New York: Routledge.

Hart, I.B. 2001. *Pesca: The History of Compania Argentina de Pesca Sociedad Anónima of Buenos Aires. An Account of the Pioneer Modern Whaling and Sealing Company in the Antarctic*. Salcombe, UK: Aidan Ellis.

Hart, I.B. 2006. *Whaling in the Falkland Islands Dependencies 1904–1931: A History of the Shore and Bay-Based Whaling in the Antarctic*. Newton St. Margarets, UK: Pequena.

Headland, R. 1984. *The Island of South Georgia*. Cambridge, UK: Cambridge University Press.

Historic Sites and Monuments Act, R.S.C., 1985, c. H-4, current to 5 June 2017, last amended on 12 December 2013. Published by the Minister of Justice at: http://laws-lois.justice.gc.ca.

Inatsisartulov nr. 11 af 19. Maj 2010 om fredning og anden kulturarvsbeskyttelse af kulturminder.

Inatsisartulov nr. 8 af 3. Juni 2015 om museumsvæsen.

Isacson, M. 2013. 'Industriarvets utmaningar. Samhällsförändringar och kulturmiljövård från 1960-tal till 2010-tal' *Bebyggelsehistorisk tidskrift* 6517–36.

Jacobsen, A.R. 2008. *Svend Foyn: fangspioner og nasjonsbygger*. Oslo: Aschehoug.

Jensen, J.F. and T. Krause 2014. 'Second world war histories and archaeology in northeast Greenland' in H.C. Gulløv (ed.) *Northern Worlds: Landscapes, Interactions and Dynamics. Research at the National Museum of Denmark*. Copenhagen: National Museum of Denmark, 491–509.

Jørgensen, A.M. 2017. *Moving Archives: Agency, Emotions and Visual Memories of Industrialization in Greenland*. Copenhagen: Københavns Universitet, Det Humanistiske Fakultet.

Josephson, P.R. 2014. *The Conquest of the Russian Arctic*. Cambridge, MA: Harvard University Press.

Keeling, A. and J. Sandlos 2009. 'Environmental justice goes underground? Historical notes from Canada's northern mining frontier' *Environmental Justice* 2(3): 117–125.

Keeling, A. and Sandlos J. 2017. 'Ghost towns and zombie mines: historical dimensions of mine abandonment, reclamation and redevelopment in the Canadian north' in B. Martin and S. Bocking (eds) *Ice Blink: Navigating Northern Environmental History*. Calgary, AB: University of Calgary Press, 377–420

Kintisch, E. 2016. 'The lost Norse' *Science* 354(6313): 696–701.

Laijus, J. 2004. 'From fishing to mining: the change of priorities in the development of the north and Russian expeditions to Spitsbergen in the early 20th century' in Jernkontoret (ed.) *Arktisk gruvdrift II. Teknik, vetenskap och historia i norr*. Stockholm: Jernkontoret, 93–106.

Laijus, J. 2013. 'In search of instructive models. The Russian state at a crossroads to conquering the north' in D. Jørgensen and S. Sörlin (eds) *Northscapes: History, Technology and the Making of Northern Environments*. Vancouver, BC: University of British Columbia Press, 110–133.

Lajus, J. 2013. 'Field stations on the coast of the Arctic ocean in the European part of Russia from the first to the second IPY' in S. Sörlin (ed.) *Science, Geopolitics and Culture in the Polar Region*. Farnham UK and Burlington, VT: Ashgate, 111–142.

Länsstyrelsen i Norrbottens Län 2000. Program för Norrbottens industriarv Luleå, Länsstyrelsen i Norrbottens län.

Le Mouël, J.F. 2004. 'Heritage in the French sub-Antarctic territory: between urgency and emergency' in S. Barr and P. Chaplin (eds.) *Cultural Heritage in the Arctic and Antarctic Regions*. Oslo: ICOMOS IPHC, 60–64.

Liljequist, G.H. 1993. *High Latitudes: A History of Swedish Polar Travels and Research*. Stockholm: Swedish Polar Research Secretariat.

LOV-1978-06-09-50. Lov om kulturminner. Oslo: Klima- og miljødepartementet.

Lüdecke, C. 2008. 'German meteorological stations in Northwest Svalbard' in S. Barr and P. Chaplin (eds) *Historical Polar Bases: Preservation and Management* Olso: ICOMOS IPHC, 18–22.

Marstrander, L. 1999. 'Svalbard cultural heritage management' in U. Wråkberg (ed.) *The Centennial of S.A. Andrée's North Pole Expedition*. Stockholm: Royal Academy of Sciences.

Martin-Nielsen, J. 2013. '"The deepest and most rewarding hole ever drilled": ice cores and the Cold War in Greenland' *Annals of Science* 70(1): 47–70.

McCannon, J. 2012. *A History of the Arctic: Nature, Exploration and Exploitation*. London, Reaktion.

Moreigneaux, N. 2012. 'A new vision of preservation: the laser programme of the whaling station at Port-Jeanne d'Arc, Kerguelen' in D. Munroe (ed.) *Managing Industrial & Cultural Heritage: South Georgia in Context*. Dundee, UK: South Georgia Heritage Trust, 108–113.

Muinaismuistolaki 295./1963. Lag om fornminnen, www.finlex.fi/sv/laki/alkup/1963/19630295.

Nielsen, H. and K.H. Nielsen 2016. 'Camp Century: Cold War city under the ice' in R.E. Doel, K. Harper and M. Heymann (eds) *Exploring Greenland: Cold War Science and Technology on Ice*. New York: Palgrave Macmillan, 195–216.

Nuttall, M. 2010. 'Narratives of history, environment and global change: expeditioner-tourists in Antarctica' in M.C. Hall and J. Saarinen (eds) *Tourism and Change in Polar Regions: Climate, Environment and Experiences*. Abingdon, UK: Routledge, 204–214.

Ojala, C-G. 2009. 'Sámi prehistories: the politics of archaeology and identity in Northernmost Europe', Occasional papers in archaeology. Uppsala, Sweden: Institutionen för arkeologi och antik historia, Uppsala universitet.

Paglia, E. 2016. 'The northward course of the Anthropocene: transformation, temporality and telecoupling in a time of environmental crisis' Stockholm: KTH-Royal Institute of Technology.

Pauketat, T.R. 2012. *The Oxford Handbook of North American Archaeology*. New York and Oxford, UK: Oxford University Press.

Piper, L. 2009. *The Industrial Transformation of Subarctic Canada: Nature, History, Society*. Vancouver, BC: University of British Columbia Press.

Pitul'ko, V.V. and E.Y. Pavlova 2016. *Geoarchaeology and Radiocarbon Chronology of Stone Age Northeast Asia*. College Station, TX: A&M University Press.

Roberts, P. 2017. 'The politics of early exploration' in K. Dodds, A.D. Hemmings and P. Roberts (eds) *Handbook on the Politics of Antarctica*. Cheltenham, UK and Northampton, MA: Edward Elgar Publishing, 318–336.

Roberts, P. and E. Paglia 2016. 'Science as national belonging: the construction of Svalbard as a Norwegian space' *Social Studies of Science* 46(6): 894–911.

Roura, R. 2008. 'Antarctic research stations: environmental and cultural heritage perspectives 1983–2008' in S. Barr and P. Chaplin P (eds) *Historical Polar Bases: Preservation and Management*. Oslo: ICOMOS IPHC, 38–52.

Roura, R. 2009. 'The polar cultural heritage as a tourism attraction: a case study of the airship mooring mast at Ny-Ålesund, Svalbard' *Téoros* 28(1): 29–38.

Roura, R. 2011. *The Footprint of Polar Tourism: Tourist Behaviour at Cultural Heritage Sites in Antarctica and Svalbard*. Groningen, the Netherlands: Barkhuis Publishing.

Roura, R. 2017. 'Antarctic cultural heritage: geopolitics and management' in K. Dodds, A.D. Hemmings and P. Roberts (eds) *Handbook on the Politics of Antarctica*. Cheltenham, UK and Northampton, MA: Edward Elgar Publishing, 468–484.

Sejersen, F. 2014. *Efterforskning og udnyttelse af råstoffer i Grønland i historisk perspektiv*. Copenhagen: University of Copenhagen & University of Greenland.

Selinger, F. 2001. *Von 'nanok' bis 'Eismitte': Meteorologische unternehmungen in der Arktis 1940–1945*. Hamburg, Germany: Convent.

Senatore, M.X. and A. Zarankin 2013. 'Tourism and the invisible historic sites in Antarctica' in F. Babics, H. Barré and A. Magnant (eds) *Heritage, a Driver of Development: Rising to the Challenge*. Paris: ICOMOS, 599–608.

SFS 1998:950. *Kulturmiljölag*. Stockholm: Kulturdepartementet.

Sörlin, S. 1988. *Framtidslandet: debatten om Norrland och naturresurserna under det industriella genombrottet*. Stockholm: Carlsson.

Sörlin, S. 2002. 'Rituals and resources of natural history: the north and the Arctic in Swedish scientific nationalism' in S. Sörlin and M. Bravo (eds) *Narrating the Arctic: A Cultural History of Nordic Scientific Practices*. Canton, MA: Science History Publications, 73–122.

Stenfoss, H.P. and J. Taagholt 2012. *Grønlands teknologihistorie, Grønlandsforskning*. Copenhagen: Gyldendal.

Sugden, D. 1982. *Arctic and Antarctic: A Modern Geographical Synthesis*. Totowa, NJ: Barnes & Noble.

Tamnes, R. 1991. *The United States and the Cold War in the High North*. Aldershot, UK: Dartmouth.

Tönnessen, J.N, and A.O. Johnsen 1982. *The History of Modern Whaling*. London: Hurst.

Tuck, J.A. and R. Grenier 1989. *Red Bay, Labrador: World Whaling Capital A.D. 1550–1600*. St. Johns, NF: Atlantic Archaeology Ltd.

Vikström, H. and P. Högselius 2017. 'From cryolite to critical metals: the scramble for Greenland's minerals' in L. Rastad Bjørst and R.C. Thomsen (eds) *Heritage and Change in the Arctic: Resources for the Present, and the Future*. Aalborg, Denmark: Aalborg University Press.

Wexelsen, E., B. Basberg and J-E. Ringstad 1993. *Whaling and History: Perspectives on the Evolution of the Industry*. Sandefjord, Norway: Sandefjordmuseene.

Wynn, G. 2007. *Canada and Arctic North America: An Environmental History*. Santa Barbara, CA: ABC-CLIO.

Zarankin, A. and M.X. Senatore 2005. 'Archaeology in Antarctica: nineteenth-century capitalism expansion strategies' *International Journal of Historical Archaeology* 9(1): 43–56.

Web published sources

www.ats.aq/e/ep.htm

www.ats.aq/documents/cep/handbook/Protocol_e.pdf

www.ats.aq/documents/recatt/Att004_e.pdf

www.ats.aq/documents/recatt/Att249_e.pdf

www.nps.gov/nr/research/index.htm accessed 2017–10–11

www.pc.gc.ca/en/lhn-nhs accessed 2017–10–11

www.polarheritage.com/content/library/50.pdf accessed on 2017–08–14

www.polarheritage.com/polarheritage/Sitelist01up/ accessed on July 30, 2017

http://whc.unesco.org/archive/websites/arctic2008/russia.html accessed 2017–08–31

http://whc.unesco.org/en/list/577 accessed 2017–10–11

http://whc.unesco.org/en/list/629 accessed 2017–10–11

http://whc.unesco.org/en/list/1412 accessed on 2017–08–17

Part II
Polar environments

10

Biodiversity in the Polar Regions in a warming world

Hans Meltofte

Introduction to polar biodiversity

In global public debates and in the popular imagination, melting ice under the feet of polar bears (*Ursus maritimus*) and emperor penguins (*Aptenodytes forsteri*) often comes up first when the theme is climate change. This is not at all a bad imagery and is used to good effect to raise awareness of the significant challenges faced by the Arctic and Antarctic, since it encompasses both the enhanced speed and marked physical effects that climate change has in the Polar Regions and the pressures that this places on the specialized polar biodiversity.

Polar bears and penguins are also good illustrations of the comprehensive differences that characterize the two Polar Regions (for the purposes of this chapter, the Arctic is defined as the land north of the tree line together with adjacent more or less ice-covered seas; the Antarctic is defined as the entire continent plus the seas south of the Antarctic Convergence). While the Arctic has a wide range of four-legged mammalian predators, there are no terrestrial mammalian predators in the Antarctic; until recently not even humans lived there until the establishment of permanent scientific stations stemming from the International Geophysical Year (1957/58). This is of decisive importance to the entire ecosystem, since it allows millions of flightless seabirds as well as marine mammals to breed and haul out on land and sea ice all along the coasts of Antarctica and adjacent islands (see Figure 10.1). In combination with an unbroken circle of highly productive marine habitat all around the Antarctic, the scene is set for a marine environment that is far richer than in the Arctic.

As an illustration of this, the most numerous seal species in the world, the Antarctic crabeater seal (*Lobodon carcinophaga*), has an estimated population in the order of 50 to 80 million individuals (Shirihai 2007) compared to an estimated world population of eight million harp seals (*Pagophilus groenlandicus*), the most abundant seal in the Arctic (Reid et al. 2013). Also, at least 24 Antarctic and sub-Antarctic seabird species number more than one million individuals, while 'only' about 13 Arctic and sub-Arctic seabirds reach this level (cf. Cramp 1983–1989; Williams 1995; Brooke 2004; Shirihai 2007; Ganter and Gaston 2013).

For the terrestrial environment, conditions are reversed. While the Antarctic is a huge and almost totally ice-covered continent with next to no genuine terrestrial wildlife, the Arctic's main feature is extensive and 'rich' tundra fringing the continents surrounding the relatively small Arctic Ocean. 'Rich' is relative to the ocean, in that more than 14,000 terrestrial species are known to

Arctic Terns telling stories from the North

Arctic Terns telling stories from the South

Figure 10.1 Arctic terns, *Sterna paradisaea*, who visit both Polar Regions during their annual migrations, may 'testify' to their differences here as fantasized by Rohan Chakravarty.

science in the Arctic proper, as compared to 7,600 marine species (Payer et al. 2013). On top of these come several thousand species of endoparasites and microorganisms in both biomes.

The terrestrial funga, flora and fauna in the Arctic evolved from species in steppe and mountainous regions of the two major continents, Eurasia and North America, but several of them have developed into highly adapted and often endemic Arctic specialists (Payer et al. 2013). An example of this is the *Calidris* sandpipers, which constitute a relatively speciose and

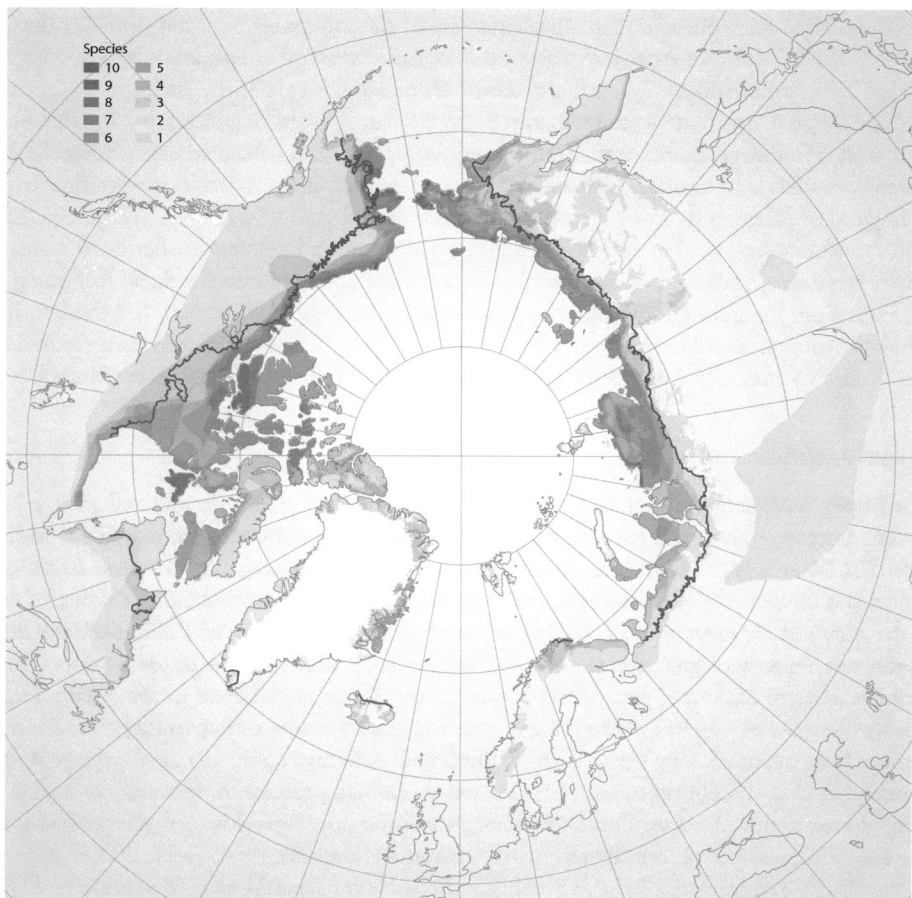

Species
10 5
9 4
8 3
7 2
6 1

Figure 10.2 Circumpolar *Calidris* sandpiper species richness with the southern limit of the terrestrial Arctic demarcated according to the CAVM Team (2003). Sandpiper distributions from Zöckler (1998), reproduced with permission from the World Conservation Monitoring Centre.

numerically dominating avifauna element on the tundra (Figure 10.2). Several of them are endemic to the Arctic during breeding but disperse to virtually all parts of the world during the non-breeding season.

This is illustrative of the extreme seasonality in the utilization by many taxa of the Polar Regions. Many birds and mammals take advantage of the short and highly productive Arctic summer, while spending the rest of the year in more clement environments. In fact, many migratory birds spend about 9–11 months of the year out of the Arctic, only moving to the Arctic for an intensive period of reproduction. At lower latitudes, concentrations of millions of shorebirds/waders utilize the rich resources on intertidal flats along sea coasts, but they cannot breed there. With such numbers of nesting birds, the eggs would be far too easy prey to a rich predator community, and their young would not be able to take advantage of the food resources exposed on the vast expanses of intertidal flats twice per day at low tide. Instead, the intense production of invertebrates on the tundra in summer is a perfect food source, particularly for precocial young (Meltofte 1996).

In the marine environment, the mere structure of the food webs is rather different between the two Polar Regions. In Antarctic waters, the well-known krill (*Euphausia superba*), which are relatively large shrimp-like euphasids, constitute an important part of the zooplankton community. Krill are to a large extent utilized directly by predatory seabirds and marine mammals such as true seals (Phocidae), southern fur seals (*Arctocephalus* spp.) and baleen whales (McGonigal and Woodworth 2001). In contrast, marine mammals primarily feed on fish in the Arctic, where a whole guild of pelagic fish constitutes a much more complex food web – like in the rest of the world's oceans. Also squid (Teuthida) constitute a direct link between smaller crustaceans and predatory seabirds, seals and toothed whales in Antarctic waters without pelagic fish being part of the equation. Demersal fish are abundant in both Polar Regions, however, but the continental shelf of Antarctica and thereby the area of relatively shallow sea is much more restricted than in the Arctic Ocean, which holds stunningly almost 30% of the global shelf area (Michel 2013).

Human impacts on wildlife through time

In the Arctic, wildlife has been the very fundament for human existence for millennia, while in the Antarctic human exploitation has a history of only a few hundred years (Shirihai 2007; Meltofte 2013). This has largely ended in Antarctica, whereas hunting and fishing remain vital to Arctic communities (Huntington 2013; Laidre et al. 2015). This meant that overexploitation of living resources was the only serious pressure on polar biodiversity until recently (Shirihai 2007; Meltofte 2013). Yet, the vast majority of species in both Polar Regions were relatively unaffected by direct human interference. This, of course, particularly applies to the Antarctic, whereas in the Arctic, humans probably contributed significantly to the extermination of the entire terrestrial megafauna and several other large mammals long before historic times with extensive consequences for the ecosystems (Normand et al. 2017). In more recent times, serious human pressure on mammals and birds was largely limited to the effects on a number of long-lived and slow reproducing species together with easily accessible colonies of seabirds and marine mammals (Krupnik 1993; Freese 2000).

Indeed, the extermination of the Steller's sea cow (*Hydrodamalis gigas*) by southern expeditions within a decade was only possible because the species had already been driven to extinction by Aleutian people at all inhabited islands long before that (Doming 1978; Turvey and Risley 2006). Most likely, also the great auk (*Pinguinus impennis*) – the original penguin – had already been driven to extinction by humans in Greenland and most of the other North Atlantic islands centuries ago, when the last individuals were killed by whalers and taxidermists in the nineteenth century (Nettleship and Evans 1985; Meldgaard 1988). Similarly, indigenous peoples exterminated several North American musk-ox (*Ovibos moschatus*) populations before any European contact (Krupnik 1993). Also, the development of reindeer husbandry in the Old World centuries ago was triggered by depletion of the wild stocks to the extent that it became necessary for local communities to monopolize and monitor the herds (Vorren 1974/75; Bjørnstad et al. 2012).

With the arrival of modern weaponry and efficient means of transport this escalated the depletion – and in some cases even extinction – of most global stocks of large whales as one of the gravest results (McGonigal and Woodworth 2001). But also, populations of walrus (*Odobenus rosmarus*), northern fur seals (*Callorhinus ursinus*) and caribou/reindeer (*Rangifer tarandus*), together with several Arctic seabird and waterfowl species were seriously depleted (Meltofte 2013). Recently, even the sustainability of the polar bear hunt has been questioned (Jørgensen 2015).

On top of harvest in the Arctic, migratory species are often also harvested when moving out of the Arctic – in some cases to such an extent that it threatens the species. Here, the gravest examples are the extinction of the New World Eskimo curlew (*Numenius borealis*) in the late nineteenth and early twentieth centuries and presently the near extinction of the spoon-billed sandpiper (*Calidris pygmeus*) of East Asia, where excessive harvests appear to play major roles in the declines of these species (Ganter and Gaston 2013).

In the Antarctic, populations of fur seals and southern elephant seals (*Mirounga leonine*) were similarly diminished, while true seals and seabirds such as penguins hardly suffered more than local declines (McGonigal and Woodworth 2001). Finally, during the second half of the twentieth century, this continued with the depletion of many fish stocks around the world, including overexploitation in the waters of the Polar Regions (McGonigal and Woodworth 2001; Christiansen and Reist 2013).

On the positive side, many of these negative trends have been reversed by improved management through national legislation and international agreements enforcing restrictions on the take of some species of fish, birds and marine mammals and the establishment of protected areas (McGonigal and Woodworth 2001; Meltofte 2013, Laidre et al. 2015; CAFF 2017; see also the chapters on northern fisheries (Chapter 30) by Hoel and conservation (Chapter 32) by Nuttall in this volume). This includes the fact that Arctic marine fisheries, which are among the largest in the world, appear, on the whole, to be reasonably well managed nowadays, although there have been management failures, and high harvest pressure continues on some stocks (Christiansen and Reist 2013; Huntington 2013). Also, bycatch of seabirds, in particular, poses serious problems to certain populations, but this problem is increasingly being solved by precautionary measures particularly in Southern Ocean long-line fisheries (Lewison et al. 2014; Mulligan and Winnard 2017).

In Antarctic waters fisheries include krill and toothfish (*Dissostichus eleginoides* and *Dissostichus mawsoni*), the take of which is under debate. Yet, many mammal and bird populations are still seriously reduced as are most of the large whales, and in some cases the negative trends have not even been overcome with thick-billed murre (*Uria lomvia*) colonies in Greenland as the most pronounced example from within the Arctic (Meltofte 2013; Merkel et al. 2014).

Due to the heavily reduced populations of many whale stocks, it is possible that other species have taken advantage of 'untapped' food resources. Besides a number of seabirds, this may apply to northern as well as southern fur seals, where the southern species now are so abundant on many Antarctic and sub-Antarctic coasts that it may cause problems to other biota (McGonigal and Woodworth 2001; Turner et al. 2009).

Climate change

While overexploitation was the single most important stressor on polar biodiversity in the past, global warming is certainly taking over that role in the future (e.g. Turner et al. 2009; Meltofte 2013). On the one hand, polar biodiversity may be better adapted to climate variability and change than biodiversity in most other biomes, but on the other, the speed and expected magnitude of the present global warming may exceed what it can sustain. Furthermore, there are indications that abrupt climate changes in the past also have caused extinctions or near extinctions of several Arctic species (Buehler and Baker 2005).

As described in several chapters of this book, global warming has resulted in a twofold increase in temperatures in the Arctic as compared to the global mean, while in the Antarctic temperature changes have so far been modest with the Antarctic Peninsula and the waters of the Antarctic Circumpolar Current having warmed most since the 1950s (ACIA 2005; Turner et al. 2009; SCAR 2017; see also Turner et al. 2016). This has already resulted in heavily reduced spring snow cover and summer sea ice in the northern hemisphere and regional sea ice loss at the Antarctic Peninsula – just to mention two of the more spectacular effects. For polar biodiversity that is highly adapted to cold, snow and ice, this inevitably results in significant changes in living conditions. Initially, earlier snow and ice melt may benefit a number of species, but in the longer term, species from lower latitudes will move pole-wards and put the specialized polar biodiversity under pressure (Turner et al. 2009; Callaghan et al. 2013; Meltofte 2013; CAFF 2017; Wheeler et al. 2017). Hence, in the north, the entire terrestrial high Arctic biome may be squeezed out between expanding ecosystems from the south and the northern coasts of the lands, although some high Arctic endemics may survive uphill in mountainous areas or on the northernmost Arctic islands. The same processes apply to the highly specialized marine biome adapted to summer sea ice and perennial coastal ice shelves, where only minor areas are expected to persist through this century north of Greenland and the northernmost Canadian islands (Turner et al. 2009; Michel 2013). For marine mammals, birds and other biota dependent on timing of spring sea ice retreat, effects – positive as well as negative – are already seen now, and more are expected in the future (Meltofte 2013; Laidre et al. 2015; CAFF 2017).

Since the late 1970s, the median speed of distributional changes in a wide range of species is 11.0 m per decade uphill in mountains and 16.9 km per decade away from the equator, tracking the concomitant temperature changes statistically significantly (Chen et al. 2011). This is also observed in the Arctic, where both scientists and northerners have seen shrub expansion over the tundra and new species appearing further north than before (Meltofte 2013). Here, also, southern populations of polar bears are declining due to reduced sea ice season length.

Pronounced regional differences exist in impact of sea ice development on Antarctic penguin populations. While winter sea ice has expanded around much of the continent followed by increasing Adélie penguin (*Pygoscelis adeliae*) populations (Southwell et al. 2015), the diminishing late winter sea ice at the Antarctic Peninsula has allowed chinstrap (*Pygoscelis Antarctica*) and gentoo (*Pygoscelis papuato*) penguins to expand southwards replacing the smaller and more ice-dependent Adélie penguins that are also under pressure from increased spring snow cover on their breeding sites (e.g. Croxall et al. 2002; Forcada and Trathan 2009; Ainley et al. 2010; Cimino et al. 2016). Likewise, several emperor penguin colonies are declining due to reduced sea ice season length (Ainley et al. 2010). The projection is that a temperature increase of just 1.3°C before the middle of this century, will put 20–40% of the world's emperor penguins and 70% of the world's Adélie penguins at risk largely because of sea ice loss (Ainley et al. 2010; Jenouvrier et al. 2014; Ballerini et al. 2015). On top of this, further regional warming could reduce biomass and abundance of krill – the basic food source both directly and indirectly for many Antarctic whales, seals and seabirds – by more than 95% across, e.g. the Scotia Sea, Western Antarctica, during the next 100 years (Murphy et al. 2007; Turner et al. 2009; Seyboth et al. 2016).

Global warming may also facilitate intrusion of invasive alien species into the Polar Regions putting endemic polar species under further pressure (Turner et al. 2009; Lassuy and Lewis 2013; CAFF and PAME 2017). On the sub-Antarctic islands major damage to ecosystems has been exerted primarily by rats (*Rattus* sp.), domestic cats (*Felis silvestris cattus*), European rabbits (*Oryctolagus cuniculus*) and reindeer introduced by humans, but so far relatively few alien species have established viable populations within the polar borders be it in the marine or terrestrial. Many more are on the 'doorstep' in several places along the borders of both the Arctic and the Antarctic, however.

One further aspect of greenhouse gas emission has received much less public attention. Due to the uptake of the atmosphere's increased CO_2 content in sea water, the acidity of the ocean's water has increased considerably since the industrial revolution (Orr et al. 2005). Furthermore, cold water takes up CO_2 more efficiently than warm water so that the polar seas are more exposed to this stressor. This causes problems particularly for animals like crustaceans and molluscs using calcium carbonate for shells and skeletons, in that it not only prevents the formation of such structures but even dissolves already formed calcium body parts exposed to sea water. The consequences that this may have for marine ecosystems are not fully understood, but it is possible that it will alter marine food webs significantly and even threaten major fisheries (Turner et al. 2009; AMAP 2013; CAFF 2017).

Taken together, the diversity of life forms that are found in the regions we now call Arctic and Antarctic will increase. But the surplus of species will mainly be those that are already common on latitudes next to the Polar Regions, while the true polar specialists are at risk (Turner et al. 2009; Meltofte 2013). However, the actual number of high latitude species at risk of extinction is relatively low as compared to several other more speciose lower-latitude biomes under pressure from climate change. Nevertheless, many of those at risk are highly charismatic Arctic and Antarctic species that are adapted to life in some of the most extreme regions of the world. The attention that polar bears and penguins receive in public media is an indication of the great importance of these creatures to human beings as iconic and charismatic species – in spite of the fact that the effects of climate change in other parts of the world will likely dwarf the already obvious problems in the Polar Regions (Marshall 2014: 135, 239).

UV-B radiation

The well-known reduction of the ozone layer particularly over the Antarctic region (see Chapter 36 by Kirkwood in this volume) has led to a 130% increase in spring-time UV-B radiation in the Antarctic and a 22% increase in the Arctic (Slanina 2009). UV-B radiation is harmful to a range of organisms including humans. For biodiversity in the Polar Regions, marine phytoplankton is considered most at risk in that the increase in UV-B may decrease net primary production by up to 10%, alter food webs and slow biogeochemical cycles.

In the terrestrial Arctic, increased UV-B radiation has negative consequences for plants (Newsham and Robinson 2009) and has been shown to affect the resource availability, productivity and trophic interactions and dynamics of freshwater organisms as well (Wrona and Reist 2013). Following international efforts, the breakdown of the ozone layer has now stopped and, provided this continues, the layer is expected to regenerate during this century (Turner et al. 2014).

Contaminants

Contamination of Arctic animals with heavy metals and other toxic substances has long been of concern, especially as humans in many regions live off marine predators that have relatively higher levels of contaminants that accumulate in humans (AMAP 2009a, 2011). However, for Arctic wildlife, we have few examples of effects on population level (Meltofte 2013). Indeed, we have not seen anything in the Arctic like the crash in the number of birds of prey populations in North America and Europe in the second half of the twentieth century.

The reason for this is probably that international efforts to curb the release of the most harmful contaminants from the industrialized world have been rather successful, but also that such effects are hard to document (Meltofte 2013). Furthermore, it does not mean that there are no problems. Recently, high levels of a variety of contaminants have been documented in three scavenging bird species in the Arctic, namely ivory gull (*Pagophila eburnean*), black-legged kittiwake (*Rissa tridactyla*) and glaucous gull (*Larus hyperboreus*), where it may be the reason for well-documented population declines (Goutte et al. 2015; Lucia et al. 2015; Petersen et al. 2015).

In the Antarctic, contaminants in wildlife originating from human activities are at much lower levels compared to the Arctic (e.g. Aubail et al. 2011), and 'only' local effects of rubbish dumps etc. are seen (Bargagli 2008). Most likely, this is because industrial activities are much smaller in the southern hemisphere than in the northern, and maybe even that the consistent air and sea circulation around the Antarctic to some extent shields the continent from influences of contaminated air and water masses from the north (see Chapter 12 by Bingham in this volume).

Mineral exploitation, extraction and transport

While the Antarctic is off limits to extractive industries at least until 2041 according to the 1991 Environmental Protocol to the Antarctic Treaty (and while the Protocol is in effect, noting that Article 25 makes room for consultative party members to request a review following its expiration after 50 years; see the chapters by Chaturvedi (Chapter 31), Dodds (Chapter 20), Dey Nuttall (Chapter 23) and Rothwell (Chapter 21) in this volume), mineral and hydrocarbon exploration and extraction activities are increasing in the Arctic (AMAP 2009b). This applies to a high extent to the huge continental shelves in the Arctic Ocean, where oil and gas appear to exist in very large quantities. This poses a number of threats to Arctic biodiversity, primarily in the form of risks of accidents with large releases of oil as the most serious (Skjoldal et al. 2009; Meltofte 2013). Whereas accidents on land usually can be contained to local environments, oil accidents at sea are particularly problematic in that the oil may disperse to entire regions and contaminate food webs and kill large numbers of seabirds and marine mammals.

Furthermore, the simple fact that a third of oil reserves, half of gas reserves and over 80% of current coal reserves should remain unused from 2010 to 2050 if a global mean temperature increase of more than 2°C from the pre-industrial level is to be avoided (McGlade and Ekins 2015), it can be argued that this most obviously should include oil finds in the most sensitive marine regions on Earth, i.e. marine Arctic regions with both sea ice and icebergs making the operations much more risky and clean up next to impossible (Lloyd's and Chatham House 2012; McGlade and Ekins 2015).

Human disturbance to wildlife

Compared with most other regions of the world, little habitat conversion has taken place in both Polar Regions. In Antarctica, the most pronounced effects are around human activity centres such as whaling and research stations. In the Arctic, mining and hydrocarbon extraction can be added to this, while ocean bottom trawling and overgrazing by semi-domestic reindeer have more geographically extensive effects (Turner et al. 2009; Meltofte 2013; CAFF 2017). Concerning the more direct effect in the form of displacement of wildlife from important habitats, the two Polar Regions are again very different. In the Antarctic, wildlife has thrived without humans or other terrestrial mammalian predators with the result that birds and mammals are next to indifferent towards the presence of humans. When observing a few simple rules, researchers and tourists alike can approach resting and breeding birds and mammals to within a few meters without serious effects. One exception is the giant petrel (*Macronectes* spp.) that is sensitive to human presence and must be protected. This is not so in the Arctic, where wildlife has had to adapt to hunters as well as other mammalian predators through millennia (Meltofte 2013). The result is that the escape distance towards humans is up to hundreds of meters in a number of hunted species. This means that even other human activities can have significant disturbance effects on these species, and that protective measures may be necessary. Tourism, in particular, is booming in both Polar Regions (see Chapter 27 by Stewart and Liggett in this volume), but tourist activities are generally easy to manage and so far no serious negative effects have been documented. Here, the most obvious threat is ship accidents with release of oil, especially because cruise ships tend to navigate in less well charted areas. On the positive side, increasing numbers of people gain insight and deep fascination with the Polar Regions and thereby become 'ambassadors' for the protection of the high latitudes.

Shipping in general, which is on the rise these days (AMSA 2009), has one more problematic effect in the form of noise. Underwater noise, in particular, may pose problems to whales, due to their dependence on long distance acoustic communication (CAFF 2017). Furthermore, seismic sampling in connection with oil and gas exploration can have direct scaring effects including displacement from important feeding areas etc.

Research needs

Due to the sheer remoteness and harsh climate of the Polar Regions, research and monitoring of species and ecosystems are lacking far behind most other biomes on Earth. This is particularly unfortunate since some of the most pronounced effects of climate change are expected on high latitudes, some of which may have global repercussions. Even for exploited species and populations this means that we have few long-term data series to document possible changes (e.g. Turner et al. 2009; Meltofte 2013). With exceedingly few exceptions, permanent research facilities have only been established since the late 1960s. The result is that only recently are we beginning to understand the processes of decisive importance to ecosystem functioning and resilience. When it comes to species richness in many groups of organisms, our knowledge is often fragmentary. Regarding the basic biology of many species, our knowledge is similarly exceedingly limited.

To be able to document changes driven by climate change and other anthropogenic as well as natural drivers, we need to be able to monitor trends in distribution, abundance and phenology. Furthermore, to be able to understand the functional causes for observed changes, we need much more research into species, population and ecosystem biology. Hence, there is a critical lack of essential data and scientific understanding necessary to improve management, planning and implementation of biodiversity conservation or monitoring strategies in the Arctic and the Antarctic (Turner et al. 2009; Meltofte 2013; CAFF 2017). In the Arctic, the Circumpolar Biodiversity Monitoring Program (CBMP) is striving to fill this gap (Barry et al. 2013), while in the Antarctic the CCAMLR Ecosystem Monitoring Program (CEMP) is doing the same for the marine environment (CCAMLR 2017).

References

ACIA 2005. *Arctic Climate Impact Assessment*. New York: Cambridge University Press.

Ainley, D., Russell, J., Jenouvrier, S., Woehler, E., Lyver, P.O., Fraser, W.R. and Kooyman, G.L. 2010. 'Antarctic penguin response to habitat change as Earth's troposphere reaches 2°C above preindustrial levels' *Ecol. Monogr.* 80: 49–66.

AMAP 2009a. *AMAP Assessment 2009: Human Health in the Arctic*. Oslo: Arctic Monitoring and Assessment Programme.

AMAP 2009b. *Oil and Gas Activities in the Arctic: Effects and Potential Effects*. Oslo: Arctic Monitoring and Assessment Programme.

AMAP 2011. *Arctic Pollution 2011*. Oslo: Arctic Monitoring and Assessment Programme.

AMAP 2013. *Arctic Ocean Acidification 2013: Summary Report*. Oslo: Arctic Monitoring and Assessment Programme.

AMSA 2009. *Arctic Marine Shipping Assessment 2009 Report*. Arctic Council.

Aubail, A., Teilmann, J., Dietz, R., Rigét, F., Harkonen, T., Karlsson, O., Rosing-Asvid, A. and Caurant, F. 2011. 'Investigation of mercury concentrations in fur of phocid seals using stable isotopes as tracers of trophic levels and geographical regions' *Polar Biol.* 34: 1411–1420.

Ballerini, T., Tavecchia, G., Pezzo, F., Jenouvrier, S. and Olmastroni, S. 2015. 'Predicting responses of the Adelie penguin population of Edmonson Point to future sea ice changes in the Ross Sea' *Front. Ecol. Evol.* 3. doi: 10.3389/fevo.2015.00008.

Bargagli, R. 2008. 'Environmental contamination in Antarctic ecosystems' *Sci. Total Environ.* 400: 212–226.

Barry, T., Christensen, T., Payne, J. and Gill, M. 2013. *Circumpolar Biodiversity Monitoring Program Strategic Plan, 2013–2017: Phase II Implementation of the CBMP*. CAFF Monitoring Series Report Nr. 8. CAFF International Secretariat. Akureyri, Iceland.

Bjørnstad, G., Flagstad, Ø. Hufthammer, A.K. and Røed, K.H. 2012. 'Ancient DNA reveals a major genetic change during the transition from hunting economy to reindeer husbandry in northern Scandinavia' *J. Archaeol. Sci.* 29: 102–108.

Brooke, M. 2004. *Albatrosses and Petrels across the World*. Oxford, UK: Oxford University Press.

Buehler, D.M. and Baker, A.J. 2005. 'Population divergence times and historical demography in Red Knots and Dunlins' *Condor* 107: 497–513.

CAFF 2017. *State of the Arctic Marine Biodiversity Report*. Akureyri, Iceland: Conservation of Arctic Flora and Fauna International Secretariat.

CAFF and PAME 2017. *Arctic Invasive Alien Species: Strategy and Action Plan*. Akureyri, Iceland: Conservation of Arctic Flora and Fauna and Protection of the Arctic Marine Environment.

Callaghan, T.V., Matveyeva, N., Chernov, Y., Schmidt, N.M., Brooker, R. and Johansson, M. 2013. 'Arctic terrestrial ecosystems' in S.A. Levin (ed.) *Encyclopedia of Biodiversity*, 2nd edition. Vol. 1. Academic Press, Waltham, MA. 227–244.

CAVM Team 2003. *Circumpolar Arctic Vegetation Map. Scale 1:7,500,000.* Conservation of Arctic Flora and Fauna (CAFF) Map No. 1. U.S. Fish and Wildlife Service, Anchorage, AK.

CCAMLR 2017. *CCAMLR Ecosystem Monitoring Program (CEMP)*. www.ccamlr.org/en/science/ ccamlr-ecosystem-monitoring-program-cemp.

Chen, I.-C., Hill, J.K., Ohlemüller, R., Roy, D.B. and Thomas, C.D. 2011. 'Rapid range shifts of species associated with high levels of climate warming' *Science* 333: 1024–1026.

Christiansen, J.S. and Reist, J.D. 2013. 'Fishes' in H. Meltofte (ed.) *Arctic Biodiversity Assessment. Status and Trends in Arctic Biodiversity*. Akureyri, Iceland: Conservation of Arctic Flora and Fauna. 192–245.

Cimino, M.A., Lynch, H.J., Saba, V.S. and Oliver, M.J. 2016. 'Projected asymmetric response of Adélie penguins to Antarctic climate change' *Sci. Rep.-UK* 6, 28785, doi: 10.1038/srep28785.

Cramp, S. (ed.) 1983–1989. *Handbook of the Birds of Europe, the Middle East and North Africa*. Vols 1–4. Oxford, UK: Oxford University Press.

Croxall, J.P., Trathan, P.N. and Murphy, E.J. 2002. 'Environmental change and Antarctic seabird populations' *Science* 297: 1510–1514.

Doming, D.P. 1978. 'Sirenian evolution in the North Pacific Ocean' *Univ. Calif. Publ. Geol. Sci.* 118: 1–176.

Forcada, J. and Trathan, P.N. 2009. 'Penguin responses to climate change in the Southern Ocean' *Glob. Change Biol.* 15: 1618–1630.

Freese, C.H. 2000. *The Consumptive Use of Wild Species in the Arctic: Challenges and Opportunities for Ecological Sustainability*. WWF Canada and WWF International Arctic Programme.

Ganter, B. and Gaston, A.J. 2013. 'Birds' in H. Meltofte (ed.) *Arctic Biodiversity Assessment. Status and Trends in Arctic Biodiversity*. Akureyri, Iceland: Conservation of Arctic Flora and Fauna. 142–180.

Goutte, A., Barbraud, C., Herzke, D., Bustamante, P., Angelier, F., Tartu, S., Clément-Chastel, C., Moe, B., Bech, C., Gabrielsen, G.W., Bustnes, J.O. and Chastel, O. 2015. 'Survival rate and breeding outputs in a high Arctic seabird exposed to legacy persistent organic pollutants and mercury' *Environ. Pollut.* 200: 1–9.

Huntington, H. 2013. 'Provisioning and cultural services' in H. Meltofte (ed.) *Arctic Biodiversity Assessment. Status and Trends in Arctic Biodiversity*. Akureyri, Iceland: Conservation of Arctic Flora and Fauna. 592–626.

Jenouvrier, S., Holland, M., Stroeve, J., Serreze, M., Barbraud, C., Weimerskirch, H. and Caswell, H. 2014. 'Projected continent-wide declines of the emperor penguin under climate change' *Nat. Clim. Change* 4: 715–718.

Jørgensen, M. 2015. *Polar Bears on the Edge*. Spitzbergen-Svalbard: NozoMojo.

Krupnik, I. 1993. *Arctic Adaptations. Native Whalers and Reindeer Herders of Northern Eurasia*. Lebanon, NH: University Press of New England.

Laidre, K.L, Stern, H., Kovacs, K.M., Lowry, L., Moore, S.E., Regehr, E.V., Ferguson, S.H., Wiig, Ø., Boveng, P., Angliss, R.P., Born, E.W., Litovka, D., Quakenbush, L., Lydersen, C., Vongraven, D. and Ugarte, F. 2015. 'Arctic marine mammal population status, sea ice habitat loss, and conservation recommendations for the 21st century' *Conserv. Biol.* 29: 724–737.

Lassuy, D.R. and Lewis, P.N. 2013. 'Invasive species: human-induced' in H. Meltofte (ed.) *Arctic Biodiversity Assessment. Status and Trends in Arctic Biodiversity*. Akureyri, Iceland: Conservation of Arctic Flora and Fauna. 558–565.

Lewison, R.L. et al. 2014. 'Global patterns of marine mammal, seabird, and sea turtle bycatch reveal taxa-specific and cumulative megafauna hotspots' *PNAS* 111: 5271–5276.

Lloyd's and Chatham House 2012. *Arctic Opening: Opportunity and Risk in the High North*. Lloyd's, London. www.lloyds.com/~/media/Files/News%20and%20Insight/360%20Risk%20Insight/Arctic_Risk_Report_20120412.pdf.

Lucia, M., Verboven, N., Strøm, H., Miljeteig, C., Gavrilo, M.V., Braune, B.M., Boertmann, D. and Gabrielsen, G.W. 2015. 'Circumpolar contamination in eggs of the high-arctic ivory gull *Pagophila eburnea*' *Environ. Toxicol. Chem.* 34: 1552–1561.

Marshall, G. 2014. *Don't Even Think About It. Why Our Brains are Wired to Ignore Climate Change*. New York: Bloomsbury.

McGlade, C. and Ekins, P. 2015. 'The geographical distribution of fossil fuels unused when limiting global warming to 2°C' *Nature* 517: 187–190.

McGonigal, D. and Woodworth, L. 2001. *The Complete Encyclopedia of Antarctica and the Arctic*. Richmond Hill, ON: Firefly Books .

Meldgaard, M. 1988. 'The great auk, *Pinguinus impennis* (L.) in Greenland' *Hist. Biol.* 1: 145–178.

Meltofte, H. 1996. 'Are African wintering waders really forced south by competition from northerly wintering conspecifics? Benefits and constraints of northern versus southern wintering and breeding in waders' *Ardea* 84: 31–44.

Meltofte, H. (ed.) 2013. *Arctic Biodiversity Assessment. Status and Trends in Arctic Biodiversity*. Akureyri, Iceland: Conservation of Arctic Flora and Fauna.

Merkel, F., Labansen, A.L., Boertmann, D., Mosbech, A., Egevang, C., Falk, K., Linnebjerg, J.F., Frederiksen, M. and Kampp, K. 2014. 'Declining trends in the majority of Greenland's thick-billed murre (Uria lomvia) colonies 1981–2011' *Polar Biol.* 37: 1061–1071.

Michel, C. 2013. 'Marine ecosystems' in H. Meltofte (ed.) *Arctic Biodiversity Assessment. Status and Trends in Arctic Biodiversity*. Akureyri, Iceland: Conservation of Arctic Flora and Fauna. 486–527.

Mulligan, B. and Winnard, S. 2017. *Towards Seabird-Safe Fisheries*. Cambridge, UK: BirdLife International and RSPB.

Murphy, E.J., Trathan, P.N., Watkins, J.L., Reid, K., Meredith, M.P., Forcada, J., Thorpe, S.E., Johnston, N.M. and Rothery, P. 2007. 'Climatically driven fluctuations in Southern Ocean ecosystems' *P. Roy. Soc. B* 274: 3057–3067.

Nettleship, D.N. and Evans, P.G.H. 1985. 'Distribution and status of the Atlantic Alcidae' in D.N. Nettleship and T.R. Birkhead (eds) *The Atlantic Alcidae*. Waltham, MA: Academic Press. 53–154.

Newsham, K.K. and Robinson, S.A. 2009. 'Responses of plants in polar regions to UVB exposure: a meta-analysis' *Glob. Change Biol.* 15: 2574–2589.

Normand, S., Høye, T.T., Forbes, B.C., Bowden, J.J., Davies, A.L., Odgaard, B.V., Riede, F., Svenning, J.-C., Treier, U.A., Willerslev, R. and Wischnewski, J. 2017. 'Legacies of historical human activities in Arctic woody plant dynamics' *Annu. Rev. Env. Resour.* 42: 541–567.

Orr, J.C., et al. 2005. 'Anthropogenic ocean acidification over the twenty-first century and its impact on calcifying organisms' *Nature* 437: 681–686.

Payer, D.C., Josefson, A.B. and Fjeldså, J. 2013. 'Species diversity in the Arctic' in H. Meltofte (ed.) *Arctic Biodiversity Assessment. Status and Trends in Arctic Biodiversity*. Akureyri, Iceland: Conservation of Arctic Flora and Fauna. 66–77.

Petersen, A., Irons, D.B., Gilchrist, H.G., Robertson, G.J., Boertmann, D., Strøm, H., Gavrilo, M., Artukhin, Y., Clausen, D.S. Kuletz, K.J. and Mallory, M.L. 2015. 'The status of glaucous gulls *Larus hyperboreus* in the circumpolar Arctic' *Arctic* 68: 107–120.

Reid, D.G., Berteaux, D. and Laidre, K.L. 2013. 'Mammals' in H. Meltofte (ed.) *Arctic Biodiversity Assessment. Status and Trends in Arctic Biodiversity*. Akureyri, Iceland: Conservation of Arctic Flora and Fauna. 78–141.

SCAR 2017. *Antarctic Climate Change and the Environment: 2017 Update*. http://epic.awi.de/44422/2/ATCM40_ip080_e.pdf.

Seyboth, E., Groch, K.R., Rosa1, L.D., Reid, K., Flores, P.A.C. and Secchi, E.R. 2016. 'Southern right whale *(Eubalaena australis)* reproductive success is influenced by krill *(Euphausia superba)* density and climate' *Sci. Rep.-UK* 6, 28205, doi: 10.1038/srep28205.

Shirihai, H. 2007. *The Complete Guide to Antarctic Wildlife: Birds and Marine Mammals of the Antarctic Continent and the Southern Ocean*, 2nd edition. Princeton, NJ: Princeton University Press.

Skjoldal, H.R., Cobb, D., Corbett, J., Gold, M., Harder, S., Low, L.L., Noblin, R., Robertson, G., Scholik-Schlomer, R., Sheard, W., Silber, G., Southall, B., Wiley, C., Wilson, B. and Winebrake, J. 2009. *Arctic Marine Shipping Assessment.* Tromsø, Norway: Protection of the Arctic Marine Environment.

Slanina, S. 2009. 'Antarctic ozone hole'. Retrieved from *The Encyclopedia of Earth* www.eoearth.org/view/article/150116.

Southwell, C., Emmerson, L., McKinlay, J., Newbery, K., Takahashi, A., Kato, A., Barbraud, C., DeLord, K. and Weimerskirch, H. 2015. 'Spatially extensive standardized surveys reveal widespread, multi-decadal increase in east Antarctic Adelie penguin populations' *Plos One* 11(10): e0165989.

Turner, J., Bindschadler, V., Convey, P., di Prisco, G., Fahrbach, E., Gutt, J., Hodgson, D., Mayewski, P. and Summerhayes, C. 2009. *Antarctic Climate Change and the Environment.* Cambridge, UK: The Scientific Committee on Antarctic Research.

Turner, J. et al. 2014. 'Antarctic climate change and the environment: an update' *Polar Record* 50: 237–259.

Turner, J., Lu, H., White, I., King, J.C., Phillips, T., Hosking, J.S., Bracegirdle, T.J., Marshall, G.J., Mulvaney, R. and Deb, P. 2016. 'Absence of 21st century warming on Antarctic Peninsula consistent with natural variability' *Nature* 535: 411–415.

Turvey, S.T. and Risley, C.L. 2006. 'Modelling the extinction of Steller's sea cow' *Biol. Lett.* 2: 94–97.

Vorren, Ø. 1974/75. 'Man and reindeer in Northern Fennoscandia: economic and social aspects' *Folk* 16–17: 243–252.

Wheeler, H.C., Høye, T.T. and Svenning, J-C. 2017. 'Wildlife species benefitting from a greener Arctic are most sensitive to shrub cover at leading range edges' *Glob. Change Biol.* doi: 10.1111/gcb.13837.

Williams, T.D. 1995. *The Penguins.* Oxford, UK: Oxford University Press.

Wrona, F.J. and Reist, J.D. 2013. 'Freshwater ecosystems' in H. Meltofte (ed.) *Arctic Biodiversity Assessment. Status and Trends in Arctic Biodiversity.* Akureyri, Iceland: Conservation of Arctic Flora and Fauna. 442–485.

Zöckler, C. 1998. *Patterns of Biodiversity in Arctic Birds.* WCMC Biodiv. Bull. No. 3.

11

Geological histories of polar environments

Tom A. Jordan

Introduction

Geology describes the varied rock strata, provinces and structures in the world around us, and the processes which created them. Geologically, the Arctic and Antarctic are very different. The Arctic is centred on a large oceanic basin surrounded by the North American and Eurasian continents. The region has been built and shaped by geological processes during which enormous landmasses have formed and shifted from southern latitudes into the northern latitudes now constituting the circumpolar Arctic, and the subsequent formation and opening of the Arctic Ocean. At the beginning of the Cambrian period, approximately 500 million years ago (Ma), for example, the lands now making up the Canadian Shield and Greenland were part of a continuous continent situated further south at the Equator. Tectonic plate movements and the shaping and formation of the geological building blocks in Arctic North America are recorded as tremendous changes in climate during past millennia, ranging from sub-equatorial climates during the Paleozoic (540–250 Ma) to Arctic conditions in the Cretaceous (~100 Ma) and the present. Some of the Earth's most ancient rock formations are found in the Arctic (for example, at Isua in Greenland's Nuuk Fjord region), but the region is also marked by geological features that are relatively young in the history of the Earth, such as the floor of the Arctic Ocean. Recent work on the geological environments of the Arctic such as the deep Arctic Ocean, including the evolution of ridges and basins, means that geological histories of this complex region are now beginning to be written (e.g. Coakley 2016), and there is an extensive literature on the geological history of other parts of the Arctic (e.g. for a popular account and survey of the history of geological research in Greenland, see Henriksen 2008).

In contrast to the Arctic, the Antarctic region is much less well understood and so this chapter focuses on outlining the geological history of Antarctic polar environments. Antarctica is dominated by a continental land mass entirely encircled by oceans. This isolation of Antarctica meant it was the last continent to be discovered and it remains one of the least explored places on our planet. Like the Arctic, it too contains some of the world's most ancient rocks. Another reason we understand less about Antarctica's geological history is because the extensive Antarctic ice sheets, which cover over 98% of the land mass with an average thickness of two kilometres, make investigation of Antarctic geology extremely challenging. Understanding the geology of

Antarctica, however, is critical. All the southern hemisphere continents, including Antarctica, were once amalgamated in a single supercontinent known as Gondwana. The formation and destruction of this landmass spans the time from the emergence of the first complex life through to the age of the dinosaurs. Antarctica linked many of the now dispersed southern hemisphere continents, and therefore provides key information about the processes which drove continental amalgamation and breakup across the southern hemisphere. The geology of the Antarctic also plays an important role in determining the existence and long-term stability of the continental ice sheets. The present isolation of Antarctica was the culmination of the tectonic process of continental breakup and was one of the key triggers for the onset of Antarctic glaciation. The deep ice-filled marine basins of West Antarctica, where the ice sheet is predicted to be the most vulnerable to warming of the adjacent oceans, were formed by tectonic stretching and thinning of the continental crust, while the high mountains in the heart of East Antarctica were the birth place of Antarctica's first ice sheets. The local geology and rock type can also directly influence how specific ice streams flow.

Techniques

Understanding the geology of Antarctica has been a priority since the first scientific expeditions to the continent at the turn of the twentieth century. Since this time the collection and analysis of rock samples has provided great insights into the antiquity of the continent, and the past and present processes that have shaped the region. For sedimentary rocks, deposited by wind or water, the study of fossils (palaeontology) has allowed scientists to correlate and date events in Antarctica with other regions around the globe. These techniques underpinned the early study of Antarctic geology and are still used today to investigate the past environments of Antarctica (Taylor and Taylor 2012). More recent development of radiometric dating techniques, which utilise the known rate of decay of naturally occurring radioactive elements, have allowed dating of a broader range of igneous (derived from liquid magma) and metamorphic (altered) rocks. These dating techniques have allowed geologists to study the tectonic evolution of Antarctica throughout its long geological history (Veevers and Saeed 2013). Most recently, techniques to date rock uplift and exposure have been used to reveal when the high Antarctic mountain ranges formed, and the extent and retreat history of the Antarctic ice sheet (Sugden et al. 2006).

Direct observation and mapping of exposed geological structures, coupled with dating techniques, has allowed geologists to begin to build a picture of the tectonic evolution of the Antarctic continent, from ancient continental collisions to chains of volcanoes and rifts. However, over 99% of the continent is blanketed by thick ice sheets so direct observation of the geology is often impossible. Geophysical sensing techniques have therefore been applied to map the sub-ice topography and underlying geology. The first geophysical exploration of Antarctica's hidden world relied on expeditions which traversed the continent using specially adapted tracked vehicles (Fuchs and Hillary 2011). These expeditions typically used seismic methods, setting off explosions and recording the time for echoes of sound waves from rocks beneath the ice to return to reveal the ice thickness. Although accurate, these ground-based techniques are time consuming, and it would take hundreds of years to provide even a preliminary survey of the buried mountains and valleys across the Antarctic continent. However, advances in radar technology since the 1970s have allowed the development of radio-echo sounding systems capable of measuring ice thickness from aircraft (Drewry 1983; Siegert, Chapter 33 of this volume). This advance has allowed the preliminary surveying of much of Antarctica, although 17% of the continent still has no ice thickness data within at least 20 km, and significant areas remain completely unexplored (Fretwell et al. 2013).

Airborne radar provides a good view of the subglacial topography but can only be used to make an inference about the geological structures below. Further information about the geological makeup of the ice-covered interior of Antarctic comes from geophysical measurements, including so-called 'potential field' observations of the strength of the local magnetic and gravity fields, and seismic techniques (Lowrie 2007). The Earth's magnetic field is created in the hot fluid iron outer core; however, its strength is locally attenuated or enhanced by the presence, absence and alignment of magnetic minerals within the shallow crust. These local variations in magnetic field strength account for as little as 0.002% of the total field but can be measured and mapped. The Earth's gravitational field reflects the downward attraction of the entire planet; however, the local strength of the gravitational field is affected by rock density. Such variations can amount to just 0.0001% of the total gravitational field but can be sensed and recorded by modern instruments. Together potential field data allows the geophysical properties (magnetism and density) of the underlying rocks of Antarctica to be mapped indirectly using aircraft (e.g. Ferraccioli et al. 2011). In addition to potential field data, advances in passive seismic techniques have helped change our understanding of Antarctica. Such techniques rely on arrays of widely spaced sensors which 'listen' for the signals from distant earthquakes. Variations in the timing and pattern of received signals are interpreted to reveal the structure and thickness of the crust and underlying mantle across broad regions (An et al. 2015). Integrating these geophysical data with radar derived sub-ice topography and geological information from sparse outcrops, allows interpretation of the past tectonic events that compressed and folded, or stretched and pulled apart, the Earth's crust in Antarctica.

Geology

The Antarctic continent can be geologically divided into three distinct parts: East Antarctica, West Antarctica and the Antarctic Peninsula (Thomson et al. 2011). These different regions have distinct tectonic histories which are reflected in the observed topographic and geological provinces. East Antarctica is the oldest part of the Antarctic continent and contains some of the oldest rocks on Earth (Fitzsimons 2000). It is dominated by elevated topography (average ~130 m above sea level), which prior to subsidence due to ice sheet loading would have been on average ~800 m high. This region includes the Transantarctic Mountains along one margin and the totally ice-covered Gamburtsev Subglacial Mountains in the continental interior. The typical rock sequence in East Antarctica consists of deformed and metamorphosed basement rocks over 500 million years old, overlain by relatively little deformed sequences of continental sediments including the very extensive Beacon Sandstone formation which are recognized from Dronning Maud Land to Northern Victoria Land. Originally, East Antarctica was considered to be a monolithic block with a relatively uniform lithology and internal structure; however, more recent studies have shown that it is likely made up of an underlying mosaic of different >500 Ma provinces (Boger 2011). The dates of most recent tectonic activity are still debated, however.

West Antarctica includes the low-lying Ross Sea, Amundsen Sea and Weddell Sea embayments and the more elevated Thurston Island, Marie Byrd Land and Haag Ellsworth Whitmore Mountains regions (Dalziel and Elliot 1982). Overall, West Antarctica has a mean elevation of 580 m below sea level, with the deepest parts of the low-lying embayments being over 1500 m below sea level. The low-lying regions are underlain by significant continental rift systems filled with thick sequences of sediments. The Weddell Sea embayment is underlain by an inferred Jurassic (~175 Ma) rift system (Jordan et al. 2017), while the Ross Sea and Amundsen Sea embayments are associated with younger Cretaceous to Cenazoic rifting (100–45 Ma) (Cande et al. 2000), and ongoing active, or recently active volcanism (LeMasurier and Thomson 1990).

However, the thick continental ice sheet and extensive floating ice shelves limit access to the rocks in these low-lying embayments, making unambiguous dating of West Antarctic rifting problematic.

The Antarctic Peninsula is often considered to be part of the West Antarctica region. However, it has a distinct geological history which warrants its discussion as a separate province. This region developed as a magmatic arc in response to subduction of oceanic crust along the Paleo Pacific margin of Gondwana (Larter and Barker 1991a). As such it is dominated by igneous intrusions and volcanic rocks similar to those observed in the South American Andean mountain range. These igneous rocks are interleaved with and intrude continental margin sediments which often contain reworked arc igneous material. Collision of separate tectonic blocks, coupled with ongoing subduction, triggered deformation and metamorphism along the Antarctic Peninsula.

Tectonic evolution

The oldest rocks in East Antarctica are Archean in age (>2.5 billion years old). These rocks outcrop in a number of distinct areas including Dronning Maud Land, Terre Adelie Land and around the Lambert Glacier (Fitzsimons 2003). These Archean provinces progressively grew over many millions of years by addition of juvenile magmatic rocks and tectonic accretion of magmatic arcs around the province margins. During this time, it is thought that the components of the East Antarctic continent may have been incorporated within a number of transient supercontinental land masses (Jacobs et al. 2008). The process of major continental accretion culminated, around a billion years ago, with an event known as the Grenville orogeny. This global event records the protracted collision and amalgamation of many of the separate earlier continental fragments into a single super continent known as Rodinia (Dalziel et al. 2000). The precise configuration of the continental fragments making up Rodinia is not well known (Li et al. 2008). It is generally agreed that much of East Antarctica and southern Australia were amalgamated. The North American continent (Laurentia) was also a single unit probably located adjacent to the present day Pacific margin of Antarctica. However, the precise location of Laurentia, the position of fragments of West Antarctic basement, and other continental fragments relative to East Antarctica remain controversial.

After the Grenville orogeny the Rodinian super continent likely rifted apart, but no clear signature of this event is recognised in Antarctica. This period of relative tectonic quiescence in East Antarctica came to an end around 500 Ma with the Pan-African and Ross orogenies. The Pan-African event reflected the amalgamation of the Gondwanan super continent by continental collision between West (South America, South Africa and Dronning Maud Land) and East (Australia, East Antarctica and India) Gondwana (Jacobs and Thomas 2004). During this event many of the older Grenville orogenic belts were reactivated and overprinted by significant deformation and magmatism. The approximately contemporary Ross orogen reflected a distinct tectonic setting associated with long-lived oceanic subduction against the paleo Pacific margin of Gondwana (Goodge 2007). This tectonic setting led to the emplacement of large volumes of arc type magmatism. The Ross orogen did not trigger the same extensive metamorphic events as the Pan-African event, although terrain accretion in Northern Victoria Land led to the suturing of a suite of distinct arc provinces against the margin of East Antarctica. Figure 11.1 provides reconstructions of East Antarctica in Pre-Cambrian times.

After the Pan-African and Ross orogenies, the geology of Antarctica is marked by deposition of thick sequences of continental sediments. In East Antarctica these rocks include Devonian to Triassic (417–200 Ma) sandstones and coals of the Beacon group which extend from Dronning Maud Land to Northern Victoria Land (Bradshaw 2013). In the Lambert Glacier region, Permian

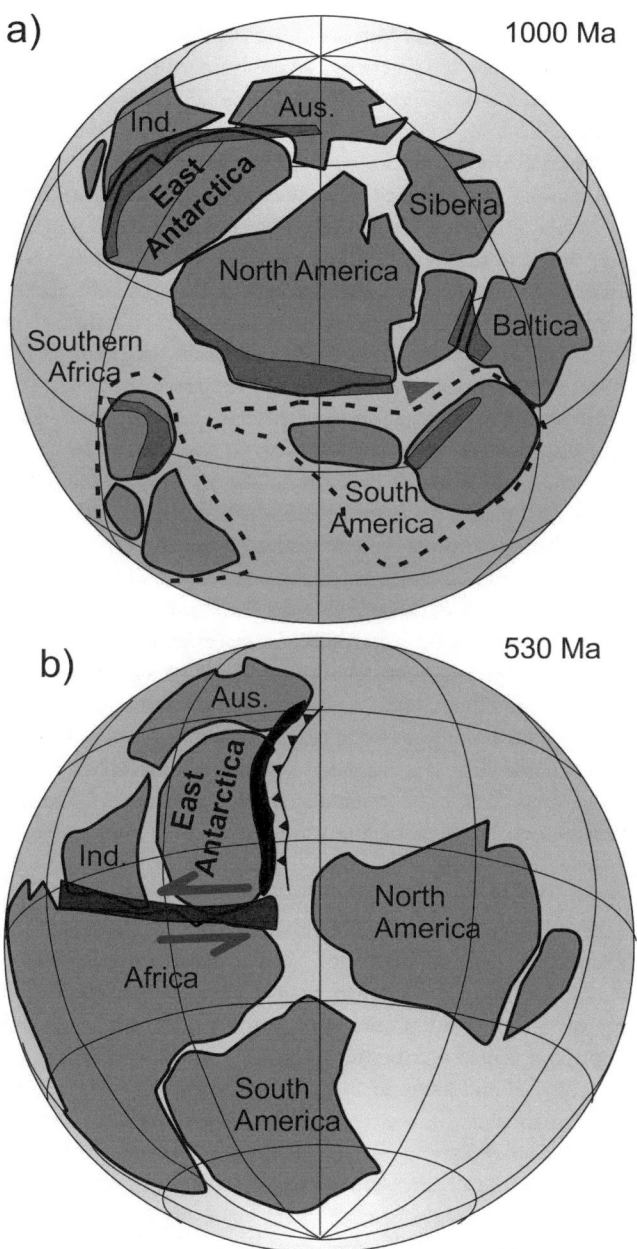

Figure 11.1   Reconstructions of East Antarctica in Pre-Cambrian times. a) Possible reconstruction of East Antarctica within Rodinia (Dalziel et al. 2000). The dark belt marks indicate the proposed Gondwanide orogen. Ind. India, Aus. Australia. b) Reconstruction of East Antarctica during Gondwana assembly (Dalziel 2014). Belt marks on the edges of East Antarctica and Australia indicate the developing Ross orogen along the continental margin (Goodge 2007), while the dark belt lines between East Antarctica, India and Africa mark the continent-continent collision zone of the Pan-African orogen (Jacobs and Thomas 2004). Figure by author.

(~275 Ma) rifting led to continental basin formation and deposition of coal measures (Harrowfield et al. 2005). This rift system is thought to cut across the heart of East Antarctica, where Permian, and possibly later Cretaceous (~100 Ma), rifting triggered uplift of the Gamburtsev Subglacial Mountains (Ferraccioli et al. 2011). Within West Antarctica significant sequences of marine sediments were laid down from Cambrian to Permian times (542–251 Ma) in the Ellsworth Whitmore Mountains region. The oldest of these West Antarctic rocks are rift-related sediments, but later Permian (~275 Ma) sequences reflect development of a proximal magmatic arc along the Pacific margin of Gondwana (Curtis 2001). These West Antarctic sediments were folded in the Permo-Triassic Gondwanide orogen (~250 Ma), which is also observed in South Africa, South America and at the margin of East Antarctica in the Shackleton Range (Veevers and Powell 1994). Elsewhere in East Antarctica these sediments remain undeformed.

The long-lived Gondwanan supercontinent began to break up in the Jurassic (~175 Ma) (Figure 11.2a). The apparent precursor to continental breakup in Antarctica was the emplacement of the mafic Ferrar Large Igneous Province (LIP). This province includes the basaltic Ferrar sills, extending along the entire length of the Transantarctic Mountains, and the Dufek Intrusion which is thought to be a deep-seated part of the same LIP (Elliot 1992). Other parts of this LIP are seen in Dronning Maud Land, although some of these rocks may reflect part of the coincident Karoo LIP, which was emplaced in South Africa. The presence of a broader Karoo/Ferrar LIP has been linked with the upwelling of a large region of hot mantle material, known as a mantle plume, beneath the Weddell Sea region (Storey 1995). Coincident with emplacement of the mafic Karoo/Ferrar LIP, the Antarctic Peninsula was subject to a period of intense silicic magmatism which has been linked with the proximity of a mantle plume, and/or changes in the subduction regime along the Pacific margin (Riley et al. 2001). During this period of magmatism, it is proposed that the West Antarctic Haag Ellsworth Whitmore province was translated from adjacent to Coats Land in East Antarctica to its present position in West Antarctica (Dalziel et al. 2013), associated with the development of a broad continental rift within the Weddell Sea Embayment.

Final continental separation, with the formation of oceanic crust, occurred between South Africa and East Antarctica about 165 Ma and continued until at least 130 Ma (Figure 11.2b). Further rifting of the other southern hemisphere continents from Antarctica progressed in an approximately clockwise manner (Seaton et al. 2012). Sri Lanka and India rifted away from Antarctica sometime between 127 and 118 Ma. Australia began to slowly rift away from Antarctica ~95 Ma until ~45 Ma when there was a significant acceleration in the rate of continental separation. Rifting between New Zealand and Antarctica began ~100 Ma, with associated significant continental extension within the Ross Sea and likely further inboard within the West Antarctic Rifts System (Figure 11.2c). This phase of extension is thought to have been a trigger for the uplift of the Transantarctic Mountains, which extend ~1500 km along the margin of the West Antarctic Rift System (Stern and ten Brink 1989). Final separation of New Zealand and the Campbell plateau likely occurred by ~72 Ma, when ocean spreading anomalies are detected. Continental extension continued within the West Antarctic Rift system into more recent times, with some authors suggesting this movement extended as far as the southern Antarctic Peninsula (Cande et al. 2000).

Throughout the Mesozoic (250–65Ma) the Antarctic Peninsula was subject to continuing subduction and arc-related magmatism (Leat et al. 1995). Accretion of exotic terrains with the Antarctic Peninsula triggered deformation and metamorphism in the Mid Cretaceous. However, both local and far travelled models have been proposed for the accreted terrains (Burton-Johnson and Riley 2015). Further subduction and arc magmatism continued after terrain accretion. However, subduction stopped along the Antarctic Peninsula from Cretaceous to recent times as the adjacent oceanic spreading ridge was consumed into the subduction zone (Larter and Barker 1991b). Only the tip of the Antarctic Peninsula, adjacent to Bransfield Strait,

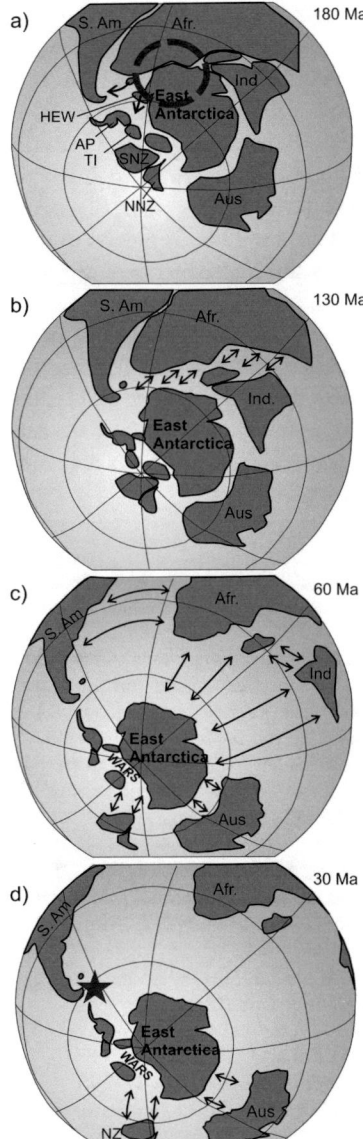

Figure 11.2 Antarctica during the breakup of Gondwana. a) Antarctica at the onset of
continental breakup. The circle marks the location of proposed mantle plume.
Arrows show proposed movement of the Falkland and Haag Ellsworth Whitmore
(HEW) microplates associated with the Jurassic Weddell Sea embayment rift.
SNZ and NNZ mark Southern and Northern New Zealand. b) Initial oblique
separation of South America and Africa from East Antarctica and India (arrows).
c) Separation of India, Australia and New Zealand. Note development of the
continental West Antarctic Rift System (WARS) at this time. Also note that no
deep seaway existed between South America and the Antarctic Peninsula at
this stage. d) Continued separation of Australia and New Zealand, and final
separation between South America and Antarctica (star) allowing development
of the cooling circumpolar current. (Figure by author).

is now the site of ongoing subduction, with associated crustal extension and active volcanism occurring within the adjacent Bransfield Strait.

Around 50 Ma, westward-directed subduction began beneath the South Sandwich Islands. Slab role back led to the development of a back arc extensional zone, and ultimately an oceanic spreading centre between South America and the Antarctic Peninsula (Eagles et al. 2005). This spreading centre led to the final separation of South America and Antarctica ~35 Ma (Figure 11.1d). This newly formed oceanic gateway led to the development of a strong circumpolar current plunging Antarctica into a deep-freeze (see Chapter 12 by Bingham in this volume).

Today the geology of Antarctica is dominated by ongoing glacial processes. In some areas slow ice flow and extreme polar desert conditions have acted to preserve the Antarctic landscape, potentially for millions of years. In other areas glacial erosion, exploiting pre-existing geological structures, has cut deep ice-filled basins over 1.5 km below sea level (Jamieson et al. 2010). This incision potentially contributed to the uplift of both the Transantarctic and Gamburtsev Subglacial Mountains and has generated and transported vast volumes of sediment onto and across the continental margins. The resulting marginal sediments preserve key records of the continent's glacial history.

References

An, M., D.A. Wiens, Y. Zhao, M. Feng, A.A. Nyblade, M. Kanao, Y. Li, A. Maggi and J. Lévêque 2015. 'S-velocity model and inferred Moho topography beneath the Antarctic Plate from Rayleigh waves' *JGR* 120: 359–383.

Boger, S.D. 2011. 'Antarctica: Before and after Gondwana' *Gondwana Research* 19: 335–371.

Bradshaw, M.A. 2013. 'The Taylor Group (Beacon Supergroup): the Devonian sediments of Antarctica. Antarctic Palaeoenvironments and earth-surface processes' in M. J. Hambrey, P. F. Barker, P. J. Barrett et al., (eds) *Geological Society London Special Publication* 381: 67–97.

Burton-Johnson, A. and T.R. Riley 2015. 'Autochthonous v. accreted terrane development of continental margins: a revised in situ tectonic history of the Antarctic Peninsula' *JGSL*, doi:10.1144/jgs2014–1110.

Cande, S.C., J.M. Stock, R.M. Müller and T. Ishihara 2000. 'Cenozoic motion between East and West Antarctica' *Nature* 404: 145–150.

Coakley, B., K. Brumley, N. Lebedeva-Ivanova and D. Mosher 2016. 'Exploring the geology of the central Arctic Ocean: understanding the basin features in place and time' *Journal of the Geological Society* 173(6): 967–987.

Curtis, M.L. 2001. 'Tectonic history of the Ellsworth Mountains, West Antarctica: Reconciling a Gondwana enigma' *GSABul* 113: 939–958.

Dalziel, I.W.D. 2014. 'Cambrian transgression and radiation linked to an Iapetus-Pacific oceanic connection?' *Geology* 42, doi:10.1130/G35886.35881.

Dalziel, I.W.D. and D.H. Elliot 1982. 'West Antarctica: Problem child of Gondwanaland' *Tectonics* 1: 3–19.

Dalziel, I.W.D., L. Lawver, I.O. Norton and L.M. Gahagan 2013. 'The Scotia Arc: Genesis, evolution, global significance' *Annu. Rev. Earth Planet. Sci.* 41: 767–793.

Dalziel, I.W.D., S. Mosher and L.M. Gahagan 2000. 'Laurentia–Kala-hari collision and the assembly of Rodinia' *The Journal of Geology* 108: 499–513.

Drewry, D.J., (ed.) 1983. *Antarctica: Glaciological and Geophysical Folio.* Cambridge, UK: University of Cambridge, Scott Polar Research Institute.

Eagles, G., R.A. Livermore, J.D. Fairhead and P. Morris 2005. 'Tectonic evolution of the west Scotia Sea' *Journal of Geophysical Research: Solid Earth* 110(B2). https://doi.org/10.1029/2004JB003154.

Elliot, D.H. 1992. 'Jurassic magmatism and tectonism associated with Gondwanaland break-up: an Antarctic perspective' in B.C. Storey and R.J. Pankhurst (eds) *Magmatism and the Causes of Continental Break-up. Geological Society London Special Publication*, 354: 165–184.

Ferraccioli, F., C. Finn, T.A. Jordan, R.E. Bell, L.M. Anderson and D. Damaske 2011. 'East Antarctic rifting triggers uplift of the Gamburtsev Mountains' *Nature* 479: 388–392.

Fitzsimons, I.C.W. 2000. 'Grenville-age basement provinces in East Antarctica: evidence for three separate collisional orogens' *Geology* 28: 879–882.

Fitzsimons, I.C.W. 2003. 'Proterozoic basement provinces of southern and southwestern Australia, and their correlation with Antarctica' in M. Yoshida, B. F. Windley and S. Dasgupta (eds) *Proterozoic East Gondwana: Supercontinent Assembly and Breakup. Geological Society of London*, 206: 93–130.

Fretwell, P., H.D. Pritchard, D.G. Vaughan, J. Bamber, N. Barrand, R. Bell, C. Bianchi, R. Bingham, D. Blankenship and G. Casassa 2013. 'Bedmap2: improved ice bed, surface and thickness datasets for Antarctica' *The Cryosphere* 7(1). http://hdl.handle.net/2152/41162.

Fuchs, V. and E. Hillary 2011. *The Crossing of Antarctica: The Commonwealth Transantarctic Expedition 1955–1958.* Whitefish, MT: Literary Licensing LLC.

Goodge, J.W. 2007. 'Metamorphism in the Ross orogen and its bearing on Gondwana margin tectonics' in M. Cloos, W.D. Carlson, M.C. Gilbert, J.G. Liou and S.S. Sorensen (eds) *Convergent Margin Terranes and Associated Regions: A Tribute to W.G. Ernst. Geological Society of America Special Paper.* 419: 185–203.

Harrowfield, M., G.R. Holdgate, C.J.L. Wilson and S. McLoughlin 2005. 'Tectonic significance of the Lambert Graben, east Antarctica: reconstructing the Gondwanan rift' *Geology* 33: 197–200.

Henriksen, N. 2008. *Geological History of Greenland: Four Billion Years of Earth Evolution.* Copenhagen: Geological Survey of Denmark and Greenland.

Jacobs, J., S.A. Pisarevsky, R.J. Thomas and M. Becker 2008. 'The Kalahari Craton during the assembly and dispersal of Rodinia' *PC* 160: 142–158.

Jacobs, J. and R.J. Thomas 2004. 'Himalayan-type indenter-escape tectonics model for the southern part of the late Neoproterozoic-early Paleozoic East African- Antarctic orogen' *Geology* 32: 721–724.

Jamieson, S.S.R., D.E. Sugden and N.R.J. Hulton 2010. 'The evolution of the sub-glacial landscape of Antarctica' *EPSL* 293: 1–27.

Jordan, T., F. Ferraccioli and P.T. Leat 2017. 'New geophysical compilations link crustal block motion to Jurassic extension and strike-slip faulting in the Weddell Sea Rift System of West Antarctica' *Gondwana Research* 42: 29–48.

Larter, R.D. and P.F. Barker 1991a. 'Effects of ridge crest-trench interaction on Antarctic-Phoenix spreading: forces on a young subducting plate' *JGR* 96(B12): 19583–19607.

Larter, R.D. and P.F. Barker 1991b. *Neogene Interaction of Tectonic and Glacial Processes at the Pacific Margin of the Antarctic Peninsula. Sedimentation, Tectonics and Eustasy: Sea-Level Changes at Active Margins. D. I. M. MacDonald.* Oxford, UK: Blackwell Publishing Ltd.

Leat, P.T., J.H. Scarrow and I.L. Millar 1995. 'On the Antarctic Peninsula batholith' *Geol. Mag* 132: 399–412.

LeMasurier, W.E. and J.W. Thomson, Eds. 1990. *Volcanoes of the Antarctic Plate and Southern Oceans. Antarctic Research Series.* Washington, DC: American Geophysical Union.

Li, Z.X., S.V. Bogdanova, A.S. Collins, A. Davidson, D. De Waele, R.E. Ernst, I.C.W. Fitzsimons, R.A. Fuck, D.P. Gladkochub, J. Jacobs, K.E. Karlstrom, S. Lu, L.M. Natapov, V. Pease, S.A. Pisarevsky, K. Thrane and V. Vernikovsky 2008. 'Assembly, configuration, and break-up history of Rodinia: A synthesis' *PC* 160: 179–210.

Lowrie, W. 2007. *Fundamentals of Geophysics.* Cambridge, UK: Cambridge University Press.

Riley, T.R., P.T. Leat, R.J. Pankhurst and C. Harris 2001. 'Origins of large volume rhyolitic volcanism in the Antarctic Peninsula and Patagonia by crustal melting'. *JP* 42: 1043–1065.

Seaton, M., Müller R.D., S. Zahirovic, C. Gaina, T. Torsvik, G. Shephard, A. Talsma, M. Gurnis, M. Turner, S. Maus and M. Chandler 2012. 'Global continental and ocean basin reconstructions since 200 Ma'. *Earth-Science Reviews* 113: 212–270.

Stern, T. and U.S. ten Brink 1989. 'Flexural uplift of the Transantarctic mountains'. *JGR* 94: 10315–10330.

Storey, B.C. 1995. 'The role of mantle plumes in continental breakup: case histories from Gondwanaland'. *Nature* 377: 301–308.

Sugden, D.E., C. Bentley and C.Ó. Cofaigh 2006. 'Geological and geomorphological insights into Antarctic ice sheet evolution'. *PTRS*: 1607–1625, DOI: 1610.1098/rsta.2006.1791.

Taylor, T.N. and E.L.E. Taylor 2012. *Antarctic Paleobiology: Its Role in the Reconstruction of Gondwana.* New York: Springer Verlag.

Thomson, M.R.A., J.A. Crame and J.W. Thomson 2011. *Geological Evolution of Antarctica.* Cambridge, UK: Cambridge University Press.

Veevers, J. and A. Saeed 2013. 'Age and composition of Antarctic sub-glacial bedrock reflected by detrital zircons, erratics, and recycled microfossils in the Ellsworth Land–Antarctic Peninsula–Weddell Sea–Dronning Maud Land sector (240°E–0°–015°E)'. *Gondwana Research* 23: 296–332.

Veevers, J.J. and C.M.E. Powell 1994. *Permian-Triassic Pangean Basins and Foldbelts Along the Panthalassan Margin of Gondwanaland.* Boulder, CO: The Geological Society of America, Inc.

12

Polar oceans and their global significance

Rory Bingham

The polar oceans and their global significance

Although occupying only a small fraction of Earth's total surface area, the polar oceans exert a powerful influence on Earth's climate on timescales that range from the seasonal up to transitions that persist over millions of years. This power owes much to the unique characteristics of the polar oceans, most notably their proximity to the polar ice sheets, their covering of sea ice, and the intense surface cooling and dense water formation that occur at high latitudes. These features give rise to a complex web of interactions and feedbacks between the polar oceans, the atmosphere, sea ice and the ice sheets that have the potential to amplify initially small amplitude changes within the climate system or its external boundary conditions. For this reason, the polar oceans will play a central role in shaping the climate system's dynamic response to the global warming we are now witnessing.

The polar oceans influence climate in numerous ways. They impact on the stability of the ice sheets, with direct implications for global temperatures and sea level. The processes that drive and sustain the global overturning circulation, which shapes Earth's climate by storing and redistributing vast quantities of heat from the sun and by drawing CO_2 out of the atmosphere, occur almost exclusively within the polar oceans. Sea ice, a distinguishing feature of the polar oceans (for now), controls the exchange of energy between the polar oceans and the atmosphere, with implications for climate and weather. These processes are closely coupled, with each having the power to affect the other. The ocean's global overturning circulation drives changes in the polar oceans that can influence ice sheet stability. The growth and retreat of ice sheets precipitate changes in the polar oceans, thereby driving changes in the global overturning circulation. Likewise, changes in sea ice cover can influence and by influenced by the overturning circulation.

In this chapter I shall describe our current understanding regarding the mechanisms by which the polar oceans influence climate. I begin with a brief overview of the worlds' oceans' overturning circulation as we find it today. This is vital to understand, as the polar oceans play a central role in this circulation. Having established the importance of the polar oceans in maintaining our present climate, I will examine their role in the dramatic changes in Earth's past climate. Finally, I consider the global significance of the polar oceans in our present and future warming world and conclude with a brief summary and outlook.

The polar oceans and the global overturning circulation

Much of the global influence of the polar oceans arises because of the central role they play in the ocean's global overturning circulation (Figure 12.1) (see, also, Chapter 11 by Jordan in this volume). This circulation, especially its Atlantic component, referred to as the Atlantic meridional overturning circulation (AMOC), is responsible for transporting vast quantities of heat from the surface layers of the Southern Hemisphere oceans to high northerly latitudes. As such, the strength of the AMOC plays a crucial role in the balance of heat between the Northern and Southern Hemispheres, with a more vigorous AMOC warming the North at the expense of the South and a weaker AMOC having the opposite effect (Broecker 1998; Stocker and Johnsen 2003). The AMOC also regulates the climate of North America and western Europe (Sutton and Hodson 2005; Jackson et al. 2015), shapes precipitation patterns across the globe (Folland et al. 1986, 2001; Zhang and Delworth 2006; Sun et al. 2012), influences the position of the inter-tropical convergence zone (ITCZ) with implications for global weather patterns (Vellinga and Wu 2004; Menary et al. 2012), impacts on the frequency and intensity of Atlantic hurricanes (Goldberg et al. 2001; Zhang and Delworth 2006) and influences CO_2 uptake by the ocean (Sabine et al. 2004).

The Northern Hemisphere polar oceans – which here we take to comprise the Arctic Ocean and Nordic and Labrador Seas and all of which lie within the 10°C July isotherm – exert a powerful influence on climate because they are crucial in sustaining and regulating the AMOC. In the upper layer of the AMOC relatively warm surface water from the Pacific, Indian and Southern Oceans converge and travel north through Atlantic eventually reaching the Nordic and Labrador Seas. Here, intense surface cooling through loss of heat to the frigid atmosphere above increases the density of this Atlantic Water until it sinks to form North Atlantic Deep Water (NADW), which returns south through the Atlantic in the lower layer of the AMOC to spread at depth through the Pacific and Indian Oceans. Eventually (as described later) this

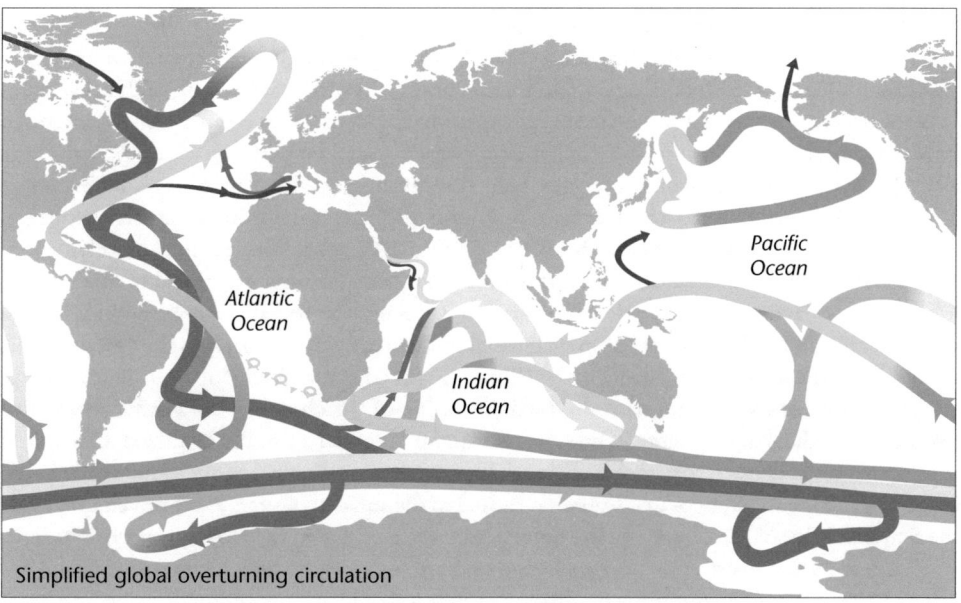

Figure 12.1 A simplified view of the ocean's global overturning circulation (Reproduced from Talley 2013).

water will reach the surface of the Pacific and Indian Oceans once more, thereby completing a circuit of the upper cell of the global overturning circulation, or the global conveyor belt as it is commonly known.

In additional to intense cooling, the formation of NADW also depends on the relatively high salinity of the Atlantic Water flowing into the Labrador and Nordic Seas. The Arctic Ocean exerts a powerful moderating influence on the AMOC by funnelling relatively fresh Pacific Water, which enters the Arctic via the Bering Strait, and freshwater from river run-off and the Greenland ice sheet into these regions of deep-water formation. This freshwater input acts as a brake on the strength of the overturning by keeping the surface waters in the regions of deep-water formation fresher and lighter than they would otherwise be, thereby inhibiting deep-water formation. Without the flow through the Bering Strait, for example, the surface of the Labrador and Nordic Seas would be more saline and consequently the overturning would be more vigorous than we find today (Shaffer and Bendtsen 1994; Goosse et al. 1997; Hasumi 2002).

The Southern Ocean and the marginal seas surrounding Antarctica also shape climate through the pivotal roles they play in the global overturning circulation. The Southern Ocean connects the major ocean basins, permitting the exchange of water masses between them and therefore a truly global overturning circulation to exist. The main pathway for this inter-basin exchange is the powerful, eastward-flowing Antarctic Circumpolar Current (ACC), which, unimpeded by meridional land boundaries, encircles the globe and dominates the dynamics of the Southern Ocean.

The Southern Ocean and ACC are also essential to the existence of the upper cell of the global overturning circulation. To complete a loop of the conveyor belt, the NADW that spreads into the deep Indian and Pacific Oceans must eventually make its way back to the surface of these oceans. While some of this will be achieved through direct upwelling in the Indian and Pacific Oceans, a greater part of it involves a more circuitous journey via the Southern Ocean. Here strong wind forcing drives surface water northwards. In doing so, Pacific and Indian deep water is sucked up from depths along the steeply sloping surfaces of constant density (isopycnals) associated with the ACC. Water that is sufficiently warm will flow into the upper layers of the Pacific and Indian Oceans and eventually back into the South Atlantic, joining water that has flowed directly from the Southern Ocean, to feed once again into the upper, northward-flowing limb of the AMOC.

The Southern Ocean is also an important location for dense water formation. The cooler fraction of the upwelled Pacific and Indian deep water, together with the NADW upwelling directly from the Atlantic, which is cooler still, flow south towards Antarctica where intense cooling and salt rejection (as sea ice forms) around the margins of Antarctica transforms it into dense Antarctic Bottom Water (AABW). The AABW feeds the second, underlying cell of the global overturning circulation that spreads northwards through the global ocean beneath the deep-water layer. In the Pacific and Indian Oceans, the AABW upwells and merges with the Pacific and Indian Ocean deep water that is eventually drawn back to the surface in the Southern Ocean through wind–driven upwelling, as already described. In the Atlantic Ocean the AAWB upwells and merges with the NADW, some of which enters the Pacific and Indian Oceans to form their deep waters, while the remainder is upwelled in the Southern Ocean.

In summary, the dense waters that feed the global overturning circulation are formed only in the polar oceans and the majority of that water is returned to the surface through wind–driven upwelling in the Southern Ocean. Clearly, then, the polar oceans are essential to the global overturning circulation. They therefore exert a powerful influence on Earth's climate system. Their role in maintaining the relatively stable overturning circulation we find today is, in a sense, the invisible influence of the polar oceans on climate. We are unaware of it as it is relatively unchanging (in our lifetimes). But, nonetheless, Earth's climate would be quite different without the polar ocean processes that sustain the present overturning circulation, as can be seen by considering Earth's past climate.

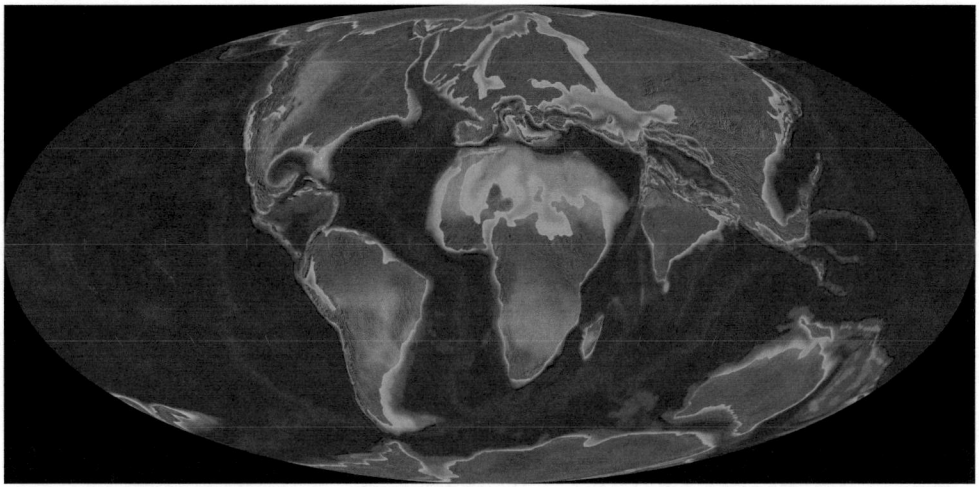

Figure 12.2 Map showing the configuration of the continents and the beginning of the Cenozoic era. (©Ron Blakey).

The polar oceans in Earth's past climate

By the beginning of the Cenozoic era (65 million years ago to present) the continents and oceans were starting to resemble their present form (Figure 12.2). During this period, Earth's climate cooled from a warm, greenhouse-like state, where temperatures were 10–15°C above those of today and the Antarctic continent, although at its present polar position, was covered in vegetation, to the much cooler semi-glaciated (icehouse) state we find today. The primary driver for this transition was carbon burial within Earth's crust, which, over millions of years, gradually reduced the amount of CO_2 in the atmosphere from 1000 to 170 parts per million (Hansen et al. 2013).

Yet, the descent towards our present ice-age climate was not smooth. Rather, it was marked by three dramatic and, in geological terms, rapid cooling events, with intervening periods of warming (Figure 12.3). Providing definitive explanations for these cooling events is challenging as the complex mechanisms – involving orbital and tectonic forces, carbon cycling and intricate climate feedbacks – through which they arose must be inferred from indirect indicators of Earth's past climate buried within the sediment that has accumulated on the sea floor over millions of years. Nonetheless, there is sufficient evidence to conclude that the polar oceans, especially the Southern Ocean, played a significant part in each of the three cooling events.

The first and largest of the Cenozoic cooling events – the Eocene-Oligocene Transition – occurred around 33.5 million years ago. Over a relatively short period (approximately 0.4 million years) Earth's climate cooled by several degrees and an ice sheet established itself over Antarctica (Miller et al. 1991, 2005; Barrett 1996; Zachos et al. 1996, 2001; Lear et al. 2000, 2008; Coxall et al. 2005). The Southern Ocean is implicated in two competing hypotheses for this transition. The "gateway hypothesis" first put forward by Kennett (1977) attributes the rapid Antarctic glaciation to the development and deepening of the Drake Passage between the West Antarctica Peninsula and South America and a deepening of the Tasman Seaway. These changes opened up the Southern Ocean allowing the ACC to develop, thermally isolating Antarctica from warm sub-tropical waters to the north and initiating the growth of an ice sheet over the continent.

161

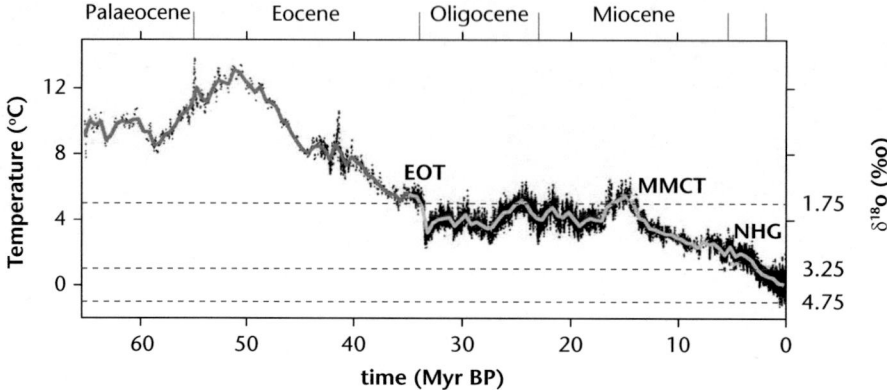

Figure 12.3   Deep ocean temperature during the Cenozoic era as estimated by Hansen et al. (2013). The Eocene-Oligocene Transition (EOT), the Mid-Miocene Climate Transition (MMCT) and intensification of the Northern Hemisphere Glaciation (NHG) appear as abrupt drops in temperature. (Adapted from Hansen et al. 2013, figure 1; Source: http://rsta.royalsocietypublishing.org/content/371/2001/20120294).

Once ice sheet growth was underway, positive feedbacks between the ice sheet and the climate system would have promoted further cooling and ice sheet growth. In addition to global cooling, due to the fact that the ice sheet with its higher albedo reflected more of the sun's energy back into space than did the vegetated surface it replaced, the developing ice sheet and associated sea ice may have switched on the production of dense AABW (Miller et al. 2009; Goldner et al. 2014). This is supported by sediment records which indicate cooling of the deep ocean at this time (Miller et al. 1987; Zachos et al. 2001; Katz et al. 2008). As this dense water sank it would have drawn CO_2 out of the atmosphere, further enhancing global cooling and ice sheet growth.

Rather than being secondary to the opening of the Southern Ocean and development of the ACC, a second, recently more favoured, hypothesis places the observed fall in CO_2 as the main driver of the cooling and ice sheet growth during the Eocene-Oligocene transition (DeConte and Pollard 2003). However, once this process was underway, it would have been reinforced by the aforementioned positive feedback mechanisms, including AABW formation and CO_2 drawdown, and possibly an invigorated AMOC that fed heat into the North Hemisphere at the expense of the Southern Hemisphere (Goldner et al. 2014).

In fact, recent explanations of the Eocene-Oligocene transition have linked the gateway and CO_2 drawdown hypotheses (Galeotti et al. 2016). From this perspective, it was not the thermal barrier provided by the birth of the ACC that led to ice sheet growth. Rather, as the ACC developed, so too did a global overturning circulation more akin to that which we find today. This drew nutrient rich deep water towards the Southern Ocean's surface, feeding an explosive growth in diatom (phytoplankton) numbers that promoted the drawdown of atmospheric CO_2. This then led to global cooling, ice sheet growth and the positive ice-ocean-atmosphere feedbacks described earlier. Further cooling may have resulted from the increased deep-ocean storage of atmospheric CO_2 associated with a more vigorous global overturning circulation (Fyke et al. 2015).

Following the Eocene-Oligocene cooling event, Earth gradually warmed again, with temperatures peaking up to 10°C higher than today at the Miocene Climate Optimum. During this period the Antarctic ice sheet receded, and the continent was once more covered in vegetation. The second major cooling event of the Cenozoic era – the Mid-Miocene Climate Transition – began approximately 15 million years ago, culminating 13.9 million years ago with a relatively short period during which the Antarctica ice sheet grew rapidly and then stabilised. Sediment records show the Southern Ocean cooling in three phases preceding the rapid ice sheet growth. The timing of these cooling events suggests a link to the relatively large variations in the eccentricity of Earth's orbit around the Sun that occurred during this period. These orbital fluctuations may have led to an intensification of the ACC, cooling the Southern Ocean and promoting ice sheet growth by increasing the thermal isolation of Antarctica (Shevenell et al. 2004, 2008).

As for the Eocene-Oligocene Transition, however, it may be that global cooling and the rapid regrowth of the Antarctic ice sheet were primarily instigated by declining atmospheric CO_2 (Vincent and Berger 1985; Raymo and Ruddiman 1992; Pagani et al. 1999; Kürschner et al. 2008), with changes in Earth's orbit playing a secondary role. The fall in CO_2 may have resulted from changes in ocean chemistry associated with weathering, perhaps of the newly formed Himalayas (Kender et al. 2014). In this case, the observed stepwise cooling of the Southern Ocean preceding the period of rapid ice sheet growth may have been due to eccentricity-paced melt water pulses from the still dynamic ice sheet (Holbourn et al. 2005). Nonetheless, it is likely that as the ice sheet growth accelerated the Southern Ocean feedbacks that featured in the Eocene-Oligocene Transition reinforced the CO_2 decline, global cooling and ice sheet growth.

The final cooling event of the Cenozoic Era occurred at the end of the Pliocene (~2.73 Ma). This transition saw a dramatic intensification of the Northern Hemisphere glaciation, and marked the beginning of the present Quaternary Period, characterised by repeated glacial cycles. There is evidence that at this time the Southern Ocean had become stably stratified, thereby inhibiting the movement of water between the surface and deep ocean (Sigman et al. 2004; Adkins 2013). This stratification may have resulted from an expanded Antarctic ice sheet which shifted Southern Hemisphere winds northward, permitting the expansion of sea ice that acted to freshen the surface of the Southern Ocean. Alternatively, an overall cooling of the water column as atmospheric CO_2 declined may have allowed salinity rather than temperature to control the stratification (Sigman et al. 2004). Whatever its cause, the stable stratification would have inhibited the upwelling of NADW in the Southern Ocean. In turn, this could have reinforced global cooling by preventing the residual heat stored in the NADW from being returned to the atmosphere as it cooled to form AABW (Woodard et al. 2014). A reduction in Southern Ocean upwelling may have further enhanced the Northern Hemisphere glaciation by weakening the AMOC and reducing the transport of heat to high northerly latitudes. Increased stratification of the Northern Hemisphere polar oceans due to global cooling may also have slowed the AMOC to similar effect. The reduction in upwelling of NADW in the Southern Ocean may have further contributed to global cooling by trapping CO_2 in the deep ocean (Sigman et al. 2004).

This brings us to the present Quaternary Period, where Earth's climate has cycled between glacial periods, characterised by extensive ice sheets covering much of the Earth's surface, and interglacial periods, such as the present time, where ice sheets have retreated to high latitudes. While the pacing of glacial cycles is determined primarily by cyclic changes in Earth's orbit – the so-called Milankovitch cycles – the polar oceans are believed to have played an important role in the much more rapid warming and cooling events which occurred during glacial and interglacial periods.

Ice core and ocean sediment records show that the last glacial period was irregularly punctuated every few thousand years by episodes of rapid warming. During these Dansgaard-Oeschger oscillations, as they are known, Northern Hemisphere temperatures rose by up to 12°C over a matter of decades, approaching those more typical of interglacial periods, before gradually and then abruptly cooling again over hundreds of years. There appears to be a strong link between Dansgaard-Oeschger events and the state of the AMOC. It is hypothesised that during stadial (cold) periods, some mechanism, such as sea ice over the Nordic Seas (Böhm et al. 2015) or a freshwater barrier in the sub-polar North Atlantic (Ganopolski and Rahmstorf 2001), restricts deep-water formation to lower latitudes in the North Atlantic. The AMOC exists in a weakened, but stable, cold state, with less heat transported to high northerly latitudes. As the ice sheet grows and becomes unstable, particularly large pulses of freshwater associated with Heinrich events can temporarily shut down the AMOC (the "off" state) further reinforcing Northern Hemisphere cooling. In contrast, during inter-stadial (warm) periods the AMOC exists in an invigorated warm state similar to that we find today.

Although the relationship between the state of the AMOC and Dansgaard-Oeschger oscil-lations and Heinrich events seems robust, there is much less certainty regarding the mecha-nisms that trigger these events and control their evolution. A popular explanation sees the AMOC as the main control on the evolution of Dansgaard-Oeschger events. During a stadial period some process, such as freshwater forcing (Manabe and Stouffer 1994; Ganopolski and Rahmstorf 2001), declining sea ice cover (Dokken et al. 2013), shifting storm tracks (Wunsch 2006; Seager and Battisti 2007), or increased wind-driven upwelling and CO_2 release to the atmosphere (Banderas et al. 2012, 2015), eventually drive the AMOC from its cold state towards its more vigorous warm state. This increases the northward transport of heat, rapidly warm-ing the Northern Hemisphere. Positive feedback loops, such as further sea ice melting due to increased northward heat transport, then act to reinforce the initial AMOC strengthening.

The interstadial persists while the AMOC remains in its warm state. The duration of the interstadial appears to grow in relation to the stability of the AMOC, with a more stable AMOC leading to longer warm periods and vice versa (Buizert and Schmittner 2015). In turn, the stabil-ity of the AMOC appears to be related to the condition of the Southern Ocean, with a warmer Southern Ocean tending to stabilise the AMOC by reducing dense AABW formation, thereby enhancing deep-water formation in the North Hemisphere, as suggested by the bipolar seesaw mechanism proposed by Broecker (1998). In a glacial climate, however, this situation is unsta-ble and eventually freshwater fluxes from the melting ice sheet force the AMOC back towards its cold state and the Northern Hemisphere begins to cool. Although the fall in temperature is initially gradual, stadial conditions return rapidly once some tipping point is reached.

However, it may well be that ocean dynamics alone are insufficient to explain the magnitude or evolution of the observed temperature changes during Dansgaard-Oeschger oscillations. Sea ice may be crucial. Declining sea ice could amplify the warming over Greenland by allowing the Nordic Seas to store and release more of the sun's energy, rather than reflecting it back into space or trapping it beneath the surface (Dokken et al. 2013). Triggered by the ocean warming at intermediate depths that occurs when the AMOC is in its cold state, the rapid collapse of an ice shelf extending into the Nordic Seas from Greenland could explain the speed with which sea ice loss and warming occurs (Petersen et al. 2013). Likewise, the slow, then rapid, cooling at the end of a Dansgaard-Oeschger event could be due to the slow regrowth of the ice shelf, which, upon reaching some threshold, causes rapid expansion of sea ice over the Nordic Seas.

Reductions in the strength of the AMOC due to freshwater forcing from the melting Laurentide ice sheet, may also explain the series of rapid cooling events, the most recent of which was the Younger Dyras, which interrupted the warming at the beginning of the present interglacial period. The relationship with the AMOC explains why these cooling events and the Northern Hemisphere warming and cooling associated with Dansgaard-Oeschger oscil-lations appear in the Antarctica ice core records as temperature fluctuations of the opposite sign, and why Heinrich events led to warming over Antarctica. During interstadial periods a more vigorous AMOC increases northward heat transport, warming the Northern Hemisphere while cooling the Southern Hemisphere, a phenomenon known as the thermal bipolar seesaw (Stocker and Johnsen 2003). By shutting off the AMOC Heinrich events have the opposite effect, warming the Southern Hemisphere at the expense of the Northern Hemisphere.

In summary, while the precise mechanisms are still uncertain, the study of Earth's past climate demonstrates the global significance of the polar oceans. Evidence suggests that the Southern Ocean played a crucial role in the three major cooling events of the Cenozoic era. The debate concerns whether changes in the Southern Ocean drove the cooling events or whether the Southern Ocean was part of a feedback loop that amplified some initial change within the climate system or its external boundary conditions. Similarly, there is good evidence that Northern Hemisphere polar oceans through their influence on the AMOC were central to the rapid climate fluctuations during glacial periods and interglacial transitions. Again, the debate centres on whether ocean dynamics were the ultimate cause of the fluctuations or one element in a larger set of interacting processes. These questions may never be definitively answered. Nonetheless, examination of Earth's past climate regimes forcefully demonstrated the global significance of the polar oceans and provides vital clues to how the climate system may respond to global warming.

The polar oceans in a warming world

While Earth's climate gradually cooled over the Cenozoic Era, the burning of fossil fuels has led to a rapid rise in atmospheric CO_2 and Earth's global mean surface temperature over the last century. The unique characteristics of the polar oceans make them especially sensitive to global warming (see Chapter 19 by Hodgkins in this volume). They are, therefore, likely to play a prominent role in shaping the dynamic response of the climate system to anthropogenic climate change. The impacts of warming on the polar oceans include surface freshening due to increased ice sheet melting, rising ocean temperatures and loss of sea ice. In this section, I examine the global significance of such changes within the polar oceans and assess their possible implications for our future climate.

The global overturning circulation

As the planet continues to warm, the Greenland and West Antarctic ice sheets are melting at an accelerating rate, freshening the surface waters of the polar oceans (Velicogna 2009; Schrama and Wouters 2011; King et al. 2012; Shepherd et al. 2012). As discussed earlier, the ocean's global overturning circulation is extremely sensitive to such changes, with increased freshening tending to reduce the strength of the overturning by inhibiting the formation of deep and bottom waters. Based on rapid fluctuations in Earth's past climate – the Younger Dryas, for example – it is plausible that accelerating mass loss from the Greenland ice sheet may eventually lead to a dramatic reduction in the strength of the AMOC, with severe consequences for Earth's climate, including, counter-intuitively, rapid cooling of Western Europe.

While there is some evidence that the AMOC has weakened over the last several decades (Bryden et al. 2005; Smeed et al. 2014; Rahmstorf et al. 2015), direct observations of AMOC strength are arguably too recent to identify a significant trend against a background of strong inter-annual variability (Cunningham et al. 2007; Kanzow et al. 2010; Roberts et al. 2014). Climate models generally show that the AMOC will weaken over this century (e.g. Jungclaus et al. 2005; Meehl et al. 2007), with warmer temperatures leading to greater weakening. However, the possibility of a total collapse of the AMOC under plausible warming scenarios is still debated (e.g. Meehl et al. 2007; Hawkins et al. 2011; Boulton et al. 2014).

In addition to mass loss from the Greenland ice sheet, an enhanced global hydrological cycle – as is expected in a warmer climate – may also result in increased surface freshening of the Nordic and Labrador Seas. An invigorated hydrological cycle would heighten the moisture transport to high latitudes, resulting in increased freshwater fluxes into the Arctic Ocean from precipitation and river run-off. At the same time, declining sea ice cover (as discussed further later; see also the chapter by Wilkinson and Stroeve (Chapter 13) in this volume) could reduce the export of freshwater, in the form of sea ice, from the Arctic Ocean. Climate model results suggest that the combined effect of these two changes will be a freshening of the Arctic Ocean *and* an increase in the freshwater flux from the Arctic (Haines et al. 2015). This could further weaken the AMOC, increasing the likelihood of global impacts on climate.

The Arctic Ocean is not only a conduit through which freshwater passively flows. Wind-driven changes in the strength of the Beaufort Gyre enable the Arctic Ocean to store and release freshwater over interannual to decadal timescales (Proshutinsky et al. 2002, 2009; Giles et al. 2012; Long et al. 2012; Stewart and Haine 2013). Changes in the circulation of the sub-polar North Atlantic (Wu and Wood 2008), variations in the sea ice export through the Fram Strait (Cox et al. 2010; Tsukernik et al. 2010; Smedsrud et al. 2011), and possibly internal dynamics (Frankcombe and Dijkstra 2011), also influence the rate at which freshwater flows from the Arctic on inter-annual to decadal timescales. Whatever their cause, such freshwater flux changes have the potential to influence climate by freshening the Nordic Seas and sub-polar North Atlantic, thereby modulating the strength of the AMOC (Cheng and Rhines 2004; Jungclaus et al. 2005; Karcher et al. 2005; Hawkins and Sutton 2007; Jahn et al. 2010; Jackson and Vellinga 2013). Moreover, with increased long-term freshening of the Arctic as described earlier, the AMOC may become more prone to the modulating influence of Arctic freshwater fluxes. Such fluctuations, important in their own right, also confound our ability to detected long-term trends.

The Southern Ocean has warmed and freshened considerably over recent decades (Bindoff et al. 1999; Jacobs et al. 2002; Böning et al. 2008; Helm et al. 2010; Azaneu et al. 2013; Purkey and Johnson 2013). The freshening may be due to increased mass loss from the Antarctic ice sheet (Jacobs et al. 2002; Jacobs and Giulivi 2010; Nakayama et al. 2014), changes in evaporation and precipitation (Jacobs et al. 2002; Durack et al. 2012; Helm et al. 2010) or variations in the wind-driven export of sea ice to the open ocean (Haumann et al. 2016). Similar to the inhibiting impact freshening from the Greenland ice sheet melting may be having on deep-water formation in the Northern Hemisphere, the observed freshening of the surface waters around the margins of Antarctica may be reducing the rate of AAWB formation and weakening the second cell of the global overturning circulation (de Lavergne et al. 2014). Indeed, the northward spread of AABW through the Atlantic is estimated to have slowed by about 1.6×10^6 m^3 s^{-1} from 1968 to 2005 at 35°S (Kouketsu et al. 2011) and by 0.9×10^6 m^3 s^{-1} between 1981 and 2010 at 24°N (Frajka-Williams et al. 2011).

Climate models indicate that through the course of the twenty-first century continued freshening and warming will intensify the stratification of the Southern Ocean (Downes et al. 2010; Meijers et al. 2012; Downes and Hogg 2013). As well as continuing to reduce the formation of AABW and weaken the lower overturning cell (de Lavergne et al. 2014), the increased stratification may inhibit the water transformation processes that eventually feed the northward flowing upper limb of the AMOC. Since these processes are responsible for a large fraction of heat and CO_2 uptake by the ocean (Frölicher et al. 2015), the impact of this could be to exacerbate the anthropogenic rise in greenhouse gasses and associated warming (Downs et al. 2010). However, models also suggest that a reduction in the rate of AABW formation may strengthening the AMOC, thereby partially compensating for the weakening due to freshening in the Labrador and Nordic Seas (Patara and Böning 2014).

Recent decades have seen an intensification and poleward shift of the vigorous westerly winds over the Southern Ocean. This has led to a strengthening of the ACC and possibly invigorated the upper cell of the global overturning circulation (Sigmond et al. 2011). Since the change in Southern Hemisphere winds has been due primarily to the depletion of ozone in the stratosphere (see Chapter 36 by Kirkwood in this volume), a weakening and equatorward shift in the zonal winds may be expected as the ozone layer recovers. However, climate models predict a similar intensification and poleward shift in zonal winds and associated changes in the Southern Ocean circulation in response to increasing atmospheric CO_2, and this is expected to become the dominant influence on the changing strength of the ACC by 2050. This may lead to a substantial increase in the strength of the AMOC (Saenko et al. 2005), again compensating to some, as yet uncertain, extent AMOC weakening due to freshening in the Labrador and Nordic Seas. However, enhanced zonal winds may promote CO_2 out-gassing by the ocean, thereby reducing the effectiveness of the Southern Ocean as a sink for atmospheric CO_2 (Le Quéré et al. 2007).

The polar oceans are, of course, not merely passive receptacles into which the Greenland and Antarctic ice sheets discharge. The polar oceans themselves may also be modulating the rate at which the ice sheets are losing mass, with warming polar oceans and their changing currents promoting the erosion of the ice sheets (Hattermann and Levermann 2010). Rising global mean sea level associated with such increased mass flux is a further means by which the polar oceans exert a global influence. Changes in the strength of the AMOC driven by the ice sheet discharge may also lead to dynamic changes in regional sea level, particularly along the US east coast (Levermann et al. 2005; Bingham and Hughes 2009; Yin et al. 2009).

Warming and sea ice loss in the Arctic

The Arctic region is warming at twice the rate of the global average, a manifestation of what is known as Arctic amplification (Manabe and Stouffer 1980; Miller et al. 2009; Screen and Simmonds 2010; Serreze and Barry 2011). This warming has contributed to a dramatic reduction in sea ice thickness and extent (Serreze et al. 2007; Stroeve et al. 2011; Wilkinson and Stroeve, Chapter 13, this volume). The summer of 2012 saw seasonal sea ice extent – the amount of ice left after summer melting – reduce to levels unprecedented in the observational record (Figure 12.4). By the middle of this century the Arctic Ocean may be ice-free (Wang and Overland 2012). This warming and sea ice loss illustrate the intricate feedbacks that exist in the polar oceans and has direct implications for Earth's climate beyond the influence it will have on the ocean's overturning circulation.

With an albedo much higher than that of seawater, sea ice reflects much of the sun's energy back into space. As sea ice has been lost, the surface waters of the Arctic Ocean have warmed as they have absorbed more of the sun's energy (see Chapter 19). In some regions, the Arctic Ocean's surface waters have warmed by up to 5°C in recent years (Polyakov et al. 2010). Such warming thins perennial sea ice, making it more likely to melt completely during summer months, and inhibits sea ice formation, allowing the Arctic Ocean to stay ice-free for longer each season (Steele et al. 2008). This increase in both the extent and duration of ice-free conditions allows the ocean to absorb even more of the sun's energy. In a positive feedback loop, the warmer surface waters then promote further sea ice loss directly and by returning more heat to the atmosphere (Stroeve et al. 2011; Screen et al. 2013)

The Arctic Ocean also receives vast quantities of heat from the inflow of relatively warm Atlantic water. Observations show that heat transfer to the Arctic from Atlantic water is higher now than at any time in the past 2,000 years (Spielhagen et al. 2011). Although most of this warm water enters and circulates around the Arctic Ocean at intermediate depths (150–800 m) beneath the cold and fresher surface layer, some heat reaches the surface through wind- and convective-driven mixing and upwelling. With the increase in heat flux into the Arctic, this may be driving additional ice loss (Polyakov et al. 2010). In a further positive feedback loop, by exposing the ocean to the influence of the wind, sea ice loss may increase ocean mixing and wind-driven upwelling of warm Atlantic Water, thereby promoting further sea ice loss. And in yet another feedback loop, changes in the atmospheric conditions due to warming from the

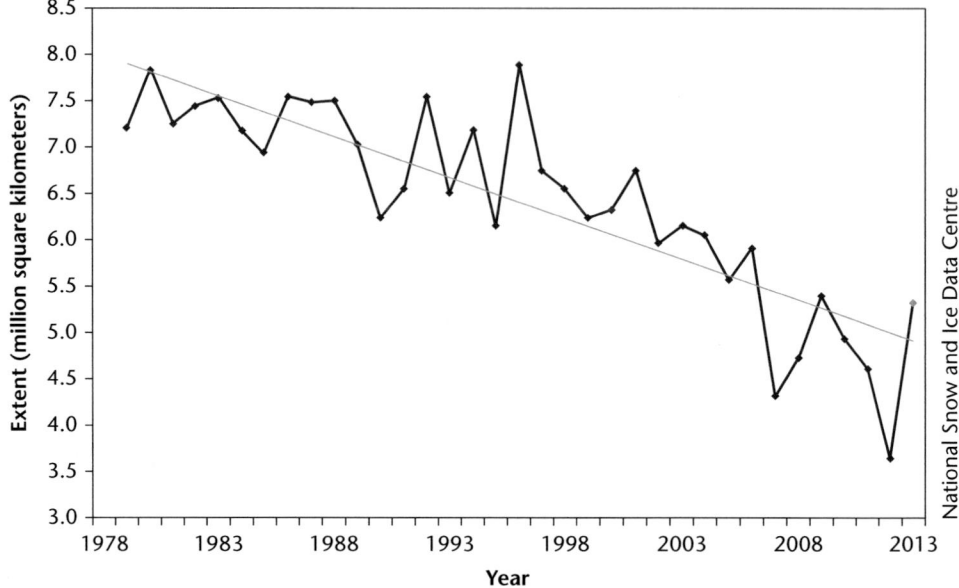

Figure 12.4   The decline in Arctic sea ice extent since 1979. (Image credit National Snow and Ice Data Centre; Source: http://nsidc.org/arcticseaicenews/2013/10/)

ice-free ocean may increase winds and storms and associated mixing, leading to further ocean warming and sea ice loss (Hakkinen et al. 2008).

Arctic Ocean warming and sea ice loss may have impacts that reach far beyond the Arctic. There is evidence that variations in heat exchange between the ocean and atmosphere associated with changing sea ice extent and sea surface temperatures can alter the large-scale structure and moisture content of the atmosphere across the Northern Hemisphere. A number of studies have linked sea ice decline with changes in the Northern Hemisphere jet streams and associated storm tracks, increasing the frequency and persistence of extreme weather events over the Northern Hemisphere, including more severe winters over North America and Europe, with colder temperatures and increased snowfall (Francis et al. 2009; Honda et al. 2009; Deser et al. 2010; Petoukhov and Semenov 2010; Cohen et al. 2012, 2014; Francis and Vavrus 2012; Liu et al. 2012; Yang and Christensen 2012). However, whether sea ice is the primary reason for the more extreme winters of recent times remains contentious (Shepherd 2016). For example, it may be that sea ice is a second order effect and modulates a first order control such as North Atlantic SST (Balmaseda et al. 2010).

Declining sea ice may also influence the exchange of CO_2 between the Arctic Ocean and the atmosphere, with implications for the global climate. The growth of phytoplankton (net primary productivity; NPP) draws CO_2 out of the atmosphere, and there is evidence that NPP in the Arctic Ocean has increased as sea ice extent has declined and nutrient rich surface waters have been exposed to more sunlight (e.g. Arrigo et al. 2008; Pabi et al. 2008; Arrigo and van Dijken 2011). The growth of phytoplankton may be further enhanced by wind-driven mixing and upwelling of nutrient rich waters that can occur under ice-free conditions. Increased exposure of the Arctic's surface waters to the atmosphere as sea ice declines also facilitates air–sea gas exchange promoting the oceanic drawdown of CO_2. Together increased NPP and air–sea gas exchange may explain the increased uptake of CO_2 by the Arctic Ocean over recent year (Bates 2006).

Lawrence et al. (2015) predict a 30% increase in NPP by the end of the century due to an ice-free Arctic. Thus, the loss of Arctic sea ice could open up a new sink for atmospheric CO_2. This sink could be further boosted by enhanced ocean mixing due to a shift in storm tracks to higher latitudes, which may occur as global temperatures rise (Hakkinen et al. 2008). However, other climate feedbacks, such as the increased outgassing of CO_2 associated with warmer surface waters (Lauderdale et al. 2016), could well offset the increased drawdown of CO_2. Or increased surface stratification could mean that the ocean rapidly equilibrates with the atmosphere such that the increased uptake of CO_2 is transitory (Cai et al. 2010). So the overall influence of Arctic sea ice loss on CO_2 and climate remains uncertain.

Summary and outlook

In this chapter I have examined the powerful influence the polar oceans exert on Earth's climate, a power that, given the diminutive dimensions of the polar oceans, is at first surprising. It arises, however, through the essential role the polar oceans play in the ocean's global overturning circulation, which shapes climate by redistributing vast quantities of heat from the Southern Hemisphere to high northerly latitudes and by storing heat and CO_2 in the deep ocean. Only in the polar oceans does water cool sufficiently to form the dense water that feeds the overturning. And only in the Southern Ocean do the conditions prevail that bring this dense water back to the surface. In the absence of these polar ocean processes our climate would be quite different to that which we experience today.

The dramatic changes in the Earth's past climate provide evidence that the global reach of the polar oceans also arises through the intimate relationship that exists between the polar oceans and the ice sheets they encircle. As well as being key to its future stability, the Southern Ocean played a crucial role in the establishment of the Antarctic ice sheet, and in so doing contributed to sea level decline and global cooling, the latter through the impact of the Antarctic ice sheet on Earth's albedo. In turn, the polar oceans furnish the ice sheets with an additional influence over Earth's climate. As ice sheets vary they can inhibit or strengthen the global overturning circulation thereby affecting the global redistribution of heat and CO_2, both of which may drive further ice sheet change.

The power of the polar oceans to affect dramatic changes in Earth's climate and weather patterns also resides in the interdependence between sea ice and ocean temperature. The polar oceans respond to sea ice loss in a way that promotes further sea ice loss, thus amplifying the initial change. In so doing, they afford themselves the opportunity to feed the atmosphere with heat and moisture, potentially altering weather patterns across the Northern Hemisphere, with increased snowfall and colder winters among the impacts.

In this chapter, I have concentrated on the global significance of the polar oceans in terms of their impact on climate. It should not be forgotten, however, that the polar oceans are globally significant in many other respects too. The retreat of sea ice is opening up new shipping routes between the west and Asia, and new opportunities for hydrocarbon and mineral exploration, with implications for the global economy (see Nuttall, Christensen and Siegert in the Introduction to this volume). International political and strategic tensions may be expected as countries compete to exert control over what they see as their territory, or, for countries not directly bordering the Arctic, international waters. Of course, changes in the polar oceans do not have to be felt globally to be of global significance. The loss of sea ice and ocean warming directly impacts on the lives of indigenous populations and unique flora and fauna, such as the iconic polar bear, as does the pollution and disruption caused by increased exploitation of the Arctic region. Loss or damage to these unique human and natural systems represent a grievous loss to humanity as a whole.

We now know that the global influence of the polar oceans arises through a complex web of interactions between the oceans, the atmosphere, sea ice and the ice sheets. Yet, this complexity, compounded by the relative inaccessibility and harsh conditions of the polar oceans, means our knowledge of the polar oceans remains far from complete. While we can be sure that the polar oceans will play a central role in shaping the climate system's dynamic response to global warming, we cannot yet be certain what this response will be. Potential changes include a significant weakening of the global overturning circulation, leading to dramatic changes to regional climate and weather patterns across the globe, increased ice sheet mass loss and global sea level rise, sea ice loss, changes in the oceanic uptake of CO_2 and more extreme winters over the Northern Hemisphere. Addressing the shortcomings in understanding the Polar Regions is made ever more pressing given the impact of such changes on humankind and by the acute sensitivity of the Polar Regions to anthropogenic climate change.

References

Adkins, J.F. 2013. 'The role of deep ocean circulation in setting glacial climates' *Paleoceanography* 28: 539–561.

Arrigo, K.R. and G.L. van Dijken 2011. 'Secular trends in Arctic Ocean net primary production' *J. Geophys. Res.* 116, C09011, doi:10.1029/2011JC007151.

Arrigo, K.R., G. van Dijken and S. Pabi 2008. 'Impact of a shrinking Arctic ice cover on marine primary production' *Geophys. Res. Lett.* 35, L19603, doi:10.1029/2008GL035028.

Azaneu, M., R. Kerr, M.M. Mata and C.A.E. Garcia 2013. 'Trends in the deep Southern Ocean (1958–2010): implications for Antarctic bottom water properties and volume export' *J. Geophys. Res. Oceans* 118: 4213–4227.

Balmaseda M.A., L. Ferranti, F. Molteni, and T. N. Palmer 2010. 'Impact of 2007 and 2008 Arctic ice anomalies on the atmospheric circulation: implications for long-range predictions' *Q. J. R. Meteorol. Soc.* 136: 1655–1664.

Banderas, R., J. Álvarez-Solas and M. Montoya 2012. 'Role of CO2 and Southern Ocean winds in glacial abrupt climate change' *Clim. Past* 8: 1011–1021.

Banderas, R., J. Álvarez-Solas, A. Robinson and M. Montoya 2015. 'An interhemispheric intervention for glacial abrupt climate change' *Clim. Dyn.* 44(9–10): 2897–2908.

Barrett, P.J. 1996. 'Antarctic paleoenvironment through Cenozoic times: a review' *Terr. Antarct.* 3: 103–119.

Bates, N.R. 2006. 'Air-sea CO_2 fluxes and the continental shelf pump of carbon in the Chukchi Sea adjacent to the Arctic Ocean' *J. Geophys. Res.* 111, C10013, doi:10.1029/2005JC003083.

Bindoff, N.L., A.P.S. Wong, and J.A. Church 1999. 'Large-scale freshening of intermediate waters in the Pacific and Indian Oceans' *Nature* 400(6743): 440–443.

Bingham, R.J. and C.W. Hughes 2009. 'Signature of the Atlantic meridional overturning circulation in sea level along the east coast of North America' *Geophys. Res. Lett.* 36, L02603, doi:10.1029/2008GL036215.

Böhm, E., J. Lippold, M. Gutjahr, M. Frank, P. Blaser, B. Antz, J. Fohlmeister, N. Frank, M.B. Andersen and M. Deininger 2015. 'Strong and deep Atlantic meridional overturning circulation during the last glacial cycle' *Nature* 517(7532): 73–76.

Böning, C.W., A. Dispert, M. Visbeck, S.R. Rintoul, and F.U. Schwarzkopf 2008. 'The response of the Antarctic circumpolar current to recent climate change' *Nature Geoscience* 1(12): 864–869.

Boulton, C.A., L.C. Allison, and T.M. Lenton 2014. 'Early warning signals of Atlantic Meridional Overturning Circulation collapse in a fully coupled climate model' *Nature Communications* 5: 5752, http://doi.org/10.1038/ncomms6752.

Broecker, W.S. 1998. 'Paleocean circulation during the last deglaciation: a bipolar seesaw?' *Paleoceanography* 13: 119–121.

Bryden, H.L., H.R. Longworth, and S.A. Cunningham 2005. 'Slowing of the Atlantic meridional overturning circulation at 25°N' *Nature* 438(7068): 655–657.

Buizert, C. and A. Schmittner 2015. 'Southern Ocean control of glacial AMOC stability and Dansgaard-Oeschger interstadial duration' *Paleoceanography* 30: 1595–1612.

Cai W.J., L. Chen, B. Chen, Z. Gao, S.H. Lee, J. Chen, D. Pierrot, K. Sullivan, Y. Wang, X. Hu, W.J. Huang, Y. Zhang, S. Xu, A. Murata, J.M. Grebmeier, E.P. Jones and H. Zhang 2012. 'Decrease in the CO2 uptake capacity in an ice-free Arctic Ocean basin' *Science* 329(5991): 556–559.

Cheng, W. and P.B. Rhines 2004. 'Response of the overturning circulation to high-latitude fresh-water perturbations in the North Atlantic' *Clim. Dyn.* 22(4): 359–372.

Cohen, J., J.A. Screen, J.C. Furtado, M. Barlow, D. Whittleston, D. Coumou, J. Francis, K. Dethloff, D. Entekhabi, J. Overland and J. Jones 2014. 'Recent Arctic amplification and extreme mid-latitude weather' *Nature Geoscience* 7(9): 627–637.

Cohen, J.L., J.C. Furtado, M.A. Barlow, V.A. Alexeev and J.E. Cherry 2012. 'Arctic warming, increasing snow cover and widespread boreal winter cooling' *Environmental Research Letters* 7(1): 014007, http://doi.org/10.1088/1748-9326/7/1/014007.

Cox, K.A., J.D. Stanford, A.J. McVicar, E.J. Rohling, K.J. Heywood, S. Bacon, M. Bolshaw, P.A. Dodd, S. De la Rosa and D. Wilkinson 2010. 'Interannual variability of Arctic sea ice export into the East Greenland Current' *J. Geophys. Res.* 115, C12063, doi:10.1029/2010JC006227.

Coxall, H.K., P.A. Wilson, H. Palike, C.H. Lear and J. Backman 2005. 'Rapid stepwise onset of Antarctic glaciation and deeper calcite compensation in the Pacific Ocean' *Nature* 433: 53–57.

Cunningham, S.A., T. Kanzow, D. Rayner, M.O. Baringer, W.E. Johns, J. Marotzke, H.R. Longworth, E.M. Grant, J.J-M Hirschi, L.M. Beal, C.S. Meinen and H.L. Bryden 2007. 'Temporal variability of the Atlantic Meridional Overturning Circulation at 26.5°N' *Science* 317(5840): 935–938.

DeConto, R.M. and D. Pollard 2003 'Rapid Cenozoic glaciation of Antarctica induced by declining atmospheric CO2' *Nature* 421: 245–249.

de Lavergne, C., J.B. Palter, E.D. Galbraith, R. Bernardello, and I. Marinov 2014. 'Cessation of deep convection in the open Southern Ocean under anthropogenic climate change' *Nature Clim. Change* 4(4): 278–282.

Deser, C., R. Tomas, M. Alexander and D. Lawrence 2010. 'The seasonal atmospheric response to projected Arctic sea ice loss in the late twenty-first century' *J. Climate* 23: 333–351.

Dokken, T.M., K.H. Nisancioglu, C. Li, D.S. Battisti and C. Kissel 2013. 'Dansgaard-Oeschger cycles: interactions between ocean and sea ice intrinsic to the Nordic seas' *Paleoceanography* 28: 491–502.

Downes, S.M., N.L. Bindoff and S.R. Rintoul 2010. 'Changes in the subduction of Southern Ocean water masses at the end of the twenty-first century in eight IPCC models' *J. Climate* 23: 6526–6541.

Downes, S.M. and A. McHugh 2013. 'Southern Ocean circulation and eddy compensation in CMIP5 models' *J. Climate* 26: 7198–7220.

Durack, P.J., S.E. Wijffels and R.J. Matear 2012. 'Ocean salinities reveal strong global water cycle intensification during 1950 to 2000' *Science* 336(6080): 455–458.

Folland, C.K., A.W. Colman, D.P. Rowell and M.K. Davey 2001. 'Predictability of Northeast Brazil rainfall and real-time forecast skill, 1987–98' *Journal of Climate*, http://doi.org/10.1175/1520-0442(2001)014<1937:PONBRA>2.0.CO;2.

Folland, C.K., D.E. Parker and T.N. Palmer 1986. 'Sahel rainfall and worldwide sea temperatures 1901–85' *Nature* 320: 602– 607.

Frajka-Williams, E., S.A. Cunningham, H. Bryden and B.A. King 2011. 'Variability of Antarctic bottom water at 24.5°N in the Atlantic' *J. Geophys. Res.* 116, C11026, doi:10.1029/2011JC007168.

Francis, J.A., W. Chan, D.J. Leathers, J.R. Miller and D.E. Veron 2009. 'Winter Northern Hemisphere weather patterns remember summer Arctic sea-ice extent' *Geophys. Res. Lett.* 36, L07503, doi:10.1029/2009GL037274.

Francis, J.A. and S.J. Vavrus 2012. 'Evidence linking Arctic amplification to extreme weather in mid-latitudes' *Geophys. Res. Lett.* 39, L06801, doi:10.1029/2012GL051000.

Frankcombe, L.M. and H.A. Dijkstra 2011. 'The role of Atlantic-Arctic exchange in North Atlantic multi-decadal climate variability' *Geophys. Res. Lett.* 38, L16603, doi:10.1029/2011GL048158.

Frölicher, T.L., J.L. Sarmiento, D.J. Paynter, J.P. Dunne, J.P. Krasting and M. Winton 2015. 'Dominance of the Southern Ocean in anthropogenic carbon and heat uptake in CMIP5 models' *J. Climate* 28: 862–886.

Fyke, J.G., M. D'Orgeville and A.J. Weaver 2015. 'Drake Passage and Central American seaway controls on the distribution of the oceanic carbon reservoir' *Global and Planetary Change* 128: 72–82.

Galeotti, S., R. DeConto, T. Naish, P. Stocchi, F. Florindo, M. Pagani, P. Barrett, S.M. Bohaty, L. Lanci, D. Pollard, S. Sandroni, F.M. Talarico and J.C. Zachos 2016. 'Antarctic ice sheet variability across the Eocene-Oligocene boundary climate transition' *Science* 1 (April): 76–80.

Ganopolski, A. and S. Rahmstorf 2001. 'Rapid changes of glacial climate simulated in a coupled climate model' *Nature* 409: 153–158.

Giles, K.A., S.W. Laxon, A.L. Ridout, D.J. Wingham and S. Bacon 2012. 'Western Arctic Ocean freshwater storage increased by wind-driven spin-up of the Beaufort Gyre' *Nature Geosci.* 5(3): 194–197.

Goldberg, S.B., C.W. Landsea, A.M. Mestas-Nunez and W.M. Gray 2001. 'The recent increase in Atlantic hurricane activity: causes and implications' *Science* 293: 474–479.

Goldner, A., N. Herold and M. Huber 2014. 'Antarctic glaciation caused ocean circulation changes at the Eocene–Oligocene transition' *Nature* 511(7511): 574–577.

Goosse, H., J. Campin, T. Fichefet and E. Deleersnijder 1997. 'Sensitivity of a global ice–ocean model to the Bering Strait throughflow' *Clim. Dyn.* 13(5): 349–358.

Haine, T.W.N., B. Curry, R. Gerdes, E. Hansen, M. Karcher, C. Lee, B. Rudels, G. Spreen, L. de Steur, K.D. Stewart and R. Woodgate 2015. 'Arctic freshwater export: status, mechanisms, and prospects' *Global and Planetary Change* 125: 13–35.

Hakkinen, S., A. Proshutinsky, and I. Ashik 2008. 'Sea ice drift in the Arctic since the 1950s' *Geophys. Res. Lett.* 35, L19704, doi:10.1029/2008GL034791.

Hansen J., M. Sato, G. Russell and P. Kharecha 2013. 'Climate sensitivity, sea level and atmospheric carbon dioxide' *Phil Trans R Soc A* 371: 20120294, http://dx.doi.org/10.1098/rsta.2012.0294.

Hasumi, H. 2002. 'Sensitivity of the global thermohaline circulation to interbasin freshwater transport by the atmosphere and the Bering Strait throughflow' *J. Climate* 15: 2516–2526.

Hattermann, T. and A. Levermann 2010. 'Response of Southern Ocean circulation to global warming may enhance basal ice shelf melting around Antarctica' *Clim. Dyn.* 35(5): 741–756.

Haumann, F.A., N. Gruber, M. Munnich, I. Frenger and S. Kern 2016. 'Sea-ice transport driving Southern Ocean salinity and its recent trends' *Nature* 537(7618): 89–92.

Hawkins, E., R.S. Smith, L.C. Allison, J.M. Gregory, T.J. Woollings, H. Pohlmann and B. de Cuevas 2011. 'Bistability of the Atlantic overturning circulation in a global climate model and links to ocean freshwater transport' *Geophys. Res. Lett.* 38, L10605, doi:10.1029/2011GL047208.

Hawkins, E. and R. Sutton 2007. 'Variability of the Atlantic thermohaline circulation described by three-dimensional empirical orthogonal functions' *Clim. Dyn.* 29(7–8): 745–762.

Helm, K.P., N.L. Bindoff and J.A. Church 2010. 'Changes in the global hydrological-cycle inferred from ocean salinity' *Geophys. Res. Lett.* 37, L18701, doi:10.1029/2010GL044222.

Holbourn, A., W. Kuhnt, M. Schulz and H. Erlenkeuser 2005. 'Impacts of orbital forcing and atmospheric carbon dioxide on Miocene ice-sheet expansion' *Nature* 438: 483–487.

Honda, M., J. Inoue and S. Yamane 2009. 'Influence of low Arctic sea-ice minima on anomalously cold Eurasian winters' *Geophys. Res. Lett.* 36, L08707, doi:10.1029/2008GL037079.

Jackson, L.C., R. Kahana, T. Graham, M.A. Ringer, T. Woollings, J.V. Mecking and R.A. Wood 2015. 'Global and European climate impacts of a slowdown of the AMOC in a high resolution GCM' *Clim. Dyn.* 45(11–12): 3299–3316.

Jackson, L.C. and M. Vellinga 2013. 'Multidecadal to centennial variability of the AMOC: HadCM3 and a perturbed physics ensemble' *J. Climate* 26: 2390–2407.

Jacobs, S.S. and C.F. Giulivi 2010. 'Large multidecadal salinity trends near the Pacific–Antarctic continental margin' *J. Climate* 23: 4508–4524.

Jacobs, S.S., C.F. Giulivi and P.A. Mele 2002. 'Freshening of the Ross Sea during the late 20th century' *Science* 297(5580): 386–389.

Jahn, A., B. Tremblay, L.A. Mysak and R. Newton 2010. 'Effect of the large-scale atmospheric circulation on the variability of the Arctic Ocean freshwater export' *Clim. Dyn.* 34: 201–222

Jungclaus, J.H., H. Haak, M. Latif and U. Mikolajewicz 2005. 'Arctic–North Atlantic interactions and multidecadal variability of the Meridional Overturning Circulation' *J. Climate* 18: 4013–4031.

Kanzow, T., S.A. Cunningham, W.E. Johns, J.J-M. Hirschi, J. Marotzke, M.O. Baringer, C.S. Meinen, M.P. Chidichimo, C. Atkinson, L.M. Beal, H.L. Bryden and J. Collins 2010. 'Seasonal variability of the Atlantic Meridional Overturning Circulation at 26.5°N' *J. Climate* 23: 5678–5698.

Karcher, M., R. Gerdes, F. Kauker, C. Koberle and I. Yashayaev 2005. 'Arctic Ocean change heralds North Atlantic freshening' *Geophys. Res. Lett.* 32, L21606, doi:10.1029/2005GL023861.

Katz, M.E., K.G. Miller, J.D. Wright, B.S. Wade, J.V. Browning, B.S. Cramer and Y. Rosenthal 2008. 'Stepwise transition from the Eocene greenhouse to the Oligocene icehouse' *Nature Geoscience* 1(5): 329–334.

Kender S., J. Yu and V.L. Peck 2014. 'Deep ocean carbonate ion increase during mid Miocene CO2 decline' *Scientific Reports* 4(4): 187, doi:10.1038/srep04187.

Kennett, J.P. 1977. 'Cenozoic evolution of Antarctic glaciation, the circum-Antarctic Ocean, and their impact on global paleoceanography' *J. Geophys. Res.* 82(27): 3843–3860.

King, M.A., R.J. Bingham, P. Moore, P.L. Whitehouse, M.J. Bentley and G.A. Milne 2012. 'Lower satellite-gravimetry estimates of Antarctic sea-level contribution' *Nature* 491: 586–589.

Kouketsu, S., T. Doi, T. Kawano, S. Masuda and N. Sugiura 2011. 'Deep ocean heat content changes estimated from observation and reanalysis product and their influence on sea level change' *J. Geophys. Res.* 116, C03012, doi:10.1029/2010JC006464.

Kürschner, W.M., Z. Kvaček and D.L. Dilcher 2008. 'The impact of Miocene atmospheric carbon dioxide fluctuations on climate and the evolution of terrestrial ecosystems' *PNAS* 105: 449–453.

Lauderdale, J.M., S. Dutkiewicz, R.G. Williams and M.J. Follows 2016. 'Quantifying the drivers of ocean-atmosphere CO_2 fluxes' *Global Biogeochem. Cycles* 30: 983–999.

Lawrence, J., E. Popova, A. Yool and M. Srokosz 2015. 'On the vertical phytoplankton response to an ice-free Arctic Ocean' *J. Geophys. Res. Oceans* 120: 8571–8582.

Lear, C.H., T.R. Bailey, P.N. Pearson, H.K. Coxall and Y. Rosenthal 2008. 'Cooling and ice growth across the Eocene–Oligocene transition' *Geology* 36: 251–254.

Lear, C.H., H. Elderfield and P.A. Wilson 2000. 'Cenozoic deep-sea temperatures and global ice volumes from Mg/Ca in benthic foraminiferal calcite' *Science* 287: 269–272.

Le Quéré, C., C. Rödenbeck, E.T. Buitenhuis, T.J. Conway, R. Langenfelds, A. Gomez, et al. 2007. 'Saturation of the Southern Ocean CO2 sink due to recent climate change' *Science* 316(5832): 1735–1738.

Levermann, A., A. Griesel, M. Hofmann, M. Montoya and S. Rahmstorf 2005. 'Dynamic sea level changes following changes in the thermohaline circulation' *Clim. Dyn.* 24(4): 347–354.

Liu, J., J.A. Curry, H. Wang, M. Song and R.M. Horton 2012. 'Impact of declining Arctic sea ice on winter snowfall' *Proceedings of the National Academy of Sciences* 109(11): 4074–4079.

Long, Z., W. Perrie, C.L. Tang, E. Dunlap and J. Wang 2012. 'Simulated interannual variations of freshwater content and sea surface height in the Beaufort Sea' *J. Climate* 25: 1079–1095.

Manabe, S. and R.J. Stouffer 1980. 'Sensitivity of a global climate model to an increase of CO2 concentration in the atmosphere' *Journal of Geophysical Research: Oceans* 85(C10): 5529–5554.

Manabe, S. and R.J. Stouffer 1994. 'Multiple-century response of a coupled ocean-atmosphere model to an increase of atmospheric carbon dioxide' *J. Climate* 7: 5–23.

Meehl, G.A. et al. 2007. 'Global climate projections' in *Climate Change 2007: The Physical Science Basis. Contribution of Working Group I to the Fourth Assessment Report of the Intergovernmental Panel on Climate Change*, edited by S. Solomon et al. Cambridge, UK: Cambridge University Press. 747–845.

Meijers, A.J.S., E. Shuckburgh, N. Bruneau, J-B. Sallee, T.J. Bracegirdle and Z. Wang 2012. 'Representation of the Antarctic Circumpolar Current in the CMIP5 climate models and future changes under warming scenarios' *J. Geophys. Res.* 117, C12008, doi:10.1029/2012JC008412.

Menary, M.B., W. Park, K. Lohmann et al. 2012. 'A multimodel comparison of centennial Atlantic meridional overturning circulation variability' *Clim. Dyn.* 38(1–2): 2377–2388.

Miller, K.G. et al. 2005. 'The phanerozoic record of global sea-level change' *Science* 310: 1293–1298.

Miller, K.G. et al. 2009. 'Climate threshold at the Eocene-Oligocene transition: Antarctic ice sheet influence on ocean circulation' in C. Koeberl and A. Montanari (eds) *The Late Eocene Earth? Hothouse, Icehouse, and Impacts* (Geol. Soc. Am. Spec. Pap. 452). 169–178.

Miller, K.G., R.G. Fairbanks and G.S. Mountain 1987. 'Tertiary oxygen isotope synthesis, sea level history, and continental margin erosion' *Paleoceanography* 2: 1–19.

Miller, K.G., J.D. Wright and R.G. Fairbanks 1991. 'Unlocking the ice house: Oligocene–Miocene oxygen isotopes, eustasy, and margin erosion' *J. Geophys. Res.* 96: 6829–6848.

Nakayama, Y., R. Timmermann, C.B. Rodehacke, M. Schröder and H.H. Hellmer 2014. 'Modeling the spreading of glacial meltwater from the Amundsen and Bellingshausen Seas' *Geophys. Res. Lett.* 41: 7942–7949.

Pabi, S., G.L. van Dijken and K.R. Arrigo 2008. 'Primary production in the Arctic Ocean, 1998–2006' *J. Geophys. Res.* 113, C08005, doi:10.1029/2007JC004578.

Pagani, M., M.A. Arthur and K.H. Freeman 1999. 'Miocene evolution of atmospheric carbon dioxide' *Paleoceanography* 14: 273–292.

Patara, L. and C.W. Böning 2014. 'Abyssal ocean warming around Antarctica strengthens the Atlantic overturning circulation' *Geophys. Res. Lett.* 41: 3972–3978.

Petersen, S.V., D.P. Schrag and P.U. Clark 2013. 'A new mechanism for Dansgaard-Oeschger cycles' *Paleoceanography* 28: 24–30.

Petoukhov, V. and V.A. Semenov 2010. 'A link between reduced Barents-Kara sea ice and cold winter extremes over northern continents' *J. Geophys. Res.* 115, D21111, doi:10.1029/2009JD013568.

Polyakov, I.V., L.A. Timokhov, V.A. Alexeev, S. Bacon, I.A. Dmitrenko, L. Fortier, I.E. Frolov, J-C. Gascard, E. Hansen, V.V. Ivanov, S. Laxon, C. Mauritzen, D. Perovich, K. Shimada, H.L. Simmons, V.T. Sokolov, M. Steele and J. Toole 2010. 'Arctic Ocean warming contributes to reduced polar ice cap' *J. Phys. Oceanogr.* 40: 2743–2756.

Proshutinsky, A., R.H. Bourke and F.A. McLaughlin 2002. 'The role of the Beaufort Gyre in Arctic climate variability: seasonal to decadal climate scales' *Geophys. Res. Lett.* 29(23): doi:10.1029/2002GL015847.

Proshutinsky, A., R. Krishfield, M-L. Timmermans, J. Toole, E. Carmack, F. McLaughlin, W.J. Williams, S. Zimmermann, M. Itoh and K. Shimada 2009. 'Beaufort Gyre freshwater reservoir: state and variability from observations' *J. Geophys. Res.* 114, C00A10, doi:10.1029/2008JC005104.

Purkey, S.G. and G.C. Johnson 2013. 'Antarctic bottom water warming and freshening: contributions to sea level rise, ocean freshwater budgets, and global heat gain' *J. Climate* 26: 6105–6122.

Rahmstorf, S., J.E. Box, G. Feulner, M.E. Mann, A. Robinson, S. Rutherford and E.J. Schaffernicht 2015. 'Exceptional twentieth-century slowdown in Atlantic Ocean overturning circulation' *Nature Clim. Change* 5(5): 475–480.

Raymo, M.E. and W.F. Ruddiman 1992. 'Tectonic forcing of late Cenozoic climate' *Nature* 359: 117–122.

Roberts, C.D., L. Jackson and D. McNeall 2014. 'Is the 2004–2012 reduction of the Atlantic Meridional Overturning Circulation significant?' *Geophys. Res. Lett.* 41: 3204–3210.

Sabine, C.L., R.A. Feely, N. Gruber, R.M. Key and K. Lee 2004. 'The oceanic sink for anthro-pogenic CO2' *Science* 305: 367–371.

Saenko, O.A., J.C. Fyfe and M.H. England 2005. 'On the response of the oceanic wind-driven circulation to atmospheric CO2 increase' *Clim. Dyn.* 25(4): 415–426.

Schrama, E.J.O. and B. Wouters 2011. 'Revisiting Greenland Ice Sheet mass loss observed by GRACE' *J. Geophys. Res.* 116, B02407, doi:10.1029/2009JB006847.

Screen, J.A. and I. Simmonds 2010. 'The central role of diminishing sea ice in recent Arctic temperature amplification' *Nature* 464(7293): 1334–1337.

Screen, J.A., I. Simmonds, C. Deser and R. Tomas 2013. 'The atmospheric response to three decades of observed Arctic sea ice loss' *J. Climate* 26: 1230–1248.

Seager, R. and D.S. Battisti 2007. 'Challenges to our understanding of the general circulation: abrupt climate change' in T. Schneider and A.H. Sobel (eds) *The Global Circulation of the Atmosphere*. Princeton, NJ: Princeton University Press. 331–371.

Serreze, M.C. and R.G. Barry 2011. 'Processes and impacts of Arctic amplification: a research synthesis' *Global and Planetary Change* 77(1–2): 85–96.

Serreze, M.C., M.M. Holland and J. Stroeve 2007. 'Perspectives on the Arctic's shrinking sea-ice cover' *Science* 315(5818): 1533–1536.

Shaffer, G. and J. Bendtsen 1994. 'Role of the Bering Strait in controlling North Atlantic Ocean circulation and climate' *Nature* 367: 354–357.

Shepherd, A., E.R. Ivins, A. Geruo, V.R. Barletta, M.J. Bentley, S. Bettadpur, et al. 2012. 'A reconciled estimate of ice-sheet mass balance' *Science* 338(6111): 1183–1189.

Shepherd, T.G. 2016. 'Effects of a warming Arctic' *Science* 353(6303): 989–990.

Shevenell, A.E., J.P. Kennett and D.W. Lea 2004. 'Middle Miocene Southern Ocean cooling and Antarctic cryosphere expansion' *Science* 305(5691): 1766–1770.

Shevenell, A.E., J.P. Kennett and D.W. Lea 2008. 'Middle Miocene ice sheet dynamics, deep-sea tem-peratures, and carbon cycling: a Southern Ocean perspective' *Geochem. Geophys. Geosyst.* 9, Q02006, doi:10.1029/2007GC001736.

Sigman, D.M., S.L. Jaccard and G.H. Haug 2004. 'Polar ocean stratification in a cold climate' *Nature* 428(6978): 59–63.

Sigmond, M., M.C. Reader, J.C. Fyfe and N.P. Gillett 2011. 'Drivers of past and future Southern Ocean change: stratospheric ozone versus greenhouse gas impacts' *Geophys. Res. Lett.* 38, L12601, doi:10.1029/2011GL047120.

Smedsrud, L.H., A. Sirevaag, K. Kloster, A. Sorteberg and S. Sandven 2011. 'Recent wind driven high sea ice area export in the Fram Strait contributes to Arctic sea ice decline' *The Cryosphere* 5: 821–829.

Smeed, D.A., G.D. McCarthy, S.A. Cunningham, E. Frajka-Williams, D. Rayner, W. E. Johns, C.S. Meinen, M.O. Baringer, B.I. Moat, A. Duchez and H.L. Bryden 2014. 'Observed decline of the Atlantic meridional overturning circulation 2004–2012' *Ocean Sci.* 10: 29–38.

Spielhagen, R.F., K. Werner, S.A. Sørensen, K. Zamelczyk, E. Kandiano, G. Budeus, K. Husum, T.M. Marchitto and M. Hald 2011. 'Enhanced modern heat transfer to the Arctic by warm Atlantic water' *Science* 331(6016): 450–453.

Steele, M., W. Ermold and J. Zhang 2008. 'Arctic Ocean surface warming trends over the past 100 years' *Geophys. Res. Lett.* 35, L02614, doi:10.1029/ 2007GL031651.

Stewart, K.D. and T.W.N. Haine 2013. 'Wind-driven Arctic freshwater anomalies' *Geophys. Res. Lett.* 40: 6196–6201.

Stocker, T.F. and S.J. Johnsen 2003. 'A minimum thermodynamic model for the bipolar seesaw' *Paleoceanography* 18(4): 1087, doi:10.1029/2003PA000920.

Stroeve, J.C., M.C. Serreze, M.M. Holland, J.E. Kay, J. Malanik and A.P. Barrett 2011. 'The Arctic's rapidly shrinking sea ice cover: a research synthesis' *Climatic Change* 110(3–4): 1005–1027.

Sun, Y., S.C. Clemens, C. Morrill, X. Lin, X. Wang and Z. An 2012. 'Influence of Atlantic meridional overturning circulation on the East Asian winter monsoon' *Nature Geosci* 5(1): 46–49.

Sutton, R.T. and D.L.R. Hodson 2005. 'Atlantic Ocean forcing of North American and European summer climate' *Science* 309: 115–118.

Talley, L.D. 2013. 'Closure of the global overturning circulation through the Indian, Pacific, and Southern Oceans: schematics and transports' *Oceanography* 26(1): 80–97.

Tsukernik, M., C. Deser, M. Alexander and R. Thomas 2010. 'Atmospheric forcing of Fram Strait sea ice export: a closer look', *Clim. Dyn.* 35: 1349–1360.

Velicogna, I. 2009. 'Increasing rates of ice mass loss from the Greenland and Antarctic Ice Sheets revealed by GRACE' *Geophys. Res. Lett.* 36, L19503. doi:10.1029/2009GL040222

Vellinga, M. and P. Wu 2004. 'Low-latitude freshwater influence on centennial variability of the Atlantic thermohaline circulation' *J. Clim.* 17(23): 4498–4511.

Vincent, E. and W.H. Berger 1985. 'Carbon dioxide and polar cooling in the Miocene' in E.T. Sundquist and W.S. Broecker (eds) *The Carbon Cycle and Atmospheric CO2: Natural Variations Archean to Present*, Geophys. Monogr, vol. 32. AGU, Washington, DC. 455–468.

Wang, M. and J.E. Overland 2012. 'A sea ice free summer Arctic within 30 years: an update from CMIP5 models' *Geophys. Res. Lett.* 39, L18501, doi:10.1029/2012GL052868.

Woodard, S.C., Y. Rosenthal, K.G. Miller, J.D. Wright, B.K. Chiu and K.T. Lawrence 2014. 'Antarctic role in Northern Hemisphere glaciation' *Science* 346(6211): 847–851.

Wu, P. and R. Wood 2008. 'Convection induced long term freshening of the subpolar North Atlantic Ocean' *Clim. Dyn.* 31(7–8): 941–956.

Wunsch, C. 2006. 'Abrupt climate change: an alternative view' *Quaternary Research* 65(2): 191–203.

Yang, S. and J.H. Christensen 2012. 'Arctic sea ice reduction and European cold winters in CMIP5 climate change experiments' *Geophys. Res. Lett.* 39, L20707, https://doi.org/10.1029/2012GL053338.

Yin, J., M.E. Schlesinger and R.J. Stouffer 2009. 'Model projections of rapid sea-level rise on the northeast coast of the United States' *Nature Geosci* 2(4): 262–266.

Zachos, J.C., M. Pagani, L. Sloan, E. Thomas and K. Billups 2001. 'Trends, rhythms, and aberrations in global climate change 65 Ma to present' *Science* 292: 686–693.

Zachos, J.C., T.M. Quinn and S. Salamy 1996. 'High resolution (104 yr) deep-sea foraminiferal stable isotope records of the earliest Oligocene climate transition' *Paleoceanography* 9: 353–387.

Zhang, R. and T.L. Delworth 2006. 'Impact of Atlantic multidecadal oscillations on India/Sahel rainfall and Atlantic hurricanes' *Geophys. Res. Lett.* 33, L17712, doi:10.1029/2006GL026267.

13

Polar sea ice as a barometer and driver of change

Jeremy Wilkinson and Julienne Stroeve

Introduction

Sea ice is the result of the freezing of the sea surface, and as such it is a phenomenon that is generally restricted to the cold, high latitude regions of the Earth. Even though sea ice is found in some of the most remote regions of the planet, it is a particularly important substance that covers about 15% of the world's oceans during part of the year.[1] Sea ice limits the transfer of heat, momentum and gases between the ocean and atmosphere. Its growth and melt process influences ocean salinity, stratification and global ocean circulation, and it provides an important ecological niche for many species, including humans. Perhaps the most important climate-related property is that the white sea ice reflects almost 80% of the incoming solar radiation, so the less sea ice there is the more sunlight reaches the ocean (Parkinson 2014). This in turn warms the ocean and amplifies the warming of the polar regions (Serreze et al. 2009; Screen and Simmonds 2010; Bingham, Chapter 12 of this volume). Since the characteristics of sea ice are significantly influenced by small changes in either the atmosphere or the ocean, sea ice is a very sensitive component of the polar environment and global climate system (Walsh 2013), and as such is known as a key climate indicator. In this chapter the various aspects of sea ice formation and melt are described, before the Arctic and Antarctic sea ice systems, and the changes to the sea ice that are occurring there, are summarised.

Sea ice formation

When heat is lost from the ocean surface to the atmosphere, the upper-most layer of the ocean cools, becomes denser and sinks, which in turn causes an upwelling of warmer water from below. This process is known as convective overturning. As more heat is lost to the atmosphere, these convective cells will continue to bring deeper, warmer water to the surface (Weeks and Ackley 1986). This will continue until the freezing point of the water column, determined by its salinity, is reached. Unlike freshwater, which has a freezing temperature of 0°C, seawater contains salt which depresses the freezing point. The exact temperature at which the sea-surface freezes depends on its salinity; the greater the salinity the lower the freezing point. Generally speaking, the sea freezes when its temperature reaches about −1.8°C. In theory, the entire water column, from top to bottom, must be cooled to the freezing temperature before ice can begin to form. This is a tricky task when the ocean may be thousands of metres deep. However, in practice, the ocean is stratified into layers of increasing density, thus it is only the upper layer, known as the mixed layer, which needs to be reduced to its freezing point.

Once the mixed layer is cooled to freezing any additional heat loss will cause the formation of small ice crystals known as frazil ice. As more frazil crystals form they produce what appears to be an oily sheen on the surface of the ocean, often referred to as grease ice. This oily-like appearance results from the frazil ice dampening the wind induced ripples on the sea, thus producing a smooth surface that resembles an oil spill on the water surface. Frazil inhibits breaking waves, which in turn decreases the turbulence in the upper ocean. If both the wave and wind effects are reduced, the agitation of the frazil ceases and the frazil crystals begin to consolidate. As the upper frazil layers are exposed to the cold atmosphere, they preferentially fuse together to form a thin, near-continuous ice sheet known as nilas. The transformation of frazil to nilas can occur within several hours (Zubov 1945). Nilas layers are transparent at first and thus appear dark, but as they thicken their colour changes to greyish-white. Due to its small thickness, when this ice type moves sections frequently raft under each other. As the cold air temperatures are conducted through the nilas, further thickening will continue with seawater freezing directly to the underside of the ice. The sea ice will continue to grow in this manner until the following summer melt period begins.

However, should wind decrease but swell persist, the frazil crystals will continue to be agitated by these waves. Under the motion of the waves this newly-formed layer of ice cannot form a continuous sheet of nilas but is broken into small rounded-shaped pieces normally a few tens of centimetres in diameter, known as pancake ice (Leonard et al. 1998). Interestingly the covering of the sea surface with pancakes and frazil ice further diminishes penetrating waves, which in turn produces less flexure on individual pancakes, and allows the freezing together of smaller ones. Once the incoming wave energy has decayed sufficiently, all pancakes will freeze together to form an extensive sheet. From then, further thickening will occur through the freezing of seawater on the bottom of the ice, as described above.

At this early stage the sea ice is particularly vulnerable to fracturing by weather. For example, wind and wave energy from a storm system could break the ice into a number of larger pieces known as floes. Between the floes will be a combination of broken ice of varying sizes, known as brash ice, and possibly newly formed pancake and frazil ice. When the storm energy subsides, this mixture of different ice types and sizes will freeze together to form a new, but lumpier, sheet of ice.

An important process of the freezing of seawater is the expulsion of salt. Sea ice is not pure ice. As it grows, salts, known as brine, are trapped in small pockets. These pockets form a network of channels through the ice that develop over time as the ice thickens and ages. In fact, most of the salt is ejected back in the sea, a process known as brine drainage (Eicken 2003). Brine drainage continues throughout the duration of the ice; young sea ice has a salinity of around 12 parts per thousand, ice older than a year may have a salinity as low as 1 or 2 parts per thousand, whereas the surrounding seawater is around 34 parts per thousand. Generally speaking, the older the ice the less salt it holds. The ejected brine increases the salinity of the surrounding sea water, making it denser and thus more likely to sink to greater depths. It is this sinking of dense water in the Polar Regions that is one of the mechanisms that drives the circulation of the world oceans (see Bingham, Chapter 12 of this volume).

Once a continuous sheet of sea ice is formed, it is collectively known as first year ice (FYI). Generally, ice formed at the start of the winter is around 1 to 2 m thick by the beginning of the following melt season and the surface appears relatively smooth. FYI that survives the summer melt then becomes known as multiyear ice (MYI). MYI can be many years old and is generally thick and heavily deformed. The morphology of the underside also differs; FYI is regarded as smooth whereas MYI is quite rugged as a result of the differential melt/growth rates throughout the season and the effects of convergence pressure.

In summary, sea ice forms when the ocean surface reaches its salinity-dependent freezing point. Once this occurs, the development of the sea ice cover can be in two ways. The first is in calm conditions, which is known as the frazil–nilas–ice sheet cycle, the second in more tempestuous conditions and is known as the frazil–pancake–ice sheet cycle.

Snow on sea ice

Once sea ice forms a stable platform on the surface of the ocean, the falling snow accumulates on its surface. On level FYI, the wind may redistribute the snow into dune-like features known as sastrugi. Surface features, such as pressure ridges (see below), trap more snow than level surfaces. Snow is an insulator, reducing the heat exchange between the atmosphere and ice–ocean interface. Depending on its thickness in winter, it reduces the rate of sea ice growth, and in summer delays the melting of the ice. A covering of snow inhibits the penetration of light to the ice surface and water below and thus plays a fundamental role in the availability of light for photosynthesis.

Sea ice movement

Sea ice is almost constantly drifting with the winds and currents (Leppäranta 2005). Even in the middle of winter it is not unusual for sea ice to drift 30 km or more in one day. However, under certain conditions sea ice is attached to the shore. This ice type is known as land fast ice. Fast ice is usually a seasonal sea ice cover that is immobilised due to the geometry of the coast or by anchoring points such as small islands, rocks, grounded sea ice ridges or icebergs. Fast ice is usually in an undeformed state, although the region between the fast ice and the drifting pack may become highly deformed.

Drifting ice may be pushed together by converging winds, causing ice pressure forming ridges or, conversely, torn apart by diverging winds to leave regions of open water (known as leads). Deformed ice is a ubiquitous feature of sea ice in the Polar Regions. Both thin and thick ice can collide to form ridges – long, linear-like features composed of blocks of ice, piled both above (known as a ridge sail) and below (known as a ridge keel) the ice surface. Ridge sails may be as much as a few metres high, while keels may reach depths of 50 m or more (Wadhams 2000). Ridge sails, as their name suggests, tend to catch the wind. Because sea ice has a greater surface roughness than the open ocean, up to 70% of the drift of ice is due to winds (Thorndike and Colony 1982). The processes affecting the morphology of sea ice continue throughout its life, and therefore morphology of sea ice is affected by the conditions under which it formed as well as the thermal (growth and melt) and dynamic (ridges and lead formation) processes. This results in a complex, highly heterogeneous sea ice cover that varies on a scale of less than a metre to hundreds of kilometres.

Summer sea ice processes

As summer approaches, the sunlit period becomes longer, the air temperatures increase and the ice steadily retreats along all longitudes. In summer, the snow and sea ice melt, and ridges become smaller and more rounded. Thinner ice types normally entirely melt during the summer, and sea ice persists only in higher latitudes, leaving behind an increasing region of open ocean. Between the remaining sea ice pack and the extending open water region is an area known as the marginal ice zone (MIZ). This region is influenced by the proximity to the open ocean, and as a result it has quite different properties to the interior pack (Wadhams 1986).

Wave and wind induced break-up leaves the ice as a mixture of broken floes and brash. The ice within the MIZ is extremely mobile (driven by winds and currents) and, consequently, the ice concentration, i.e. the amount of open water surrounding the floes, is constantly varying.

In the Arctic the melting snow cover forms melt ponds, and these play a critical role in enhancing the melt of sea ice by changing its albedo (Polashenski et al. 2012). The impact of melt ponds on the ice albedo feedback is central to the seasonal evolution and predictability of ice decay in the Arctic (Schroeder et al. 2014), whereas in the Antarctic they are not a common feature because the melt processes are different. There is a natural buffer to summer sea ice processes because as the summer progresses into autumn, the Sun's elevation declines until it reaches an angle where incoming solar radiation does not produce a significant warming of the ocean surface, after which the formation of sea ice begins again and eventually the sun dips below the horizon.

The Arctic sea ice

The most common definition of the Arctic is the region above the Arctic Circle (66° 34' N), an imaginary line that marks the latitude above which the sun does not rise on the winter solstice. However, Arctic sea ice in winter extends well beyond this boundary (Figure 13.1(a)). Generally, Arctic sea ice covers about 14–16 million km² of the ocean surface during the northern hemisphere winter, extending as far south as Newfoundland, Canada (50°N), and to Bohai Bay, China (38°N). In western Europe, however, warm ocean currents help keep the northern coast of Norway at 70°N generally ice free. During summer, the sea ice retreats to within the Arctic Ocean, or to about 6 and 7 million km² in summer (Figure 13.1(b)).

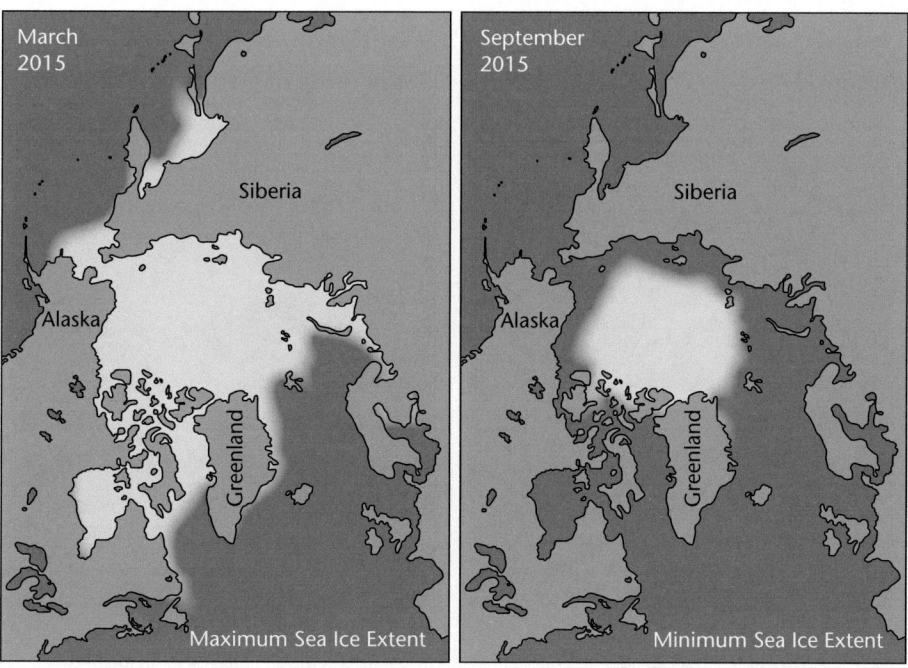

Figure 13.1   Climatological average extent of Arctic sea ice during March (a) and September (b). Figure by Julienne Streove drawing on data provided by the National Snow and Ice Data Center.

Our knowledge of sea ice variability and change largely comes from the modern passive microwave satellite data record, which began in the late 1970s. Given the large dielectric contrast between open water and sea ice, polar regions can be mapped even during polar night and in the presence of cloud cover using frequencies in the microwave portion of the electromagnetic spectrum. Based on this data record, it has become clear that the Arctic is undergoing a period of profound sea ice loss. The changes have been most prominent in summer, with trends during September (the end of the melt season), in the order of $-86,000$ km^2/yr, or a rate of -13.2% per decade from 1979 to 2017. Since 2007, the minimum sea ice extent has fallen below 5 million km^2 each year, with the lowest extent reached on 16 September 2012 of 3.41 million km^2. Thus, today the Arctic has about 40–50% less sea ice in summer than it did in the 1980s and 1990s. Spatially the summer ice losses have been dominated by large reductions within the Beaufort, Chukchi, and East Siberian seas (Figure 13.2).

Changes in winter have been much smaller, dominated by reductions within the Barents Sea and the Sea of Okhotsk. While the trend in the maximum winter extent is about half that for the September minimum (~43,000 km^2/yr or -2.7% per decade), the last three winters (2014–17) have seen consecutive record low values for the maximum extent, dropping below 14 million km^2 (Figure 13.3).

While these changes have received lots of media attention, the data record is still relatively short. Considerable effort has gone into extending this data record back in time, by incorporating earlier satellite data, aircraft and ship observations, whaling records, and ice chart records. A new data set released by the National Snow and Ice Data Center provides mid-monthly estimates of sea ice concentration back to 1850. These data show how the changes during the 2010s are unprecedented in at least 150 years and now encompass all calendar months of the year.

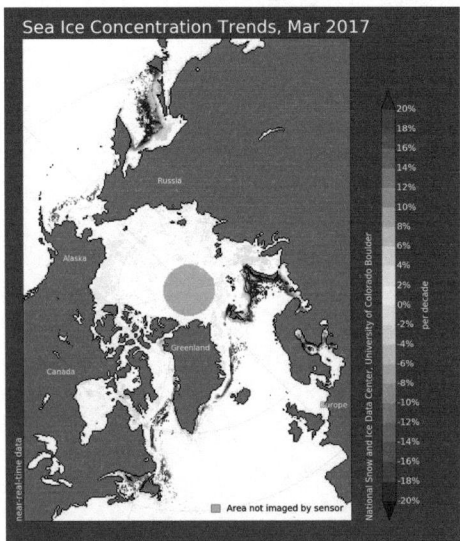 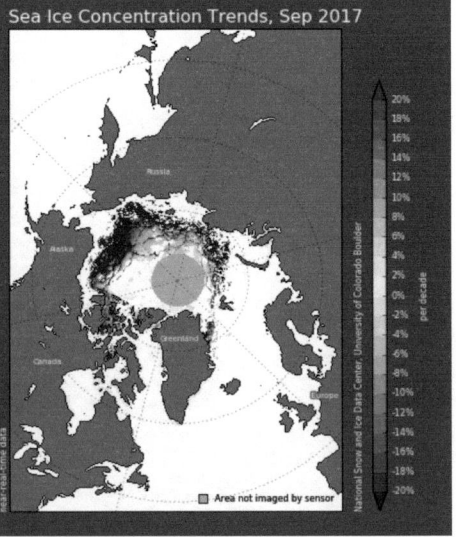

Figure 13.2 Spatial trends in sea ice concentration during March (left) and September (right). Trends are computed from 1979 through 2017. Sea Ice Index image, courtesy of the National Snow and Ice Data Center, University of Colorado Boulder (Fetterer et al. 2017).

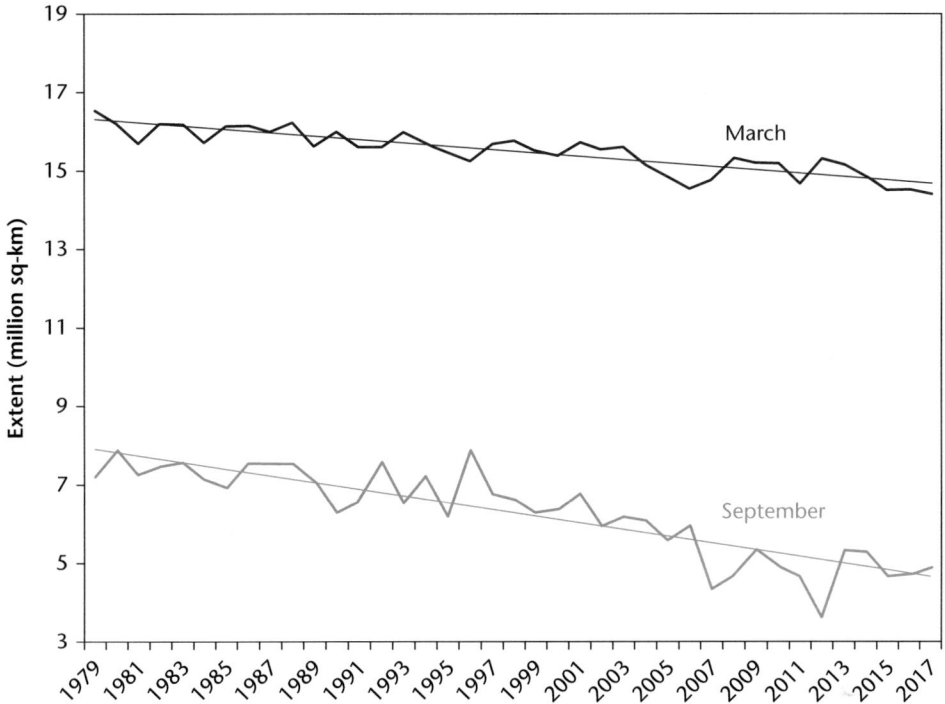

Figure 13.3 Satellite derived sea ice extent for the March (winter maximum) and September (summer minimum) for the years 1979 to 2017. Figure by Julienne Stroeve. Sea ice extent is based on data distributed by the National Snow and Ice Data Center Sea Ice Index (http://nsidc.org/data/seaice_index).

Sea ice extent and/or concentration is just one part of the story. While similarly long-term observations on sea ice thickness are not available on a pan-Arctic scale, early submarine observations, combined with radar and laser satellite altimetry and airborne campaigns, have shown that the MYI has thinned significantly since the late 2000s (Lindsay and Schweiger 2015). This thinning has been accompanied by a decline in the amount of the Arctic Ocean consisting of MYI (Maslanik et al. 2011), which has reduced from 70% of the Arctic Ocean to about 30%. More importantly, ice five years or older only makes up about 3% of the sea ice cover compared to 20% in the 1980s.

Numerous studies have reported on the drivers of observed Arctic sea ice loss, including studies on winter atmospheric circulation (Rigor et al. 2002), summer atmospheric circulation (e.g. Serreze et al. 2016), changes in ocean circulation (Polyakov et al. 2017; Woodgate et al. 2005a, 2005b), and the role of atmospheric greenhouse gases (Notz and Stroeve 2016). While variable atmospheric and oceanic circulation play key roles for individual years, the long-term decline shows a clear link with atmospheric CO_2 concentrations. Notz and Stroeve (2016) found a linear relationship between September sea ice extent and cumulative CO_2 emissions, showing a rate of sea ice loss of $3m^2$ per metric ton of CO_2. These results provide a clear limit on the amount of additional CO_2 that can be put into the atmosphere before the Arctic Ocean becomes ice-free in summer (700 Gt of Carbon). At current emission rates of 35–40 Gt of Carbon each year, this will happen within the next 20 years, and will have profound implications for climate, not only within the Arctic but also for the rest of the planet. Arctic amplification refers to the outsized warming in the Arctic compared to mid-latitudes and is largely driven by warming in expanded open water areas.

The Antarctic sea ice

The Antarctic, which comprises land surrounded by the ocean, is very different from the Arctic, which is an ocean almost completely surrounded by land. Because of this, Antarctic sea ice is exposed to significantly different land, ocean, cryospheric, and atmospheric processes and feedbacks. For example, the persistence of certain atmospheric patterns, such as the southern annular mode, or SAM, is thought to play a crucial role in determining the seasonal extent and distribution of sea ice around Antarctica (Schroeter et al. 2017). Consequently, we should consider Antarctic sea ice, and the processes that influence it, as being considerably different to Arctic sea ice.

Being in the southern hemisphere the Antarctic seasons are reversed, with the maximum sea ice extent occurring in September or October (about 18 to 19 million km²), and the minimum extent reached in February or March (about 3 to 4 million km²). In winter the entire continent is surrounded by sea ice, after which ice retreat is rapid with many regions of the continent being ice free in summer (Gloersen et al. 1992); see Figure 13.4. The seasonal retreat of the Antarctic ice cover in spring is one of the greatest seasonal events on earth; almost 7 million km² of ice melts in a single month. In summary, Antarctica possesses a largely seasonal ice cover which has a greater winter extent than Arctic sea ice, and because the winter ice extent exceeds that of the Arctic, the maximum global sea ice extent coincides with the Antarctic winter.

Until recently the long-term trend in Antarctic sea ice extent has been a small increase of approximately 1.5% per decade, since satellite records began in the late 1970s (Turner et al. 2015). This modest increase is surprising when put within the context of a warming global climate, the dramatic reductions in Arctic sea ice, and that the output of the Coupled Model Intercomparison Project Phase 5 (CMIP5 – now known as the Climate Model Intercomparison

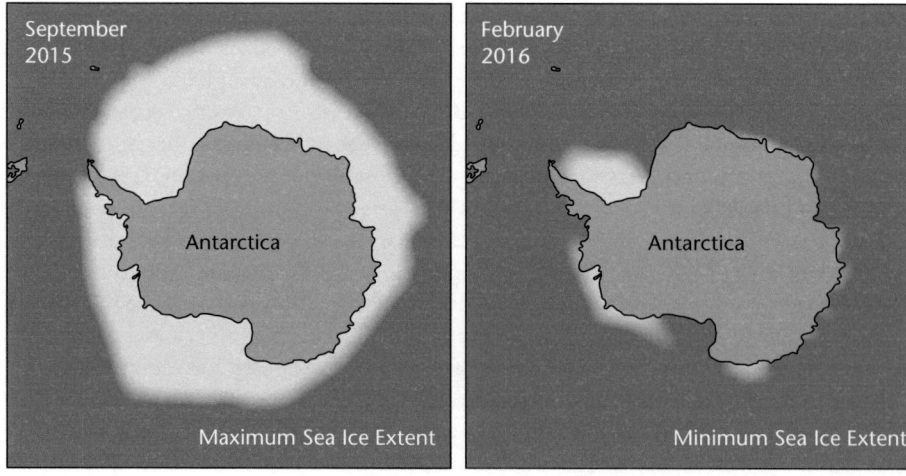

Figure 13.4 Satellite images of maximum (left) and minimum (right) Antarctic sea ice extent. Antarctic sea ice reaches its maximum extent in September or October and reaches a minimum extent in February. Source: https://earthobservatory.nasa.gov/Features/SeaIce/page4.php.

Project) show a decreasing Antarctic sea ice extent over recent decades (Turner et al. 2013). Thus the behaviour of Antarctic sea ice has presented a conundrum for global climate change science (National Academies of Sciences, Engineering, and Medicine 2017), and the reasons for this the disparity between observed and modelled trends are not yet well understood (Hobbs et al. 2016). Globally, the sea ice increases in Antarctica sea ice do not make up for the decreases in Arctic sea ice loss over the last decades (Parkinson 2014).

Sea ice extent is not the full story, and one has to look at how other sea ice properties are varying, especially sea ice thickness. For the Antarctic this is tricky as estimating sea ice thickness from satellite data remains challenging, and while military submarines have been critical in the monitoring of Arctic sea ice thickness no similar data set exists for the Antarctic. Our knowledge of Antarctic sea ice thickness is therefore derived almost entirely from limited drilling and coring campaigns on a small number of cruises, sporadic electromagnetic measurements, and visual estimates done during the passage of icebreakers (Williams et al. 2014). Thus establishing trends in ice thickness is presently not possible. The situation has been stressed in the WCRP-SCAR summary report on the International Workshop on Antarctic Sea Ice Thickness: "Sea ice thickness remains arguably the largest single gap in our knowledge of the climate system".

It is worth noting that the long-term increase in sea ice extent came to a halt in March 2017, when the summer sea ice recorded a new minimum extent (about 2 million km^2). This was almost 200,000 km^2 below the previous lowest minimum extent in 1997. There are many local, regional, and global processes that influence sea ice growth and melt, and therefore it is too early to conclude that the 2017 minimum is the start of an Arctic-like retreat for Antarctic sea ice, but it has prompted a call for significant investment in Antarctic sea ice science (Turner and Comiso 2017).

Note

1 https://oceanservice.noaa.gov/facts/sea-ice-climate.html.

References

Eicken, H. 2003. 'From the microscopic to the macroscopic to the regional scale: Growth, microstructure and properties of sea ice' in D. Thomas and G.S. Dieckmanni (eds) *Sea Ice: An Introduction to Its Physics, Chemistry, Biology and Geology*. Malden, MA: Blackwell. 22–81.

Fetterer, F., K. Knowles, W. Meier, M. Savoie and A.K. Windnagel. 2017. Updated daily. Sea Ice Index, Version 3. Boulder, CO. *NSIDC: National Snow and Ice Data Center*. doi: http://dx.doi.org/10.7265/N5K072F8.

Gloersen P., W.J. Campbell, D.J. Cavalieri, J.C. Comiso, C.L. Parkinson and H.J. Zwally. 1992. *Arctic and Antarctic Sea Ice, 1978–1987: Satellite Passive Microwave Observations and Analysis*. NASA Special Publication 511. Washington, DC: National Aeronautics and Space Administration.

Hobbs, W.R. et al. 2016. 'A review of recent changes in Southern Ocean sea ice, their drivers and forcings' *Global Planetary Change* 143: 228–250.

Leonard, G.H., H.H. Shen and S.F. Ackley 1998. 'Initiation and evolution of pancake ice in a wave field' *Antarct. J. U. S.*, 33: 53–55.

Leppäranta, M. 2005. *The Drift of Sea Ice*. Berlin: Springer-Verlag.

Lindsay, R.W and A. Schweiger 2015. 'Arctic sea ice thickness loss determined using subsurface, aircraft, and satellite observations' *The Cryosphere* 9: 269–283.

Maslanik, J., J. Stroeve, C. Fowler and W. Emery 2011. 'Distribution and trends in Arctic sea ice age through spring 2011' *Geophys. Res. Lett.*, 38, L13502, doi:10.1029/2011GL047735.

National Academies of Sciences, Engineering, and Medicine 2017. *Antarctic Sea Ice Variability in the Southern Ocean-Climate System*. Washington, DC: The National Academies Press.

Notz, D. and J. Stroeve 2016. 'Arctic sea-ice loss directly follows cumulative anthropogenic $CO2$ emissions' *Science*, doi:.10.1126/science.aag2345.

Parkinson, C.L. 2014. 'Global sea ice coverage from satellite data: annual cycle and 35-yr trends' *J. Climate* 27: 9377–9382.

Polashenski, C., D. Perovich, and Z. Courville 2012. 'The mechanisms of sea ice melt pond formation and evolution' *J. Geophys. Res.*, 117, C01001, doi:10.1029/2011JC007231.

Polyakov, I.V., A.V. Pnushkov, M.B. Alkire, I.M. Ashik, T.M. Baumann, E.C. Carmack, I. Goszczko, J. Guthrie, V.V. Ivanov, T. Kanzow, R. Krishfield, R. Kwok, A. Sundfjord, J. Morison, R. Rember and A. Yulin 2017. 'Greater role for Atlantic inflows on sea-ice loss in the Eurasian Basin of the Arctic Ocean' *Science* 356(6335), doi:10.1126/science.aai8204.

Rigor, I.G., J.M. Wallace and R.L. Colony 2002. 'Response of sea ice to the Arctic Oscillation' *J. Climate*, doi:10.1175/1520-0442(2002)015<2648:ROSITT>2.0.CO;2

Schroeder D., D. Feltham, D. Flocco and M. Tsamados 2014. 'September Arctic sea-ice minimum predicted by spring melt-pond fraction' *Nat. Clim. Change* 4: 353–357.

Schroeter, S., W. Hobbs and N.L. Bindoff 2017. 'Interactions between Antarctic sea ice and large-scale atmospheric modes in CMIP5 models' *The Cryosphere* 11: 789–803.

Screen, J.A. and I. Simmonds 2010. 'The central role of diminishing sea ice in recent Arctic temperature amplification' *Nature* 464: 1334–1337.

Serreze, M.C., A.P. Barrett, J.C. Stroeve, D.N. Kindig, and M.M. Holland 2009. 'The emergence of surface-based Arctic amplification' *The Cryosphere* 3: 11–19.

Serreze, M.C., J. Stroeve, A.P. Barrett, and L. Boisvert 2016. 'Summer atmospheric circulation anomalies over the Arctic Ocean and their influences on September sea ice extent: a cautionary tale' *J. Geophys. Res.*, doi:10.1002/2016JD025161.

Thorndike, A.S. and R. Colony 1982 'Sea ice motion in response to geostrophic winds' *J. Geophys. Res.* 87(C8): 5845–5852.

Turner, J., T.J. Bracegirdle, T. Phillips, G.J. Marshall, and J. Scott Hosking 2013. 'An initial assessment of Antarctic sea ice extent in the CMIP5 models' *Journal of Climate* 26: 1473–1484.

Turner J. and J. Comiso 2017. 'Solve Antarctica's sea-ice puzzle' *Nature* 547(7663): 275–277.

Turner, J., J.S. Hosking, T.J. Bracegirdle, G.J. Marshall, and T. Phillips 2015. 'Recent changes in Antarctic sea ice' *Philosophical 15 Transactions of the Royal Society of London A: Mathematical, Physical and Engineering Sciences*, 373, 20140163, 2015.

Wadhams, P. 1986. 'The seasonal ice zone' *The Geophysics of Sea Ice*, NATO ASI 820 Series, 825–991, 1986. DOI:10.1007/978-1-4899-5352-0_15.

Wadhams, P. 2000. *Ice in the Ocean*. New York: Taylor and Francis.

Walsh, J.E. 2013. 'Melting ice: what is happening to Arctic sea ice, and what does it mean for us?' *Oceanography* 26: 171–181.

Weeks, W.F. and S.F. Ackley 1986. 'The growth, structure and properties of sea ice' in N. Untersteiner (ed.) *The Geophysics of Sea Ice*. New York: Plenum Press (NATO ASI B146). 9–164.

Williams, G., T. Maksym, J. Wilkinson, C. Kunz, P. Kimball and H. Singh 2014. 'Thick and deformed Antarctic sea ice mapped with autonomous underwater vehicles' *Nature Geoscience* 8: 61–67.

Woodgate, R.A., K. Aagaard, and T.J. Weingartner 2005a. 'A year in the physical oceanography of the Chukchi Sea: moored measurements from autumn 1990–1991' *Deep Sea Res.*, Part II, 52(24–26): 3116–3149.

Woodgate, R.A., K. Aagaard, and T.J. Weingartner 2005b. 'Monthly temperature, salinity and transport variability of the Bering Strait throughflow' *Geophys. Res. Lett.*, 32, L04601, doi:10.1029/2004GL021880.

Zubov, N.N. 1945. *L'dy Arktiki [Arctic Ice]*. Izdatel'stvo Glavsermorputi, Moscow. Engl. transl. 1963 by U.S. Naval Oceanogr. Office and Amer. Meteorol. Soc., San Diego, CA.

14

The current health of polar ice sheets and implications for sea level

Mal McMillan

Introduction

Earth's two Polar ice sheets are the largest reservoirs of fresh water on the planet; greater by two orders of magnitude than the sum total of all glaciers and ice caps. Currently stored within the Greenland and Antarctic ice sheets is the equivalent of almost 66 meters of global sea level; approximately one-tenth is held within Greenland and the remainder within Antarctica (Fretwell et al. 2013; Vaughan et al. 2013). The size of these great ice sheets, and the volumes of water that they hold, is of global significance. Their health, whether they are contributing or removing water from the oceans, directly impacts upon present rates of sea level rise, with far-reaching implications for many coastal communities. Beyond their direct contribution to sea level, ice sheets have the capacity both to affect, and to be affected by, changes to the Earth system. Climatic changes, for example, have the capacity to modify ice sheet topography and ice discharge into the ocean, which in turn may alter atmospheric (Toniazzo et al. 2004; Junge et al. 2005) and oceanic (Fichefet et al. 2003; Hu et al. 2013) circulation patterns (see Bingham, Chapter 12 in this volume).

Systematically observing the current health of ice sheets forms part of wider efforts to monitor Earth's Polar Regions, and aids projections of the future evolution of the Earth system. Historically, analyses of ice sheet behaviour at the continental scale were limited by the implausibility of making observations across such a vast, remote, and inhospitable environment. Since the early 1990s, however, the launch of polar orbiting satellites has brought the opportunity to make measurements that cover the entire ice sheet. This chapter reviews these satellite observations, along with other contemporary measurements that together describe the current health of the Greenland and Antarctic ice sheets. More specifically, the chapter focuses upon estimates of recent ice sheet mass change or mass imbalance; that is, quantifying the net ice mass that is gained or lost each year, and projections of ice sheet evolution during the next century.

Methods for monitoring the health of an ice sheet

Most estimates of present day mass loss from the Greenland and Antarctic ice sheets are derived from one of three techniques – commonly referred to as the mass budget, altimetry and gravimetry methods. These three approaches all utilise satellite observations, which provide regular sampling and comprehensive coverage, complemented in some cases with model and airborne data. An overview of each method is given below.

Mass budget method

The mass budget method uses a combination of satellite and model data to compute the difference, or imbalance, between the mass input and output from an ice sheet (Rignot and Kanagaratnam 2006; Rignot 2008; Rignot et al. 2008; van den Broeke et al. 2009). If the mass imbalance is positive, then more mass has been accumulated than has been lost; if it is negative then more mass has been lost. Total mass input to the ice sheet is derived from its surface mass balance, which accounts for the net contribution from surface processes, including precipitation (snow and rain), melt-water run-off, wind-redistribution and sublimation. Such estimates can be calculated at the continental-scale using regional atmospheric climate models (Fettweis et al. 2013; van Angelen et al. 2013), which simulate the physical atmospheric processes acting at the ice surface, typically with a spatial resolution of 10–30 km (Figure 14.1a). Mass output due to ice discharge is estimated using satellite-derived observations of ice surface flow, together with estimates of ice thickness derived from airborne or satellite measurements (Rignot and Kanagaratnam 2006; Depoorter et al. 2013; Rignot et al. 2013). The principal challenges associated with this technique are that of acquiring comprehensive ice thickness datasets, making regular ice sheet wide velocity measurements and accurately modelling and validating the temporal evolution of surface mass balance. Because the net imbalance is much smaller than the total mass flux, both the mass input and output terms must be measured with a high degree of accuracy. This is particularly true in Antarctica where the mass imbalance (~150 Gt/yr) is less than 10% of the annual mass flux (~2000 Gt/yr) (van den Broeke et al. 2011).

Altimetry

Repeated satellite and airborne altimetry measurements chart temporal changes in ice sheet surface elevation (Wingham et al. 1998; Zwally et al. 2005; Shepherd and Wingham 2007; Pritchard et al. 2009; McMillan et al. 2014). These observations can be used to identify regions where surface uplift is occurring, as a result of accumulation of ice or snow, or of water at the ice sheet base. Conversely, a lowering of surface elevation indicates ice, snow or subglacial water loss (Figure 14.1b). At the continental scale, measurements of elevation change can be used to show whether the volume of the ice sheet is growing or shrinking. These, in turn, can be converted into estimates of ice mass imbalance by using a model to define the material density of the observed volume change (Zwally et al. 2005; Sørensen et al. 2011; Shepherd et al. 2012). The principal challenges associated with the altimetry technique are that of determining the appropriate density to use, because the altimeter cannot distinguish between elevation changes due to snow (e.g., increased snowfall) or ice (e.g., from changing ice dynamics), of detecting artefacts relating to the varying electromagnetic properties of the snowpack (Scott et al. 2006; Nilsson et al. 2015), and of achieving comprehensive spatial sampling of the entire ice sheet.

Gravimetry

Satellite gravimetry measures temporal variations in Earth's gravity field, from which changes in ice sheet mass can be inferred (Horwath and Dietrich 2006; Luthcke et al. 2006; Velicogna and Wahr 2006; Bouman et al. 2014; Schrama et al. 2014). This method, which was pioneered with the launch of the Gravity Recovery and Climate Experiment (GRACE) satellite mission in 2002, has provided approximately monthly estimates of ice sheet mass change since then (Figure 14.1c). Unlike other methods, gravimetry directly measures the movement of mass within the Earth system, including transfers between ice sheet and ocean, albeit at a relatively coarse spatial resolution. The major challenges associated with this technique relate to mapping mass changes at the scale of individual glacier basins, and to the isolation of ice mass change signals. The latter issue arises because the GRACE measurements are sensitive to all mass change within the Earth system, which includes the movement of mass within the underlying crust and mantle, in response to the reduction in ice load since the last glacial maximum. This process, termed Glacial Isostatic Adjustment, has a relatively large impact on GRACE estimates of Antarctic mass imbalance, and is commonly accounted for using model simulations (King et al. 2012; Whitehouse et al. 2012).

The health of the Greenland Ice Sheet

During the era of systematic polar satellite observations, which began in the early 1990s, the Greenland Ice Sheet has shifted from a state of near balance, to one of significant mass loss (Rignot et al. 2011; van den Broeke et al. 2011; Shepherd et al. 2012; Vaughan et al. 2013). Recent estimates by the Intergovernmental Panel on Climate Change (IPCC) found that between 1992 and 2001, the average rate of ice loss from Greenland was 34 ± 40 Gt yr^{-1} (Vaughan et al. 2013). Over the following decade, the rate of ice loss accelerated substantially, to reach an average of 215 ± 59 Gt yr^{-1} between 2002 and 2011 (Vaughan et al. 2013), and increasing further since then (Schrama et al. 2014). Total ice losses from Greenland equate to approximately 10% of the measured global sea level rise since the late 1990s (Church et al. 2013).

The rapidly increasing ice loss from Greenland since the late 2000s has provoked significant scientific attention, aimed at better understanding the processes responsible for these changes (Figure14.1). Fluctuations in ice sheet mass are primarily linked to two processes: changing surface mass balance, for example from increased melting or accumulation, and variable glacier flow, which affects both the redistribution of ice and the quantity of ice lost to the ocean. In the

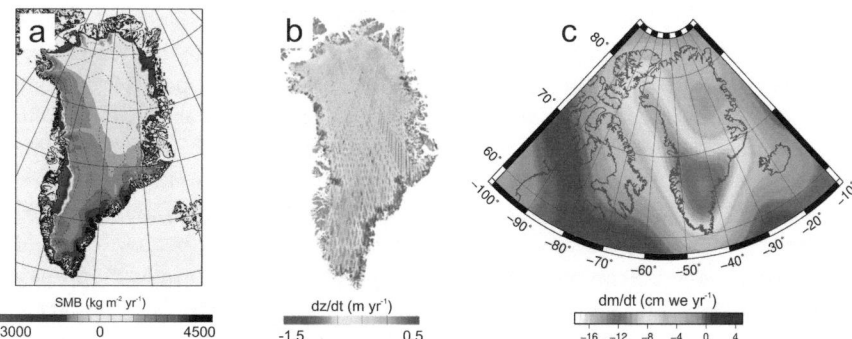

Figure 14.1   Health of the Greenland Ice Sheet. a. Mean surface mass balance for the period 1989–2004 from regional atmospheric climate modelling (van den Broeke et al. 2011; reproduced with permission Springer Publishing); b. Average rate of ice sheet elevation change between 2011 and 2014 from CryoSat-2 radar altimetry (McMillan et al. 2016); c. Average rate of ice sheet mass imbalance between 2003 and 2013 from GRACE gravimetry (Schrama et al. 2014; reproduced with permission John Wiley and Sons).

case of Greenland, the mass imbalance observed during the 2000s (van den Broeke et al. 2011; Shepherd et al. 2012; Vaughan et al. 2013) was caused, in roughly equal parts, by decreased surface mass balance and increased ice discharge (van den Broeke et al. 2009). The decreased surface mass balance was primarily due to higher rates of surface melting (van den Broeke et al. 2009) caused by warmer summer temperatures over the ice sheet. This resulted from a combination of global temperature increases and shifting patterns of regional atmospheric circulation (Hanna et al. 2008; Fettweis et al. 2011; Bindoff et al. 2013).

Alongside the observed increase in ice sheet melting, significant changes in ice flow have also been detected (Joughin, Smith, Howat and Moon 2010; Moon et al. 2012; Enderlin et al. 2014; Joughin 2014). Ice velocity, and hence ice discharge, vary over a wide range of timescales, as a result of changing subglacial conditions, ocean and atmospheric forcing, ice sheet rheology and topography. Here the discussion is limited to annual to decadal timescales, to assess the impact of changing ice dynamics on ice sheet mass balance over the same period. Much of the recent work to document and understand the changing nature of ice flow has focused on two primary mechanisms. Firstly, the dynamic response of relatively slow-flowing (typically 100 m yr^{-1}), land-terminating sectors of the ice sheet to observed increases in surface melting, and secondly, changes to relatively fast-flowing (typically 1–10 km yr^{-1}), marine-terminating glaciers in response to changes in ocean- and atmospheric-driven melting at their termini.

In the case of the first mechanism, observations have shown that surface melt water can penetrate to the base of the ice sheet, where it may act as a lubricant at the ice-bed interface and produce seasonal increases in ice flow (Zwally et al. 2002; Joughin et al. 2008; Bartholomew et al. 2011). This initially raised concerns that under a warmer climate, increasingly larger areas of the ice sheet may begin to flow faster. However, several recent studies have shown that the subglacial hydrological system responds to enhanced water input by increasing its drainage capacity, therefore reducing the longer-term impact that this process may have on ice velocity across these land-terminating glacier systems (Sole et al. 2013; Tedstone et al. 2013).

In the case of the second mechanism, satellite surveys have shown that marine-terminating outlet glacier dynamics are highly variable in both space and time (Joughin, Smith, Howat and Moon 2010; Joughin et al. 2012; Moon et al. 2012; Enderlin et al. 2014). Although this complexity makes generalisations difficult, some broad observations can be made. At the regional scale, glaciers in both the northwest and southeast sectors of the ice sheet accelerated between 2000 and 2010, with speeds on average increasing by around 30% (Moon et al. 2012). The widespread glacier acceleration occurring in these two regions was responsible for almost 90% of the total increase in ice discharge from the ice sheet (Enderlin et al. 2014). The timing of glacier acceleration has, however, differed between these two regions. In the northwest, glaciers tended to accelerate throughout the decade, whereas in the southeast the acceleration primarily occurred during the first half of this period (Moon et al. 2012; Enderlin et al. 2014). Understanding exactly what has caused these changes is challenging because of the complex nature of ice-ocean interactions and the numerous processes that can induce and modulate a dynamic response (Joughin et al. 2012). There is, however, evidence that a greater influx of warm ocean water occurred in some regions at the time of glacier acceleration (Holland et al. 2008; Rignot et al. 2010), although a direct causal link has yet to be established.

Since 2010, the rate of ice loss from Greenland has increased still further (Enderlin et al. 2014; Schrama et al. 2014) and the ice sheet has experienced episodes of rare and extreme surface melt (Nghiem et al. 2012). This period has been characterised by exceptionally high summer melting (Fettweis et al. 2013), particularly along Greenland's western coast, and increased winter temperatures (Hanna et al. 2012). The unusually high summer temperatures have been linked to the increased frequency of persistent atmospheric circulation patterns, which have drawn warm air over the ice sheet (Fettweis et al. 2011). As a result, the contribution of surface mass balance to total ice loss has become increasingly dominant since 2010, making up almost 70% of the observed ice sheet imbalance (Enderlin et al. 2014).

The health of the Antarctic Ice Sheet

Satellite data acquired since the late 1990s show that the Antarctic Ice Sheet lost mass at an average rate of 88 ± 35 Gt yr^{-1} during that period (Vaughan et al. 2013). Like Greenland, Antarctic mass losses have increased with time, from a rate of 30 ± 67 Gt yr^{-1} (1992–2001) to 147 ± 75 Gt yr^{-1} (2002–2011). Because of its cooler climate, the ice sheet experiences relatively little surface melting by the atmosphere, with the exception of parts of the Antarctic Peninsula. Instead, its mass balance is dominated by ice loss into the ocean, which is offset by inland snowfall accumulation. Satellite observations have revealed a wide range of behaviour across this vast continent (Figure 14.2), which is summarised in the following sections.

West Antarctica

The West Antarctic Ice Sheet has attracted considerable scientific attention during recent years. Long-held concerns about the stability of this last remaining marine-based ice sheet (Weertman 1974; Mercer 1978) have been reinforced by satellite data showing rapid loss of ice from some regions (Rignot 1998; Shepherd et al. 2002; Thomas et al. 2004; Zwally et al. 2005; Pritchard et al. 2009; McMillan et al. 2014). One of the most extensively studied areas is the Amundsen Sea Sector, which drains around one-third of West Antarctica and has a comparable ice flux to that of the entire Greenland Ice Sheet (Rignot and Kanagaratnam 2006; Rignot et al. 2008). Between 2005 and 2010, the mass imbalance of this sector alone contributed annually 0.28 ± 0.05 mm to sea level (Shepherd et al. 2012), equivalent to almost 10% of the observed rate of global sea level rise (Church et al. 2013).

In this region, ice is transported from the interior of West Antarctica to the Amundsen Sea coastline by several fast-flowing ice streams (Figure 14.3). The largest of these are the Pine Island and Thwaites Glaciers, which have undergone substantial flow acceleration in recent decades, and now reach maximum speeds of around 4 km yr^{-1} (Mouginot et al. 2014). As a result of increased glacier flow, ice discharge into the ocean along this coastline has increased by 77% since the 1970s, with half of the recorded acceleration occurring between 2003 and 2009 (Mouginot et al. 2014). The increased discharge has led to a significant ice mass imbalance in this region, which has manifested itself as widespread ice sheet thinning (Shepherd et al. 2002; Pritchard et al. 2009; Zwally et al. 2005; Helm et al. 2014; McMillan et al. 2014) and grounding line retreat (Rignot 1998; Park et al. 2013; Rignot et al. 2014). Compared to the early 1990s, the grounding lines of these glaciers have retreated up to 35 km inland (Joughin, Smith and Holland 2010; Park et al. 2013; Rignot et al. 2014). This retreat, across a broadly inward sloping bed (Holt et al. 2006; Vaughan et al. 2006; Fretwell et al. 2013), has renewed questions about the stability of this region and its potential to contribute relatively rapid rates of sea level rise in the future (Joughin 2014; Rignot et al. 2014).

The coherent pattern of mass loss from glaciers draining into the Amundsen Sea suggests that they may be responding to a common forcing mechanism. More specifically, it has been proposed that shifts in wind patterns that have occurred since the late 1990s have increased the ocean heat flux onto the continental shelf and into the cavities underlying the coastal ice shelves (Thoma et al. 2008). These warmer waters have increased rates of ice shelf basal melting (Jacobs et al. 2011), produced widespread ice shelf thinning (Shepherd et al. 2004, 2010; Pritchard

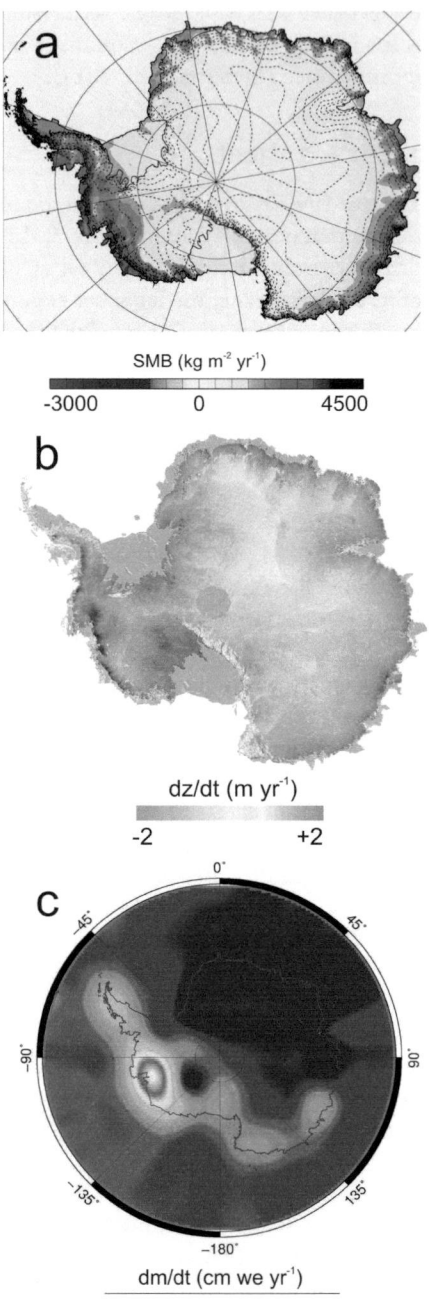

SMB (kg m^{-2} yr^{-1})

-3000   0   4500

dz/dt (m yr^{-1})

-2   +2

dm/dt (cm we yr^{-1})

Figure 14.2   Health of the Antarctic Ice Sheet. a. Mean surface mass balance for the period 1989–2004 from regional atmospheric climate modelling (van den Broeke et al. 2011; reproduced with permission Springer Publishing); b. Rate of ice sheet elevation change between 2010 and 2013 from CryoSat radar altimetry (McMillan et al. 2014); c. Average rate of ice sheet mass imbalance between 2003 and 2013 from GRACE gravimetry (Schrama et al. 2014; reproduced with permission John Wiley and Sons).

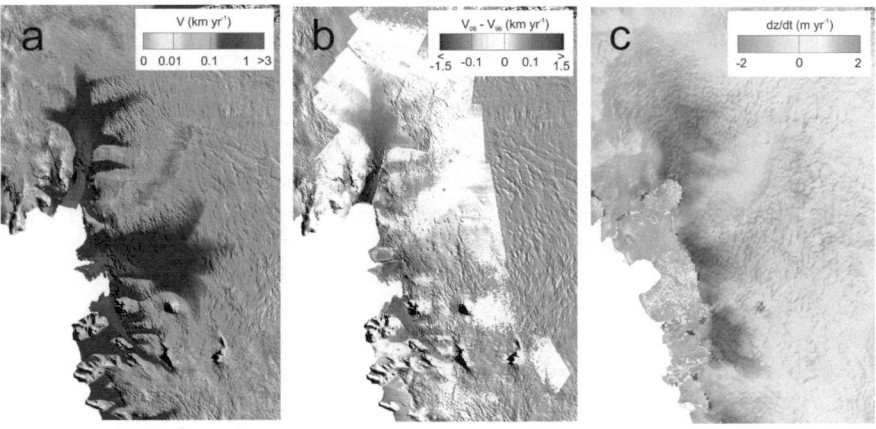

Figure 14.3 Satellite observations of ice loss in the Amundsen Sea sector of West Antarctica. a. Ice flow velocity (Mouginot et al. 2014); b. Ice flow acceleration between 1996 and 2008 (Mouginot et al. 2014); c. Rate of ice sheet thinning between 2010 and 2013 (McMillan et al. 2014; reproduced with permission John Wiley and Sons).

et al. 2012), grounding line retreat (Joughin, Smith and Holland 2010; Park et al. 2013; Rignot et al. 2014), and an associated acceleration of inland ice (Dupont and Alley 2005; Rignot 2008; Joughin and Alley 2011). Such a response mechanism, albeit modulated by the geometrical configuration of each glacier, could broadly explain the pattern of ice mass loss observed. Understanding the drivers of shifting wind patterns in this region and the associated delivery of warm water onto the continental shelf remains an area of ongoing research.

East Antarctica

The East Antarctic Ice Sheet has been broadly in balance since the late 1990s (Shepherd et al. 2012). Although there is some evidence of modest mass gains in recent years, the large annual variability caused by snowfall fluctuations, and high uncertainty associated with estimates of imbalance in this region, restrict any identification of a longer-term trend. In this region, further extension of the current observational record is required to determine any long-term imbalance with confidence.

Although the East Antarctic Ice Sheet is broadly in balance, at a smaller spatial and temporal scale there are some notable exceptions (Boening et al. 2012; Shepherd et al. 2012; Lenaerts et al. 2013; McMillan et al. 2014). The Totten Glacier, for example, discharges the largest volume of ice of all glaciers in East Antarctica (Rignot and Thomas 2002). During the past 20 years, there has been a persistent signal of ice thinning across the fast-flowing area close to the glacier's grounding line (Davis et al. 2005; Pritchard et al. 2009; Flament and Rémy 2012; McMillan et al. 2014). The coincidence of ice thinning and fast flow suggests that, like glaciers in the Amundsen Sea Sector of West Antarctica, Totten Glacier might too be responding to ice shelf thinning, caused by high rates of ocean–driven melting at the ice shelf base (Pritchard et al. 2012; Shepherd et al. 2012; Depoorter et al. 2013; Khazendar et al. 2013; Rignot et al. 2013). Other significant signals of mass imbalance in East Antarctica appear to be meteorological in origin. One example in recent years has been a series of extreme accumulation events that deposited several hundred gigatonnes of snow in Dronning Maud Land between 2009 and 2011 (Boening et al. 2012; Shepherd et al. 2012; Lenaerts et al. 2013; McMillan et al. 2014). Events such as these may be linked to prolonged changes in atmospheric pressure patterns, which have increased atmospheric flows inland from the coast (Boening et al. 2012).

Antarctic Peninsula

In the latter half of the twentieth century, parts of the Antarctic Peninsula have been among the fastest warming regions of Earth. Alongside rising air temperatures (Turner et al. 2005), there is also evidence of increasing ocean temperatures in recent decades (Robertson et al. 2002; Martinson et al. 2008). During this period, substantial glaciological changes have occurred across this region, particularly near to the coast, where fast-flowing glaciers feed numerous small ice shelves. Many of these ice shelves have undergone thinning and retreat (Shepherd et al. 2003; Cook and Vaughan 2010; Shepherd et al. 2010; Pritchard et al. 2012), and in some cases have disintegrated entirely (Rott et al. 1996; Scambos et al. 2000; Humbert et al. 2010), triggering sustained acceleration and mass loss from their tributary glaciers (Rignot et al. 2004, 2005; Rott et al. 2011).

Limits in detecting long-term ice sheet mass imbalance

Robust estimates of ice sheet mass imbalance are essential for understanding the future implications of current ice sheet evolution, yet they are hampered by the relatively short period over which satellite observations have been made. For example, data have shown increased rates of ice loss in recent years from both ice sheets; determining whether or not these increases represent part of a sustained acceleration will affect the long-term significance that is attributed to these observations. Such an assessment is complicated because ice sheets respond to factors which operate over a large range of timescales; from daily fluctuations in temperature and precipitation, through longer-lasting atmospheric events, such as those linked to variability in atmospheric circulation, to long-term changes, which includes those linked to the anthropogenic alteration of Earth's climate system and ice retreat since the Last Glacial Maximum.

Identifying signals of long-term imbalance using the current observational record requires that they be distinguished from the short-term variability that is inherent within the climate system. This is particularly important when future sea level projections are based upon the extrapolation of current observations. As an example, sea level rise projections at 2100 vary by several tens of centimetres, depending upon whether it is assumed that the rate of mass loss will continue to accelerate at its present rate, or remains constant (Meier et al. 2007; Rignot et al. 2011; Wouters et al. 2013). The longer the observational record of ice mass fluctuations, the more certain the identification of any underlying long-term trends. In this regard, assessments have found that observations spanning 10–20 years are required to determine long-term acceleration in ice sheet mass loss with reasonable certainty (Wouters et al. 2013), highlighting the importance of building long-term, multi-decadal records.

Total ice sheet contribution to sea level rise

The observed changes in ice sheet mass can be converted into an equivalent contribution to global sea level rise, given knowledge of the global ocean area and the mean ocean water density. Commonly, global ocean area is taken to be $\sim 362.5 \times 10^6$ km^2 and the density of seawater to be 1000 kg m^{-3} (Cogley 2012; Tian et al. 2015). Between 1992 and 2011, the Antarctic and Greenland ice sheets lost a total of 4260 $\pm$ 1200 Gt of ice, which is equivalent to a rise in global sea level of 11.7 $\pm$ 3.3 mm (Vaughan et al. 2013). Of this, approximately 60% has come from Greenland, and the remainder from Antarctica. During these two decades, the rate of mass loss from both ice sheets has increased with time, so that their recent contribution to sea level – estimated to be 1.2 ± 0.4 mm yr^{-1} (Vaughan et al. 2013) – now stands substantially above the 20-year mean.

The contribution of ice sheets to future sea level rise

The satellite datasets compiled since the 1990s have highlighted the complexity of ice sheet evolution and the challenges associated with making projections of their contribution to future sea level rise. Any projection must account for the varying factors which affect ice sheet mass loss or gain, that is, they must address how both surface mass balance and ice discharge will evolve given anticipated changes in climate. Based on current observations and projections of future climate evolution, the relative importance of these two processes is expected to differ significantly between the two ice sheets of Greenland and Antarctica.

Greenland is primarily a terrestrial-based ice sheet, grounded on bedrock above sea level. During summer, atmospheric surface temperatures rise above freezing and so, in a warming climate, more ice will melt. The Greenland Ice Sheet is expected to be particularly sensitive to future global temperature increases, due to amplified warming in the Polar Regions (see Marshall, Chapter 15 of this volume) and the non-linear response of ice melting to rising temperatures, as the bare ice zone expands inland, surface albedo reduces and the ice sheet surface elevation lowers (Serreze et al. 2009; Fettweis et al. 2013; Edwards et al. 2014). The dominant process responsible for ice loss from Greenland over the next century is expected to be changing surface mass balance (Goelzer et al. 2013; Enderlin et al. 2014), although changes in ice discharge may contribute additional mass loss over shorter timescales (Howat et al. 2010; Moon et al. 2012). In contrast, Antarctica sits within a much cooler climate and, with the exception of the Antarctic Peninsula, air temperatures are not expected to rise by enough during the next century to cause significant surface melting (Fyke et al. 2010). Consequently, changing surface mass balance is anticipated to exert only a moderate positive influence on overall mass imbalance, and rather it is varying ice discharge, driven by changes to ocean circulation and temperatures, that is likely to dominate.

The two ice sheets therefore place different demands on attempts to make projections of their future contribution to sea level. In Antarctica, the primary need is to understand the drivers of variability within the surrounding ocean, and the influence that these changes may have upon ice dynamics, through complex interactions at the ice-ocean interface. In Greenland, there is a similar requirement to understand the influence of oceanic processes on outlet glacier dynamics. In addition, however, it is critical to assess the likely future evolution of atmospheric conditions, their impact upon surface mass balance, and the strength of several feedback mechanisms that may exert a large influence upon the rate of ice loss.

To develop quantitative estimates of future sea level rise, several approaches have been taken. In the absence of comprehensive observational datasets, or where the underlying processes driving ice sheet change are not sufficiently well understood, techniques based upon empirical arguments and physical intuition have been used to place bounds on future sea level rise (Meier et al. 2007; Pfeffer et al. 2008). These, by their very nature, do not provide a precise estimate of future sea level rise, but rather aim to identify what is physically possible or improbable, given current knowledge of how glacier systems evolve.

Alternatively, where data and understanding are sufficiently advanced, sea level projections can be derived using process-based modelling. This approach aims to develop models that simulate the underlying physical processes that govern an ice sheet's response to its surrounding climate. For example, process-based models can be used to simulate future ice sheet surface mass balance given projected changes in global climate (Church et al. 2013). This approach offers the important capacity to couple together ice, ocean, and atmospheric models to enable each component of the Earth system to interact with each other.

Developing process-based models that are capable of yielding a realistic representation of evolving ice dynamics remains highly challenging. Furthermore, developing these models at a continental scale, while maintaining the complexity and resolution required to simulate dynamical changes across individual glacier basins, is theoretically and computationally demanding. As a result, although significant progress has been made in recent years (Nick et al. 2009; Price et al. 2011; Gladstone et al. 2012; Nick et al. 2013), process-based ice sheet models are not yet incorporated within global climate models. Instead, projections of future dynamical changes commonly rely on models of individual glaciers, with the results upscaled to provide estimates of ice imbalance at the continental scale (Price et al. 2011; Nick et al. 2013). Due to the complexity of this approach, such process-based, continental-scale dynamical projections have, to date, focused upon Greenland rather than the Antarctic Ice Sheet.

A recent synthesis report summarised current projections of the ice sheets' contribution to sea level rise by the end of this century, drawn primarily from process-based modelling (Church et al. 2013). In Greenland, changing surface mass balance is projected to produce a sea level rise in the range 10 mm to 160 mm, while increased glacier discharge is anticipated to add a further 10 mm to 70 mm. In Antarctica, increased snowfall is projected to lower sea level by between 0 mm and 70 mm, and changes in ice discharge to cause between a 20 mm fall and a 185 mm rise in sea level. These range of values do not account for the possibility of more widespread deglaciation across the Amundsen Sea Sector of West Antarctica. At present, the likelihood of this occurring remains uncertain, although it holds the capacity to raise sea levels by an additional several tens of centimetres. Continued monitoring of the Polar Regions, together with the further development of process-based models, will help to improve our understanding of how Earth's ice sheets may evolve in the future.

References

Bartholomew, I.D. et al. 2011. 'Seasonal variations in Greenland Ice Sheet motion: Inland extent and behaviour at higher elevations' *Earth and Planetary Science Letters* 307: 271–278.

Bindoff, N. et al. 2013. 'Detection and attribution of climate change: from global to regional' in *Climate Change 2013: The Physical Science Basis. Contribution of Working Group I to the Fifth Assessment Report of the Intergovernmental Panel on Climate Change*. Cambridge, UK: Cambridge University Press. 867–952.

Boening, C. et al. 2012. 'Snowfall-driven mass change on the East Antarctic Ice Sheet' *Geophysical Research Letters* 39, p. L21501.

Bouman, J. et al. 2014. 'Antarctic outlet glacier mass change resolved at basin scale from satellite gravity gradiometry' *Geophysical Research Letters* 41: 1–8.

Church, J.A. et al. 2013. 'Sea level change' in Stocker, T.F., D. Qin, G.-K. Plattner, M. Tignor, S.K. Allen, J. Boschung, A. Nauels, Y. Xia (eds) *Climate Change 2013: The Physical Science Basis. Contribution of Working Group I to the Fifth Assessment Report of the Intergovernmental Panel on Climate Change*. Cambridge, UK: Cambridge University Press.

Cogley, J.G. 2012. 'Area of the ocean' *Marine Geodesy* 35(4): 379–388.

Cook, A.J. and Vaughan, D.G. 2010. 'Overview of areal changes of the ice shelves on the Antarctic Peninsula over the past 50 years' *The Cryosphere* 4(1): 77–98.

Davis, C.H. et al. 2005. 'Snowfall-driven growth in East Antarctic ice sheet mitigates recent sea-level rise' *Science* 308(5730): 1898–1901.

Depoorter, M.A. et al. 2013. 'Calving fluxes and basal melt rates of Antarctic ice shelves' *Nature* 502(7469): 89–92.

Dupont, T.K. and Alley, R.B. 2005. 'Assessment of the importance of ice-shelf buttressing to ice-sheet flow' *Geophysical Research Letters* 32(4), p. L04503.

Edwards, T.L. et al. 2014. 'Effect of uncertainty in surface mass balance-elevation feedback on projections of the future sea level contribution of the Greenland ice sheet' *The Cryosphere* 8: 195–208.

Enderlin, E. et al. 2014. 'An improved mass budget for the Greenland ice sheet' *Geophysical Research Letters* 41: 1–7.

Fettweis, X. et al. 2011. 'The 1958–2009 Greenland ice sheet surface melt and the mid-tropospheric atmospheric circulation' *Climate Dynamics* 36: 139–159.

Fettweis, X. et al. 2013. 'Estimating the Greenland ice sheet surface mass balance contribution to future sea level rise using the regional atmospheric climate model MAR' *The Cryosphere* 7: 469–489.

Fichefet, T. et al. 2003. 'Implications of changes in freshwater flux from the Greenland ice sheet for the climate of the 21st century' *Geophysical Research Letters* 30(17): 8–11.

Flament, T. and Rémy, F. 2012. 'Dynamic thinning of Antarctic glaciers from along-track repeat radar altimetry' *Journal of Glaciology* 58(211): 830–840.

Fretwell, P. et al. 2013. 'Bedmap2: improved ice bed, surface and thickness datasets for Antarctica' *The Cryosphere* 7(1): 375–393.

Fyke, J.G. et al. 2010. 'Surface melting over ice shelves and ice sheets as assessed from modeled surface air temperatures' *Journal of Climate* 23: 1929–1936.

Gladstone, R.M. et al. 2012. 'Calibrated prediction of Pine Island Glacier retreat during the 21st and 22nd centuries with a coupled flowline model' *Earth and Planetary Science Letters* 333–334: 191–199.

Goelzer, H. et al. 2013. 'Sensitivity of Greenland ice sheet projections to model formulations' *Journal of Glaciology* 59(216): 733–749.

Hanna, E. et al. 2008. 'Increased runoff from melt from the Greenland Ice Sheet: A response to global warming' *Journal of Climate* 21(2): 331–341.

Hanna, E. et al. 2012. 'Recent warming in Greenland in a climatic context: I. Evaluation of surface air temperature records' *Environmental Research Letters* 7.

Helm, V., Humbert, A. and Miller, H. 2014. 'Elevation and elevation change of Greenland and Antarctica derived from CryoSat-2' *The Cryosphere* 8(4): 1539–1559.

Holland, D.M. et al. 2008. 'Acceleration of Jakobshavn Isbræ triggered by warm subsurface ocean waters' *Nature Geoscience* 1: 659–664.

Holt, J. et al. 2006. 'New boundary conditions for the West Antarctic Ice Sheet: Subglacial topography of the Thwaites and Smith glacier catchments' *Geophysical Research Letters* 33, p. L09502.

Horwath, M. and Dietrich, R. 2006. 'Errors of regional mass variations inferred from GRACE monthly solutions' *Geophysical Research Letters* 33, p. L07502.

Howat, I.M. et al. 2010. 'Seasonal variability in the dynamics of marine-terminating outlet glaciers in Greenland' *Journal of Glaciology* 56(198): 601–613.

Hu, A. et al. 2013. 'Influence of continental ice retreat on future global climate' *Journal of Climate* 26: 3087–3111.

Humbert, A. et al. 2010. 'Deformation and failure of the ice bridge on the Wilkins Ice Shelf, Antarctica' *Annals of Glaciology* 51(55): 49–55.

Jacobs, S.S. et al. 2011. 'Stronger ocean circulation and increased melting under Pine Island Glacier ice shelf' *Nature Geoscience* 4(8): 519–523.

Joughin, I. 2014. 'Marine ice sheet collapse potentially under way for the Thwaites Glacier Basin, West Antarctica' *Science* 344(6185): 735–738.

Joughin, I. et al. 2008. 'Seasonal speedup along the western flank of the Greenland Ice Sheet' *Science* 320(5877): 781–783.

Joughin, I. and R.B. Alley 2011. 'Stability of the West Antarctic ice sheet in a warming world' *Nature Geoscience* 4: 506–513.

Joughin, I., R.B. Alley and D. Holland 2012. 'Ice-sheet response to oceanic forcing' *Science* 338(6111): 1172–1176.

Joughin, I., B.E. Smith and D.M. Holland 2010. 'Sensitivity of 21st century sea level to ocean-induced thinning of Pine Island Glacier, Antarctica' *Geophysical Research Letters*, 37, p. L20502. https://doi.org/10.1029/2010GL044819.

Joughin, I., B.E. Smith, I. Howat and T. Moon 2010. 'Greenland flow variability from ice-sheet-wide velocity mapping' *Journal of Glaciology* 56(197): 415–430.

Junge, M.M. et al. 2005. 'A world without Greenland: Impacts on the Northern Hemisphere winter circulation in low- and high-resolution models' *Climate Dynamics* 24: 297–307.

Khazendar, A. et al. 2013. 'Observed thinning of Totten Glacier is linked to coastal polynya variability' *Nature Communications* 4, 2857. doi:10.1038/ncomms3857.

King, M.A. et al. 2012. 'Lower satellite-gravimetry estimates of Antarctic sea-level contribution' *Nature* 491(7425): 586–589.

Lenaerts, J.T.M. et al. 2013. 'Recent snowfall anomalies in Dronning Maud Land, East Antarctica, in a historical and future climate perspective' *Geophysical Research Letters* 40: 2684–2688.

Luthcke, S.B. et al. 2006. 'Recent Greenland ice mass loss by drainage system from satellite gravity observations' *Science*, 314(November): 1286–1289.

Martinson, D.G. et al. 2008. 'Western Antarctic Peninsula physical oceanography and spatio-temporal variability' *Deep-Sea Research II* 55: 1964–1987.

McMillan, M. et al. 2014. 'Increased ice losses from Antarctica detected by CryoSat-2' *Geophysical Research Letters* 41: 1–7.

McMillan M. et al. 2016. 'A high-resolution record of Greenland mass balance' *Geophysical Research Letters* 43: 7002–7010.

Meier, M.F. et al. 2007. 'Glaciers dominate eustatic sea-level rise in the 21st century' *Science* 317(5841): 1064–1067.

Mercer, J. 1978. 'West Antarctic Ice Sheet and CO2 greenhouse effect: a threat of disaster' *Nature* 271: 321–325.

Moon, T. et al. 2012. '21st-century evolution of Greenland outlet glacier velocities' *Science* 336(6081): 576–578.

Mouginot, J., E. Rignot and B. Scheuchl 2014. Sustained increase in ice discharge from the Amundsen Sea Embayment, West Antarctica, from 1973 to 2013. *Geophysical Research Letters* 41(5): 1576–1584.

Nghiem, S.V. et al. 2012. 'The extreme melt across the Greenland ice sheet in 2012' *Geophysical Research Letters* 39, p.L20502. https://doi.org/10.1029/2012GL053611.

Nick, F.M. et al. 2009. 'Large-scale changes in Greenland outlet glacier dynamics triggered at the terminus' *Nature Geoscience* 2: 110–114.

Nick, F.M. et al. 2013. 'Future sea-level rise from Greenland's main outlet glaciers in a warming climate' *Nature* 497: 235–238.

Nilsson, J. et al. 2015. 'Greenland 2012 melt event effects on CryoSat-2 radar altimetry' *Geophysical Research Letters* 42(10): 3919–3926.

Park, J.W. et al. 2013. 'Sustained retreat of the Pine Island Glacier' *Geophysical Research Letters* 40: 1–6.

Pfeffer, W.T., J.T. Harper and S. O'Neel 2008. 'Kinematic constraints on glacier contributions to 21st-century sea-level rise' *Science* 321(5894): 1340–1343.

Price, S.F. et al. 2011. 'Committed sea-level rise for the next century from Greenland ice sheet dynamics during the past decade' *Proceedings of the National Academy of Sciences of the United States of America* 108(22): 8978–8983.

Pritchard, H.D. et al. 2009. 'Extensive dynamic thinning on the margins of the Greenland and Antarctic ice sheets' *Nature* 461(7266): 971–975.

Pritchard, H.D. et al. 2012. 'Antarctic ice-sheet loss driven by basal melting of ice shelves' *Nature* 484(7395): 502–505.

Rignot, E. 1998. 'Fast recession of a West Antarctic glacier' *Science* 281(5376): 549–551.

Rignot, E. 2008. 'Changes in West Antarctic ice stream dynamics observed with ALOS PALSAR data' *Geophysical Research Letters* 35, p. L12505. https://doi.org/10.1029/2008GL033365.

Rignot, E. et al. 2004. 'Accelerated ice discharge from the Antarctic Peninsula following the collapse of Larsen B ice shelf' *Geophysical Research Letters* 31, p. L18401. https://doi.org/10.1029/2004GL020697.

Rignot, E. et al. 2005. 'Recent ice loss from the Fleming and other glaciers, Wordie Bay, West Antarctic Peninsula' *Geophysical Research Letters* 32, p. L07502. https://doi.org/10.1029/2004GL021947.

Rignot, E. et al. 2008. 'Recent Antarctic ice mass loss from radar interferometry and regional climate modelling' *Nature Geoscience* 1(2): 106–110.

Rignot, E. et al. 2011. 'Acceleration of the contribution of the Greenland and Antarctic ice sheets to sea level rise' *Geophysical Research Letters* 38. https://doi.org/10.1029/2011GL046583.

Rignot, E. et al. 2013. 'Ice-shelf melting around Antarctica' *Science* 341(6143): 266–270.

Rignot, E. et al. 2014. 'Widespread, rapid grounding line retreat of Pine Island, Thwaites, Smith, and Kohler glaciers, West Antarctica, from 1992 to 2011' *Geophysical Research Letters* 41: 3502–3509.

Rignot, E. and Kanagaratnam, P. 2006. 'Changes in the velocity structure of the Greenland Ice Sheet' *Science* 311(5763): 986–990.

Rignot, E., Koppes, M. and Velicogna, I. 2010. 'Rapid submarine melting of the calving faces of West Greenland glaciers' *Nature Geoscience* 3: 187–191.

Rignot, E. and Thomas, R.H. 2002. 'Mass balance of polar ice sheets' *Science* 297(5586): 1502–1506.

Robertson, R. et al. 2002. 'Long-term temperature trends in the deep waters of the Weddell Sea' *Deep-Sea Research II* 49: 4791–4806.

Rott, H. et al. 2011. 'The imbalance of glaciers after disintegration of Larsen-B ice shelf, Antarctic Peninsula' *The Cryosphere* 5: 125–134.

Rott, H., P. Skvarca and T. Nagler 1996. 'Rapid collapse of northern Larsen Ice Shelf, Antarctica' *Science* 271(5250): 788–792.

Scambos, T.A. et al. 2000. 'The link between climate warming and break-up of ice shelves in the Antarctic Peninsula' *Journal of Glaciology* 46(154): 516–530.

Schrama, E.J.O., B. Wouters and R. Rietbroek 2014. 'A mascon approach to assess ice sheet and glacier mass balances and their uncertainties from GRACE data' *Journal of Geophysical Research* 119(7): 6048–6066.

Scott, J.B.T. et al. 2006. 'Importance of seasonal and annual layers in controlling backscatter to radar altimeters across the percolation zone of an ice sheet' *Geophysical Research Letters* 33(24), p. L24502. https://doi.org/10.1029/2006GL027974.

Serreze, M.C. et al. 2009. 'The emergence of surface-based Arctic amplification" *The Cryosphere* 3: 11–19.

Shepherd, A. et al. 2003. 'Larsen Ice Shelf has progressively thinned' *Science* 302(856): 856–859.

Shepherd, A. et al. 2010. 'Recent loss of floating ice and the consequent sea level contribution' *Geophysical Research Letters* 37, p. L13503. https://doi.org/10.1029/2010GL042496.

Shepherd, A. et al., 2012. 'A reconciled estimate of ice-sheet mass balance' *Science* 338(6111): 1183–1189.

Shepherd, A. and Wingham, D. 2007. 'Recent sea-level contributions of the Antarctic and Greenland Ice Sheets' *Science* 315(5818): 1529–1532.

Shepherd, A., Wingham, D. and Mansley, J. 2002. 'Inland thinning of the Amundsen Sea sector, West Antarctica' *Geophysical Research Letters* 29(10), p. 1364. https://doi.org/10.1029/2001GL014183.

Shepherd, A., Wingham, D. and Rignot, E. 2004. 'Warm ocean is eroding West Antarctic Ice Sheet' *Geophysical Research Letters* 31, p. L23402. https://doi.org/10.1029/2004GL021106.

Sole, A. et al. 2013. 'Winter motion mediates dynamic response of the Greenland Ice Sheet to warmer summers' *Geophysical Research Letters* 40: 3940–3944.

Sørensen, L.S. et al. 2011. 'Mass balance of the Greenland ice sheet (2003–2008) from ICESat data: the impact of interpolation, sampling and firn density' *The Cryosphere* 5(1): 173–186.

Tedstone, A.J. et al. 2013. 'Greenland ice sheet motion insensitive to exceptional meltwater forcing' *Proceedings of the National Academy of Sciences of the United States of America* 110(49): 19719–19724.

Thoma, M. et al. 2008. 'Modelling circumpolar deep water intrusions on the Amundsen Sea continental shelf, Antarctica' *Geophysical Research Letters* 35(L18602). https://doi.org/10.1029/2008GL034939.

Thomas, R. et al. 2004. 'Accelerated sea-level rise from West Antarctica' *Science* 306(5694): 255–258.

Tian, Y. et al. 2015. 'On the conversion of Antarctic ice-mass change to sea level equivalent' *Marine Geodesy* 38(1): 89–97.

Toniazzo, T., Gregory, J.M. and Huybrechts, P. 2004. 'Climatic impact of a Greenland deglaciation and its possible irreversibility' *Journal of Climate* 17: 21–33.

Turner, J. et al. 2005. 'Antarctic climate change during the last 50 years' *International Journal of Climatology* 25(February): 279–294.

Van Angelen, J.H. et al. 2013. 'Rapid loss of firn pore space accelerates 21st century Greenland mass loss' *Geophysical Research Letters* 40(10): 2109–2113.

Van den Broeke, M.R. et al. 2009. 'Partitioning recent Greenland mass loss' *Science* 326(5955): 984–986.

Van den Broeke, M.R. et al. 2011. 'Ice sheets and sea level: thinking outside the box' *Surveys of Geophysics* 32: 495–505.

Vaughan, D.G. et al. 2006. 'New boundary conditions for the West Antarctic ice sheet: subglacial topography beneath Pine Island Glacier' *Geophysical Research Letters* 33, p. L09501. https://doi.org/10.1029/2005GL025588.

Vaughan, D.G. et al. 2013. 'Observations: cryosphere' in Stocker, T.F., D. Qin, G.-K. Plattner, M. Tignor, S.K. Allen, J. Boschung, A. Nauels, Y. Xia (eds) *Climate Change 2013: Physical Science Basis. Contribution of Working Group I to the Fifth Assessment Report of the Intergovernmental Panel on Climate Change.* Cambridge, UK: Cambridge University Press.

Velicogna, I. and J. Wahr 2006. 'Measurements of time-variable gravity show mass loss in Antarctica' *Science* 311(5768): 1754–1756.

Weertman, J. 1974. 'Stability of the junction of an ice sheet and an ice shelf' *Journal of Glaciology* 13: 3–11.

Whitehouse, P.L. et al. 2012. 'A new glacial isostatic adjustment model for Antarctica: calibrated and tested using observations of relative sea-level change and present-day uplift rates' *Geophysical Journal International* 190(3): 1464–1482.

Wingham, D. et al. 1998. 'Antarctic elevation change from 1992 to 1996' *Science* 282(5388): 456–458.

Wouters, B. et al. 2013. 'Limits in detecting acceleration of ice sheet mass loss due to climate variability' *Nature Geoscience* 6(8): 613–616.

Zwally, H.J. et al. 2002. 'Surface melt-induced acceleration of Greenland Ice-Sheet flow' *Science* 297(July): 218–223.

Zwally, H.J. et al. 2005. 'Mass changes of the Greenland and Antarctic Ice Sheets and shelves and contributions to sea-level rise: 1992–2002' *Journal of Glaciology* 51(175): 509–527.

15

Polar climate and evidence for anthropogenically-driven climate change

Gareth Marshall

Introduction

Both Polar Regions have seen significant changes in their climate over the periods of available observations although such changes can be divergent (e.g. Turner and Overland 2009). The most striking contrast is between the marked reduction in Arctic sea ice and the much smaller but nevertheless statistically significant increase in the Antarctic (see the Chapter 13 in this volume by Wilkinson and Stroeve). Moreover, while there has been warming across most of the Arctic associated with reductions in snow cover and melting permafrost, with the exception of the western Antarctic Peninsula the Antarctic has experienced relatively little temperature change and even cooling in some regions. The contrast between the two Polar Regions can be partially explained by their differences in land and sea distribution and orography. The Arctic Ocean, surrounded by land masses, is located at the highest latitudes and thus receives the greatest levels of solar radiation during summer, allowing the ice-albedo feedback mechanism (e.g. Flanner et al. 2011) to operate efficiently as temperatures are close to freezing. In contrast the Antarctic continent is surrounded by ocean and because much of it is elevated and covered in highly reflective snow and ice the ice-albedo feedback mechanism has relatively little effect. Furthermore, one of the results of the Antarctic ozone hole (see Kirkwood, Chapter 36 of this volume) has been to modify the atmospheric circulation such that the circumpolar westerlies around Antarctica have increased, diminishing the advection of warmer air masses into the continent. It has also acted to reduce the speed of the density-driven katabatic winds over much of Antarctica, causing a reduction of turbulent heat flow towards the surface, again leading to cooling.

In this chapter I examine whether and with what level of certainty such changes in polar climate may be directly attributed to human activity. Note that climate change related to sea ice, and to glaciers and ice sheets, are discussed elsewhere in this volume, in Chapters 13 and 14, respectively. Much of the discussion in this chapter is based on the findings of the recent Fifth Assessment Report of the Intergovernmental Panel on Climate Change (IPCC), in particular chapter 10 of The Physical Science Basis contribution of Working Group I (Bindoff et al. 2013). This used a series of model experiments to analyse the results from many different climate models, known as the fifth Climate Model Intercomparison Project (hereinafter CMIP5).

Detecting anthropogenic change

Before considering the question of whether anthropogenically-driven climate change can be observed in the Polar Regions, I describe the general methodology behind such attribution studies. The causes responsible for such changes are determined by examining the observational record for spatial and temporal 'fingerprints' related to individual climate forcings. These can include anthropogenic forcings, such as greenhouse gas (GHG) increases, ozone depletion and aerosols, and natural forcings external to the basic climate system, such as solar variability and volcanic eruptions.

One methodology is to use structural model analysis, as employed by Chylek et al. (2014), to examine Arctic warming (see later in this chapter). This technique uses a model, often linear, that assumes the observations are the product of a combination of individual explanatory variables. The coefficients for the predictors are determined by minimising the difference between the observed and modelled dependent variable. Comparing the percentage of the dependent variable explained using different combinations of the explanatory variables reveals which of the latter are likely to be the most important, whether they be anthropogenic and/or natural, in driving the observations.

More commonly, climate models, run with one or more forcings, are used to derive simulated 'fingerprints', which are then compared to the available observation datasets to determine which one or combination of the forcings best matches reality. As an example, Arblaster and Meehl (2006) used an ensemble of model simulations run with the observed time-series of five different forcings to separate their contribution to the marked trend in the Southern Annular Mode (SAM) in the late twentieth century. Such analyses have the advantage over structural models in that any significant spatial aspects of climate change can be examined and, importantly, that process studies using model diagnostics can be utilised to prove causation.

In addition, for us to be certain that the 'fingerprint' of any forcing is responsible for the observed change it needs to be statistically separable from internal climate variability. This unforced variability is a consequence of the interaction of the many different processes within the climate system, which operate across a range of spatial and temporal timescales. In their work for the IPCC, Bindoff et al. (2013) provide an example of such an analysis for historical global surface air temperature (SAT); in addition to the time-series of the average, they show how the spatial fingerprints of SAT trends in model runs with natural and combined forcings, providing further evidence for an anthropogenic contribution to 'global warming'. Climate models can be run over long time-periods as 'control runs' in which no external forcings are applied, either natural or anthropogenic, in order to establish their internal variability. By comparing the output of a forced run with a control run we can determine whether the apparent 'fingerprint' of an external forcing is statistically different from internal climate variability. An alternative method of establishing whether an observed trend is outside the range of internal variability is to derive an estimate of the likelihood of it occurring based on the probability distribution function of that parameter, as calculated from a long control run; if the likelihood is very small then we can assume that the trend has occurred as a response to external forcing.

Thus, a climate model must reliably simulate both the 'fingerprint' patterns associated with individual forcings and the patterns of unforced internal variability in order to produce a valid attribution assessment. Although far from perfect, modern state-of-the-art climate models are now sufficiently accurate to undertake such assessments at a regional as well as global scale. However, there are several issues that make these attribution studies problematic in the Polar Regions, especially the Antarctic. The small number, sparse distribution and short timescale of the observational records, in conjunction with the high magnitude of climate variability as compared to other regions of the Earth, means that it can be difficult to isolate any externally-forced 'fingerprint' within the observations and also to determine whether it is statistically significantly different from internal climate variability. Moreover, uncertainty between climate models has a clear maximum at high latitudes (Hawkins and Sutton 2009), reflecting differences in the representation of the various climate feedbacks that lead to amplification of climate change signals, such as the ice-albedo feedback mechanism (see Hodgkins, Chapter 19 of this volume).

The Arctic

Arctic modes of atmospheric circulation variability

In addition to the presence of the aforementioned amplified 'global warming' signal in the Arctic, changes in the surface climate there have occurred as a result of dynamical modifications in both the atmosphere and the ocean (see Bingham, Chapter 12 in this volume for discussion of the latter). There are several large-scale modes of atmospheric circulation variability that influence climate across all or part of the Arctic region and such sources of climate variability need to be accounted for when attempting to attribute regional climate change to an anthropogenic source.

The North Atlantic Oscillation (NAO) (and the related Northern Annular Mode (NAM), here considered to be the same for simplicity) is the major mode of extra-tropical atmospheric circulation variability in the Northern Hemisphere and plays a key role in determining the Arctic climate (e.g. Hurrell et al. 2003). The NAO describes co-variability in sea level pressure (SLP) between the Icelandic Low and Azores High and is most apparent in winter. When it is positive, with negative (positive) SLP anomalies near Greenland (the Azores), SAT is generally warmer than average over the Eurasian Arctic and cooler over the North American sector. There was a generally positive trend in the winter NAO from the 1960s up to 2000. Given that climate models generally show a more positive NAO with increasing greenhouse gases, human activity was considered as a likely driver for the warming Arctic. However, in the twenty-first century, there has been a strongly negative trend in the NAO, which included the most negative value in winter 2009/10 in an observation-based index starting in 1823. This recent trend is not reflected in climate models, suggesting that it is a manifestation of internal variability, or possibly a mechanism not resolved in the models (Gillett and Fyfe 2013).

Climatic conditions across Alaska and northwest Canada are strongly influenced by the Aleutian Low, which is a climatological feature of the North Pacific atmosphere. A deep Aleutian Low results in predominantly southerly air masses affecting the region giving higher temperatures, and vice versa. The depth of the Aleutian Low in turn co-varies with the phase of the Pacific Decadal Oscillation (PDO), which is a major cycle of sea surface temperature (SST) variability of the north Pacific region. In the mid-1970s there was a shift from negative to positive PDO that resulted in a deeper Aleutian Low and warmer conditions across Alaska, while during the twenty-first century the phase of the PDO has returned to being primarily negative. Models reveal that stochastic (or random) atmospheric forcing, such as the passage of storms, can induce longer-term SST variability such as a change in the phase of the PDO. However, modelling studies have also suggested an influence from anthropogenic aerosols on the PDO (Allen et al. 2014).

The Aleutian Low also comprises one of four 'centres of action' associated with the Pacific North American (PNA) atmospheric pattern of co-varying pressure anomalies; positive PNA events are associated with an intensified Aleutian Low (similar to the PDO), and therefore greater than average storminess in the Gulf of Alaska, which, together with ridging over western Canada and Alaska, drives warm winds and increased moisture northward over these regions, the so-called 'Arctic Express'. During negative PNA periods significant atmospheric blocking occurs in the northern Pacific leading to broadly opposite temperature and precipitation anomalies. Both the PDO and PNA are influenced by tropical variability through the phase of the El Niño-Southern Oscillation (ENSO), although there is debate in the literature about the nature and magnitude of such co-variance. Abatzoglou and Redmond (2007) showed that seasonal asymmetry in trends in the PNA can lead to an enhanced and masked anthropogenic warming signal in western North America in spring and autumn, respectively.

The Atlantic Multidecadal Oscillation (AMO) is a mode of internal climate variability that occurs in the North Atlantic. It is expressed primarily as changes in SSTs, and its signal can also be observed in Arctic sea ice variability and SAT (Miles et al. 2013; Chylek et al. 2014). The second of these studies demonstrated that it was very important to include the AMO as an explanatory variable when trying to isolate an anthropogenic forcing for warming SATs over the Arctic.

Arctic surface air temperature

Although the number of continuous SAT records in the Arctic is still relatively sparse compared to most of the Earth's landmasses, it is significantly larger than Antarctica. There are ~60 that begin around 1930–1940 and three that commenced in the middle of the nineteenth century. Turner and Overland (2009) report on the locations of the 59 stations with long records in the Arctic and SAT anomaly values for the observational records from these stations for January, relative to the 1961–1990 mean for each station.

The average Arctic temperature warmed in the early twentieth century, followed by a significant cooling from 1940 to 1965 after which the current warming began, which is approximately twice the global mean. Northwestern North America and central Siberia have experienced the greatest temperature rises over the last 50 years, in the range of 2–3°C. This recent warming, especially from the 1990s, has been relatively widespread whereas previously SAT changes were neither spatially nor temporally uniform. For example, the sudden warming across Alaska and northern Canada in the late 1970s has been linked to a deepening of the Aleutian Low whereas the Siberian warming is related to the positive trend in the NAO (see the previous section). The former warming has resulted in extensive permafrost melting and deforestation caused by higher survival rates of spruce bark beetle larvae during winter. However, large-scale cooling has occurred during boreal winter over much of the Eurasian Arctic. Cohen et al. (2012) related this seasonal asymmetry in SAT trends to a dynamical process whereby a warmer summer/autumn leads to enhanced snow cover that in turn forces a negative NAO/NAM in the following winter.

Gillett et al. (2008) undertook the first formal attribution study of Arctic land-surface SAT increases and found a clear anthropogenic influence, distinguishable from natural forcings, with a consistent magnitude in simulations and observations. Based on a structural model analysis, Chylek et al. (2014) calculated that about half of the recent warming has anthropogenic causes. However, the early twentieth century warming could not be reproduced so was ascribed to unforced variability. Shindell and Faluvegi (2009) inferred a large contribution to the subsequent mid-century Arctic cooling from changes in aerosol forcing and suggested that a reduction in the emission of anthropogenic sulphates in the late twentieth century may well have contributed significantly to the observed SAT increase.

For the IPCC, Bindoff et al. (2013) concluded that despite the uncertainties introduced by limited observational coverage, high internal variability, poorly understood local forcings and modelling uncertainties, there is sufficient evidence for it to be likely (66–100% confidence) that there has been an anthropogenic contribution to the Arctic land surface temperatures over the past 50 years.

Arctic precipitation

There is significant uncertainty regarding even mean precipitation totals in the Arctic (e.g. Serreze and Barry 2005). Imperfect adjustment procedures for the significant gauge undercatch of solid precipitation are compounded both by the variety of gauge types and reporting practices used by different Arctic countries and by a switch to automated stations at some locations. The already sparse precipitation network has declined since the 1990s, with observations now especially sparse over northern Canada and Siberia. While some historical observations exist in the central Arctic Ocean, from the Russian floating North Pole stations, there is no current systematic observing program for precipitation across this large region. Increases in Arctic precipitation since the 1960s have been observed in several different observational data sets although high latitude trends vary significantly with available coverage (Polson et al. 2013).

Analysis of precipitation trends from October to May (the snowfall season) reveals increases across much of the Arctic (e.g. Callaghan et al. 2011). Climate model simulations show clear global and regional scale changes in precipitation associated with anthropogenic forcing, with the largest differences between models with and without anthropogenic forcing found in the Arctic, where increases in precipitation are a robust feature of CMIP5 model projections and predominantly forced by GHG increases. An earlier detection and attribution study by Min et al. (2008) focussed solely on precipitation in the Arctic and found an attributable human influence unrelated to changes in circulation associated with the NAO. Observed changes are significantly larger than the model simulated changes, although Polson et al. (2013) noted that the difference between models and observations decreases if changes are expressed as a percentage of climatological precipitation. Furthermore, observed and simulated changes are largely consistent between CMIP5 models and observations given data uncertainty. However, the latter mean that regional-scale attribution of Arctic precipitation change remains problematic.

Arctic snow cover

Northern Hemisphere snow cover extent (SCE) is among the most important indicators of global climate change because of the threshold response of snow formation and melt to the 0°C isotherm. However, an increase in winter precipitation could be sufficient to offset this response. Available in situ and satellite data reveal different regional snow cover responses to the widespread warming and increasing winter precipitation observed across the Arctic in recent decades (e.g. Callaghan et al. 2011). The largest and most rapid decreases in snow water equivalent and snow cover duration are observed over maritime regions of the Arctic. Here, increases in air temperatures, which are close to freezing point in spring, are most effective at reducing snow accumulation, increasing snowmelt, or both (Brown and Mote 2009). In the North American sector of the Arctic, snow cover has been decreasing since the 1950s whereas in Eurasia widespread losses did not occur until the 1980s and more recently increases have been observed in autumn. Cohen et al. (2012) suggested that the latter impacts the broadscale atmospheric circulation in the following winter.

A formal detection and attribution study of these observed springtime changes in Northern Hemisphere SCE was undertaken by Rupp et al. (2013), who compared CMIP5 model runs that included both natural and anthropogenic forcings with those with only the former. They demonstrated that the former runs could largely explain the observed decrease in SCE while the latter were inconsistent with these observations. However, the model runs with both types of forcing under-predicted the response in SCE by a factor of two, either because the models were not sufficiently sensitive to the forcing and/or because they under-represented internal climate variability. Nevertheless, the IPCC concluded with high confidence that the decrease in Northern Hemisphere SCE since the 1970s is likely to be caused by external forcings with an anthropogenic contribution (Bindoff et al. 2013).

Arctic haze

One easily identifiable aspect of anthropogenic influence on the Arctic climate is the significant air pollution that is most noticeable during winter and spring, the so-called 'Arctic haze'. This comprises aerosols transported poleward from mid-latitude industrial source regions, particularly during east-west pressure gradients. Modelling studies by Hu et al. (2005) indicated that radiative cooling is associated with most aerosol types – through the enhanced scattering of solar radiation and increasing cloud reflectivity and cloud lifetimes – but the low reflectivity of soot actually has a heating effect through the absorption of sunlight. However, spatially comprehensive surveys of impurities in Arctic snow suggested that impurities actually decreased between the mid-1980s and late 2000s and hence changes in surface reflectivity have probably not contributed significantly to recent reductions in Arctic ice and snow (Vaughan et al. 2013).

Lubin and Vogelmann (2006) utilised multisensory radiometric data to show that enhanced aerosol concentrations over the North Slope of Alaska altered the microphysical properties of clouds in such a way as to cause an average increase of 3.4 Wm^{-2} in surface longwave fluxes under clouds and, hence, a net warming. More recently, using four years of observations from Barrow, Alaska, Zhao and Garrett (2015) demonstrated that the cloud radiative impact on surface temperature is a net warming between October and May and a net cooling in summer. During episodes of high surface haze aerosol concentrations and cloudy skies, both the net warming and net cooling are amplified. Averaged over a calendar year the warming and cooling effects are approximately in balance, but the seasonality of the net effect may be to exert a control on the amplitude and timing of sea ice melt. In addition, changes in aerosols are likely to have had an effect on longer-term trends in Arctic climate (Shindell and Faluvegi 2009).

The Antarctic

The ozone hole

Stratospheric ozone depletion over Antarctica, the 'ozone hole', has been described as the first unequivocal demonstration of anthropogenic climate change (see also Kirkwood, Chapter 36 of this volume). The primary cause has been traced to the emission of chloroflurocarbons (CFCs) and other halogen source gases containing chlorine and bromine. These unreactive gases accumulate in the troposphere and are then transported into the stratosphere above Antarctica, where strong circumpolar winds that form a polar vortex during the polar night isolate this air. Within the vortex the gases descend to altitudes at which polar stratospheric clouds (PSCs) form, at temperatures of about −78°C, from nitric acid and sulphur-containing gases condensing with water vapour. It is the heterogeneous chemical reactions on the surfaces of the PSC particles, where inert chlorine and bromine compounds are converted into reactive forms, which lead to ozone depletion. Ultraviolet (UV) sunlight is necessary for these reactions to take place so maximum ozone loss occurs during austral spring at high latitudes but starts at the sunlit edge of the polar vortex and propagates poleward. When temperatures rise in late spring PSC formation ceases and the polar vortex breaks down allowing ozone-rich air into the Antarctic stratosphere once again. As the absorption of incoming solar UV radiation by stratospheric ozone heats the atmosphere, ozone depletion has led to a cooler stratosphere and thus an increase in the meridional temperature gradient. This in turn causes a stronger and more persistent polar vortex (e.g. Previdi and Polvani 2014).

Observational and modelling studies have revealed that these changes in stratospheric circulation are dynamically coupled to changes in the troposphere with a lag of one to two months: thus the impact of spring ozone loss on Southern Hemisphere high latitude climate is predominantly felt during austral summer. The primary effect has been to enhance and shift poleward the mean westerly flow around Antarctica.

The Southern Annular Mode

The Southern Annular Mode (SAM) is the principal mode of atmospheric circulation variability in the Southern Hemisphere extra-tropics and has a huge influence on Antarctic climate. It is essentially an index of the strength of the circumpolar westerlies around Antarctica; the ozone hole has caused a positive trend in the SAM (Figure 15.1) that is most pronounced in austral summer and autumn. The effects of the SAM on Antarctic SAT and precipitation are described later; here I discuss the attribution of the changes in the SAM itself.

There have been several types of model experiment used to demonstrate the impact of the ozone hole on the SAM with most showing a clear link between the two, suggesting the response is robust. One example, updated from Gillett et al. (2013), is shown in Figure 15.1, which compares observed and simulated seasonal trends in the SAM for 1955–2011. The simulated trends are from CMIP5 climate models forced with one or all of greenhouse gases, aerosols, ozone changes and natural (solar and volcanic) forcings. It can be seen that the observed

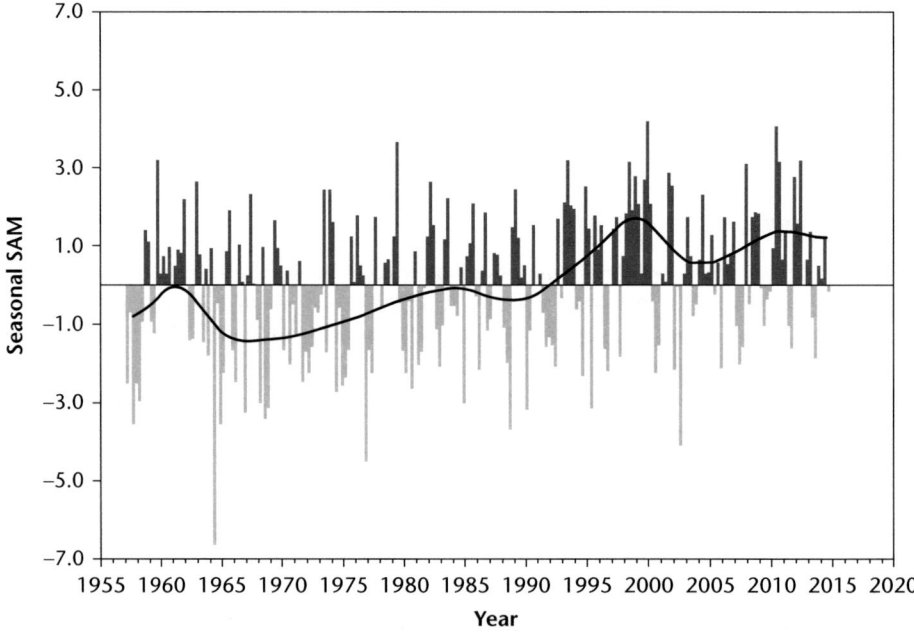

Figure 15.1   Seasonal and observed trends in the SAM index. Black lines show observed SAM trends derived from the stations used by Marshall (2003). Adapted from figure S3b from Gillett and Fyfe (2013) and Figure 10.13b from Bindoff et al. (2013).

trends are outside the range of the model control runs in all seasons except spring but are especially so in summer. Comparing the SAM trend for the individual forcings in this season indicates clearly that ozone loss has had the biggest impact with GHGs also contributing and aerosols having a negative effect. Lee and Feldstein (2013) showed that ozone contributed 50% more than GHGs based on analysis of the westerly winds themselves.

Fogt et al. (2009) compared observed SAM trends with those in a statistically reconstructed SAM dataset going back to the nineteenth century and climate models. Their study indicated that while the recent summer trend is unprecedented in the past 150 years, a negative autumn trend after 1930 in the reconstruction is larger than the recent positive trend. Moreover, the seasonal model trends in autumn during 1957–2005 are the most different from observations. The authors concluded that the recent autumn trend is most likely natural climate variability and thus, only the summer SAT changes induced by the positive SAM trends can be considered likely to be anthropogenic in origin. Analysing the findings of the various model experiments, the IPCC conservatively stated that the positive summer SAM trend is likely (66–100% probability) to be due in part to stratospheric ozone depletion.

Antarctic temperatures

Undertaking attribution studies in the southern high latitudes is made particularly difficult by the high inter-annual variability and the short time-series of available data. Antarctic SAT observations are both few and far between prior to the International Geophysical Year (IGY) of 1957–1958. Some earlier data exist, such as the Orcadas record from Signy Island, which begins in 1903, and the Faraday/Vernadsky record from the Antarctic Peninsula, starting in 1947. Even following the IGY there are fewer than 20 records with which to identify long-term temperature trends (Turner et al. 2005). Furthermore, the vast majority of the stations are located in the Antarctic coastal region or sub-Antarctic islands of the Southern Ocean, with only two providing long-term meteorological data for the interior of the continent. Averaging over all observed locations, Antarctica has warmed over the 1950–2008 period although some individual locations have cooled in certain seasons.

The recent acceleration of glaciers in West Antarctica and their significant contribution to sea level rise (McMillan, Chapter 14 of this volume) has made this region a focus for recent climate change studies. West Antarctica has a particular paucity of observations although a 50-year SAT series has recently been reconstructed for Byrd station (Bromwich et al. 2013). This record, together with several statistical reconstructions of West Antarctic SATs, indicates that West Antarctica is actually one of the most rapidly warming regions on Earth.

The key influence on Antarctic SATs in the past few decades has been the trend towards a more positive phase of the SAM in summer and autumn, as noted previously. The SAM is generally positively correlated with SAT over the Antarctic Peninsula and negatively correlated with SAT over the majority of the rest of the continent (e.g. Thompson et al. 2011). Marshall et al. (2006) demonstrated a regional mechanism linking the more positive summer SAM to the warming and subsequent disintegration of the Larsen B ice shelf in the northeast Antarctic Peninsula, thereby suggesting an anthropogenic driver for the latter. Nicolas and Bromwich (2014) showed that the SAM trends have had a statistically significant cooling effect in East Antarctica in both summer and autumn and also West Antarctica in the latter season only. This cooling may be interpreted as having 'shielded' these parts of the continent from 'global warming' or as an additional warming that might occur if the SAM reverts to its average pre-1980 state as some models predict will happen with the recovery of the ozone hole.

By removing the temperature trends congruent with the SAM in annual data, Gillett et al. (2008) revealed a residual warming across most of the Antarctic continent in which a clear anthropogenic influence could be detected. However, the IPCC concluded that there is only low confidence that anthropogenic influence has contributed to the overall warming due to the issues with the observational datasets as described, and with the Gillett et al. (2008) study being the only formal attribution study of Antarctic SAT to date (Bindoff et al. 2013). The situation is further complicated by the discovery of decadal-scale reversals in the sign of the SAM-SAT relationship due to changes in the longwave pattern around Antarctica, as occurred in the first decade of the twenty-first century (Marshall et al. 2013).

Antarctic precipitation

Austral summer precipitation change over the Southern Ocean has been attributed to anthropogenic forcing by Fyfe et al. (2012). In the Southern Hemisphere zonally averaged precipitation has declined around 45°S and increased around 60°S since 1957, consistent with CMIP5 historical simulations, with the magnitude of the half-century trend outside the range of simulated natural variability. Model simulations with individual forcings reveal that the observed pattern of precipitation change is substantially forced by anthropogenic GHG and ozone changes, with an opposing influence from aerosols. Confidence in this result is enhanced by its consistency with the trends in the SAM. Such changes in precipitation may have contributed to the observed increase in Antarctic sea ice, with greater snowfall on the ice increasing both its albedo, making it less susceptible to melting, and thickness, as the snow depresses the ice surface below sea level and the consequent flooding leads to rapid transformation of snow to ice (Wilkinson and Stroeve, Chapter 13 of this volume). In addition, any further changes in southern high latitude precipitation have the potential to affect ocean stratification and thus the uptake of heat and carbon into the Southern Ocean, which could have a profound impact on future global climate change.

Concluding remarks

There is now major interest in climate change in the Polar Regions, in particular with the suggestion that the amplified warming in the Arctic has had a large impact on the weather of the populated Northern Hemisphere by altering the speed and amplitude of the meanders of the mid-latitude jet stream (e.g. Francis and Vavrus 2012; Screen and Simmonds 2014). A growing number of studies have already demonstrated an anthropogenic contribution to both the substantial Arctic warming since the middle of the twentieth century and other aspects of the regional climate such as the hydrological cycle. Moreover, the IPCC model projections of climate change through the coming century, based on a range of different possible GHG emission scenarios, all predict that the greatest future warming is also likely to occur in northern high latitudes.

Directly attributing Antarctic climate change to anthropogenic forcing has proved more difficult due to the shorter timescales of observations and high climate variability, which includes SAT trends of opposite sign in different areas of the continent. Thus, attribution studies have focused on the impact of the ozone hole on driving stronger circumpolar westerlies (a more positive SAM), which has had a significant influence on recent Antarctic climate. In a future with increasing GHGs there is likely to be a more positive SAM in all seasons except for the austral summer. In this season the effects of greater atmospheric GHG concentrations and ozone recovery, which has likely begun already (e.g. Shepherd et al. 2014), will act to 'push' the SAM in opposite directions. Modelling studies to date suggest that the two will be approximately in balance (e.g. Polvani et al. 2011; Gillett and Fyfe 2013) although this is clearly dependent on the amount of GHGs emitted. Nevertheless, when considering both Polar Regions across all seasons, it seems highly likely that existing anthropogenically-driven polar climate change will become even stronger in the future.

References

Abatzoglou, J.T. and K.T. Redmond 2007. 'Asymmetry between trends in spring and autumn temperature and circulation regimes over western North America' *Geophysical Research Letters* 34, L18808, doi:10.1029/2007GL030891.

Allen, R.J., J.R. Norris and M. Kovilakam 2014. 'Influence of anthropogenic aerosols and the Pacific Decadal Oscillation on tropical belt width' *Nature Geoscience* 7: 271–274.

Arblaster, J.M. and G.A. Meehl 2006. 'Contributions of external forcings to Southern Annular Mode trends' *Journal of Climate* 19: 2896–2905.

Bindoff, N.L. and 14 others 2013. 'Detection and attribution of climate change: from global to regions' in Stocker, T.F. and 9 others (eds) *Climate Change 2013: The Physical Science Basis. Contribution of Working Group I to the Fifth Assessment Report of the Intergovernmental Panel on Climate Change.* Cambridge, UK and New York: Cambridge University Press.

Bromwich, D.H., J.P. Nicolas, A.J. Monaghan, M.A. Lazzara, L.M. Keller, G.A. Weidner and A.B. Wilson 2013. 'Central West Antarctica among the most rapidly warming regions on Earth' *Nature Geoscience* 6: 139–145.

Brown, R.D. and P.W. Mote 2009. 'The response of Northern Hemisphere snow cover to a changing climate' *Journal of Climate* 22: 2124–2145.

Callaghan, T.V. and 20 others 2011. 'The changing face of Arctic snow cover: a synthesis of observed and projected changes' *Ambio* 40: 17–31.

Chylek, P., N. Hengartner, G. Lesins, J.D. Klett, O. Humlum, M. Wyatt and M.K. Dubey 2014. 'Isolating the anthropogenic component of Arctic warming' *Geophysical Research Letters* 41: 3569–3576.

Cohen, J.L., J.C. Furtado, M.A. Barlow, V.A. Alexeev and J.E. Cherry 2012. 'Arctic warming, increasing snow cover and widespread boreal winter cooling' *Environmental Research Letters* 7, doi:10.1088/1748-9326/7/1/014007.

Flanner, M.G., K.M. Shell, M. Barlage, D.K. Perovich and M.A. Tschudi 2011. 'Radiative forcing and albedo feedback from the Northern Hemisphere cryosphere between 1979 and 2008' *Nature Geoscience* 4: 151–155.

Fogt, R.L., J. Perlwitz, A.J. Monaghan, D.H. Bromwich, J.M. Jones and G.J. Marshall 2009. 'Historical SAM variability. Part II: twentieth-century variability and trends from reconstructions, observations, and the IPCC AR4 models' *Journal of Climate* 22: 5346–5365.

Francis, J.A. and S.J. Vavrus 2012. 'Evidence linking Arctic amplification to extreme weather in mid-latitudes' *Geophysical Research Letters* 39, L06801, doi:10.1029/2012GL051000.

Fyfe, J.C., N.P. Gillett and G.J. Marshall 2012. 'Human influence on extratropical Southern Hemisphere summer precipitation' *Geophysical Research Letters* 39, L23711, doi:10.1029/2012GL054199.

Gillett, N.P. and J.C. Fyfe 2013. 'Annular mode changes in the CMIP5 simulations' *Geophysical Research Letters* 40: 1–5.

Gillett, N.P., J.C. Fyfe and D.E. Parker 2013. 'Attribution of observed sea level pressure trends to greenhouse gas, aerosol, and ozone changes' *Geophysical Research Letters* 40: 2302–2306.

Gillett, N P., D.A. Stone, P.A. Stott, T. Nozawa, A.Y. Karpechko, G.C. Hegerl, M.F. Wehner and P.D. Jones 2008. 'Attribution of polar warming to human influence' *Nature Geoscience* 2: 750–754.

Hawkins, E. and R. Sutton 2009. 'The potential to narrow uncertainty in regional climate predictions' *Bulletin of the American Meteorological Society* 9: 1095–1107.

Hu, R.M., J.P. Blanchet and E. Girard 2005. 'Evaluation of the direct and indirect radiative and climate effects of aerosols over the western Arctic' *Journal of Geophysical Research* 110, doi:10.1029/2004JD005043.

Hurrell, J.W., Y. Kushnir, G. Ottersen and M. Visbeck 2003. 'An overview of the North Atlantic Oscillation' in Hurrell, J.W., Y. Kushnir, G. Ottersen and M. Visbeck (eds) *The North Atlantic Oscillation: Climatic Significance and Environmental Impact. Geophysical Monograph* 134: 1–35. Washington, DC: American Geophysical Union.

Lee, S. and S.B. Feldstein 2013. 'Detecting ozone- and greenhouse gas-driven wind trends with observational data' *Science* 339: 563–567.

Lubin, D. and M. Vogelmann 2006. 'A climatologically significant aerosol longwave indirect effect in the Arctic' *Nature* 439: 453–456.

Marshall, G.J. 2003. 'Trends in the Southern Annular Mode from reanalyses' *Journal of Climate* 16: 4134–4143.

Marshall, G.J., A. Orr, N.P.M. van Lipzig and J.C. King 2006. 'The impact of a changing Southern Hemisphere Annular Mode on Antarctic Peninsula summer temperatures' *Journal of Climate* 19: 5388–5404.

Marshall, G.J., A. Orr and J. Turner 2013. 'A predominant reversal in the relationship between the SAM and East Antarctic temperatures during the twenty-first century' *Journal of Climate* 26: 5196–5204.

Miles, M.W., D.V. Divine, T. Furevik, E. Jansen, M. Moros and A.E.J. Ogilvie 2013. 'A signal of persistent Atlantic multidecadal variability in Arctic sea ice' *Geophysical Research Letters* 41: 463–469.

Min, S-K., X. Zhang and F. Zwiers 2008. 'Human-induced Arctic moistening' *Science* 320, 518–520.

Nicolas, J.P. and D.H. Bromwich 2014. 'New reconstruction of Antarctic near-surface temperatures: multidecadal trends and reliability of global reanalyses' *Journal of Climate* 27: 8070–8093.

Polson, D., G.C. Hegerl, X. Zhang and T.J. Osborn 2013. 'Causes of robust seasonal land precipitation changes' *Journal of Climate* 26: 6679–6697.

Polvani, L.M., M. Previdi and C. Deser 2011. 'Large cancellation, due to ozone recovery, of future Southern Hemisphere atmospheric circulation trends' *Geophysical Research Letters* 38, L04707, doi:10.1029/2011GL046712.

Previdi, M. and L.M. Polvani 2014. 'Climate system response to stratospheric ozone depletion' *Quarterly Journal of the Royal Meteorological Society* doi:10.1002/qj.2330.

Rupp, D.E., P.W. Mote, N.L. Bindoff, P.A. Stott and D.A. Robinson 2013. 'Detection and attribution of observed changes in Northern Hemisphere spring snow cover' *Journal of Climate* 26: 6904–6914.

Screen, J.A. and I. Simmonds 2014. 'Amplified mid-latitude planetary waves favour particular regional weather extremes' *Nature Climate Change* 4: 704–709.

Serreze, M. and R. Barry 2005. *The Arctic Climate System.* Cambridge, UK and New York: Cambridge University Press.

Shepherd, T.G., D.A. Plummer, J.F. Scinocca, M.I. Hegglin, V.E. Fioletov, M.C. Reader, E. Rembsberg, T. Von Clarmann and H.J. Wang 2014. 'Reconciliation of halogen-induced ozone loss with the total-column ozone record' *Nature Geoscience* 7: 443–449.

Shindell, D. and G. Faluvegi 2009. 'Climate response to regional radiative forcing during the twentieth century' *Nature Geoscience* 2: 294–300.

Thompson, D.W.J., S. Solomon, P.J. Kushner, M.H. England, K.M. Grise and D.J. Karoly 2011. 'Signatures of the Antarctic ozone hole in Southern Hemisphere surface climate change' *Nature Geoscience* 4: 741–749.

Turner, J. and J. Overland 2009. 'Contrasting climate change in the two Polar Regions' *Polar Research* 28: 146–164.

Turner, J. and 8 others 2005. 'Antarctic climate change during the last 50 years' *International Journal of Climatology* 25: 279–294.

Vaughan, D.G. and 13 others 2013. 'Observations: cryosphere' in Stocker, T.F. and 9 others (eds) *Climate Change 2013: The Physical Science Basis. Contribution of Working Group I to the Fifth Assessment Report of the Intergovernmental Panel on Climate Change.* Cambridge, UK and New York: Cambridge University Press.

Zhao, C. and T.J. Garrett 2015. 'Effects of Arctic haze on surface cloud radiative forcing' *Geophysical Research Letters* doi:1002/2014GL062015.

16

Post Last Glacial Maximum processes in the Polar Regions

Pippa Whitehouse

Introduction

Over the last ~20,000 years, since the peak of the last glaciation, the Polar Regions have seen dramatic changes in the extent of the polar ice sheets, the state of the polar oceans, and even the shape of the land. The underlying cause of these changes can be traced to changes in the global climate system, which is largely driven by periodic variations in Earth's orbit around the sun (Hays et al. 1976). Over the last million years Earth has passed through a series of glacial cycles, each one roughly 100,000 years long (Shackleton 2000). However, complex feedbacks between the ice sheets, oceans, atmosphere, ecosystems and solid Earth all play a role in determining the details of each glacial cycle.

The changes that have taken place since the peak of the last glaciation – known as the Last Glacial Maximum, or LGM – include the melting of ~52 million cubic kilometres of ice, enough to raise mean sea level by ~130 m (Lambeck et al. 2014), an increase in CO_2 levels from ~190 ppm to pre-industrial levels of ~280 ppm (Petit et al. 1999), and an increase in global mean air temperatures of ~3.5°C (Shakun et al. 2012). Projections for future climatic changes include a large degree of uncertainty (IPCC 2013); it is therefore useful to explore the response of the Polar Regions to past climatic changes, and to consider the degree to which those changes may potentially be mirrored in future.

A degree of caution is necessary when comparing past changes with the present because conditions during the last deglaciation – for example, the geometry of the ice sheets, Earth's orbital state, and the state of the ocean-atmosphere system – may have placed the Polar Regions near a tipping point for change, pre-conditioning them to undergo periods of rapid ice loss. However, although the conditions today may be different, many of the underlying processes that were active during the post-LGM period continue to shape the present evolution of the Arctic and Antarctic. We begin by briefly describing conditions across the Polar Regions during the LGM.

The Polar Regions during the LGM

Global ice volume was at a maximum between ~26,500 and ~19,000 years ago (Clark et al. 2009), but on a regional scale maximum ice extents and thicknesses were not necessarily reached contemporaneously. For example, analysis of marine sediment cores from the seafloor around Antarctica suggest that some sectors of that ice sheet had already begun to retreat prior to the time of the global LGM (Livingstone et al. 2012), while ice thicknesses in other areas only reached a maximum ~17,000 years ago (Todd et al. 2010). In the Canadian Arctic ice coverage did peak during the global LGM (Dyke et al. 2002), but in the Eurasian Arctic maximum ice extents were reached much earlier, between 90,000 and 80,000 years ago (Svendsen et al. 2004 and references therein). During this early glacial expansion, ice extended across north-west Russia but there was limited ice in Scandinavia. This situation was reversed by the time of the global LGM, when a thick ice sheet was centred on Scandinavia, but north-west Russia remained ice free (Lambeck et al. 2010 and references therein).

The reason for these regional variations is that conditions for maximal ice extent and thickness were not favourable in all locations at the same time. Some settings required a change in snowfall distribution or local seasonality, e.g. an increase in the duration or severity of winter conditions, for substantial ice-sheet growth, while others required a drop in ocean temperatures or even sea level itself for existing ice sheets to advance. Many of the polar ice sheets were marine-terminating during the LGM, meaning that grounded ice extended into the ocean until an increase in the depth of the seafloor meant that the ice began to float and eventually break off as icebergs. Marine-terminating ice sheets are likely to have responded

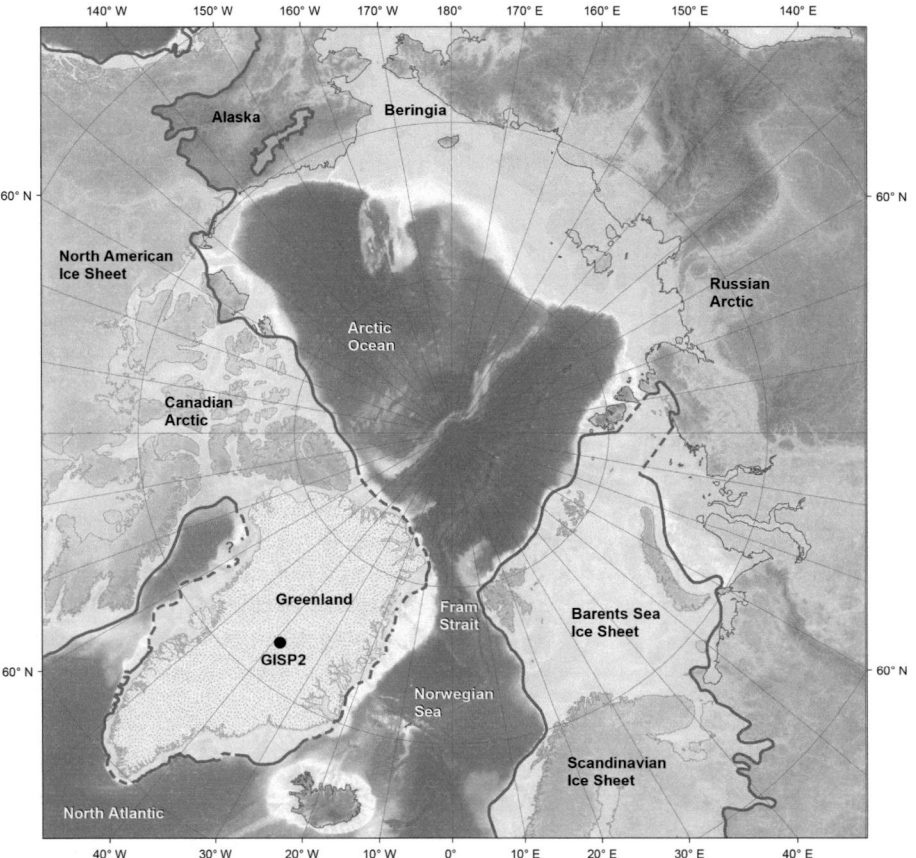

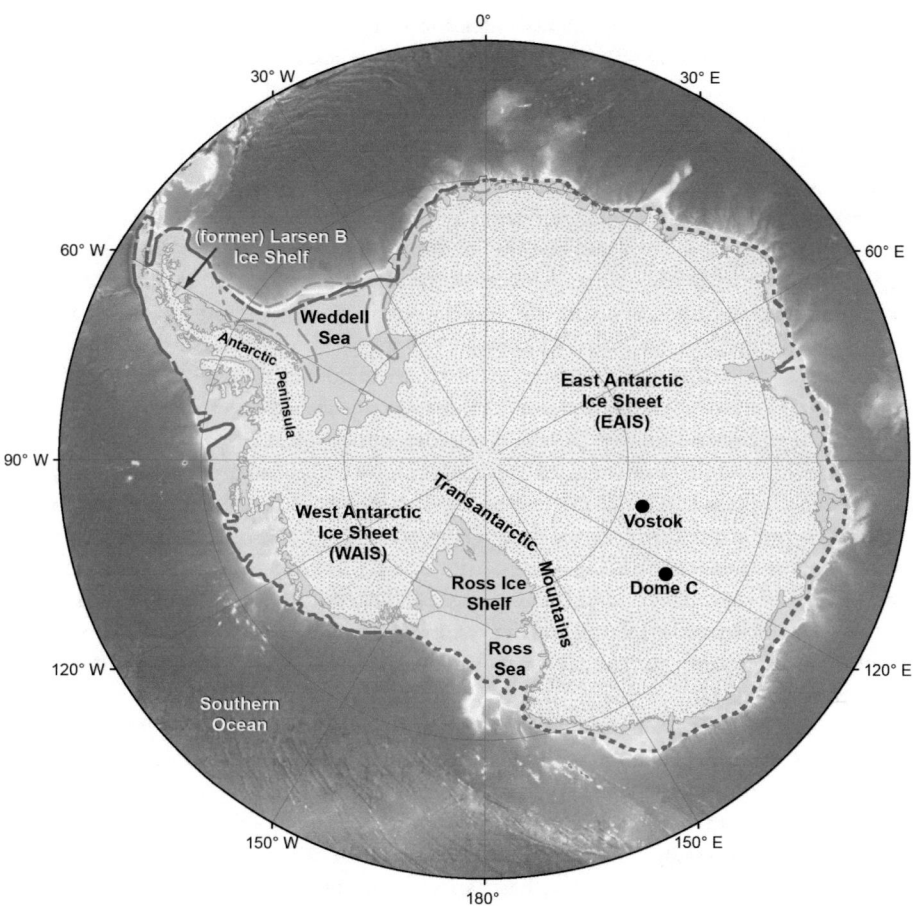

Figure 16.1 Locations mentioned in the text for (a) the Arctic and (b) the Antarctic. Ice core locations are shown as a black dot. LGM ice extents are indicated by a transparent grey screen surrounded by a line; areas where former ice extent is less well constrained are indicated by dashed (inferred ice margin) and dotted (speculative ice margin) lines. Ice-sheet margins are drawn by the author after Kleman et al. (2010), Manley and Kaufman (2002), Funder et al. (2011), Ó Cofaigh et al. (2013), Svendsen et al. (2004) and Bentley et al. (2014). Arctic bathymetry: IBCAO (Jakobsson et al., 2012). Antarctic bathymetry: IBCSO (Arndt et al., 2013).

more strongly to changes in ocean temperatures than atmospheric temperatures, with the latter playing an important role in governing the behaviour of land-terminating ice sheets.

The ice sheets will also have influenced each other as they evolved. For example, the expansion of the West Antarctic Ice Sheet (WAIS) into the Ross Sea will have dammed the flow of ice from the East Antarctic Ice Sheet (EAIS) as it flowed through the Transantarctic Mountains (Figure 16.1), while the presence of the Scandinavian ice sheet during the LGM meant that the Russian Arctic was in a rain shadow, and hence it received very little snowfall. The growth of the North American ice sheet will have similarly perturbed atmospheric and oceanic conditions around Greenland. These differences in geographical setting, local climate, and internal ice dynamics led to large variations in the behaviour of the polar ice sheets during and after the LGM.

Our knowledge of the past extent and thickness of the polar ice sheets is based on a diverse range of marine and terrestrial observations. Onshore evidence for past ice extent and flow direction includes features such as moraines, trim lines, drumlins, and mega-scale glacial lineations (Benn and Evans 2010), while techniques such as cosmogenic exposure dating (Balco 2011) are used to determine the time at which a site was last covered by ice. Geomorphological evidence must be carefully interpreted, bearing in mind that different processes prevail in the cold, dry environment of Antarctica and the relatively warmer, wetter environments of the terrestrial Arctic. Every passage of ice will reshape the landscape, and while the destructive nature of ice means that the majority of the evidence relates to the most recent ice retreat, there can be a more complex picture to decipher; it is therefore crucial to be able to date the age of each landform. This can be challenging in the Polar Regions due to a lack of 'dateable' organic material or the biasing presence of 'old' recycled carbon, while non-uniqueness is a problem when trying to reconstruct local ice cover history via cosmogenic exposure dating (Alexanderson et al. 2014).

Offshore, bathymetric surveys reveal the geomorphological fingerprint of past ice flow across the seafloor, while additional information can be deduced from the analysis of marine sediment (e.g. Cofaigh 2012). If the age of the sediment can be determined then it is possible to work out from the type of material deposited when that location was variously covered by grounded ice, floating ice (an 'ice shelf'), sea ice, or open ocean. However, the acquisition of such data is expensive and time-consuming, and there are many regions where we lack information relating to the maximum marine extent of the major ice sheets during the LGM, in both the Arctic (e.g. Funder et al. 2011) and Antarctic (Bentley et al. 2014) (see Figure 16.1).

Onshore ice extents are better known but determining past ice *thickness* is difficult in regions where the topography is very flat, e.g. across the Canadian and Eurasian Arctic, because there are no mountains to act as a 'dipstick' and record evidence of past ice thickening and thinning. In regions where ice is still present it is possible to use evidence from ice cores to determine past changes in ice elevation. Ice core data are particularly useful for understanding changes across the vast East Antarctic plateau, where evidence suggests that, contrary to all other polar ice masses, the centre of the EAIS was actually thinner during the LGM (Parrenin et al. 2007). This surprising result is linked to the fact that the colder temperatures decreased the ability of the atmosphere to hold and transport moisture, and that the source of that moisture, i.e. the open ocean, was further away due to increased sea ice coverage (Gersonde et al. 2005).

Several features of the polar environment during the LGM can be attributed to the fact that atmospheric and oceanic circulation were likely different to the present. The details of these differences are still poorly known due to sparse data records, but the growth of the polar ice sheets, up to several kilometres thick, will have affected atmospheric circulation patterns and hence snowfall patterns. Changes in wind strength and direction will also have had an impact on ocean circulation, which is driven by a combination of wind stress at the surface of the ocean and density variations within the ocean. Atmospheric temperatures, the freshwater input to the ocean, and sea ice extent will all have played a role in governing the density structure of the ocean and hence ocean circulation during the LGM.

The growth of the ice sheets will also have affected the shape of the solid Earth, with land beneath the kilometre-scale ice sheets being depressed by several hundred metres and the surrounding regions being elevated due to mantle material being squeezed out sideways from beneath the ice sheets (Milne and Shennan 2013). An additional consequence of the growth of the polar ice sheets was a drop in mean sea level, which altered the shape of the polar coastlines, most notably resulting in the emergence of Beringia (Figure 16.1). The presence of this land bridge between Alaska and Kamchatka will not only have influenced local climate and ocean circulation, but it will have had long-lasting effects on biodiversity across the Polar Regions and on human migration patterns.

Post-LGM changes across the Polar Regions

The dramatic changes that have occurred across the Polar Regions since the LGM, and the processes that have brought about these changes, were triggered by a combination of (i) changes to Earth's orbit and (ii) dynamic feedbacks between the ice sheets, the ocean, the atmosphere, and the solid Earth. The former plays a fundamental role in regulating Earth's climate and varies with known periodicity; it is the less well-understood feedbacks that govern the details. In this section we briefly outline the changes that have occurred across the Polar Regions since the LGM, and then go on to discuss the processes that brought about these changes.

The rate and magnitude of climatic changes in the Polar Regions typically exceed global mean changes; a phenomenon referred to as 'polar amplification'. For example, while global mean temperatures have increased by ~3.5°C since the LGM, evidence from ice cores suggests that temperatures in Antarctica and Greenland increased by 10–15°C over this period. The reason for this amplification is linked to the way in which heat is transferred between different components of the Earth system (Miller et al. 2010a; Serreze and Barry 2011); in particular, changes in sea ice extent produced enhanced temperature changes due to feedbacks associated with albedo change. A second surprising observation is that the Polar Regions have not only experienced warming since the LGM, but there have also been periods of cooling, for up to several thousand years at a time. Furthermore, these periods of cooling and warming have often been asynchronous in the two hemispheres. A summary of the major post-LGM climatic changes in each region is given in Table 16.1.

Timeline of climatic changes in the Polar Regions since the LGM

Early post-LGM

The North American and Scandinavian ice sheets began to melt soon after 20 ka BP, with warming also recorded in the Southern Ocean (Alley 2000) and Antarctica (Fudge et al. 2013) at this time. However, Greenland remained cold until ~14.6 ka BP, likely due to reduced ocean circulation in the North Atlantic (see later).

Bølling-Allerød/Antarctic Cold Reversal

Around 14.6 ka BP there was an abrupt change in conditions in both Polar Regions, which coincided with a ~400-year period of rapid sea-level rise, known as Meltwater Pulse 1a (Fairbanks 1989). The source of ice melt that led to this rapid sea-level rise remains the source of much debate (Gomez et al. 2015; Weaver et al. 2003), but it is clear that around this time temperatures across Greenland abruptly increased, while Antarctica experienced a decrease in the rate of warming or a return to cooling. In the northern hemisphere, this period of warming is known as the Bølling-Allerød, while in the southern hemisphere, the return to cooling is known as the Antarctic Cold Reversal.

Younger Dryas

Following the Bølling-Allerød/Antarctic Cold Reversal, there was a return to cold, dry conditions in the northern hemisphere between 12.8–11.5 ka BP (Alley 2000) during a period known

Table 16.1 Climatic changes in the Polar Regions since the LGM; # = warming; * = cooling; ** = no clear signal. Dates prior to AD 0 are expressed in 'ka BP', which represents 'thousands of years before present'. Table by author.

Name	Approx. dates	Arctic	Antarctic
LGM	Until ~19 ka BP	*Cold; more extensive ice than present; lower precipitation	
Early post-LGM	~19 ka BP – ~14.6 ka BP	**Insolation increase; cold conditions persist in Greenland; warming and ice retreat outside Greenland; meltwater flux suppresses North Atlantic circulation	#Insolation decrease; warming recorded across Antarctica and the Southern Ocean
Bølling-Allerød/Antarctic Cold Reversal	~14.6 ka BP – ~12.8 ka BP	#Abrupt warming; coincides with a sudden increase in the rate of sea-level rise	*Reduced warming or a return to cooling
Younger Dryas	~12.8 ka BP – ~11.5 ka BP	*Cold, dry conditions; regional ice re-advance; ends with abrupt warming	#Warming to present temperatures; increase in snowfall rates
Holocene Thermal Maximum	~11.5 ka BP – ~5.5 ka BP	#Maximum temperatures and minimum ice extents for the post-LGM period; brief return to cold, dry conditions at ~8.2 ka BP	***Temperatures peak at ~0.4°C above present; no response to 8.2 ka event
Mid-to-Late Holocene/ Neoglacial	~5.5 ka BP – ~AD 950	*Decreasing summer insolation; cooling; ice advance	**Increasing summer insolation; both advance and retreat of ice extent
Medieval Warm Period	~AD 950 – ~AD 1250	#Warming; ice retreat behind present in Greenland	**No clear signal
Little Ice Age	~AD 1250 – ~AD 1850	*Cooling; ice advance	**No clear signal
Twentieth century	AD 1900 onwards	#Warming; reducing sea ice extent; net ice loss	#Warming; ice shelf collapse; net ice loss

as the Younger Dryas. Temperatures were ~15°C colder than present, and there was an increase in ice extent in many regions of the northern hemisphere. Once again, changes in Antarctica seem to have been almost opposite to those in the northern hemisphere, with warming resuming across Antarctica shortly after the beginning of the Younger Dryas, and temperatures reaching present-day values by ~12 ka BP. The Younger Dryas terminated abruptly with a period of rapid warming in the northern hemisphere (Alley 2000).

Early Holocene and Holocene Thermal Maximum

Since the beginning of the Holocene period, around 11.5 ka BP, the climate across the Polar Regions has been much more stable, with average temperatures in both regions typically lying within a few degrees of present. Temperature reconstructions, as determined from changes in the position of the treeline, changes in glacier extents, and via the analysis of pollen, tree-ring, speleothem, ice core, lake and marine records (Kaufman et al. 2004; Marcott et al. 2013; Nesje et al. 2005), imply that northern hemisphere summer temperatures were above present during the Holocene Thermal Maximum (~11.5 to 5.5 ka BP). Glacier extents across much of the Arctic were smaller than present during this period (Solomina et al. 2015), although this warm period was briefly interrupted by a return to cold, dry conditions around 8.2 ka BP (Alley et al. 1997).

Neoglacial, Medieval Warm Period, Little Ice Age

During the last 5000 years the Arctic experienced gradual cooling – the 'Neoglacial' – which culminated in the Little Ice Age and an increase in ice extent in many northern hemisphere regions. There were short-lived local variations to this pattern, e.g. Greenland experienced a temporary return to warmer conditions during the Medieval Warm Period (Miller et al. 2010b). These fluctuations in climate had a direct impact on the distribution of human settlements around the Arctic during this period (D'Andrea et al. 2011). In contrast, evidence from Antarctica shows a more complicated pattern, with advance and retreat occurring at the same time in different regions (Hall 2009). There is evidence for a return to cooler conditions during the last 1000 years (Orsi et al. 2012), but there is not yet consensus as to whether a clear signature of the Little Ice Age can be identified across Antarctica.

Improved data resolution for the recent past reveals that the dominant feature of polar climates during the last few hundred years has been one of regional and temporal variability (e.g. Overpeck et al. 1997), likely driven by variations in solar activity, volcanism, and internal feedbacks within the climate system. However, in both regions, warming during the twentieth century is unequivocal.

Processes associated with climatic changes in the Polar Regions

The dramatic climate reversals that occurred during the last deglaciation reflect the strong sensitivity of the Polar Regions to external forcing during this period. The trigger for each change is the subject of ongoing debate, and in many cases a combination of factors likely played a role. It is therefore crucial to be able to accurately date any event in order to attribute cause and effect. The main processes that have contributed to changing Polar conditions since the LGM are discussed below.

Insolation effects

Gradual changes to the seasonal distribution of incoming solar radiation ('insolation') – arising as a result of changes in Earth's orbit – determine when in the past each hemisphere had the hottest summers or the coldest winters. Summer insolation dramatically increased in the northern hemisphere between 24 and 12 ka BP, which helped to trigger the melting of the North American and Scandinavian ice sheets. The resulting flux of meltwater into the North Atlantic from these *subpolar* ice sheets played an important role in governing climate changes during the deglaciation. Conversely, although increases in summer insolation did trigger some warming in the *polar* ice sheets (Greenland and Antarctica) early in the deglaciation, air temperatures at the surface of these higher-latitude, higher-elevation ice sheets were so low that a few degrees warming was insufficient to trigger large scale melting. The mechanism by which these ice sheets predominantly lost mass during this period was not by melting at their surface, but via an increase in the rate at which ice flowed into the ocean.

In general, insolation variations in the two hemispheres have been out of phase since the LGM, and this is one of the reasons that the Arctic and Antarctic have experienced different post-LGM climatic histories. However, feedbacks associated with changes in ocean circulation also played an important role, and the very different setting of the two regions – with the Antarctic being a continent surrounded by ocean and the Arctic being an ocean surrounded by continents – means that different processes are important in each region.

Oceanic processes

Ocean circulation and heat transport

Ocean circulation plays an important role in determining how heat is distributed across the surface of Earth (e.g. Rahmstorf 2002: Bingham, Chapter 12 of this volume). Incoming solar radiation warms the surface of the ocean; this heat is stored, and then redistributed throughout the ocean via global-scale currents. These currents are driven by wind and by density variations, with the latter arising due to variations in temperature and salinity.

In the North Atlantic Ocean, wind-driven surface currents flow from south to north. During this journey the water cools and becomes more saline, with the salinity increasing initially due to evaporation and later due to sea ice formation. As this increasingly dense water reaches the Norwegian Sea it sinks, displacing the water beneath and setting up an ocean-scale circulation system which has been likened to a large conveyor belt (see Bingham, Chapter 12 of this volume). When the speed of this conveyor belt is fast, warm tropical water from the south is quickly translated north towards the Arctic, and the land around the North Atlantic experiences anomalously warm conditions. However, if the conveyor belt slows or stops for any reason then less heat is delivered to the Arctic and temperatures plummet.

Northern hemisphere cooling – freshwater pulses

One way to slow the conveyor belt is to introduce a large volume of cold, freshwater into the North Atlantic. Such a water mass would be less dense than the surrounding ocean (even though it is cold, its freshness wins out) and therefore it would not sink; without sinking there can be no circulation.

Early during the last deglaciation Greenland remained cold despite an increase in northern hemisphere insolation. This is thought to be associated with a reduction in the strength of North Atlantic circulation due to an influx of meltwater (Broecker and Denton 1990), but where did the meltwater come from if Greenland itself was not melting? The answer can be found by studying the marine sediment of the North Atlantic, which contains distinct layers of iceberg-transported debris. It is thought that rapid ice loss from North America soon after the LGM resulted in a series of massive iceberg-discharge events – 'Heinrich' events (Heinrich 1988). This sudden influx of freshwater caused a reduction in North Atlantic circulation which helped maintain cold, dry conditions across Greenland until ~14.6 ka BP. It remains to be determined whether these ice discharge events were driven by natural climatic cycles, or oscillations associated with internal ice-sheet dynamics (e.g. Macayeal 1993).

Later in the deglaciation, gradual warming in the northern hemisphere was twice interrupted by a return to colder, drier conditions; during the Younger Dryas and the '8.2 ka BP event' (see Table 16.1). These episodes may also have been triggered by iceberg discharge, but an alternative hypothesis revolves around the catastrophic drainage of water from ice-dammed lakes (Alley et al. 1997). During the final stages of deglaciation in North America, meltwater from the disintegrating ice sheet was trapped in a giant lake adjacent to the ice sheet – Lake Agassiz – which, at its peak, stretched from Manitoba to Quebec. Episodically, the lake breached the ice that was holding it in place, leading to large flooding events that are recorded in sedimentary deposits across North America (Murton et al. 2010) and in the chemical composition of the ocean at the time (Carlson et al. 2007).

Hemispheric differences

During prolonged periods of cold Arctic conditions, e.g. between the LGM and 14.6 ka BP, and during the Younger Dryas, Antarctic ice cores conversely indicate warming (Table 16.1). A potential mechanism to explain this is that during periods of weak North Atlantic circulation heat builds up in the tropics rather than being transported to the northern hemisphere, leading to warming in the South Atlantic and Antarctica (Alley 2000 and references therein).

Northern hemisphere warming

The abrupt transition to warmer Arctic conditions at the beginning of the Bølling-Allerød and the end of the Younger Dryas can similarly be explained as being due to the redistribution of heat, this time from the southern hemisphere to the northern hemisphere as ocean circulation resumes in the North Atlantic. This phenomenon is referred to as the 'bipolar seesaw' effect (Broecker 1998), and importantly it does not require an abrupt change to the global energy budget. Hypotheses to explain the sudden resumption of circulation in the North Atlantic include a decrease in the rate of meltwater from North America (Alley 2000) or a sudden flux of meltwater from Antarctica into the Southern Ocean (Weaver et al. 2003).

The Arctic Ocean and sea ice

The Arctic Ocean was almost isolated from the rest of the global ocean during the LGM because ice cover or newly-exposed land bridges blocked all of the marine passages except the Fram Strait (see Figure 16.1). There was still water exchange through this passage, and the flow of relatively warm, salty water from the Atlantic helped to maintain seasonal ice-free conditions along the Norwegian margin, assisted by katabatic winds from the Barents Sea Ice Sheet (Nørgaard-Pedersen et al. 2003). This open water would have been an important moisture source for snowfall across the ice sheet.

North of 85°N, low seafloor sedimentation rates and a lack of evidence for biological productivity suggest that the Arctic Ocean was permanently ice-covered during the LGM, with the ice being underlain by a layer of cold, low-salinity water. This low-salinity layer persists today, fuelled by high runoff rates from the surrounding Arctic rivers, but its origin during the much drier LGM period is probably linked to the lower input of high-salinity water from the Atlantic and minimal ocean circulation, with the thick ice cover preventing any wind-driven mixing.

Increased meltwater input, the gradual opening of additional marine gateways, and increasing summer insolation all contributed to a change in temperature and salinity conditions across the Arctic Ocean in the early Holocene. Sea ice is thought to have been at a minimum at this time (Polyak et al. 2010). This assumption is largely based on coastal evidence for ice-free conditions; however, once sea ice coverage begins to decrease this can lead to a runaway scenario because the darker ocean surface results in less solar energy being reflected back into space, i.e. the *albedo* is lower. This leads to the increased absorption of heat into the ocean and hence further sea ice melt. This feedback process is one of the explanations for the polar amplification effect.

Following the relatively ice-free conditions of the early Holocene, Arctic sea ice coverage increased again during northern hemisphere cooling in the Mid-to-Late Holocene (Polyak et al. 2010), potentially assisted by another feedback effect whereby increasing sea ice coverage can enhance local *atmospheric* cooling by insulating it from the relatively warm ocean beneath. The increase in sea ice cover will also have minimised wind-induced mixing of the cold upper ocean with warmer water beneath, with the wind instead potentially helping to compress and thicken the ice, thus increasing its chances of persisting for multiple years. The increase in coastal 'fast ice' at this time will have reduced coastal erosion rates and impacted on the migration and hunting methods adopted by both humans and other members of the Arctic food chain.

Atmospheric processes

Warm air is able to hold more moisture than cold air, and hence, counter-intuitively, the increase in polar air temperatures since the LGM contributed to an increase in snowfall in the Polar Regions, as recorded in ice cores from Greenland and Antarctica (e.g. Alley et al. 1993), although both regions are still classed as a polar desert. The post-LGM increase in snowfall was facilitated by a decrease in sea ice extent, which meant that the distance to the open ocean – the moisture source for precipitation – decreased. Reduced sea ice is also implicated in causing a change in atmospheric circulation patterns, which led to storm tracks being shifted towards Greenland (Kapsner et al. 1995), thus further enhancing snowfall rates.

Despite the increase in snowfall, which potentially prolonged the existence of some ice caps well into the Holocene, rising air temperatures also resulted in an increase in surface melt, and this eventually led to net ice loss and the retreat of the snowline, particularly in the Arctic. As the ice receded this will have decreased surface albedo, reinforcing local temperature increases and hence ice loss.

Surface melt is typically less pervasive in the Antarctic, where air temperatures at the surface of the ice sheet rarely rise above freezing. However, at the low elevation margins of the Antarctic Peninsula, increasing air temperatures have been implicated in causing the collapse of floating ice shelves (Mercer 1978). Changes in atmospheric circulation can enhance this process by driving an increase in ocean circulation, causing warm water to be drawn up beneath the ice shelves, thus increasing melt at their base as well as their surface.

Changes in atmospheric temperatures will also have affected ground temperatures, and hence permafrost extent throughout the Arctic (Hinzman et al. 2005; Overpeck et al. 1997). Permafrost forms an impermeable layer that allows wetlands to persist on a seasonal basis, and its degradation since the LGM has led to changes in the hydrology and ecology of the Arctic (Rouse et al. 1997), changes to the global carbon cycle due to the release of methane (Zimov et al. 2006), and localised subsidence.

Ice-sheet processes

Global mean sea level began to rise sharply between 20,000 and 19,000 years ago following the onset of ice-sheet melt (Clark et al. 2009), but the timing and style of polar deglaciation varied due to differences in the forcing mechanisms required to trigger retreat.

Ice sheets that extend into the ocean are typically faster flowing and thinner than ice sheets confined to land because the presence of water and deformable marine sediments at the base of the ice reduces the friction between the ice and the seafloor (Cuffey and Paterson 2010). In regions where this relatively thin ice is close to flotation, i.e. the ice is only just thick enough to prevent it from floating, this leads to a potentially unstable situation because small triggers such as an increase in sea level or an increase in melting can lead to the ice beginning to float. This further reduces friction at the base of the ice sheet, potentially leading to ice acceleration and net ice loss.

This scenario is exacerbated where the seafloor deepens towards the centre of the ice sheet, as is the case beneath West Antarctica, because the flux of ice into the ocean depends on the thickness of ice at the point where it begins to float, the 'grounding line' (Schoof 2007). Retreat of the grounding line into deeper water results in an increase in the flux of ice; if this is not matched by an increase in snowfall then the system enters a positive feedback loop – known as the marine-ice-sheet instability (Joughin and Alley 2011) – with the net result being accelerated ice loss.

The presence of a floating ice shelf in front of a grounded ice sheet, e.g. the Ross Ice Shelf, will reduce the flux of ice across the grounding line due to the resistive force provided by friction between the ice shelf and the adjacent land. If the thickness or extent of an ice shelf is reduced, either by increased melting at the base of the ice shelf following an influx of warm water, or an increase in the rate of iceberg calving from the front of the ice shelf, this will lead to accelerated ice flow and net ice loss, as has been observed following the rapid disintegration of the Larsen B ice shelf in the Antarctic Peninsula in early 2002 (Scambos et al. 2004).

These processes outline the main mechanisms by which the extensive marine-grounded ice sheets of the LGM retreated, either to their present extent (e.g. Antarctica and some regions of Greenland), or to a configuration where they became land-based ice sheets (e.g. the North American and Eurasian ice sheets). Once an ice sheet becomes land-based the ocean no longer has a direct influence on ice dynamics, so the style of ice flow will change, and the rate of ice loss will be dependent on atmospheric conditions; specifically, the balance between snowfall and ice melt. However, feedback processes can still play a role here, via a runaway process known as 'saddle collapse' (Gregoire et al. 2012). As the atmosphere warmed during the last deglaciation, melting will have lowered the ice surface, but as the ice surface lowered it will have come into contact with warmer temperatures, thus accelerating the rate of ice melt. This process, along with the marine-ice-sheet instability mechanism, has been implicated in driving the most rapid periods of sea-level rise during the last deglaciation.

Water also plays a role in governing the behaviour of a land-based ice sheet. If the melt generated at the surface of an ice sheet is able to make its way to the bed of the ice sheet, it will act as a lubricant and can cause temporary accelerations in ice flow. The timing of these episodic accelerations may be related to the seasonal production of meltwater or the catastrophic drainage of meltwater lakes which form on the surface of the ice sheet (Schoof 2010).

Aside from obvious differences between the processes acting on land-based and marine-based ice sheets, there has been much regional variability in the rate and style of ice-sheet change since the LGM due to variations in topography, geology, and local climate (Solomina et al. 2015). And the picture has not just been one of ice loss: many Arctic regions experienced ice-sheet regrowth during the cooler Younger Dryas and Neoglacial periods, while in Antarctica, ice-sheet growth has been associated with both warmer periods, due to snowfall increase, and cooler periods, when losses to the ocean were mitigated by colder ocean temperatures.

Solid Earth processes and sea-level change

During the LGM, the surface of Earth was depressed beneath regions of more extensive ice cover, and – to a lesser extent – elevated in the surrounding regions. As the ice sheets thinned and retreated the land beneath began to rebound and is still rebounding today at over 1 cm/yr in parts of Scandinavia and Canada (Lidberg et al. 2010; Sella et al. 2007). The reason for the ongoing nature of this rebound is that Earth deforms *viscoelastically*, comprising an instantaneous elastic response to unloading plus a transient viscous response that continues today.

Over the past 20,000 years, the changing shape of the solid Earth in the Polar Regions, in combination with contemporaneous changes in sea level, has influenced the shape of the coast-line, migration pathways (Elias et al. 1996), ocean circulation (Nørgaard-Pedersen et al. 2003), ice dynamics (Gomez et al. 2010), coastal processes (Whitehouse et al. 2007), and river trajectories (Mangerud et al. 2004). The depressed land also enabled meltwater from the shrinking ice sheets to become trapped in vast lakes (Mangerud et al. 2004), which will have altered the local climate. If these lakes drained catastrophically the resulting floodwaters will have dramatically altered the landscape, eroding and re-depositing vast volumes of sediment. The shape of the land will also have been altered by the direct erosional activity of ice sheets, but only if the ice was thick enough. Erosion occurs as ice flows across the land, but if the ice is too thin then the cold air temperatures above mean that the ice will be frozen to the bed and no erosion can occur.

In many regions, the local ice-sheet history has actually been determined by interpreting field-based reconstructions of local sea-level change, which reflect both global changes in ocean volume and local deformation of the solid Earth due to ice loading (e.g. Simpson et al. 2009). This technique has been successfully used in many regions of Greenland, but it is harder to apply in Antarctica, where much of the coastline has been continuously covered by ice throughout the post-LGM period.

The most dramatic changes to the geography of the Arctic since the LGM can be attributed not to solid Earth deformation, but to the ~130 m global mean sea-level rise that accompanied the melting of the major ice sheets (Lambeck et al. 2014). This rise flooded the land bridge between North America and Asia, and consequently much of this region experienced a shift from a continental interior climate to a milder, wetter maritime climate (Mann and Hamilton 1995). The accompanying expansion of vegetation will also have altered the Arctic climate. Vegetation has a lower albedo and higher heat capacity than ice, thus as vegetation cover increased more heat would have been retained and the associated warming would have further reduced ice cover and enhanced vegetation expansion.

Summary

Many gaps remain with regard to the details of past change across the Polar Regions, with more data needed to better constrain changes in ice extent and thickness, snowfall patterns, ecological changes, oceanic conditions, and sea ice extent. However, more importantly, inter-disciplinary collaboration is enabling dramatic improvements in our understanding of the processes responsible for change. Many feedbacks exist between the ice sheets, the solid Earth, the oceans, the atmosphere, and the biosphere; ongoing work to understand these feedbacks, and the response time of each component to external forcing, will enable us to better constrain potential future rates of change.

References

Alexanderson, H. et al. 2014. 'An Arctic perspective on dating Mid-Late Pleistocene environmental history' *Quat. Sci. Rev.* 92: 9–31.

Alley, R.B. 2000. 'The Younger Dryas cold interval as viewed from central Greenland' *Quat. Sci. Rev.* 19(1–5): 213–226.

Alley, R.B. et al. 1993. 'Abrupt increase in Greenland snow accumulation at the end of the Younger Dryas event' *Nature* 362(6420): 527–529.

Alley, R.B. et al. 1997. 'Holocene climatic instability: a prominent, widespread event 8200 yr ago' *Geology* 25(6): 483–486.

Arndt, J.E. et al. 2013. 'The International Bathymetric Chart of the Southern Ocean (IBCSO) Version 1.0: a new bathymetric compilation covering circum-Antarctic waters' *Geophys Res Lett* 40(12): 3111–3117.

Balco, G. 2011. 'Contributions and unrealized potential contributions of cosmogenic-nuclide exposure dating to glacier chronology, 1990–2010' *Quat. Sci. Rev.* 30(1–2): 3–27.

Benn, D.I. and D.J.A. Evans 2010. *Glaciers and Glaciation*, 2nd edition. London: Hodder Education.

Bentley, M.J. et al. 2014. 'A community-based geological reconstruction of Antarctic Ice Sheet deglaciation since the Last Glacial Maximum' *Quat. Sci. Rev.* 100, 1–9.

Broecker, W.S. 1998. 'Paleocean circulation during the last deglaciation: a bipolar seesaw?' *Paleoceanography* 13(2): 119–121.

Broecker, W.S. and G.H. Denton 1990. 'The role of ocean-atmosphere reorganizations in glacial cycles' *Quat. Sci. Rev.* 9(4): 305–341.

Carlson, A.E. et al. 2007. 'Geochemical proxies of North American freshwater routing during the Younger Dryas cold event' *Proc. Natl. Acad. Sci. U.S.A.* 104(16): 6556–6561.

Clark, P.U. et al. 2009. 'The Last Glacial Maximum' *Science* 325(5941): 710–714.

Cofaigh, C.O. 2012. 'Ice sheets viewed from the ocean: the contribution of marine science to understanding modern and past ice sheets' *Philos T R Soc A* 370(1980): 5512–5539.

Cofaigh, C.O. et al. 2013. 'An extensive and dynamic ice sheet on the West Greenland shelf during the last glacial cycle' *Geology* 41(2): 219–222.

Cuffey, K.M. and W.S.B. Paterson 2010. *The Physics of Glaciers*, 4th edition. Oxford, UK: Butterworth-Heinemann.

D'Andrea, W.J. et al. 2011. 'Abrupt Holocene climate change as an important factor for human migration in West Greenland' *Proc. Natl. Acad. Sci. U.S.A.* 108(24): 9765–9769.

Dyke, A.S. et al. 2002. 'The Laurentide and Innuitian ice sheets during the Last Glacial Maximum' *Quat. Sci. Rev.* 21(1–3): 9–31.

Elias, S.A. et al. 1996. 'Life and times of the Bering land bridge' *Nature* 382(6586): 60–63.

Fairbanks, R.G. 1989. 'A 17,000-year glacio-eustatic sea level record: influence of glacial melting rates on the Younger Dryas event and deep-ocean circulation' *Nature* 342(6250): 637–642.

Fudge, T.J. et al. 2013. 'Onset of deglacial warming in West Antarctica driven by local orbital forcing' *Nature* 500(7463): 440–444.

Funder, S. et al. 2011. 'The Greenland Ice Sheet during the past 300,000 years: a review' in J. Ehlers, P.L. Gibbard and P.D. Hughers (eds) *Quaternary Glaciations. Extent and Chronology, Developments in Quaternary Science*, volume 15. Amsterdam: Elsevier. 699–713.

Gersonde, R. et al. 2005. 'Sea-surface temperature and sea ice distribution of the Southern Ocean at the EPILOG Last Glacial Maximum: a circum-Antarctic view based on siliceous microfossil records' *Quat. Sci. Rev.* 24(7–9): 869–896.

Gomez, N. et al. 2010. 'Sea level as a stabilizing factor for marine-ice-sheet grounding lines' *Nat Geosci* 3(12): 850–853.

Gomez, N. et al. 2015. 'Laurentide-Cordilleran Ice Sheet saddle collapse as a contribution to meltwater pulse 1A' *Geophys Res Lett* 42(10): 3954–3962.

Gregoire, L.J. et al. 2012. 'Deglacial rapid sea level rises caused by ice-sheet saddle collapses' *Nature* 487, 219–222.

Hall, B.L. 2009. 'Holocene glacial history of Antarctica and the sub-Antarctic islands' *Quat. Sci. Rev.* 28(21–22): 2213–2230.

Hays, J.D. et al. 1976. 'Variations in earth's orbit: pacemaker of ice ages' *Science* 194(4270): 1121–1132.

Heinrich, H. 1988. 'Origin and consequences of cyclic ice rafting in the Northeast Atlantic-Ocean during the past 130,000 years' *Quaternary Res* 29(2): 142–152.

Hinzman, L.D. et al. 2005. 'Evidence and implications of recent climate change in northern Alaska and other arctic regions' *Climatic Change* 72(3): 251–298.

IPCC 2013. Climate change 2013. 'The physical science basis. Contribution of working group I to the fifth assessment report of the intergovernmental panel on climate change' *Rep.* 1535 pp. www.ipcc.ch/report/ar5/wg1/.

Jakobsson, M. et al. 2012. 'The International Bathymetric Chart of the Arctic Ocean (IBCAO) Version 3.0' *Geophys Res Lett* 39.

Joughin, I. and R.B. Alley 2011. 'Stability of the West Antarctic Ice Sheet in a warming world' *Nat Geosci* 4(8): 506–513.

Kapsner, W.R. et al. 1995. 'Dominant influence of atmospheric circulation on snow accumulation in Greenland over the past 18,000 years' *Nature* 373(6509): 52–54.

Kaufman, D.S. et al. 2004. 'Holocene thermal maximum in the western Arctic (0–180 degrees W)' *Quat. Sci. Rev.* 23(5–6): 529–560.

Kleman, J. et al. 2010. 'North American Ice Sheet build-up during the last glacial cycle, 115–21 kyr' *Quat. Sci. Rev.* 29(17–18): 2036–2051.

Lambeck, K. et al. 2010. 'The Scandinavian Ice Sheet: from MIS 4 to the end of the Last Glacial Maximum' *Boreas* 39(2): 410–435.

Lambeck, K. et al. 2014. 'Sea level and global ice volumes from the Last Glacial Maximum to the Holocene' *Proc. Natl. Acad. Sci. U.S.A.* 111(43): 15296–15303.

Lidberg, M. et al. 2010. 'Recent results based on continuous GPS observations of the GIA process in Fennoscandia from BIFROST' *J. Geodyn.* 50(1): 8–18.

Livingstone, S.J. et al. 2012. 'Antarctic palaeo-ice streams' *Earth-Sci Rev* 111(1–2): 90–128.

Macayeal, D.R. 1993. 'Binge/purge oscillations of the Laurentide Ice-Sheet as a cause of the North-Atlantic Heinrich events' *Paleoceanography* 8(6): 775–784.

Mangerud, J. et al. 2004. 'Ice-dammed lakes and rerouting of the drainage of northern Eurasia during the Last Glaciation' *Quat. Sci. Rev.* 23(11–13): 1313–1332.

Manley, W.F. and D.S. Kaufman 2002. *Alaska PaleoGlacier Atlas: Institute of Arctic and Alpine Research (INSTAAR)*, edited by i. c. e. Q. a. p. a. University of Colorado, v. 1. Boulder, CO: INSTAAR.

Mann, D.H. and T.D. Hamilton 1995. 'Late Pleistocene and Holocene Paleoenvironments of the North Pacific Coast' *Quat. Sci. Rev.* 14(5): 449–471.

Marcott, S.A. et al. 2013. 'A reconstruction of regional and global temperature for the past 11,300 years' *Science* 339(6124): 1198–1201.

Mercer, J.H. 1978. 'West Antarctic Ice Sheet and CO2 greenhouse effect: Threat of disaster' *Nature* 271(5643): 321–325.

Miller, G.H. et al. 2010a. 'Arctic amplification: can the past constrain the future?' *Quat. Sci. Rev.* 29 (15–16): 1779–1790.

Miller, G.H. et al. 2010b. 'Temperature and precipitation history of the Arctic' *Quat. Sci. Rev.* 29(15–16): 1679–1715.

Milne, G.A. and I. Shennan 2013. 'Isostasy: glaciation-induced sea-level change' in S.A. Elias and C J. Mock (eds) *Encyclopedia of Quaternary Science*. Oxford, UK: Elsevier. 452–459.

Murton, J.B. et al. 2010. 'Identification of Younger Dryas outburst flood path from Lake Agassiz to the Arctic Ocean' *Nature* 464(7289): 740–743.

Nesje, A. et al. 2005. 'Holocene climate variability in the northern North Atlantic region: A review of terrestrial and marine evidence' in H. Drange, T. Dokken, T. Furevik, R. Gerdes and W. Berger (eds) *The Nordic Sea: An Integrated Perspective. Geophysical Monograph Series*. Washington, DC: American Geophysical Union. 289–322.

Nørgaard-Pedersen, N. et al. 2003. 'Arctic Ocean during the Last Glacial Maximum: Atlantic and polar domains of surface water mass distribution and ice cover' *Paleoceanography* 18(3). https://doi.org/10.1029/2002PA000781.

Orsi, A.J. et al. 2012. 'Little Ice Age cold interval in West Antarctica: evidence from borehole temperature at the West Antarctic Ice Sheet (WAIS) Divide' *Geophys Res Lett* 39. https://doi.org/10.1029/2012GL051260.

Overpeck, J. et al. 1997. Arctic environmental change of the last four centuries' *Science* 278(5341): 1251–1256.

Parrenin, F. et al. 2007. '1-D-ice flow modelling at EPICA Dome C and Dome Fuji, East Antarctica' *Clim Past* 3(2): 243–259.

Petit, J.R. et al. 1999. 'Climate and atmospheric history of the past 420,000 years from the Vostok ice core, Antarctica' *Nature* 399(6735): 429–436.

Polyak, L. et al. 2010. 'History of sea ice in the Arctic' *Quat. Sci. Rev.* 29(15–16): 1757–1778.

Rahmstorf, S. 2002. 'Ocean circulation and climate during the past 120,000 years' *Nature* 419(6903): 207–214.

Rouse, W.R. et al. 1997. 'Effects of climate change on the freshwaters of Arctic and subarctic North America' *Hydrol Process* 11(8): 873–902.

Scambos, T.A. et al. 2004. 'Glacier acceleration and thinning after ice shelf collapse in the Larsen B embayment, Antarctica' *Geophys Res Lett* 31(18). https://doi.org/10.1029/2004GL020670.

Schoof, C. 2007. 'Ice sheet grounding line dynamics: steady states, stability, and hysteresis' *J Geophys Res-Earth* 112(F3). https://doi.org/10.1029/2006JF000664.

Schoof, C. 2010. 'Ice-sheet acceleration driven by melt supply variability' *Nature* 468(7325): 803–806.

Sella, G.F. et al. 2007. 'Observation of glacial isostatic adjustment in "stable" North America with GPS' *Geophys Res Lett* 34(2). https://doi.org/10.1029/2006GL027081.

Serreze, M.C. and R.G. Barry 2011. 'Processes and impacts of Arctic amplification: a research synthesis' *Global and Planetary Change* 77(1–2): 85–96.

Shackleton, N.J. 2000. 'The 100,000-year ice-age cycle identified and found to lag temperature, carbon dioxide, and orbital eccentricity' *Science* 289(5486): 1897–1902.

Shakun, J.D. et al. 2012. 'Global warming preceded by increasing carbon dioxide concentrations during the last deglaciation' *Nature* 484(7392): 49–54.

Simpson, M.J.R. et al. 2009. 'Calibrating a glaciological model of the Greenland Ice Sheet from the Last Glacial Maximum to present-day using field observations of relative sea level and ice extent' *Quat. Sci. Rev.* 28(17–18): 1631–1657.

Solomina, O.N. et al. 2015. 'Holocene glacier fluctuations' *Quat. Sci. Rev.* 111, 9–34.

Svendsen, J.I. et al. 2004. 'Late Quaternary ice sheet history of northern Eurasia' *Quat. Sci. Rev.* 23(11–13): 1229–1271.

Todd, C. et al. 2010. 'Late Quaternary evolution of Reedy Glacier, Antarctica' *Quat. Sci. Rev.* 29(11–12): 1328–1341.

Weaver, A.J. et al. 2003. 'Meltwater Pulse 1A from Antarctica as a trigger of the Bølling-Allerød Warm Interval' *Science* 299(5613): 1709–1713.

Whitehouse, P.L. et al. 2007. 'Glacial isostatic adjustment as a control on coastal processes: an example from the Siberian Arctic' *Geology* 35(8): 747–750.

Zimov, S.A. et al. 2006. 'Permafrost and the global carbon budget' *Science* 312(5780): 1612–1613.

17

Biogeochemical cycling in glacial environments

Elizabeth A. Bagshaw

Introduction

Meltwaters from the Polar Regions enter the ocean each summer, contributing some $0.3^{-1} \times 10^{12}$ m^3 of freshwater to the world's oceans (Jones et al. 2002; Mernild and Hasholt 2009; Pattyn 2010). Exported in these meltwaters is a distinct array of dissolved and particulate material evacuated from glacial environments. These waters have a unique signature, which has a lasting impact on global biogeochemical cycles. This chapter explores how glacial meltwaters acquire solute, how this evolves as meltwaters travel through the glacial environment and investigates the impact of the exported meltwaters on local and global biogeochemical processes downstream.

Geochemical weathering in glacial environments

Low temperatures are traditionally associated with low weathering potential (Drever 1997). However, polar environments are the location of some extremely effective weathering mechanisms: freeze-thaw shattering exposes fresh mineral surfaces and glacial grinding releases large quantities of finely ground sediment. Because of this, chemical weathering processes in glacial environments are extremely effective (Tranter and Wadham 2014). Once large quantities of dilute icemelt are flushed through these fresh materials, chemical weathering can occur at rates comparable to temperate environments (Anderson et al. 1997).

Chemical weathering operates in different regions of the glacial environment: at the margins, in lateral moraines and sidewall debris-covered slopes; in front of the glacier, in the proglacial zone, where meltwater streams incise through unconsolidated debris and moraines; on the surface of glaciers, in the supraglacial zone, where materials deposited on the ice surface by winds or debris avalanches react with icemelt; or in the englacial zone, where dust and debris trapped in the ice matrix come into contact with meltwater. By far the most important environment is the region beneath the ice, the subglacial environment. Here, meltwaters are concentrated into channels which interact with bedrock, rock flour and sediments. The nature of the subglacial environment thus broadly controls the potential for release of solute from glaciated regions, and we will consider how changes in this region control chemical weathering processes.

Thermal regime

The quantity of solute entrained by the meltwater depends on the ratio and contact time between rock/sediment and water. This is governed by the type of glacier and the type of bed material (Tranter and Wadham 2014). Firstly, we will consider how glacier thermal regime impacts the weathering environment. Thermal regime is a balance between factors that heat basal ice, most notably the geothermal heat flux, energy produced by basal friction, plastic deformation of ice and heat released by freezing water, and those that cool basal ice, including conduction into cold overlying ice or the frozen substrate beneath, which acts as a heat sink (Benn and Evans 1996).

Glaciers which are entirely frozen to their beds (so-called 'cold-based glaciers') experience very limited chemical weathering, since little meltwater reaches the ice-bed interface. Instead, weathering is concentrated in other environments (see supraglacial habitats, below). This thermal regime occurs when ice at the bed does not reach the pressure melting point and is most common in ice masses at very high altitudes, or those which are very thin. Indeed, until the 1990s, it was considered that the majority of the Greenland and Antarctic ice sheets were also frozen to their beds, and hence experienced little chemical weathering (we will later find that this is no longer considered to be the case). When energy produced by deformation, friction or geothermal heating balances that released by freezing, the bed of a glacier reaches the 'pressure melting point', and the glacier is said to be 'warm-based' or temperate. Here, at least a thin layer of water is present at the interface between the ice and the bedrock, and frequently there are large meltwater channels which route water through the bed material to discharge at the meltwater 'portal' at the glacier front. This means that there is great potential for interaction between rocks and water, and hence opportunity for chemical weathering.

Many ice masses exhibit zones of both 'warm' and 'cold' ice; these are known as 'polythermal' glaciers. In high latitude environments such as Svalbard, large glaciers frequently have a warm core, surrounded by cold, thin margins (Hodgkins 1997; Irvine-Fynn et al. 2011). Beneath the polar ice sheets, geothermal hotspots and frictional heating caused by high pressures and undulating bed topography produce regions where ice is at the pressure melting point. Water can exist in a thin film, as in temperate glaciers, and there is the potential for meltwater generated at the ice sheet surface to penetrate to the bed. Chemical weathering processes can therefore occur beneath large swathes of polar ice sheets which were previously considered geochemically inactive.

Bed material

The second parameter to consider is the characteristics of the bed itself. The mineralogy of the bedrock of course has an impact on the solute released by chemical weathering processes, although as we will discover below, the bedrock composition does not have as much influence as one might expect. Instead, it is the form of the bed material which exerts the major control. The surface area to water volume ratio governs the ease with which meltwaters can access mineral surfaces for weathering. In regions where ice overrides smooth, hard bedrock, only the top surface is available for weathering. By contrast, where ice erodes sections of unconsolidated sediment or deforming till, waters can penetrate into the bed material and weather freshly ground sediments. Here, meltwaters have ready access to freshly comminuted rock flour, which is typically silt-sized and coated with microparticles, and adsorbed organic matter or surface precipitates that may otherwise hinder water–rock interactions are largely absent (Tranter and Wadham 2014). An extreme example can be found in ice stream pore waters beneath Antarctica, where saturated deforming till contains waters in an order of magnitude more concentrated in major ions than typical subglacial waters (Skidmore et al. 2010).

Glacier hydrology

The final controlling factor which governs the potential for contact between meltwater and material available for weathering is the length of time that water spends at the bed. Meltwater generated at the glacier surface flows down through the ice to discharge into the proglacial environment. The water finds its way through cracks, crevasses and defects in the ice. When the pressure of water pooled in a crevasse exceeds the ice overburden pressure, the water pressure can force the crevasse to open further downwards. This forms a direct chute to the bed, known as a moulin because of the 'milling' action of the water swirling into the hole. Moulins transport water deep into the ice, sometimes to englacial passageways (Gulley et al. 2009), which link up further down the glacier, and in some cases all the way to the bed (Das et al. 2008; Catania and Neumann 2010). Once water reaches the bed, it is routed downglacier via a number of different flow configurations:

1   Thin film
    Here the water is forced into a very layer, a few mm thick, between the ice and the bed. This is unlikely to be stable for prolonged time periods (Weertman 1957).
2   Linked cavities
    Water forced to the bed is stored in cavities formed by separation between the ice and the bed, generally between 1 and 10 m high. These are scattered across the glacier bed and linked by narrow connections (<0.1 m high) through which water can pass (Kamb 1987).
3   R channel
    Röthlisberger channels incise upwards into the ice through viscous heating as water flows. The channel size is proportional to the balance between water pressure, which keeps the channels open, and ice deformation, which acts to close them (Röthlisberger 1972).
4   N channel
    Nye channels are the opposite of R channels, cutting passages in the bedrock beneath the ice. They can remain stable for prolonged periods and remain etched into the bedrock once the ice above has retreated (Nye 1976).
5   Sediment canals
    If the ice flows over unconsolidated sediment, meltwaters can cut broad, shallow canals into the sediment (Walder and Fowler 1994).
6   Groundwater
    If water pressure is sufficient and the glacier bed is porous, meltwaters can penetrate the bed and flow as groundwater. Occasionally they emerge in front of the glacier as a spectacular upwelling or artesian fountain (Hodgkins 1997).

Each of these drainage system morphologies has different consequences for rock–water contact times. Structures where water transits rapidly, including R and N channels, are known as 'efficient' or 'low pressure' systems, since they rapidly convey water to the glacier front. They only form when there is sufficient water pressure to maintain the balance of frictional heating and creep closure; once the water flow decreases, the channels close and the meltwater is forced into an alternative configuration, usually a form of linked cavity system. These 'high pressure' or 'inefficient' drainage system morphologies transport water more slowly, and as such, provide increased opportunity for interaction between meltwaters and the bed and promote long rock–water contact times.

Chemical weathering reactions

Once freshly eroded material comes into contact with dilute meltwaters, a number of chemical reactions occur. Supraglacial ice melt is usually quite dilute (ionic strengths <40 mmol l^{-1}) (Tranter and Wadham 2014). The first reactions to occur once this dilute water comes into contact with eroded material are carbonate and silicate hydrolysis (Brown et al. 1996; Tranter et al. 2002; Equations 1 and 2). These reactions raise the pH to high values (>9) and lower the pCO$_2$, maximising the potential of the water to absorb CO$_2$. The high pH in turn enhances the dissolution of aluminosilicates (Equation 3), since Al and Si become more soluble at elevated pH.

$$Ca_{1-x}(Mg_x)CO_{3(s)} + H_2O_{(l)} \leftrightarrow (1-x)Ca^{2+}_{(aq)} + xMg^{2+}_{(aq)} + HCO^{-}_{3(aq)} + OH^{-}_{(aq)} \tag{1}$$

$$KAlSi_3O_{8(s)} + H_2O_{(l)} \leftrightarrow HAlSi_3O_8 + K^{+}_{(aq)} + OH^{-}_{(aq)} \tag{2}$$

$$CaAl_2Si_2O_{8(s)} + 2CO_{2(aq)} + 2H_2O_{(l)} \leftrightarrow Ca^{2+}_{(aq)} + 2HCO^{-}_{3(aq)} + H_2Al_2Si_2O_{8(s)} \tag{3}$$

Gases dissolved in the water, sourced from the atmosphere, from gas bubbles in basal ice, or from microbial respiration lower the pH and saturation with respect to carbonates, enabling continuous carbonate dissolution. Carbonate dissolution is extremely important for meltwater geochemistry, since favourable reaction kinetics prompt rapid dissolution, even if carbonates are present in only trace quantities in the bedrock.

In debris-rich environments, sulphide oxidation can occur. This is an important reaction for providing protons to solution, which lowers the pH further and enables more carbonate dissolution. The reactions are said to proceed in a 'coupled' fashion (Equation 4), where carbonate dissolution continues in tandem with sulphide oxidation. Evidence for coupling of reactions can be found in the composition of borehole waters sampled directly from the bed of the Haut Glacier d'Arolla: water displays a 2:1 relationship between Ca^{2+} + Mg^{2+} and SO$_4^{2-}$ (Tranter et al. 2002). The oxidation of sulfides preferentially dissolves carbonates, rather than silicates, because the rate of carbonate dissolution is orders of magnitude faster (Tranter and Wadham 2014).

$$4FeS_{2(s)} + 16Ca_{1-x}(Mg_x)CO_{3(s)} + 15O_{2(aq)} + 14H_2O_{(l)} \leftrightarrow$$
$$16(1-x)Ca^{2+}_{(aq)} + 16xMg^{2+}_{(aq)} 16HCO^{-}_{3(aq)} + 8SO^{2-}_{4(aq)} + 4Fe(OH)_{3(s)} \tag{4}$$

The concentration of SO$_4^{2-}$ present in glacial runoff suggests that either oxygen concentrations in glacial melt must be in excess of 250% air saturation, or that other oxidising agents are employed (Tranter et al. 2002). Trace quantities of Fe^{3+} in the bedrock or sourced from sulphide oxidation (Equation 4) may be used to oxidise sulphides under anoxic conditions (Equation 5). Kinetically, this reaction proceeds slowly, unless it is mediated by microbial activity.

$$FeS_{2(s)} + 14Fe^{3+}_{(aq)} + 8H_2O_{(l)} \leftrightarrow 15Fe^{2+}_{(aq)} + 2SO^{2-}_{4(aq)} + 16H^{+}_{(aq)} \tag{5}$$

Biogeochemical weathering in glacial environments

Until relatively recently, glacial environments were presumed to be largely sterile, with very little in situ microbial activity. However, mounting chemical evidence, and eventually in situ observations, demonstrated that there are widespread and active microbial populations throughout the glacial environment, and that glaciers and ice sheets may be considered as a distinct biome (Anesio and Laybourn-Parry 2012). Considering Equation 5, this process is likely to be catalysed by microbial activity, which may also oxidise organic carbon (Equation 6). Equations 5 and 6 lower the partial pressure of O_2 in solution, promoting anoxic conditions and forcing Fe^{3+} to act as an alternative oxidising agent (Tranter et al. 2002). When oxygen is exhausted, alternative oxidising agents may be employed, including Mn(IV), NO_3^- and SO_4^{2-} (Cockell et al. 2013). The order in which these agents are utilised depends on a strict hierarchy of energy expenditure known as a 'redox tower': microorganisms will utilise the substance that allows the most efficient release of energy and will usually only utilise alternative oxidants (>Fe(III)>Mn(IV)>NO_3^- >SO_4^{2-}) when oxygen supplies are exhausted.

$$C_{org(s)} + O_{2(aq)} + H_2O_{(l)} \leftrightarrow CO_{2(aq)} + H_2O_{(l)} \leftrightarrow H^+_{(aq)} + HCO^-_{3(aq)} \tag{6}$$

The access to oxidising agents is governed by the location of the habitat within the glacial environment. Microbial habitats are ubiquitous throughout the glacial environment; wherever there is liquid water and a source of energy, microbes can be found (Anesio and Laybourn-Parry 2012). Whilst there is some evidence for viable habitats in the ice matrix which exploit microscopic veins of water between ice crystals (Mader et al. 2006) and are potentially associated with sediment within englacial passageways (Laybourn-Parry et al. 2012), the primary habitats are the supraglacial and subglacial zones.

Supraglacial habitats

Debris deposited on the glacier surface by wind, avalanching from the glacier sidewalls, and via uplift of bed materials, supports a diverse microbial community which includes photosynthetic organisms. The presence of dark debris on the light, high albedo glacier surface causes local regions of enhanced melting and the formation of water-filled depressions. These depressions, known as cryoconite holes or cryolakes, together with more dispersed debris accumulations, constitute supraglacial habitats for microbial life in the Antarctic and the Arctic. They cover varying amounts of the glacier surface, typically between 3 and 15% (Fountain et al. 2004; Hodson et al. 2008). The cryoconite holes are home to a variety of organisms but are typically dominated by cyanobacteria. These organisms can fix carbon, and may be responsible for uptake of CO_2 from the atmosphere and accumulation of labile organic carbon over time (Anesio et al. 2009), as evidenced by in situ measurements (Hodson et al. 2010; Telling et al. 2010; Bagshaw et al. 2011), and the fact that the organic carbon content of supraglacial debris far exceeds the typical organic carbon concentration of debris from basal ice and ice-marginal moraines (1–5%, vs. 0.5%; Kastovska et al. 2005). There is also evidence for active nutrient cycling: N can be fixed from the atmosphere (Telling et al. 2011), and P extracted from dead organisms and sediment by enzymes (Stibal et al. 2009). These processes recycle nutrients to more labile forms, which are easily utilised by other microorganisms which receive glacier melt. Meltwaters periodically flush supraglacial habitats, redistributing the now bioavailable forms of C, N and P to downstream environments (see later). The newly produced organic material is readily available for microbes and is an important resource in nutrient-limited ecosystems (Bagshaw et al. 2013). Recent investigations have also revealed that ice sheet surfaces support extensive populations of dispersed ice algae, which are capable of photosynthesising during the summer months and hence generate organic carbon over widespread areas of the ablation zone (Benning et al. 2014). As the algae produce organic matter, they darken the ice surface, which has important implications for ice surface reflectivity (Yallop et al. 2012).

Subglacial habitats

Although the supraglacial environment is perhaps the most obvious glacial habitat, the habitable zone beneath the ice is as significant in size. Here, there is access to freshly eroded material as an energy source, and liquid water from melting. Overridden soils and organic material may be an important resource: old organic carbon may be oxidised (Equation 5), and there is evidence that microbes can also utilise kerogen in bedrock (Wadham et al. 2010). The presence of liquid water again determines the habitability. Beneath temperate glaciers where ice is at the pressure melting point, significant concentrations of microorganisms have been reported (104–107 cells/ml; Sharp et al. 1999; Skidmore et al. 2000). Beneath colder ice masses, the presence of liquid water depends on ice thickness and dynamics, or the existence of geothermal hotspots.

Ice streams are regions of fast-flowing ice, driven by topography or the presence of deforming till, which drain ice from the interior of ice sheets (Skidmore et al. 2010). The fast-flowing ice deforms rapidly, promoting basal melting and frequently causing saturation of till. This environment, which includes liquid water and large quantities of freshly eroded debris, is an ideal habitat for subglacial microbes. Here, the supply of oxygen from melting ice is generally less than the demand from reducing agents in the water-saturated till, so the waters become anoxic and the biomass is dominated by communities which can utilise less thermodynamically favourable energy sources (Wadham et al. 2004; Cockell et al. 2013). For example, culturable cells have been extracted from sediments beneath the Kamb Ice Stream in West Antarctica, with an abundance of ~107 cells g^{-1}, and genetic evidence suggests that the community was a simplified version of that found in subglacial alpine and Arctic sediments and water (Foght et al. 2004).

In regions where ice overrides significant reserves of organic carbon, for example ancient forests, there is the potential for methanogenesis. Here, microorganisms known as methanogens oxidise carbon under anoxic conditions and generate methane. Methane is an extremely effective greenhouse gas, thus release of methane has the potential to act as a significant climate amplifier. The potential for methane release from melting Arctic permafrost is well established (Shakhova et al. 2010), but comparatively little attention has been paid to the potential for a similar scenario occurring as glaciers and ice sheets retreat. Measurements of methanogenesis in subglacial sediments beneath Arctic and Antarctic sediments reveal that there is up to 63 Pg of organic carbon which could be metabolised by methanogens and that the reservoir of methane hydrates beneath Antarctica could be of a comparable magnitude to that stored in Arctic permafrost (Wadham et al. 2012).

Subglacial lakes also exist in areas of topographic lows but are generally maintained by geothermal heating and constrained by bed topography. These extraordinary oases of liquid water exist beneath hundreds of meters of ice, and likely maintain an active microbial community. There are over 400 such lakes in Antarctica (Siegert et al. 2016), and new evidence is emerging for their existence beneath the Greenland Ice Sheet (Palmer et al. 2013; Willis et al. 2015). Large, hydrologically closed lakes, such as Lake Vostok, may have oxic regions, since input of reducing glacial debris is balanced by melting of meteoric ice from above which contains oxygen (Siegert et al. 2003). In debris-rich lakes, particularly shallow, connected water bodies, low oxygen conditions may prevail. Recent analysis of rigorously clean samples from subglacial Lake Whillans, a shallow, sediment rich water body, revealed an active community of diverse heterotrophic and autotrophic microorganisms present in the water column and in surface sediments (Christner et al. 2014). Analysis of DNA from sediments at the base of Lake Whillans shows that sulphate transformations (sulphate reduction and sulphur oxidation) play an important role in community metabolism (Purcell et al. 2014). This shows that microorganisms can extract energy from chemical compounds even under dark, likely anoxic conditions.

The largest and most enigmatic subglacial lake is Vostok, located at the centre of the East Antarctic Ice Sheet. The lake is likely hydrologically isolated, meaning that there is very little flow in and out of the lake beyond melting of ice from above, although the duration of isolation is presently disputed (Royston-Bishop et al. 2004). To date, the lake has been sampled twice: on the first occasion samples were contaminated by drilling fluid, but the second sampling mission (austral summer 2014/15) promises microbially clean results (Lukin and Vasiliev 2014). The majority of information about the water in the lake has to date been obtained from so-called 'accretion ice', which is refrozen lake water accreted onto the ice above the lake (Priscu et al. 1999). Weak geothermal heating drives thermal circulation within the lake, causing melting at one end and refreezing at the other. The accretion ice therefore contains mineral inclusions and microorganisms from within the lake itself. Examination of these fragments suggests that the lake does support chemotrophic microbial life, although its form and function will remain disputed until analysis of clean samples from the lake waters.

Impacts of microbial activity

The presence of active populations of microorganisms in glacial environments has profound impacts on the composition of runoff. We have seen how microbes may catalyse chemical weathering reactions, but microbial activity can also change the phase of carbon and nutrients in glacial regions. Carbon, nitrogen, phosphorus and other trace elements are required for cell maintenance, reproduction and sustenance in microbial communities. These species are present in the physical environment in glacial regions but not necessarily in forms which are accessible to all microorganisms. Snow and ice melt provide limited quantities of N (Tranter and Wadham 2014), and P is sourced from ground-up rock debris (Hodson et al. 2004). Much of this P is present in fractions which are unavailable to microorganisms on biological timescales, so the organisms must employ strategies to exploit the available resources, such as utilising enzymatic processes to extract P from debris (Stibal et al. 2009). A similar situation occurs for N species: whilst snow and ice melt can provide easily 'digestible' forms of nitrogen in NO_3^- and NH_4^+, this is rapidly utilised and exhausted, so microbial communities must source N from elsewhere. Nitrogen fixation, where organisms use enzymes to convert inert atmospheric N_2 to reactive NH_4^+, has been observed in glacier surface communities in Svalbard and Antarctica (Telling et al. 2011, 2015), and catalysis of nitrification, which cycles NH_4^+ to NO_2^- and NO_3^- is an important pathway in both supraglacial and subglacial communities (Wynn et al. 2006; Cameron et al. 2012).

Supraglacial carbon cycling

Numerous studies have demonstrated the cycling of carbon in cryoconite hole ecosystems on glaciers across the world (Bagshaw et al. 2012). Measurements of primary production and respiration in cryoconite holes show that communities can utilise inorganic carbon, sourced from ice and snow melt or respiration by heterotrophic organisms, to manufacture organic carbon through photosynthesis. The balance between these two processes – uptake of organic carbon through respiration vs generation of organic carbon via photosynthesis – is disputed, with some studies demonstrating that cryoconite holes generate significant quantities of carbon (Anesio et al. 2009), others showing that additional sources of carbon are required to sustain production (Stibal and Tranter 2007), and others showing that the processes are broadly balanced (Hodson et al. 2010; Telling et al. 2010). It is most likely that the communities vary over time, and that the balance between P and R is controlled by external forcings, such as temperature and solar radiation, and microbial community dynamics. Yet while the magnitude of the total flux may be subject to discussion, all the research clearly demonstrates the potential for glacier surface ecosystems to recycle carbon into more easily available forms (Barker et al. 2006).

Subglacial carbon cycling

Subglacial ecosystems also recycle carbon. However, the lack of sunlight in these environments means that the mechanism differs from that occurring on glacier surfaces. In the subglacial environment, organisms are able to oxidise ancient organic matter (Equation 6) that has been overridden by the glacier in more recent times. Isotopic studies have shown that old sources of organic carbon are recycled to more bioavailable forms which are then flushed out from beneath the glacier in seasonal runoff (Hood et al. 2009). These processes are common to glaciers around the world, including the Alps (Singer et al. 2012), the Arctic (Hood et al. 2009; Bhatia et al. 2013; Lawson et al. 2014) and the Antarctic (Hood et al. 2015). Evidence for methane cycling (see earlier) beneath the Antarctic and Greenland ice sheets (Wadham et al. 2012; Dieser et al. 2014) show that microorganisms can degrade organic carbon even under anoxic conditions.

Nutrient export

A consequence of these processes is that meltwater flushed from glaciers contains C, N and P in converted or recycled forms, which are readily available to organisms in downstream ecosystems. Absolute concentrations are relatively low, but the organic carbon contains a high percentage of protein-like compounds which are easily utilised in biological processes, and the C:N ratios are relatively low (Barker et al. 2006; Hood et al. 2009). This means that dissolved organic carbon (DOC) derived from glacial environments is extremely labile (bioavailable) in comparison to DOC from other terrestrial environments. Laboratory experiments where microorganisms were fed DOC exported from glaciers demonstrates that 25–95% of the flux may be readily metabolised by heterotrophic organisms (Hood et al. 2015). The same is true for N and P, which are cycled into readily available forms by glacial microbial processes (see earlier).

Biogeochemical cycling in polar regions

The impact of exported material on downstream ecosystems is, at present, poorly quantified. However, a number of studies have shown that nutrients produced by microbial processes in glacial ecosystems have a positive influence on downstream microbial communities. Glacier forefields are frequently nutrient-poor, since there are few vascular plants and soil development is limited. Nutrient content tends to increase with soil development, which is correlated with the time since the land was last covered in ice (Bradley et al. 2014). Input of nutrients, even in relatively low concentrations, is thus crucial for supporting biological activity in otherwise oligotrophic environments. Studies in the McMurdo Dry Valleys of Antarctica, where glaciers are frozen to their beds and all runoff originates from the supraglacial environment, surmise that nutrients generated in cryoconite holes can stimulate primary production in the ice-covered lakes which receive the meltwaters (Foreman et al. 2004; Bagshaw et al. 2013). These systems are generally P-limited since the majority is locked up in forms which are unavailable to organisms on biological timescales, thus an influx of bioavailable P in glacier melt is an important contributor to regional processes.

In the Dry Valleys, processes are simple to model since glacier melt is the only source of liquid water and the ice-covered lakes represent the end-point of the local hydrological cycle.

In areas where glacier melt travels long distances along streams, rivers, through fjords and eventually into the ocean, the impact of glacial runoff is more complex to interpret. Nevertheless, a number of studies have shown that glacier melt does stimulate microorganisms which live in streams and fjords (Singer et al. 2012; Hood et al. 2015), and that some is transported to the oceans. This is particularly important for trace elements such as iron, which are severely limited in some areas of the ocean. Primary production in many of the oceans adjacent to glaciated regions is iron-limited, since they are far distant from other sources of iron such as terrestrial dust and riverine discharge (Statham et al. 2008). It is therefore hypothesised that iron sourced from microbial activity in subglacial environments is delivered to the ocean via meltwaters and iceberg rafting of debris. Crucially, the iron is recently cycled by glacial microbial activity, so the majority is bioavailable. Measurements show that iron export from Greenland and Antarctica is a potentially significant process, and that the flux could be of a similar size to that from aeolian dust (Hawkings et al. 2014).

Total flux

Modelling studies have been undertaken to understand the impact of export of nutrients from glacial environments on local, regional and global biogeochemical cycles. Field observations are used as inputs to biogeochemical models which can estimate the likely response of coastal ecosystems to an influx of nutrients. Modelling of iron flux from the Antarctic ice sheet, via subglacial runoff and iceberg discharge, predicts that the flux can increase primary production in the Southern Ocean by up to 40% (Death et al. 2014). This provides an explanation for the observed seasonal highs of production observed in the near-coastal zone.

There have been varying efforts to quantify export of organic carbon from the Greenland and Antarctic ice sheets to the oceans. Some estimates of the flux of DOC suggest that the Greenland ice sheet is comparable with that from a small Arctic river (0.08 Tg a^{-1}) and that particulate DOC is equivalent to a large Arctic river (0.9 Tg a^{-1}) (Bhatia et al. 2013); but values scaled up from a large catchment (Leverett Glacier, SW Greenland) demonstrate that the DOC is likely 0.13–0.17 Tg a^{-1} and POC of the order of 0.36–1.52 Tg a^{-1} (Lawson et al. 2014). This is equivalent to a large river system and hence represents an important carbon input to the world's oceans.

Values are not as well constrained for exports from Antarctica, primarily because the flux of subglacial meltwater and suspended sediment is not well quantified over the vast area. Estimates are based on maximum, mean and minimum estimates of total runoff from the entire ice sheet and a variety of measured concentrations across the continent, and range from 1.5 to 253 Gg DOC a^{-1} (Wadham et al. 2013). A total combined estimate of DOC export in global glacier runoff, including the Greenland and Antarctic ice sheets and all mountain glaciers, is of the order of 1.04 Tg a^{-1} (Hood et al. 2015). Hood et al. (2015) argue that the input from mountain glaciers is the larger contributor at present (0.58 Tg a^{-1} compared to 0.22 for Greenland and 0.24 for Antarctica) because of the larger turnover of mass in these systems, but acknowledge that this is likely to change in coming decades.

Impact on Earth's biogeochemical cycles

The flux of carbon, nutrients and trace elements from glaciers and ice sheets to the world's oceans is significant because (a) the flux is generally bioavailable, and (b) it is released in otherwise nutrient limited regions. For example, high nutrient, low chlorophyll regions of the ocean have limited phytoplankton productivity because of a lack of micronutrients for growth. One of these regions lies just off the coast of Antarctica (Statham et al. 2008) and is limited by a lack of iron. Input of dissolved iron from subglacial meltwaters and iceberg rafted debris could have a positive impact on phytoplankton productivity. Plankton blooms in these regions occur in the summer melt season, and are often associated with low salinity waters, suggesting that iron delivered by glacial melt is an important mechanism for supporting ocean productivity.

The role of glaciers and ice sheets in global biogeochemical cycles will very likely change in coming decades as ice melt increases. Hood et al. (2015) estimate that 15.34 Tg of DOC will be released from storage within and beneath glaciers by 2050. The impact of this release, and the associated N, P and micronutrients, will alter local, regional and global biogeochemical processes. Glaciologists, biogeochemists and oceanographers are seeking to quantify these effects now.

References

Anderson, S.P., J.I. Drever and N.F. Humphrey 1997. 'Chemical weathering in glaciated environments' *Geology* 25: 399–402.

Anesio, A.M., A.J. Hodson, A. Fritz, R. Psenner and B. Sattler 2009. 'High microbial activity on glaciers: Importance to the global carbon cycle' *Global Change Biology* 15(4): 955–960.

Anesio, A.M. and J. Laybourn-Parry 2012. 'Glaciers and ice sheets as a biome' *Trends in Ecology & Evolution* 27(4): 219–225.

Bagshaw, E.A., M. Stibal, A.M. Anesio, C. Bellas, M. Tranter, J. Telling and J.L. Wadham 2012. *Glacier Surface Habitats. Life at Extremes: Environments, Organisms and Strategies for Survival*. Wallingford, UK: E.M. Bell and CABI.

Bagshaw, E.A., M. Tranter, A.G. Fountain, K. Welch, H.J. Basagic and W.B. Lyons 2013. 'Do cryoconite holes have the potential to be significant sources of C, N and P to downstream depauperate ecosystems of Taylor Valley, Antarctica?' *Arctic Antarctic and Alpine Research* 45(4): 440–454.

Bagshaw, E.A., M. Tranter, J.L. Wadham, A.G. Fountain and M. Mowlem 2011. 'High-resolution monitoring reveals dissolved oxygen dynamics in an Antarctic cryoconite hole' *Hydrological Processes* 25(18): 2868–2877.

Barker, J.D., M.J. Sharp, S.J. Fitzsimons and R.J. Turner 2006. 'Abundance and dynamics of dissolved organic carbon in glacier systems' *Arctic Antarctic and Alpine Research* 38(2): 163–172.

Benn, D.I. and D.J.A. Evans 1996. 'The interpretation and classification of subglacially-deformed materials' *Quaternary Science Reviews* 15(1): 23–52.

Benning, L.G., A.M. Anesio, S. Lutz and M. Tranter 2014. 'Biological impact on Greenland's albedo' *Nature Geosci* 7(10): 691–691.

Bhatia, M.P., S.B. Das, L. Xu, M.A. Charette, J.L. Wadham and E.B. Kujawinski 2013. 'Organic carbon export from the Greenland ice sheet' *Geochimica Et Cosmochimica Acta* 109: 329–344.

Bradley, J.A., J.S. Singarayer and A.M. Anesio 2014. 'Microbial community dynamics in the forefield of glaciers' *Proceedings of the Royal Society B-Biological Sciences* B 281: 20140882. http://dx.doi.org/10.1098/rspb.2014.0882.

Brown, G., M. Tranter and M. Sharp 1996. 'Experimental investigation of suspended sediment by Alpine glacial meltwater' *Hydrological Processes* 10: 579–597.

Cameron, K.A., A.J. Hodson and A.M. Osborn 2012. 'Carbon and nitrogen biogeochemical cycling potentials of supraglacial cryoconite communities' *Polar Biology* 35(9): 1375–1393.

Catania, G.A. and T.A. Neumann 2010. 'Persistent englacial drainage features in the Greenland Ice Sheet' *Geophysical Research Letters* 37, L02501, doi:10.1029/2009GL041108.

Christner, B.C., J.C. Priscu, A.M. Achberger, C. Barbante, S.P. Carter, K. Christianson, A.B. Michaud, J.A. Mikucki, A.C. Mitchell, M.L. Skidmore, T.J. Vick-Majors and W.S. Team 2014. 'A microbial ecosystem beneath the West Antarctic Ice Sheet' *Nature* 512(7514): 310–313.

Cockell, C.S., E. Bagshaw, M. Balme, P. Doran, C.P. McKay, K. Miljkovic, D. Pearce, M.J. Siegert, M. Tranter, M. Voytek and J. Wadham 2013. *Subglacial Environments and the Search for Life Beyond the Earth*. Antarctic Subglacial Aquatic Environments, American Geophysical Union. 129–148.

Das, S.B., I. Joughin, M.D. Behn, I.M. Howat, M.A. King, D. Lizarralde and M.P. Bhatia 2008. 'Fracture propagation to the base of the Greenland Ice Sheet during supraglacial lake drainage' *Science* 320(5877): 778–781.

Death, R., J.L. Wadham, F. Monteiro, A.M. Le Brocq, M. Tranter, A. Ridgwell, S. Dutkiewicz and R. Raiswell 2014. 'Antarctic Ice Sheet fertilises the Southern Ocean' *Biogeosciences* 11(10): 2635–2643.

Dieser, M., E.L.J.E. Broemsen, K.A. Cameron, G.M. King, A. Achberger, K. Choquette, B. Hagedorn, R. Sletten, K. Junge and B.C. Christner 2014. 'Molecular and biogeochemical evidence for methane cycling beneath the western margin of the Greenland Ice Sheet' *Isme Journal* 8(11): 2305–2316.

Drever, J.I. 1997. *The Geochemistry of Natural Waters: Surface and Groundwater Environments*. Englewood Cliffs, NJ: Prentice Hall.

Foght, J., J. Aislabie, S. Turner, C.E. Brown, J. Ryburn, D.J. Saul and W. Lawson 2004. 'Culturable bacteria in subglacial sediments and ice from two Southern Hemisphere glaciers' *Microbial Ecology* 47(4): 329–340.

Foreman, C.M., C.F. Wolf and J.C. Priscu 2004. 'Impact of episodic warming events on the physical, chemical and biological relationships of lakes in the McMurdo Dry Valleys, Antarctica' *Aquatic Geochemistry* 10(3): 239–268.

Fountain, A. G., M. Tranter, T. H. Nylen, K. J. Lewis and D. R. Mueller 2004. 'Evolution of cryoconite holes and their contribution to meltwater runoff from glaciers in the McMurdo Dry Valleys, Antarctica' *Journal of Glaciology* 50(168): 35–45.

Gulley, J.D., D.I. Benn, E. Screaton and J. Martin 2009. 'Mechanisms of englacial conduit formation and their implications for subglacial recharge' *Quaternary Science Reviews* 28(19–20): 1984–1999.

Hawkings, J.R., J.L. Wadham, M. Tranter, R. Raiswell, L.G. Benning, P J. Statham, A. Tedstone, P. Nienow, K. Lee and J. Telling 2014. 'Ice sheets as a significant source of highly reactive nanoparticulate iron to the oceans' *Nature Communications* 5, 3939. 1–8. ISSN 2041–1723.

Hodgkins, R. 1997. 'Glacier hydrology in Svalbard, Norwegian High Arctic' *Quaternary Science Reviews* 16(9): 957–973.

Hodson, A., A.M. Anesio, M. Tranter, A. Fountain, M. Osborn, J. Priscu, J. Laybourn-Parry and B. Sattler 2008. 'Glacial ecosystems' *Ecological Monographs* 78(1): 41–67.

Hodson, A., K. Cameron, C. Boggild, T. Irvine-Fynn, H. Langford, D. Pearce and S. Banwart 2010. 'The structure, biological activity and biogeochemistry of cryoconite aggregates upon an Arctic valley glacier: Longyearbreen, Svalbard' *Journal of Glaciology* 56(196): 349–362.

Hodson, A., P. Murnford and D. Lister 2004. 'Suspended sediment and phosphorus in proglacial rivers: bioavailability and potential impacts upon the P status of ice-marginal receiving waters' *Hydrological Processes* 18(13): 2409–2422.

Hood, E., T.J. Battin, J. Fellman, S. O'Neel and R.G.M. Spencer 2015. 'Storage and release of organic carbon from glaciers and ice sheets' *Nature Geoscience* 8(2): 91–96.

Hood, E., J. Fellman, R.G.M. Spencer, P.J. Hernes, R. Edwards, D. D'Amore and D. Scott 2009. 'Glaciers as a source of ancient and labile organic matter to the marine environment' *Nature* 462(7276): 1044–1100.

Irvine-Fynn, T.D.L., A.J. Hodson, B.J. Moorman, G. Vatne and A.L. Hubbard 2011. 'Polythermal glacier hydrology: a review' *Reviews of Geophysics* 49, RG4002 / 2011.

Jones, I.W., G. Munhoven, M. Tranter, P. Huybrechts and M.J. Sharp 2002. 'Modelled glacial and non-glacial HCO_3-, Si and Ge fluxes since the LGM: little potential for impact on atmospheric CO_2 concentrations and a potential proxy of continental chemical erosion, the marine Ge/Si ratio' *Global and Planetary Change* 33(1–2): 139–153.

Kamb, B. 1987. 'Glacier surge mechanism based on linked cavity configuration of the basal water conduit system' *Journal of Geophysical Research-Solid Earth and Planets* 92(B9): 9083–9100.

Kastovska, K., J. Elster, M. Stibal and H. Santruckova 2005. 'Microbial assemblages in soil microbial succession after glacial retreat in Svalbard (High Arctic)' *Microbial Ecology* 50(3): 396–407.

Lawson, E.C., J.L. Wadham, M. Tranter, M. Stibal, G.P. Lis, C.E.H. Butler, J. Laybourn-Parry, P. Nienow, D. Chandler and P. Dewsbury 2014. 'Greenland Ice Sheet exports labile organic carbon to the Arctic oceans' *Biogeosciences* 11(14): 4015–4028.

Laybourn-Parry, J., M. Tranter and A.J. Hodson 2012. *The Ecology of Snow and Ice Environments*. Oxford, UK: Oxford University Press.

Lukin, V.V. and N.I. Vasiliev 2014. 'Technological aspects of the final phase of drilling borehole 5G and unsealing Vostok Subglacial Lake, East Antarctica' *Annals of Glaciology* 55(65): 83–89.

Mader, H.M., M.E. Pettitt, J.L. Wadham, E.W. Wolff and R.J. Parkes 2006. 'Subsurface ice as a microbial habitat' *Geology* 34(3): 169–172.

Mernild, S.H. and B. Hasholt 2009. 'Observed runoff, jokulhlaups and suspended sediment load from the Greenland Ice Sheet at Kangerlussuaq, West Greenland, 2007 and 2008' *Journal of Glaciology* 55(193): 855–858.

Nye, J.F. 1976. 'Water flow in glaciers: Jokulhaups, tunnels and veins' *Journal of Glaciology* 17: 181–207.

Palmer, S.J., J.A. Dowdeswell, P. Christoffersen, D.A. Young, D.D. Blankenship, J.S. Greenbaum, T. Benham, J. Bamber and M.J. Siegert 2013. 'Greenland subglacial lakes detected by radar' *Geophysical Research Letters* 40(23): 6154–6159.

Pattyn, F. 2010. 'Antarctic subglacial conditions inferred from a hybrid ice sheet/ice stream model' *Earth and Planetary Science Letters* 295(3–4): 451–461.

Priscu, J.C., E.E. Adams, W.B. Lyons, M.A. Voytek, D.W. Mogk, R.L. Brown, C.P. McKay, C.D. Takacs, K.A. Welch, C.F. Wolf, J.D. Kirshtein and R. Avci 1999. 'Geomicrobiology of subglacial ice above Lake Vostok, Antarctica' *Science* 286(5447): 2141–2144.

Purcell, A.M., J.A. Mikucki, A.M. Achberger, I.A. Alekhina, C. Barbante, B.C. Christner, D. Ghosh, A.B. Michaud, A.C. Mitchell, J.C. Priscu, R. Scherer, M.L. Skidmore, T.J. Vick-Majors and W.S. Team 2014. 'Microbial sulfur transformations in sediments from Subglacial Lake Whillans' *Frontiers in Microbiology* 5. doi: 10.3389/fmicb.2014.00594.

Rothlisberger, H. 1972. 'Water pressure in intra- and subglacial channels' *Journal of Glaciology* 11: 117–203.

Royston-Bishop, G., M. Tranter, M.J. Siegert, V. Lee and P.D. Bates 2004. Is Vostok Lake in steady state? *Annals of Glaciology* 39: 490–494.

Shakhova, N., I. Semiletov, A. Salyuk, V. Yusupov, D. Kosmach and O. Gustafsson 2010. 'Extensive methane venting to the atmosphere from sediments of the East Siberian Arctic shelf' *Science* 327(5970): 1246–1250.

Sharp, M., J. Parkes, B. Cragg, I.J. Fairchild, H. Lamb and M. Tranter 1999. 'Widespread bacterial populations at glacier beds and their relationship to rock weathering and carbon cycling' *Geology* 27(2): 107–110.

Siegert, M.J., M. Tranter, J.C. Ellis-Evans, J.C. Priscu and W.B. Lyons 2003. 'The hydrochemistry of Lake Vostok and the potential for life in Antarctic subglacial lakes' *Hydrological Processes* 17(4): 795–814.

Siegert, M.J., N. Ross, and A. Le Brocq 2016. 'Recent advances in understanding Antarctic subglacial lakes and hydrology.' *Philosophical Transactions of the Royal Society of London, A.* 374, 20140306.

Singer, G.A., C. Fasching, L. Wilhelm, J. Niggemann, P. Steier, T. Dittmar and T.J. Battin 2012. 'Biogeochemically diverse organic matter in Alpine glaciers and its downstream fate' *Nature Geoscience* 5: 710–714.

Skidmore, M., M. Tranter, S. Tulaczyk and B. Lanoil 2010. 'Hydrochemistry of ice stream beds: evaporitic or microbial effects?' *Hydrological Processes* 24(4): 517–523.

Skidmore, M.L., J.M. Foght and M.J. Sharp 2000. 'Microbial life beneath a high Arctic glacier' *Applied and Environmental Microbiology* 66(8): 3214–3220.

Statham, P.J., M. Skidmore and M. Tranter 2008. 'Inputs of glacially derived dissolved and colloidal iron to the coastal ocean and implications for primary productivity' *Global Biogeochemical Cycles* 22(3), GB3013, doi:10.1029/2007GB003106

Stibal, M., A.M. Anesio, C.J.D. Blues and M. Tranter 2009. 'Phosphatase activity and organic phosphorus turnover on a high Arctic glacier' *Biogeosciences* 6(5): 913–922.

Stibal, M. and M. Tranter 2007. 'Laboratory investigation of inorganic carbon uptake by cryoconite debris from Werenskioldbreen, Svalbard' *Journal of Geophysical Research-Biogeosciences* 112(G4): 9. G04S33, doi:10.1029/2007JG000429.

Telling, J., A.M. Anesio, J. Hawkings, M. Tranter, J.L. Wadham, A.J. Hodson, T. Irvine-Fynn and M.L. Yallop 2010. 'Measuring rates of gross photosynthesis and net community production in cryoconite holes: a comparison of field methods' *Annals of Glaciology* 51(56): 153–162.

Telling, J., A.M. Anesio, M. Tranter, A.G. Fountain, T.H. Nylen, J. Hawkings, V.B. Singh, P. Kaur, M. Musilova and J. Wadham 2015. 'Spring thaw ionic pulses boost nutrient availability and microbial growth in entombed Antarctic Dry Valley cryoconite holes' *Frontiers in Microbiology* 5: 694. doi: 10.3389/fmicb.2014.00694.

Telling, J., A.M. Anesio, M. Tranter, T. Irvine-Fynn, A. Hodson, C. Butler and J. Wadham 2011. 'Nitrogen fixation on Arctic glaciers, Svalbard' *Journal of Geophysical Research-Biogeosciences* 116. G03039, doi:10.1029/2010JG001632.

Tranter, M., M.J. Sharp, H.R. Lamb, G.H. Brown, B.P. Hubbard and I.C. Willis 2002. 'Geochemical weathering at the bed of Haut Glacier d'Arolla, Switzerland: a new model' *Hydrological Processes* 16(5): 959–993.

Tranter, M. and J.L. Wadham 2014. *Geochemical Weathering in Glacial and Proglacial Environments. Treatise on Geochemistry.* Oxford, UK: Elsevier. 157–173.

Wadham, J.L., S. Arndt, S. Tulaczyk, M. Stibal, M. Tranter, J. Telling, G.P. Lis, E. Lawson, A. Ridgwell, A. Dubnick, M.J. Sharp, A.M. Anesio and C.E.H. Butler 2012. 'Potential methane reservoirs beneath Antarctica' *Nature* 488(7413): 633–637.

Wadham, J.L., S. Bottrell, M. Tranter and R. Raiswell 2004. 'Stable isotope evidence for microbial sulphate reduction at the bed of a polythermal high Arctic glacier' *Earth and Planetary Science Letters* 219(3–4): 341–355.

Wadham, J.L., R. De'ath, F.M. Monteiro, M. Tranter, A. Ridgwell, R. Raiswell and S. Tulaczyk 2013. 'The potential role of the Antarctic Ice Sheet in global biogeochemical cycles' *Earth and Environmental Science Transactions of the Royal Society of Edinburgh* 104(1): 55–67.

Wadham, J.L., M. Tranter, M. Skidmore, A.J. Hodson, J. Priscu, W.B. Lyons, M. Sharp, P. Wynn and M. Jackson 2010. 'Biogeochemical weathering under ice: size matters' *Global Biogeochemical Cycles* 24. GB3025, doi:10.1029/2009GB003688

Walder, J.S. and A. Fowler (1994. 'Channelized subglacial drainage over a deformable bed' *Journal of Glaciology* 40(134): 3–15.

Weertman, J. 1957. 'On the sliding of glaciers' *Journal of Glaciology* 3: 33–38.

Willis, M.J., B.G. Herried, M.G. Bevis and R.E. Bell 2015. 'Recharge of a subglacial lake by surface meltwater in northeast Greenland' *Nature* 518(7538): 223–227.

Wynn, P.M., A. Hodson and T. Heaton 2006. 'Chemical and isotopic switching within the subglacial environment of a High Arctic glacier' *Biogeochemistry* 78(2): 173–193.

Yallop, M.L., A.M. Anesio, R.G. Perkins, J. Cook, J. Telling, D. Fagan, J. MacFarlane, M. Stibal, G. Barker, C. Bellas, A. Hodson, M. Tranter, J. Wadham and N.W. Roberts 2012. 'Photophysiology and albedo-changing potential of the ice algal community on the surface of the Greenland ice sheet' *Isme Journal* 6(12): 2302–2313.

18

Permafrost dynamics

Margareta Johansson

Introduction

Permafrost, perennial frozen ground, is defined as any material (soil, rock, sediment or other earth material) that stays at or below 0°C for two or more consecutive years (Brown and Péwé 1973). Approximately 17% of the global land masses is presently underlain by permafrost and it is widespread in the northern Polar Region (French 2017), while in the southern Polar Region it is mainly found in the non-glaciated areas of Antarctica and in high altitude areas (Table 18.1; Gruber 2012; Bockheim et al. 2013). In addition to terrestrial permafrost, subsea permafrost is also widespread in the continental shelves in the Arctic Ocean. The distribution of permafrost is usually defined in four zones based on the percentage of land surface that is underlain by permafrost: (1) continuous permafrost (90–100%), (2) discontinuous permafrost (50–90%), (3) sporadic permafrost (10–50%) and (4) isolated patches of permafrost (0–10%) (Brown et al. 1998). In general, the proportion of landscape underlain by permafrost increases with increasing latitude from the southern/northern limits of the permafrost zone to the High Arctic/Antarctic (Figure 18.1). Permafrost can be anything from a few dm thick at the limit of the permafrost zone to about 1500 metres (French 2007). On top of the permafrost is an active layer that thaws and refreezes on an annual basis, below which is a transient layer that can remain frozen in some summers (Shur et al. 2005), and below the permafrost is unfrozen material. The active layer can

Table 18.1 Summary of global permafrost distribution (adapted from Gruber 2012).

Region	Permafrost region	
	10^6 km^2	*% of exposed area*
Global, north of 60°S	21.7 (18.8–24.4)	17 (14–19)
Northern Hemisphere	21.7 (18.7–24.3)	22 (19–25)
Southern Hemisphere, north of 60°S	0.05 (0.01–0.15)	0.14 (0.03–0.43)
Antarctica, ice-free[1]	0.28	100
Global, exposed surface	22.0 (19.1–24.7)	17 (14–19)

1 Data from Bockheim (1995).

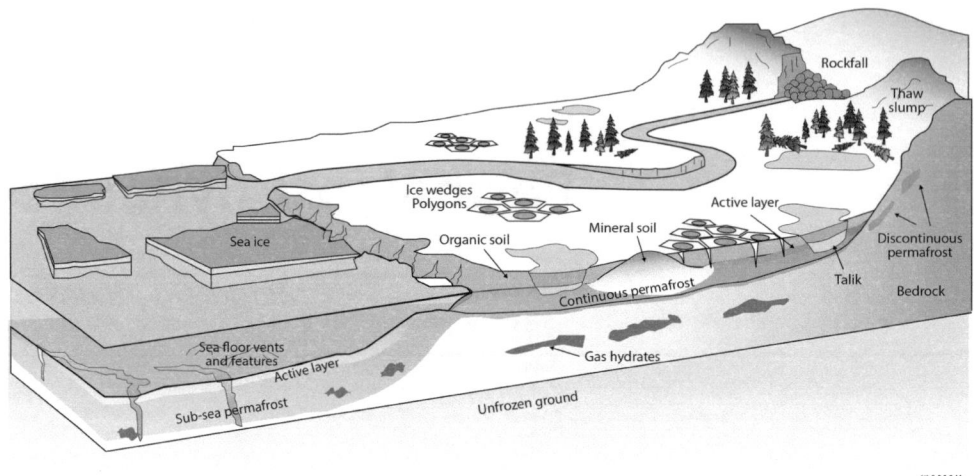

Figure 18.1   Distribution of permafrost along a conceptual transect from the sub-Arctic to the continental shelves (AMAP 2011).

be anything from less than few tens of cm in vegetated, organic terrain to several metres in areas of exposed bedrock (Callaghan et al. 2011).

Permafrost forms when temperatures drop and has especially developed during the colder periods (glacial and stadial periods) of Earth history. Past air temperatures have hence been important for the formation of permafrost; current air temperature is an influential parameter for permafrost exist-ence; and the future air temperature will determine the fate of permafrost in the Polar Regions. Air temperatures can be used to indicate the presence of permafrost at a continental scale (McGuire et al. 2016). It is however not only air temperature that affects the presence and dynamics of permafrost. At a local scale, there are many factors that are important for the energy balance at the ground surface and for the response of permafrost to changes in climate (Williams and Smith 1989; French 2017).

Parameters affecting permafrost dynamics

In addition to air temperatures, snow cover is another important climatic parameter that affects permafrost (Park et al. 2015). Snow insulates the ground and an increase in snow cover can hence increase the speed of permafrost degradation (e.g. Johansson et al. 2013). Wind is also an important factor as it distributes snow especially in areas with low vegetation. Other factors that are important for permafrost dynamics include the thermal properties of the earth material, soil organic layer thickness, soil moisture/ice content and drainage conditions, vegetation and large water bodies.

As the organic layer thickness decreases, the depth of thaw generally increases as heat pen-etrates much less rapidly in organic soils (Yi et al. 2007) than in mineral soils (Waelbrock 1993) due to the soil texture and the water content. The thermal properties do not only differ between soil materials but also for any one material over a year. The heat transfer by conduc-tion varies between frozen and unfrozen states, as the thermal conductivity of ice is four times that of water (Smith and Riseborough 2002).

Vegetation cover affects ground temperatures mainly in three ways: (1) by forming a snow trap in winter which increases the snow accumulation and persistence of snow cover resulting in warmer permafrost (Palmer et al. 2012), (2) through reducing the amount of solar radiation reaching the ground surface in summer resulting in a cooling effect (Rouse 1984) and (3) as an insulating layer, e.g. mosses (Blok et al. 2011). There are examples from both Polar Regions of how vegetation affects the thermal regime of the ground. From the northern Polar Region, Sturm et al. (2001) reported that shrub canopies increased winter soil temperature by 2°C relative to adjacent shrub-free tundra. A study from King George Island, in the maritime Antarctic, located at the edge of the climatic limit of Antarctic permafrost, showed that sites with similar soil conditions located only 10 m apart experienced significant differences in mean annual ground temperature due to different vegetation cover. One site that was dominated by lichen had a higher ground temperature amplitude compared to the other site dominated by mosses which acted as an insulating layer (Almeida et al. 2014).

Next to climate, large water bodies have the greatest local effect on ground temperatures (Williams and Smith 1989). Water bodies that do not freeze to the bottom in winter, act as a heat source and have a marked effect on ground temperatures and the local distribution of permafrost and it is common to find taliks (unfrozen soil layers; Figure 18.1) underneath water bodies in areas of discontinuous permafrost.

Past trends in permafrost dynamics

To understand the ongoing changes in permafrost and its impacts better, it is important to link the sensitivity of permafrost to climate change in the past (DeConto et al. 2011). Evidence of the former extent of permafrost and climate can also be used when predicting future permafrost dynamics. Vandenberghe et al. (2014) concluded that the impacts of climatic changes on permafrost development could best be discussed starting at the time when permafrost had its maximum extent during the last glacial period. This was when the maximum cold-climate conditions occurred, at the end of the last ice age, and Vandenberghe and colleagues named this period the "Last Permafrost Maximum – LPM" (note that it does not necessary coincide in time with the Last Glacial Maximum). By using both published and unpublished literature about permafrost indicators between 25,000 and 17,000 years ago, they reconstructed the permafrost distribution during the LPM for the northern hemisphere. The permafrost distribution was much more widespread during the LPM than at present, extending south of 45°N with the southern limit of permafrost in, e.g. southern Europe. Also, land area that is currently under sea (Bering Land Bridge) expanded the terrestrial permafrost distribution northward (Vandenberghe et al. 2014). Lindgren et al. (2016) concluded that the permafrost extent during the last glacial maximum in the northern circumpolar region was about 33% larger than at present.

The climatic transition from LPM to the current interglacial period (Whitehouse, Chapter 16 of this volume) was associated with a rapid thaw of permafrost both from the top and bottom at the southernmost limits of its maximum distribution. In areas of coastal uplift, there was on the contrary some permafrost formation occurring (Callaghan et al. 2011). Permafrost had completely disappeared from most of Europe, northern Kazakhstan and western Siberia by the time of the Holocene optimum (5000 to 9000 years BP; Velichko and Faustova 2009; Velichko and Nechaev 2009), but in areas with continuous permafrost there was in general no widespread thaw. Where near surface permafrost was ice-rich, many thermokarst lakes developed as a consequence of increasing active layer thickness (Walter et al. 2007).

During more recent stadials (cold periods in the Middle and Late Holocene), new fairly shallow areas of permafrost formed in the northern Polar Region. The latest example is during the so-called Little Ice Age (from the seventeenth to the nineteenth century) when shallow permafrost (15 to 25 m, e.g. Romanovsky et al. 1992) was formed in sediments that had been predominantly unfrozen during most of the Holocene. Current climate warming has initiated the Little Ice Age permafrost thawing that is ongoing at present (Callaghan et al. 2011).

Current trends in permafrost dynamics

To determine the current trends in permafrost dynamics, the two most direct indicators of change are monitored, namely the thickness of the active layer and ground temperatures. Monitoring of the active layer thickness is mainly done through probing either throughout the thaw season (summer) or at the end of the thaw season (Figure 18.2). Active layer thickness can also be measured through thaw tubes (e.g. Smith et al. 2009b). The Circumpolar Active Layer Monitoring (CALM) project (www2.gwu.edu/~calm/; Nelson et al. 2004; Shiklomanov et al. 2008, 2012) has developed a standardized protocol that has been used since 1990. Sporadic measurement of active layer thickness exists at least from the 1930s (e.g. Ekman 1957), but continuous monitoring started in the 1970s in the northern Polar Region (e.g. Akerman and Johansson 2008).

Ground temperatures are recorded at different depths in boreholes in both Polar Regions (Figure 18.3). The best indicator for long-term change is permafrost temperature at a depth where there is practically no annual fluctuation in ground temperatures (ranging from a few metres to more than 20 metres; Romanovsky et al. 2017). During the International Polar Year (2007–2008), the Thermal State of Permafrost (TSP) project developed a standardized monitoring scheme and many new boreholes were made in both Polar Regions (Romanovsky et al. 2010a; Vieira et al. 2010). Data on both active layer thickness and ground temperatures can be found in the GTN-P database (http://gtnpdatabase.org/; Biskaborn et al. 2015).

Trends in active layer thickness

There are CALM sites in both the northern and southern Polar Regions (Figure 18.4). In the northern Polar Region, there are 164 sites registered in the CALM database (located >60°N; CALM database 2016). The results from those sites show in general substantial inter-annual

Figure 18.2 Active layer thickness is monitored in both Polar Regions using the standardized monitoring scheme from the Circumpolar Active Layer Monitoring (CALM) programme (Photo: J. Åkerman).

Figure 18.3   Ground temperatures are monitored in boreholes to record the thermal state of permafrost. Here is an example from the northern Polar Region (Photo: M. Johansson).

variability in active layer thickness mainly due to variations in summer air temperature (e.g. Popova and Shmakin 2009; Smith et al. 2009a) and the decadal trends in the active layer thickness differ by region (Shiklomanov et al. 2012). The active layer thickness reported from the northern Polar Region in 2016 ranges from 30 to more than 200 cm at the end of the thawing season. In a warming climate the active layer is in general increasing in thickness over time (Romanovsky et al. 2015). This has been reported for the last decades from many sites in the northern Polar Region. Some examples are from the Russian European North sites (Romanovsky et al. 2017), some Nordic sites (e.g. Akerman and Johansson 2008) and some sites in northeast Greenland (Elberling et al. 2013). Insignificant increase, and even decrease, in active layer thickness has been observed in some parts of the northern hemisphere, e.g. parts of Alaska and Canada (Duchesne et al. 2015; Luo et al. 2016). However, in northern Alaska, the average date of freeze-up of the active layer has become almost two months later during the last 30 years (Romanovsky et al. 2017).

In the southern Polar Region there are 29 sites that are registered in the CALM database (located >60°S; CALM database 2016). The monitoring began in the end of the 1990s and there are very few continuous monitoring sites extending more than ten years. At the end of

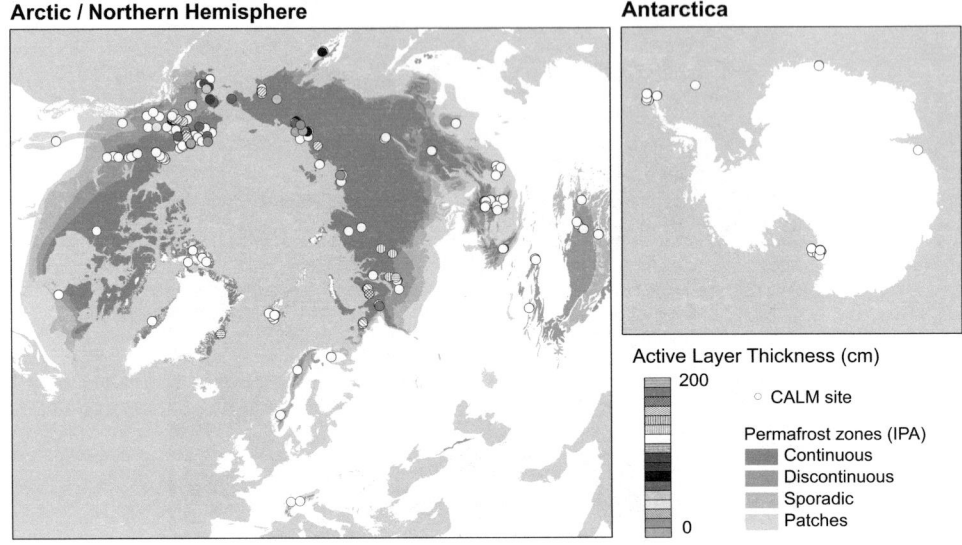

Figure 18.4   Location of Circumpolar Active Layer Monitoring (CALM) sites in (a) the Northern (>60°N) and (b) the Southern (>60°S) Polar Regions (Illustration: Boris Biskaborn, Alfred Wegener Institute – Helmholtz Zentrum für Polar und Meeresforschung, Potsdam, Germany).

the thawing season, the active layer thickness ranges from 15 cm to more than 130 cm (CALM database 2016). Guglielmin and Vieira (2014) reported an increase in active layer thickness in continental Antarctica, despite the absence of air warming, since the early 2000s (see Marshall, Chapter 15 in this volume). In other areas with shorter monitoring history it is difficult to talk about trends in active layer thickness, but large interannual variability in maximum thaw depth has been recorded from e.g. Adelaide Island (ranging from 76–140 cm; Guglielmin et al. 2014) and from Livingston Island (44–92 cm; de Pablo et al. 2013).

Trends in ground temperatures

Data compiled from the TSP project concluded that the range of ground temperatures is greater in the Antarctic than in the Arctic. In the Antarctic, ground temperatures ranged from just below 0°C in the South Shetland Islands (Bockheim et al. 2013) to almost −24°C on Mount Fleming (Ross Island) (Vieira et al. 2010). In the Arctic, ground temperatures ranged from just below 0°C in the southern boundary of permafrost existence to −15°C in the northernmost continuous permafrost zone (e.g. Romanovsky et al. 2010a; Smith et al. 2013).

There has been a general warming trend in ground temperatures in the northern Polar Region since the 1980s (Brown and Romanovsky 2008; Marshall, Chapter 15 of this volume). The magnitude of increase in ground temperatures varies regionally (Christiansen et al. 2010; Romanovsky et al. 2010b; Smith et al. 2010), but are in general between 0.5 to 2°C at the depth of zero annual amplitude since the late 1970s (Callaghan et al. 2011). In Russia, the monitored increase in ground temperatures has been especially pronounced since the late 1990s compared with the previous three decades (Streletskiy et al. 2015). The greatest increases in ground temperatures in the northern Polar Region have been found in cold permafrost (< −2°C) or bedrock. Less warming has been observed at warm permafrost sites with temperatures close to 0°C as those are affected by latent heat release when a large portion of the additional heat flux from increasing air temperatures into the deeper ground is spent on melting of the pore ice in the ground (Romanovsky et al. 2017). As a result, warm permafrost can persist for some time under rising air temperatures (James et al. 2013).

Movement of permafrost boundaries

Movement of permafrost boundaries has already been detected in the northern Polar Region. By combining many different methodologies such as active layer thickness probing, manual excavations, ground temperature monitoring using boreholes and geophysical techniques, James et al. (2013) concluded that the southern limit of permafrost appears to have retreated northward by at least 25 km along the Alaska Highway (in southern Yukon and northern BC) since the late 1960s.

Permafrost boundaries can also shift vertically. An example is from the East Siberian Arctic Shelf area where the upper subsea permafrost boundary was moved from 6–8 m down to 16–18 m as a consequence of increased near bottom water temperature from 1985 until present (Romanovsky et al. 2017).

Future trends in permafrost dynamics

Assessing the response of permafrost to future climate warming is difficult as there are many different factors in addition to climate that affect permafrost dynamics. For example, vegetation and soil processes can make the permafrost more resilient to climate warming, and on the contrary, some of the changes in surface stability and hydrology can make permafrost more vulnerable to warming (Jorgenson et al. 2010). Given the projected increase in GHG concentrations by the late 2030s, global circulation models project that average autumn and winter Arctic air temperatures will increase by as much as 4°C by 2040 (Overland et al. 2017). In combination with increases in snow depth, this will significantly affect permafrost dynamics. Projections for the twenty-first century suggest an enhanced thaw rate and an increase in active layer thickness (Slater and Lawrence 2013; Figure 18.5). Hence, by the end of the twenty-first century, late-Holocene permafrost may be actively thawing at the southern boundary of the permafrost domain and Late Pleistocene permafrost could start to thaw in some locations (Callaghan et al. 2011). Romanovsky et al. (2017) conclude that the sensitivity of the modelled area of permafrost (with an active layer thickness more than three metres) to climate change was 0.8 to 2.3 million km² per degree C increase in local air temperature.

Impacts of permafrost dynamics

As permafrost determines the living conditions for flora, fauna and humans in many areas, there are a number of good reasons why permafrost dynamics should be studied and monitored, as changes can result in large impacts at different local, regional and global scales. Figure 18.6 gives an overview of some examples of such impacts from warming and thawing permafrost, and these are further discussed below.

A) Impacts on infrastructure

Ongoing climate change in the Polar Regions, especially changes in the amount and type of precipitation, its timing and the rate of snow melt will affect permafrost, which represents great challenges to municipalities and transportation networks in areas underlain by permafrost (Instanes

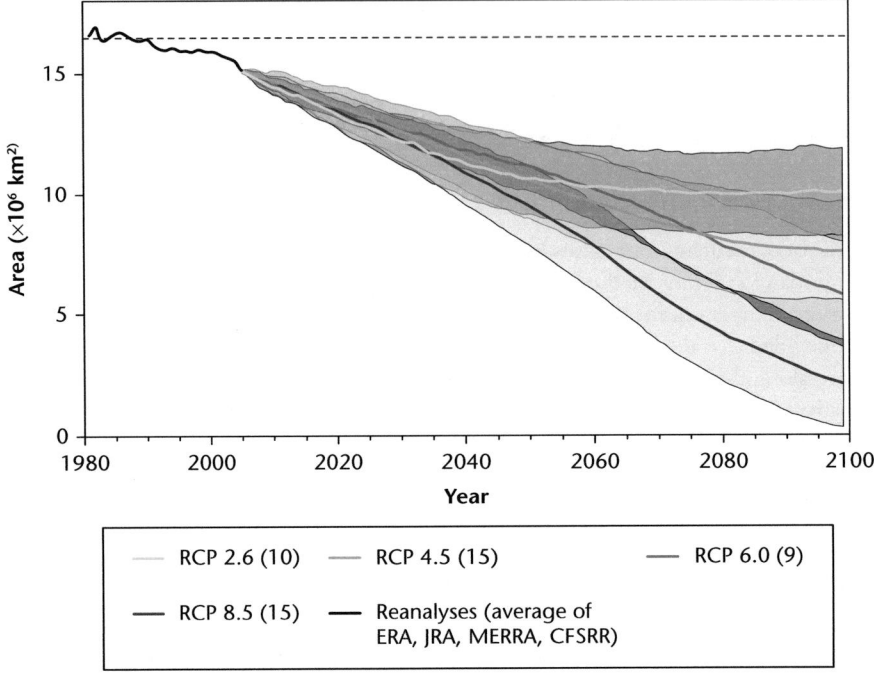

Figure 18.5   Projected changes in sustainable permafrost area. Shaded areas represent one standard deviation across the CMIP5 models and the dashed black line is the model equivalent present-day total area of continuous and discontinuous permafrost (Slater and Lawrence, 2013 © American Meteorological Society. Used with permission).

et al. 2016). The stability and strength of frozen ground is strongly linked to ground temperatures. Ongoing changes in permafrost ground temperatures and the active layer thickness have resulted in decreases in ground-bearing capacity (Streletskiy et al. 2012). It is especially a problem in areas of ice-rich permafrost where the stability can decrease as ground ice melts. This has major impacts on infrastructure performance (Instanes et al. 2005, 2016). The impacts of changing permafrost dynamics on infrastructures are a much bigger concern in the northern Polar Region compared to the southern Polar Region. This is especially true in Russia where large cities are built on permafrost and information on future projections of permafrost conditions was not available at the time of construction. Streletskiy et al. (2012) found a 40% decrease in bearing capacity in the city of Nadym and a 20% decrease for Yakutsk and Salekhard over the past 40 years. Warming permafrost will provide great challenges in the future for local and regional infrastructure.

B) Impacts on coastal areas

Permafrost coasts in the northern Polar Region account for 34% of Earth's coasts (Lantuit et al. 2012). As a consequence of warming permafrost and decreasing sea ice, coastal erosion has been observed in the Arctic (Gunther et al. 2015). This results in release of organic carbon into the nearshore zone, with potential impact on global carbon fluxes and their climate feedbacks, on nearshore food webs and on local communities (Fritz et al. 2017). Local communities are likely to be affected in two main ways: (1) their food availability is likely to decrease as many communities still rely on marine biological resources and (2) the coastal erosion poses a threat

Figure 18.6 An overview of local impacts from changing permafrost dynamics A) on infrastructure B) on coastal erosion C) on carbon budget D) on hydrology E) on mountain slopes F) on vegetation and G) on human health (Illustration: Susanna Olsson, Lund University, Sweden).

to infrastructure close to the shore line. As a consequence, communities in the northern Polar Region have and will experience the need for relocation (Hamilton et al. 2016).

C) Impacts on carbon fluxes

Permafrost is like a big freezer where dead plants are being preserved and a large carbon pool has built up in permafrost areas. The new northern permafrost zone carbon inventory reported that the known pool of terrestrial permafrost carbon in the northern permafrost zone is 1,330–1,580 Pg carbon (Hugelius et al. 2014; Schuur et al. 2015). When permafrost thaws, carbon can be decomposed, and this results in an enhanced carbon cycling process (Schuur et al. 2015; also see Chapter 17 in this volume on biogeochemical cycling by Liz Bagshaw, and Chapter 19 on polar feedbacks by Richard Hodgkins). The impacts of the potential release of methane (CH_4) and carbon dioxide (CO_2) from warming and thawing permafrost may be both regional and global (Koven et al. 2015).

D) Impacts on hydrology

As a result of changing permafrost dynamics, especially increasing active layer thickness, changes in hydrology have been reported from both Polar Regions (e.g. Smith et al. 2005; Bockheim et al. 2013; Liljedahl et al. 2016). In ice-rich permafrost areas, ice lenses at the top of the permafrost table melt when the active layer thickness increases. As a consequence, the ground subsides (Nelson et al. 2001) and new lakes – so-called thermokarst lakes – are formed. In other

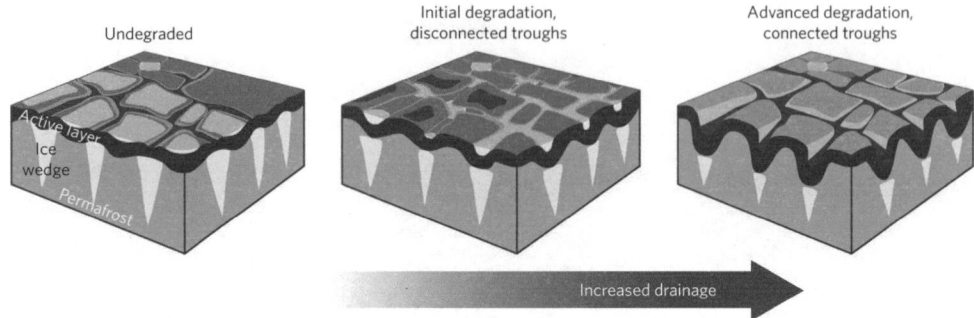

Figure 18.7 Impact on hydrology from degrading permafrost in an area with ice wedges (Figure reprinted by permission from Macmillan Publishers Ltd: Nature Geoscience (Liljedahl et al., 2016), copyright 2016).

areas, increased active layer thickness leads to drying of the landscape, for example in areas with ice wedges (e.g. Liljedahl et al. 2016; Figure 18.7) as the water finds new drainage pathways in the newly unfrozen material. The impact on hydrology is also important for the carbon fluxes (Olefeldt et al. 2016).

E) Impacts on slope instability

The impacts of warming and thawing mountain permafrost can be severe and can affect, for example, the stability of slopes and rock walls that in turn can affect vegetation, people and infrastructure. In many high-altitude areas, rock fall events and slope instabilities are related to thawing permafrost (e.g. Gruber and Haeberli 2007; Krautblatter et al. 2013). In addition to the long-term changes in permafrost temperatures that affect slope stability when the permafrost starts to thaw (Callaghan et al. 2011), it is also extreme events such as unusual warm periods and heavy rain (especially in the so-called shoulder seasons: spring and fall) that have great impact on slope stability (Stoffel et al. 2014).

F) Impacts on vegetation

No vegetation is dependent on permafrost and on a Polar Region scale no ecosystems are restricted by the presence or absence of permafrost. However, at the local scale, the presence of permafrost strongly influences the plant species' composition and the types of plant that can grow (Callaghan et al. 2011). Changes in permafrost dynamics are also likely to have great impact on vegetation; for example, where slumping of the land surface occurs, trees become unstable and fall.

G) Impacts on human health

Many of the above-mentioned impacts can affect human health. A potential "threat" from thawing permafrost is the release of ancient diseases. People have been buried in permafrost for several centuries and there is a possibility that viruses within the bodies of people that died in epidemics could again become prevalent under current and projected permafrost thawing and expose modern ecosystems and environments to relic life, with largely unknown consequences (Callaghan et al. 2011). An example that occurred in 2016 was an outbreak of anthrax in the

Yamalsky District in Siberia, Russia. More than 1,200 reindeer were killed, and families were evacuated from the tundra.

Conclusions

Permafrost dynamics in both Polar Regions are affected by the ongoing climate change. This is very likely to continue in the future and is expected to pose many challenges at a local scale (to communities that are living in areas underlain by permafrost), at a regional scale (for example, by affecting infrastructure that connects large areas) and also at a global scale (through feedback effects, e.g. the additional GHG release as a consequence of thawing permafrost). Continuous monitoring of permafrost dynamics is required in both Polar Regions to provide information needed for adaptation to future changes in permafrost dynamics.

Acknowledgement

The author is grateful for financial support from the Nordic Center of Excellence's project DEFROST under the Nordic Top-Level Research Initiative and the Adsimnor project funded by the Swedish Research Council VR. The author is also very grateful to Boris Biskaborn at Alfred Wegener Institute, Potsdam, Germany for kindly providing Figure 18.4; to Susanna Olsson, Lund University, Sweden for providing Figure 18.7; and to Jonas Åkerman, Lund University, Sweden for providing Figure 18.2.

References

Akerman H.J. and Johansson, M. 2008. 'Thawing permafrost and thicker active layers in Sub-arctic Sweden' *Permafrost and Periglacial Processes* 19(3): 279–292.
Almeida, I.C.C., et al. 2014. 'Active layer thermal regime at different vegetation covers at Lions Rump, King George Island, Maritime Antarctica' *Geomorphology* 225: 36–46.
AMAP 2011. 'Snow, Water, Ice and Permafrost in the Arctic (SWIPA): climate change and the cryosphere' Arctic Monitoring and Assessment Programme (AMAP), Oslo, Norway. xii + 538 pp.
Biskaborn, B.K. et al. 2015. 'The new database of the Global Terrestrial Network for Permafrost (GTN-P)' *Earth System Science Data* 7(2): 245–259.
Blok, D. et al. 2011. 'The cooling capacity of mosses: controls on water and energy fluxes in a Siberian tundra site' *Ecosystems* 14: 1055–1065.
Bockheim, J. G. 1995. 'Permafrost distribution in the southern circumpolar region and its relation to the environment: a review and recommendations for further research' *Permafrost and Periglacial Processes* 6(1): 27–45.
Bockheim, J. et al. 2013. 'Climate warming and permafrost dynamics in the Antarctic Peninsula region' *Global and Planetary Change* 100: 215–223.
Brown, J., Ferrians, Jr. O.J., Heginbottom, J.A. and Melnikov, E.S. 1998. revised February 2001. *Circum-Arctic Map of Permafrost and Ground Ice Conditions.* Boulder, CO: National Snow and Ice Data Center. Digital media.
Brown, J. and Romanovsky, V.E. 2008. 'Report from the International Permafrost Association: state of permafrost in the first decade of the 21st century' *Permafrost and Periglacial Processes* 19: 255–260.
Brown, R.J.E and Péwé, T.L. 1973. 'Distribution of permafrost in North America and its relationship to the environment: a review, 1963–1973' in *Permafrost: North American Contribution, Second International Conference*, pp. 71–100. Washington, DC: National Academy of Sciences.
Callaghan, T.V. et al. 2011. 'Changing permafrost and its impacts' in *Snow, Water, Ice and Permafrost in the Arctic (SWIPA) 2011*. Arctic Monitoring and Assessment Programme (AMAP), Oslo.
Christiansen, H.H. et al. 2010. 'The thermal state of permafrost in the Nordic area during the International Polar Year 2007–2009' *Permafrost and Periglacial Processes* 21: 156–181.
Circumpolar Active Layer Monitoring (CALM) website, 2016. www2.gwu.edu/~calm/data/data-links. html.

DeConto, R.M. et al. 2011. 'Past extreme warming events linked to massive carbon release from thawing permafrost' *Nature* 484, 87–91.

de Pablo, M.A. et al. 2013. 'Interannual active layer variability at the Limnopolar Lake CALM site on Byers Peninsula, Livingston Island, Antarctica' *Antarctic Science* 25(2): 167–180.

Duchesne, C. et al. 2015. *20 Years of Active Layer Monitoring in the Mackenzie Valley, Northwest Territories*. Geological Survey of Canada, Scientific Presentation SP31.

Ekman, S. 1957. Die Gewässer des Abisko-Gebietes und Ihre Bedingungen. Kungliga Svensk Vetenskapsakademiens Handling 4. 172 pp (In German).

Elberling, B., Michelsen, A., Schädel, C., Schuur, E.A., Christiansen, H.H., Berg, L., Tamstorf, M.P. and Sigsgaard, C. 2013. 'Long-term CO_2 production following permafrost thaw' *Nature Climate Change* 3(10): 890–894.

French, H.M. 2007. *The Periglacial Environment*. 3rd edition. Chichester, UK: John Wiley & Sons.

French, H.M. 2017. *The Periglacial Environment*. 4th edition. Hoboken, NJ: Wiley-Blackwell.

Fritz, M., Vonk, J.E. and Lantuit, H. 2017. 'Collapsing Arctic coastlines' *Nature Climate Change* 7: 6–7.

Gruber, S. 2012. 'Derivation and analysis of a high-resolution estimate of global permafrost zonation' *Cryosphere* 6(1): 221–233.

Gruber, S. and Haeberli, W. 2007. 'Permafrost in steep bedrock slopes and its temperature-related destabilization following climate change' *Journal of Geophysical Research – Earth Surface* 112 (F2): F02S18. DOI: 10.1029/2006JF000547.

Guglielmin, M. and Vieira, G. 2014. 'Permafrost and periglacial research in Antarctica: new results and perspectives' *Geomorphology* 225: 1–3.

Guglielmin, M. et al. 2014. 'Permafrost and snow monitoring at Rothera Point (Adelaide Island, Maritime Antarctica): implications for rock weathering in cryotic conditions' *Geomorphology* 225: 47–56.

Gunther, F. et al. 2015. 'Observing Muostakh disappear: permafrost thaw subsidence and erosion of a ground-ice-rich island in response to arctic summer warming and sea ice reduction' *Cryosphere* 9(1): 151–178.

Hamilton, L.C. et al. 2016. 'Climigration? Population and climate change in Arctic Alaska' *Population and Environment* 38(2): 115–133.

Hugelius, G. et al. 2014. 'Estimated stocks of circumpolar permafrost carbon with quantified uncertainty ranges and identified data gaps' *Biogeosciences* 11: 6573–6593.

Instanes, A. et al. 2005. 'Infrastructure: buildings, support systems, and industrial facilities' in C. Symon, L. Arris and B. Heal (eds) *Arctic Climate Impact Assessment*, ACIA. Cambridge, UK: Cambridge University Press 907–944.

Instanes, A. et al. 2016. 'Changes to freshwater systems affecting Arctic infrastructure and natural resources' *Journal of Geophysical Research – Biogeosciences* 121(3): 567–585.

James, M. et al. 2013. 'Multi-decadal degradation and persistence of permafrost in the Alaska Highway corridor, northwest Canada' *Environmental Research Letters*, 8: 045013. DOI: 10.1088/1748-9326/8/4/045013.

Johansson, M. et al. 2013. 'Rapid responses of permafrost and vegetation to experimentally increased snow cover in sub-arctic Sweden' *Environmental Research Letters* 8: 035025. DOI: 10.1088/1748-9326/8/3/035025.

Jorgenson, M.T. et al. 2010. 'Resilience and vulnerability of permafrost to climate change' *Canadian Journal of Forest Research-Revue Canadienne de Recherche Forestiere* 40(7): 1219–1236.

Koven, C.D. et al. 2015. 'A simplified, data-constrained approach to estimate the permafrost carbon-climate feedback' *Philosophical Transactions of the Royal Society A: Mathematical Physical and Engineering Sciences* 373(2054): 20140423. DOI: 10.1098/rsta.2014.0423.

Krautblatter, M., Funk, D. and Gunzel, F.K. 2013. 'Why permafrost rocks become unstable: a rock-ice-mechanical model in time and space' *Earth Surface Processes and Landforms* 38(8): 876–887.

Lantuit, H. et al. 2012. 'The Arctic Coastal Dynamics Database: a new classification scheme and statistics on Arctic permafrost coastlines' *Estuaries and Coasts* 35(2): 383–400.

Liljedahl, A.K. et al. 2016. 'Pan-Arctic ice-wedge degradation in warming permafrost and its influence on tundra hydrology' *Nature Geoscience* 9(4): 312–318.

Lindgren, A. et al. 2016. 'GIS-based maps and area estimates of Northern Hemisphere permafrost extent during the Last Glacial Maximum' *Permafrost and Periglacial Processes* 27(1): 6–16.

Luo, D.L. et al. 2016. 'Recent changes in the active layer thickness across the Northern Hemisphere' *Environmental Earth Sciences* 75(7). 555, DOI: 10.1007/s12665-015-5229-2.

McGuire, A.D. et al. 2016. 'Variability in the sensitivity among model simulations of permafrost and carbon dynamics in the permafrost region between 1960 and 2009' *Global Biogeochemical Cycles* 30(7): 1015–1037.

Nelson, F.E. et al. 2004. 'The circumpolar-active-layer-monitoring (CALM) Workshop: introduction' *Permafrost and Periglacial Processes* 15(2): 99–101.

Nelson, F.E., Anisimov, O.A. and Shiklomanov, N.I. 2001. 'Subsidence risk from thawing permafrost: the threat to man-made structures across regions in the far north can be monitored' *Nature* 410(6831): 889–890.

Olefeldt, D. et al. 2016. 'Circumpolar distribution and carbon storage of thermokarst landscapes' *Nature Communications* 7: 13043. DOI: 10.1038/ncomms13043.

Overland, J.E., Walsh, J. and Kattsov, V. 2017. 'Trends and feedbacks' in *Snow, Water, Ice and Permafrost in the Arctic (SWIPA) 2017*. Arctic Monitoring and Assessment Programme (AMAP), Oslo, Norway. 9–23.

Palmer, M.J., Burn, C.R. and Kokelj, S.V. 2012. 'Factors influencing permafrost temperatures across tree line in the uplands east of the Mackenzie Delta, 2004–2010' *Canadian Journal of Earth Sciences* 49(8): 877–894.

Park, H. et al. 2015. 'Effect of snow cover on pan-Arctic permafrost thermal regimes' *Climate Dynamics* 44(9–10): 2873–2895.

Popova, V.V. and Shmakin, A.B. 2009. 'The influence of seasonal climatic parameters on the permafrost thermal regime, West Siberia, Russia' *Permafrost and Periglacial Processes* 20: 41–56.

Romanovsky, V.E., Garagula, L.S. and Seregina, N.V. 1992. 'Freezing and thawing of soils under the influence of 300- and 90-year periods of temperature fluctuation' in *Proceedings of the International Conference on the Role of Polar Regions in Global Change*, University of Alaska Fairbanks, 2: 543–548.

Romanovsky, V.E., Smith, S.L. and Christiansen, H.H. 2010a. 'Permafrost thermal state in the polar Northern Hemisphere during the International Polar Year 2007–2009: a synthesis' *Permafrost and Periglacial Processes* 21: 106–116.

Romanovsky, V.E. et al. 2010b. 'Thermal state of permafrost in Russia' *Permafrost and Periglacial Processes* 21: 136–155.

Romanovsky, V.E. et al. 2015. 'Terrestrial permafrost (in "State of the climate in 2014")' *Bulletin of the American Meteorological Society* 96(7). S139–S141.

Romanovsky, V.E. et al. 2017. 'Changing permafrost and its impacts' in *Snow, Water, Ice and Permafrost in the Arctic (SWIPA) 2017*. Arctic Monitoring and Assessment Programme (AMAP), Oslo, Norway. 65–102.

Rouse, W. 1984. 'Microclimate of Arctic tree line. 2: Soil microclimate of tundra and forest' *Water Resources Research* 20: 67–73.

Schuur, E.A.G. et al. 2015. 'Climate change and the permafrost carbon feedback' *Nature* 520 (7546): 171–179.

Shiklomanov, N.I. et al. 2008. 'The Circumpolar Active Layer Monitoring (CALM) Program: data collection, management, and dissemination strategies' *9th International Conference on Permafrost*, 2: 1647–1652.

Shiklomanov, N.I., Streletskiy, D.A. and Nelson, F.E. 2012. 'Northern Hemisphere component of the global Circumpolar Active Layer Monitoring (CALM) program' *Proceedings of the Tenth International Conference on Permafrost*, 25–29 June, Salekhard, Russia, 1: 377–382.

Shur, Y., Hinkel, K.M. and Nelson, F.E. 2005. 'The transient layer: implications for geocryology and climate-change science' *Permafrost and Periglacial Processes* 16: 5–17.

Slater, A.G. and Lawrence, D.M. 2013. 'Diagnosing present and future permafrost from climate models' *Journal of Climate* 26: 5608–5623.

Smith, L.C., et al., 2005. 'Disappearing Arctic lakes' *Science* 308: 1429–1429.

Smith, M.W. and Riseborough, D.W. 2002. 'Climate and the limits of permafrost: a zonal analysis' *Permafrost and Periglacial Processes* 13(1): 1–15.

Smith, S.L. et al. 2009a. 'Active-layer characteristics and summer climatic indices, Mackenzie Valley, Northwest Territories, Canada' *Permafrost and Periglacial Processes* 20: 201–220.

Smith, S.L. et al. 2009b. 'Data for geological survey of Canada active layer monitoring sites in the Mackenzie Valley, NWT' Geological Survey of Canada Open File 6287.

Smith, S.L. et al. 2010. 'Thermal state of permafrost in North America: a contribution to the International Polar Year' *Permafrost and Periglacial Processes* 21: 117–135.

Smith, S.L. et al. 2013. 'A map and summary database of permafrost temperatures in Nunavut, Canada. Geological Survey of Canada Open File 7393' https://geoscan.nrcan.gc.ca/cgi-bin/starfinder/0?path=geoscan.fl&id=fastlink&pass=&search=R%3D292615&format=FLFULL.

Stoffel, M., Tiranti, D. and Huggel, C. 2014. 'Climate change impacts on mass movements: case studies from the European Alps' *Science of the Total Environment* 493: 1255–1266.

Streletskiy, D.A., Shiklomanov, N.I. and Nelson, F.E. 2012. 'Permafrost, infrastructure, and climate change: a GIS-based landscape approach to geotechnical modeling' *Arctic, Antarctic and Alpine Research* 44(3): 368–380.

Streletskiy, D.A. et al. 2015. 'Changes in the 1963–2013 shallow ground thermal regime in Russian perma-frost regions' *Environmental Research Letters* 10(12): 125005. DOI: 10.1088/1748–9326/10/12/125005.

Sturm, M. et al. 2001. 'Snow-shrub interactions in Arctic tundra: a hypothesis with climatic implications' *Journal of Climate* 14: 336–344.

Vandenberghe, J. et al. 2014. 'The Last Permafrost Maximum (LPM) map of the Northern Hemisphere: permafrost extent and mean annual air temperatures, 25–17 ka BP' *Boreas* 43(3): 652–666.

Velichko, A.A. and Faustova, M.A. 2009. 'Glaciation during the Late Pleistocene' in A.A. Velichko (ed.) *Paleoclimates and Paleoenvironments of Extra-Tropical Area of the Northern Hemisphere. Late Pleistocene–Holocene.* GEOS, Moscow. 32–42.

Velichko, A.A. and Nechaev, V.P. 2009. 'Subaerial cryolithozone of the Northern Hemisphere during the Late Pleistocene and Holocene' in A.A. Velichko (ed.) *Paleoclimates and Paleoenvironments of Extra-Tropical Area of the Northern Hemisphere. Late Pleistocene–Holocene.* GEOS, Moscow. 42–49.

Vieira, G. et al. 2010. 'Thermal state of permafrost and active-layer monitoring in the Antarctic: advances during the International Polar Year 2007–2009' *Permafrost and Periglacial Processes* 21(2): 182–197.

Waelbrock, C. 1993. 'Climate–soil processes in the presence of permafrost: a systems modelling approach' *Ecological Modelling* 69: 185–225.

Walter, K.M. et al. 2007. 'Thermokarst lakes as a source of atmospheric CH_4 during the last deglaciation' *Science* 318(5850): 633–636.

Williams, P. J. and Smith, M.W. 1989. *The Frozen Earth: Fundamentals of Geocryology.* Cambridge, UK: Cambridge University Press.

Yi, S.H., Woo, M.K. and Arain, M.A. 2007. 'Impacts of peat and vegetation on permafrost degradation under climate warming' *Geophysical Research Letters* 34 DOI:10.1029/2007gl030550.

19

Polar feedbacks in a changing climate

Richard Hodgkins

Introduction

Feedbacks in the climate system are processes in which an initial atmospheric, oceanic or – as the Polar Regions are being considered here – cryospheric perturbation results in a response which either amplifies (positive) or dampens (negative) that perturbation. Positive feedbacks therefore increase the rate and/or magnitude of change, while negative feedbacks decrease and/or stabilize it. In reality, multiple environmental processes, stores and fluxes interact to give rise to a wide range of feedbacks on varying spatial and temporal scales, generating complex climate responses to any particular perturbation.

Despite their similarities as the Earth's extreme cold environments, the Polar Regions are vast and heterogeneous, therefore complex and variable processes, responses and feedbacks should be anticipated. It is not possible to speak of a single or a characteristic polar response, because the Arctic and Antarctic, fundamental coldness aside, are so contrasting in terms of topography, land and ocean distribution, oceanic and atmospheric circulations, ecology and the nature of human intervention, to name just a few of the more obvious characteristics. Moreover, each region has experienced markedly different climatic changes in recent decades.

Climate and environmental variability are of course highly complex in any part of the world, characterized by large-scale, non-linear climate dynamics and regional feedbacks. The drivers of change may be intrinsic/internal variability within atmospheric and oceanic circulation systems (natural change), or external forcing from both natural and anthropogenic sources (e.g. solar variability and volcanic emissions, or Greenhouse Gas (GHG) and aerosol emissions, respectively). Either of these may be modified by positive or negative feedbacks through atmospheric, oceanic or terrestrial processes. Environmental processes in both the Arctic and Antarctic interact with the atmospheric and oceanic circulations of lower latitudes, but they also exhibit distinctive characteristics that generate feedbacks which modulate these interactions in varying ways, many of which are not fully understood, partly because of their inherent physical complexity, and partly because of the challenge of deconvoluting their influence from the broader nexus of climate and environmental change – and in the Polar Regions specifically – because observational data are typically sparse and instrumental time-series are characteristically short.

Recent polar climate change and what it tells us about feedbacks

Changes in the direction and strength of feedbacks have been very important in the climate evolution of the Arctic over the past few decades. The autumn formation of sea ice formerly provided a dominant negative feedback on Arctic air temperature variability by raising the regional surface reflectivity (albedo) and insulating the relatively warm ocean from the overlying atmosphere. Decreased ice extent and thickness from the 1990s, in combination with increased heat storage in ice-free regions of the Arctic Ocean, however, have promoted a shift towards the positive feedback processes known collectively as *Arctic amplification* (Serreze and Francis 2006; Serreze and Barry 2011; Marshall, Chapter 15 of this volume). The *polar amplification* of atmospheric warming is a term originally coined by Manabe and Stouffer (1980) and is applicable to both hemispheres, but as will become clear later, contemporary amplification is significantly stronger in the North. Current, important amplification processes are dominated by the ice-albedo feedback, but the autumn, positive, ice-insulation feedback as a consequence of the additional oceanic heat storage in formerly ice-covered ocean is also significant (Jackson et al. 2010).

A number of studies have now indicated that the observed increase in Arctic near-surface air temperatures in the late twentieth and early twenty-first centuries, unlike the warm interval of the 1930s and 1940s, is inconsistent with simulated internal variability (e.g. Wang et al. 2007; Gillett et al. 2008; Shindell and Faluvegi 2009). The state of the science is such that, in the Fifth Assessment Report of the Intergovernmental Panel on Climate Change (IPCC), Working Group I (WGI) were able to conclude that, despite the uncertainties associated with limited observational coverage, high internal variability, modelling uncertainties and insufficiently understood regional processes and forcings – it is likely there has been an anthropogenic contribution to the substantial Arctic land-surface warming since the 1960s (IPCC 2013).

However, it has proved more difficult to detect signals of climate change (natural or anthropogenic) in Antarctica: the low concentration of weather stations, their predominant location near the coast, the short instrumental records, the strong dependence of near-surface temperature on wind speed and cloudiness, and significant natural climate variability all seem to have combined – outside of the Antarctic Peninsula – to obscure the statistical significance of temperature trends (van den Broeke et al. 2004).

The major mode of atmospheric variability across the mid- and high-latitude areas of the southern hemisphere is the Southern Annular Mode (SAM), also known as the Antarctic Oscillation (AAO) or the high-latitude mode. This is defined as the difference in mean sea-level pressure between 45–60°S and is associated with synchronous, inverse anomalies in surface pressure, geopotential height, zonal wind, surface temperature and other climatological variables between Antarctica and mid-latitudes. Since the mid-twentieth century, the SAM shifted into a positive phase, increasing the sea-level pressure gradient, yielding a strengthening of the westerly wind field. Such a shift may be consistent with trends in the lower-stratospheric polar vortex at least partially driven by photochemical ozone destruction (Thompson and Solomon 2002; Kirkwood, Chapter 36 of this volume), the occurrence of which is a significant, anthropogenically-driven point of contrast between the Arctic and Antarctic. It has been suggested that the absence of recent warming over interior Antarctica is related to this SAM intensification. However, the SAM is also believed to be responsible for the recent, marked warming on the eastern side of the Antarctic Peninsula (Marshall et al. 2004): the stronger westerlies reduce the orographic blocking effect of the Peninsula, advecting more air masses eastward over the barrier. A combination of climatological temperature gradient across the Peninsula and the formation of Föhn winds on its eastern side subsequently yield a summer SAM temperature sensitivity three times greater on the eastern side of the Peninsula than the west.

When temperature changes associated with the SAM are removed statistically, both observations and simulations indicate warming at all observed locations except the South Pole in the second half of the twentieth century, with separate natural and anthropogenic responses of consistent magnitude (Gillett et al. 2008). Nevertheless, the evidence for human influence on Antarctic temperature remains weaker than for other parts of the world, particularly as a consequence of observational uncertainties. In contrast to the Arctic therefore, the IPCC WGI (2013) were only able to express low confidence in the identified Antarctic near-surface air temperatures changes and conclude likewise that there is low confidence in an anthropogenic influence on observed warming (IPCC 2013).

Evidence for accelerating rates of climate feedback is therefore clearer and stronger in the Arctic than in the Antarctic. The rapid depletion of Arctic sea ice in the late twentieth and early twenty-first centuries, in particular, cannot be replicated in modelled, internal climate variability, with natural processes appearing to be capable of accounting for no more than 50% of the observed loss (Kay et al. 2011; Schweiger et al. 2011; Day et al. 2012; Jahn et al. 2012). On the other hand, the recent trend in Antarctic sea ice is one of a slight increase of 1.3–1.7% decade^{-1} over the satellite instrumental record (1979–2012; Bindoff et al. 2013) though with regional variability. At first sight, this is at odds with the expected effects of a globally-warming atmosphere and ocean, and with the potential for attendant, strong, polar feedbacks. This increase is also at odds with most climate simulations (Turner et al. 2013; Zunz et al. 2013), though the relatively brief observational record and significant uncertainties in simulated and observed internal variability limit confidence in current model results. Initial studies suggested that stratospheric ozone loss may have driven atmospheric cooling and the increasing trend in Antarctic sea ice (Goosse et al. 2009; Turner et al. 2009; WMO 2011), but subsequent simulations (Sigmond and Fyfe 2010; Bitz and Polvani 2012) have questioned these. One hypothesis is that sub-surface ocean warming and enhanced freshwater input from sub-ice-shelf melting have freshened the Southern Ocean adjacent to continental Antarctica, strengthening stratification, decreasing the upward heat flux and favouring greater sea ice formation (Zhang 2007; Goosse et al. 2009; Bintanja et al. 2013) in a somewhat counter-intuitive negative feedback loop. However, scientific understanding of this trend is currently low.

Feedback processes of the Polar Regions

Table 19.1 summarizes the range of feedback types which are particularly important in either or both Polar Regions, or which could become so as climate evolves over the twenty-first century and beyond. These broadly include: (1) changes to the surface albedo, which may increase or decrease reflectivity and therefore promote either positive or negative feedback, respectively; (2) changes induced in the atmospheric circulation and near-surface heat fluxes with changes in sea ice and cloud covers (Turner et al. 2007); (3) potential release into the atmosphere of large stores of carbon and other GHGs from terrestrial and shallow coastal ocean areas; (4) release into the oceans of freshwater from a perturbed hydrological cycle generally, and from melting ice in particular, affecting the thermohaline circulation; and (5) decisions by governments and businesses that influence future rates of GHG emissions.

Both the Arctic and Antarctic exhibit high overall reflectivity as a result of their extensive snow and ice covers, and also as a result of their low vegetation covers; a major factor in low surface-radiation receipts. A further factor contributing to low radiation receipts is the relative prevalence of cloud-free skies, as a consequence of predominantly descending polar air with low humidity, which permits the reflected radiation to escape Earth's atmosphere effectively (Turner and Marshall 2011). This importance of reflectivity in polar climates imparts to them

Table 19.1 Positive (amplifying) and negative (dampening) feedbacks which are, or could become, important in the Arctic and Antarctic. Table by author.

Positive feedbacks	Negative feedbacks
Albedo decrease from decreasing permanent and seasonal snow and ice covers, increasing shrub tundra cover, evergreen forest expansion, and greater black carbon deposition on snow and ice from more frequent fires.	Albedo increase from greater deciduous vs. coniferous forest cover following more frequent disturbance, enhanced levels of aerosols from marine phytoplankton DMS production and vegetation degradation from more frequent fires.
Increased non-summer cloud formation, preventing long-wave radiation loss and insulating the near surface.	Increased summer cloud formation, increasing albedo.
Carbon release from enhanced permafrost thaw, organic material decomposition, more frequent disturbance and enhanced coastal erosion.	Carbon storage in more abundant terrestrial plants and increasing carbon uptake by marine plants.
Increase in volume of freshwater from rivers draining into the Arctic Ocean, and thawing of sea ice and glaciers, depressing Arctic halocline, making ice formation more difficult.	Increase in volume of freshwater from rivers draining into the Arctic Ocean, or from Antarctic ice shelves, and thawing of sea ice and glaciers, lowering ocean surface temperature.
Reduced latent heat absorption with diminishing availability of snow and ice for melting.	Sea ice insulation effect, in which the ice cover prevents ocean heat from interacting with the overlying atmosphere.
Accelerated resource development demanded by Arctic residents.	Greater efforts to control GHG emissions demanded by Arctic residents.
Failure to achieve multinational agreement to reduce GHG emissions.	Success in achieving multinational agreement to reduce GHG emissions.

considerable sensitivity to variations in the extent of oceanic (in the case of both the Arctic and Antarctic) and terrestrial (in the case of the Arctic) snow and ice covers. Consequently, the albedo feedback – the changing rate of reflection of solar radiation with the surface properties of high latitudes (Table 19.1) – is a critical feature of polar climates. This is particularly so in the Arctic, where the seasonal and multi-decadal variability of sea ice and terrestrial snow covers affects proportionally greater areas than in the generally more stable Antarctic.

Given their locations at the Earth's highest latitudes, both Polar Regions experience significant annual solar radiation deficits, therefore they constitute large-scale heat sinks, which are important in driving global atmospheric and oceanic circulations. Cai (2005) found that increased poleward atmospheric heat transport, as a consequence of enhanced, anthropogenically-driven radiative forcing, would contribute to the polar amplification of atmospheric warming by redistributing part of the excess energy from low to high latitudes (even in the absence of the ice-albedo feedback; Table 19.1). This effect would then tend to strengthen the high-latitude water vapour feedback – a positive feedback in which higher temperatures allow the atmosphere to hold a greater amount of water vapour, reducing longwave radiation losses to space and so increasing near-surface air temperatures (Table 19.1). This *dynamical amplifier* was found to contribute c. 25% of high-latitude surface warming in winter, compared with a global average of c. 10%. How such an effect might evolve as the faster-than-average pace of Arctic warming ultimately reduces the equator-pole temperature gradient is as yet unclear.

For the case of the Arctic specifically, Turner et al. (2007) indicate that, irrespective of the main drivers of warming, the pervasive changes in the ocean, sea ice, and terrestrial and marine ecosystems introduce a strong inertia to the Arctic climate system by promoting the development of strong, currently predominantly positive, feedback loops, thereby making it difficult for any future natural atmospheric variability to reverse recent changes. The persistence of large, negative sea ice anomalies and the continued, steep decline in spring northern hemisphere snow cover through years of unexceptional air temperatures potentially attest to the development of this inertia.

The particular case of the Arctic: warming amplification through strong, positive feedbacks

The Arctic is warming at least twice as fast as the global average (ACIA 2005; Jeffries and Richter-Menge 2013). Amplified atmospheric warming in high northern latitudes has been described as an inherent characteristic of the behaviour of the Earth's climate system, operating over a wide range of spatial and temporal scales (Serreze and Barry 2011).

An initial point is that it takes less energy to achieve a given amount of warming in the Arctic than in lower latitudes. The Arctic troposphere (the lowermost layer of the atmosphere, where weather systems are concentrated) is relatively thin: only about half as deep as the equatorial troposphere. Therefore, a given amount of energy will be more effective in warming the Arctic atmosphere, other factors being equal. Next, the cover of perennial and seasonal snow and ice on the land and ocean surfaces of the Arctic acts as an energy sink: inputs of atmospheric energy are consumed in first raising this snow and ice to the melting temperature, and then in converting it from its solid to its liquid phase (latent heat absorption; Table 19.1). Only when the snow and ice cover has been removed, with the consumption of large amounts of atmospheric energy, can the temperature of the underlying surface and its overlying air be raised. As the extent of snow and ice cover decreases, the efficiency of this energy sink decreases: northern-hemisphere June snow cover extent (a month in which snow is largely confined to the Arctic) is now decreasing at a faster rate (18% per decade) than September sea ice extent (13% per decade) (Derksen and Brown 2012; Jeffries and Richter-Menge 2013). Less energy is therefore being consumed in melting, so more is available to raise surface temperatures. In a similar way but to a lesser extent, the presence of impermeable permafrost beneath the Arctic land surface leads to poor sub-surface drainage, meaning that soils are often waterlogged in summer, keeping the supply of water for evaporation high, which also consumes atmospheric energy. This is also an effect which will diminish as permafrost degrades and Arctic soils become more freely draining.

Then, most importantly and as noted earlier, the loss of snow and ice, particularly over the ocean, greatly reduces the albedo of the Earth's surface. Over fresh snow, the reflection of solar radiation is extremely efficient, with 90% or more of incoming energy returned back out to space. Over open ocean water, however, this figure falls to around 10%. Therefore, as snow and ice diminish in area, the albedo of the surface decreases, more atmospheric energy is absorbed, snow and ice cover is further diminished, and albedo is further decreased: a classic, and highly effective, positive feedback on atmospheric warming. Pistone et al. (2014) quantified the effectiveness of this feedback in an analysis of 30 years of satellite microwave data, finding that the decline in Arctic albedo yielded atmospheric forcing equivalent to 25% of that due to CO_2 emissions in the same period. Snow-free Arctic land surfaces are generally more efficient reflectors than the ocean, but they too are becoming less efficient as warming contributes to vegetation growth, and bare ground or tundra vegetation gives way to shrubs and trees (Epstein et al. 2013; Xu et al. 2013).

Black carbon in the atmosphere (soot, derived from various sources including forest fires, diesel or wood combustion, oil and gas flaring) absorbs solar radiation and contributes to warming (Table 19.1), although its effects are short-lived. However, the accumulation of black carbon particles on snow or ice surfaces contributes to the lowering of albedo and reduces the efficiency with which such surfaces reflect solar radiation back to space, a further positive feedback which is thought to be particularly effective in the Arctic (AMAP 2011a): the fallout of black carbon is at a maximum there in late winter, and models suggest that this may be increasing spring snowmelt rates by 20–30% (Flanner et al. 2007; AMAP 2011b), though understanding here too is still at an early stage.

Differences between the Arctic and Antarctic

It is already clear, therefore, that there are some significant differences in the types and importance of feedbacks between the North and South Polar Regions. Indeed, observed polar climate change from the instrumental record is quite different between the Arctic and Antarctic. Significant atmospheric warming in Antarctica, for instance, is largely confined to the Antarctic Peninsula, whereas it is widespread in the Arctic (Bindoff et al. 2013). Furthermore, the sea ice extent in the Southern Ocean is currently increasing slowly (+1.1% decade^{-1}; NSIDC, 2015), whereas it is declining rapidly (−13% decade^{-1}; NSIDC, 2015) in the Arctic Ocean. Antarctic sea ice growth is likely driven by a combination of wind field and ocean-circulation changes, potentially augmented by freshening associated with sub-ice-shelf melt (Table 19.1): wind field changes have probably caused ice compaction and ridging, making ice more resistant to melt (Zhang 2012). Arctic sea ice decline is likely a response to the interaction of atmospheric warming with shifts in the dominant mode of north polar climate variability, the Arctic Oscillation (AO). The AO is a pattern of alternating atmospheric pressure between Arctic and mid-latitudes, the positive phase of which produces a strong polar vortex, with the polar-front jet stream shifted northward (and conversely, the negative phase of which produces a weak vortex with a southward shift in the jet stream). The AO entered a strong positive phase in the early 1990s, with strong circumpolar winds flushing older, thicker ice out of the Arctic through the Fram Strait (Rigor et al. 2002; Kwok 2004; Rigor and Wallace 2004), leaving proportionally more, younger, less resilient ice which melted more readily.

Unlike the Arctic, many climate models show only limited polar amplification over the Southern Ocean and Antarctica for the last century. Indeed, there seem to be some important negative feedbacks within the Antarctic climate system. The extremely cold, high-elevation ice sheet experiences barely any melt to support any albedo feedback, and projected temperature increases of the twenty-first century are unlikely to change this; the majority of sea ice is in contact with a near-surface, cold-water layer formed by the cold, continental katabatic wind flow (Stroeve et al. 2007), which may be strengthening rather than weakening with the addition of sub-ice-shelf melt (Bintanja et al. 2013). The Southern Ocean appears to be a particularly important negative feedback on Antarctic atmospheric warming: the deep circulation component increases ocean heat uptake as the climate warms (Gregory 2000), yielding an asymmetry in the polar response to atmospheric warming. Importantly, this asymmetry does not result from a difference in the actual efficacy of feedback processes: when climate models are run to equilibrium, such that the ocean heat uptake effect is removed, the hemispheres have nearly equal polar amplification (Bitz 2006). Furthermore, the considerable warming on the Antarctic Peninsula, contrasting with the modest or absent warming elsewhere on the continent, has been attributed to reduced stratospheric ozone levels in the past few decades (Shindell and Schmidt 2004). However, as a consequence of the Montreal Protocol (UNEP 2012) to reduce ozone-depleting chlorofluorocarbon (CFC) emissions, these ozone levels have stabilized and are expected to recover by about mid-century or slightly later, after which continental Antarctica is expected to warm somewhat. Therefore, while Arctic amplification has already become apparent (Serreze et al. 2009), its Antarctic equivalent appears to be suppressed so far by negative feedbacks associated with heat uptake in the Southern Ocean and trends in stratospheric ozone.

Twenty-first-century prospects

Over the Arctic Ocean in the course of the twenty-first century, the greatest atmospheric warming is anticipated for autumn and winter. The seasonality of this warming is accounted for mainly by the positive feedbacks of albedo decrease and insulation loss with the ongoing decrease in sea ice extent and thickness, coupled with the delay in the post-summer re-formation of sea ice and the initiation of the winter snow cover (Hinzman et al. 2013). The increased extent of open water and thin ice at the end of summer absorb more solar radiation when skies are clear or contribute to more low cloud and long-wave radiation absorption (McGuire et al. 2006), leading to delayed and/or reduced ice growth the following winter. However, while twenty-first-century model projections indicate the greatest changes in autumn and winter, and over the ocean, recent Arctic change has been most marked during the spring and over the land surface, mainly driven by the positive feedback of albedo decrease associated with the steep decline in northern hemisphere spring snow cover (Hinzman et al. 2013).

Given the substantial decline in Arctic sea ice extent in recent decades, Turner and Overland (2009) ask whether there are any significant negative feedbacks on north polar climate and environmental change. It is suggested that polar amplification of global warming may slow the poleward transport of sensible heat, given the reduced thermal gradient between high and low latitudes, but the transport of latent heat may increase. Arctic cloud cover is changing seasonally, yielding positive feedback outside summer (Table 19.1), by decreasing long-wave radiation losses, but potentially contributing to negative feedback during the summer melt season by increasing albedo (Table 19.1; Najafi et al. 2015). However, the impact of clouds on albedo is likely to be small because of their low contrast with snow and ice surfaces. An intensified hydrological cycle, with increased precipitation and evaporation accompanying the increased temperatures of a slightly more positive AO (Stocker et al. 2013), may slow the thermohaline circulation, but only to an extent that more or less compensates for the atmospheric warming driving the process; model simulations generally indicate a constant or even increasing flow of warm waters into the Arctic Ocean during the twenty-first century.

In the Antarctic, on the other hand, current, predominantly negative, feedbacks are expected to give way increasingly to positive feedbacks during the course of the current century. The recent increase in sea ice extent is expected to peak before mid-century, as GHG levels rise and stratospheric ozone levels recover; by the century's end there is expected to be an annual reduction in Antarctic sea ice extent of about 25% (Bracegirdle et al. 2008). Similarly, the anticipated recovery of stratospheric ozone is expected to weaken the positive trend in the SAM, likewise weakening the negative feedback on GHG warming and polar amplification across the continent and the sea ice zone (Perlwitz et al. 2008). The corresponding temperature response is projected to be one of greater warming in the continental interior ($0.34\pm0.10°C$), which may also weaken katabatic winds, particularly in summer (Mayewski et al. 2009). Increased precipitation is also projected for Antarctic coastal regions, due to a poleward shift of storm tracks (Stocker et al. 2013).

Significant, additional feedbacks may become important in the course of the twenty-first century

Other, potential, strong, positive feedbacks on warming may become important during the later decades of the twenty-first century, and again these are likely to be most apparent in the Arctic. Permafrost is believed to contain around 1400–1700 Gt of carbon, which is about four times more than all the carbon emitted by human activity in modern times. The upper 3 m of permafrost is estimated to hold as much carbon as all known coal reserves (Tarnocai et al. 2009). If that permafrost were to thaw and soils become waterlogged, soil microbes could convert the carbon into methane (CH_4). If the soils instead drained, the carbon would be respired into the atmosphere as carbon dioxide (CO_2). Much uncertainty currently surrounds the potential for these processes to occur and at what rates, but it is clear that permafrost is currently thawing over wide areas. On the North Slope of Alaska, the highest temperatures since measurements began in the late 1970s were recorded at 20 m depth in the ground in 2011. Since the 1990s, the thickness of the active layer (the surface ground layer that thaws seasonally) has increased in the Eurasian Arctic, Siberia, Chukotka, Svalbard and Greenland (Romanovsky et al. 2012).

Another perspective is that permafrost soils bordering the Arctic Ocean contain as much shallow carbon as all the world's temperate and tropical forests, grass and shrubland ecosystems and agricultural land combined. Reduced sea ice cover is now exposing Arctic coastlines to greater wave action and accelerated coastal erosion compared to previous decades. Coastal erosion at Yedoma, northern Siberia, is currently believed to destabilize about 44 Mt of permafrost carbon each year, of which about 67% becomes atmospheric CO_2 (Vonk et al. 2012). As yet, there is no direct atmospheric evidence that either emissions of CH_4, or the net carbon balance, are changing over the Arctic as a whole (Bruhwiler and Dlugokencky 2012), but reducing the uncertainty surrounding these processes is a key scientific priority, given the potential magnitude of the associated feedback. IPCC (2013) states that it is virtually certain that Arctic permafrost extent will decrease as atmospheric temperature increases through the twenty-first century, with the area of near-surface permafrost projected to decrease by about 80% for a business-as-usual emissions scenario. Lawrence et al. (2008) predict that annual permafrost carbon emissions could eventually equate to 15–35% of today's annual emissions from human activity.

A potentially even more significant feedback is associated with methane clathrates (a form of water ice that contains large amounts of methane within its crystal structure) in the seabed of the Arctic Ocean, notably off the coast of Siberia, which are estimated to contain 1400 Gt of methane, which is a much more powerful GHG than CO_2. These clathrates remain stable under a combination of high pressure and low temperature and are thought to become vulnerable as sea ice retreats and ocean temperatures increase. Even greater uncertainty surrounds the potential for large-scale methane release compared to carbon, but the potential for positive feedback is sufficient to warrant further attention.

As a region, the Arctic is particularly sensitive to the effects of black carbon because of its large snow and ice cover; the deposition of soot on snow and ice has a much greater impact than its presence in the atmosphere, which is a more important effect in mid-latitudes. For the same reason, the Arctic is especially sensitive to localized emissions which stay at low altitudes (Sand et al. 2013). Greater controls on air pollution have recently contributed to reduced black carbon emissions from industrialized northern countries (AMAP 2011a), but this trend may not continue. There are currently relatively few sources of black carbon emissions within the Arctic itself, although this may also change as human activities increase in the northern circumpolar regions, particularly if significant oil and gas extraction goes ahead. It has further been suggested that black carbon emissions from increased Arctic shipping may increase five-fold by 2030 (Corbett et al. 2010).

Conclusions

The changing twenty-first-century climate undoubtedly has the potential to produce considerable – and in human terms, unprecedented – environmental change in both Polar Regions. Such change is already clearly apparent in the Arctic, in terms of air temperature change, sea ice and snow cover decline, permafrost thawing, and related effects. Similar changes are becoming apparent in the Antarctic, but on the whole are not so far as readily identifiable as their northern equivalents. Feedback processes contingent on the distinctive characteristics of the polar environments, such as the effect of a decreasing albedo with diminishing snow cover and greening land surfaces, or modified, seasonal ocean-atmosphere interactions with declining sea ice extent, are making an important contribution to this apparent or impending change. Positive feedbacks are currently dominating Arctic climate systems, whereas negative feedbacks still largely characterize the Antarctic, although this appears set to change during the course of the current century. But it is important to recall that the Polar Regions are vast and heterogeneous, therefore diversity in the occurrence, direction and magnitude of feedbacks should be anticipated: it is not possible to speak of a typical polar response to global warming, as the two regions are so contrasting. Polar observational data are still sparse and instrumental time-series short, which is clearly an impediment to progress in understanding climate systems and the likely outcomes of feedback processes; therefore a key scientific priority is to maintain and expand fundamental monitoring of climate and associated environmental processes. A final point is that polar change is not limited to the Polar Regions themselves: Arctic changes are transmitted to lower latitudes largely via atmospheric influence on the behaviour of the polar-front jet stream; Antarctic changes are transmitted to lower latitudes largely via oceanic influence on Antarctic Bottom Water formation and circulation. Therefore, the effective reach of polar change extends well south of 66°N and far north of 66°S and matters to us all.

References

ACIA (Arctic Climate Impact Assessment). 2005. *Arctic Climate Impact Assessment*. Cambridge, UK: Cambridge University Press.

AMAP (Arctic Monitoring and Assessment Programme). 2011a. *The Impact of Black Carbon on Arctic Climate*. Arctic Monitoring and Assessment Programme, Oslo, Norway.

AMAP (Arctic Monitoring and Assessment Programme). 2011b. *Snow, Water, Ice and Permafrost in the Arctic (SWIPA). Climate Change and the Cryosphere*. Arctic Monitoring and Assessment Programme, Oslo, Norway.

Bindoff, N.L., P.A. Stott, K.M. AchutaRao, M.R. Allen, N. Gillett, D. Gutzler, K. Hansingo, G. Hegerl, Y. Hu, S. Jain, I.I. Mokhov, J. Overland, J. Perlwitz, R. Sebbari and X. Zhang. 2013. *Detection and Attribution of Climate Change: From Global to Regional*. In Stocker, T.F., D. Qin, G.-K. Plattner, M. Tignor, S.K. Allen, J. Boschung, A. Nauels, Y. Xia, V. Bex and P.M. Midgley (eds) *Climate Change 2013: The Physical Science Basis. Contribution of Working Group I to the Fifth Assessment Report of the Intergovernmental Panel on Climate Change*. Cambridge, UK and New York: Cambridge University Press.

Bintanja, R., G.J. van Oldenborgh, S.S. Drijfhout, B. Wouters and C.A. Katsman. 2013. 'Important role for ocean warming and increased ice-shelf melt in Antarctic sea-ice expansion' *Nature Geosci.* 6: 376–379.

Bitz, C.M. 2006. *Polar Amplification*. www.realclimate.org/index.php?p=234.

Bitz, C.M. and L.M. Polvani. 2012. 'Antarctic climate response to stratospheric ozone depletion in a fine resolution ocean climate model' *Geophys. Res. Lett.* 39: L20705, doi:10.1029/2012GL053393.

Bracegirdle, T.J., W.M. Connolley and J. Turner. 2008. 'Antarctic climate change over the twenty first century' *J. Geophys. Res.* 113, D03103, doi:10.1029/2007JD008933.

Bruhwiler, L. and E. Dlugokencky. 2012. 'Carbon dioxide (CO_2) and methane (CH_4)' [in *Arctic Report Card 2012*], www.arctic.noaa.gov/reportcard.

Cai, M. 2005. 'Dynamical amplification of polar warming' *Geophys. Res. Lett.* 32(L22710), doi:10.1029/2005GL024481.

Corbett, J.J., D.A. Lack, J.J. Winebrake, S. Harder, J.A. Silberman and M. Gold. 2010. 'Arctic shipping emissions inventories and future scenarios' *Atmos. Chem. Phys.* 10: 9689–9704.

Day, J.J., J.C. Hargreaves, J.D. Annan and A. Abe-Ouchi. 2012. 'Sources of multi-decadal variability in Arctic sea ice extent' *Environ. Res. Lett.* 7, 034011, doi:10.1088/1748-9326/7/3/034011.

Derksen, C. and R. Brown. 2012. 'Spring snow cover extent reductions in the 2008-2012 period exceeding climate model projections' *Geophys. Res. Lett.* 39: L19504, doi:10.1029/2012GL053387.

Epstein, H.E. and 23 others. 2013. *Vegetation* [in *Arctic Report Card 2012*], www.arctic.noaa.gov/reportcard.

Flanner, M.G., C.S. Zender, J.T. Randerson and P.J. Rasch. 2007. 'Present-day climate forcing and response from black carbon in snow' *J. Geophys. Res: Atmos.* 112(D11202), doi: 10.1029/2006JD008003.

Gillett, N.P., D.A. Stone, P.A. Stott, T. Nozawa, A.Y. Karpechko, G.C. Hegerl, M.F. Wehner and P.D. Jones. 2008. 'Attribution of polar warming to human influence' *Nature Geosci.* 1: 750–754.

Goosse, H., W. Lefebvre, A. de Montety, E. Crespin and A.H. Orsi. 2009. 'Consistent past half-century trends in the atmosphere, the sea ice and the ocean at high southern latitudes' *Clim. Dyn.* 33: 999–1016.

Gregory, J.M. 2000. 'Vertical heat transports in the ocean and their effect on time-dependent climate change' *Clim. Dyn.* 16: 501–515.

Hinzman, L.D., C.J. Deal, D.A. McGuire, S.H. Mernild, I.V. Polyakov and J.E. Walsh. 2013. 'Trajectory of the Arctic as an integrated system' *Ecol. App.* 23(8): 1837–1868.

IPCC (Intergovernmental Panel on Climate Change). 2013. *Summary for Policymakers*. In *Climate Change 2013: The Physical Science Basis. Contribution of Working Group I to the Fifth Assessment Report of the Intergovernmental Panel on Climate Change*, www.climatechange2013.org/.

Jackson, J.M., E.C. Carmack, F.A. McLaughlin, S.E. Allen and R.G. Ingram. 2010. 'Identification, characterization, and change of the near-surface temperature maximum in the Canada Basin, 1993–2008' *J. Geophys. Res.* 115(C05021), doi:10.1029/2009JC005265.

Jahn, A., K. Sterling, M.M. Holland, J.E. Kay, J.A. Maslanik, C.M. Bitz, D.A. Bailey, J. Stroeve, E.C. Hunke, W.H. Lipscomb and D.A. Pollak. 2012. 'Late-twentieth-century simulation of Arctic sea-ice and ocean properties in the CCSM4' *J. Clim.* 25: 1431–1452.

Jeffries, M.O. and J. Richter-Menge, eds. 2013. 'Arctic [in "State of the climate in 2012"]' *Bull. Amer. Meteor. Soc.* 94(8): S111–S146.

Kay, J.E., M.M. Holland and A. Jahn. 2011. 'Inter-annual to multi-decadal Arctic sea ice extent trends in a warming world' *Geophys. Res. Lett.* 38(L15708), doi: 10.1029/2011GL048008.

Kwok, R. 2004. 'Annual cycles of multiyear sea ice coverage of the Arctic Ocean: 1999–2003' *J. Geophys. Res. Oceans* 109(C11004), doi: 10.1029/2003JC002238.

Lawrence, D.M., A.G. Slater, R.A. Tomas, M.M. Holland and C. Deser. 2008. 'Accelerated Arctic land warming and permafrost degradation during rapid sea ice loss' *Geophys. Res. Lett.* 35(L11506), doi:10.1029/2008GL033985.

Manabe, S. and R.J. Stouffer. 1980. 'Sensitivity of a global climate model to an increase of CO_2 concentration in the atmosphere' *J. Geophys. Res.* 85(C10): 5529–5554.

Marshall, G.J., P.A. Stott, J. Turner, W.M. Connolley, J.C. King and T.A. Lachlan-Cope. 2004. 'Causes of exceptional atmospheric circulation changes in the Southern Hemisphere' *Geophys. Res. Lett.* 31(L14205), doi:10.1029/2004 GL019952.

Mayewski, P.A., M.P. Meredith, C.P. Summerhayes, J. Turner, A. Worby, P.J. Barrett, G. Casassa, N.A.N. Bertler, T. Bracegirdle, A.C. Naveira Garabato, D. Bromwich, H. Campbell, G.S. Hamilton, W.B. Lyons, K.A. Maasch, S. Aoki, C. Xiao and T. van Ommen. 2009. 'State of the Antarctic and Southern Ocean climate system' *Rev. Geophys.* 47(RG1003), doi:10.1029/2007RG000231.

McGuire, A.D., F.S. Chapin III, J.E. Walsh and C. Wirth. 2006. 'Integrated regional changes in Arctic climate feedbacks: implications for the global climate system' *Annu. Rev. Environ. Resour.* 31: 61–91.

Najafi, M.R., F.W. Zwiers and N.P. Gillett. 2015. 'Attribution of Arctic temperature change to greenhouse-gas and aerosol influences' *Nature Climate Change* doi:10.1038/nclimate2524.

National Snow and Ice Data Center (NSIDC). 2015. *State of the Cryosphere: Sea Ice.* http://nsidc.org/cryosphere/sotc/sea_ice.html.

Perlwitz, J., S. Pawson, R.L. Fogt, J.E. Nielsen and W.D. Neff. 2008. 'Impact of stratospheric ozone hole recovery on Antarctic climate' *Geophys. Res. Lett.* 35(L08714), doi:10.1029/2008GL033317.

Pistone, K., I. Eisenman and V. Ramanathan. 2014. 'Observational determination of albedo decrease caused by vanishing Arctic sea ice' *PNAS* 111(9): 3322–3326.

Rigor, I.G. and J.M. Wallace. 2004. 'Variations in the age of Arctic sea-ice and summer sea-ice extent' *Geophys. Res. Lett.* 31(L09401), doi:10.1029/2004GL019492.

Rigor, I.G., J.M. Wallace and R.L. Colony. 2002. 'Response of sea-ice to the Arctic Oscillation' *J. Clim.* 15: 2648–2663.

Romanovsky, V.E., S.L. Smith, H.H. Christiansen, N.I. Shiklomanov, D.S. Drozdov, N.G. Oberman, A.L. Kholodov and S.S. Marchenko. 2012. *Permafrost* [in *Arctic Report Card 2012*], www.arctic.noaa.gov/reportcard.

Sand, M., T.K. Berntsen, O. Seland and J.E. Kristjánsson, 2013. 'Arctic surface temperature change to emissions of black carbon within Arctic or midlatitudes' *J. Geophys. Res. Atmos.* 118(14): 7788–7798.

Schweiger, A., R. Lindsay, J. Zhang, M. Steele, H. Stern and R. Kwok. 2011. 'Uncertainty in modeled Arctic sea ice volume' *J. Geophys. Res. Oceans* 116: C00D06, doi: 10.1029/2011JC007084.

Serreze, M.C., A.P. Barrett, J.C. Stroeve, D.M. Kindig and M.M. Holland. 2009. 'The emergence of surface-based Arctic amplification' *The Cryosphere* 3: 11–19.

Serreze, M.C. and R.G. Barry. 2011. 'Processes and impacts of Arctic amplification: a research synthesis' *Global and Planetary Change* 77(1–2): 85–96.

Serreze, M.C. and J.A. Francis. 2006. 'The Arctic amplification debate' *Climatic Change* 76: 241–264.

Shindell, D. and G. Faluvegi. 2009. 'Climate response to regional radiative forcing during the twentieth century' *Nature Geosci.* 2: 294–300.

Shindell, D.T. and G.A. Schmidt. 2004. 'Southern Hemisphere climate response to ozone changes and greenhouse gas increases' *Geophys. Res. Lett.* 31, L18209, doi:10.1029/2004GL020724.

Sigmond, M. and J C. Fyfe. 2010. 'Has the ozone hole contributed to increased Antarctic sea ice extent?' *Geophys. Res. Lett.* 37(L18502), doi: 10.1029/2010GL044301.

Stocker, T.F., D. Qin, G-K. Plattner, L.V. Alexander, S.K. Allen, N.L. Bindoff, F-M. Bréon, J.A. Church, U. Cubasch, S. Emori, P. Forster, P. Friedlingstein, N. Gillett, J.M. Gregory, D.L. Hartmann, E. Jansen, B. Kirtman, R. Knutti, K. Krishna Kumar, P. Lemke, J. Marotzke, V. Masson-Delmotte, G.A. Meehl, I.I. Mokhov, S. Piao, V. Ramaswamy, D. Randall, M. Rhein, M. Rojas, C. Sabine, D. Shindell, L.D. Talley, D.G. Vaughan and S-P. Xie. 2013. 'Technical summary' in Stocker, T.F., D. Qin, G.-K. Plattner, M. Tignor, S.K. Allen, J. Boschung, A. Nauels, Y. Xia, V. Bex and P.M. Midgley (eds) *Climate Change 2013: The Physical Science Basis. Contribution of Working Group I to the Fifth Assessment Report of the Intergovernmental Panel on Climate Change.* Cambridge, UK and New York: Cambridge University Press.

Stroeve, J., M.M. Holland, W. Meier, T. Scambos and M. Serreze. 2007. 'Arctic sea ice decline: faster than forecast' *Geophys. Res. Lett.* 34(L09501), doi:10.1029/2007GL029703.

Tarnocai, C., J.G. Canadell, E.A.G. Schuur, P. Kuhry, G. Mazhitova and S. Zimov. 2009. 'Soil organic carbon pools in the northern circumpolar permafrost region' *Global Biogeochem. Cycles* 23(GB2023), doi:10.1029/2008GB003327.

Thompson, D.W.J. and S. Solomon, 2002. 'Interpretation of recent southern hemisphere climate change' *Science* 296(5569): 895–899.

Turner, J., T.J. Bracegirdle, T. Phillips, G.J. Marshall and J.S. Hosking. 2013. 'An initial assessment of Antarctic sea ice extent in the CMIP5 models' *J. Clim.* 26: 1473–1484.

Turner, J., J.C. Comiso, G.J. Marshall, T.A. Lachlan-Cope, T.J. Bracegirdle, T. Maksym, M.P. Meredith, Z. Wang and A. Orr. 2009. 'Non-annular atmospheric circulation change induced by stratospheric ozone depletion and its role in the recent increase of Antarctic sea ice extent' *Geophys. Res. Lett.* 36(L08502), doi: 10.1029/2009GL037524.

Turner, J. and G.J. Marshall. 2011. *Climate Change in the Polar Regions.* Cambridge, UK: Cambridge University Press.

Turner, J. and J. Overland, 2009. 'Contrasting climate change in the two polar regions' *Polar Res.* 28: 146–164.

Turner, J., J.E. Overland and J.E. Walsh. 2007. 'An Arctic and Antarctic perspective on recent climate change' *Int. J. Climatol.* 27: 277–293.

United Nations Environment Program (UNEP) Ozone Secretariat. 2012. *Handbook for the Montreal Protocol on Substances that Deplete the Ozone Layer. Nairobi, Secretariat for The Vienna Convention for the Protection of the Ozone Layer and The Montreal Protocol on Substances that Deplete the Ozone Layer.* http://ozone.unep.org/Publications/MP_Handbook/MP-Handbook-2012.pdf.

van den Broeke, M., N. van Lipzig and G. Marshall. 2004. 'On Antarctic climate and change' *Weather* 59: 3–7.

261

Vonk, J.E., L. Sánchez-García, B.E. van Dongen, V. Alling, D. Kosmach, A. Charkin, I.P. Semiletov, O.V. Dudarev, N. Shakhova, P. Roos, T.I. Eglinton, A. Andersson and Ö. Gustafsson. 2012. 'Activation of old carbon by erosion of coastal and subsea permafrost in Arctic Siberia' *Nature* 489: 137–140.

Wang, M., J.E. Overland, V. Kattsov, J.E. Walsh, X. Zhang and T. Pavlova. 2007. 'Intrinsic versus forced variation in coupled climate model simulations over the Arctic during the 20th century' *J. Clim.* 20: 1093–1107.

WMO (World Meteorological Organization). 2011. *Scientific Assessment of Ozone Depletion: 2010. Global Ozone Research and Monitoring Project – Report No. 52, World Meteorological Organization*, Geneva, Switzerland.

Xu, L., R.B. Myneni, F.S. Chapin III, T.V. Callaghan, J.E. Pinzon, C.J. Tucker, Z. Zhu, J. Bi, P. Ciais, H. Tømmervik, E.S. Euskirchen, B.C Forbes, S.L. Piao, B.T. Anderson, S. Ganguly, R.R. Nemani, S.J. Goetz, P.S.A. Beck, A.G. Bunn, C. Cao and J.C. Stroeve. 2013. 'Temperature and vegetation seasonality diminishment over northern lands' *Nature Climate Change* 3: 581–586.

Zhang, J. 2012. 'Modeling the impact of wind intensification on Antarctic sea ice volume' *J. Clim.* 27: 202–214.

Zhang, J.L. 2007. 'Increasing Antarctic sea ice under warming atmospheric and oceanic conditions' *J. Clim.* 20: 2515–2529.

Zunz, V., H. Goosse and F. Massonnet. 2013. 'How does internal variability influence the ability of CMIP5 models to reproduce the recent trend in Southern Ocean sea ice extent?' *The Cryosphere* 7: 451–468.

Part III

Polar politics and resource futures

20

The Antarctic Treaty, territorial claims and a continent for science

Klaus Dodds

Introduction

The Antarctic Treaty, which was adopted in 1959 and entered into force in 1961, is a remarkable treaty with a geopolitical history that is perhaps not as well appreciated as it might be. One reason why could be a common assumption that the Antarctic is somehow divorced from global political, cultural and economic histories and geographies (Dodds, Hemmings and Roberts 2017). It is still commonplace to read that the Antarctic is a 'pole apart', as if to suggest that it is an outlier because of its relative geographical remoteness and absence of an indigenous human population. While humans have imported their ideas, practices and objects to Antarctica, the southern circumpolar polar region has always been more than simply a hub for national and international scientific investigation and resource exploitation.

As we shall note, the Antarctic remains an important site for experimentation in human governance, which continues to influence the politics of other parts of Earth and beyond. Examples would include the replication of nuclear-free zones of peace and the adoption of ideas about how to govern Areas Beyond National Jurisdiction (ABNJ) including the seabed and the Moon/outer space. But what often captures the attention of international legal and political commentators is the negotiation of the 1959 Antarctic Treaty itself and the manner in which the so-called seven claimant states (Argentina, Australia, Chile, France, New Zealand, Norway and the United Kingdom) agreed to put aside their territorial claims (under Article IV of the Treaty) and dedicate themselves alongside non-claimant and semi-claimant signatories, such as the United States and Soviet Union/Russia, to co-operation and harmony. Semi-claimant in this context refers to the fact that both the then Cold War superpowers held on to the right to press a territorial claim in the future.

The Antarctic Treaty is not the only treaty applicable to the governance of Antarctica but it is the most significant. The Treaty's articles helped transform the area of application legally and politically (south of 60° South) into something quite distinct in terms of its promotion and support of science, international co-operation and demilitarization. Under Article IV, all signatories agree to defer on the question of the legal status of the continent and surrounding ocean for the duration of the Treaty. The geopolitical circumstances and the dramatis personae enrolled in the process of treaty negotiation and its entry into force ensured that this was not an inevitable outcome. It was quite possible, as this chapter explains, to imagine there to be a failure in agreement in 1959 and 1960. For its supporters, however, the Treaty's eventual entry into force appeared to enshrine further a near perfect union of diplomacy and science. When connected to the International Geophysical Year (IGY) of 1957–8, including its Antarctic programme, the securing of the Antarctic Treaty nourishes the mythologies of polar science and governance as products of co-operation and goodwill.

What this chapter does, is to challenge a near-dominant view that the Antarctic Treaty's genesis lies with the IGY and the altruistic scientists and diplomats responsible (for example, Walton 2013). The moniker a 'continent for science', was popularized by journalist Richard Lewis in his book published after two personal visits to Antarctica in the late 1950s and early 1960s (Lewis 1965). In his accounting of the politics of Antarctica, Lewis draws attention to how multi-national collaboration and scientific knowledge-sharing produced an inspiring modus operandi for politics and government. The Antarctic as a 'continent for science' stood in stark contrast to other continents and their multiple burdens, complexities and conflicts. The Antarctic functions as a proverbial spatial container for a distinct polar politics largely insulated from elsewhere. What such a geographical and technical framing underestimates, however, is the contested geopolitics of the polar continent and surrounding Southern Ocean.

It is important to resist the idea that Antarctica should be treated as geopolitically exceptional. To be fair, the Antarctic and Southern Ocean Science Horizon Scan report, released by the Scientific Committee on Antarctic Research (SCAR) in 2013, posed the following: "How will external pressures and changes in geopolitical configurations of power affect Antarctic governance and science?" Even to pose such a question would have been unthinkable some years earlier, such was the belief that the Antarctic Treaty helped to 'seal off' Antarctica from mainstream geopolitical machinations. Distinctly modern activities and practices such as mobilizing rival sovereignty claims, mapping and charting, cultural and historical commemoration, scientific base construction and the intensifying politicization of science reveal a more complex entanglement with the wider world. The techniques of measurement and base settlement used in Antarctica would have been familiar to anyone working for colonial powers such as Britain and France. The map, the chart and the base/outpost were indispensable in the colonization of Africa, Asia and Latin America (Dodds and Collis 2017). All of which matters because the seven claimant states believe that they exercise jurisdictional rights over their own citizens in their territories as well as accruing resource rights and capacities to regulate activity. However, claimant states like Norway and New Zealand have also expressed their support for the principles underlying the Antarctic Treaty and associated legal instruments.

Territorial claims and the division of Antarctica

The claiming of Antarctica as distinct territory has an intriguing political history and offers interesting insights into how sovereign nation-states present and implement claims to distinct national jurisdictions. For the last 400 years, international law has placed emphasis on the occupation, settlement and administration of territory. When thought of as a 'bounded space', territory brings to the fore the image of the border as opposed to a frontier, a zone or a periphery. Lines were drawn on maps and charts and the world divided up in a way that meant rivers, mountains, and coastlines are very often significant markers of the external limits of nation-states. In remoter areas of the world, including the Arctic and Antarctic, lines of latitude have been instrumental in the demarcation of territorial space (Dodds and Nuttall 2016).

When nation-states press their claims to territory they do so as part of the development of ideas concerning an attachment to place. As political philosophers note, there are what are termed connection-based theories, which explore how states lay claim to territory and invest in it so that attachment is strengthened (Nine 2012). This might come about from measurement, settlement, and integration into public cultures. Claimant states argue that important connections exist between their polar and non-polar national territories through a shared history (including exploration and resource use), geography and occupancy. Jack Child, for example, was a pioneer of work about Antarctica and the geopolitics of South America, writing on how South American states such as Argentina and Chile invested heavily in territorial forms of nationalism, informed by geopolitical thinking and geographical education (for a summary, see Child 2008).

Claimant states such as Australia and the UK act as explicitly territorial agents, and as part of their attachment to place, they put considerable emphasis on how their actions and responsibilities are performed in those territories. This might include publicizing scientific expeditions, issuing postage stamps, performing legal-judicial duties and/or preserving evidence of earlier inhabitation. A good example would be the work of the UK Antarctic Heritage Trust which preserves and populates (in the summer months with volunteers) the restored Port Lockroy base located in the Antarctic Peninsula region.

This approach is termed functional because it explores and recognises that states need geographical space to secure land and population, regulate affairs and protect property rights. Antarctica is particularly interesting because it is counter-intuitive to our taken-for-granted political-legal world. While we are familiar with territorial disputes between states, the Antarctic is the only continent where its entire ownership is disputed between parties. Without an indigenous human population, it is also claimed by outsiders to enjoy global significance, especially in terms of its importance for science, conservation, ecosystem management and debates about environmental protection and sustainable development, because of its intimate connections to global climate change and its relatively untapped resource potential including minerals and freshwater reserves. Marine resources have, historically, been at the forefront of resource exploitation in the southern polar region and have proven controversial, as disputes over whaling (involving today, most notably, Japan and 'scientific whaling') and, most recently, the introduction of the Ross Sea marine protection area in 2016 would exemplify (Brooks et al. 2016).

Reinforcing the prevailing division of the world by nation-states are an abundance of ideologies, practices, objects and discourses dedicated to its endurance and reproduction. Popular and official nationalisms reinforce that sense of how the world is divided up into national territories. Citizens, often through public culture including education, learn about the geographical extent of their nation. Argentine and Chilean students, for instance, learn earlier on in their lives that their national territory extends all the way to the South Pole. As claimant states, the Argentine and Chilean governments have invested greatly in ensuring that national maps, charts and postage stamps depict the respective countries as having polar territories. Geographical location and division often help determine citizenship, and the sovereign rights and the territorial reach and scope of the state. In the late 1970s, Argentina flew pregnant women down south so that they could have their babies in Antarctica. States and governments invest heavily and regularly in reminding their citizens where national territories begin and end. It is, historically speaking, an artificial and arbitrary way of dividing the world into distinct spaces but it remains enduring.

With the development of this modern political system came a distinctly modern understanding of sovereignty as being tied to exclusive sovereignty over a bounded territory (Elden 2013). European nation-states, and various thinkers who expounded on the idea of sovereignty, such as seventeenth-century theorist Thomas Hobbes, who wrote of the unlimited power of the sovereign, found formal expression in the expansion of European imperialism. What has been described as being fundamentally European in terms of ideas of territory and forms of administration and governance, evolving from a patchwork of local and regional authoritative structures including city-states and imperial unions predominant in the Middle Ages, ultimately became a global export. New territories were identified, demarcated and delimited through practices such as legal proclamations, settlement building, infrastructure planning, and investment in mapping and surveying. Between 1500–1900, European empires encompassed the Americas, Asia, Africa and both polar regions. Shirley Scott (2011, 2017) has argued that Antarctica underwent three distinct waves of imperialism involving South American, European and Australian and American actors and interests. Was the Antarctic, therefore, outside this dominant Euro-American model for appropriating territory at a distance or merely the logical extension of a colonial-administrative legacy that bestowed territorial legacies on post-colonial and independent states such as Argentina and Chile?

In the first wave, Scott (2017) draws attention to the peculiar position of Argentina and Chile, who argued that their territorial claims to Antarctica were part of a post-independence inheritance in the early part of the nineteenth century. Utilizing the doctrine of *uti possidetis*, literally meaning from the Latin 'what one possesses', the modern boundaries of South American countries owe their origins to the papal division between the Spanish and Portuguese empires in the fifteenth century and subsequent internal divisions within their respective imperial territories.

While the newly independent South American countries of the nineteenth century were caught up in border conflicts over common boundaries along the Andes, the Chaco and the Amazon; Argentina and Chile contended that their imperial inheritance enabled them to extend further south and encapsulate the Antarctic Peninsula and parts of the South Atlantic Ocean and Pacific Ocean. Adrian Howkins argues that these claims were at their heart a form of 'environmental nationalism', grounded literally in the intersection of ice, rock, air and sea. Argentina and Chile were entitled by nature to imagine themselves as 'Antarctic nations' and their polar territories were southerly elemental extensions (Howkins 2016).

Geopolitically, this notion of natural inheritance mattered to both South American nations and their enduring investment in polar nationalism. Educationally, children were taught under the 'patriotic education' curricula designed to inculcate a sense of how the country was geographically and geologically connected to its remote island territories (e.g. Easter Island in Chile's case and the Falkland Islands and South Georgia in Argentina's case) and the Antarctic continent (Escude 1992). Geographers and the discipline of geography were essential accomplices in the support of this public consciousness by highlighting the shared geology, ice, weather and fauna between southern Patagonia and the Antarctic Peninsula (Dodds 2002). By the mid-twentieth century, military and civilian governments were committing investment to polar education and ensuring that popular culture including maps, magazines and postage stamps represented their respective countries as 'Antarctic nations' (for example, see Child 2008). In 1948, President Gabriel Videla of Chile visited the Antarctic and declared that Chile had to be willing, "to defend the sovereignty and unity of our nation, from Arica [in the far north] to the South Pole" (cited in Howkins 2016: 9). Videla was the first head of state to visit Antarctica.

The British spearheaded the second wave of Antarctic imperialism at the turn of the twentieth century. Having been at the heart of exploratory activity and subsequently Southern Ocean resource exploitation since the eighteenth century, the first formal claim via Letters Patent to the Antarctic was made by Britain in 1908 and refined further in 1917. The earliest Letters Patent of 1843 and 1876 were, however, instrumental in articulating British arrangements for their resource interests in the South West Atlantic. Sealing was the strategically significant activity in the nineteenth century. The regulation of whaling was the main economic driver of the early twentieth century Antarctic claim and Norway, as the main whaler nation operating in the South West Atlantic and Antarctic waters, helpfully recognized this nascent imperial authority.

The British established what was termed the Falkland Islands Dependencies (FID), and the Falkland Islands acted as a strategic gateway. The Letters Patent of 1908 affirmed the enlarged territorial scope of the FID to include the South Orkney, South Shetland and South Sandwich Islands, and the Antarctic Peninsula region (called Graham Land by British administrators) with local administrative activity in the whaling command and control mission of South Georgia. The modified 1917 Letters Patent adopted the so-called sector principle (which was adopted in both the Canadian Arctic and Russian Arctic) to extend the geographical parameters of the FID to include the seas surrounding island chains and the Antarctic Peninsula. Renamed the British Antarctic Territory in 1962, after the entry into force of the Antarctic Treaty in 1961, the area in question remains substantial, encompassing over 1.7 million square kilometres. South Georgia, lying north of the 60° South line of latitude, was excluded from the revised boundaries of the British Antarctic Territory.

Prior to the Antarctic Treaty negotiations, however, whaling was a lucrative business, as a provider of whale oil and baleen used in the modern clothing industry. Whale oil was also used in the production of margarine and explosives. It was, in the first half of the twentieth century, a major resource. Between 1904 and 1962, the South Georgia settlement of Grytviken hosted the Norwegian-dominated whaling industry. The whaling station there handled over 50,000 slaughtered whales and produced over 450,000 metric tons of whale oil and over 190,000 metric tons of whale meat for European and North American markets (Howkins 2016). Millions of pounds of revenue were generated by companies such as the Edinburgh-based Christian Salvesen and the Norwegian-Argentine operation, Compañía Argentina de Pesca.

As the British became more involved with the regulation of whaling, so interest in mapping and surveying the Antarctic Peninsula and islands such as the South Orkneys and Shetlands became more pressing. Obtaining reliable weather information was also considered useful given that the primary activity in the FID was whaling. By the 1930s and 1940s, the work of British surveyors, oceanographers, sailors and scientists helped to fix and mark boundary points, enhance understanding of the marine biology of the Southern Ocean and establish a network of base huts in order to exercise control over the Antarctic Peninsula region (Dodds 2002). Establishing and maintaining what was termed 'effective occupation' meant that interested parties such as the United Kingdom were concerned with activities and gathering knowledge that helped consolidate their onshore presence. Under prevailing norms of international law, the modes of acquiring territory were two-fold – a formal expression of intention to occupy (a territory not already under the sovereign authority of another recognized nation-state) and a demonstration of continuous 'effective occupation' thereafter. Antarctica was considered to be *a terra nullius* and thus legally capable of being occupied.

By the late 1940s, Antarctic territory was increasingly owned, distributed, mapped and bordered. But it was also deeply contested. What's important to recognize is that while the British were seeking to cement further their legal, resource and geopolitical presence in Antarctica, other interested parties such as New Zealand, France and Australia were articulating their own territorial claims to territory in 1923, 1924 and 1933 respectively. Australia and New Zealand were working alongside Britain in extending a UK-Dominion arc of territorial claiming and administration. The French claim to Adélie Land was predicated on, like their Anglo-Saxon counterparts, a public history of polar exploration, discovery and exploitation. Their collective approach was quite different from the South American states of Argentina and Chile who believe, as already indicated in this chapter, that their claims to Antarctica were inherited rather than something that had to be formally claimed and occupied.

As Shirley Scott (2011: 55–56) usefully reminds us:

> For the European states the issue was that of who owned Antarctica; for the South Americans, the issue was that of delimiting the mutual boundary, on the Antarctic portion of their territory. Once the United Kingdom began to challenge the Antarctic rights of Argentina and Chile they felt compelled to justify their position in terms of the contemporary European international law of colonialism – hence their belated Antarctic "claims". The "homogenous claims" interpretation of pre-Treaty Antarctic politics, by which all seven states were involved in an equivalent process of territorial claim-making during the twentieth century, ignores the distinction between two pre-Treaty waves of Antarctic colonialism.

President Videla's visit to Chilean Antarctic Territory in 1948, as we noted earlier, was a belated response to four decades of European Antarctic colonization. For Argentina and Chile, what was fundamentally at stake was how best to establish a mutual boundary between themselves, extending from the Andes to the South Pole. Both countries viewed the UK's presence on the Antarctic Peninsula and surrounding islands as an unwelcome and unjustified expression of Western imperialism. Britain stood accused of calumniating Antarctica, spreading falsehoods about its unclaimed status.

The final wave identified was embodied in the approach taken by the United States (and to a lesser extent the Soviet Union), which was an active participant in Antarctic exploration, exploitation and settlement. American airmen, sailors and scientists were pivotal, including the legendary naval officer and explorer, Richard Byrd, who presided over American Antarctic activities from the late 1920s until the 1950s. Notably in the 1940s, Byrd and the US Navy participated in the large-scale Operation High Jump (1946–7) involving some 4000 personnel, and later Operation Deep Freeze (1955–7), which provided logistical support for the US IGY program. Eschewing a formal claim to Antarctic territory, the United States in conjunction with the Soviet Union (which reactivated its Antarctic interests in the late 1940s) reserved the right to make a claim in the future. For now, in the immediate aftermath of World War II, both countries took the view that they were also not going to be limited in terms of where they might operate in Antarctica.

In effect, the third wave of Antarctic imperialism was, geographically speaking, the most ambitious. Byrd, like other American explorers and geographers such as Laurence Gould, believed passionately in an American polar claim. But their political masters feared that a formal American claim might provoke the Soviets to issue their own claim (on the basis of a long history of exploration and exploitation in the Antarctic) and inadvertently escalate Cold War tension. What they proposed instead was a temporary solution to the knotty issue of Antarctic ownership. In 1948, a proposal was circulated to the seven claimant states (and later after formal protests the Soviet Union) proposing a condominium whereby there would be a commitment to collectively govern the Antarctic. At the heart of the condominium was a conviction by US authorities that there should be freedom of access across the continent, and that claimant states were not going to be recognized as sovereign authorities.

What compounded matters still further was growing evidence that the United Kingdom, Argentina and Chile as counter-claimants were embroiled in a polar 'great game', with investment in rival mapping and surveying projects and a collective determination to strengthen their 'presence' in the region. British diplomat William Hunter Christie coined the term 'Antarctic Problem' as shorthand for what was at stake – three rival countries armed with their particular attachments to place (Hunter Christie 1951). Fearing the outbreak of conflict in Antarctica, the

United States seized an opportunity to marshal debate about how the polar continent should be governed. What was apparent in the late 1940s and 1950s, however, was that no argument was being offered up in support of common heritage or universal administration of the region. What US officials had in mind was a far more limited vision for the Antarctic, a select group of states acting on behalf of humanity but not desirous of close scrutiny of their individual and collective action.

For all the scientific achievements of the International Geophysical Year of 1957–8, including the Antarctic programme, the preparations leading up to it were mindful of potential complications regarding polar sovereignty. The Arctic was largely absent from the IGY because of Cold War tension between the Soviet Union and the United States, with some research eventually carried out by drifting ice stations in the Arctic Ocean. In the Antarctic, an agreement in Paris in 1955 established the principle that IGY parties would have freedom of access and thus the establishment of research priorities would be grounded in scientific reasoning rather than explicit territorial design. In other words, countries such as the United States and the Soviet Union were adamant that their research plans were not going to be vetted by claimant states. For the largest claimant state, Australia, this meant that they had to accept the Soviets were going to situate their research contribution (through the establishment of a number of bases) in Australian Antarctic Territory. The United States established a South Pole station and worked collaboratively with others including New Zealand. Both superpowers, in other words, were not going to be spatially contained by claimant states.

The IGY hard-wired the principle that scientific investigation in Antarctica was open-ended in principle and not artificially constrained by territorial claims even if claimant states such as the United Kingdom based their IGY contribution in the Falkland Islands Dependencies. Claimant states were rooted in their 'areas' while non-claimant states worked across the polar continent and offshore. Notably, US glaciological studies worked across the ice sheet in their concerted efforts to discover more about the thickness of the ice and the underlying morphology of the continent. Through their geographical mobility and scientific probing, the traversing of the Antarctic ice sheet serves as a powerful reminder of what was at stake: the American snow-cats moving over the polar continent unimpeded by others and indifferent to lines on the map (Barr and Lüdecke 2010).

Recognising that there was an opportunity to re-wire the geopolitics of Antarctica for their benefit, the Eisenhower administration convened a series of preliminary meetings with the eleven other Antarctic IGY parties (Argentina, Australia, Belgium, Chile, France, Japan, New Zealand, Norway, South Africa, Soviet Union and the United Kingdom). When the parties did gather to negotiate an Antarctic Treaty in October 1959, agreement on its form and content was not guaranteed. There were knotty issues to resolve including how to manage conflicting positions on territorial sovereignty. For six weeks, details were thrashed out and arguments raged about how to accommodate the negotiating positions of claimant states such as Australia, France and Argentina and the interests and wishes of the Soviet Union and the United States. Even after agreement was secured, the entry into force of the Treaty in June 1961 depended on its ratification by national legislatures and in Argentina there was serious resistance to its content, which was seen by some nationalists as weakening Argentina's sovereignty.

Antarctica as a continent for science

The 1959 Antarctic Treaty scrambled some of the received wisdom of the previous 500 years. It established the continent and surrounding ocean as exceptional in the sense of being unique in encouraging the seven claimant states to defer their territorial claims in support of scientific investigation, peace and wider co-operation. Other signatories were also obligated to restrain from making any claims to territorial sovereignty. As the relevant part of Article IV stipulated,

> no acts or activities taking place while the present Treaty is in force shall constitute a basis for asserting, supporting or denying a claim to territorial sovereignty in Antarctica or create any rights of sovereignty in Antarctica. No new claim, or enlargement of an existing claim, to territorial sovereignty in Antarctica shall be asserted while the present Treaty is in force.

As Article I of the Treaty notes,

> Antarctica shall be used for peaceful purposes only. There shall be prohibited, *inter alia*, any measure of a military nature, such as the establishment of military bases and fortifications, the carrying out of military manoeuvres, as well as the testing of any type of weapon.

After establishing the principle that the Antarctic should be a zone of peace and co-operation, Article II posits the following, "Freedom of scientific investigation in Antarctica and co-operation toward that end, as applied during the International Geophysical Year, shall continue, subject to the provisions of the present Treaty". Article III reiterates that importance by stipulating that international co-operation is essential in order to build confidence and "permit maximum economy of and efficiency of operations" given the expense of operating in the Antarctic and the logistical challenges posed by remoteness, weather and distance.

The Treaty's area of application actively encouraged a containment model of scientific politics. Additional legal instruments such as the Convention on the Conservation of Antarctic Seals (CCAS 1972) and the Convention on the Conservation of Antarctic Marine Living Resources (CCAMLR 1981) either used the designation of 60° South or extended the area of application to the Antarctic Convergence on the grounds that it made more sense to incorporate more of the Southern Ocean for the development of conservation measures. The terms and conditions of the Antarctic Treaty 'containerized' the Southern Ocean and polar continent, conveying inadvertently that the region itself was isolated and capable of being isolated repeatedly from the interactions of other parties, interests and flows.

Social scientists such as Aant Elzinga (1993, 2009) have been at the forefront of critical scholarship exploring how the Antarctic Treaty and associated legal instruments, including the 1991 Protocol on Environmental Protection (as part of an assemblage of practices and values embodied in the Antarctic Treaty System, or ATS), were complicit in framing the region as a 'continent by and for science'. The epistemic dominance of science and scientists was written into the DNA of the ATS. Scientists and polar science-educated diplomats responsible for the workings of the Treaty were dominant and largely unchallenged for much of the 1960s and 1970s. This changed in the late 1970s and throughout the 1980s as interest in Antarctica was globalized. Non-governmental organizations such as Greenpeace and the Antarctic and Southern Ocean Coalition became more active in polar environmental matters, often accusing the ATS of being too secretive and isolated from global auditing and scrutiny. The resource potential of the region was attracting a new body of interest from historically marginalized states, many of whom were still under imperial administration when the Antarctic Treaty was negotiated. Led by the former British colony of Malaysia, members of the Global South used the United Nations General Assembly to consider the 'Question of Antarctica' – an issue first raised in the UN by Malaysia's prime minister Mahathir Bin Mohamad in September 1982, and promoted further by Malaysia at the signing of the UN Convention on the Law of the Sea (UNCLOS) in Jamaica three months later, and at a summit of non-aligned countries in New Delhi in March 1983.

Other countries such as Brazil, India and China were making their presence felt as they joined the ATS in the 1980s and brought their own interests to bear on a group that was dominated by a Euro-American cluster led by the United States and its allies including Norway, Australia and New Zealand. While the emphasis on science and scientific knowledge retained its valence, it was no longer unchallenged. The framing of Antarctica as a 'continent by and for science' was being openly challenged by newer parties to the ATS and outsiders as self-interested rhetoric designed to entrench the power of the original signatories. It also contributed to the legitimacy of pariah states such as apartheid South Africa, which was allowed for years to participate in the ATS because fellow members were complicit in ensuring that global politics should not 'contaminate' Antarctic science and politics. For all the high-minded rhetoric, however, Antarctic science continued to serve the interests of claimant and non-claimant states alike. Politics was to be found everywhere in Antarctica and beyond, ranging from decisions to inspect other scientific bases to the production of research papers that were being used to champion particular nation-states as *primus inter pares*. If Antarctic science was the prevailing currency of status and influence then all the parties, it was noted, were adept at adjusting their political ideologies, outlooks and practices.

The politics of Antarctica, and in particular the framing of the idea of a 'continent for science' also revealed other commitments that were part of the taken-for-granted world of the late 1950s. As a raft of feminist, post-colonial and critical race studies have shown, the Antarctic was assumed to be a space for white men in the main. Women were largely absent, and the world beyond Europe, Oceania and Latin America was marginal with the exception of Japan and South Africa. The Antarctic Treaty in its earliest incarnation was born out of privilege and the terms and conditions of the Treaty were in favour of restricting membership. Article IX of the Treaty notes that any contracting party could send representatives to participate in meetings with other parties to discuss matters of interests to Antarctica as long as it "demonstrates its interest in Antarctica by conducting substantial research activity there, such as the establishment of a scientific station or the dispatch of a scientific expedition". Little detail was offered, at the time, about how such activities would be evaluated by the existing membership and until the 1980s and 1990s the membership of the ATS was largely unchanged.

When considering the contemporary challenges facing the Antarctic, the position of polar science remains significant but co-exists with other actors, interests and regimes. The fate of Antarctica's resources, including fish and, in the future, minerals, reveals the competing claims that parties do make to scientific authority and knowledge. There are disagreements over what constitutes sustainable fishing in the Southern Ocean. There are challenges for the ATS that extend beyond the political-scientific nexus and include overlapping legal regimes, which mean the exploitation, management and protection of an array of issues, including biodiversity, heritage, tourism, marine pollution and climate change, and involves the intersection of regional and global governance. Finally, the ATS itself relies on individual and collective restraint as all parties recognize that the sovereignty of Antarctica remains unsettled. Arguably, the very durability of the ATS rests on investing in the epistemic authority of scientific knowledge and practice while also finding ways of defusing or even deflecting the more destructive potential of conflict over resources.

The politics of Antarctica will become increasingly complicated and even controversial in future years. The struggle for the mastery of Antarctic futures will be emblematic of wider earthly politics regarding what we value, where and what we protect, and who decides on such matters. The role of non-human actors and forces will also make itself felt in the intervening period – warming oceans, acidification, underground volcanic eruptions, melting ice, alien species invasion to name but a few. In a warming world, with a population some projections suggest will approach ever closer to 10 billion around the 2050s, it might mean that the provisions of the Antarctic Treaty and associated legal instruments like the Protocol on Environmental Protection (and its prohibition on mining and mineral resource exploitation) are stress-tested in ways that were unimaginable in the late 1950s. For now, the living resources of the Southern Ocean will remain in the proverbial frontline of struggles to reconcile human-led conservation with exploitation.

References

Barr, S. and C. Lüdecke 2010. *The History of the International Polar Years (IPYs)*. Berlin: Springer-Verlag.

Brooks, C., L. Crowder, L. Curran, R. Dunbar, D. Ainley, K. Dodds, K. Gjerde and U. Sumaila 2016. 'Science-based management in decline in the Southern Ocean' *Science* 354(6309): 185–187.

Child, J. 2008. *Miniature Messages: The Semiotics and Politics of Latin American Postage Stamps*. Durham, NC: Duke University Press.

Dodds, K. 2002. *Pink Ice: Britain and the South Atlantic Empire*. London: I.B. Tauris.

Dodds, K. and C. Collis 2017. 'Post-colonial Antarctica' in K. Dodds, A.D. Hemmings and P. Roberts (eds) *Handbook of the Politics of Antarctica*. Cheltenham, UK: Edward Elgar. 50–68.

Dodds, K., A. Hemmings and P. Roberts (eds) 2017. *Handbook on the Politics of the Antarctic*. Cheltenham, UK: Edward Elgar.

Dodds, K. and M. Nuttall 2016. *The Scramble for the Poles: The Geopolitics of the Arctic and Antarctic*. Cambridge, UK: Polity.

Elden, S. 2013. *The Birth of Territory*. Chicago, IL and London: University of Chicago Press.

Elzinga, A. 1993. 'Science as the continuation of politics by other means' in T. Brante, S. Fuller and W. Lynch (eds) *Controversial Science: From Content to Contention*. Albany, NY: State University of New York Press. 127–152.

Elzinga, A. 2009. 'Through the lens of the polar years: changing characteristics of polar research in historical perspective' *Polar Record* 45: 313–336.

Escude, C. 1992. *Education, Political Culture and Foreign Policy: The Case of Argentina*. Durham, NC: Duke – UNC Working Paper Series on Latin America.

Howkins, A. 2016. *The Polar Regions: An Environmental History*. Cambridge, UK: Polity.

Hunter, C.W. 1951. *The Antarctic Problem*. London: George Unwin.

Lewis, R. 1965. *A Continent for Science: Antarctic Adventure*. New York: Viking Books.

Nine, C. 2012. *Global Justice and Territory*. Oxford, UK: Oxford University Press.

SCAR 2013. *1st SCAR Antarctic and Southern Ocean Science Horizon Scan*. Cambridge, UK: SCAR.

Scott, S. 2011. 'Ingenious and innocuous? Article IV of the Antarctic Treaty as imperialism' *The Polar Journal* 1: 49–60.

Scott, S. 2017. 'Three waves of Antarctic imperialism' in K. Dodds, A. Hemmings and P. Roberts (eds) *Handbook on the Politics of the Antarctic*. Cheltenham, UK: Edward Elgar, 37–49.

Walton, D. 2013. *Antarctica: Global Science from a Frozen Continent*. Cambridge, UK: Cambridge University Press.

21

The Polar Regions and the law of the sea

Current controversies

Donald R. Rothwell

Introduction

During the first decade of the twenty-first century there has been a renewed focus on the polar oceans. This has been partly driven by the attention generated by claims to an outer continental shelf made in both the Arctic Ocean and Southern Ocean by a number of countries. These claims have been the subject of review by the Commission on the Limits of the Continental Shelf (CLCS) and that process remains ongoing. It has also been driven by renewed interest in the Polar Regions as a result of the impact of climate change supposedly making both the Arctic and Antarctic more accessible to a range of activities, including commercial shipping, fishing operations, and seabed exploration and development (Rayfuse 2007). The polar oceans have also been the scene of clashes over contentious environmental issues such as whaling, which resulted in Australia commencing in 2010 a case before the International Court of Justice over the legitimacy of Japan's Southern Ocean scientific whaling programme (Fitzmaurice and Tamada 2016). While the polar oceans are governed by a legal regime founded upon the 1982 United Nations Convention on the Law of the Sea (UNCLOS) (1833 United Nations Treaty Series 397), different regional approaches apply. In Antarctica large parts of the Southern Ocean are subject to the Antarctic Treaty (402 United Nations Treaty Series 71) and associated international legal instruments that regulate fisheries and marine environmental protection. In the Arctic there is no equivalent regional legal regime, though the Arctic Council is increasingly paying attention to Arctic Ocean issues and in 2011 and 2013 sponsored agreements on search and rescue and marine pollution response. This chapter commences with a discussion of the international law of the sea and then progresses to consider current controversies that exist with respect to the law of the sea in the polar oceans.

The international law of the sea

International law recognizes states as having sovereignty over their territory, which extends to all of the land territory of the state. However, that sovereignty must be recognized for the state to be able to exercise all of the traditional attributes that go with state sovereignty. The recognition of state sovereignty has been a significant issue in the Polar Regions, and in the case of Antarctica the ongoing debates over state sovereignty ultimately proved to be a catalyst for the negotiation of the 1959 Antarctic Treaty (Rothwell 1996: 63–68). With respect to the maritime dimensions of the state, the traditional view was that the limits of the state ended at the low-water mark around the coast, and that the waters beyond that were free for all states to enjoy as part of the so-called 'freedom of the seas' over what were referred to as the 'high seas' (Rothwell and Stephens 2016: 1–5). This was the dominant position with respect to the seas and oceans up until the beginning of the nineteenth century, when there began to emerge a very gradual change in state practice with initial claims of sovereign rights and jurisdiction over seas adjacent to the coast. By the commencement of the twentieth century, there was recognition in customary international law of the territorial sea as a maritime zone over which all coastal states had a right to claim. This right was further recognized at the First United Nations Conference on the Law of the Sea which adopted the four 1958 Geneva Conventions, one of which specifically dealt with the territorial sea. These developments in the law of the sea which had been occurring over a number of centuries culminated with the adoption of the 1982 United Nations Convention on the Law of the Sea (UNCLOS) which gave clear recognition to a number of maritime zones (Rothwell and Stephens 2016: 10–20). UNCLOS is also known as the Law of the Sea Convention (LOSC), or the Law of the Sea treaty; for the purposes of the discussion in this chapter, it is referred to as LOSC. The effect of these developments has been to expand considerably the breadth and scope of coastal state sovereignty and jurisdiction over the sea, while also guaranteeing certain fundamental freedoms of the seas for all states. This legal framework is one that revolves around the provisions of the LOSC, supplemented by additional instruments, including regional and bilateral arrangements, in addition to state practice. All of these developments have had considerable implications for the Polar Regions and the law of the sea (Rothwell 2012a).

The LOSC identifies the limits and parameters of six maritime zones. They are:

- a 12 nautical mile territorial sea (LOSC, Article 3);
- a 24 nautical mile contiguous zone (LOSC, Article 33);
- a 200 nautical mile exclusive economic zone (EEZ) (LOSC, Article 57);
- a minimum 200 nautical mile continental shelf (LOSC, Article 76);
- the deep seabed (LOSC, Part XI); and
- the high seas (LOSC, Part VII).

In addition, the LOSC refers to the internal waters of the state (LOSC, Article 8) which are those waters which are on the landward side of any baselines the coastal state may have proclaimed.

The law of the sea and the Polar Regions

There has been ongoing debate as to whether the law of the sea – as represented by the LOSC – applied to the Southern Ocean and the Arctic Ocean (Rothwell 2012a). In the case of Antarctica, that the Antarctic Treaty predated the LOSC raised issues as to whether it created a *lex specialis* which is so comprehensive and distinctive that all other international law is excluded. However, the fact that the Antarctic Treaty made express reference to the 'high seas' (Antarctic Treaty, Article VI) at a minimum suggested that the two regimes – that dealing with Antarctica up to the limits of 60° S and that dealing with the sea – did apply within the Antarctic Treaty area.

With respect to the Arctic, the LOSC addressed environmental protection over ice-covered waters in Article 234. For Canadians this was a particularly symbolic development as it provided some international recognition for the legitimacy of its actions in adopting in 1970 the Arctic

Waters Pollution Prevention Act which proclaimed the enforcement of environmental shipping laws in waters 100 nm from the islands of the Canadian Arctic Archipelago (Huebert 2001). However, with this albeit important exception, the LOSC made no specific reference to environmental management of polar oceans and seas.

Nevertheless, that the LOSC made so little allowance for the polar oceans did have consequences as the 'new' law of the sea was operationalized in both the Southern Ocean and Arctic Ocean. One is that some of the Antarctic claimants sought to assert maritime claims, consistent with their status as 'coastal states' under the law of the sea. Yet doubt remains as to whether the category of 'coastal states' is relevant for Antarctica, given that each of the seven territorial claims to the continent remain contested and, in any event, the active assertion of claims has been effectively suspended during the life of the Antarctic Treaty (Hemmings and Stephens 2009). Nevertheless, some of the Antarctic claimants have continued to assert some semblance of traditional maritime claims. For example, Australia claims a 200 nm 'Australian Whale Sanctuary' offshore the Australian Antarctic Territory (Mossop 2008).

In the case of the Arctic, any possible ambiguity or uncertainty with respect to the status of the LOSC was unambiguously removed in 2008 when representatives of Canada, Russia, Denmark, Norway and the US (the so-called Arctic Five, referring to their status as Arctic Ocean coastal states) issued the Ilulissat Declaration which noted that:

> [t]he law of the sea provides for important rights and obligations concerning the delineation of the outer limits of the continental shelf, the protection of the marine environment, including ice-covered areas, freedom of navigation, marine scientific research, and other uses of the sea.
>
> *(Ilulissat Declaration 2008)*

The Declaration continued that the five Arctic coastal states "remain committed to this legal framework and to the orderly settlement of any possible overlapping claims". This makes clear that for the key Arctic Ocean states the law of the sea is an all encompassing regime with respect to the rights and obligations which exist within that maritime space.

Current controversies

Marine living resource management

Marine living resource management has been a longstanding issue in the Southern Ocean. Adopted in 1980, the Convention on the Conservation of Antarctic Marine Living Resources (CCAMLR) (1329 United Nations Treaty Series 48) was a ground-breaking instrument that adopted both the ecosystem and precautionary approach. The Convention did not rely upon the geographic outer limits of the Antarctic Treaty as its area of operation, but instead relied upon the 'Antarctic Convergence' as the biological boundary between polar waters and more temperate waters to the north (CCAMLR, Article 1). CCAMLR was negotiated in an effort to ensure that unregulated harvesting of krill did not take place in the Southern Ocean, and since the 1980s has been able to evolve into a sophisticated management regime for marine living resources. In doing so it was also at the forefront of developing responses to compliance and monitoring, and also Illegal, Unreported and Unregulated (IUU) fishing (Kaye 2001: 399–442). Reflecting the challenges associated with maritime regulation and enforcement in remote waters, some Southern Ocean states have also adopted innovative co-operative maritime surveillance and enforcement arrangements such as the Australia-France Agreements in the sub-Antarctic between Heard and McDonald Islands (Antarctica) and Kerguelen (France) (Rothwell 2012b: 142–143).

With respect to Arctic fisheries there exist a number of bilateral fisheries agreements and frameworks, which to date have particular application at lower latitudes. Many of these have a longstanding historical basis, such as agreements for Icelandic fisheries and the Barents Sea. With the greater acceptance of the LOSC, and the mutual benefit of joint EEZ and continental shelf boundaries, some momentum is building for the resolution of new boundaries which have implications for fisheries management. An example is the 2010 Norway-Russia Barents Sea and Arctic Ocean Treaty which delimits the EEZ/continental shelf between the two countries and resolves several outstanding issues, such as the 1978 Grey Zone Agreement (Neumann 2010). However, in all other respects the bilateral management of Arctic fisheries is no more distinctive than equivalent arrangements elsewhere.

Non-living seabed resource management

Non-living seabed resource management in the Polar Regions in the form of issues related to offshore oil and gas exploration and development has proven to be particularly contentious and will most likely remain so for some time to come. In the Southern Ocean, Article 7 of the Madrid Protocol on Environmental Protection to the Antarctic Treaty (30 ILM 1461) placed a prohibition on all mineral resource activities. The effect of this has been to remove all oil and gas development from Antarctic discourse for the time being. However, recent interest in continental shelf claims by the seven Antarctic territorial claimants has revived interest in Southern Ocean oil and gas (Leary 2014).

In the Southern Ocean, a series of CLCS submissions consistent with LOSC Article 76 have also been made though there has been no uniformity amongst the seven Antarctic claimants as to the approach adopted. Australia, for example, gathered all of the data necessary for a claim offshore the Australian Antarctic Territory and its sub-Antarctic possessions, but ultimately asked the Commission effectively to set aside for the time being its claim offshore the continent. Six states – Germany, India, Japan, the Netherlands, the Russian Federation, and the United States – all parties to the Antarctic Treaty – made it clear that in their view Article IV of the Treaty placed constraints on the capacity of Treaty parties to assert rights or claims over the seabed offshore Antarctica. New Zealand, on the other hand, indicated to the CLCS that it reserved its right to make a future claim to the continental shelf offshore the Ross Dependency (Rothwell 2008). To date Australia (2004), New Zealand (2006), Argentina (2009), Norway (2009) and the United Kingdom (2009) have all made CLCS submissions that relate to their Southern Ocean territories, however the Commission has made no recommendations which relate to those territories (Hemmings and Stephens 2009).

This is also an issue that has been the focus of considerable attention in the Arctic. Much of the Arctic Ocean remains beyond the limits of existing 200 nm continental shelf claims, including the seabed at the North Pole. In theory, therefore, a great deal of the Arctic Ocean is potentially susceptible to outer continental shelf (OCS) claims that extend beyond 200 nm. Russia was the first to make a CLCS submission in 2001 (revised and resubmitted in 2015), followed in 2006 by Norway, and Denmark with three submissions relating to Greenland in 2012, 2013 and 2014. Canada provided Preliminary Information regarding its Arctic submission in 2013. The US, once it accedes to the LOSC, would also become eligible to assert a claim following a CLCS submission. It can therefore be anticipated that over the course of the next few years the CLCS will face a number of claims to an Arctic OCS, a number of which will overlap.

Marine environmental management

A common theme in polar oceans governance has been a focus on marine environmental management. This has been very evident in the Southern Ocean through the adoption of CCAMLR and the Madrid Protocol, all of which sought to manage a variety of environmental issues in the Southern Ocean, notwithstanding the absence in most instances of recognized state sovereignty. Likewise, though equivalent Arctic initiatives have been slower to develop, the Arctic Environmental Protection Strategy (AEPS) and Arctic Council process has also given attention to marine environmental management issues since the 1990s. This is particularly reflected in a number of the Arctic Council Working Groups, including the Arctic Monitoring and Assessment Programme (AMAP), Conservation of Arctic Flora and Fauna (CAFF), and particularly the Protection of the Arctic Marine Environment (PAME) working group. A number of recent initiatives have highlighted the capacity of these mechanisms, such as the development of Arctic marine biodiversity monitoring plans, and the Arctic Marine Strategic Plan.

CCAMLR has continued to play a pivotal role in Southern Ocean marine environmental protection and has been able to evolve through the adoption of Conservation Measures, and innovative mechanisms for maritime regulation and enforcement. In 2016 the CCAMLR Commission endorsed a proposal for a Marine Protected Area (MPA) encompassing 1.55 million km^2 in the Ross Sea. This proposal had first been promoted by New Zealand and the United States in 2011 and had been discussed over six years before finally being adopted by consensus. The MPA, which became effective in December 2017, is designed to limit or entirely prohibit "certain activities in order to meet specific conservation, habitat protection, ecosystem monitoring and fisheries management objectives" (CCAMLR 2016). While the establishment of the MPA is a singular achievement for CCAMLR, the length of time taken to reach consensus on the proposal highlighted differences of views amongst member states and it remains to be seen whether CCAMLR members will be supportive of similar initiatives in other parts of the Southern Ocean.

The polar oceans are also subject to general marine environmental instruments that have global application, some of which have particular relevance for the law of the sea. These include instruments such as the 1946 International Convention for the Regulation of Whaling (ICRW) (161 United Nations Treaty Series 72), and the 1992 Convention on Biological Diversity (CBD) (1760 United Nations Treaty Series 79). The ICRW has particular application because of the historical abundance of whale stocks in the polar oceans, the cultural and economic significance of whales for Arctic indigenous peoples, and evolution of whaling from a resource management to an environmental management issue. Antarctica and the Southern Ocean has recently been the scene of discord with respect to whaling. Since the 1985/86 season there has been a global moratorium on whaling; however, Japan continued its whaling activities in the Southern Ocean in reliance upon Article VIII of the ICRW which allows 'special permit' whaling for the purpose of scientific research. Australia and New Zealand have been particularly critical of Japan's conduct of these whaling programmes and in 2010 Australia commenced proceedings against Japan in the International Court of Justice seeking to halt Japan's 'scientific whaling'. The court handed down its judgment in 2014 (*Whaling in the Antarctic* (Australia v. Japan; New Zealand Intervening) Judgment of 31 March 2014 [2014] ICJ Reports). Finding in favour of Australia, the court ruled that Japan's conduct was inconsistent with Article VIII of the ICRW and ordered the country to cease its whaling activities. Japan accepted the court's judgement and undertook no whaling during the 2014/15 season but resumed whaling in 2015/16 under the guise of a new whaling program named NEWREP-A. Australia and other states parties to the ICRW continue to protest against Japan's Southern Ocean whaling program at the meetings of the International Whaling Commission, which is the forum for the meeting of the states' parties to the ICRW, while Japan asserts that it has now modified its behaviour and is in compliance with Article VIII (Rothwell 2016).

Shipping

Perhaps unique amongst the world's oceans, the polar oceans are not major trade routes and therefore commercial shipping is significantly less in these oceans than elsewhere. A number of factors contribute to this, including the absence of major population centres, the difficulties posed by navigation through ice covered waters, and the generally unfavourable weather for shipping. While both the Southern and Arctic Oceans have been the scene of some of the more remarkable ocean voyages, they have not been the subject of very particular attention with respect to shipping in general. This is not to suggest that shipping issues are not alive in the polar oceans – far from it – and with the onset of the effects of climate change, which are increasingly apparent in terms of diminishing sea ice, there has been an upsurge of interest in polar shipping, particularly in the Arctic (Chircop 2009).

Nevertheless, two Arctic shipping routes have been a particular focus of attention. The first – the so-called Northern Sea Route – runs along Russia's northern coast. As the Northern Sea Route falls primarily within Russian coastal waters there has been little dispute over Russia's capacity to control substantially navigation within these waters consistently with the law of the sea (Tymchenko 2001). This has not been the case with respect to the Northwest Passage. A variety of interconnected sea routes which pass between the islands that make up the Canadian Arctic Archipelago, the status of the Northwest Passage has been the subject of ongoing dispute between Canada and the US ever since the 1969 voyage of the SS *Manhattan*. In response to the *Manhattan* voyage Canada adopted the Arctic Waters Pollution Prevention Act, placing significant constraints on the passage of vessels through its Arctic waters on environmental grounds. Canada took further steps in the 1980s to bolster its control of its Arctic waters through the declaration of straight baselines around the outer limits of its Arctic islands with the effect that all vessels passing through the Northwest Passage would be within Canadian internal waters. The significance of Canada's initiatives was to convert waters that may at one time have been a part of the territorial sea or EEZ, within which certain navigational rights existed for foreign ships, into waters over which Canada had complete sovereignty and the capacity to regulate all shipping – including the right to deny entry to foreign vessels (Pharand 1988).

Related to these measures, in August 2008 Canada announced that it intended to make what was previously a voluntary Vessel Traffic System (VTS) within Canadian Arctic waters mandatory. The amendments became effective in 2009 and remain operative as the 'Northern Canada Vessel Traffic Services Zone Regulations' (NORDREG) (SOR/2010–127 (10 June 2010) Canada Shipping Act, 2001, SC 2001, c 26.), made under the framework of the Canada Shipping Act, 2001. NORDREG applies to essentially two types of Canadian Arctic waters: waters prescribed under the Shipping Safety Control Zones Order, which encompass waters that extend to a distance of 200 nm from the territorial sea baseline, and various internal waters within Canada's Arctic baselines, such as the waters of Hudson Bay and James Bay. The relevant waters commence in places as far south as 60°N and encompass the Canadian EEZ as it surrounds the islands of the Canadian Arctic Archipelago to the limits of agreed or respected maritime boundaries with Denmark and the United States. Vessels entering the NORDREG controlled area are to provide a sailing plan report, position reports upon entry into the NORDREG area and thereafter on a 24-hour basis, and as soon as feasible once a vessel's master becomes aware of another vessel in difficulty, an obstruction to navigation, hazardous navigational conditions, or a pollutant in the water.

The legitimacy of the NORDREG regulations were the subject of diplomatic exchange between Canada and the United States in 2010, and between Canada, the United States and other states at the International Maritime Organization (IMO) at the same time (Kraska 2015: 243–247). The US critique of NORDREG is framed around the view that they are not supportable under LOSC Article 234 and represent an infringement of the freedom of navigation. Notwithstanding these criticisms, Canada has proceeded to give effect to NORDREG within its declared waters that make up the Canadian Arctic. For commercial shipping, compliance is the "easiest way to ensure hassle-free transit" (Kraska 2015: 247) and because of the exemptions for sovereign immune vessels no issues to date have arisen with respect to US government vessels operating in the Canadian Arctic beyond the Northwest Passage. Much of the Canadian/US disagreement with respect to NORDREG comes down to Canada's history of unilateral Arctic marine environmental protection measures against continued US assertions of the freedom of navigation via diplomatic channels and the actions of the US military. Nevertheless, Canada's reliance upon Article 234 as justification for NORDREG may increasingly be on a weak legal foundation given the need for the waters subject to such measures to be ice-covered for 'most of the year'. As the Arctic sea ice cover therefore recedes from parts of the Canadian EEZ, so too may the legal basis for NORDREG. Canada may therefore have to confront the fact that given its express reliance upon Article 234 as providing an international law foundation for the enactment of NORDREG, and given the intent of that provision in providing a coastal State with an exceptional right to adopt certain EEZ marine pollution measures that go beyond the general capacity of a coastal State to adopt equivalent measures in ice-free waters, the right would cease if the waters no longer possessed the relevant ice-covered characteristics. The United States, and other potential user States of Canadian Arctic waters, may therefore be carefully monitoring Canada's position with respect to NORDREG if climate change continues to impact upon the extent of ice in the Canadian EEZ.

One of the more significant areas of recent controversy with respect to shipping and marine environmental protection in the polar oceans has been with respect to the negotiation of a Polar Code (Rayfuse 2014). The IMO has coordinated discussions over the development of a Polar Code since 2009. These culminated in decisions of two IMO Committees in 2014 and 2015 that adopted an 'International Code for Ships Operating in Polar Waters' which entered into force on 1 January 2017 (Chircop 2016: 275). The Polar Code will have particular implications in the Arctic, raising issues as to whether existing national legislation is appropriate for regulating shipping and the compliance obligations upon Arctic states (Chircop 2016). The IMO has also responded to the particular circumstances of the polar oceans through instruments such as 1973/1978 International Convention for the Prevention of Pollution from Ships (MARPOL) (1340 UNTS 62). The Southern Ocean is listed as a 'Special Area' under MARPOL, Annex I, II, and V. However, there is no equivalent listing for the Arctic Ocean. The potential for greater numbers of vessels to navigate within and through the polar oceans raises for consideration whether MARPOL and the regime for its implementation is adequate. For example, under MARPOL Annex VI neither the Southern Ocean or the Arctic Ocean has a designated emission control area. In 2009 the Arctic Marine Shipping Assessment (AMSA), carried out under the auspices of the Arctic Council, recommended that Arctic states support the development of improved practices and innovative technologies for ships so as to reduce emissions (Arctic Council 2009: 7, Recommendation II: H). In that regard, MARPOL has given increasing attention to coastal and port state implementation. However, in the Southern Ocean, with the exception of sub-Antarctic islands, coastal states are not recognized as having sovereignty or jurisdiction and are therefore unable to exercise traditional coastal state jurisdiction with respect to marine pollution. Likewise, port states may be some considerable distance from the area where a pollution incident occurs, which due to its isolation may never have been identified in the first instance. While coastal state jurisdiction is recognized in the Arctic, the central Arctic Ocean is beyond national jurisdiction, and extremely remote. Issues arise here also with respect to the effectiveness of port state jurisdiction. These factors suggest the need for MARPOL to be modified to reflect the particular issues that arise in regard to marine pollution in the Polar Regions.

Concluding remarks

The Polar Regions have significant maritime domains and accordingly are subject to the law of the sea as reflected in the LOSC and its associated instruments, and the multiple additional legal instruments that make up the modern law of the sea. Yet for much of the post-World War II period the Polar Regions have not been the focus of the law of the sea. However, as the Polar Regions came more into the mainstream of international affairs they have likewise become a focus for the law of the sea. There have been two drivers for this. The first was concern over state sovereignty and environmental protection which in the case of the Arctic was reflected in LOSC Article 234, and the ongoing disagreements between Canada and the United States over the status of the Northwest Passage. In the Southern Ocean these aspects were reflected in the 1991 Madrid Protocol and more recently over Australia taking Japan to the International Court of Justice in 2010 over its whaling programme. The second driver has been increased interest in continental shelf resources and the subsea mapping activities of states wishing to delimit their outer continental shelf beyond 200 nm, the catalyst of which has been the need to assert OCS claims consistently with LOSC Article 76 via CLCS submissions. In the Arctic this process has developed in a similar fashion to other oceans and seas, with the exception of the United States not yet having asserted an OCS claim offshore Alaska due to it not yet having acceded to the LOSC. In the Southern Ocean the constraints created by the Antarctic Treaty have acted as a check on the assertion of OCS claims, but have nonetheless created some legal tensions between the claimant states. The polar oceans have increasingly entered into the mainstream of the law of the sea and the current controversies that have been highlighted indicate their immediate significance for Antarctica and the Arctic and most likely well into the future.

References

Arctic Council 2009. *Arctic Marine Shipping Assessment 2009 Report*. Tromsø, Norway: Arctic Council.

CCAMLR 2016. 'CCAMLR to create world's largest Marine Protected Area' (media release, 28 October 2016). ccamlr.org.

Chircop, A. 2009. 'The growth of international shipping in the Arctic: is a regulatory review timely?' *International Journal of Marine and Coastal Law* 24: 355–380.

Chircop, A. 2016. 'Jurisdiction over ice-covered areas and the Polar Code: an emerging symbiotic relationship' *Journal of International Maritime Law* 22: 275–290.

Fitzmaurice, M. and D. Tamada (eds) 2016. *Whaling in the Antarctic: Significance and Implications of the ICJ Judgment*. Leiden, Netherlands: Brill Nijhoff.

Hemmings, A.D. and T. Stephens 2009. 'Australia's extended continental shelf: what implications for Antarctica?' *Public Law Review* 20: 9–16.

Huebert, R. 2001. 'Article 234 and marine pollution jurisdiction in the Arctic' in A. G. Oude Elferink and D. R. Rothwell (eds) *The Law of the Sea and Polar Maritime Delimitation and Jurisdiction*. The Hague, Netherlands: Martinus Nijhoff, 249–267.

Ilulissat Declaration 2008. Oceans Conference, Nuuk, Greenland, www.oceanlaw.org/downloads/arctic/Ilulissat_Declaration.pdf.

Kaye, S. 2001. *International Fisheries Management*. Dordrecht, Netherlands: Kluwer.

Kraska, J. 2015. 'The Northern Canada Vessel Traffic Services Zone Regulations (NORDREG) and the Law of the Sea' *International Journal of Marine and Coastal Law* 30: 225–254.

Leary, D. 2014. 'From hydrocarbons to psychrophiles: the "scramble" for Antarctic and Arctic resources' in T. Stephens and D.L. VanderZwaag (eds) *Polar Oceans Governance in an Era of Environmental Change*. Cheltenham, UK: Edward Elgar, 125–145.

Mossop, J. 2008. 'Opposing Japanese whaling in the Southern Ocean: the international law implications of contrasting approaches' *Asia Pacific Journal of Environmental Law* 11: 221–234.

Neumann, T. 2010. 'Norway and Russia agree on maritime boundary in the Barents Sea and the Arctic Ocean' *ASIL Insights* 14(34): 1–4.

Pharand, D. 1988. *Canada's Arctic Waters in International Law*. Cambridge, UK: Cambridge University Press.

Rayfuse, R. 2007. 'Melting moments: the future of polar oceans governance in a warming world' *Review of European Community and International Environmental Law* 16(2): 196–216.

Rayfuse, R. 2014. 'Coastal state jurisdiction and the Polar Code: a test case for Arctic oceans governance?' in T. Stephens and D. L. VanderZwaag (eds) *Polar Oceans Governance in an Era of Environmental Change*. Cheltenham, UK: Edward Elgar, 235–252.

Rothwell, D.R. 1996. *The Polar Regions and the Development of International Law*. Cambridge, UK: Cambridge University Press.

Rothwell, D.R. 2008. 'Issues and strategies for outer continental shelf claims' *International Journal of Marine and Coastal Law* 23: 185–211.

Rothwell, D.R. 2012a. 'Polar oceans governance in the twenty-first century' *Ocean Yearbook* 26: 343–360.

Rothwell, D.R. 2012b. 'Law enforcement in Antarctica' in A. Hemmings, D.R. Rothwell and K.N. Scott (eds) *Antarctic Security in the Twenty-First Century*. London: Routledge. 135–153.

Rothwell, D.R. 2016. 'The whaling case: an Australian perspective' in M. Fitzmaurice and D. Tamada (eds) *Whaling in the Antarctic: Significance and Implications of the ICJ Judgment*. Leiden, Netherlands: Brill Nijhoff. 271–307.

Rothwell, D.R. and T. Stephens 2016. *The International Law of the Sea*, 2nd edition. Oxford, UK: Hart.

Tymchenko, L. 2001. 'The Northern Sea Route: Russian management and jurisdiction over navigation in Arctic Seas' in A.G. Oude Elferink and D.R. Rothwell (eds) *The Law of the Sea and Polar Maritime Delimitation and Jurisdiction*. The Hague, Netherlands: Martinus Nijhoff. 269–291.

22

The Arctic Council

An intergovernmental forum facing constraints and utilizing opportunities

Timo Koivurova

The current period of Arctic-wide co-operation between the states of the northern circumpolar region can be said to have commenced with the 1991 Arctic Environmental Protection Strategy (AEPS). The activities of the AEPS, which were carried out under the auspices of several working groups concerned with conservation, pollution issues, and protection of the Arctic environment, were subsumed by the Arctic Council in 1996. While the Arctic Council continues to be the predominant Arctic inter-governmental forum focused on the region today, it is essentially the same type of arrangement for circumpolar co-operation on the environment and sustainability that the AEPS had set out to be. There have obviously been some changes in the nature of this co-operation, but the fundamental principles of how it is enacted and operationalized have remained in place and guide the Arctic Council's work. At the same time, the Council has been able to find a distinctive role in the ever-changing multilevel governance environment in which it operates.

The argument in this chapter is two-fold. First, it will be demonstrated that there are various constraints as to how much it can change. While this is true for any institution in today's world, and so the Arctic Council is no exception, this is nonetheless a novel perspective in Arctic scholarship. Most analysis of the Arctic Council starts from the premise that it can develop – and needs to develop, because of the enormous challenges the region faces – and move towards whatever direction is desirable. An especially good example of this discourse is whether member states should negotiate, conclude and sign an Arctic treaty (and how it should be operationalized), or consolidate the Arctic Council in some other manner (Koivurova 2008; Koivurova and Molenaar 2009; Young 2009, 2011).[1]

Secondly, broader geopolitical changes, the dense multilevel governance systems influencing the Arctic, and certain strategies of resilience, which all institutions tend to follow as they continue their work, despite significant environmental changes that throw up challenges, are shown as reasons why the Arctic Council has retained its fundamental characteristics throughout the period of Arctic co-operation that has developed since the end of the Cold War. Yet, it will also be shown that the Council has been able to carve itself a niche in a very dense institutional environment that characterizes the contemporary Arctic. In particular, attention will be drawn to how the Council has been able to transform itself from a decision-shaping body to a decision-making forum. Before demonstrating how the Arctic Council has been able to change or operate within the constraints with which it is able to function, it is imperative to examine how Arctic-wide co-operation has changed over the years.

A brief history of Arctic-wide inter-governmental co-operation

The initial idea for Arctic-wide co-operation was laid out by the former Soviet Secretary-General Mikhail Gorbachev in a speech in Murmansk in 1987. The Soviet leader proposed that the Arctic states could and should initiate co-operation in various areas, one being protection of the Arctic environment. This idea was concretized when Finland convened a conference of the eight Arctic states – Canada, Denmark, Finland, Iceland, Norway, the then-Soviet Union, Sweden, and the United States – in Rovaniemi in 1989 to discuss the issue. On 14 June 1991, after two additional preparatory meetings, the national delegations, which were led mostly by the ministers of environment – with the participation of other actors – signed the Declaration on the Protection of the Arctic Environment in Rovaniemi (and sometimes referred to as the Rovaniemi Declaration), thereby adopting and initiating the Arctic Environmental Protection Strategy (Tennberg 1998).

The AEPS identified six priority environmental problems threatening the Arctic in particular (persistent organic contaminants, radioactivity, heavy metals, noise, acidification and oil pollution). It also outlined international environmental protection treaties that apply in the Arctic and specified additional actions to counter the identified environmental threats. Four working groups were established: Conservation of Arctic Flora and Fauna (CAFF), Protection of the Arctic Marine Environment (PAME), Emergency Prevention, Preparedness and Response (EPPR) and the Arctic Monitoring and Assessment Programme (AMAP). Three ministerial meetings followed in this first phase of Arctic co-operation, which was also referred to as the Rovaniemi Process. Senior Arctic Affairs Officials (which since 1996 have been known as Senior Arctic Officials, or SAOs), who were representatives of the ministries of foreign affairs of the eight states (including the Russian Federation following the dissolution of the Soviet Union in December 1991), co-ordinated the co-operation in-between the ministerial meetings, while the experts in the working groups represented the environmental sector and science. The final AEPS ministerial meeting was held in Alta, Norway in 1997 and it focused on integrating the AEPS into the structure of the newly established Arctic Council.

The Arctic Council was established as a high-level intergovernmental forum on the basis of a declaration (known as the Ottawa Declaration) signed by the ministers of foreign affairs of the Arctic states in Ottawa, Canada on 19 September 1996. Initially, the founding of the Arctic Council brought only minor modifications to the format and procedures of Arctic co-operation developed under the AEPS, and extended slightly the terms of reference beyond the previous focus on environmental protection to include more priority given to sustainable development than previously. It is important to note that there was not much change in the specific kinds of Arctic co-operation practices following the transition to the Arctic Council from 1996[2] and its fundamental elements – soft-law legal status, institutional structure, and no permanent funding mechanism – remained the same as in the AEPS. Of particular importance is that the Arctic Council – as an intergovernmental forum – cannot enact any legally-binding rules.

The Council was empowered to deal with "common Arctic issues, in particular issues of sustainable development and environmental protection in the Arctic".[3] This potentially yielded a very broad mandate, since "common issues" can include almost any facet of international policy apart from "matters related to military security".[4] Environmental co-operation is now included as a principal focus within the mandate of the Arctic Council, with the four working groups that had been established by the AEPS co-operation continuing under the umbrella of the Council.[5]

The second 'pillar' of the Council's mandate is co-operation on sustainable development, which has a working group of its own (the Sustainable Development Working Group, or SDWG). The Arctic Council has also adopted new programmes related to environmental protection, such as the Arctic Council Action Plan to Eliminate Pollution in the Arctic (ACAP), for which a sixth working group has been established.[6]

What is unique about the Arctic Council is the role it has given to the region's indigenous peoples' organizations. They are defined as Permanent Participants, a distinct category between the Council's members (which are exclusively nation states) and observers, which have to be consulted before any decision-making. The group of observers is relatively large and consists of inter-governmental and non-governmental organizations, as well as non-Arctic states with active interests in the region (Graczyk 2011; Graczyk and Koivurova 2014).

Before 2007, international Arctic co-operation evolved in relative isolation from similar co-operative initiatives that worked within a global context, and its work, progress and initiatives were largely unnoticed. This relative isolation has been challenged since 2007. The scientific outlook for Arctic climate change and in particular the melting sea ice of the Arctic Ocean – a development which opened up speculation about new economic and security threats and opportunities in the region – led to change in the international perception of the region. This intensified in summer 2007, when a Russian group planted a Russian flag on the sea bed underneath the North Pole. This act was interpreted by many in the media as a claim on the sea bed and its resources for Russia, triggering an international discussion that an all-out scramble for resources had begun among the Arctic Ocean coastal states. It is now broadly acknowledged that this was a grave misunderstanding: it was no more than an instance of the coastal states following international legal guidelines under the United Nations Convention on the Law of the Sea (UNCLOS), for which purpose they have been actively mapping the continental shelf of the Arctic Ocean (Baker 2012; Koivurova 2011a, 2011b; see also Donald Rothwell's chapter in this handbook on the Polar Regions and the law of the sea). The five Arctic coastal states of the United States, Canada, Russia, Norway, and the Kingdom of Denmark argued as much in their Ilulissat Declaration in 2008.

However, because of the perceived scramble for resources, and the resulting international and public attention given to Arctic issues in recent years, states began to pay more attention to the Arctic Ocean sea ice melting and to possible ways to exploit its hydrocarbon riches, its navigational highways, and other potential (Anderson 2009; Koivurova 2011b [n 28]). This has also partly influenced the demands placed on the Arctic Council, given that during 2008–2010 it appeared that intensified co-operation had emerged among an inner circle consisting of the five coastal states (Koivurova 2011a, 2011b).

The global attention to the Arctic has translated into expressions of the interests of various actors and non-Arctic states to take part in the work of the Arctic Council. In its 2013 Kiruna ministerial meeting, and in the resulting Kiruna Declaration, the Council accepted China, India, Japan, South Korea, Singapore and Italy as observers.[7] The acceptance of these new observers meant that the matters dealt with within the Arctic Council are increasingly taken into consideration worldwide.

More recent structural developments in the work of the Arctic Council have strengthened its capacity and role. With the 2011 Nuuk Declaration, which emerged from the ministerial meeting in Greenland's capital that marked the end of Denmark's chairmanship, the Council's jointly-funded permanent secretariat was established in Tromsø, Norway. Another new development is that the Arctic Council has acted as a catalyst for negotiating three international agreements: for responding to marine oil spills, co-operation in marine and aeronautical search and rescue operations and for enhancing international scientific co-operation (SAR Agreement; Oil Spills Agreement; Scientific Cooperation Agreement).[8]

Overall, even if the Arctic Council has been strengthened over time, each new incremental step has been built firmly on the same fundamentals that have remained in place from its beginnings. Even if there have been occasional suggestions from non-governmental organizations to change these fundamentals via an international treaty,[9] this has not been accepted by most states and other actors as an appropriate institutional arrangement for the Arctic (Koivurova 2008). A greater challenge has been the Arctic Ocean coastal state co-operation that was not received well by the three states that were not invited to attend meetings (Iceland, Finland and Sweden), or by the permanent participants. Yet, even if the initial signs of this co-operation seemed to suggest that Arctic Ocean coastal states are establishing a competing forum of inter-governmental co-operation, this did not happen – only the negotiations surrounding Arctic fisheries have been undertaken with the lead of the Arctic Ocean coastal states, primarily because the Arctic Council does not have any competence concerning fisheries. All the three Agreements – on search and rescue, oil spills, and scientific co-operation – have been negotiated under the Arctic Council, between all eight Arctic states.[10]

What are the constraints on how the Arctic Council can evolve?

It is evident, as the short review above reveals, that the Arctic Council has changed in both its scope and shape (at least in terms of the wider range of observers) in the last few years. But the fundamental premises of circum-Arctic co-operation – the legal nature and status of such co-operation, the funding system, the operating format, and other issues concerning procedure – have remained almost the same from the beginning of the AEPS in 1991 to the current day activities and operations of the Arctic Council. What can explain this?

Taking a broader and long-standing perspective, it seems obvious that the geopolitical situation has determined to a considerable degree what it is possible to do in the Arctic in terms of international co-operation, given that the two formerly competing superpowers of the Cold War era are members in the Arctic Council. The AEPS would not have been possible without the end of the Cold War. During the Cold War, the Arctic emerged as a major strategic region of confrontation between the coalitions led by the United States and the Soviet Union. The future of the Arctic Council depends on the evolving relations between the Russian Federation and the Western powers. It is significant that the Ukraine crisis, which in so many ways has become the major focal point of tension and possible confrontation between Russia and the Western powers in recent years, has not – at least yet – had significant consequences for the work of the Arctic Council, largely because the form of co-operation explicitly avoids touching on military issues but also because most of the work is low-ambition level, civil servant and scientist-driven down to earth co-operation.

The second constraining reason for the limitations in the evolution of the Arctic Council relates to the very dense multilevel governance that already exists in the Arctic. As was recognized in the AEPS, there are many international environmental treaties that already regulate how environmental protection must be done in the Arctic by the Arctic states. In the Arctic,[11] all levels of law – international, European, national and sub-national, the customary law of indigenous peoples – come into play, together with many policy areas, from environmental protection to maritime and indigenous issues and trade, to name a few. Of the eight states, three are federal in structure (the United States, Canada and Russia), with varying division of powers between the regional and federal levels: the state of Alaska in the United States, the three northern territories of Canada and the various federal subjects of the Russian Federation, all of which are areas where indigenous peoples have also been granted different powers and rights. Moreover, northern municipalities are key actors in governance and developments occurring in the region. The EU is an important actor in Arctic governance both via its own regulations and through participation in international normative processes (Koivurova et al. 2012; Pedersen 2008; Svalbard Treaty).[12]

The eight Arctic states are parties to a large number of international environmental treaties and other normative instruments and are bound by customary international law. And, even further, northern regions have a number of their own co-operative structures across borders, such as the Barents Regional Council in northern Fennoscandia and northwest Russia, or different trans-boundary water commissions, or co-management arrangements such as the conservation agreement for the Porcupine caribou herd between Canada and the US, which involves co-operation between federal, territorial, state, and indigenous organizations and communities in the Northwest Territories, Yukon Territory and Alaska. These levels of co-operation contribute to the complexity of international governance in the circumpolar North. Also, much of the Arctic is already within the sovereignty, and so subject to the sovereign rights, of the Arctic states. All of the land area is firmly under the sovereignty of the Arctic states,[13] and most Arctic waters fall under their respective exclusive maritime jurisdictions. It is obvious that within this ever-changing dense multilevel governance system, the Arctic Council can develop only to a limited extent beyond its fundamental remit (Byers 2013).

Finally, and importantly, international institutions also have their internal dynamics that diminish the possibilities for any major development jumps. At one level, we can see that this incremental change – each change built on the basis of an earlier one – has its explanation in what we nowadays refer to as imprinting. As Marquis and Tilcsik define it, it is "a process whereby, during a brief period of susceptibility, a focal entity develops characteristics that reflect prominent features of the environment, and these characteristics continue to persist despite significant environmental changes in subsequent periods" (Marquis and Tilcsik 2013). Hence, the basic characteristics of an institution remain unchanged even in the face of significant change in its exterior environment. This seems well in tune with how Arctic-wide co-operation has evolved over almost thirty years.

Hence, there seem to be clear limits as to how and to where the Arctic Council can develop, both because of internal factors, but also because of the constraints imposed on it by the region's geopolitical fundamentals, multilevel governance system, and those international agreements and arrangements which prevail also in the Arctic. Yet, since it is located in this complex international and national governance setting, it has to search continually for its own unique niche in order not to become redundant. It is here that the Arctic Council has demonstrated that it indeed can evolve.

How the Arctic Council has found its niche

Regional organizations such as the Arctic Council are faced with clear constraints as to how far and, indeed, into what they can evolve and the reasons for this have been reviewed above. But within those constraints, the Arctic Council has been able to find its niche – a relevant role within the dense governance framework applicable in the Arctic. The following characteristics have enabled the organization to do this:

1   A flexible unique soft-law structure, which also accords the Arctic Council's indigenous peoples a unique status in an inter-governmental forum.
2   The strong commitment shown by those who have been working for the Arctic Council – what we could also call an epistemic community (Haas 1997).

It seems clear that with this structure, the Council has been able to perform activities that cannot be done in other levels of governance. This transformation from decision-shaping body to decision-making body has happened gradually but with success.

The Council has gradually focused more of its efforts on carrying out large-scale scientific assessments, identifying environmental problems in the region and the ways they have been created, in particular how climate change transforms the region at twice the rate as compared to the rest of the world. The 2005 Arctic Climate Impact Assessment (ACIA)[14] has been followed up by various other large-scale scientific assessments that have demonstrated how climate change influences economic activities in the region or how it changes its biodiversity (ACIA; Graczyk 2011; Nilsson 2007). Moreover, these assessments also include recommendations, some of which are now carefully followed-up as to whether they have been considered or even implemented. Because of these large-scale scientific assessments, the Council has also been able to influence broader frameworks of environmental governance, even global ones.

A well-known example is the way the Arctic Council actors were able to influence the negotiations of what became the 2001 Stockholm Convention on Persistent Organic Pollutants (POPs), which became effective in 2004. The background to this process was the making of the 1997 AMAP assessment, which provided scientific information that enabled a concrete outcome for international environmental protection. POPs, defined in the assessment as a threat, end up in the Arctic from southern industrial regions via prevailing northerly winds and ocean circulation. Therefore, in order to address the POPs issue, the eight nation-states and permanent participants had to try and influence global levels of governance. As has been demonstrated by Downie and Fenge, it was the science that AMAP compiled that prompted joint action by the member states of the Arctic Council and its permanent participants (Downie and Fenge 2003).

The role played in the process by the permanent participants, such as the Inuit Circumpolar Conference (ICC – since 2006 renamed as the Inuit Circumpolar Council) was particularly compelling. They were able to concretize the impacts of POPs: although Arctic peoples do not use POPs, the substances are found in the large marine mammals which constitute for them a vital source of food. An especially convincing argument in encouraging the progress of negotiations was put forward that highlighted the scientific evidence for large concentrations of POPs in pregnant Inuit women, potentially damaging the foetus and having long-term adverse effects on human health, and thus being harmful for future generations (Downie and Fenge 2003). This relevant finding, the scientific information compiled by AMAP in general, and the coalitions between Arctic Council member states, influenced the negotiations on a protocol for the United Nations Economic Commission for Europe (UNECE) Convention on the Long-Range Transboundary Air Pollution (CLRTAP) on POPs (Geneva Convention; Aarhus Protocol). Even more importantly, these developments also influenced the successful conclusion of the 2001 Stockholm POPs Convention (Stockholm Convention). Here the role of actors such as ICC was crucial. The risks to the Arctic are highlighted in the Convention's preamble:

> <u>Acknowledging</u> that the Arctic ecosystems and indigenous communities are particularly at risk because of the biomagnification of persistent organic pollutants and that contamination of their traditional foods is a public health issue.[15] (underlining in original)

A final example of how the Arctic Council has been able to transform itself from a decision-shaping to a decision-making body is how it has been able to catalyse legally-binding agreements, even if it is a soft-law body established via a declaration. Under the auspices of the Council, three legally-binding agreements have been negotiated, namely, the 2011 agreement on search and rescue co-operation (one of the key Arctic Marine Shipping Assessment [AMSA] recommendations)[16], the agreement on marine oil pollution preparedness and response concluded in 2013, and the recently concluded agreement on enhancing international scientific co-operation at the 10th Ministerial Meeting in May 2017 at Fairbanks, Alaska (SAR Agreement; Oil Spills

Agreement; Scientific Cooperation Agreement). Given the possibility of a major accident or oil spill, a legal action going beyond a soft-law approach was needed. Moreover, the Arctic states had also been pushing to make the 2009 non-binding Polar Code for shipping a mandatory International Maritime Organization (IMO) instrument – the action recommended in AMSA, and this came into force on 1 January 2017 (see, also, the chapter by Donald Rothwell in this volume).[17] Currently, there is preliminary work being done within the Arctic Council on a possible oil spills prevention agreement or another type of instrument to address the issue, which was, again, recommended by AMSA.[18] In fact, the Arctic appears to counter the general trend of states seemingly being more reluctant than before to bring about and conclude treaties (Pauwelyn, Wessel and Wouters 2014).[19] This is primarily a consequence of the attention currently given to Arctic climate change and its impacts, highlighted in ACIA's findings, especially the anticipated increase in various human activities in the region. As a result, there is a heightened focus on the adaptation of Arctic governance to a new climate change-driven reality in such areas as Arctic maritime navigation, oil spills or fisheries.

Conclusions

It seems obvious that whatever region of the globe we nowadays focus our attention on, it will be densely regulated by various layers of governance. Consequently, all regional or global regimes or institutions need to search continually for, and evaluate, the rationale for their existence. This will be influenced by more long-term changes in geopolitics, the current multilevel governance system that we have and internal factors that make it difficult to change the way the institution develops, even if clear changes take place in its operating environment.

All this is clear as regards the Arctic Council. It was the end of the Cold War that made the current state of Arctic-wide co-operation possible, but it is also entirely possible that the worsening of relations between the Russian Federation and the Western powers will have a significant influence on how this co-operation can continue into the future (related to this, see the chapter in this handbook by Andreas Østhagen for a discussion of Arctic geopolitics and security). It is also obvious that the dense multilevel governance framework imposes constraints on what can be done and on what can be achieved within the Arctic Council. As this chapter has outlined, even if there are changes in the Arctic Council, these have been built on the foundations that were laid down already at the beginning of the current period of Arctic-wide co-operation, with the initiation of the AEPS in 1991.

Yet, even with all these constraints, it is also clear that the Arctic Council has been able to find its niche in, for instance, influencing policy at various levels of governance. As this chapter has also shown, the focus of the Council on producing large-scale scientific assessments on the environmental threats facing the Arctic have been taken up by Arctic Council actors in influencing global environmental negotiations successfully, such as the persistent organic pollutants negotiations and more recently the Minamata Convention negotiations over protecting human health and the environment from the emissions from mercury and mercury compounds (Minamata Convention).[20]

Perhaps the most radical change that has taken place in the Arctic Council is that it has been able to act as a catalyst for negotiating legally-binding agreements between the eight Arctic states. The motivation for this treaty-making comes from climate change and its dramatic transformational consequences for the Arctic, especially the new possibilities this change entails such as economic activities entering into the region. Evidently, with offshore oil and gas development as well as various types of shipping accessing the region, search and rescue and oil spill preparedness and response are needed to tackle the possibility of accidents in this region where distances are vast, and the infrastructure needed for responding to accidents is scarce. Given the rapidly thinning and retreating sea ice in the Arctic Ocean, legal rules responding to shipping and possible fisheries are needed. The Arctic Council has been able to work successfully within the limits it faces. Whether this will hold true for the future depends most importantly from how things will evolve in Ukraine as well as Russia's power projections in other parts of the world.

Notes

1 See 6.2. International Co-operation in the Arctic, p.44 in Finland's Strategy, which can be downloaded from the Arctic Council website at www.arctic-council.org/index.php/en/about-us/member-states/finland.

2 Also, the status of the Arctic Council was made stronger than that of the AEPS, insofar as the Council met at the level of foreign ministries instead of ministries of the environment.

3 Art 1(a) of the 1996 Ottawa Declaration.

4 Ibid at p.3.

5 Ibid art 1(b) reads: "The Arctic Council is established as a high level forum to [. . .] b. oversee and co-ordinate the programs established under the AEPS on the Arctic Monitoring and Assessment Program (AMAP); Conservation of Arctic Flora and Fauna (CAFF); Protection of the Arctic Marine Environment (PAME); and Emergency Prevention, Preparedness and Response (EPPR)".

6 Recently, the Arctic Council has been implementing several projects outside of the working groups, which are monitored directly by the SAOs (e.g. the Arctic Resilience Report, which was later renamed the Arctic Resilience Assessment). While the environmental work is carried out in five working groups with their respective secretariats, the SDWG is responsible for all the other sectors of governance but has just a single secretariat. The Arctic Council's work focuses clearly, then, on environmental issues.

7 From autumn 2014, that also includes the European Union. The EU was earlier considered an "observer in principle". At the Arctic Council's Kiruna Ministerial Meeting in May 2013, the Council "received the application of the EU for observer status affirmatively", with a final decision awaiting "implementation", but with being invited to observe Council proceedings as any other observer.

8 The third Agreement on Enhancing International Arctic Scientific Cooperation entered into at the 2017 Ministerial Meeting in Fairbanks, Alaska https://oaarchive.arctic-council.org/handle/11374/1916.

9 It is important to note that it is not only non-governmental organizations in this regard, but a comprehensive Arctic treaty has also been suggested by the European Parliament in its resolutions of 2008 and 2009, and by Finland in its 2013 Strategy.

10 This required some extra measures to get Finland and Sweden in, include aviation emergencies with respect to search and rescue agreement, and even including the Gulf of Bothnia region of the Baltic Sea in the case of the oil spills convention.

11 The Arctic constitutes ocean and land areas around the North Pole, but there is no universally-agreed definition for its southern boundary. Tree line, 10° centigrade July isotherm, or Arctic Circle are often used in natural sciences and Arctic Council working groups and particular assessment processes (such as those undertaken by CAFF and PAME, or the 2004 Arctic Human Development Report) have adopted their own definitions.

12 Finland and Sweden are member states of the EU, while Iceland and Norway (with the exception of Svalbard) adopt much EU legislation (including environmental law) owing to the European Economic Area Agreement; Greenland, which itself left the European Communities in 1985, possesses extensive autonomous powers. The Svalbard archipelago has a unique status, established through the international Svalbard Treaty in 1920 (Svalbard Treaty). See Timo Koivurova, Kai Kokko, Sebastien Duyck, Nikolas Sellheim and Adam Stepien (2012), "The present and future competence of the European Union in the Arctic" *Polar Record* 48(4): 361–371, and Torbjørn Pedersen (2008) "The dynamics of Svalbard diplomacy" *Diplomacy and Statecraft* 19(2): 236–262.

13 The only exception as regards sovereignty over land territory is Hans Island, a barren islet located in the Kennedy Channel portion of Nares Strait between Ellesmere Island (Canada) and Greenland (Kingdom of Denmark). See Byers (2013: 10–16).

14 ACIA (2004) *Impacts of a Warming Arctic: ACIA Overview Report*. See also ACIA Scientific Report (2005) Scientific report of the Arctic Climate Impact Assessment. Cambridge University Press.

15 Ibid, Preamble, 3rd para.

16 PAME (AMSA) (n 39) 6, Recommendation IE.
17 PAME (Arctic Marine Shipping Assessment) 6, Recommendation IB.
18 Ibid, Recommendation IIF.
19 The American Society of International Law recently invited papers on "The end of treaties? An online agora", identifying a trend on the decreasing importance of treaties. www.asil.org/blogs/call-papers-end-treaties-online-agora
20 A fairly similar process took place when the Arctic Council actors – based themselves on AMAP compiled knowledge on problems caused by mercury – were able to influence the mercury negotiations, the convention, which states in its preamble that "[n]oting the particular vulnerabilities of Arctic ecosystems and indigenous communities because of the biomagnification of mercury and contamination of traditional foods, and concerned about indigenous communities more generally with respect to the effects of mercury". See Minamata Convention on Mercury (adopted 10 October 2013 at Kumamoto). www.mercuryconvention.org.

References

ACIA 2004. *Impacts of a Warming Arctic: ACIA Overview Report*. Cambridge, UK: Cambridge University Press, www.amap.no/documents/doc/impacts-of-a-warming-arctic-2004/786.
ACIA 2005. *Arctic Climate Impact Assessment: Scientific Report*. Cambridge, UK: Cambridge University Press.
Arctic Environmental Protection Strategy (AEPS) *Declaration on the Protection of Arctic Environment*. Rovaniemi, Finland, June 1991. http://library.arcticportal.org/1542/1/artic_environment.pdf.
Arctic Monitoring and Assessment Programme (AMAP) 1997. *Arctic Pollution Issues*. Oslo: AMAP https://oaarchive.arctic-council.org/handle/11374/924.
Agreement on Cooperation on Aeronautical and Maritime Search and Rescue in the Arctic (adopted on 12 May 2011 in Nuuk, entered into force on 19 January 2013) 50 I.L.M. 1119 (2011) (SAR Agreement). www.ifrc.org/docs/idrl/N813EN.pdf.
Agreement on Cooperation on Marine Oil Pollution, Preparedness and Response in the Arctic (adopted on 15 May 2013 in Kiruna) (Oil Spills Agreement) www.arctic-council.org/eppr/agreement-on-cooperation-on-marine-oil-pollution-preparedness-and-response-in-the-arctic/.
Agreement on Enhancing International Arctic Scientific Cooperation entered into at the 2017 Ministerial Meeting in Fairbanks, Alaska (Scientific Cooperation Agreement) https://oaarchive.arctic-council.org/handle/11374/1916.
Anderson, A. 2009. *After the Ice: Life, Death, and Geopolitics in the New Arctic*. New York: Smithsonian.
Baker, B. 2012. 'Oil, gas and the Arctic Continental Shelf: what conflict?' *Oil, Gas and Energy Law* 2. www.ogel.org/article.asp?key=3251.
Byers, M. 2013. *International Law and the Arctic*. Cambridge, UK: Cambridge University Press.
Downie, D.L. and T. Fenge (eds) 2003. *Northern Lights Against POPs: Combatting Toxic Threats in the Arctic*. Montreal, QC: McGill-Queen's University Press.
Finland's Strategy for the Arctic Region 2013. *6.2: International Cooperation in the Arctic*. Prime Minister's Office Publications 44: 16/2013 http://vnk.fi/documents/10616/334509/Arktinen+strategia+2013+en.pdf/6b6fb723-40ec-4c17-b286-5b5910fbecf4.
Graczyk, P. 2011. 'Observers in the Arctic Council: evolution and prospects' *The Yearbook of Polar Law* 3: 575–633.
Graczyk, P. and T. Koivurova 2014. 'A new era in the Arctic Council's external relations? Broader consequences of the Nuuk observer rules for Arctic governance' *Polar Record* 50: 225–236.
Haas, P.M. 1997. 'Introduction: epistemic communities and international policy co-ordination' in P.M. Haas (ed.) *Knowledge, Power, and International Policy Co-ordination*. Columbia, SC: University of South Carolina Press.
Ilulissat Declaration (28 May 2008). Arctic Ocean Conference (Ilulissat Declaration).
Kiruna Declaration (15 May 2013). 8th Ministerial Meeting of the Arctic Council (Kiruna Declaration).
Koivurova, T. 2008. 'Alternatives for an Arctic Treaty: evaluation and a new proposal' *Review of European, Comparative & International Environmental Law* 17(1): 14–26.
Koivurova, T. 2011a. 'Power politics or orderly development? Why are states "claiming" large areas of the Arctic seabed' in S.R. Silverburg (ed.) *International Law: Contemporary Issue and Future Developments*. Boulder, CO: Westview Press, 362–375.
Koivurova, T. 2011b. 'The actions of the Arctic states respecting the Continental Shelf: a reflective essay' *Ocean Development & International Law* 42(3): 211–226.
Koivurova, T., K. Kokko, S. Duyck, N. Sellheim and A. Stepien 2012 'The present and future competence of the European Union in the Arctic' *Polar Record* 48(4): 361–371.

Koivurova, T. and E.J. Molenaar 2009. *International Governance and Regulation of the Marine Arctic: Overview and Gap Analysis*. Oslo: WWF International Arctic Programme.

Marquis, C. and A. Tilcsik 2013. 'Imprinting: toward a multilevel theory' *The Academy of Management Annals* 7(1): 193–243.

Minamata Convention on Mercury (adopted on 10 October 2013 in Kumamoto) (Minamata Convention).

Nilsson, A.E. 2007. 'A changing Arctic climate: science and policy in the Arctic climate impact assessment'. PhD thesis, Linköping University Electronic Press, Linköping University, Sweden.

Nuuk Declaration (12 May 2011). 7th Meeting of the Arctic Council (Nuuk Declaration).

Pauwelyn, J., R.A. Wessel and J. Wouters 2014. 'When structures become shackles: stagnation and dynamics in international lawmaking' *European Journal of International Law* 25(3): 733–763.

Pedersen, T. 2008. 'The dynamics of Svalbard diplomacy' *Diplomacy and Statecraft* 19(2): 236–262.

Polar Code: IMO's International Code for Ships Operating in Polar Waters www.imo.org/en/MediaCentre/HotTopics/polar/Documents/POLAR%20CODE%20TEXT%20AS%20ADOPTED.pdf.

Protection of the Arctic Marine Environment (PAME) – Arctic Marine Shipping Assessment (AMSA) Recommendations IB, IE 6.

Tennberg, M. 1998. *The Arctic Council. A Study in Governmentality* PhD thesis Acta Universitatis Lapponiensis 19, University of Lapland, Finland.

The Convention on the Long-Range Transboundary Air Pollution (adopted on 13 November 1979 in Geneva, entered into force on 16 March 1983) (Geneva Convention).

The Convention on Persistent Organic Pollutants (adopted 22 May 2001 in Stockholm, entered into force 17 May 2004) 2256 UNTS 119 (Stockholm Convention).

The Declaration on the Establishment of the Arctic Council, 35 ILM 1385–1390 (1996) (Ottawa Declaration).

The Protocol on Persistent Organic Pollutants (POPs) (adopted on 24 June 1998 in Aarhus, entered into force on 23 October 2003) (Aarhus Protocol).

Treaty Concerning the Archipelago of Spitsbergen (adopted on 9 February 1920 in Paris, entered into force on 14 August 1925) 2 LNTS 8; UKTS (1924) 18 (Svalbard Treaty).

Young, O.R. 2009. 'Whither the Arctic? Conflict or cooperation in the circumpolar North' *Polar Record* 45: 73–82.

Young, O.R. 2011. 'If an Arctic ocean treaty is not the solution, what is the alternative?' *Polar Record* 47: 327–334.

23

National Antarctic programmes
The politics-science interface

Anita Dey Nuttall

Introduction

Until the mid-twentieth century, the interests a small number of nation states, such as the United Kingdom, Norway, France, Belgium, Russia, the United States, Australia and New Zealand, had in Antarctica were expressed mainly through sporadic, but large-scale and often long-term expeditions of geographical discovery and scientific exploration. Since then, and notably following the end of World War II and especially after the International Geophysical Year (IGY) of 1957–58, international political interest in the continent has been affirmed, consolidated and instituted with the establishment of permanent national Antarctic science programmes. This chapter examines the origins and development of national Antarctic programmes within a context of understanding this in relation to the interface of politics and science in Antarctica. It further describes the attributes of national Antarctic programmes and the general organizational framework within which countries operate and sustain their long-term science activities and the associated logistics and infrastructure needed to maintain a scientific (and political) presence in Antarctica.

The discovery and exploration of Antarctica in the nineteenth century and into the early twentieth century was the beginning of a process which often resulted in asserting and laying claims to parts of the continent, including the sub-Antarctic islands. It was a product of European expansion in commerce and mercantile activities in which the search for resources and associated empire building led to the dispatch of national expeditions of discovery to the southern circumpolar regions. Economic motives played an important role in sustaining political interests in the Antarctic, especially when the resources of the southern waters were seen to be highly lucrative with the start of the sealing and whaling industries in the nineteenth century. In this context, a significant development was the use of science to support an economic activity – the protection of the whaling industry (Heap 1990: 183), which in turn also contributed to geographical exploration. The first long-term state-sponsored Antarctic scientific programme, the *Discovery* Investigations, was initiated by Britain and arose as a result of its whaling activities in Antarctic waters.

Participation in the discovery and subsequent exploration of the Antarctic by Britain, France and Russia, later joined by Norway, Germany, Belgium, the USA, Sweden, Australia,

New Zealand and Japan, laid the foundation for their future political claims in Antarctica (Kirwan 1959). For some of these countries, a relatively continuous historical involvement, including sealing, whaling and exploration, and indeed their competence in organizing and implementing polar operations, justified their permanent interest in the continent. For others, a less continuous involvement, but the existence nevertheless of even a single historical association, gave them the same prerogative but not nearly the same competence.

Between 1939 and the start of IGY, the contest for territorial claims in the Antarctic prompted the establishment of permanent national operations and the administrative machinery required to run and maintain them. The protection and consolidation of these claims led to a range of national Antarctic operations and activities which later provided the conditions for the establishment of permanent national Antarctic science programmes. These operations consisted of establishing bases for permanent occupation in Antarctica (Quigg 1983). Establishing a base, rather than a temporary seasonal scientific field camp, was important for countries if they wanted to fulfil the condition of effective occupation – something no longer inconceivable in the southern circumpolar regions, where settlement in its conventional sense was restricted by geographical and climatic constraints, as well as the difficulties of reaching Antarctica in the first place. These bases for occupation were later to serve another purpose: providing part of an infrastructure integral to the logistics of national Antarctic programmes. In the case of some of these national operations, the administrative machinery a country established at the time was the forerunner of their current national Antarctic operating agency.

Since the signing of the Antarctic Treaty in 1959 by twelve IGY countries (which had all maintained Antarctic research programmes during IGY – but this criterion for qualifying a country to being a part of the negotiations for the Antarctic Treaty was controversial and debated at the time [e.g., see Bos 1991]),[1] the operations of national Antarctic programmes have inscribed and maintained a diverse range of scientific activities on the continent, with research activities clustered in and around a number of permanent and seasonal bases. Over the last 50 years, the Treaty has been acceded by 41 other countries, 17 of whom have acquired consultative status, thus joining the ranks of the original 12 signatories as influential decision-makers in Antarctic matters; 24 of whom acceded as Non-Consultative Parties with no decision-making powers, but who attend the Antarctic Treaty Consultative Party (ATCP) meetings as observers. Scientific activities in the Treaty area are presently carried out in 41 year-round and 39 seasonal scientific stations with a presence of over 4,000 scientific and logistic personnel during peak season work (COMNAP 2017).

Antarctica: a continent for science

It has long been a commonplace remark that Antarctica is a continent for science (see also, the chapters by Klaus Dodds and Sanjay Chaturvedi in this handbook). From the time of the discovery of Antarctica and the earliest expeditions to explore and map its coastlines and mountain ranges, investigate its vast ice sheet, chart its seas and ice extent, and study and classify its wildlife, the economic importance of the continent has been recognized as minimal compared to the contributions scientific research can make to understanding regional and global geophysical, climatic and biological systems.

Although the immediate post-World War II period in international relations came to be quickly dominated by the beginnings of the Cold War and influenced how governments saw the nature and scope of science (and the reasons for doing it in the first place) over the next few decades, the 1950s in particular were marked by a surge of internationalism which was especially evident in the realm of science. The Pugwash Conferences beginning from 1957 emphasized the supranational character of science and created opportunities for international co-operation (Heap 1983: 105). It was against this background that the IGY was able to exploit the desire for international co-operation in science in Antarctica. In the midst of political tensions (not just because of the Cold War, but in a context of post-colonial movements), the twelve nations recognized that peaceful and constructive co-operation between states was possible at the scientific level; the IGY established principles of conduct in international relations which emphasized the pre-eminence of science over sovereignty in Antarctica, marking the beginning of a new era (Beck 1986; Sullivan 1961).

The original signatories of the Antarctic Treaty suspended the issue of territorial claims and turned to the next closest issue of ownership: the management of Antarctic resources, which set the scene for the future course of Antarctic science and politics. Conservation measures were not built into the Treaty when it was signed, but it was agreed that assigning priority to the preservation and conservation of Antarctic resources would need to be done and so the development of what was seen initially as an environmental sub-regime (Elliott 1994) paved the way for the adoption by Treaty members of such legal instruments as the Agreed Measures for the Conservation of the Antarctic Fauna and Flora (often just referred to as the Agreed Measures; 1964), the Convention for the Conservation of Seals (CCAS 1972), the Convention for the Conservation of Marine Living Resources (CCAMLR 1982), the Convention for the Regulation of Antarctic Mineral Resource Activities (opened for signature in 1988 but not ratified) and the Protocol on Environmental Protection (1991; also just known as the Environmental Protocol). These instruments complement the Treaty and together comprise the Antarctic Treaty System (ATS).[2]

Notwithstanding the effects and legacies of historic commercial sealing and whaling operations, current discussion over bioprospecting and its implications, and the possibilities of future activities related to minerals and fisheries, Antarctica's economic potential is limited by strict regulatory agreements and regimes. Mining, for example, is prohibited under the Protocol on Environmental Protection. Further efforts to protect the region from economic activities such as commercial fishing have been articulated recently by designating the Ross Sea as the world's largest marine protected area (MPA). The new MPA, which came into force in December 2017, established 72% of the area as a 'no-take' zone forbidding all fishing, but permitting in some sections some harvesting of fish and krill for scientific research (CCAMLR 2017). Currently, discussion is also taking place over similar measures to protect the Weddell Sea.

Towards sustained science activities in Antarctica

National operations changed in nature and extent in response to changing images and perceptions of Antarctica. With the outbreak of World War II and the wartime activities of the Axis powers in the South Atlantic and South Pacific, Antarctica assumed strategic importance, aroused US interest in the idea of a pan-American Antarctic and fanned geopolitical aspirations, particularly those of Chile and Argentina (Sullivan 1957). United States, Argentine and Chilean claims and activities stimulated the expansion of British activity. To protect its claims in Antarctica, Britain established bases for occupation that would perfect title to the territory claimed, which was mainly in the Antarctic Peninsula and a few sub-Antarctic islands (Fuchs 1982). The need arose for an administrative organization to run the bases, reinforcing government participation in Antarctic science, but also in international Antarctic dialogues. In turn, Australia, France, Argentina and Chile perceived similar needs. It was the start of nation states beginning to combine occupation, to one degree or another, with science.

The IGY transformed national activities in Antarctica from primarily political to primarily scientific operations and it initiated the concept of a unified international scientific programme. Preparations for the IGY required each nation co-operating in research in Antarctica to establish infrastructure, both at home and in the field, to support a scientific programme. Such infrastructure entailed coordination and collaboration between a number of national research agencies to draw up a scientific programme, dispatch scientific expeditions to the continent, and organize the logistics. The bringing together of scientific resources and logistics available to each country is the feature which continues to dominate each present-day Antarctic programme. Although the IGY was for a limited period, it nonetheless set in motion a process of organized, concerted and sustained scientific investigation in Antarctica which the participating nations did not wish to disturb. Moreover, results of the investigations carried out on the continent established links between Antarctica and Earth system processes as a whole that justified the long-term reasons for continued research. The first step towards the continuation of national operations in Antarctica was the extension of the IGY by the International Geophysical Cooperation Year (IGCY) in 1959. It led to the replacement of the Comité Speciale de l'Année Geophysique Internationale (CSAGI) by a Special Committee for Antarctic Research, later renamed the Scientific Committee on Antarctic Research (SCAR) which had the task of bringing together scientists of different disciplines and of coordinating programmes and exchanging information (Walton and Clarkson 2011).

The considerable costs of Antarctic science required political incentive to support and sustain scientific endeavours. The presence of such powers as the UK, USA and, later, the then Soviet Union, each to some degree politically at variance with the other, paved the way for permanent activities in the continent. With the conclusion of the IGY, the Antarctic Treaty provided the political framework to continue scientific research. Although the Treaty does not stipulate permanent national activity, the original parties to it each gave definition to their political, economic and scientific interests on the continent by establishing a permanent presence there through the conduct of long-term national Antarctic programmes.

The Treaty set the scene for future patterns of activity in Antarctica. While the 'claims of sovereignty' of seven of the IGY countries and the 'historic association' of the other five gave them a permanent place in Antarctica, for those countries which were not part of that history, the Treaty provided an interpretation of new terms for establishing their permanent interests in the continent. Acceding countries needed to fulfil a set of conditions in order to acquire powers equal to those of the original signatories of the Treaty. For new members to gain decision-making power or consultative status, Article IX of the treaty requires contracting parties to 'demonstrate [their] interest in Antarctica by conducting substantial scientific research activity there, such as the establishment of a scientific station or the dispatch of a scientific expedition'.[3] Acceding countries needed to display continuing interest in the region if they were to maintain consultative status. This obligation did not apply to the original signatories to the Treaty. More recently, however, and as a result of an unsuccessful application for consultative status by a non-Consultative Party, discussions at the 39th Antarctic Treaty Consultative Meeting in 2016 raised the issue of improving the mechanisms currently available to the parties in considering applications by Contracting Parties wishing to be granted Consultative Party status pursuant to Article IX, paragraph 2, of the Antarctic Treaty. It resulted in the establishment of an intersessional contact group to develop clearly defined criteria for achieving consultative status (Gray and Hughes 2016: 3)

Most of the countries that later acceded to the Treaty, particularly in the 1980s, sought membership with the sole objective of achieving consultative status (for example, Sweden, Brazil, Uruguay, South Korea, China and India). This is reflected in the sequence of events that took place in these countries soon after their accession to the Treaty. In keeping with the requirements of the Treaty to achieve consultative status, the pattern of the route taken by the countries toward achieving consultative status consisted of them first acceding to the Treaty, secondly sending expeditions to the continent, and then establishing a scientific presence through construction of a refuge hut, weather station, a summer field station, or a year-round scientific station. The mix of such different types of what constitutes a 'scientific station' reflected the room available to stretch what comprised 'substantial scientific research interest'. Long-term involvement in Antarctica required development of the appropriate infrastructure, which contributed to the maintenance of long-term national Antarctic programmes.

In scientific terms, the need for the continuity of scientific programmes was recognized by SCAR in order to provide a permanent machinery whose function was to facilitate future scientific co-operation. SCAR established a minimum requirement of four years of involvement to pursue science in Antarctica (SCAR Manual 1987: 7). In economic terms, the establishment of scientific stations and the dispatch and running of annual scientific expeditions amounted to considerable commitment and investment in resources. This contributed to making national Antarctic programmes permanent and continuous. In political terms, the unresolved territorial claims and the probationary nature of Treaty membership made continuity and permanence of national Antarctic activity necessary.

In the 1980s, the surge of nations establishing scientific stations in the Antarctic (which was occurring almost annually) and concerns about environmental practices in Antarctica led the Treaty nations to recognize the need for enhanced stewardship of Antarctica. At a time of increasing global and environmental awareness, the perceived plight of Antarctica generated intense interest in both the use and regulation of scientific activities in the continent (Elliott 1994). The adverse environmental effects as a result of stations crowded into the few easily accessible coastal areas of Antarctica, and of mutual interference among the different research programmes or logistics activities, became major concerns.

The formation of the Council of Managers of National Antarctic Programs (COMNAP) in 1988 was an acknowledgement of the increasing importance of coordinating and managing the logistics and operational aspects of a growing number of national Antarctic science programmes, while ensuring the protection of the Antarctic environment. The establishment of COMNAP evolved from a SCAR working group on Antarctic logistics and was an important mechanism for co-operation of the national operators (Retamales and Rogan-Finnemore 2011: 232). This was also the time when there was a major departure from the past practice of nations acquiring consultative status with the Netherlands setting an example in 1990 by successfully making a case for gaining it without the construction of a permanent station of its own and instead choosing to use existing infrastructure of other Antarctic Treaty Consultative Parties (ATCPs) with whom it collaborated (Pannatier 1994: 127). With the ATCPs agreeing that establishing a station is not a precondition for conducting the substantial scientific research in Antarctica necessary for obtaining consultative status, this paved the way for national applications to be judged on the basis of their past record of scientific contributions and future plans for continuing such work in the continent. Science undertaken in co-operation with other countries became an acceptable basis for attaining full membership (Walton 2017: 1).

In 1991, following a series of negotiations, an international consensus for stewardship and protection of the Antarctic environment emerged in the form of the Protocol on Environmental Protection which bans mineral activities for a period of 50 years from when it came into force (which was in 1998). The Environmental Protocol underscored the principle that the Antarctic must be regarded as a protected area, a natural reserve, to be utilized only for peaceful purposes, for science and for its unique aesthetic value as a wilderness. The Protocol's Article 22.4 placed a new condition for states seeking to acquire consultative status. In order for a state to be acknowledged as a Consultative Party, it must have first ratified, accepted, approved, or acceded to the Protocol (Pannatier 1994: 125).

Aspects of national Antarctic programmes

A national Antarctic programme is a permanent, long-term plan for a country's activities in the continent. It involves sending regular scientific expeditions or teams of scientists to Antarctica, providing the infrastructure necessary for its successful operation within the country concerned and in the Antarctic, and generating research of sound scientific quality.

Science and logistics are the two key components of a national Antarctic programme. Each is a function of – and is vital to – the other. Success in Antarctic operations depends upon their combined effectiveness. Yet they are expensive and often complex initiatives, and different organizations, institutions or departments may manage various aspects of the two components. Implementing a national Antarctic programme comprises all the activities relating to the conduct of scientific research in the Antarctic Treaty area. These activities have objectives that fall into the categories of science, logistics and administration.

The 29 Antarctic Treaty Consultative Parties have established their national Antarctic programmes in circumstances that were unique to each of them, and the kinds of organization selected to assume responsibility for operating individual national Antarctic programmes have varied accordingly. The vast differences in the organizational structures and scientific capabilities of the ATCP countries result in different national styles and standards in the implementation of the various Antarctic programmes. Since each country is responsible for the content and emphasis of its own programmes, national Antarctic operating agencies also have the task of translating general scientific priorities into national programmes that are compatible with the scientific priorities and capabilities of the participating countries.

Since all the ATCPs are members of SCAR, the categories under which SCAR places them based on the stage of development of their Antarctic research programmes is informative, given the length of time some of these individual programmes have had to develop.[4] The USA and Russia are the only two countries that fall under the level of 'special contributors'. SCAR regards countries at this level as those that demonstrate the importance of the Antarctic region aligned to their national priorities, despite the size of their programme. The second category of 'well-developed programmes' includes only 16 (Argentina, Australia, Brazil, China, France, Germany, India, Italy, Japan, South Korea, the Netherlands, New Zealand, Norway, South Africa, Spain, and the UK) of the 29 ATCPs that 'have a multi-disciplinary and productive Antarctic research community. This can include having a base in Antarctica, logistical resources, and an established community of scientists working together with the international community' (SCAR, www.scar.org/about-us/members/detailed-information/). The third category of 'initial-stage programmes' includes nine of the ATCPs (Belgium, Bulgaria, Chile, Ecuador, Finland, Peru, Poland, Sweden, Uruguay) that are still growing their national programmes and developing resources needed for sustained activities. The goal of each nation state in this category is to establish well-developed programmes over time. It is interesting to note that Belgium and Chile, two of the IGY countries and original signatories of the Antarctic Treaty, have been full members of SCAR since 1958 but are still considered as not having well-developed programmes. Non-IGY countries in this group received full membership of SCAR between 1978 and 2004, and yet their programmes continue to be at their initial stages. The remaining two ATCPs – Ukraine and the Czech Republic – remain 'associate members' of SCAR even though they gained consultative status in 2004 and 2014 respectively. Countries at this level acknowledge their interest in establishing an Antarctic research programme, but are seen not as having a large community of national Antarctic researchers for all areas of science. 'The goal for associate members is to move up to Initial-Stage Programmes in five or six years' (SCAR, www.scar.org/about-us/members/detailed-information/). As the Treaty does not provide a formal mechanism for reviewing whether an existing Consultative Party continues to meet the criterion of 'substantial scientific research activity', and since decision-making at the ATS is consensus-based, the possibility of a consultative party member consenting to its own exit from consultative status is improbable (Dudeney and Walton 2012). The role of science as a currency of influence in Antarctica in this instance perhaps plays a lesser role, thereby underscoring the overriding influence and power of politics over science.

Scope of national Antarctic programmes

Scientific research activities in Antarctica – not just during expeditions and field programmes, but which include and encompass the everyday life and running of an Antarctic base – must be self-contained. The particular geographical and climatic conditions in which programmes are implemented require appropriate logistics support – transportation must be readily available, while communication must be rapid, effective and dependable. Governments supporting Antarctic research must provide an entire community infrastructure – accommodation, laboratory facilities, food, clothing, recreation, transport, and support staff all need to be brought in from outside the southern circumpolar regions and maintained at considerable cost. The Antarctic climate and the remoteness of the continent from centres of population and transportation have a significant impact on the choice and cost of the logistic components (Hulsey et al. 1991: 3). This feature 'of total dependence by science on an imported, expensive logistic support' (Quilty 1988: 66) is particular to Antarctic science more than any other scientific activity. Commonly, logistics take up more than half the share of the funding of an Antarctic programme and so a critically important part of logistics capabilities is that they are enabled by policy and funding decisions made at the national level. This makes the interface between science and politics in the Antarctic profound due to the dependence of its science on the kind of enormous financial support that can only be agreed upon at government level and sustained through continued funding.

Different organizations may manage the different components of a country's Antarctic programme. This aspect makes national Antarctic operations multifaceted and complex systems. The systems employed by individual ATCPs for organizing and delivering on their programmes are influenced by each country's politics, science, culture and economy, which in turn influence scientific priorities and the availability of funds and equipment (Dey Nuttall 1997). Governments, however, delegate responsibility to a single national organization which assumes responsibility for a country's overall activity in the Antarctic.

A national Antarctic operating agency is invariably part of a much larger government organization or institution. Some have responsibilities in other areas, for example those that are region-specific like the Arctic, or science-specific such as oceans, environment or earth sciences, or politically-specific such as foreign affairs. The list of such operating agencies therefore includes a considerable range of government departments, ministries, national research institutes, and national funding bodies, all with widely differing terms of reference. The degree of executive control and oversight of Antarctic scientific activities in these organizations varies from one country to another.

The type of a national Antarctic agency can be determined by its range of activities. They could be an Antarctic-specific or bipolar or a non-polar agency. Antarctic-specific agencies, as the name would imply, are solely responsible for conducting Antarctic programmes (for example British Antarctic Survey, Australian Antarctic Division, French Polar Institute). Bipolar agencies such as the Norwegian Polar Institute, Alfred Wegener Institute for Polar and Marine Research (Germany), and National Institute of Polar Research (Japan) are active in both the Arctic and the Antarctic. Non-polar agencies are those that may have very broad mandates with responsibilities that go beyond Antarctic science, for example, the Netherlands Organisation for Scientific Research (NWO Dutch: Nederlandse Organisatie voor Wetenschappelijk Onderzoek) is the national research council of the Netherlands responsible for the country's Antarctic programme. These agencies can be either research organizations conducting in-house Antarctic research, or administrative organizations that coordinate the science and logistics arms of Antarctic programmes, or they can be responsible for both the research and administrative functions as well as being the source of funding of a national Antarctic programme.

Some ATCP countries describe their Antarctic organization or agency as a "programme" or "expedition". For example, Brazil calls its organization the Programa Antarctica (PROANTAR) or Brazilian Antarctic Program and Russia calls theirs the Russian Antarctic Expedition (see COMNAP members listed in Table 23.1). Such names should not be confused with similar ones adopted by other ATCPs to describe their Antarctic science activities rather than their organization that manages these activities. However, COMNAP refers to 'National Antarctic Programs' as those organizations that have responsibility for delivering and supporting scientific research in the Antarctic Treaty Area on behalf of their respective governments and in the spirit of the Antarctic Treaty. The complexity of the structure and composition of individual national agencies is reflected in COMNAP's constitution that further describes 'National Antarctic Program' members as such:

> Whenever this lead agency has a broader mission, only those parts of the organisation that have this national responsibility are considered part of the 'National Antarctic Program' member of COMNAP . . . Whenever this national responsibility is divided between several national organisations, the lead agency will, as appropriate, organise for relevant parts of the other national organisations to participate in the work of COMNAP under its authority and responsibility.[5]

Among the recommendations and other actions agreed by the Consultative Parties, the 13th Antarctic Treaty Consultative Parties Meeting's Recommendation XIII-1 (Brussels 1985)[6] distinguishes two kinds of contact points for the institutions or entities designated as national leads for their Antarctic programmes: one dealing with Treaty matters and one dealing with scientific data. In some countries the same organization has responsibility for both Treaty matters and scientific and operational matters. A majority of the countries active and operating in Antarctica identify their foreign ministries as the organization that is responsible for dealing with national obligations relating to Antarctic Treaty matters. Table 23.1 draws from the list available from the Antarctic Treaty Secretariat and COMNAP and illustrates the range of organizations that are responsible for the country's national Antarctic programme in the area of 'Treaty matters', 'Scientific and operational matters', 'National competent authorities' and those that are COMNAP members who are the officials responsible for carrying out national activity in the Antarctic on behalf of their governments all Parties to the Antarctic Treaty.

Table 23.1 Agencies responsible for national Antarctic programmes.

	ATCPs	Treaty matters*	Scientific and Operational matters*	National competent authorities*	COMNAP members
1	Argentina	Dirección Nacional de Política Exterior Antártica, Ministerio de Relaciones Exteriores y Culto	Dirección Nacional del Antártico, Ministerio de Relaciones Exteriores y Culto	Ministerio de Relaciones Exteriores y Culto	Dirección Nacional del Antártico
2	Australia	Department of Foreign Affairs and Trade of Australia/ Australian Antarctic Division	Australian Antarctic Division	Australian Antarctic Division	Australian Antarctic Division
3	Belgium	FPS Foreign Affairs, Foreign Trade and Development Cooperation	FPS Foreign Affairs, Foreign Trade and Development Cooperation/Federal Public Planning Service Science Policy	FPS Health, Food Chain Safety & Environment, DG Environment	Belgian Science Policy Office (BELSPO) and the Polar Secretariat
4	Brazil	Division for Sea, Antarctic and Outer Space Affairs, Ministry of Foreign Affairs	Division for Sea, Antarctic and Outer Space Affairs, Ministry of Foreign Affairs	Division for Sea, Antarctic and Outer Space Affairs, Ministry of Foreign Affairs	Brazilian Antarctic Program (PROANTAR)
5	Bulgaria	Ministry of Foreign Affairs of Bulgaria	Bulgarian Antarctic Institute	Contact point not identified.	Bulgarian Antarctic Institute
6	Chile	Dirección de Antártica, Ministry of Foreign Affairs	Instituto Antártico Chileno (INACH)	Dirección de Antártica, Ministry of Foreign Affairs	Instituto Antártico Chileno (INACH)
7	China	Chinese Arctic and Antarctic Administration/Ministry of Foreign Affairs of China	Chinese Arctic and Antarctic Administration	Chinese Arctic and Antarctic Administration	Chinese Arctic and Antarctic Administration; Polar Research Institute of China
8	Czech Republic	Ministry of Foreign Affairs of the Czech Republic	Ministry of Foreign Affairs of the Czech Republic	Ministry of Foreign Affairs of the Czech Republic	Masaryk University (a lead institution of the Czech Polar Research Centre)
9	Ecuador	Ministerio de Relaciones Exteriores y Movilidad Humana	Instituto Antártico Ecuatoriano (INAE)	Ministerio de Relaciones Exteriores y Movilidad Humana	Instituto Antártico Ecuatoriano (INAE)

10	Finland	Ministry for Foreign Affairs of Finland	Finnish Institute of Marine Research/ Ministry for Foreign Affairs of Finland	*Contact point not identified.*	Finnish Antarctic Research Program (FINNARP) is operated under the Finnish Meteorological Institute
11	France	Ministry of Foreign Affairs of France; French Southern and Antarctic Lands (TAAF); Ministry of Ecology, Sustainable Development, Transportation and Housing	French Polar Institute Paul Emile Victor (IPEV); Ministry of Foreign Affairs of France;	French Southern and Antarctic Lands (TAAF)	French Polar Institute Paul Emile Victor (IPEV)
12	Germany	Federal Foreign Office	Alfred Wegener Institute for Polar and Marine Research	Federal Environmental Agency of Germany	Alfred Wegener Institute for Polar and Marine Research
13	India	Ministry of Earth Sciences, Government of India	National Centre for Antarctic & Ocean Research (Ministry of Earth Sciences, Government of India)	National Centre for Antarctic & Ocean Research (Ministry of Earth Sciences, Government of India)	National Centre for Antarctic & Ocean Research (Ministry of Earth Sciences, Government of India)
14	Italy	Ministry of Foreign Affairs of Italy	Italian National Agency for New Technologies, Energy and the Environment (ENEA)	Ministry of Foreign Affairs of Italy	Programma Nazionale di Ricerche in Antartide (PNRA)
15	Japan	Ministry of Foreign Affairs of Japan; Ministry of Education, Culture, Sports, Science and Technology; Fisheries Agency of Japan; Ministry of the Environment of Japan	National Institute of Polar Research; Ministry of Foreign Affairs of Japan; Ministry of Education, Culture, Sports, Science and Technology; Fisheries Agency of Japan; Ministry of the Environment of Japan	Ministry of the Environment of Japan	National Institute of Polar Research
16	Republic of Korea	Ministry of Foreign Affairs of the Republic of Korea; Korea Polar Research Institute (KOPRI)	Korea Polar Research Institute (KOPRI); Ministry of Foreign Affairs of the Republic of Korea	*Contact point not identified.*	Korea Polar Research Institute (KOPRI)

(continued)

Table 23.1 (continued)

	ATCPs	Treaty matters*	Scientific and Operational matters*	National competent authorities*	COMNAP members
17	The Netherlands	Ministry of Foreign Affairs of the Kingdom of the Netherlands; Ministry of Infrastructure and Environment	Netherlands' Polar Research Programme, Nederlandse Organisatie voor Wetenschappelijk Onderzoek (NWO)	*Contact point not identified.*	Nederlandse Organisatie voor Wetenschappelijk Onderzoek (NWO)
18	New Zealand	Ministry of Foreign Affairs and Trade of New Zealand	Ministry of Foreign Affairs and Trade of New Zealand	Ministry of Foreign Affairs and Trade of New Zealand	Antarctic New Zealand
19	Norway	Royal Ministry of Foreign Affairs of Norway	Norwegian Polar Institute	Norwegian Polar Institute	Norwegian Polar Institute
20	Peru	Dirección de Asuntos Antárticos del Ministerio de RREE (Sovereignty Limits and Antarctic Affairs of the Ministry of Foreign Affairs)	Dirección de Asuntos Antárticos del Ministerio de RREE (Sovereignty Limits and Antarctic Affairs of the Ministry of Foreign Affairs)	Dirección de Asuntos Antárticos del Ministerio de RREE	Division of Antarctic Affairs under the Ministry of Foreign Affairs
21	Poland	Ministry of Foreign Affairs of the Republic of Poland	Institute of Biochemistry and Biophysics Polish Academy of Sciences	*Contact point not identified.*	Institute of Biochemistry and Biophysics Polish Academy of Sciences
22	Russian Federation	Ministry of Foreign Affairs of the Russian Federation; Russian Antarctic Expedition, Arctic and Antarctic Research Institute of Russia	Polar and Marine Division, Federal Service for Hydrometeorology and Environmental Monitoring (Roshydromet); Russian Antarctic Expedition, Arctic and Antarctic Research Institute of Russia	Polar and Marine Division, Federal Service for Hydrometeorology and Environmental Monitoring	Arctic and Antarctic Research Institute of Russia (AARI)/ Russian Antarctic Expedition (RAE)
23	South Africa	Department of Environmental Affairs; Ministerio de Economía, Industria y Competitividad	Department of Environmental Affairs	Department of Environmental Affairs	South African National Antarctic Programme (SANAP)

24 Spain	Ministry of Foreign Affairs and Cooperation; Spanish Ministry of Economy, Industry and Competitiveness	Spanish Ministry of Economy, Industry and Competitiveness; The Mediterranean Center for Marine and Environmental Research (CMIMA), Spanish National Research Council (CSIC)	Comité Polar Español; Spanish Ministry of Economy, Industry and Competitiveness	Comité Polar Español
25 Sweden	Ministry of Education and Research	Ministry of Education and Research	Ministry of Education and Research	Swedish Polar Research Secretariat under the Ministry of Education and Research
26 Ukraine	Ministry of Education and Science; Ukrainian National Antarctic Scientific Center	Ministry of Education and Science; Ukrainian National Antarctic Scientific Center	Ministry of Education and Science; Ukrainian National Antarctic Scientific Center	National Antarctic Scientific Center of Ukraine
27 UK	Polar Regions Department, Foreign and Commonwealth Office	British Antarctic Survey	Polar Regions Department, Foreign and Commonwealth Office	British Antarctic Survey
28 USA	Office of Ocean and Polar Affairs (OES/OPA), United States Department of State	Office of Ocean and Polar Affairs (OES/OPA), United States Department of State	Office of Ocean and Polar Affairs (OES/OPA), United States Department of State	US National Science Foundation (NSF) Office of Polar Programs
29 Uruguay	Ministry of Foreign Affairs of Uruguay; Antarctic Institute of Uruguay	Antarctic Institute of Uruguay; Ministry of Foreign Affairs of Uruguay	Antarctic Institute of Uruguay; Ministry of Foreign Affairs of Uruguay	Antarctic Institute of Uruguay

(*Source:* produced by Anita Dey Nuttall from a diverse range of information available from the Antarctic Treaty Secretariat* and the Council of Managers of National Antarctic Programs).

How has the science-politics interface in Antarctica changed?

While political factors continue to be essential to sustain national Antarctic programmes, particularly because of the large amount of public funding required for their operations, there appears to be a shift from the political importance of having a presence in Antarctica to defend territorial claims, or the non-recognition of such claims, to the political importance of ensuring that the quality of science in Antarctica is at a sufficiently high level to enable countries to remain at the forefront of decision-making. The influence of politics on science in Antarctica has intensified as a result of environmental pressures and the increasingly noticeable effects of climate change and marine and airborne contaminants. The function of science in the Antarctic has changed and evolved beyond being used only to support political ends like permanent occupation or economic activity, such as it did for whaling in the past. It is now a necessary tool to influence decisions on managing the protection of the environment and, indeed, to bring about measures for the conservation of ecosystems, wildlife and resources (the Ross Sea marine protected area being a notable recent example of an ecosystem-based approach). The current definition of political interests thus aligns more with the scientific agenda in the Antarctic now that environmental and climate change concerns are high on the global political agenda.

The mechanisms developed within the Antarctic Treaty System have underscored the need and demand for collaboration between countries, and nations look increasingly to ways of pooling resources and sharing logistics. By its very nature, the global relevance of Antarctic research encourages international co-operation and collaboration among countries with comparable economic and technological capabilities for launching and running joint research programmes. Other states may think there are more strategic reasons for co-operation, which arise from common resource and security issues. For example, Asia-Pacific states such as Japan, India, China and South Korea consider the Indian and Pacific oceans as strategically significant for living and non-living resources and sea-borne trade routes, and so the reasons for Antarctic research may be framed in this context. Yet another group of co-operating nations may emerge among the southern hemisphere nations – i.e., Argentina, Chile, Australia, New Zealand – which are directly affected by activities in Antarctica. National agencies will increasingly have to balance their national scientific priorities to fit with the expectations and demands of the three key bodies – the Committee for Environmental Protection of the Antarctic Treaty (CEP), the Scientific Committee on Antarctic Research (SCAR) and the Council of Managers of National Antarctic Programs (COMNAP).

Concluding remarks

Research activity in Antarctica since the IGY has resulted in considerable and sustained effort in scientific discovery, observation and monitoring (e.g., Summerhayes 2008). Long-term Antarctic programmes, though, have distinct phases with characteristics and programmatic priorities which change over time and which reflect changing political and scientific perspectives on Antarctica. While science continues to be seen, in some respects, as the political currency for countries to safeguard and indeed advance their strategic interests in Antarctica, the geopolitical significance of the continent has increased in recent decades (e.g., Dodds 2010; Hemmings 2009). Yet, despite discussions of the resource potential of Antarctica, the continent is unique in being protected by international agreements that privilege science, conservation and environmental governance over development.

The Antarctic Treaty has developed into a complex system of protocols and conventions which aim to limit the effects of human activities and regulate the extent of human impacts on the continent, yet Antarctica is an increasingly busy place with more scientific activity carried out each year and greater numbers of tourists visiting the region. Furthermore, the Antarctic Treaty and its associated instruments have not been concerned specifically with biprospecting, for example, which has been increasing in recent years (Hemmings 2009; Hughes and Bridge 2010; see also, the chapter by Sanjay Chaturvedi in this volume), or necessarily tested significantly in terms of how effective its institutions concerned with ecosystem-based management actually are (e.g., in the case of CCAMLR and Antarctic krill fishing; see Österblom and Olsson 2017). All of this increased interest in Antarctica poses challenges to how science is not only conducted within the Antarctic Treaty System through national Antarctic programmes, but how it is managed and what its priorities should be (Hemmings 2009). The history of the development of national Antarctic programmes illustrates how governments have responded in organizational terms, through their national Antarctic operating agencies, to the shifting physical and political challenges Antarctica presents. Nations may have different reasons for investing in and maintaining research programmes in Antarctica – there could be dominant political and strategic interests, or concerns with security and environmental protections – but science for the sake of conducting science and discovering new knowledge about the continent and its significance for the rest of the globe is still often the primary inspiration. Nonetheless, urgent global challenges such as climate change, rising sea levels, and threats to marine ecosystems will require nations active in Antarctic science to respond to them in new and innovative ways that transform how Antarctic science is organized, managed and delivered in the global interest.

Notes

1 Argentina, Australia, Belgium, Chile, France, Japan, New Zealand, Norway, South Africa, the UK, Russia (formerly USSR) and the USA. Seven signatory states (the UK, Argentina, Chile, Norway, France, Australia and New Zealand) of the Antarctic Treaty of 1959 claim territory there. Within their domestic jurisdictions these claimant states treat their territorial claims in Antarctica as part of their overall national territory, even though the governance of Antarctica stipulated by the Treaty does not recognize such claims nor deny them.
2 For more information on these instruments see: Secretariat of the Antarctic Treaty *Compilation of key documents of the Antarctic Treaty System* 2014 Secretariat of the Antarctic Treaty, Buenos Aires.
3 www.ats.aq/documents/ats/treaty_original.pdf Accessed 21 March 2017.
4 www.scar.org/about-us/members/detailed-information/ Accessed 1 November 2017.
5 www.comnap.aq/Shared%20Documents/comnap-constitution-adopted-04-july-2008.pdf Accessed15 August 2017.
6 www.ats.aq/devAS/info_measures_listitem.aspx?lang=e&id=144 Accessed 16 February 2017.

References

Beck, P.J. 1986. *The International Politics of Antarctica*. London and Sydney: Croom Helm.
Bos, A. 1991. 'Consultative status under the Antarctic Treaty: redefining the criteria?' in A. Jørgensen-Dahl and W. Østreng (eds) *The Antarctic Treaty System in World Politics*. New York: Palgrave Macmillan.
CCAMLR 2017. 'CCAMLR to create world's largest Marine Protected Area' www.ccamlr.org/en/news/2016/ccamlr-create-worlds-largest-marine-protected-area.
COMNAP 2017. *Antarctic Station Guide*. Christchurch, New Zealand: COMNAP Secretariat.
Dey Nuttall, A. 1997. 'Profiles of national Antarctic operating agencies of Antarctic Treaty Consultative Parties: an introductory study' *Polar Record* 33(185): 133–144.
Dodds, K. 2010. 'Governing Antarctica: contemporary challenges and the enduring legacy of the 1959 Antarctic Treaty' *Global Policy* 1(1): 108–115.
Dudeney, J.R. and D.W.H. Walton 2012. 'Leadership in politics and science within the Antarctic Treaty' *Polar Research* 31(1): 11075, DOI: 10.3402/polar.v31i0.11075.
Elliott, L.M. 1994. *International Environmental Politics: Protecting the Antarctic*. London: St. Martin's Press.

Fuchs, V.E. 1982. *Of Ice and Men: The Story of British Antarctic Survey* 1943–73. London: Anthony Nelson.

Gray A.D. and K.A. Hughes 2016. 'Demonstration of "substantial research activity" to acquire consultative status under the Antarctic Treaty' *Polar Research* 35:134061, DOI: 10.3402/polar.v35.34061.

Heap, J.A. 1983. 'Cooperation in the Antarctic: a quarter of a century's experience' in F.O. Vicuna (ed.) *Antarctic Resources Policy*. Cambridge, UK: Cambridge University Press.

Heap, J.A. 1990. 'Sovereignty as a source of stress' in R.A. Herr . (eds) *Antarctica's Future: Continuity or Change*. Hobart, Tasmania: Tasmanian Government Printing Office for the Australian Institute of International Affairs.

Hemmings, A.D. 2009. 'From the new geopolitics of resources to nanotechnology: emerging challenges of globalism in Antarctica' *Yearbook of Polar Law* 1: 55–72.

Hughes, K.A. and P.D. Bridge 2010. 'Potential impacts of Antarctic bioprospecting and associated commercial activities upon Antarctic science and scientists' *Ethics in Science and Environmental Politics* 10: 13–18.

Hulsey, J.L., P.A. Koushki, F.L. Bennett and J. Kelly 1991. 'Cold region logistics planning and management' *Journal of Cold Regions Engineering* 6(1): 1–11.

Kirwan, L.P. 1959. *The White Road: A Survey of Polar Exploration*. London: Hollis & Carter.

Österblom, H. and O. Olsson 2017. 'CCAMLR: an ecosystem approach to the Southern Ocean in the Anthropocene' in K. Dodds, A.D. Hemmings and P. Roberts (eds) *Handbook on the Politics of Antarctica*. Cheltenham, UK: Edward Elgar.

Pannatier, S. 1994. 'Acquisition of consultative status under the Antarctic Treaty' *Polar Record* 30: 123–129.

Quigg, P.W. 1983. *A Pole Apart*. London: McGraw-Hill Book Company.

Quilty, P.G. 1988. 'Cooperation in Antarctica in scientific and logistic matters: status and means of improvement' in Wolfrum, R. (ed.) *Antarctic Challenge III. Conflicting Interest, Co-operation, Environmental Protection, Economic Development*. Berlin: Duncker & Humblot.

Retamales, J. and M. Rogan-Finnemore 2011. 'The role of the Council of Managers of national Antarctic programs' in P.A. Berkman, M.A. Lang, D.W.H. Walton and O. Young (eds) *Science Diplomacy: Antarctica, Science, and the Governance of International Spaces*. Washington, DC: Smithsonian Institution Scholarly Press. 231–239.

SCAR Manual 1987. Cambridge, UK: Scientific Committee on Antarctic Research.

Sullivan, W. 1957. *Quest for a Continent*. New York: McGraw-Hill.

Sullivan, W. 1961. *Assault on the Unknown: The International Geophysical Year*. New York: McGraw-Hill

Summerhayes, C.P. 2008. 'International collaboration in Antarctica: the International Polar Years, the International Geophysical Year, and the Scientific Committee on Antarctic Research' *Polar Record* 44(231): 321–334.

Walton, D.W H. 2017. 'Achieving consultative party status' *Antarctic Science* 29(1): 1–1.

Walton, D.W.H. and Clarkson, P.D. (eds) 2011. *Science in the Snow*. Cambridge, UK: Scientific Committee on Antarctic Research.

24

Sustainable development and sustainability in Arctic political discourses[1]

Birger Poppel

The terms 'sustainable development' and 'sustainability'[2] have, since the World Commission on Environment and Development in 1987 published the report *Our Common Future*,[3] been a prominent part of the vocabulary used in political declarations, reports and other public documents, internationally and nationally, as well as in company strategies and prospects for economic activities – not least in relation to the Arctic region. More recent examples can be found in the World Economic Forum's Arctic Investment Protocol that "aspires to promote sustainable and equitable economic growth in the region" (WEF 2015) and the Arctic Economic Council[4] having the mission "To facilitate sustainable Arctic economic and business development" (Arctic Economic Council 2014). The frequent use in different arenas does not necessarily mean that there is a common understanding of these terms – far from it. Some of the different meanings and understandings will be recited and examined below.

This chapter will focus on the concepts 'sustainable development', 'sustainable' and 'sustainability',[5] the background, the significance and the actual application of these concepts in the Arctic strategies and policies of Arctic states and in the co-operation between Arctic states, primarily in the Arctic Council. Sustainable development has been a significant concept both as an overall vision and a strategic goal globally (most prominently in the joint efforts within the United Nations), regionally (for instance in the Arctic and in the Nordic region) as well as nationally (as the member states of the United Nations have agreed to implement the UN's 2030 sustainable development goals into sector- and geography-specific goals) (UN 2015a). This chapter will further search for potential inspiration, co-operation, and synergy between the different levels.

The ambition of the following analysis is not to make an in-depth and all-encompassing investigation into all aspects of the individual Arctic states' contributions to the Arctic Council's and the United Nations' sustainable development initiatives, but rather to get an overall idea of whether there is solid ground for an assumption that the Arctic states' participation in various international sustainable development initiatives inspires the Arctic countries' individual strategies and the jointly-developed Arctic Council strategies, and vice versa.

Before elaborating on the contemporary use of the terms and some actual contexts in which they are applied, it might, however, be useful to review the understanding and use of 'sustainable development' in the Brundtland Report, not least because 'sustainable development' is used frequently and in still more paradoxical connections such as when companies operating in the mineral, oil and gas sectors characterize their economic activities as sustainable,[6] which is of course contradicting the original meaning of sustainability, as the raw material is not replaced when extracted.[7]

Sustainable development and sustainability – origins and definitions of the concepts

As pointed out by both Owens (2003), Gad et al. (2016) and Søndergaard (2017), the concept of sustainability was not an invention of the World Commission on Environment and Development (most often named after the chairperson of the commission, the former Premier of Norway, Gro Harlem Brundtland) (WCED 1987). The concept had been part of both national and international agendas since the first international conference on the environment and development in Stockholm 1972 and the publication of *Limits to Growth* (Meadows et al. 1972), and was also used in the World Conservation Strategy, published in 1980 by the International Union of Conservation of Nature (IUCN 1980). It was, however, through the work, the focus and the recommendations of the commission in *Our Common Future* (WCED 1987) that 'sustainable development' more permanently entered the political discourse about economic, social and environmental development, and since the late 1980s has taken centre stage – not least in the work of the United Nations.

When a definition of 'sustainable development' is presented it is most often the following: "Sustainable development is development that meets the needs of the present without compromising the ability of future generations to meet their own needs" (WCED 1987, Chapter 2.IV.1, 40). Usually, however, the remainder of the paragraph in the Brundtland Report is omitted. It contains within it two key concepts:

- the concept of 'needs', in particular the essential needs of the world's poor, to which overriding priority should be given; and
- the idea of limitations imposed by the state of technology and social organization on the environment's ability to meet present and future needs. (ibid.)

This passage, which is usually not quoted, includes a focus on poverty and the distribution of income and wealth, as well as on key preconditions for sustainable development. Furthermore, the different preconditions for sustainable development listed and explained in the WCED report seldom – or at best implicitly – occur in Arctic debates and documents when 'sustainable development' is claimed.

A more thorough analysis of sustainable development as a strategic goal presupposes that the system of concepts and the preconditions listed in the Brundtland Report are applied.

The Brundtland Report sums up seven systemic processes of change required to pursue sustainable development:

> In its broadest sense, the strategy for sustainable development aims to promote harmony among human beings and between humanity and nature. In the specific context of the development and environment crises of the 1980s, which current national and international political and economic institutions have not and perhaps cannot overcome, the pursuit of sustainable development requires:
>
> - a *political system* that secures effective citizen participation in decision making
> - an *economic system* that is able to generate surpluses and technical knowledge on a self-reliant and sustained basis

- a *social system* that provides for solutions for the tensions arising from disharmonious development.
- a *production system* that respects the obligation to preserve the ecological base for development,
- a *technological system* that can search continuously for new solutions,
- an *international system* that fosters sustainable patterns of trade and finance,
- and
- an *administrative system* that is flexible and has the capacity for self-correction.

(WCED 1987, Chapter 2.IV.81, 74–75)

Furthermore, "These requirements are more in the nature of goals that should underlie national and international action on development. What matters is the sincerity with which these goals are pursued and the effectiveness with which departures from them are corrected" (WCED 1987, Chapter 2.IV.82, 75). One of the objectives of this chapter is to probe whether 'sustainable development' when applied in different Arctic contexts has this wider sense.

From environmental concerns to a broader understanding of sustainable development in the Arctic

Concern for the Arctic environment and the "special importance to the cooperation of the northern countries in environmental protection" was among the initiatives highlighted in Mikhail Gorbachev's (then Secretary of the Soviet Communist Party) 1987 Murmansk speech supporting the overall goal "Let the North Pole be a pole of peace" (Gorbachev 1987). Arctic environmental concerns also shaped the agenda for the so-called Rovaniemi Process that started in 1989 when Finland hosted a meeting in Rovaniemi between the Arctic countries to initiate international co-operation to protect the Arctic environment. The process gained momentum and was followed up by a conference in Rovaniemi in 1991, the first on the Arctic environment to include ministers of the Arctic countries as well as representatives of the Arctic indigenous peoples. The ministerial conference agreed upon an Arctic Environmental Protection Strategy (AEPS 1991) and thus paved the way for more permanent co-operation to protect the Arctic environment. The concerted efforts implementing the AEPS led to the development of a number of environmental working groups that are now working groups of the Arctic Council: Arctic Monitoring and Assessment Programme (AMAP), Conservation of Arctic Flora and Fauna (CAFF), Emergency Prevention, Preparedness and Response (EPPR), Protection of the Arctic Marine Environment (PAME).

The second ministerial conference was held in Nuuk, Greenland, in September 1993. The conference agreed upon the 'Declaration on Environment and Development in the Arctic' and noted "that in order to achieve sustainable development, environmental protection shall constitute an integral part of the development process and cannot be considered in isolation from it" (Ministry of Foreign Affairs 1993: 3). The Nuuk declaration recognized "the importance of applying the results of the United Nations Conference on the Environment and Development to the Arctic region" (ibid.) and further referred to the Arctic states' implementation of "relevant provisions of the Rio Declaration, Agenda 21 and the Forest Principles" (ibid.) through the AEPS. The foundation of the AEPS and the implementation of the strategy was thus seen as closely linked to the work of the United Nations.

The declaration further established the precursor for the Sustainable Development Working Group (SDWG) of the Arctic Council "reaffirming the commitment to sustainable development, including the sustainable use of renewable resources by indigenous peoples, and to that end agreeing to establish a Task Force for this purpose" (ibid.: 4).

Common interests among the Arctic states, an increasing international focus on environmental concerns, and the need for timely political initiatives – not least in the circumpolar region – seems to be a unifying principle from Gorbachev's Murmansk speech via the Rovaniemi Process, and the Arctic co-operation in the AEPS, to the inauguration of the high-level intergovernmental forum, the Arctic Council in Ottawa in 1996 (see also the chapter by Koivurova in this volume). In the first paragraph of the Declaration of the Establishment of the Arctic Council, it is stated that the Arctic Council (AC) is established to:

> [p]rovide a means for promoting cooperation, coordination and interaction among the Arctic States, with the involvement of Arctic indigenous communities and other Arctic inhabitants on common Arctic issues [a footnote explicitly excluded matters related to 'military security' as AC agenda items – BP], in particular issues of sustainable development and environmental protection in the Arctic.
>
> *(Arctic Council 1996)*

The focus on sustainable development was further emphasized by the AC at the Ministerial meeting in Iqaluit in Nunavut in 1998, which established the Arctic Council's Sustainable Development Program, agreed the Terms of Reference (ToR) as the founding document for the programme, and set up the SDWG. These decisions were followed up by the Ministerial meeting in Barrow, Alaska in 2000 as the ministers agreed to a Sustainable Development Framework Document (SDFD) (Arctic Council 2000). As stated in the Terms of Reference, the goal of the sustainable development programme was to:

> [p]ropose and adopt steps to be taken by the Arctic States to advance sustainable development in the Arctic, including opportunities to protect and enhance the environment and the economies, culture and health of indigenous communities and of other inhabitants of the Arctic, as well as to improve the environmental, economic and social conditions of Arctic communities as a whole.
>
> *(Arctic Council 1998)*

The SDFD linked the Sustainable Development Program of the Arctic Council to the 1992 UN Rio Conference on Environment and Development, stating that "The Council's Sustainable Development Program has the objective of addressing the special circumstances of the Arctic in that context" (Arctic Council 1998).

In the Sustainable Development Framework Document, the Arctic Council thus defined sustainable development taking its starting point as the conventional understanding of the concept – i.e. the short definition in the Brundtland Report – focusing on the environmental, economic and social aspects, but it went a little further, explicitly expanding the definition and stating that "Economic, social and *cultural* [author's emphasis] developments are, along with environmental protection, interdependent and mutually reinforcing aspects of Sustainable Development and are all part of the Council's focus in this regard" (ibid.). The Framework Document also included capacity building as "a necessary element for achievement of Sustainable Development" and a number of other 'subject areas' such as

- Health issues and the well-being of peoples in the Arctic. Prevention and control of disease and injuries, as well as the long-term monitoring of the impact of pollution and climate change
- Sustainable economic activities and increasing community prosperity

- Education and cultural heritage
- Children and youth
- Management of natural, including living, resources
- Infrastructure development

(ibid.)

Incorporating these elements in the Framework Document emphasizes an understanding of sustainable development that – like the Brundtland Report – also focused on preconditions and included, for instance, cultural heritage. While the Brundtland Report argued for sustainable development on a systemic level (see above) the SDFD primarily focuses on broader policy areas/domains.

'Sustainable development' in national Arctic strategies and the international efforts of the Arctic states

All Arctic states have developed national Arctic strategies, and these all include strategic considerations, reflections and goals related to sustainable development. Furthermore, when it comes to establishing strategies, goals and measures/indicators to assess sustainable development, all eight Arctic Council member states have, individually, been engaged at the national level, as well as regionally[8] and internationally in processes such as those guided and managed by the United Nations. Most recently, the UN General Assembly on 25 September 2015 unanimously agreed upon the 2030 sustainable indicators (United Nations 2015b)[9] and almost all of the Arctic states have developed national indicators complying with the overall UN goals (see Table 24.1 below).

Given the contents of the Arctic Council's Sustainable Development Framework Document, not least the explicit reference to UN initiatives, a fair assumption seems to be that these often parallel initiatives and processes focussing on environmental concerns might have created synergy and inspired the Arctic countries' efforts under the auspices of the Arctic Council in developing sustainable development mechanisms, indicators and implementation plans for the Arctic region as has been done in the UN since the Rio Declaration (United Nations 1992).

Furthermore, as all Arctic states have agreed that one of the top priorities of the Arctic Council as such and the overall objective of one of the Arctic Council's working groups is sustainable development, it might also be expected that sustainable development in various aspects (including preconditions for sustainable development) would be high on the agenda of the national Arctic strategies.

To make an in-depth, comparative analysis of sustainable development visions, strategies, implementation plans and assessments of the eight Arctic states, and how they include the Arctic regional co-operation in national, other regional or global collaborative efforts, including not least the United Nations development strategies, would demand thorough studies into, primarily but not only, the documents below and the processes they are products of:

- Arctic Council member states' Arctic and Northern strategies and policies (Table 24.1 below lists documents in this category);
- Arctic Council Ministerial Meeting resolutions as well as Senior Arctic Officials' reports to the Ministerial Meetings;
- United Nations development goals (including annual reports as well as summary and evaluation reports) such as 'Agenda 21 goals' (United Nations 1992), 'Millennium Development Goals' (United Nations 2000) and the UN 'the 2030 Agenda for Sustainable Development'

(United Nations 2015a) also called 'the 2030 Goals' and the processes applied – as these goals and indicators condenses and represents the UN understanding of sustainable development;

- Arctic states' reports on handling, implementation, and following up on the UN goals (Table 24.1 below lists the documents in this category).
- Sustainable development strategies and strategies focussing on the Arctic developed by regional associations such as the Nordic Council of Ministers that developed the council's first set of sustainable development indicators in 2001 as a follow up to the council's Nordic Strategy – Sustainable Development – New Bearings for the Nordic Region to assess the Nordic countries' performance (see e.g. Nordic Council of Ministers 2002; 2013), and the European Union (European Union 2008 and 2016).

Such an analysis is beyond the scope of this chapter. It has, however, been the ambition of the research behind this chapter to probe potential continuity and synergy between sustainable development initiatives in different fora including the Arctic Council and wording and initiatives in key documents focusing on the Arctic. For that reason, the Arctic strategies of the eight Arctic states as well as a number of national, sustainable development documents including the most recent national follow-up documents to the UN 2030 agenda have been selected and compared. Finally, the most recent Arctic Council declarations from the Ministerial Meetings in Kiruna, 2013 (Arctic Council 2013a, 2013b, 2013c); Iqaluit, 2015 (Arctic Council 2015) and Fairbanks 2017 (Arctic Council 2017a) are included in the comparisons.

The Arctic Council's focus on sustainable development

In 2016, under the heading '20 Years of Arctic Council', the Arctic Council official website (www.arctic-council.org) presented key achievements of its work over two decades, including the following:

> During its first 20 years, the Arctic Council focused much of its work on issues of sustainable development and environmental protection in the Arctic. Since its establishment, it has produced many landmark studies on topics important in this unique region, including climate change, environmental pollutants, shipping, tourism, safety and search-and-rescue, biodiversity of flora and fauna, oil pollution response, human health, indigenous languages, and much more.
>
> *(Arctic Council 2016)*[10]

More than 300 research-based reports and publications have been published by the working groups mandated by the Arctic Council (ibid.)[11] and several of these have not only provided new insights but have also been agenda setting outside the Arctic. The Arctic Climate Impact Assessment report (ACIA 2005) and the Arctic Biodiversity Assessment (CAFF 2013) are prominent examples.

Every Arctic Council Ministerial concludes in a declaration (named by the city hosting the Ministerial) reporting progress, stating consensus about key Arctic challenges and the role of the Arctic Council, including working group tasks for the next chairmanship's period. Environmental protection and sustainable development have been addressed in all declarations. Arctic Council decisions are based on the principle of consensus, and when it comes to legally-binding agreements, the achievements are modest. The first binding agreement among the Arctic states – on search and rescue – was signed in 2011[12] and the second, addressing oil pollution preparedness (signed in 2013),[13] was the first legally-binding agreement addressing environmental protection and sustainable development.[14]

Table 24.1 Arctic/Northern strategies and policies; United Nations' global initiatives and national reports developing the UN initiatives in the individual Arctic countries (table drawn up by author).

	Denmark/Greenland/Faroe Islands	Iceland	Norway	Sweden	Finland	Russian Federation[1]	USA/Alaska	Canada	European Union
Arctic/Northern strategy(ies)/policy(ies)	2011: Kingdom of Denmark Strategy for the Arctic 2011–2020	2011: A Parliamentary Resolution on Iceland's Arctic Policy	2006: The Norwegian Government's High North Strategy; 2014: Norway's Arctic Policy for 2014 and Beyond	2011: Sweden's Strategy for the Arctic Region	2010: Finland's Strategy for the Arctic Region; 2013: Finland's Strategy for the Arctic Region 2013	2008: Policy of the Russian Federation the Arctic to 2020; 2013: Strategy for the Development of the Arctic Zone of the Russian Federation	2013: National Strategy for the Arctic Region; 2016: US-Canada Joint Statement on Climate, Energy and Arctic Leadership	2009: Canada's Northern Strategy; 2016: US-Canada Joint Statement on Climate, Energy and Arctic Leadership	2008: The European Union and the Arctic; 2016: An Integrated European Union Policy for the Arctic
UN initiatives: Agenda 21[2] 2000–2015: Millennium Development Goals 2016–2030: Sustainable Development Goals National strategies, plans, assessments & evaluations	2013/14: Et bæredygtigt Danmark – Udvikling i balance[3] (A sustainable Denmark – A balanced development) 2017: The world 2030. Denmark's strategy for development cooperation and humanitarian action	2002: Welfare for the Future: Iceland's National Strategy for Sustainable Development 2005 & 2009: (next four year priorities); 2010–2013	2016: The 2030 Agenda: A Roadmap for National Action and Global Partnership 2016: Norway's follow-up of Agenda 2030 and the Sustainable Development Goals	2015: Sustainable Development Goals for Sweden: Insights on Setting a National Agenda (Stockholm Environment Institute)	2016: The Finland we want by 2050 – Society's commitment to Sustainable Development	2012: Report on implementing the principles of sustainable development in the Russian Federation. . . . Preparing for 'Rio +20'[4]	2010: The United States of America. National Report[5]	2015: Progress Report of the Federal Sustainable Development Strategy 2015: Planning for a Sustainable Future: A Federal Sustainable Development Strategy for Canada 2016–2019[6]	

Notes:

1 On 30 November 2016 the Russian Federation released a 'Foreign Policy Concept of the Russian Federation' aiming at a large number of themes such as sustainable development and the regions of the world including the Arctic (The Ministry of Foreign Affairs of the Russian Federation 2016).

2 Follow-up of the 1992 UNCED (United Nations Conference on Environment and Development) in Rio de Janeiro (United Nations 1992).

3 Denmark's strategy for sustainability including 24 goals (Danish Government 2014).

4 The report refers to an 'Arctic clean-up programme' (Russian Federation 2012).

5 This very detailed report with numerous links to relevant North American and international web sites on environmental and sustainability issues has no references to the Arctic Council's working groups' activities (Government of the United States 2010).

6 The report (Government of Canada 2015b), covering a broad variety of environmental and sustainability issues, has just one Arctic related reference, Health Canada's website focusing on 'First Nations and Inuit Health': www.hc-sc.gc.ca/fniah-spnia/promotion/public-publique/water-eau-eng.php, accessed July 7, 2017.

The Arctic Council Ministerial in Kiruna, Sweden in 2013 concluded the Swedish Arctic Council chairmanship. At this Ministerial the Arctic states and the indigenous peoples' organizations 'set out a vision for the future' of the Arctic region. In seven brief paragraphs, key messages for future Arctic co-operation were presented. One of the paragraphs titled 'a prosperous Arctic' links "sustainable development almost entirely to the success of economic activities" (Arctic Council 2013b: 3). Another paragraph recognizes "the critical importance of healthy environments to sustainable communities" (ibid.: 4). Finally, the Arctic states confirm that they "remain committed to managing the region with an ecosystem-based approach which balances conservation and sustainable use of the environment" (ibid.). The 'Vision for the Arctic' does not refer to other international initiatives related to sustainable development. Only a general statement – "we are committed to demonstrating leadership in regional and global forums to address challenges affecting our homes" (Arctic Council 2013b: 3) indicates future outreach for co-operation as a possibility.

At the Ministerial meeting in Iqaluit in 2015, which concluded the Canadian AC-chairmanship, the Arctic Council reaffirmed:

> [o]ur commitment to sustainable development in the Arctic region, including economic and social development, improved health conditions and cultural well-being, and our commitment to the protection of the Arctic environment, including the health of Arctic ecosystems, conservation of biodiversity in the Arctic and sustainable use of natural resources, as stated in the Ottawa Declaration of 1996.
>
> *(Arctic Council 2015: 4)*

The Iqaluit Declaration also reaffirms the "Arctic States' commitment to work together and with partners . . . towards . . . climate agreement in Paris in December 2015" and "to work within and beyond the United Nations Framework Convention on Climate Change" (ibid.: 10), and urge to ratify the Minamata Convention on Mercury (ibid.: 9), and includes a positive mention of the International Maritime Organization's "progress made on the Polar Code" (ibid.: 10). The Iqaluit Declaration, however, includes no references to other international collaborative efforts such as the UN Millennium Development Goals 2001–2015 or the UN Sustainable Development Goals (that would be decided later in 2015) for the period 2016–2030. This might make one wonder why, since a number of the challenges and topics included in UN endeavours also face the Arctic region and its people (such as poverty, and lack of access to drinking water in some regions).

Contrary to the Iqaluit Declaration, that made no mention of the global sustainable development initiatives under the auspices of the UN, the 2017 Fairbanks Declaration does relate to the UN 2030 goals by "Reaffirming the United Nations Sustainable Development Goals and the need for their realization by 2030", (Arctic Council 2017a: 3). The Fairbanks Declaration further emphasizes its commitment to sustainable development by "Reaffirming our commitment to the well-being of the inhabitants of the Arctic, to sustainable development and to the protection of the Arctic environment" (ibid.: 2) and to "Reaffirm the role of the Arctic Council in promoting sustainable development through harmonizing its three core pillars in an integrated way: economic development, social development and environmental protection" (ibid.: 6).

Arctic Council member states' Arctic and Northern strategies and policies and the application of 'sustainable development', 'sustainability' and 'variations of the concepts' to the strategies/policies – some examples

Whereas almost all Arctic strategies and Northern policies[15] contain paragraphs focusing particularly on the environment and environmental protection (Government of Canada

2009: 24; Norwegian Ministry of Foreign Affairs 2009: 45; Prime Minister's Office, Finland 2010: 13; Althingi 2011: 9; Ministry of Foreign Affairs (Denmark) 2011: 43; Regeringskansliet 2011: 23, 27; The White House 2013: 9) 'sustainable development' is, however, not applied as an overarching concept in any of the strategies/policies. Neither are the concepts defined in the strategies, but which seem to be used interchangeably. Most of the countries subscribe, however, to a more general use of the concept as in the paragraph 'General Provisions' of the Russian Federation's strategy: "The strategy defines the basic mechanisms, ways and means to achieve the strategic goals and priorities for the sustainable development of the arctic Zone of the Russian Federation and the national security" (Putin 2013).[16]

The most common applications of the concepts 'sustainable development', 'sustainability' and 'sustainable' are not in their broader sense, including the cultural component, and also not in the sense that is generally used in the Arctic Council,[17] but are most often used in connection with the utilization of resources as also pointed to by Heininen: "Indeed, in many cases the rhetoric generally indicates that economic development, including activities, means "sustainable use of natural resources" (Heininen 2012:73; 80). All Arctic states and not least the Kingdom of Denmark, Norway, Sweden and Finland include 'sustainable' in the characterization of the preferred use of both renewable and non-renewable resources. A few examples might be illustrative (author's emphasis):

- "The huge economic potential in the Arctic must be realized while appreciating its human impact, i.e. the economic and social integrations of the population and with sensitivity to environmental concerns, thereby creating a healthy productive and *self-sustaining community*" (Ministry of Foreign Affairs (Denmark) 2011: 24);
- "Ecosystem based *management of marine resources* based on the principle of conservation and *sustainable use*" (Regeringskansliet 2011: 28);
- "Further extraction of petroleum should be done sustainably. Sweden is also striving for *environmentally sustainable use of the forest* in the Arctic" (Regeringskansliet 2011: 32); as well as "*sustainable* land transport" (Regeringskansliet 2011: 33);
- "Management of the living marine resources is to be based on the rights and duties that follow from the Law of the Sea, and the principle of *optimal utilisation of these resources within a sustainable framework*" (Norwegian Ministry of Foreign Affairs 2009: 32);
- "The resources in the Barents Sea could provide long-term secure *energy supply* to the markets in Europe and the US *within an environmentally sustainable framework*" (Norwegian Ministry of Foreign Affairs 2009: 55);
- "In addition to on-shore exploration and development there is renewed interest in the off-shore, including a new era of oil and gas exploration in the deeper waters of the Beaufort Sea. Canada will continue to support the *sustainable development of these strategic resource endowments*" (Government of Canada 2009: 16);
- "Use the renewable natural resources in the Arctic *in a sustainable way*" (Prime Minister's Office, Finland 2013: 53);
- "Develop infrastructure and services using the resources of Finnish companies to support mining operations *consistent with sustainable development*" (Prime Minister's Office, Finland 2013: 54);
- "Promote a *sustainable mining industry* in accordance with the action plan for a sustainable mining industry" (Prime Minister's Office, Finland 2013: 54);
- "Increase and renew the tourist industry to bring well-being to the region *in accordance with the principles of sustainability*" (Prime Minister's Office, Finland 2013: 55).

'Sustainability' and 'sustainable development' are further used in a number of relations including 'ecosystems':

- "Ecosystem based *management of marine resources* based on the principle of conservation and *sustainable use*" (Regeringskansliet 2011: 28);
- "Supporting actions will promote healthy, *sustainable* and resilient[18] *ecosystems* over the long term" (The White House 2013: 7);

and 'indigenous peoples':

- "*sustainable Development of Indigenous Peoples of the North, Siberia and Far East of the Russian Federation*" (Putin 2013: 18);
- "*sustainable socioeconomic development*" (Putin 2013: 16).

The Arctic states' references to international co-operation – especially United Nations global efforts towards a sustainable development

All Arctic states refer in their Arctic strategies and policies to international co-operation in general and most of the states refer also to the Arctic Council as the key instrument and vehicle for Arctic co-operation (see for instance Government of Canada 2009: 33; Regeringskansliet 2011: 18; Putin 2013: 4; The White House 2013: 9; Russian Federation 2016: 22). Most of the states include the United Nations Convention on the Law of the Sea, or UNCLOS (e.g. Government of Canada 2009: 33; Ministry of Foreign Affairs (Denmark) 2011: 13; Prime Minister's Office, Finland 2013: 44, 46), the International Maritime Organization, IMO (e.g. Althingi 2011: 9; Regeringskansliet 2011: 30) as well as the Kyoto Protocol and the UN's Climate Change Convention, UNFCCC (e.g. Norwegian Ministry of Foreign Affairs 2006: 46; Althingi 2011: 9; Ministry of Foreign Affairs (Denmark) 2011: 49); Indigenous peoples' rights and the United Nations Permanent Forum on Indigenous issues are also mentioned in some of the strategies as important issues in international co-operation (e.g. Prime Minister's Office, Finland 2010: 8, 2013: 50; Ministry of Foreign Affairs (Denmark) 2011: 50).

All the Nordic states emphasize Nordic collaboration (see for instance Prime Minister's Office, Finland 2010: 9, 19; Regeringskansliet 2011: 20), and it is important to recollect and stress the collective experience acquired in the endeavours of the Nordic states to develop Nordic visions and strategies concerning sustainable development,[19] including developing sustainable development indicators (Nordic Council of Ministers 2013) and backing up the efforts with substantial funding.[20]

The eight Arctic states have developed their Arctic strategies and policies very differently – methodologically, on issues and themes, and how Arctic objectives are prioritised.[21] This also applies to the way international collaboration is included in the strategies. The Russian Federation represents one approach in its strategy as it emphasizes international co-operation but only in general terms. Finland, on the other hand, in both the 2010 strategy (Prime Minister's Office, Finland 2010) and the 2013 strategy (Prime Minister's Office, Finland 2013), meticulously reviews Arctic relevant international cooperative organizations, institutions and key fora, such as the Arctic Council, the Barents Euro-Arctic Council, the Barents Regional Council, the Nordic Council of Ministers and the European Union, as well as issues of relevance for Arctic development, such as conventions on the environment (e.g. Prime Minister's Office, Finland 2010: 76). Finland's strategy also mentions the United Nations and UN activities including efforts in relation to sustainable development:

The United Nations (UN) and various UN bodies promote international cooperation for instance in the following sectors important for the Arctic Region: maritime law; human rights and the rights of indigenous peoples; sustainable development; environmental issues; and climate change.

(Prime Minister's Office, Finland 2010:10)

In the most recent Finnish Arctic Strategy from 2013 the "international cooperation in the Arctic" is again emphasised and declared as one of "the four pillars of policy outlined by the Government" (Prime Minister's Office, Finland 2013: 7).

United Nations sustainable development activities and the Arctic states' involvement in UN global efforts towards sustainable development

In 1992 the United Nations held a Conference on Environment and Development (the Earth Summit) in Rio de Janeiro in Brazil. The conference resulted in a number of far-reaching agreements (such as the UN Framework Convention on Climate Change, UNFCCC) and a United Nations action plan connected to sustainable development. The Agenda 21 action plan was non-binding but engaged governments, municipalities, and civil society over the next decennia in activities focusing on its implementation. Follow-up activities at a global level were arranged in 1997 (the so-called Rio +5); in 2002 with the World Summit on Sustainable Development in Johannesburg (Rio +10); and in 2012 in Rio de Janeiro with the United Nations Conference on Sustainable Development (Rio +20) (Dodds et al. 2012). Based on the experiences gathered and shared at a number of conferences on environmental and human development, the UN launched eight Millennium Development Goals (MDG) in 2000:

Goal 1: Eradicating extreme poverty and hunger

Goal 2: Achieving universal primary education

Goal 3: Promoting gender equality and empower women

Goal 4: Reducing child mortality

Goal 5: Improving maternal health

Goal 6: Combating HIV/AIDS, malaria and other diseases

Goal 7: Ensuring environmental sustainability

Goal 8: Develop a global partnership for development

(www.undp.org)

Indicators were set for each of the goals and an assessment procedure was established to assess the progress and to finally review the results in 2015. The United Nations Millennium Development Goals Report (2015) documents that improvements have been made in all goal areas and not least in reducing the number of people living in extreme poverty (United Nations 2015b). The UN MDG Report also concluded that despite the success and the achievements of the fifteen years of work with the Millennium Development Goals a large number of people still live in poverty and huge health problems and other challenges remain. This awareness resulted in a UN General Assembly decision to build on the MDG momentum and continue with a post-2015 development agenda to improve the lives of people – quoting the UN General Secretary Ban-Ki Moon from the introduction to the MDG Report:

> We need to tackle root causes and do more to integrate the economic, social and environmental dimensions of sustainable development. The emerging post-2015 development agenda, including the set of Sustainable Development Goals, strives to reflect these lessons, build on our successes and put all countries, together, firmly on track towards a more prosperous, sustainable and equitable world.
>
> *(United Nations 2015b: 3)*

The post-2015 development agenda includes the following seventeen Sustainable Development Goals (SDG) (encompassing a total of 169 target indicators):

1 No poverty
2 Zero Hunger and food security
3 Good health and well-being
4 Quality education
5 Gender equality and women's empowerment
6 Clean water and sanitation
7 Affordable and clean energy
8 Decent work and economic growth
9 Industry, innovation and infrastructure
10 Reduced inequalities
11 Sustainable cities and communities
12 Responsible consumption and production
13 Climate action
14 Life below water
15 Life on land
16 Peace, justice and strong institutions
17 Partnerships for the goals

(www.undp.org)

The goals address both the most basic needs embedded in economic, social and environmental aspects of sustainability, and systemic preconditions such as political, economic, social and administrative systems and cultural dimensions of sustainable development.

Strategies and policies of the Arctic states – economic (and sustainable?) development

All Arctic states have, as UN member states, participated in UN conferences and processes as well as in the decisions on the United Nations development goals: Agenda 21 goals, Millennium Development Goals and the UN Sustainable Development Goals. Furthermore, all Arctic states have, though in somewhat different ways, implemented UN plans by developing national sustainable development strategies and goal-settings as well as national follow-up initiatives, including reporting procedures. As reflected in Table 24.1, the documents may have different titles, but they are all focused on sustainable development goals in national settings.

To realise the ambition of getting a rough idea (and sketch potential research questions for further analyses) about the overall coherence in the Arctic states' approach to 'sustainable development' as well as actual and potential synergy between sustainable development initiatives in different fora, a number of sustainable development documents have been selected and reviewed to probe both 'overall coherence' as well as 'actual and potential synergy between sustainable development initiatives in different fora'.[22] The documents on national sustainable development goals (including implementation plans of UN SDG) as well as a number of Arctic states reports to the United Nations Commission on Sustainable Development have been reviewed to examine actual and potential synergy between sustainable development initiatives in different fora.[23]

The overall impression from reviewing the documents and reports is that, despite differences in some of the approaches, experiences based on the Arctic states activities in the Arctic, including Arctic co-operation on sustainable development, are absent. The report from Sweden to the UN Commission on Sustainable Development serves as an example. It presents a thorough historical review focussing on Sweden's role nationally, regionally (through the Nordic Council of Ministers and Baltic Sea co-operation) and globally (through the United Nations) but does not refer to Arctic Council co-operation or initiatives through its working groups (Swedish Government 2010).

Another – and more recent – example is 'A sustainable Denmark – a balanced development', a Danish government report from 2014 on Denmark's international co-operation with a sustainable development focus. The European Union, OECD and United Nations are mentioned whereas there is no mention on Denmark's activities focusing on co-operation related to sustainable development within the Arctic Council.

Summing up: Arctic and global frameworks for developing and implementing sustainable development initiatives

The Arctic Council was established with an aspiration towards developing co-operation between the Arctic states, focusing primarily on economic, social and environmental aspects of sustainable development, seemingly with a unifying principle from Gorbachev's Murmansk speech in 1987 via the Rovaniemi Process and the Arctic Environmental Protection Strategy of 2001, to the establishment of the Arctic Council in 1996, and the activities of the Arctic Council working groups including the SDWG.

It has been a common feature in the work of the Arctic Council, especially that of its working groups and not least the SDWG, that the concept of sustainable development has been applied in a broad sense, often including cultural aspects. Further, and in continuation of the SDWG Terms of References, the focus has been on projects rather than on establishing goals and selecting indicators (neither for the entire circumpolar Arctic nor the individual Arctic states and Arctic regions) to apply developmental assessments.

The increased focus on sustainable development in the Arctic has developed parallel to the endeavours of the United Nations to gather the global community's environmental and other concerns related to sustainable development. Using the work of the Brundtland Report *Our Common Future* as a point of departure, the UN has created platforms and joint initiatives such as Agenda 21; the Millennium Development Goals (MDG) 2000–2015 and, most recently, the Sustainable Development Goals (SDG) 2016–2030. Targets and sets of indicators were part of the MDG and SDG initiatives from the very beginning. The rationale has been to form the basis for measuring both total global developments and assessments between countries and of individual countries' development within the period of the initiatives. At the end of the MDG-period, in 2015, it was thus possible to draw the overall conclusion that the targets were met and also, make more detailed evaluations of differences between and within countries. Furthermore, the final levels reached in 2015 became the point of departure for the next global campaign of the United Nations, the Sustainable Development Goals 2016–2030.

The eight Arctic states – with the representatives of the indigenous peoples of the Arctic – debate and decide jointly in the Arctic Council on sustainable development issues and a number of other issues of mutual interest. The Arctic states also operate as individual nation states in the UN and in other fora, not least at a regional level such as the Nordic Council of Ministers and the Barents Regional Council. At the same time, all Arctic states have established national profiles on Arctic issues in national strategies and policies including issues such as sustainable development.

Some of the Arctic/Northern strategies and policies of the eight Arctic states have been analysed thoroughly from a comparative perspective (see e.g. Heininen 2012) whereas there has been less focus on the individual Arctic countries' policies and specific policy issues in different international fora and arenas, and to which degree these policies inform other policy areas in the country in question, and whether they are consistent with and include perceptions and experiences from other, relevant fields. To which degree, as an example, are a country's environmental concerns and commitment to sustainable development reflected in its economic strategies (including exploitation of renewable and non-renewable resources such as oil, gas and minerals) and how are they linked to the systemic preconditions of the Brundtland Report? And further, how are the Arctic states applying the goals of the United Nations to their own Arctic regions? Are there 'Sustainable Development Goals' for the different parts of the Circumpolar North and the Arctic as such? If not, why? Are the efforts of the UN in relation to sustainable development reflected in the work of the Arctic Council? If not, why?

Attempting to answer some of these questions, key documents focusing on the terms 'sustainable development' and 'sustainability' have been assessed for the purposes of this chapter. The documents include reports and decisions of the Arctic Council – including the declarations of the biannual Ministerials, the Arctic countries' Arctic strategies and policies, and Arctic states national reports related to activities following up on UN sustainable development initiatives. Based on a tentative assessment there does not seem to be a close coordination between the sustainable development policies of the Arctic states in relation to the Arctic region nor in relation to the United Nations. Further and more in-depth research is needed to further investigate this apparent lack of coherence and synergy and thus answer 'why?' – including the obvious question to which degree 'silo-thinking' in the administrations of the Arctic states when 'Arctic' is involved plays a role.

Summing up: how the Arctic Council copes with sustainable development issues

The establishment of the Arctic Council and the co-operation within it, the inclusion of the indigenous peoples of the Arctic and the openness to observers, is in many ways unique. Through the council's support to and organization of projects in the SDWG and other Arctic Council working groups (not least AMAP and CAFF) the Arctic states have contributed to agenda setting and debates – both in and outside the Arctic – about climate change, pollution and other environmental problems in the Arctic, as well as the health and well-being of indigenous peoples and other Arctic residents. Because of the Arctic Council's primary focus on projects as a key instrument to deal with important issues and challenges, documenting changes based on focused political initiatives is much harder compared to, for example, the way the United Nations has been working with goals, targets and indicators in the Millennium Development Goals programme and has set up the programme for Sustainable Development Goals. The lack of quantitatively-defined political goals not only characterizes the Arctic Council's declarations and visions but also the strategies and policies of the individual Arctic states. It might, however, be considered ironic – or at least somewhat surprising – that the Arctic Council has not set measurable goals. Not least because all Arctic states as UN members are part of UN initiatives as the abovementioned and because a number of Artic Council supported projects such as the Arctic Human Development Report (AHDR I and II) (AHDR 2004, 2014), the Arctic Social Indicators (ASI I and II) (ASI 2010, 2015) and the Survey of Living Conditions in the Arctic, SLiCA (Poppel et al. 2007; Kruse et al. 2008; Poppel 2015) have developed and applied a number of overall goals and indicators in analyses of the Arctic and its societies, peoples and cultures. Some are closely related to the Arctic and Arctic ways of life but most of them are generally used all over the world (as for example most of the health indicators). The Arctic Council has welcomed the projects mentioned, their development and application of indicators but so far neither the Arctic Council's sustainable development initiatives nor the Arctic states' national strategies and policies for the Arctic have applied the indicators systematically.

. . . and some future perspectives

In the resolution 'Transforming our world: the 2030 Agenda for Sustainable Development' adopted by the United Nations in September 2015, paragraph 55 states that:

> Targets are defined as aspirational and global, with each Government setting its own national targets guided by the global level of ambition but taking into account national circumstances. Each Government will also decide how these aspirational and global targets should be incorporated into national planning processes, policies and strategies. It is important to recognize the link between sustainable development and other relevant on-going processes in the economic, social and environmental fields.
>
> *(UN 2015a: 13)*[24]

This excerpt is part of the guidelines that the Arctic states individually implement when setting up national goals, assessment plans and policies to meet the UN SDGs. As the Fairbanks Declaration explicitly refers to the UN SDG "Reaffirming the United Nations Sustainable Development Goals and the need for their realization by 2030" (Arctic Council 2017a: 3), an obvious next step in developing and qualifying the Arctic Council's efforts in relation to sustainable development might be to probe whether indicators and targets for the Arctic could jointly be agreed upon.[25]

Another indication that the sustainable development focus of the Arctic Council might develop more in alignment with the UN SDGs can be found in the Arctic Council's SDWG's 2017 Strategic Framework, 'The Human Face of the Arctic'. The following quotation illustrates that:

> The linkages between SDWG's vision for the Arctic region and the global set of 17 United Nations Development Goals (UN SDGs), their sub-targets and indicators provides an opportunity for the work of the SDWG to contribute to the implementation of the Agenda 2030. (Sustainable Development Working Group 2017)

If the Arctic Council agreed to include the UN SDGs in its political agenda and to base Arctic regional goals on the individual national goals for the Arctic region, and further qualify and focus the work by the experience and insights Arctic research and numerous Arctic Council projects over the years have provided, by adding some of the indicators developed in Arctic

Council projects it would also contribute to a joint Arctic profile and to implement the vision stated in the Arctic Vision from 2013 that the Arctic Council "will continue our work to strengthen the Arctic Council to meet new challenges and opportunities for cooperation, and pursue opportunities to expand the Arctic Council's roles from policy-shaping into policy-making" (Arctic Council 2013b).

Notes

1 Acknowledgements to Jørgen Søndergaard for ongoing thought provoking discussions and useful advice during the development of the chapter and to Uffe Jakobsen, Rasmus Leander and Jens Vendel Hansens for helpful input to improve the manuscript.
2 The terms were, however, used earlier. A few examples are provided below.
3 Most often referred to as the 'Brundtland Report' after the chair of the commission, the former Norwegian Prime minister Gro Harlem Brundtland (The World Commission on Environment and Development 1987).
4 The Arctic Economic Council was established under the Canadian Arctic Council chairmanship 2013–2015. https://arcticeconomiccouncil.com/about-us/.
5 For more thorough conceptual deliberations see for example: WCED 1987; Rasmussen 2002; Owens 2003; Nuttall 2010; Pram Gad et al. 2016; Søndergaard 2017.
6 Sustainability as a concept in relation to non-renewable resources is for instance discussed in the AMAP Report on Oil and Gas Activities: "The question of sustainability usually relates to whether the use of the natural environment can be sustained indefinitely. In the case of natural resources where reserves are limited, as the case is for petroleum, this is not possible.

"The sustainability issue in such cases therefore has more to do with whether operations happen without irreversible harm to the environment, and how use of the resources may harm the environment. In a wider sense, sustainability may also be thought of as including issues relating to sustaining economic and social viability of societies, including how societies carry out contingency planning for accidents, and liabilities of responsible parties" (AMAP 2007: 3.58).
7 One example is the redefinition of the terms 'sustainability' and 'sustainable development' applied by the magazine Corporate Knights. The magazine annually ranks "the world's most sustainable companies" including energy companies. The ranking is based on a number of key performance indicators such as management of resources, employees and finances, and supplier performance www.corporateknights.com/wp-content/uploads/2017/01/2017-Global-100_Methodology-Final.pdf. Whereas, for instance, producing in an economical, social and environmental responsible way is preferable to producing without these goals, meeting these goals alone does not necessarily mean living up to the core meaning of sustainable development.
8 The Nordic countries, as an example, have since 2001 within the framework of the Nordic Council of Ministers developed goals and indicators to assess sustainable development.
9 www.un.org/ga/search/view_doc.asp?symbol=A/RES/70/1&referer=/english/&Lang=E.
10 www.arctic-council.org/index.php/en/about-us/arctic-council/20-year-anniversary.
11 https://oaarchive.arctic-council.org/.
12 The first binding Arctic Council (2011) agreement was on search-and-rescue in the Arctic. https://oaarchive.arctic-council.org/bitstream/handle/11374/531/EDOCS-3661-v1-ACMMDK07_Nuuk_2011_SAR_Search_and_Rescue_Agreement_signed_EN_FR_RU.PDF?sequence=5&isAllowed=y.
13 The second binding Arctic Council agreement was on oil pollution preparedness. https://oaarchive.arctic-council.org/bitstream/handle/11374/529/EDOCS-2068-v1-ACMMSE08_KIRUNA_2013_agreement_on_oil_pollution_preparedness_and_response_signedAppendices_Original_130510.PDF?sequence=6&isAllowed=y.
14 A third binding agreement on Arctic research collaboration was signed in 2017 at the Arctic Council Ministerial meeting in Fairbanks, Alaska (Arctic Council 2017b). https://oaarchive.arctic-council.org/handle/11374/1916.
15 The Russian strategy from 2013 (Putin 2013) is organised differently from the other strategies and policies as the structure 'follows' the process from identifying "main risks and threats" to "Monitoring the implementation of the Strategy" and thus does not approach the challenges thematically.
16 See also Government of Canada 2009: 15; Norwegian Ministry of Foreign Affairs 2009: 63; Prime Minister's Office, Finland 2010: 13; Regeringskansliet 2011: 23; White House 2013: 9 for similar use of the term.

17 The Swedish strategy is the only one that explicitly departs in the AC definition stating that "Sweden will promote economically, socially and environmentally sustainable development in the entire Arctic region" (Regeringskansliet 2011: 30) but also the strategy of the Danish Kingdom refers to all three components when mentioning preconditions for 'a healthy, productive and self-sustaining community (Ministry of Foreign Affairs (Denmark) 2011: 24).

18 One of many aspects in relation to the study of the use of the term 'sustainable development' and a topic for further studies might be when 'sustainable development' started being used side by side with 'resilient' and the potential implications thereof.

19 In 1998 the Nordic prime ministers and the political leaders of the self-governing regions agreed upon a declaration on a sustainable Nordic region. The declaration was followed by the agreement on a strategy for a sustainable Nordic region in 2000. The original strategy covering the period 2001–2004 was followed by a first revised strategy for 2005–2008 and a 2009–2012 second revised strategy called 'New Bearings for the Nordic Countries'. With a point of departure in this strategy the Nordic Council of Ministers decided 'Nordic Sustainable Indicators 2013–2018' based not least on the Nordic welfare model (www.norden.org/en/search?SearchableText=sustainable+development).

20 One example is the Nordic Council of Ministers' Arctic Co-operation Program 2015–2017 focusing on 'sustainable development' with a programme budget of DKK 8.8 million in 2016 (www.norden.org/en/nordic-council-of-ministers/ministers-for-co-operation-mr-sam/the-arctic/the-nordic-council-of-ministers-arctic-co-operation-programme-2015-2017).

21 See Bailes and Heininen 2012, and Heininen 2012, for more thorough comparative research on the strategies and policies of the eight Arctic states.

22 As several Arctic countries at a given time (three-year terms) participate in United Nations' Commission on Sustainable Development, national reports to a given session (session 18 in 2010) were selected. Full national reports (on chemicals, waste management, sustainable consumption and production, transport and mining) were available from the USA, Iceland and Sweden and reports on mining were available from Finland and Canada. To include the three remaining Arctic states, a 2012 Report from the Russian Federation on 'Implementing the Principles of Sustainable Development in the Russian Federation . . . Preparing For "Rio + 20"'; 'A sustainable Denmark – a balanced development' a Danish government report from 2014; and 'Initial steps towards the implementation of the 2030 Agenda' by the Norwegian government, 2016a were selected and included in the review process.

23 For further references to documents applied, see entries under the heading "Arctic states' reporting to the UN Commission on sustainable development 2010, session 18 and UN documentation.

24 "The Sustainable Development Goals and targets are integrated and indivisible, global in nature and universally applicable, taking into account different national realities, capacities and levels of development and respecting national policies and priorities. Targets are defined as aspirational and global, with each Government setting its own national targets guided by the global level of ambition but taking into account national circumstances. Each Government will also decide how these aspirational and global targets should be incorporated into national planning processes, policies and strategies. It is important to recognize the link between sustainable development and other relevant on-going processes in the economic, social and environmental fields" (UN 2015: 13).

25 It is worth mentioning that a joint and focused Arctic effort seems to be supported by the recently approved Russian Foreign Policy Concept stating, inter alia, that "Russia intends to proactively contribute to the creation of . . . the determination of global sustainable development guidelines and achievement of the UN sustainable development guidelines and achievement of the UN sustainable development goals" (Russian Federation 2016: 13) and that "Russia considers that the Arctic states bear special responsibility for the sustainable development of the region, and in this connection advocates enhanced cooperation in the Arctic Council, the coastal Arctic Five and the Barents Euro-Arctic Council" (Russian Federation 2016: 22).

References

ACIA 2005. *Impacts of a Warming Arctic: Arctic Climate Impact Assessment* Cambridge, UK: Cambridge University Press (www.acia.uaf.edu).

AHDR 2004. Einarsson N, Larsen J N Nilsson A and Young O R eds 2004 *Arctic Human Development Report* Prepared by the Stefansson Arctic Institute, under the auspices of the Icelandic Chairmanship of the Arctic Council 2002–2004, Akureyri, Iceland (https://oaarchive.arctic-council.org/handle/11374/51).

AHDR 2014. Larsen. J.N. and Fondahl, G. eds 2014 *Arctic Human Development Report. Regional Processes and Global Linkages* TemaNord 2014:567 Nordic Council of Ministers, Copenhagen (http://norden.diva-portal.org/smash/record.jsf?pid=diva2%3A788965).

ASI 2010. Larsen, J.N., Schweitzer, P. and Fondahl, G. eds. *Arctic Social Indicators: A Follow-up to the Arctic Human Development Report* TemaNord 2010:519 Nordic Council of Ministers, Copenhagen (www.norden.org/sv/publikationer/publikationer/2010-519).

ASI 2015. *Arctic Social Indicators II: Implementation* Larsen J N Schweitzer P and Petrov A eds 2015 TemaNord 2014:568 Nordic Council of Ministers, Copenhagen (http://sdwg.org/wp-content/uploads/2015/02/ASI-II.pdf).

Bailes, A.J.K. and Heininen, L. 2012. *Strategy Papers on the Arctic or High North: A Comparative Study and Analysis.* Reykjavik: Centre for Small State Studies, Institute of International Affairs, University of Iceland (https://rafhladan.is/bitstream/handle/10802/5104/arctic_strategies_innsidur.pdf?sequence=1).

CAFF 2013. *Arctic Biodiversity Assessment: Status and Trends in Arctic Biodiversity* Akureyri, Iceland: Conservation of Arctic Flora and Fauna (www.arcticbiodiversity.is/).

Commission of the European Communities 2008. Communication from the Commission to the European Parliament and the Council *The European Union and the Arctic Region* 20.11.2008, COM (2008) 763 final, Brussels (http://eeas.europa.eu/archives/docs/arctic_region/docs/com_08_763_en.pdf).

Corporate Knights 2017. *Most Sustainable Corporations in the World: The 2017 Global 100: Overview of Methodology* www.corporateknights.com/wp-content/uploads/2017/01/2017-Global-100_Methodology-Final.pdf.

Dodds, F. and Strauss, M. with Strong, M. 2012. *Only One Earth: The Long Road via Rio to Sustainable Development.* London: Routledge.

European Commission, High Representative of the Union for Foreign Affairs and Security Policy 2016. Joint Communication to the European Parliament and the Council *An Integrated European Union Policy for the Arctic,* 27.4.2016, JOIN (2016) 21 final, Brussels (www.eeas.europa.eu/archives/docs/arctic_region/docs/160427_joint-communication-an-integrated-european-union-policy-for-the-arctic_en.pdf).

Gad, U.P., Jakobsen, U. and Strandsbjerg, J. 2015. 'Politics of sustainability in the Arctic – a research agenda' in Fondahl, G. and Wilson, G. eds. *Northern Sustainabilities.* Dordrecht, Netherlands: Springer.

Gorbachev, M. 1987. The speech in Murmansk at the ceremonial meeting on the occasion of the presentation of the Order of Lenin and the Gold Star to the city of Murmansk October 1, 1987 Moscow: Novosti Press Agency. 23–31.

Heininen, L. 2012. *Arctic Strategies and Policies – Inventory and Comparative Study* Akureyri, Iceland: The Northern Research Forum and The University of Lapland.

International Union for Conservation of Nature and Natural Resources (IUCN) 1980 *World Conservation Strategy. Living Resource Conservation for Sustainable Development.* IUCN-UNEP-WWF.

Kruse, J., Poppel, B., Abryutina, L., Duhaime, G., Martin, S., Poppel, M., Kruse, M., Ward, E., Cochran, P. and Hanna, V. 2008. 'A Survey of Living Conditions in the Arctic' in Møller V., Huschka D. and Michalos A.C. eds. *Barometers of Quality of Life around the Globe.* Dordrecht, Netherlands: Springer. 107–134.

Meadows, D.H., Meadows, D.L., Randers, J. & Behrens III, W.W. 1972. *The Limits to Growth: A Report for the Club of Rome's Project on the Predicament of Mankind.* New York: Universe Books.

Nordic Council of Ministers 2002. *Will We Achieve Our Objectives? A Nordic Set of Indicators* APN 2002:737 Nordic Council of Ministers, Copenhagen ISBN 92-893-0790-0.

Nordic Council of Ministers 2013. *Nordic Sustainable Development Indicators 2013* APN 2013:757 Nordic Council of Ministers, Copenhagen ISBN 978-92-893-2600-1 (http://norden.diva-portal.org/smash/record.jsf?pid=diva2%3A702860&dswid=-9671).

Nordic Council of Ministers' Arctic Co-operation Program 2015–17. (http://www.norden.org/en/nordic-council-of-ministers/ministers-for-co-operation-mr-sam/the-arctic/arctic-co-operation-programme-201520132017).

Nuttall, M. 2010. *Pipeline Dreams, People, Environment, and the Arctic Energy Frontier* IWGIA Document 126, Copenhagen: IWGIA.

Owens, S. 2003. 'Is there a meaningful definition of sustainability?' *Plant Genetic Resources 1(1)*; 5–9. *NIAB 2003* ISSN 1479-2621.

Poppel, B. ed. 2015. SLiCA: Arctic living conditions. Living conditions and quality of life among Inuit, Sami and indigenous peoples of Chukotka and the Kola Peninsula TemaNord 2015: 501 Nordic Council of Ministers, Copenhagen DOI:10.6027/TN2015-501.

Poppel, B., Kruse, J., Abryutina, L. and Duhaime, G. 2007. *Survey of Living Conditions in the Arctic: SLiCA Results* Anchorage, AK: Institute of Social and Economic Research, University of Alaska.

Rasmussen, R.O. 2002. 'Om bæredygtig udvikling i Grønland' ['On sustainable development in Greenland'] in Rasmussen R O & Hansen K G eds *Aspekter af bæredygtig udvikling i Grønland [Aspects of sustainable development in Greenland]* Sisimiut, Greenland: Sisimiut Museum and NORS.

Søndergaard, J.S. 2017. 'Noget om aktuelle muligheder for anvendelse af begrebet "bæredygtig udvikling" når det gælder udviklingen af en "Bæredygtig udvikling for et arktisk velfærdssamfund"' ['On possibilities to apply the concept of "sustainable development" in relation to the formation of a "sustainable development of an Arctic welfare society"'] *Tidsskriftet Grønland 2/2017*, Copenhagen. 118–124.

World Commission on Environment and Development 1987. *Development and International Cooperation Report of the World Commission on Environment and Development* United Nations General Assembly 42 session item 83; *(A/42/427)* 4 August 1987.

World Economic Forum Global Agenda Council on the Arctic 2015. *Arctic Investment Protocol. Guidelines for Responsible Investment in the Arctic* (www3.weforum.org/docs/WEF_Arctic_Investment_Protocol.pdf).

AEPS & Arctic Council documents (listed chronologically)

AEPS 1991. Declaration on the Protection of Arctic Environment Arctic Environmental Protection Strategy Signed by the Eight Arctic Nations June 14 1991 Rovaniemi, Finland.

Ministry of Foreign Affairs 1993. The Arctic Environment Second Ministerial Conference September 1993 – Nuuk, Greenland Ministry of Foreign Affairs, Copenhagen.

Arctic Council 1996. The Declaration of the Establishment of the Arctic Council September 19 1996 Ottawa, Canada.(https://oaarchive.arctic-council.org/bitstream/handle/11374/85/EDOCS-1752-v2-ACMMCA00_Ottawa_1996_Founding_Declaration.PDF?sequence=5&isAllowed=y).

Arctic Council 1998. *Terms of Reference for the Sustainable Development Program* (https://oaarchive.arctic-council.org/bitstream/handle/11374/1658/EDOCS-930-v1-ACMMUS02_BARROW_2000_5b_SDWG_Terms_of_Reference.PDF?sequence=1&isAllowed=y).

Arctic Council 2000. The Sustainable Development Framework Document (SDFD) (https://oaarchive.arctic-council.org/bitstream/handle/11374/1657/EDOCS-931-v1-ACMMUS02_BARROW_2000_5b_Sustainable_Development_Progam.PDF?sequence=1&isAllowed=y).

Arctic Council 2011. Agreement on Cooperation on Aeronautical Maritime Search and Rescue in the Arctic Nuuk, Greenland.(https://oaarchive.arctic-council.org/bitstream/handle/11374/531/EDOCS-3661-v1-ACMMDK07_Nuuk_2011_SAR_Search_and_Rescue_Agreement_signed_EN_FR_RU.PDF?sequence=5&isAllowed=y).

Arctic Council 2013a. Kiruna Declaration. The Eight Ministerial Meeting of the Arctic Council May 15 2015 Kiruna, Sweden (https://oaarchive.arctic-council.org/bitstream/handle/11374/93/MM08_Kiruna_Declaration_final_formatted.pdf?sequence=5&isAllowed=y).

Arctic Council 2013b. Vision for the Arctic. The Eighth Ministerial Meeting of the Arctic Council May 15 2015 Kiruna, Sweden (https://oaarchive.arctic-council.org/bitstream/handle/11374/287/MM08_Kiruna_Vision_for_the_Arctic_Final_formatted%20%281%29.pdf?sequence=1&isAllowed=y).

Arctic Council 2013c. Agreement on Cooperation on Marine Oil Pollution Preparedness and Response Kiruna, Sweden (https://oaarchive.arctic-council.org/bitstream/handle/11374/529/EDOCS-2068-v1-ACMMSE08_KIRUNA_2013_agreement_on_oil_pollution_preparedness_and_response_signedAppendices_Original_130510.PDF?sequence=6&isAllowed=y).

Arctic Economic Council (AEC) 2014. (https://arcticeconomiccouncil.com/about-us/)

Arctic Council 2015. Iqaluit Declaration. The Ninth Ministerial Meeting of the Arctic Council. April 24 2015 Iqaluit, YT. (https://oaarchive.arctic-council.org/bitstream/handle/11374/662/EDOCS-3431-v1-ACMMCA09_Iqaluit_2015_Iqaluit_Declaration_original_scanned_signed_version.PDF?sequence=7&isAllowed=y).

Arctic Council 2016. *20 years of Arctic Council* (www.arctic-council.org).

The Sustainable Development Working Group 2017. *The Human Face of the Arctic*. Strategic Framework 2017. (https://oaarchive.arctic-council.org/bitstream/handle/11374/1940/SDWG-Framework-2017-Final-Print-version.pdf?sequence=1&isAllowed=y).

Arctic Council 2017a. Fairbanks Declaration. *The Tenth Ministerial Meeting of the Arctic Council*. May 11 2017 Fairbanks, AK. (https://oaarchive.arctic-council.org/bitstream/handle/11374/1910/EDOCS-4072-v5-ACMMUS10_FAIRBANKS_2017_Fairbanks_Declaration-2017.pdf?sequence=2&isAllowed=y).

Arctic Council 2017b. *Agreement on Research Collaboration* (https://oaarchive.arctic-council.org/handle/11374/1916).

Arctic states' Arctic and northern strategies (organized geographically)

Ministry of Foreign Affairs (Denmark) 2011. *Denmark, Greenland and the Faroe Islands: Kingdom of Denmark Strategy for the Arctic 2011–2020* Ministry of Foreign Affairs (Denmark), Copenhagen (http://library. arcticportal.org/1890/1/DENMARK.pdf).

Althingi 2011. *A Parliamentary Resolution on Iceland's Arctic Policy* Approved by Althingi at the 139th legislative session 28 March 2011, Reykjavik (http://library.arcticportal.org/1861/).

Norwegian Ministry of Foreign Affairs 2006. *The Norwegian Government's High North Strategy*, Oslo (www. regjeringen.no/globalassets/upload/UD/Vedlegg/strategien.pdf).

Norwegian Ministry of Foreign Affairs 2009. *New Building Blocks in the North – the Next Step in the Governments High North Strategy* 12 March 2009 Oslo/Tromsø, Norway (www.regjeringen.no/ globalassets/upload/UD/Vedlegg/strategien.pdf).

Norwegian Ministry of Foreign Affairs 2014. *Norway's Arctic Policy* (www.regjeringen.no/globalassets/ departementene/ud/vedlegg/nord/nordkloden_en.pdf).

Regeringskansliet (Ministry of Foreign Affairs) 2011. *Sweden's strategy for the Arctic region* Stockholm (www. openaid.se/wp-content/uploads/2014/04/Swedens-Strategy-for-the-Arctic-Region.pdf).

Prime Minister's Office (Finland) 2010. *Finland's Strategy for the Arctic Region* Prime Minister's Office Publication 8/2010, Helsinki (.http://vnk.fi/documents/10616/622962/J0810_Finland's+Strategy+f or+the+Arctic+Region.pdf/8c6c2c20-f0c2-4cd8-9227-3aba6d2db405?version=1).

Prime Minister's Office, Finland 2013. *Finland's Strategy for the Arctic Region 2013. Government resolution on 23 August 2013* Prime Minister's Office Publication 16/2013, Helsinki (http://vnk.fi/documents/10616/334509/Arktinen+strategia+2013+en.pdf/6b6fb723-40ec-4c17-b286-5b5910fbecf4).

Putin, V. 2013. *Russian Strategy of the Development of the Arctic Zone and the Provision of National Security Until 2020* (adopted by the President of the Russian Federation 8 February 2013, no. Pr-232) (translated from Russian) (www.iecca.ru/en/legislation/strategies/item/99-the-development-strategy-of-the-arctic-zone-of-the-russian-federation).

Russian Federation's Policy for the Arctic to 2020 (adopted by the President of the Russian Federation D Medvedev 18 September 2008). Promulgated: 30 March 2009 publication of the official governmental newspaper "Rossiyskaya Gazeta" (translated from Russian) (www.arctis-search.com/Russian+Federati on+Policy+for+the+Arctic+to+2020).

The Ministry of Foreign Affairs of the Russian Federation 2016. *Foreign Policy Concept of the Russian Federation* (approved by President of the Russian Federation Vladimir Putin 30 November 2016) (www.mid.ru/en/foreign_policy/official_documents/-/asset_publisher/CptICkB6BZ29/content/ id/2542248).

The White House 2013. *National Strategy for the Arctic Region* Governmental Printing Office, Washington, DC (https://obamawhitehouse.archives.gov/sites/default/files/docs/nat_arctic_strategy.pdf).

Government of Canada 2009. *Canada's NORTHERN STRATEGY Our North, Our Heritage, Our Future* Minister of Indian Affairs and Northern Development and Federal Interlocutor for Metis and Non-Status Indians, Ottawa, ON (www.northernstrategy.gc.ca/cns/cns-eng.asp).

Arctic states' sustainable development strategies and implementation plans related to UN sustainable development initiatives

Danish Government 2014. *Et bæredygtigt Danmark - Udvikling i balance* [A sustainable Denmark – a balanced development] (www.fm.dk/publikationer/2014/et-baeredygtigt-danmark-udvikling-i-balance) Ministry of Finance, Copenhagen.

Government of Canada 2015a. *2015 Progress Report of the Federal Sustainable Development Strategy* Environment and Climate Change Canada, Gatineau, QC (www.canada.ca/en/services/environ-ment/conservation/sustainability/federal-sustainable-development-strategy/2015-progress-report. html).

Government of Canada 2015b. *Planning for a Sustainable Future: A Federal Sustainable Development Strategy for Canada 2016–2019* Sustainable Development Office, Ottawa, ON (www.ec.gc.ca/dd-sd/default. asp?lang=En&n=F93CD795-1).

Ministry for the Environment in Iceland 2002. Welfare for the Future: Iceland's National Strategy for Sustainable Development 2002–2020 (https://eng.umhverfisraduneyti.is/media/PDF_skrar/Sjalfbar__roun_enska.pdf).

Ministry for the Environment in Iceland 2009. Welfare for the Future: Iceland's National Strategy for Sustainable Development Priorities 2010–2013 (www.government.is/media/umhverfisraduneyti-media/media/PDF_skrar/Welfare-for-the-Future-Priorities-2010-2013.pdf).

Norwegian Government 2016a. *Initial steps towards the implementation of the 2030 Agenda* Voluntary national review presented at the high-level political forum on sustainable development (hlpf). (https://sustainabledevelopment.un.org/content/documents/10692NORWAY%20HLPF%20REPORT%20-%20full%20version.pdf).

Norwegian Government 2016b. *Norway's follow-up Agenda 2030 and the Sustainable Development Goals* (www.regjeringen.no/en/dokumenter/follow-up-sdg2/id2507259/).

Russian Federation 2012. Report on Implementing the Principles of Sustainable Development in the Russian Federation. Russian Outlook on the New Paradigm for Sustainable Development. Preparing For "Rio + 20" Interagency Working Group of experts on Russia's participation in the UN Conference on Sustainable Development (Rio + 20), Moscow (https://sustainabledevelopment.un.org/content/documents/1043natrepeng.pdf).

Stockholm Environment Institute 2015. Sustainable Development Goals for Sweden: Insights on Setting a National Agenda Stockholm Environment Institute, Stockholm (www.ym.fi/download/noname/{45A196E9-03BC-41E3-A7E8-15A1273B3FE8}/112274).

Arctic states' reporting to the UN Commission on Sustainable Development 2010, session 18 and UN documentation

Finnish Government 2010. Finland National Reporting to the UN Commission on Sustainable Development CSD18–19. Thematic Profile on Mining (www.un.org/esa/dsd/dsd_aofw_ni/ni_pdfs/NationalReports/finland/Mining.pdf).

Government of Canada 2010 National Reporting to CSD-18/19. Thematic Profile on Mining (www.un.org/esa/dsd/dsd_aofw_ni/ni_pdfs/NationalReports/canada/Mining.pdf).

Government of the United States 2010. United States National Reporting to the UN Commission on Sustainable Development CSD18–19 (www.un.org/esa/dsd/dsd_aofw_ni/ni_pdfs/NationalReports/usa/Full_text.pdf).

Icelandic Government 2010. Iceland National Reporting to the UN Commission on Sustainable Development CSD18–19 (www.un.org/esa/dsd/dsd_aofw_ni/ni_pdfs/NationalReports/iceland/full_text.pdf).

Swedish Government 2010. Sweden National Reporting to the UN Commission on Sustainable Development CSD18–19 (www.un.org/esa/dsd/dsd_aofw_ni/ni_pdfs/NationalReports/sweden/Full_text.pdf).

UN Commission on Sustainable Development 2010. *Report on the 18th Session* E/CN.17/20/10/15 (www.un.org/ga/search/view_doc.asp?symbol=E/CN.17/2010/15%20(SUPP)&Lang=E).

United Nations reports and initiatives

United Nations 1982. United Nations Convention on the Law of the Sea, Dec. 10, 1982, 1833 U.N.T.S. 397. (www.un.org/Depts/los/convention_agreements/texts/unclos/unclos_e.pdf).

United Nations 1992. United Nations Conference on Environment and Development. Rio de Janeiro, Brazil, 3 to 14 June 1992. Agenda 21. (https://sustainabledevelopment.un.org/content/documents/Agenda21.pdf).

United Nations 2000. Resolution adopted by the General Assembly on 18 September 2000: 55/2. ()

United Nations Millennium Declaration (http://www.un.org/millennium/declaration/ares552e.htm).

United Nations 2015a. Resolution adopted by the General Assembly on 25 September 2015 70/1. Transforming our world: the 2030 Agenda for Sustainable Development. (www.un.org/en/ga/search/view_doc.asp?symbol=A/RES/70/1).

United Nations 2015b. *Millennium Development Goal Report 2015* (www.undp.org/content/dam/undp/library/MDG/english/UNDP_MDG_Report_2015.pdf).

Websites:

Arctic Council: (www.arctic-council.org).

Arctic Council documents archive: (https://oaarchive.arctic-council.org/).

Nordic Council of Ministers: (www.norden.org).

United Nations sustainable development knowledge platform: (https://sustainabledevelopment.un.org/hlpf/2016/).

UN Commission on Sustainable Development 2010 – session 18: (https://sustainabledevelopment.un.org/index.php?menu=1135).

UN Millennium Development Goals: (www.undp.org/content/undp/en/home/sdgoverview/mdg_goals.html).

UN Sustainable Development Goals: (www.undp.org/content/undp/en/home/sustainable-development-goals.html).

25

Indigeneity, sovereignty, and Arctic indigenous internationalism

Jessica M. Shadian

Over the past decade global interest in the Arctic has grown exponentially. So has its regional role in world politics. Accompanying this interest is the increasing number of non–Arctic states, substates (from Scotland to Maine in the United States), suprastates (such as the EU), and indigenous non–governmental organizations (INGOs) which believe they have a vested interest in its future. In light of this greater interest, it seems appropriate to reflect on the role that indigenous diplomacies have played historically and continue to play in Arctic and global politics. In the context of international relations and particularly from a historical perspective, indigenous diplomacies have been constitutive of global political transformations since the emergence of the Westphalian political system. The history of indigenous peoples, according to Marshall Beier (2009: 26) "can be neither left out nor brought in; their knowledges can no more be used to inform theory in International Relations than knowledges about them can be divorced from it".

This chapter, as such, places Arctic indigenous politics within the broader historical context of Arctic and global politics in order to highlight how the diplomacy of the region's indigenous peoples are both a product of the history of the Westphalian political system while also a harbinger of its changes (Beier 2009: 10). Specifically, this chapter addresses some of the main themes in International Relations (IR) including sovereignty, the state, and non-state diplomacy with a particular focus on Inuit and Sámi politics. It should not go unacknowledged, however, that there are a wide range of indigenous peoples in the Arctic, all of whom have varying degrees of sovereignty and autonomy as well as differing politics, size, and relations with their respective states. This chapter cannot provide coverage of indigenous politics across the circumpolar North – as such it does not set out to downplay these realities nor to assume that all Arctic indigenous peoples have similar histories or want the same futures. Rather, the point is to focus on Inuit and, to a lesser effect, Sámi politics as a means to offer a particular window into the larger discussions of the ways in which indigeneity challenges conventional thinking about IR theory. The chapter begins with a brief introduction to indigeneity in IR before bringing this thinking to bear on recent debates over who owns the Arctic. The following two sections then turn to modern international law and the means by which international law has increasingly accommodated indigenous diplomacies, giving rise to indigenous internationalism – particular attention is given to the relationship between international discourse and aims for greater indigenous self-determination at home. Rather than assuming that international law has driven these possibilities, these two sections view local indigenous efforts for greater self-determination and changing international law as constitutive practices. The chapter then turns to the Arctic Council and finally takes a look at indigeneity in a global Arctic.

IR theory and indigeneity: cosmopolitan indigenous diplomacies

In 1979, thirty years after the founding of the modern international legal system, International Relations (IR) scholar Hedley Bull wrote of the possible unravelling of the beliefs underpinning the very institutions comprising the Westphalian political system:

> One may imagine that if nationalist separatist groups were content to reject the sovereignty of the states to which they are at present subject, but at the same time refrained from advancing any claims to sovereign statehood themselves, some genuine innovation in the structure of the world political system might take place. . . [creating actual doubt as to] whether sovereignty lay with national governments or with the organs of the community. . . [and granting] cultural differences the political recognition that had been withheld in the past.
>
> *(Bull 1977, in Linklater 1998: 115–116)*

Since writing this piece, these prospects have gone from a hypothetical discussion to a reality in myriad contexts around the world. The end of the Cold War, in particular (alongside the telecommunications revolution) gave rise to a host of alternative politics marked, neither by a waning of nationalism and nationalist movements nor the demise of the nation–state system, but by what James Rosenau referred to as *fragmegration* or "resistances to boundary-spanning activities" (1997: 243). For Rosenau, fragmentation acts simultaneously with the rise of new orders and institutions or *integration* (ibid.). Though nationalisms appear to be growing in force, the self-determination desired by national groups is not necessarily conceived as a matter of territorial integrity (1997: 243). In line with Rosenau's thinking, Dittmer and McConnell look at alternative cultures and spaces of diplomacy which disrupt the national/international binary that is central to conventional theorizations (Dittmer and McConnell 2015) of diplomacy in IR. According to those authors, diplomacy is defined as 'translocal network[s] of practices' which are a function of social life that negotiates self/other relations in plural and quotidian ways. Diplomatic cultures, further, simply cannot be reduced to a single, cosmopolitan culture, but instead are viewed as plural cultures which are fundamentally rooted in the times and spaces that compose them (Dittmer and McConnell 2015: 6). From this historical and non-state centric framework, indigenous diplomacies are not viewed as 'new' diplomatic actors. Rather, their diplomatic practices and cultures pre-exist the Westphalian inter-state system and continue to co-exist along conventional state diplomacy (Dittmer and McConnell 2015:18). This relational understanding of diplomacy de-centres the practices of state diplomats and highlights the vast cultural and political infrastructure that makes state-based diplomacies meaningful (Dittmer and McConnell 2015:6).

In the Arctic, indigenous diplomacies pre-exist European discovery (followed later by colonization) and consequently, operate below, above, and through the boundaries of individual states. In contemporary politics, many of the Arctic's indigenous peoples seek political legitimacy through aspirations to attain *cultural* integrity, or the right to be political (Shadian 2010), rather than by means of becoming a state. According to Broderstad and Dahl (2002), many indigenous groups have, in response to past assimilation policies, reinvented the concept of nation-building in their own terms, perceiving it as "efforts . . . to increase their capacities for a self-rule and for self-determined sustainable community and economic development. It also involves building institutions of self-government" (2002: 2). In essence, many indigenous groups aspire to create alternative sovereignties.

These moves towards an alternative politics of self-determination have emerged interdependent of broader changes taking place at the international institutional level; in particular, international changes brought about during the 1970s with the emergence of indigenous internationalism. In the particular field of IR, however, indigenous internationalism failed at that time to affect mainstream theory. Where indigenous diplomacies did find a place in IR, the theoretical assumptions regarding state sovereignty and the inter-state system largely remained intact (e.g. Keck and Sikkink 1998). This early literature was either critical of indigenous peoples' advancements (Corntassel 2007) or focused on the influence they now have on state behaviour (Keck and Sikkink 1998), particularly around international policy-making.

More recently, a number of social scientists (e.g. Shaw 2002; Beier 2009; Shadian 2010) have begun to engage with the international politics of indigenous diplomacies in the context of the 'grand debates' in IR theory. Marshall Beier first coined the notion of indigenous diplomacies in an effort to create an analytical space to examine the role of indigenous people in global politics – a space that recognizes their constitutive political agency rather than rendering them as apolitical objects of study. Indigenous diplomacies, according to Beier, are often a means of entering the dominant political discourses rather than necessarily aiming for secession. Their goal, according to Castree, is in many instances to attain political control over various aspects related to place, as opposed to simply seeking ownership of physical territory (Castree 2004). These alternative political aims often relate to land and resource rights, control over indigenous knowledge and ideas, and often are rooted in cosmologies not found in European philosophy (Beier 2007a: 121, 2007b: 9. See also Castree 2004; Shadian 2015). Such indigenous diplomacies, according to Beier, are not 'new' but rather 'newly noticed' in the field of IR. The consequence has been the opening of a conceptual space within which to analyse the intersections between indigenous diplomacies and the foreign policies of states (Beier 2007b: 9).

Along these lines, Castree (2004) has also looked at how indigenous groups have constructed new ways of thinking about political relationships to land that go beyond traditional considerations of state sovereignty. Shadian (2010, 2015), likewise, has looked at the ways in which transnational indigenous groups are changing traditional assumptions of sovereignty and carving out spaces for alternative sovereignties. For the Arctic's indigenous peoples, this includes attaining *cultural* sovereignty, rather than state sovereignty, affording indigenous groups a distinct kind of diplomatic authority to participate formally in Arctic regional and global politics (ibid.). Such cultural sovereignties have been institutionalized in the Arctic through land claims agreements and other forms of self-determination at home, as well as the creation of transnational indigenous groups, such as the Inuit Circumpolar Council (ICC) and the Saami Council, who, as Permanent Participants, have a seat at the negotiating table of the Arctic Council. Internationally, they are also members of the United Nations Economic and Social Council (ECOSOC). Despite these inroads that indigeneity theorists have finally made into the broader mainstream debates of IR theory, when it comes to the dominant debates over who owns the Arctic and who decides, indigeneity studies is nowhere to be found.

Who owns the Arctic: indigenous governance in Arctic history

The Russian flag planting by Artur Chilingarov in August 2007 led many international pundits[1] to believe that the Arctic was a newly emerging significant geopolitical region. Global Arctic politics, likewise, particularly the politics relating to the Arctic's resources, has a long history beginning with its indigenous peoples who lived in and across the Arctic for thousands of years prior to European attempts to explore the region. This includes the ancestors of the Inuit, who migrated from Siberia across to Greenland more than a thousand years ago, or the Sámi who have lived in northern Fennoscandia and Russia's Kola Peninsula since time immemorial. This was followed centuries later by European fur traders, whalers, prospectors, and eventually mining and oil companies who have all set their sights on exploiting the Arctic's resources.

In tandem with this history is the Westphalian political history of exploration, colonization, state-building, and, eventually, indigenous political agency (Shadian 2014). The 1648 Peace of Westphalia provided a basis for European co-operation which was followed by state building and, several centuries later, codification of an international system of states. Throughout, the Arctic played a vital rather than a peripheral political role because of its wealth of resources. As Jonathan Greenberg (2009: 1315) has written:

> Even the smallest northern aboriginal nation and the most remote polar micro-environment have been impacted by foreign intervention. . . No community or environment in the Arctic remained untouched and unchanged by the larger historical forces, ideologies, and economic engines that have shaped the international system as it evolved since the beginning of the European nation-state system.

For Inuit, encounters such as those helped to strengthen ties between neighbouring Inuit communities and eventually those encounters led to the evolution of a pan-Inuit polity from the 1970s. Indigenous politics and diplomacy throughout the Arctic, as such, not only predates the Arctic Council but it also predates the creation of the Arctic states themselves. Early Inuit diplomacy can be found in the historical encounters between Inuit and explorers such as Jacques Cartier as well as the Basque, Scottish, Dutch, American, and Russian fishers and whalers that followed. Those encounters were eventually formalized through the permanent presence of the Hudson's Bay Company (Canada Heirloom Series 1988; Mitchell 1996: 50; Canadian Museum of Civilisation Corporation 2001; Alaska Humanities Forum 2004–2013) and other foreign government outposts and eventually by newly made states. Rhoda Innuksuk, an Inuk diplomat, writes that:

> Inuit welcomed the first foreign visitors. They were exotic and rich – loaded down with valuable materials like wood and metal, and equipped with highly useful devices like firearms – but in other ways they were unbelievably poor and incompetent, ill-equipped for arctic conditions, and unable to survive without Inuit help. Since their behavior was unpredictable and sometimes uncivilized, they could also be frightening.
>
> *(Innuksuk 1994: 3)*

The processes of colonization of the Arctic's indigenous peoples which followed the making of the eight Arctic states were part of these larger and ongoing global processes of nation-state building. As the Westphalian nation-state system spread throughout the world, European states embraced a moral imperative to take on the 'White Man's burden' in order to justify their acts of territorial expansion. The belief in Europe at that time was that by conquering every part of the globe, Europeans would civilize the uncivilized world (Rudolph 2005: 7). For the Arctic's indigenous peoples who had inhabited the region for thousands of years prior to European contact, the result was the relinquishment of all sovereignty over their own affairs.

By the 1960s, the colonized regions of the world outside the Arctic had begun to assert their rights to self-determination. At that time, self-determination was based on territorial integrity (i.e. statehood); a pre-existing political architecture, according to which sovereignty arose from territory and national identity, and an independent state required both (ibid.: 6).

Decolonization in Africa and Asia had effectively divided the world into a global north and global south; internal colonization in the Arctic had led to an internal inverse of these events. Southern capitals eventually controlled the political affairs of their northern territories. The unique situation that indigenous peoples shared across the Arctic soon became a point of common ground. While relations between the Arctic states and their indigenous populations differed from one another and from colonial relations in other parts of the world, they were also similar in that they all constituted internal colonization.

When Inuit and Sámi, for instance, began to assert their political rights within the countries that had colonized them and act as single transnational polities in the international arena, creating states or a state of their own was not their central aim. The international political architecture was already undergoing a process of change through expanding practices of Westphalian politics, including changing notions of sovereignty in IR theory.

By the 1970s, existing policies governing the Arctic's indigenous peoples began to shift away from paternalism and assimilation towards an entirely new set of political ideals. Those ideals were themselves part of global changes taking place at the international legal level regarding international human rights, indigenous rights, and indigenous ideas of self-determination. The manner in which many indigenous groups pursued their own paths towards self-determination reflected those global political changes in international law including shifting ideas of land and resource ownership and new notions of international development. While Inuit land claims and Sámi aims for greater cultural self-determination were very much a product of local circumstances and domestic politics, equally so, the political awareness that affected the aims for greater autonomy was rooted in changing understandings of the links between resource development, territory, and sovereignty at the global level.

Expanding the domain of international law

The founding of the United Nations in October 1945 marked a new era in international law. Since that time, international law, while always grounded in liberal theory, has undergone continuous change. These changes have rearticulated who is included and who is excluded from international politics. During colonial times, indigenous peoples were part of a system whereby political boundaries were established between 'legitimate' international actors and those deemed outside of the system. After World War II, decolonization in much of Africa and Asia greatly increased the number of 'inside' actors. In the 1970s, the Arctic's indigenous peoples began to attain their own political rights. Rather than formally moving 'inside' the system, they reconceptualized the political space of global politics altogether.

Whereas, decolonization in Africa and Asia beginning in the 1950s expanded and reaffirmed the Westphalian political architecture with the addendum of new states, a growing number of international treaties which followed highlighted the relations between cultural *rights* and land through an alternative discourse on territorial integrity. This relationship between culture and land (not necessarily attached to statehood) opened up a space for indigenous peoples to begin to articulate their own exclusion and demand autonomy based on their collective rights.

The expansion of legal discourse regarding the right to individual existence to include the right to cultural existence became central to these aims; a concept which has evolved in relation to the Westphalian notion of territorial integrity. Under international law, nation-states are highly discouraged from promoting or supporting secessionist movements in other states (the principle of territorial integrity). However, the 1960 UN Resolution 1514, on the "Declaration on the Granting of Independence to Colonial Countries and Peoples", created an initial crack in the bounded nature of this legal thinking. The affirmation of the political relation between land and the state (territorial integrity) which was reaffirmed with Resolution 1514 soon thereafter began to weaken with the advent of new legal references to land. Rather than conceptualising land solely as a matter of property, cultural meanings concerning land began to make their way into international law. In 1997, the UN Human Rights Committee International Covenant on Civil and Political Rights passed article 27 which recognized that, with regard to indigenous communities, the "traditional land tenure is an aspect of the enjoyment of culture". The Inter-American Commission on Human Rights goes further, defining property as a *facet of* cultural integrity (Inter-American Commission on Human Rights 1997). The recognition that cultural integrity is a facet of indigenous self-determination was a significant departure from early modern international law, under which only *states* had international legitimacy with regard to territory and resource ownership.

Alongside these changes, the human rights regime has expanded from being limited to states' rights to rights and responsibilities for states, individuals, and non-state collective groups. Modern international law eventually extended into the environmental arena, bringing together the environment, development, and human rights under a grand international policy tool of sustainable development. For many Arctic indigenous organizations such as the ICC and the Saami Council, the emergence of an international discourse of sustainable development validated the long held international concept of indigenous stewardship. Through effective diplomatic efforts, international indigenous groups reshaped indigenous ideas about stewardship to fit within the discourse of sustainable development and accompanying international policy instruments.

Over time, through the UN and other intergovernmental and non-governmental bodies, international indigenous groups rearticulated traditional ideas – such as subsistence hunting and whaling – so that they impart a modern approach to the stewardship of land and resources. This, in turn, has shifted the image of indigenous peoples from being ungovernable, pre-modern, and backward into possessors of a unique knowledge that qualifies them as stewards over the lands and seas they inhabit (Shadian 2014: 125). The discourse of *indigenous stewardship* has, in turn, become part of a larger framework of *sustainable development* – one that has merged human (including indigenous) rights with international development policy (see also the chapter by Birger Poppel in this volume). Essentially, indigenous peoples came to symbolize sustainable development *in practice*. The international legal discourse surrounding stewardship has been subsumed under the broader legal umbrella of sustainable development.

The biggest success of international indigenous diplomacy and greatest challenge to traditional Westphalian state sovereignty at the international level came 200 years after the Maori delegation visited Britain to see the Crown, when a UN Declaration officially affirmed the rights of indigenous peoples as legitimate political actors at the international level. In September 2007, after two decades of work, the United Nations Declaration on the Rights of Indigenous Peoples (UNDRIP) was formally adopted by the General Assembly. The recognition of indigenous rights was unprecedented in international human rights law and remains "the clearest indication yet that the international community is committing itself to the protection of the individual and collective rights of indigenous peoples" (UN Permanent Forum on Indigenous Issues).[2] The incorporation of indigenous rights into formal international legal structures has transformed international law. Rights, including resource rights, that were once held by sovereign states are now seen as held by political collectives as well. The Rio Earth Summit then created a new basis for non-state political participation, and state sovereignty has since become diluted through the participation of new political actors from above, below, and across state borders.

UNDRIP offers some of the most substantial pieces of international law relating to the right to cultural self-determination (e.g. see Articles 3, 4, 5 and 46). Article 26 specifically refers to the idea of cultural integrity as it relates directly to land and resource ownership. According to that article, "Indigenous peoples have the right to the lands, territories and resources which they have traditionally owned, occupied or otherwise used or acquired". This includes requiring states to give legal recognition to these territories. (URL: http://indigenousfoundations.arts.ubc.ca/home/ global-indigenous-issues/un-declaration-on-the-rights-of-indigenous-peoples.html). This international politics has, in turn, had significant ramifications for indigenous self-determination at home.

Constructing a global narrative of indigenous self-determination at home: a case of Inuit and Sámi sovereignties

The early Alaskan, Canadian, and Greenlandic aims for self-determination highlight the international legal relationship between stewardship and indigenous self-determination. According to the Arctic Council's Arctic Human Development Report (Larsen and Fondahl 2004), indigenous self-determination is more than territorial or *jus solis* rights. Inuit land claims, for instance, have been asymmetrical, in the sense that individuals living in particular regions, based on distinctive identity differences – or *jus sanguinus* – possess more rights to autonomy than individuals living in other regions (Broderstad and Dahl 2002: 93).

Inuit land claims processes in Canada, Alaska, and Greenland played out in relation to the expansion of international rights, including changing notions of what it means to be *indigenous* and *Inuit*. Throughout the colonial era, Europeans justified their expansion into the Arctic on the grounds that while Inuit lived off the land, they did not own it. Thus, by claiming the Arctic, Europeans undermined existing forms of Inuit self-governance. Since this time, the old Inuit principle of stewardship – of living off the land without actually owning it – has resurfaced in the discourse of the Arctic indigenous land claims. International law no longer considers stewardship as a means to ignore or override Inuit autonomy; instead, it has become the means which Inuit leaders justify their claims to self-determination.

The first land claims settlement in the North American Arctic was the Alaska Native Claims Settlement Act (ANCSA) of 1971. A major impetus to settle Alaska's indigenous land claims was the discovery of a major oil field in the North Slope Borough. When the negotiations were all said and done ANCSA helped to create local governments as well as a corporate structure to deal with financial transfers and tax revenues from resource development. In total, ANSCA created twelve regional corporations and 200 village corporations which received title to surface and subsurface lands (a thirteenth corporation was created for Alaska Native people living elsewhere in the United States). Seventy percent of the revenue from natural resources was then redistributed equally among the regional corporations.

The Canadian land claims which followed gave Canadian Inuit more extensive political sovereignty than in Alaska, though much like in Alaska, the Canadian Inuit gave up title to their land in exchange. In Canada, Inuit sovereignty has largely become embedded in the idea that strong Inuit sovereignty helps to bolster Canadian sovereignty over the Canadian Arctic; particularly the Northwest Passage which Canada regards as internal waters. To help buttress these sentiments, in 2008, Inuit Tapiriit Kanatami (ITK), the national Canadian organization which represents Canada's Inuit, decided on a change of their terminology from 'Inuit Nunaat' – meaning Inuit homeland – to 'Inuit Nunangat':

> 'Inuit Nunaat' is a Greenlandic term that describes land but does not include water or ice. The term 'Inuit Nunangat' is a Canadian Inuktitut term that includes land, water, and ice.

> As Canadian Inuit consider the land, water and ice, of our homeland to be integral to our culture and our way of life it was felt that 'Inuit Nunangat' is a more inclusive and appropriate term to use when describing our lands.
>
> *(www.itk.ca/publication/maps-inuit-nunangat-inuit-regions-canada)*

The relinquishment of title in the United States and Canada is distinct from self-determination in Greenland. Though Greenland Home Rule, like Alaska and Canada, was also driven by resource politics, the Greenlandic government has slowly gone through a process from devolution to re-centralization in Nuuk (Nuttall 2017). Since Greenland Home Rule was established in 1979, successive governments have gone through several Commissions, each of which concluded with greater self-autonomy, and in June 2009 Self-Rule was passed giving Greenlanders total control over all surface and subsurface rights. Under Self-Rule, all resource revenues go towards paying against the Danish block grant allocated to Greenland every year. Today, the Greenlandic government has control over almost all of its affairs including international affairs except those issues that implicate Danish security (e.g. exporting uranium) (Olsen and Shadian 2016).

The Sámi in the Nordic Arctic have acquired a significantly distinct form of self-determination from Inuit in North America and Greenland. Most notably, the biggest difference between Sámi governance and Inuit governance is that in the Nordic countries indigenous governance is managed by the state. However, much like North America, Sámi issues became most publicly politicized surrounding issues of resource extraction and development on Sámi lands. In Norway, Sámi concerns came to a head when the government wanted to build a hydroelectric dam on what was argued a Sámi watershed. There is no clearly defined Sámi homeland within Norway and it was only when the country's perceived need for energy came into conflict with Sámi rights and issues relating to the environment that the wider issue of the 'Sámi question' became an issue of political concern. The debates that followed led to the *Sámi Act* of 1987 which obligated State authorities to create the conditions necessary for the Sámi to protect and develop their language, their culture, and their society. This was done through a Constitutional guarantee which states that the Sámi are an indigenous people and have a legal and political guarantee for the protection and development of the Sámi language, culture, and society (Nettheim, Meyers, and Craig 2002).

Two years later, in 1989, the Sámi Parliament in Norway was established under the provisions of the *Sámi Act*. It required all national, regional, and local authorities to consult with the Sámi Parliament before making any decisions which may affect the Sámi people. The Sámi Parliament primarily acts as an advisory body to the Norwegian Parliament, which bears the financial responsibility for activities of the Sámi Parliament and its subsidiary bodies. Sufficient funds must be made available by the Norwegian Parliament in its annual budget to meet these purposes (ibid.).

While Norwegian Sámi live throughout all of Norway, a large majority of the Sámi live in Finnmark County and in 2005 the *Finnmak Act* was passed. In an attempt to strengthen Sámi rights, the *Finnmark Act* gave the Sámi greater influence over the property of Finnmark County. A major reason for this was to address issues of reindeer herding. The Act does not, however, cover fishing rights in saltwater, mining, or oil rights (ibid.).

In Sweden, like Norway, there is no specific Sámi territory. The Swedish government, rather than increasing Sámi rights as indigenous peoples, views the protection of the Sámi under the larger umbrella of equal rights for all Swedish citizens. Nevertheless, in 1992, the *Sámi Act* was passed creating the Sámi Parliament. The Saami Council of Sweden is a state authority and the members are elected by Swedish Sámi citizens. In 1993, the Swedish Parliament passed another law which allowed for small game hunting above the cultivation line and in reindeer grazing mountains. Effectively, the law opened traditional Sámi hunting grounds to all Swedish citizens. It was a controversial decision in that it undermined the previous exclusive Sámi privileges to certain hunting and fishing areas. Together, both the Swedish and the Norwegian Sámi parliaments do not have formal legal positions or authority with respect to the use and management of traditional Sámi land. This is a major divergence from the co-management structures set up in Canada as well as co-management regimes in Alaska and Greenland Self-Rule (ibid.).

In Finland, the Finnish Sámi Parliament was established in 1973. In 1995, the Finnish Constitution and the passage of its *Sámi Act* recognized 35,000 square kilometres of Northern Finland as the 'Sámi Homeland'. Within the Sámi homeland, the Sámi have a right to cultural autonomy. According to the Act: "The Sámi as an indigenous people shall, according to the provisions in the law, be ensured cultural autonomy within their Homeland area, in relation to their language and culture". The Act also states that all national authorities in Finland are obligated to negotiate with the Sámi Parliament in matters of concern and relevance to the Sámi. Among other things this includes: management, use, leasing and designation of state lands; conservation and wilderness areas as well as applications for mining licences. The obligation does not give the Sámi Parliament a power of veto but it does require that State authorities negotiate with the Sámi Parliament to resolve any disagreements. There is nothing similar to this obligation to negotiate with the Sámi in either Norway or Sweden (ibid.).

In Russia, the Sámi have attained far fewer rights than in the Nordic countries. In 2010 the Kola Sámi Assembly was created. It is comprised of a body of elected officials though its origins reach back to a 1992 Sámi Conference. The united Sámi of all four countries created a common 'Sámi People's Day' (Berg-Nordlie 2011: 57). From that conference, the Kola Sámi determined to begin to work towards a Kola Sámi Parliament (ibid). In 2008, the initial Russian Sámi Congress took place. The aim was to create a Russian Sámi Parliament that is elected by local Sámi. It was also decided that there would be a Council of Representatives that would work to establish the Parliament and represent the Russian Sámi. In 2010, a second Congress of the Russian Sámi took place in Murmansk. At that time, the Kola Sámi Assembly (Kuelnegk Soamet Sobbar) was elected. It was also decided that the Congress would be held every four years. The aims of the Assembly are to represent the Sámi people at the domestic and international levels and to work towards a recognized Russian Sámi Parliament (Berg-Nordlie 2011: 70). Presently, the Russian Sámi Council is not formally recognized by the Russian Government. They are, however, represented on the international Saami Council as the Kola Sámi Association (AKS) and the Sámi Association of the Murmansk Region (OOSMO).

Arctic regime building: indigenous diplomacies and the Arctic Council

In the context of Arctic regional regime building, Arctic indigenous groups were actively involved from the outset. In 1988, the Finnish government led consultations with the seven other Arctic states.[3] These consultations eventually led to the creation of the Arctic Environmental Protection Strategy (AEPS) which was a non-binding strategy signed by the eight Arctic states in 1991. During the discussions leading to the AEPS, the ICC called on the Arctic states to create an "effective cooperative process" which would ensure direct indigenous participation in all matters of Arctic co-operation (Simon 1990: 8). At that time, the ICC argued that "relations of state governments with indigenous peoples are required to be based on principles of co-operation and respect, rather than on unilateral state action" (ibid.: 8). According to one former ICC leader, Mary Simon, "involvement solely as members of state government delegations or as 'observers' would not suffice and would not meet even minimal international standards concerning indigenous participation" (ibid.: 7–8, 13).

By the last meeting leading up to the AEPS (1991) the ICC, Saami Council, and the Russian Association of Indigenous Peoples of the North (RAIPON) had become full participants in the processes of Arctic regime building. It was the first time that indigenous organizations had their representatives working alongside state diplomats in the crafting of intergovernmental policies. The AEPS was followed by the establishment of the Arctic Council. In September 1996, the Arctic states and three indigenous organizations (the ICC, the Saami Council, and RAIPON) met in Ottawa where the eight states signed the Declaration creating a new Arctic political collaboration. The Arctic Council became a consensus-based body, instituted through political declaration rather than a legally-binding charter. The central mandate of the Arctic Council was to help facilitate sustainable economic and social development in the Arctic and to date, it remains the only fully circumpolar intergovernmental institution. It was also decided at that time that the three indigenous groups would become something entirely unique in the realm of traditional international relations. They were officially designated as Permanent Participants (PPs). The three PPs were mandated with helping in "articulating the consensus" at the Senior Arctic Officials (SAO) and ministerial meetings. Since this time three other indigenous organizations have also become PPs (making a total of six today): the Arctic Athabaskan Council (AAC), the Aleut International Association (AIA), the Gwich'in Council International (GCI), the Inuit Circumpolar Council (ICC), the Russian Association of Indigenous Peoples of the North (RAIPON), and the Saami Council.

While the role of PPs was not entirely satisfactory to the indigenous representatives present at the negotiations (they were not invited to formally sign the Ottawa Declaration) the position of PP was, nonetheless, a historic shift in international governance. The six PPs sit at the table as ministers, debate the issues with the other ministers, and are recognized by the Chair in all matters. While their disapproval cannot technically break consensus (a point of dissatisfaction during the negotiations) their level of influence will rarely allow a decision on which they disagree to move forward without a high level of consideration by the Arctic states. During the two-year period between ministerial meetings, PPs are also full partners in all working groups including having the ability to submit projects and activities.

Up until 2007, the Arctic Council focused its attention to issues of low politics, namely sustainable development and environmental conservation and protection. At the first Arctic Council Ministerial meeting in 2009, following Artur Chilingarov's flag planting at the North Pole (which unfolded alongside new data on all-time lows for summer ice extent), it became apparent that Arctic politics and thus the Arctic Council was on the cusp of major change. By the 2009 Ministerial meeting world interest in the Arctic had grown exponentially and by the sheer numbers attending the meeting in Tromsø, Norway, it was apparent that the Arctic Council had outgrown a politics constrained to a discourse of sustainable development.

Recognising the impacts of this global interest that was already underway in the two years leading up to this meeting, the ICC took the opportunity at that time to clarify what and who the ICC represents in Arctic and global governance. At that meeting the ICC released its 'Circumpolar Inuit Declaration on Sovereignty in the Arctic'. The Declaration asserts that, not only are Inuit essential to the debates over Arctic governance, but even further, Inuit have sovereign rights to contribute to Arctic policy-making by virtue of their status as 'rights holders' – something entirely unique and distinct from other stakeholders. This includes being party to all discussions regarding who owns the Arctic and who has the right to develop its resources (Shadian 2014).

Since that 2009 Arctic Council Ministerial meeting, global interest in the politics of the Arctic has continued to grow considerably and a central facet of this interest is the question over what type of legal and political power do the PPs maintain and should continue to hold in Arctic governance. For many of the new observers which joined the Artic Council in 2011 including China, South Korea and Japan, the PPs continue to be viewed as 'stakeholders'. Though the ICC asserts that it is a 'rights holder' rather than a stakeholder this concept has very little international legal clarity for countries such as China or South Korea which orbit in a global politics where states operate under the traditional umbrella of state sovereignty.

While the UN Declaration on the Rights of Indigenous Peoples can very well be argued to have transformed the boundaries of the UN as well as international law by bringing non-state indigenous collectivities directly into the formal realm of the UN in addition to providing a minimal set of obligations by which states must consult with indigenous peoples, it does not provide guarantees for the future political power of the PPs on the Arctic Council. Though it is very often argued by many, from non-state observers to industry, that the PPs play an important role on the Arctic Council and that they will consult with indigenous peoples when it is an issue that concerns them, the ongoing challenge that IR and legal theorists are faced with is what specific rights and governing power can and do the Arctic's indigenous peoples – who have rights and sometimes ownership over Artic land, seas, and resources – expect to have in Arctic governance going into the future? This quasi-legal space in which the PPs operate poses questions that are not reserved for the Arctic Council but include the proliferation of additional Arctic and global governance arrangements which sit outside the formal parameters of conventional interstate relations altogether and is the focus of the following section.

Indigenous self-determination in a global Arctic

Over the past several decades, global politics has seen a shift from international relations or interstate co-operation which leads to formal treaties to global governance where a myriad of non-state actors participate in the creation of new legal norms, best practices, hybrid, and other multilayered governance solutions. These include integrated management, ecosystem management, and marine spatial planning, to name a few. Central to these governance frameworks is the participation of a wide variety of non-state actors from scientists and consultants to indigenous collectives and private companies (Karkkainen 2004).

Yet, who is invited, to what extent non-state entities participate, and who decides is not always clearly defined or understood. In effect, political power has become decentralized away from the state towards multiple, new centres of authority including local, regional, and international policies and law including, in the Arctic, indigenous governments and corporations as well as non-state (non-binding) compliance measures from the International Standards Organization (ISO) and the International Association of Oil & Gas Producers (IOGP) to Moody's. Taking note of the complex world in which governance now operates, a group of legal experts have written about ways in which conventional regulatory models are being challenged by the rise of and incorporation of various scales and levels of governance and non-government mandates in regional and global governance.

Though interdisciplinary and often concerned with one particular issue, what these theories share in common is the emphasis on incorporating decentralized, multi-stakeholder, informal arrangements into ideas of governance. Taking these commonalities into consideration Osofsky, Shadian and Fechtelkotter (2016) look at emerging forms of "hybrid cooperation" in the Arctic. Hybrid cooperation, according to these authors, focuses on how a variety of public and private stakeholders in the Arctic – including indigenous peoples and corporations, among others – can and are currently actively contributing to the making and operation of multilevel law and governance (Osofsky, Shadian and Fechtelkotter 2016). According to them, governance includes:

[t]hose systems which not only directly include non-state actors and varying levels of governments but also systems which directly include or incorporate rules, regulations, and standards of other entities (e.g. standard setting organizations, regional regimes, indigenous legal infrastructures, domestic law).

(Osofsky, Shadian and Fechtelkotter 2016)

Using this theoretical construct as a point of departure, the following two sub-sections focus on the role of indigeneity in the creation of what is viewed as two types of Arctic governance constructs. The first is what can be considered as 'traditional' forms of Arctic co-management which are based on formal policy instruments. The second includes newly emerging and other existing forms of hybrid cooperation which are driven by soft-law mechanisms.

Traditional co-management

In North America, early forms of hybrid structures in the Arctic emerged through indigenous aims for greater self-determination beginning in the early 1970s. Many of these aims were brought about by expectations of resource development on indigenous inhabited lands. In North America, the land claims were quite often in reaction to proposed resource developments. In many circumstances indigenous communities were not against resource development but instead were focused on how they could control and benefit from the development of resources on their land. For many indigenous communities, resource development has been viewed as one means to improve standards of living and gain further cultural-economic autonomy. Along those lines, in May 2011, the ICC released a Circumpolar Inuit Declaration on Resource Development Principles in Inuit Nunaat. The Declaration argues that:

[r]esponsible non-renewable resource development an also make an important and durable contribution to the well-being of current and future generations of Inuit. Managed under Inuit Nunaat governance structures, non-renewable resource development can contribute to Inuit economic and social development through both private sector channels (employment, incomes, businesses) and public sector channels (revenues from publicly owned lands, tax revenues, infrastructure) . . . Inuit welcome the opportunity to work in full partnership with resource developers, governments and local communities in the sustainable development of resources of Inuit Nunaat, including related policy-making, to the long-lasting benefit of Inuit and with respect for baseline environmental and social responsibilities.

(ICC 2009)[4]

Indigenous aims for self-determination in North America culminated with varying forms of political rights and autonomy including public governments. Yet, they also include specific indigenous rights to and sometimes ownership over those lands. Moreover, while the land claims agreements are often co-governance arrangements which include substate and state governments, the resources being managed and the revenues generated are often controlled by local indigenous corporations.

Looking specifically at Inuit governance, although Alaskan boroughs such as the North Slope Borough, the Canadian Inuit land claims agreements, and Self-Rule in Greenland are fundamentally different from one another in terms of taxation, compensation, rights, royalties, and resource management they all share a governance structure based on Inuit notions of stewardship. They have also established unique approaches to financing Inuit governments and the economic development of their territories through a combination of public governments, co-management resource regimes (which account for subsistence activities, such as whaling, fishing, and hunting), traditional knowledge (called Inuit Qaujimajatuqangit, or IQ in Nunavut), and Inuit corporations to determine how governments should create and deliver their programmes and services and manage their resources.

Inuit corporations such as Makivik (the legal representative of Inuit in Northern Quebec) or the Ukpeavik Inupiat Corporation (Inupiat is the term for the Inuit of northern Alaska), for instance, earn over 300 million USD in annual revenues. The Kuukpik Corporation in Alaska earns close to 5 million USD per year (Larsen and Fondahl 20042004: 133) and in 2014 the Arctic Slope Regional Corporation earned revenues exceeding 2.7 billion USD.[5] Inuit corporations very often rely directly on the forces of the global economy. Subsequently, the development and exportation of hydrocarbons and other non-renewable resources, such as diamonds, and shipping or flying them to markets around the world (with Inuit shipping, airline and other transportation companies) therefore effectively places Inuit and Inuit controlled resources at the heart of the 'global Arctic' (Shadian 2017).

More than forty years have now passed since the first Arctic indigenous land claim was settled in Alaska. As global interest in the Arctic expands and debates about what is 'best' for the Arctic continues, indigenous co-management practices which have been in place for the past four decades require a central place in these debates. Yet, beyond these conventional governance structures the growing numbers of alternative forms of non-state led hybrid cooperation are occupying a growing space in the world of Arctic politics.

Hybrid institutions

Beyond 'conventional' Arctic co-management governance structures are the wide range and growing numbers of public/private collaborations which are working together to monitor, make recommendations, set standards, and even divide industry profits. This section focuses on a number of such corporations in Alaska. Alaskan politics since the 1970s can largely be understood through a prism of oil politics. Eben Hopson, the founder of the ICC, stated at that time that "the politics of the Arctic are no longer the politics of the people, but they are the politics of oil" (Hopson 1976). More recently, Alaskan politics has expanded to include increasing shipping in the region. These realities have helped to initiate a number of unique forms of hybrid cooperation which aim to mitigate the potential negative impacts that these activities can create.

One example is the Open Water Season Conflict Avoidance Agreement (CAA) which was created by the Alaska Eskimo Whaling Commission (AEWC) in 1996 to manage offshore oil and gas impacts. The AEWC is a collaboration between eleven Inupiat subsistence whaling associations from St. Lawrence Island in the Bering Sea to Kaktovik in the Beaufort Sea (Aron 2000 in Shadian 2013). The AEWC came into being as a consequence of the shifting direction of the International Whaling Commission (IWC) from the conservation of bowhead whales to eliminating all bowhead whaling as well as a growing concern over the increase of oil exploration and development in Alaska (Shadian 2013) and its aim is to protect the bowhead whale.

Since its founding, the AEWC through its CAA has been involved with "offshore oil and gas exploration and development companies, including oil majors, in an annual process of collaboration and negotiation to create mitigation measures capable of avoiding adverse impacts to bowhead whales, habitat, and hunting opportunities". This includes an insurance agreement which provides logistical support to subsistence hunters and compensation should an oil spill occur (Shadian 2013). The CAA is based on an annually revised agreement comprised of a collaboration between the subsistence hunters and offshore oil and gas operators who meet "face-to-face" every year (Lefevre 2013: 10894). According the legal representative Jessica Lefevre, "the annual process allows the stakeholders to refine management techniques over time, based on experience, so that they provide the necessary mitigation of impacts with the least disruption to planned activities" (ibid.). Its work is based on sound science which is informed directly from the observations by Inupiat subsistence hunters (Brower 2009 in Shadian 2013). Through participation in the Open Water Season Peer Review Meeting, the CAA is able to expand its practices of "blending our Traditional Knowledge of the arctic ecosystem with western science and research" (Lefevre 2013).

Despite it being an independent institution (it is not a government entity), the CAA's work has resulted in "amendments to federal law, creating a role for the CAA in Arctic offshore development planning and making AEWC and the North Slope Borough part of the scientific review process for offshore development". The National Marine Fisheries Service Alaska (NMFS) relies on the CAA as does the Bureau of Ocean Energy Management, Regulation and Enforcement in order to meet requirements for its permits (Brower 2009 in Shadian 2013).

In a related example, the Alaska Beluga Whale Committee, the Eskimo Walrus Commission, the Ice Seal Committee, the Alaska Nanuuq Commission, and the Alaska Eskimo Whaling Associations formed the Arctic Marine Mammal Coalition (AMMC) in September 2001 to begin to address the effects of increased shipping in and around that region from Barrow through the Aleutian Islands (Chukchi and Beaufort Seas). The AMMC aims to "represent the Arctic marine mammal hunters' interests relating to the potential adverse impacts from commercial ship traffic on marine mammals".

More recently, the Arctic Waterways Safety Committee (AWSC) was created in October 2014. The AWSC is a self-governing multi-stakeholder group focused on creating and documenting best practices to ensure a safe, efficient, and predictable operating environment for all users of the Arctic waterways. This includes aims to balance the diverse interests in the Arctic waterways in the face of increased maritime travel, due in part to Arctic offshore drilling. The committee members consist of three categories: Subsistence Hunters, Industry, and Other representatives. Ultimately, the committee is an opportunity to give various stakeholders a forum to solve differences in the Arctic waterways without involving regulatory intervention from federal authorities, therefore avoiding a drawn out bureaucratic process (Osofsky, Shadian and Fechtelkotter 2016).

One last example is Arctic Inupiat Offshore, LLC. In July 2014, the Arctic Slope Regional Corporation (ASRC 2014) and six North Slope village corporations joined together to create a new company known as the Arctic Inupiat Offshore, LLC (AIO). AIO and Shell Gulf of Mexico Inc. (Shell) then entered into a binding agreement that would allow AIO the option to acquire an interest in Shell's acreage and activities on its Chukchi Sea leases. This interest would be managed by AIO (ASRC Press Release).

According to the agreement, Shell would assign AIO an overriding royalty interest in oil and gas produced from specific Chukchi Sea leases. AIO also would have the option to participate in project activities by acquiring a working interest at the time Shell makes the decision to proceed with development and production. Shell and AIO would also hold quarterly meetings to exchange information and address regional and development issues (ibid.) Shortly following this collaboration, however, Shell Oil opted not to pursue Chukchi Sea drilling for the foreseeable future. Despite limiting the current practical impact of the agreement, it continues to serve as an interesting example of hybrid cooperation (Osofsky, Shadian and Fechtelkotter 2016). Though not comprehensive and only focusing on Alaska, the examples of hybrid cooperation discussed here further bring into focus the additional facet of public/private partnerships in which indigenous corporations, associations, and governments play a central role and which need to be studied and taken seriously in the ongoing debates over Arctic governance.

Conclusion

The history of Arctic politics has been inexorably tied to the making and expansion of the Westphalian political system. Global interest in the Arctic extends back to the early days of commercial whaling and the founding of the Muscovy, Hudson's Bay, and other companies that set out to exploit the region's riches. The colonization of the region's indigenous peoples by Europeans (and later by Canadians and Americans) and Russians, the Cold War, the further discovery of new resources, and varying forms of indigenous autonomy that followed were all major steps in transforming the ways in which the Arctic was governed and by whom. The end of the Cold War paved the way for the creation of the Arctic Council, which added yet another building block to the Arctic's governing architecture (Shadian 2014; see the chapter by Timo Koivurova in this volume). Arctic indigenous diplomacies have been central to this history. The creation of the wide ranging forms of indigenous governance from land claims agreements and co-management to public/private hybrid forms of cooperation are the culmination of this longstanding relationship between the Arctic's indigenous peoples and the states which came to occupy their homelands.

The growing interest in the Arctic in recent years has opened the door to an extensive debate over who owns the Arctic and who should govern the region. Lacking in these debates, however, is an adequate discussion or understanding of the history and role of indigenous diplomacies. The political aims that indigenous groups have made in recent decades, and present diplomacies between indigenous and non-indigenous peoples, cannot be viewed as a break from the past, but as an extension of that past, and examples such as Greenlandic Self-Rule, the land claims agreements, ongoing devolution in Canada and Alaska, cultural autonomy for the Sámi, and the broader issue of who owns the Arctic are all institutional renewals of an ongoing relationship between aboriginal and non-aboriginal societies. This historical relationship is the foundation upon which the Arctic's indigenous peoples' legitimacy is grounded, and upon which their subsequent diplomatic successes in helping to define the Arctic as a political region, alongside broader contributions to a new international discourse of human rights (especially as it relates to sustainable development), rest.

Beyond the governance mechanisms which were put into place through land claims agreements at the domestic level, and the creation of a minimal set of standards for indigenous people's rights to land and resources through international law (specifically UNDRIP), is the role of the private sector. The now long-standing co-management arrangements which were made between subnational and state governments and Arctic indigenous peoples also included corporate entities which were established through those negotiations. While the indigenous corporations vary in the extent to which indigenous peoples have shared authority over land and on and offshore resources, they have in certain places included rights and ownership, as well as revenue sharing and revenue ownership over the Arctic's resources, land, and marine areas. Such forms of property rights in the Arctic have become critical forces allowing Arctic indigenous peoples not only a voice in decision-making but also control in international, regional, and national processes of Arctic resource development and management (Osofsky, Shadian and Fechtelkotter 2016).

Combined, the varying forms of indigenous diplomacies and governance mechanisms discussed in this chapter are symbolic of the changes that conventional Westphalian sovereignty and politics are undergoing. Indigeneity studies not only brings into question conventional assumptions about these concepts, but further indigenous diplomacies have forced IR scholars to rethink how global politics can effectively account for and accommodate the rising numbers of non-state actors which maintain varying degrees of political power above, below and across formal state borders. The current debates over who owns the Arctic, and who should govern thus, cannot be fully informed if they do not similarly take the history of indigenous politics and the changing landscape in which global politics operates seriously.

Notes

1 See endnote 3. Also see e.g. European Union 2008. 'The European Union and the Arctic Region'. Communication from the Commission to the European Union Parliament and the Council, 20 November, Brussels. Available at: http://eur- lex.europa.eu/LexUriServ/LexUriServ. do?uri=COM:2 008:0763:FIN:EN:PDF (accessed 9 September 2013); U.S. Department of State. May 2013. 'National Strategy for the Arctic Region'. The White House. Available at: www.whitehouse.gov/sites/default/files/docs/nat_ arctic_strategy.pdf (accessed 10 September 2013); Norwegian Ministry of Foreign Affairs. 2009c. 'New Building Blocks in the North: The Next Step in the Government's High North Strategy'. Norwegian Ministry of Foreign Affairs. Available at: www.regjeringen.no/upload/UD/Vedlegg/Nordomr%C3%A5dene/new_building_blocks_in_the_north.pdf (accessed 10 September 2013); Government of Canada. N.d. 'Canada's Northern Strategy'. Available at: www.northernstrategy.ca/cns/cns-eng.asp (accessed 16 September 2013).
2 www.un.org/development/desa/indigenouspeoples/declaration-on-the-rights-of-indigenous-peoples.html.
3 These are the US, Norway, Sweden, Russia, Iceland, Denmark, and Canada.
4 www.inuitcircumpolar.com/resource-development-principles-in-inuit-nunaat.html.
5 Forbes. www.forbes.com/companies/arctic-slope-regional-corporation/.

References

Alaska Humanities Forum 2004–2013. 'Alaska history and cultural studies', www.akhistorycourse.org/articles/article.php?artID=315.
Aron, W. 2000. 'The International Whaling Commission: a case of malignant neglect' in R.S. Johnston (ed.) and Shriver, A.L. (compiler) *Microbehavior and Macroresults*. International Institute of Fisheries Economics and Trade (IIFET) (Proceedings of the Tenth Biennial Conference of the International Institute of Fisheries Economics and Trade, 10–14 July 2000, Corvallis, Oregon, USA).
ASRC 2014. 'North Slope Village Corporations and Shell announce historic venture' Press release, 31 July 2014, www.asrc.com/PressReleases/Pages/Arctic-Inupiat-Offshore.aspx.
Beier, J.M. 2007a. 'Inter-national affairs: indigeneity, globality and the Canadian state' *Canadian Foreign Policy Journal* 13(3): 121–131.
Beier, J.M. 2007b. 'Introduction: indigenous diplomacies' *Canadian Foreign Policy Journal* 13(3): 9–11.
Beier, J.M. 2009. 'Forgetting, remembering, and finding indigenous peoples in international relations' in J.M. Beier (ed.) *Indigenous Diplomacies*. New York: Palgrave Macmillan, 11–28.
Berg-Nordlie, M. 2011. 'Striving to unite. The Russian Sámi and the Nordic Sámi Parliament model' *Arctic Review on Law and Politics* 2(1): 52–76.
Broderstad, E.G. and J. Dahl 2002. 'Political systems' *Tromsø, Norway: Commissioned by the Standing Committee of Parliamentarians of the Arctic Region (SCPAR) in preparation for a discussion of the future of Arctic governance*.
Canadian Museum of Civilization Corporation 2001. 'Canadian Inuit history: a thousand-year odyssey'. www.civilization.ca/educat/oracle/module.dmorrison/page01_e.html.
Castree, N. 2004. 'Differential geographies: place, indigenous rights and "local" resources' *Political Geography* 23(2): 133–167.
Corntassel, J. 2007. 'Towards a new partnership? Indigenous political mobilization and co-optation during the first UN indigenous decade (1995–2004)' *Human Rights Quarterly* 29(1): 137–166.
Dittmer, J. and F. McConnell (eds) 2015. *Diplomatic Cultures and International Politics: Translations, Spaces and Alternatives*. London: Routledge.
Greenberg, J. 2009. 'The Arctic in world environmental history' *Vanderbilt Journal of Transnational Law* 42: 1307–1392.
Hopson, E. 1976. 'Testimony before the Berger inquiry on the experience of the Arctic Slope Inupiat with oil and gas development in the Arctic' *Eben Hopson Memorial Archives*. www.ebenhopson.com/papers/1976/BergerSpeech.html.

Innuksuk, R. 1994. 'Inuit self-determination: the role of the Inuit circumpolar conference' Presentation at the *Seminar on Self-Determination by Indigenous Peoples*, 8–10 February 1994, Amsterdam, Netherlands.

Inter-American Commission on Human Rights 1997. 'Proposed American Declaration on the Rights of Indigenous Peoples' (Approved by the Inter-American Commission on Human Rights on 26 February 1997, at its 1333rd Session, 95th Regular Session). Human Rights Library, University of Minnesota. Online. www1.umn.edu/humanrts/instree/indigenousdecl.html.

Inuit Circumpolar Council 2011. 'A circumpolar Inuit declaration on resource development principles in Inuit Nunaat'. www.inuitcircumpolar.com/uploads/3/0/5/4/30542564/declaration_on_resource_development_a3_final.pdf.

Karkkainen, B. 2004. 'Post-sovereign environmental governance' *Global Environmental Politics* 4(1): 72–96.

Keck, M.E. and K. Sikkink 1998. *Activists Beyond Borders: Advocacy Networks in International Politics*. New York: Cornell University Press.

Larsen, J.N. and G. Fondahl (eds) 2004. *Arctic Human Development Report: Regional Processes and Global Linkages*. Copenhagen: Nordic Council of Ministers.

Lefevre, J.S. 2013. 'A pioneering effort in the design of process and law supporting integrated Arctic ocean management' *Environmental Law Reporter* 43(10): 10893–10908.

Linklater, A. 1998. *The Transformation of Political Community: Ethical Foundations of the Post-Westphalian Era*. Columbia, SC: University of South Carolina Press.

Mitchell, M. 1996. *From Talking Chiefs to a Native Corporate Elite: The Birth of Class and Nationalism Among Canadian Inuit*. Montreal, QC and Kingston, ON: McGill-Queen's University Press.

Nettheim, G., G.D. Meyers and D. Craig 2002. *Indigenous Peoples and Governance Structures: A Comparative Analysis of Land and Resource Management Rights*. Canberra: Aboriginal Studies Press.

Nuttall, M. 2017. *Climate, Society and Subsurface Politics in Greenland: Under the Great Ice*. London and New York: Routledge.

Olsen, I.H. and J.M. Shadian 2016. 'Greenland and the Arctic Council: subnational regions in a time of Arctic Westphalianisation' *Arctic Yearbook 2016*. 229–250.

Osofsky, H.M., J.M. Shadian and S.L. Fechtelkotter 2016. 'Preventing and responding to Arctic offshore drilling disasters: the role of hybrid cooperation' in J. Peel and D. Fisher (eds) *Role of International Environmental Law in Disaster Risk Reduction*. Leiden, Netherlands: Brill Publishers.

Rosenau, J.N. 1997. *Along the Domestic-Foreign Frontier: Exploring Governance in a Turbulent World*. Cambridge, UK: Cambridge University Press.

Rudolph, C. 2005. 'Sovereignty and territorial borders in a global age' *International Studies Review* 7(1): 1–20.

Shadian, J.M. 2010. 'From states to polities: re-conceptualizing sovereignty through Inuit governance' *European Journal of International Relations* 16(3): 35–62.

Shadian, J.M. 2013. 'Of whales and oil: learning from indigenous resource governance' *Polar Record* 49(4): 392–405.

Shadian, J.M. 2015. 'Not seeing like a state: Inuit diplomacies meet state sovereignty' in J. Dittmer and F. McConnell (eds) *Diplomatic Cultures and International Politics: Translations, Spaces and Alternatives*. London: Routledge.

Shadian, J.M. 2017. 'Navigating political borders old and new: the territoriality of indigenous Inuit governance' *Journal of Borderlands Studies*, 1–16.

Shaw, K. 2002. 'Indigeneity and the international' *Millennium: Journal of International Relations Studies* 31(1): 55–81.

Simon, M. 1990. 'Towards an Arctic sustainable and equitable development strategy: some preliminary views' in *Protecting the Arctic Environment: Report on the Yellowknife Preparatory Meeting*, 18–23 April, Yellowknife, Ottawa.

26

Geopolitics and security in the Arctic

Andreas Østhagen

As the international community rediscovered the Arctic as a geographic area of environmental and strategic interest at the start of the new millennium, researchers, media and policy-makers alike began launching and asserting a range of claims about the trajectory of the region. It was quickly heralded as the world's "new energy frontier", and the "next arena for geopolitical conflict".[1] When Russia's relationship with the other Arctic states began to deteriorate in 2014, stemming from Russia's annexation of Crimea and military intervention in Ukraine, headlines warning about imminent confrontation in the north re-appeared in the media (Dougherty 2015). Proving that such quick conceptions do not provide accuracy, a number of studies have outlined a more nuanced and far more complex situation than projected by some narratives and discourses that focus on the "new North".

This chapter builds on these studies and their claims, asking: what are the layers of potential inter-state conflict in the Arctic? Subsequently, how sufficient are one-sentence conclusions regarding the Arctic security environment? To contend, as some do, that the Arctic is "completely uninteresting geopolitically", neglects the role it serves for some of the Arctic states in their security considerations (Welch 2013: 2). The potential for conflict in the Arctic should not be overstated, but neither should it be ignored, nor should concerns about such potential be discarded, as parts of the region still stand as theatres for potential clashes with Russia. Yet, this has arguably little to do with symbolic quarrels over the North Pole, but everything to do with the relationship between Russia and the other Arctic states – both regionally and globally.

There is an additional need to differentiate between the various forms of 'security' in the region. Conceptualizations framed under the umbrella of human and environmental security will not be dealt with here.[2] Yet, it is important to recognize that they cannot be completely separated from the narrower concept of 'state security' or 'military security'. We should have an awareness of the various dimensions of security and what they entail in a changing Arctic. Military assets in particular perform a multitude of civilian tasks in the Arctic. In this chapter, however, the emphasis will be on how Arctic security has been dealt with from a state perspective, and how we can conceptualize and differentiate the Arctic as a so-called 'security region'.

Imminent conflict in the Arctic

When Russia annexed Crimea in 2014, media and political attention turned towards the Arctic. The north was depicted as yet another region where Russia is bolstering its military capabilities and flexing muscle. Military activity in the Russian parts of the Arctic have kept increasing after the region was neglected as the Cold War came to an end. From 2007 onwards, Russian bombers have increased in numbers, traversing along the north Norwegian coast or across the North Pole from the Kola Peninsula. Russian investment in military infrastructure in the Arctic has similarly been growing (Hilde 2013; Conley and Rohloff 2015; Expert Commission 2015: 17, 20). Canadian investment in a new Arctic Offshore Patrol Ships (AOSP) project, which will provide armed sea-borne surveillance of Arctic waters, and Norwegian acquisition of F-35 fighter jets, most of which will be stationed at airfields near Trondheim in south-central Norway and Narvik in the northern part of the country (Norway has given projected Russian activities in northern European airspace and Arctic waters as a reason for the purchase of the aircraft) have been placed in the same context. The argument is thus that the Arctic is being militarized, as littoral states are placing pieces on the chessboard in advance of an imminent geopolitical conflict. Subsequently, the 'Western' media's rhetoric and concern over activity in the Arctic has taken on a particularly demagogic language, with headlines such as "Cold War Echoes Under the Arctic Ice" (*Wall Street Journal*) and "Russia prepares for ice-cold war with show of military force in the Arctic" (*The Guardian*) (Barnes 2014; Mandraud 2014).

Yet, in an Arctic context, such headlines should not trigger surprise. The same spike in aggressive rhetoric from both 'Western' and Russian media came after the Russian planting of a titanium flag at the seabed of the North Pole in 2007. A seminal symbolic moment in the new Arctic era, it marked the purported scramble northwards for resources and territory (Grindheim 2009; Sale and Potapov 2010). In the aftermath, a number of scholars portrayed the Arctic as a region where empirical analysis could feed predictions of imminent conflict. I will turn to the attributes of potential conflict – and its diffusive mechanisms – later in the chapter. The goal in this context is to highlight how the upsurge in hype concerning Arctic conflict amongst media and scholars alike is not something that rose out of the Ukraine conflict in 2014. Instead, the possibility of 'geopolitical conflict' has been a reoccurring theme in discussions over the new Arctic security environment for a decade.

A particular dimension of these claims has been the use of 'geopolitical' or 'Arctic geopolitics' (Welch 2013; Tamnes and Offerdal 2014). The popular conception of what this involves varies. Inherently it says something about politics concerning a specific geographic area or geographic units. In an Arctic context, it can be interpreted as the emergence of new interests bound to territorial units, prompting a geographical conflict as other actors set their sights on the same units. Yet, the elusiveness of 'geopolitics' and what it truly entails, seems to have prompted widespread use of the word in Arctic descriptions (Haverluk et al. 2014). As such, it ties into the trend showcased above: describing the Arctic as the location for immediate conflict tied to power politics and profitable resource extraction (against the backdrop of climate change and a rapidly melting Arctic), in tandem with civilian and military activity increasing across the region.

Conflict *over* the Arctic?

Increased activity does not imply that a standoff is imminent. The foremost argument for why there would be a conflict over the Arctic has been the region's resource abundance. When examining the location and accessibility of these resources, however, the facts fail to bolster the argument. The resources – both onshore and offshore – are more or less located in what are already the economic zones or territories of the Arctic states themselves. Estimates vary since the totality of the resources is yet somewhat uncertain. Still, approximately 90 percent of the oil and gas resources of the circumpolar North are already under the control of the littoral states. This does not fuel a race northward to grab unclaimed resources. Instead it fuels a desire for stable operating environments to extract costly resources far away from their presumed markets (Claes and Moe 2014; Keil 2014).

This correlates with another point – namely the lack of economic profitability and slow pace of offshore resource development in the Arctic. Here, again, there has been a tendency to generalize across the region and discuss 'Arctic resources'. The truth is that the resource potential and accessibility greatly vary across the region, depending on which part of the Arctic is under examination (Harsem et al. 2011: 8040–8042; Østhagen 2013a). Oil and gas production is already taking place in the Barents Sea, an offshore area where more than 100 exploratory wells have been drilled. Climatic conditions are different in waters around Greenland or Alaska, and so are the infrastructural set-ups, prompting a different economic reality for operating companies. Thus Arctic resource development cannot be sufficiently generalized (and generalizing about Arctic ecosystems, societies and cultures similarly ignores the diversity of the region), and the lack of infrastructure, geographic distances and the harsh climate have made it questionable whether the Arctic littoral states can even exploit their own natural resource potential in their respective economic zones.

The Arctic riches are also more or less divided, with the largest maritime border dispute – between Norway and Russia in the Barents Sea – settled in 2010. Minor border disputes exist between the United States and Canada in the Beaufort Sea and between Canada and Greenland (Denmark) in the Lincoln Strait, but these disagreements will arguably not inspire conflict of any scale (Hoel 2009). The oft-cited dispute between Canada, Denmark/Greenland and Russia over who can claim the North Pole seabed is moreover unlikely to become anything but a diplomatic conflict, at worst. As argued by Michael Byers, the North Pole is a distraction (Byers 2014). The Arctic states have neither the economic nor the strategic incentive to undertake any significant operation to assert and establish further claims over the seabed of the North Pole. Symbolism is undoubtedly of great value, but the cost of North Pole operations does not match the Pole's perceived gains (Byers 2013: 281–283). Even if we do see a spike in worldwide commodity prices in the next decade, Arctic resource extraction will still remain a specialized, localized and costly affair.

Consequently, the Arctic states are mutually dependent on creating a favourable political environment for investments and economic development. The expectations of a scramble were thus founded on thin ice. Instead, Arctic relations in the twenty-first century have turned out to be surprisingly peaceful. Artur Chilingarov, the leader of the Russian flag-planting expedition in 2007, has likened the endeavour to the American flag-planting on the Moon. The goal of the North Pole venture was to showcase capabilities and interest, not incite conflict in a contested geopolitical region (BBC News 2007).[3] The five Arctic coastal states also convened in Ilulissat, Greenland, in 2008, to declare the Arctic as a region of co-operation and to affirm their intentions to work within established international arrangements and agreements, particularly the United Nations Convention on the Law of the Sea (UNCLOS – see, also, Donald Rothwell's chapter in this handbook on the Polar Regions and the law of the sea).[4] After Ilulissat, the mantra by *all* Arctic states has been co-operation (Heininen 2012). What is perhaps more surprising is exactly how streamlined the emphasis on co-operation across the region appears to be.

The Arctic Council's emergence as the primary forum for regional affairs in the Arctic plays into this, serving as an arena from which the states can portray themselves harmoniously and working towards common goals. Related is the convergence of state interests in the region. Whether or not this is a result of joint collaborative schemes like the Arctic Council,[5] or a coincidental interest alignment due to each states' preference in maximizing its relative power position vis-à-vis its neighbours, is an interesting debate on its own. Regardless, the Arctic states have shown preference for a stable political environment in which they maintain dominance in the region. This is supported by the importance of the UNCLOS and issue-specific agreements signed under the auspices of the Arctic Council.[6] These developments benefit the northern countries more than anyone else, while they ensure that Arctic issues overall remain an affair where the Arctic states are in the lead.

In sum, military conflict with other states to claim a limited number of out-of-bounds off-shore resources, many of which look likely to remain unexplored for the next couple of decades at least, is neither economically nor politically profitable. The resource argument for an outright conflict *over* the Arctic does not hold. Academics and experts examining the region's conflict potential have by and large been in agreement in recent years, concluding that direct conflict over the Arctic in itself is unlikely. As Tamnes and Offerdal argue, "[d]iscord does exist, but the main characteristics of the region are cooperation, stability and peace" (2014: 167).

Conflict *in* the Arctic?

The Arctic states themselves are, on the other hand, not exempt from conflict and instability. Although struggle over the Arctic is not cause for immediate concern, the regional relationships with Russia in the Arctic cannot be sheltered from the overarching deterioration of the relationship between Russia and the other Arctic states. Despite the emphasis on governance and regime building by governance scholars and Arctic foreign ministries, an amicable regional order is not an inherent and unchallenged trait of the region. The Arctic states have a multitude of interests at play simultaneously, some of which are in opposition to the idea of a peaceful region. In particular, this revolves around the balancing act between sovereignty enforcement and amicable bilateral relations. In this context, the occurrence of small-scale disputes should not be underestimated.

When operating in an Arctic maritime environment with limited communications and long distances, the potential for small-scale scenarios escalating is considerable. Take the case of the Russian trawler *Elektron*, which was operating in the Fisheries Protection Zone around Svalbard in October 2005. When it became clear it had conducted legal offenses, the vessel fled toward Russian waters with two Norwegian Fisheries Inspectors on-board, in effect kidnapping the officers. Only bad weather hindered Norwegian Special Forces from boarding the vessel. The situation was eventually solved through diplomacy and the involvement of the Russian Navy (Kosmo 2010: 47–55, Forsvarsdepartementet 2007). In another case in the Bering Sea in 1999, the Russian fishing vessel *Gissar* was boarded by crew from the United States Coast Guard cutter *Hamilton* on charges of illegal fishing in US waters.[7] With the boarding crew on the Russian trawler, both vessels were surrounded by a number of other Russian fishing vessels intent on blocking the *Hamilton* from taking the trawler to a port in Alaska. The incident was resolved with the arrival of the Russian Coast Guard (Kaczynski 2007).

These examples highlight the volatility of maritime situations taking place in the Arctic. Both cases were solved by diplomacy and interaction by coast guards across borders, although the incidents had potential to escalate into a further conflict. This is especially the case for situations involving resource management – with an emphasis on fisheries – where a coastal state is protecting its sovereign rights in areas that have been subject to dispute (Till 2005: 330–347; Kidmose et al. 2015: 23–27). Moreover, it should not be forgotten that such incidents are not without recent historical precedent, as skirmishes during the three Cod Wars between Iceland and Britain in the second half of the twentieth century illustrate, when Icelandic coast guard vessels and British gunships rammed one another and took shots across each other's bows. On the one hand, as the Cod Wars teach us and as the recent events around Svalbard and in the Bering Sea warn us, a diplomatic incident could escalate due to unintended actions taken by one of the parties. In an unpredictable international environment, such issues are volatile and have the potential to escalate beyond their status as minor disagreements. On the other hand, an Arctic state could feel the need to set precedence by turning an Arctic incident into an example for future interactions. This is particularly relevant because the Arctic plays well in domestic politics. Protecting the nation's sovereign rights and showing strength in the north have been particularly frequent mantras in Russia, Canada and Norway (Lackenbauer 2010: 881, 896–897; Jensen and Hønneland 2011: 44–47). Yet, we should recognize the inherent difference between outright 'conflict' between Arctic states, and small-scale conflictual incidents that may or may not flare up.

At the same time, the role the Artic plays in national defence considerations cannot be neglected. This, however, varies across the Arctic, as there is vast differentiation in what emphasis and focus each country has on its northern areas in terms of national security and defence. Russia in particular signals a desire to keep Arctic co-operation unharmed, while simultaneously expanding their military posturing in the Arctic for both symbolic and strategic purposes. As Katarzyna Zysk argues, Russian activity and rhetoric with regards to the Arctic may seem contradictory (2011, 2013: 294). Russia continues to emphasize co-operation and low-tension, coinciding with military investments and actions aimed at protecting their Arctic interests. Alexander Sergunin (2014) argues that most of Russia's increased military activity in the Arctic is a consequence of replacing out-dated and ineffective equipment, in preparation for the increased activity that Russia foresees in their northern areas. Much of this military activity is not linked to Arctic developments per se but comes as a consequence of Russia being an Arctic country with essential military bases located in the region. These bases are imperative to Russia's access to the North Atlantic, and its status as a nuclear power through its strategic submarines (Hilde 2014: 153–155).

For the Nordic countries, the Arctic is equally integral to national defence policies. In a Norwegian context, the 'High North' (nordområdene) constitutes the primary security concern for any government in Oslo. Through a two-track relationship with Russia, Norway aims both to cultivate a friendly neighbouring relationship *and* to showcase defence and sovereignty enforcing capabilities along its northern border (Flikke 2011; Jensen and Hønneland 2011; Jensen 2014). This is, however, not generally framed as an 'Arctic security' issue. Instead it is placed in the wider context of Norwegian national security and defence, related to Norway's relationship with Russia and as its role as a northern member of NATO. Similarly, Sweden and Finland place emphasis on their proximity to Russia, albeit as non-aligned states with ties to NATO through 'partnerships' and the Nordic Defence Cooperation (NORDEFCO). Iceland, as the midway point between North America and Europe, holds a highly strategic role in what was termed the G-I-UK-gap during the Cold War.[8] NATO and the Nordic countries share the responsibility of air policing exercises two to three times per year, after the US departed Keflavik air base in 2006. Beyond that, Iceland lacks a dedicated defence of its own and is subsequently dependent on its allies for assistance.

In North America,[9] the Arctic arguably does not hold the same seminal role in national security. Although rhetoric might suggest otherwise, the Arctic has primarily been the location of missile defence, surveillance infrastructure and a limited amount of strategic forces. Greenland is protected by the Danish Armed Forces (Arctic Command has its headquarters in

Nuuk, with a liaison office based at the Faroese capital Thorshavn), in addition to the existence of the American Thule Air Base. Yet, it has not held an integral role in Danish defence policy in modern times (Petersen 2011: 159–161; Government of Denmark 2013: 3, 10; Rasmussen and Struwe 2013: 25). Canada's vast Arctic territories are important from a strategic point of view, as well as a departure point for lofty symbolic statements concerning sovereignty and national identity made by politicians in Ottawa. Beyond that, however, recent Canadian governments have not prioritized Arctic military investment, as the threat from the north has been minimal. Commentators tend to argue that the most immediate concern for the Canadian Arctic is the social and health situation of indigenous people and the poor economic development of northern communities, in contrast to defence capabilities (Griffiths et al. 2011; Byers 2012; Lackenbauer 2013). Yet, in the Arctic, these two tend to go hand in hand, as the military performs a whole range of tasks crucial to Arctic inhabitants. Finally, Alaska has a somewhat more important role in US defence policy, bordering the Russian region of Chukotka across the Bering Strait. Both traditional forces and missile defence are located in the northern state. For the US Navy and the Coast Guard the region also holds importance, although the US has yet to invest in considerable Arctic capabilities and infrastructure (Zellen 2013: 228; Hilde 2014: 149).

In sum, the Arctic's importance in the various national security and defence policies varies quite considerably. The dividing line appears to fall between the European Arctic and the North American Arctic, in tandem with variations in climatic conditions. Whereas the north Norwegian and the northwest Russian coastlines are ice-free during winter, the ice – albeit melting – is an ever-constant factor in the Alaskan, Canadian and Greenlandic Arctic. Due to the sheer size and inaccessibility of the region, the spillover of security issues between the various parts of the Arctic is in turn relatively limited. Despite the overflow of rhetoric suggesting otherwise, Russian investment in Arctic troops and infrastructure has very little impact on the Canadian security outlook. Indeed, Russian overtures with bomber and fighter planes may cause alarm, but the real threat for the North American states in the Arctic is limited.

This is in contrast to the perception of an Arctic that generates its own hostile security environment. For that, the Arctic Ocean is too vast and remote. Had the Arctic Ocean been as frequently traversed (and as ice-free) as the Indian Ocean, for example, the dynamics would have been quite different. For the next generation, however, the security dynamics in the Arctic are mainly kept at a sub-regional level, i.e. in the Barents area, the Bering Sea/Strait area, and at a stretch, the Baltic Sea-region. It is thus somewhat futile to generalize security across the whole northern circumpolar region, as it makes more sense to discuss security in the specific parts of the Arctic rather than in the Arctic at large.

Conclusion

Hype concerning outright conflict over the Arctic is largely inaccurate. It does not disseminate the essence of more nuanced discussion of this vast region and the differing roles it plays in the security outlook of the Arctic states. Arguably, when thinking of Arctic security, it is more relevant to divide the area into sub-regions – separated roughly into spaces encompassing the North American and the Eurasian Arctic. It is also not possible to boil the dynamics of this region down to binary options of conflict or non-conflict scenarios. Small-scale conflict in parts of the Arctic is plausible, under certain conditions. In such circumstances, the relationship between Russia and the other Arctic states determines the parameters for the security environment. This is especially the case for incidents involving resource management – with an emphasis on fisheries – where coastal states are safeguarding sovereign rights.

Although conflict over the Arctic itself seems unlikely, a deteriorating relationship between Russia and the other Arctic countries could also pose problems for the Arctic security situation, in particular for those states bordering Russia. As Kristian Åtland (2014) argues, the effect of the conflict over Ukraine has thus been the development of an Arctic 'security dilemma'. As trust deteriorates and traditional avenues of military co-operation are disbanded, the other Arctic states begin to observe regional Russian troop movements and exercises with greater scepticism, and vice versa. Co-operating with Russia, however, is not a dichotomous choice. There are ways to sustain dialogue while not being at odds with a coherent response to Russian actions elsewhere.

An Arctic security environment, or the Arctic as a 'security region', must be understood – at a bare minimum – along two dimensions. The first is the role the Arctic has in the various countries, for security purposes. The second is the separation over outright cross-regional conflict, and smaller, more localized disputes that have the potential to derail amicable relations. Adding these arguments and dimensions to the debate can, at best, provide more clarity to our understanding of the Arctic, at the expense of one-sentence conclusions.

Notes

1 See for example Borgerson 2008; Skogrand 2008; Ingimundarson 2011 and Østhagen 2013b.
2 See Hoogensen Gjørv et al. 2014 for a conceptualization of the various security dimensions in the Arctic.
3 Notwithstanding there is a difference between the Moon and the North Pole seabed, as the United States did not claim parts of the Moon as an extension of US territory, whereas Russia did exactly that with its re-submission of hydrographic data in August 2015, to the UN Commission on the Limits of the Continental Shelf (CLCS).
4 By the five littoral states (nicknamed the Arctic Five) as a response to the hype after the Russian flag planting in 2007. The document can be found here: www.oceanlaw.org/downloads/arctic/Ilulissat_Declaration.pdf
5 Or in other arenas such as the Barents Euro Arctic Council (BEAR), the Nordic Council of Ministers, the various bilateral forums for dialogue, and NATO (which includes five out of eight Arctic states).
6 The Search and Rescue (SAR) agreement from 2011, the oil spill preparedness and response agreement from 2013, and the agreement on limiting black carbon from 2015. See www.arctic-council.org/index.php/en/ for more information.
7 Breaching the Baker-Shevardnadze agreement from 1990, which delimitates the maritime border in the Bering Sea between Russia (then USSR) and the United States.
8 Greenland, Iceland and the United Kingdom.
9 Including Greenland as geographically part of North America, although politically part of the Realm of Denmark (Europe).

References

Åtland, K. 2014. 'Interstate relations in the Arctic: an emerging security dilemma' *Comparative Strategy* 33(2): 145–166.
Barnes, J.E. 2014. 'Cold War echoes under the Arctic ice' *The Wall Street Journal*. Available at: www.wsj.com/articles/SB10001424052702304679404579461630946609454.
BBC News, 2007. 'Russia plants flag under N Pole' *World*. Available at: http://news.bbc.co.uk/1/hi/world/europe/6927395.stm.
Borgerson, S. 2008. 'Arctic meltdown: the economic and security implications of global warming' *Foreign Affairs* 87(2): 63–77.
Byers, M. 2012. 'You can't replace real icebreakers' *The Globe and Mail*. Available at: www.theglobeandmail.com/globe-debate/you-cant-replace-real-icebreakers/article534351/.
Byers, M. 2013. *International Law and the Arctic*. New York: Cambridge University Press.
Byers, M. 2014. 'The North Pole is a distraction' *The Globe and Mail*. Available at: www.theglobeandmail.com/globe-debate/the-north-pole-is-a-distraction/article20126915/.
Claes, D.H. and A. Moe 2014. 'Arctic petroleum resources in a regional and global perspective' in R. Tamnes and K. Offerdal (eds) *Geopolitics and Security in the Arctic: Regional Dynamics in a Global World*. London: Routledge. 97–120.

Conley, H.A. and C. Rohloff 2015. *The New Ice Curtain*. Washington, DC: Center for Strategic & International Studies (CSIS).

Dougherty, J.E. 2015. 'The next region where the U.S. and Russia could clash is the Arctic' *National Security News*. Available at: www.nationalsecurity.news/2015-10-09-the-next-region-where-the-u-s-and-russia-could-clash-is-the-arctic.html.

Expert Commission 2015. *Unified Effort*. Oslo: Norwegian Ministry of Defence.

Flikke, G. 2011. 'Norway and the Arctic: between multilateral governance and geopolitics' in J. Kraska (ed.) *Arctic Security in an Age of Climate Change*. New York: Cambridge University Press. 64–85.

Forsvarsdepartementet 2007. Arbeidsgruppen for utredning av oppgave- og myndighetsfordelingen mellom Kystvakten, Politiet og påtalemyndigheten Forsvarsdepartementet, ed., Oslo: Forsvarsdepartementet.

Government of Denmark 2013. *Danish Defence Agreement 2013–2017*. Copenhagen. Available at: www.fmn.dk/eng/allabout/Documents/TheDanishDefenceAgrement2013-2017english-version.pdf.

Griffiths, F., R. Huebert and W.P. Lackenbauer 2011. *Canada and the Changing Arctic: Sovereignty, Security and Stewardship*. Waterloo, ON: Wilfrid Laurier University Press.

Grindheim, A. 2009. *The Scramble for the Arctic? A Discourse Analysis of Norway and the EU's Strategies Towards the European Arctic*. Oslo: Fridtjof Nansen Institute.

Harsem, Ø., A. Eide and K. Heen 2011. 'Factors influencing future oil and gas prospects in the Arctic' *Energy Policy* 39(12): 8037–8045.

Haverluk, T.W., K.M. Beauchemin and B. Mueller 2014. 'The three critical flaws of critical geopolitics: towards a neo-classical geopolitics' *Geopolitics* 19(1): 19–39. Available at: www.tandfonline.com/doi/abs/10.1080/14650045.2013.803192.

Heininen, L. 2012. *Arctic Strategies and Policies: Inventory and Comparative Study*, (April), 1–97. Available at: www.rha.is/static/files/NRF/Publications/arctic_strategies_7th_draft_new_20120428.pdf.

Hilde, P.S. 2013. 'The "new" Arctic: the military dimension' *Journal of Military and Strategic Studies* 15(2): 130–153.

Hilde, P.S. 2014. 'Armed forces and security challenges in the Arctic' in R. Tamnes and K. Offerdal (eds) *Geopolitics and Security in the Arctic: Regional Dynamics in a Global World*. London: Routledge. 147–165.

Hoel, A.H. 2009. 'Do we need a new legal regime for the Arctic Ocean?' *International Journal of Marine and Coastal Law* 24(2): 443–456.

Hoogensen Gjørv, G., D. Bazely, M. Goloviznina and A. Tanentzap (eds) 2014. *Environmental and Human Security in the Arctic*. Abingdon, UK: Routledge.

Ingimundarson, V. 2011. 'Territorial discourses and identity politics' in J. Kraska (ed.) *Arctic Security in an Age of Climate Change*. New York: Cambridge University Press. 174–189.

Jensen, L.C. 2014 '"The times they are a-changin": 'Norsk sikkerhet og usikkerhet i nordområdene' *Internasjonal Politikk* 72(1): 7–29.

Jensen, L.C. and Hønneland, G. 2011. 'Framing the High North: public discourses in Norway after 2000' *Acta Borealia* 28(1), 37–54.

Kaczynski, V.M. 2007. 'US-Russian Bering Sea marine border dispute: conflict over strategic assets, fisheries and energy resources' *Russian Analytical Digest* 20 (May 1, 2007): 2–6.

Keil, K., 2014. 'The Arctic: a new region of conflict? The case of oil and gas' *Cooperation and Conflict* 49(2): 162–190.

Kidmose, J., K.S. Kristenensen and L.B. Struwe 2015. *Maritim sikkerhed i Arktis : Magtanvendelse og myndighedsudøvelse*. Copenhagen, Denmark.

Kosmo, S. 2010. Kystvaktsamarbeidet Norge-Russland. En fortsettelse av politikken med andre midler? Oslo: Norwegian Joint Staff College.

Lackenbauer, W.P. 2010. 'Mirror images? Canada, Russia, and the circumpolar world' *International Journal* 65(4): 879–897.

Lackenbauer, W.P. 2013. 'Harper's Arctic evolution' *The Globe and Mail*. Available at: www.theglobeandmail.com/globe-debate/harpers-arctic-evolution/article13852195/.

Mandraud, I. 2014. 'Russia prepares for Ice-Cold War with show of military force in the Arctic' *The Guardian*. Available at: www.theguardian.com/world/2014/oct/21/russia-arctic-military-oil-gas-putin.

Østhagen, A. 2013a. *Arctic Oil and Gas. The Role of Regions IFS*, ed. Oslo: Norwegian Institute for Defence Studies (IFS). Available at: https://brage.bibsys.no/xmlui/handle/11250/99831.

Østhagen, A. 2013b. 'Arctic oil and gas: hype or reality?' *The Fletcher Forum of World Affairs*, April 9. Available at: www.fletcherforum.org/home/2016/8/22/arctic-oil-and-gas-hype-or-reality.

Petersen, N. 2011. 'The Arctic challenge to Danish foreign and security policy' in J. Kraska (ed.) *Arctic Security in an Age of Climate Change*. New York: Cambridge University Press. 145–165.

Rasmussen, M.V. and L.B. Struwe 2013. *Megatrends i dansk sikkerheds- og forsvarspolitik*. Copenhagen, Denmark. Available at: www.ft.dk/samling/20131/almdel/upn/bilag/108/1335378.pdf.

Sale, R. and E. Potapov 2010. *The Scramble for the Arctic: Ownership, Exploitation and Conflict in the Far North*. London: Frances Lincoln.

Sergunin, A. 2014. 'Four dangerous myths about Russia's plans for the arctic' *Russia Direct*. Available at: www.russia-direct.org/analysis/four-dangerous-myths-about-russias-plans-arctic.

Skogrand, K. 2008. 'Emerging from the frost: security in the 21st century Arctic' Available at: http://brage.bibsys.no/xmlui/handle/11250/99775.

Tamnes, R. and K. Offerdal (eds) 2014. *Geopolitics and Security in the Arctic: Regional Dynamics in a Global World*. London: Routledge.

Till, G. 2005. *Seapower: A Guide for the Twenty-First Century*, e-book. London: Frank Cass Publishers.

Welch, D.A. 2013. 'The Arctic and geopolitics' *East Asia-Arctic Relations: Boundary, Security and International Politics*, 6 (December 2013): 1–14.

Zellen, B.S. 2013. 'U.S. defense policy and the North: the emergent Arctic power' in B.S. Zellen (ed.) *The Fast-Changing Arctic: Rethinking Arctic Security for a Warmer World*. Calgary, AB: Calgary University Press. 227–256.

Zysk, K. 2011. 'Military aspects of Russia's Arctic policy: hard power and natural resources' in J. Kraska (ed.) *Arctic Security in an Age of Climate Change*. New York: Cambridge University Press. 85–106.

Zysk, K. 2013. 'Russia's Arctic strategy: ambitions and restraints' in B.S. Zellen (ed.) *The Fast-Changing Arctic: Rethinking Arctic Security for a Warmer World*. Calgary, AB: Calgary University Press. 281–296.

27

Polar tourism

Status, trends, futures

Emma J. Stewart and Daniela Liggett

Introduction

Research on the phenomenon of tourism in the Polar Regions can be traced to a period of growth in the sector in the early 1990s (Stewart, Draper and Johnston 2005). At this time, cruise ship tourism was developing in both the Arctic and Antarctic. While the Antarctic was witnessing sustained growth as retro-fitted icebreakers came into service (Crosbie and Splettoesser 2011), communities in the Arctic were also experiencing the arrival of ship-based seasonal visitors, but at the same time many were embracing the development opportunities and potential of nature-based, cultural and business tourism (Johnston 2011). It is not surprising, therefore, that allied to the growth of the sector came the first tentative forays into studying tourism in these regions. Research tended to cluster around key scholars and their students, such as the well-known polar ecologist Bernard Stonehouse at the Scott Polar Research Institute (SPRI) in the UK, the eminent tourism anthropologist Valene Smith from the USA, polar enthusiast and geographer Margaret Johnston from Canada, and New Zealand-based renowned tourism scholar Michael Hall. This early and pioneering work established polar tourism as a legitimate research domain drawing attention to definitions, geographic boundaries and issues related to social, cultural and environmental effects (Maher and Stewart 2007). Publications arising from it appeared in polar-related and tourism studies-specific journals and other formats, including a special issue of the journal *Annals of Tourism Research* (Smith 1994), an edited book (Hall and Johnston 1995) and various symposia and conferences. Not surprisingly, later work emerged and continued to build an understanding of polar tourism as a developing and growing industry (see: Humphreys et al. 1998; Marsh 2000; Bauer 2001; Stonehouse 2001). However, akin to most emerging fields of study, these foundational research efforts were often descriptive in nature, were often concerned with tourist numbers and tourism trends, and were lacking in theoretical, methodological and empirical rigour (Stewart, Draper and Johnston 2005).

In this chapter, we illustrate how polar tourism research has matured in recent years, how polar tourism researchers engage with questions of global relevance to the study of tourism, and we identify current and future trajectories of research in the study of polar tourism. In order to achieve these tasks, we update findings presented in the first review of tourism research in the Polar Regions conducted by Stewart, Draper and Johnston in 2005. This review was framed by a model developed by Grano (1981), which examined external and internal changes in the discipline of geography and was later adapted by Hall and Page (2002) to assess the nature and status of the geography of tourism. The model provided a valuable organizational framework within which to situate polar tourism research, addressing three interrelated areas: (1) knowledge – the available information and content of the study of polar tourism; (2) action – polar tourism research in the context of research praxis; and (3) culture – academics, students and other researchers within the context of influences on the research community.

This chapter will follow broadly the same structure. The data that form the basis of this chapter were generated through a keyword search of two online scholarly databases (Scopus and Google Scholar) for any scholarly publications that appeared after 2005 (up until 2015) and had a focus on polar tourism. A list of journal articles, book chapters and books (including edited volumes) resulting from the keyword search was compiled in a spreadsheet for analysis. In an iterative process, entries were thematically coded jointly by Stewart and Liggett, including a 'key theme' and 'sub-theme' for each entry with some entries also requiring a second sub-theme. The emerging clusters were reviewed for consistency of thematic and sub-thematic coding and were, if necessary, adjusted to ensure that they were meaningful. We note that an exercise of this nature cannot capture every piece of research conducted on polar tourism, and we caution that the clusters described in this chapter tend to reflect a bias toward North American and European literature and that which is available in the English language through the traditional means of research communications. This is clearly a limitation of the survey of literature and identifies the research culture within which the authors work.

A comparative approach across the Polar Regions is taken in this chapter. Despite their obvious differences (in terms of sovereignty, use of scientific bases, management regimes, indigenous peoples, legislative control, access, cultural heritage and physical geography), with regard to tourism the major issues are largely similar and relate to the "regulation of tourists, protection of the environmental and cultural heritage, management of trans-national space, and effects on local populations" (Hall and Johnston 1995: 3–4) as well as the fact that many cruise operators operate throughout the year in both the southern and northern polar summers.

Knowledge and action

Stewart, Draper and Johnston's initial review identified four existing and growing clusters of research focusing on: tourist demand and behaviour, their effects, the policies and management that surround tourism, and the development issues that tourism presents. The review concluded that, despite emerging research clusters, polar tourism research was dominated by description, with very little empirical evidence offered to help understand polar tourism phenomena. Stewart, Draper and Johnston emphasised the need for further empirical research fashioned in a more coordinated and focused manner. This call for high-quality empirical research built on an earlier case made by Harman (2003) for high-impact scholarship that would satisfy deep curiosity as well as solve practical problems and contribute to public policy. The review also outlined a research agenda that would develop the research activity within the existing clusters as well as adding two new complementary research areas: the tourist experience and the implications of global and large-scale change on tourism.

This first review was undertaken during buoyant times for the polar tourism sector. For example, in the Canadian Arctic, some communities located on the Northwest Passage regarded 2005 as a watershed year with cruises to the region doubling from the previous season (Stewart, Draper and Dawson 2010). Soon after 2005, Arctic tourism experienced a short downturn related to the global economic crisis, but growth returned in 2010 and intensified throughout the decade, with the 2015 season recorded as the busiest on record with thirty planned cruises across Arctic Canada. The catalyst for this growth was thought to be, in part, related to climate change and significant decreases in sea ice across the Arctic allowing for greater access to northern waters (Lamers and Amelung 2010). Similarly in the Antarctic, tourism in 2005 was in a sustained period of growth, which eventually peaked in the 2007/08 season when Antarctic tourists numbered in excess of 45,000. As in the Arctic, decline followed, before a return to growth in the 2011/12 season. Forecasts by the International Association of Antarctica Tour Operators (IAATO) for the 2015/16 season indicate that the numbers of Antarctic tourists will rise to approximately 40,000 individuals (Liggett and Stewart 2017).

While there has been some variability in the patterns of growth since 2005, both temporally and spatially, the overall trajectory has been one of growth. This has been matched by an intensification of scholarly focus on polar tourism and a proliferation of polar tourism research. Our current review reveals that in the period from 2005 to 2015 at least ten dedicated books or edited books have been published; numerous book chapters on polar-tourism-related matters have appeared in either tourism or polar edited texts; at least two special issues of journals have been published; 36 post-graduate theses have been written and an astonishing 146 peer-reviewed journal articles have been published in a wide variety of journals.

A number of the books resulted from specific conferences on polar tourism, or have been based on the particular interests of the editors, and therefore can be regarded as thematic collections on topics such as global change, cruise ship tourism and regional development (e.g. Hall and Saarinen 2010; Lück, Maher and Stewart 2010; Grenier and Müller 2011; Müller, Lundmark and Lemelin 2013). Other texts are more generic, encompassing a broad range of topics and usually trans-polar in nature (e.g. Baldacchino 2006; Müller and Jansson 2006; Snyder and Stonehouse 2007; Stonehouse and Snyder, 2010; Maher, Stewart, and Lück 2011). Notable contributions in edited books elsewhere include Jabour's chapter on regulations of Antarctic tourism in *Antarctic Futures* (Jabour 2014); Liggett's chapter on policy challenges associated with Antarctic tourism in *Tourism Policy and Planning* (Liggett 2013); and Lamers, Eijgelaar and Amelung's analysis of last chance tourism in Antarctica in *Last Chance Tourism: Adapting tourism opportunities in a changing world* (Lamers, Eijgelaar and Amelung 2012).

Carefully positioned to showcase the emerging range of empirically-based polar tourism research during the International Polar Year (IPY), a special issue of the journal *Tourism in Marine Environments* on Polar Tourism was published by editors Maher and Stewart (2007). Articles in that volume addressed diverse issues such as climate change, law and policy, management, gateway ports and wildlife tourism. More recently, a dedicated edition of *The Polar Journal* edited by Liggett and Stewart (2015) brings together a collection of papers arising from a polar tourism conference on polar gateways. Interestingly, many of these manuscript-based contributions (books, edited books and special issues) augment the pre-existing clusters identified in Stewart, Draper and Johnston's initial review, e.g. by centring on issues related to tourism development, impacts of tourism activity, and matters of tourism policy, but they also extend into topics identified in that review as needing attention, such as the implications of global change for polar tourism.

Adding to the growing community of polar tourism scholars have been successful completions of post-graduate theses directly relevant to polar tourism. Our review reveals 36 such contributions with approximately two-thirds being at the doctoral level (*n*=22) and the remaining at the master's level (*n*=14). Reflecting the strong scholarly contribution from North America, 16 students completed their theses in Canada and nine in the USA. Elsewhere in the northern hemisphere, theses were successfully completed in the UK, Finland and in the Netherlands during the period 2005–2015. In the southern hemisphere, one thesis was completed in Brazil while New Zealand dominated with six completions. Considering the northern-hemisphere bias in theses completions, it is surprising that over 40% of these theses had an Antarctic focus. Thematically, while the majority of the theses could be included in the clusters identify in the 2005 review by Stewart et al., a number of theses crossed disciplinary boundaries (e.g. integration of social and physical sciences), employed novel methodological approaches (e.g. the analysis of blogs or photo diaries) or ventured into topics far behind those traditionally covered (e.g. space tourism law with lessons from Antarctica). Considering the very narrow geographic focus of polar tourism research and the relatively small numbers of tourists involved in tourism to the Arctic and Antarctic, when compared to tourism activities in other parts of the world, 36 theses completed in 10 years is not an insignificant effort and is illustrative of the building of capacity among the polar tourism community. It is noteworthy that many of those students completing research degrees during the 2005–2015 period are now actively publishing as well as contributing to the wider polar tourism research environment.

Reflecting the core currency of researchers today it is not surprising that polar tourism scholars are targeting their publishing energies and activities at international peer-reviewed journals. What is surprising though is the volume of articles produced over a relatively short period of time, for what is arguably regarded as a niche tourism destination. During the period 2005–2015, we calculate that 146 journal articles have been published in 49 different journals. Further analysis reveals that just over one half (51%) were published in tourism-specific journals, with *Tourism in Marine Environments* publishing the most papers (*n*=16), followed by the *Journal of Sustainable Tourism* and the *Scandinavian Journal of Hospitality and Tourism* (both *n*=7). Journal articles published in generic polar journals (i.e. non-tourism specific) accounted for 35% of all polar tourism papers with *Polar Record* (*n*=15), *Polar Geography* (*n*=8) and *Arctic* (*n*=7) being the most popular publishing outlets. Interestingly, the remainder of articles (14%) were published in non-tourism and non-polar specific journals (16 of them in total) giving an indication that understanding polar tourism is a cross-cutting issue and one that is emerging as a multi-disciplinary area of research. The disciplinary bases of these journals vary widely and include fields as far apart as law, marine policy, arts, ornithology and biology. The journal, *Human Dimensions of Wildlife* received the most polar-tourism-focused papers (*n*=3) out of all non-tourism and non-polar journals targeted by tourism researchers.

Perhaps reflecting the longer history of tourism in the Arctic, the reach of Arctic tourism into indigenous communities and its potential for economic development, the greater number of articles (58%) are focused specifically in the Arctic region (including the sub-Arctic). Only 29% of the articles can be classified as Antarctic or sub-Antarctic, with the remainder (13%) examining issues across the Polar Regions. A thematic analysis of the articles (see Fig. 1) showed that the pre-existing categories identified in Stewart et al.'s 2005 review remained strong, particularly in relation to tourism development and management. However, the two key areas of research needs identified by Stewart et al. – tourist experience and the implications of global and large-scale change on tourism – have, in recent years, become critical areas of scholarship, amassing significant and welcome research attention. Indeed, these two areas appear to have taken over the study of 'impacts' and 'patterns' that were so prevalent in the early days of polar tourism scholarship. Research related to governance also has emerged strongly since the late 2000s (Figure 27.1).

Further analysis of our 2005–2015 database reveals that the sub-themes (Figure 27.2) of the articles are strongly skewed towards cruise tourism which is not entirely surprising given that this is the most usual mode of transport for polar tourists. Perhaps as a reflection of the growing empirical strength of the research in this field many of the articles were classified as contributing methodologically. The prevalence of research focused on wildlife is also not surprising, given the largely nature-based approach to tourism in the Polar Regions. However, this analysis does point to the wide variety of themes addressed by researchers from heritage, economy and the climate.

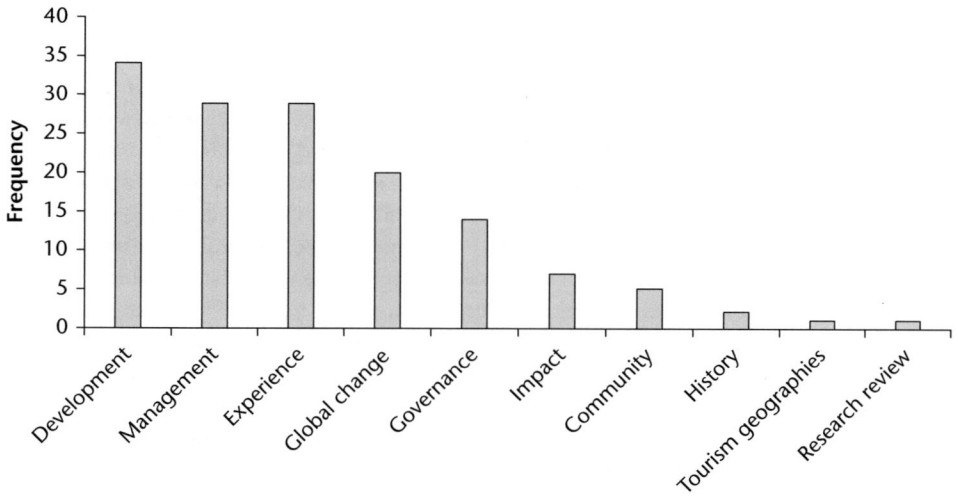

Figure 27.1   Key themes of journal articles published on polar tourism post 2005.

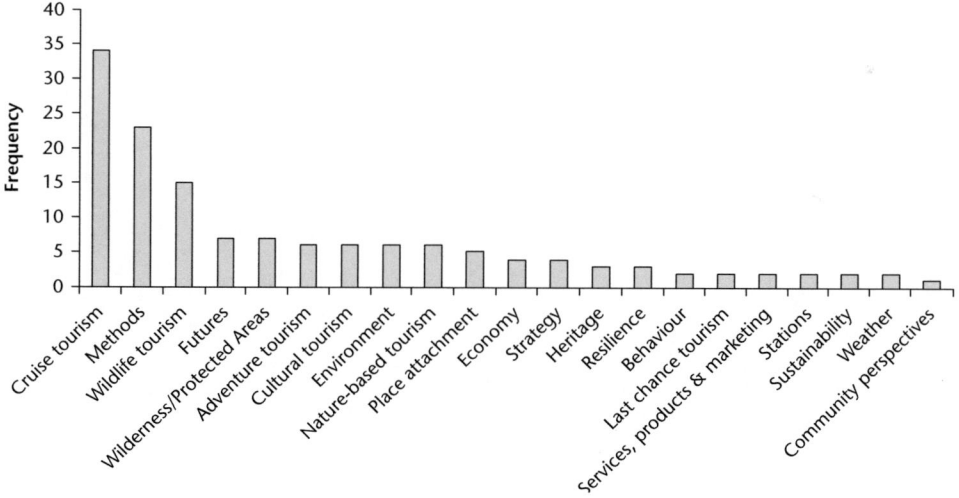

Figure 27.2   Sub-themes of journal articles published on polar tourism post 2005.

In addition to running frequencies of the themes and sub-themes, radar charts were created to illustrate the key relationship between key overarching themes and the associated sub-themes of the articles. The points on the radar charts refer to the number of articles published on the sub-theme within the key theme. As the following charts illustrate, many of the key themes show a strong relationship to the sub-theme of cruise ship tourism, particularly 'management', 'global change' and 'governance' (see Figures 27.3, 27.4 and 27.5). Not surprisingly, the attention of researchers has turned to the pressing issues of managing an increasing number of cruise ships and cruise passengers visiting the Polar Regions as illustrated in Figure 27.3.

The strong relationship between the 'global change' theme and the cruise sector (Figure 27.4) reflects the links between climate change in particular and important issues for the cruise sector, related to accessibility and safety. Especially changes in sea ice extent and thickness, which are greatly influenced by atmospheric and hydrologic conditions, have important implications with regard to access to landing sites or navigational safety, as has been highlighted by a number of incidents and accidents of cruise vessels in polar waters over the past ten years (see e.g. Swanson et al. 2015).

The considerable increase in cruise tourism activity raised questions and concerns with regard to the sufficiency of existing governance structures for polar cruise tourism. Consequently, tourism researchers chose to devote some of their energies to helping understand, and in some cases, develop governance regimes which could effectively meet the needs of local communities (in the Arctic) or research stations and networks (in the Antarctic) as well as the cruise sector in an endeavour to offer regulatory frameworks that would ensure safe and responsible visits to remote polar locations (see Figure 27.5).

Out of those articles related to tourism development, the majority appear to favour methodological innovation, while others are, unsurprisingly, focused on economic development, notions of place (including protected areas, wilderness and nature-based activities), and future tourism development (see Figure 27.6).

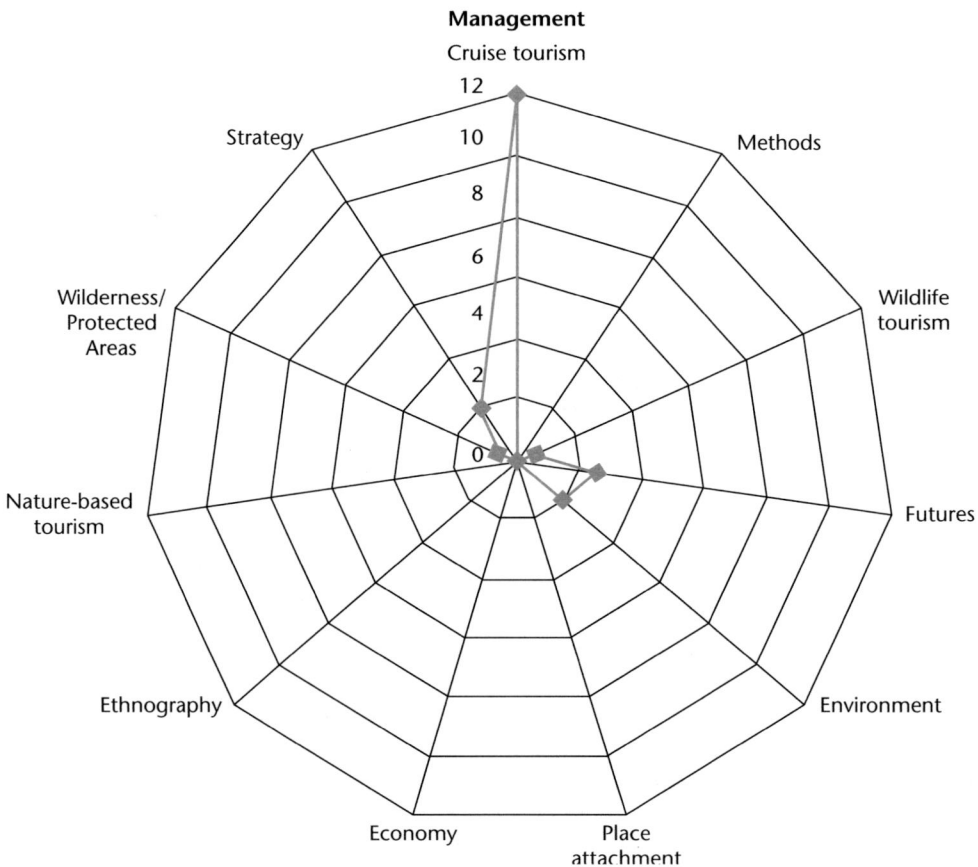

Figure 27.3   Management theme and associated sub-themes

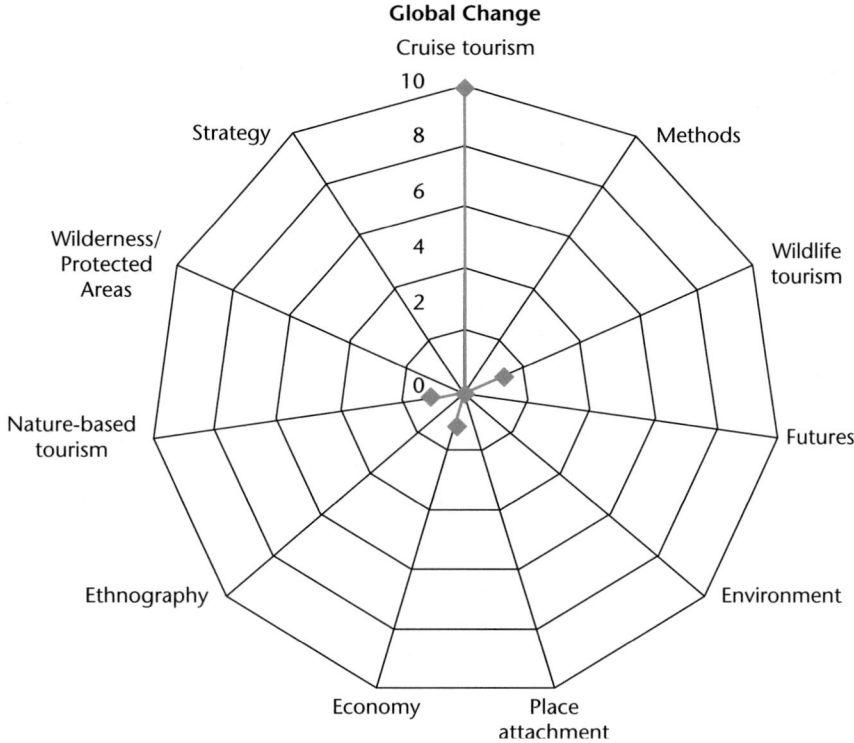

Figure 27.4   Global change theme and associated sub-themes

Stewart et al.'s 2005 review highlighted the area of visitor experience as one of the areas that required more research attention. With roughly 20% of all journal articles on polar tourism over the last decade addressing multiple facets of visitor experience, Stewart, Draper and Johnston's call for action seems to have been heard. As this area of polar tourism research was relatively undeveloped until 2005, it is not surprising that many of the papers addressing this topic over the last decade explored methodological opportunities and offered innovative approaches to understanding experiences of polar tourists. At the same time, as highlighted in Figure 27.7, the study of 'visitorexperience' is closely connected to one of the key motives for polar tourism; that of wildlife viewing.

Culture

The recent boost in academic attention on the issues related to tourism in the Polar Regions is not accidental, and as Grenier and Müller (2011) point out, a complex network of factors has coalesced and stimulated the maturing of polar tourism research. These factors include political shifts (such as the ending of the Cold War and a new impetus to develop polar resources); the emancipation and empowerment of indigenous peoples in the north; the emergence of the Polar Regions as entities and jurisdictions in their own right; developing economies; the fortification of the green movement; and also the media devoting attention to climate change matters and the fear of disappearing polar species and landscapes. Furthermore, broad changes to the spectrum of demand for polar tourism and the tourists' tendency to be more active, seeking out alternative and remote destinations and the desire to avoid politically sensitive destinations, are presenting additional interesting lines of inquiry for the research community. Finally, proliferated through online technologies, we now have information on tourism practice at our fingertips. For example, through blogs or online diaries by polar tourists themselves, through detailed media stories on incidents or accidents in the Polar Regions, through the possibility of accessing satellite data tracking vessels in polar waters, and through greater transparency related to reports by tour operators about their visits to the Polar Regions (especially as Antarctic itineraries are concerned). A number of recent doctoral projects on polar tourism matters have already tapped into these resources, which remain underutilised and offer opportunities to tourism researchers to use these resources more fully in future years (Roura 2011; Jellum 2012).

The polar tourism research community is now much broader than it has been in the past, perhaps as a result of exposure through the International Polar Year (IPY) 2007–2009; the emergence of a dedicated network of polar tourism researchers; the engagement of researchers with key industry partners such as the Association of Arctic Expedition Cruise Operators (AECO) and IAATO; and more recently the links made to the wider polar community, e.g. through participation in multi-disciplinary polar conferences such as the Scientific Committee on Antarctic Research's biennial Open Science Conferences, the biennial International Congress of Arctic Social Sciences (ICASS), or the Arctic Science Summits. One of the key achievements of the 2007–2009 IPY was the recognition of the contribution made by polar social scientists, along with an opportunity for polar social scientists to obtain IPY funding for their research (Krupnik et al. 2011). While there were few tourism-specific projects supported directly through the IPY, the momentum generated served as an important catalyst for bringing together diverse

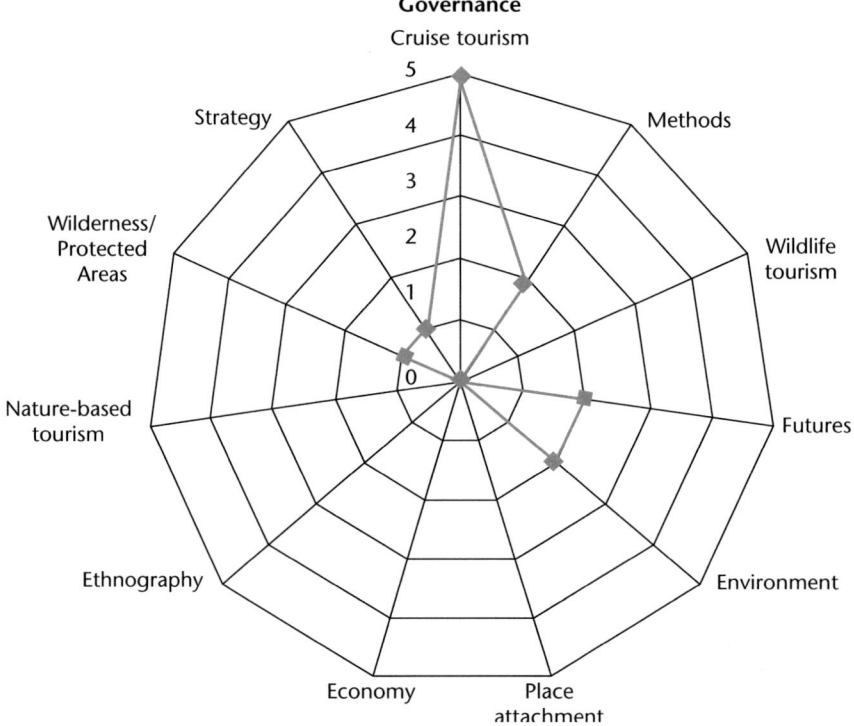

Figure 27.5 Governance theme and associated sub-themes

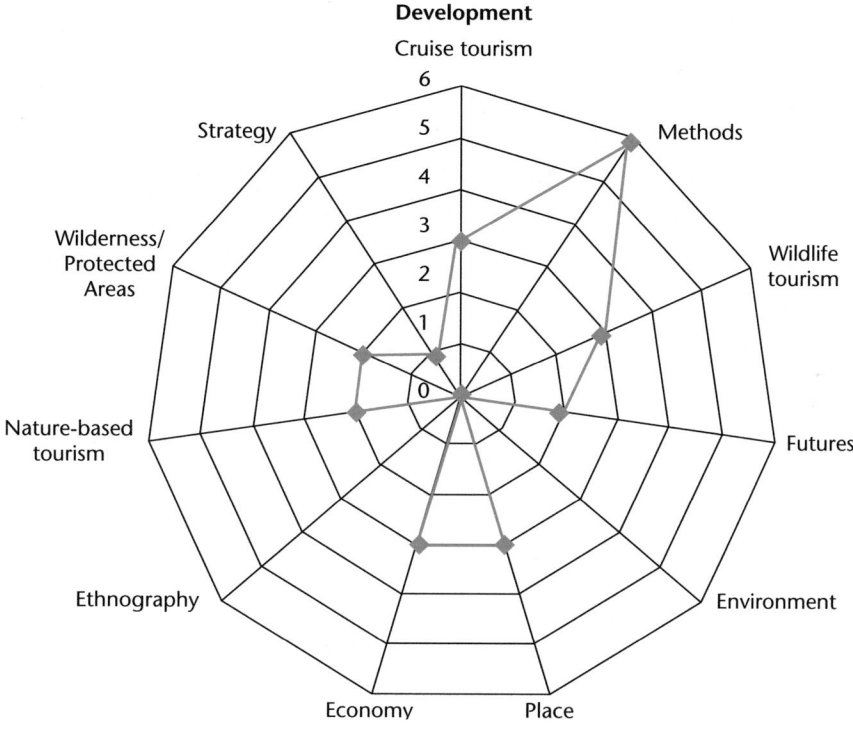

Development

Figure 27.6 Development theme and associated sub-themes

groups with interests in polar research including northern communities, the science community, tourism researchers, granting bodies, non-governmental organizations, regulatory bodies, and national communities. The various IPY fora gave polar tourism researchers the opportunity to connect, present results (e.g. at the IPY flagship conferences in Oslo and later in Montreal) and develop new collaborations.

However, it was a meeting outside of IPY that created one of the watershed moments in the development of polar tourism scholarship. At the annual conference of the Canadian Association of Geographers (CAG) held at Lakehead University in 2006, a group of scholars interested in polar tourism, spearheaded by Canada's Patrick Maher, gathered at the conference and came up with the idea for a research network, which was later to become known as the International Polar Tourism Research Network (IPTRN) (Maher 2007). With funding support from the Université du Québec à Montréal, the new network was born with the launch of its original website and Alain Grenier hosted the first of IPTRN's biennial conferences in Nunavik, Arctic Quebec in 2008. At that first gathering, 16 researchers in addition to consultants and government delegates discussed some of the most pressing tourism issues facing Arctic communities (Grenier and Müller 2011). From these modest beginnings, IPTRN has grown in membership, formalized an executive committee and hosted four further conferences in the following polar locations or gateways: Abisko, Sweden in 2010; Nain, Canada in 2012; Christchurch, New Zealand in 2014 and Raufarhöfn, Iceland in 2016. The fact that plans are already in place to host the sixth conference in the Canadian Yukon in 2018, and the seventh conference in Ushuaia, Argentina in 2020 is evidence of the strong and growing momentum in the polar tourism research community.

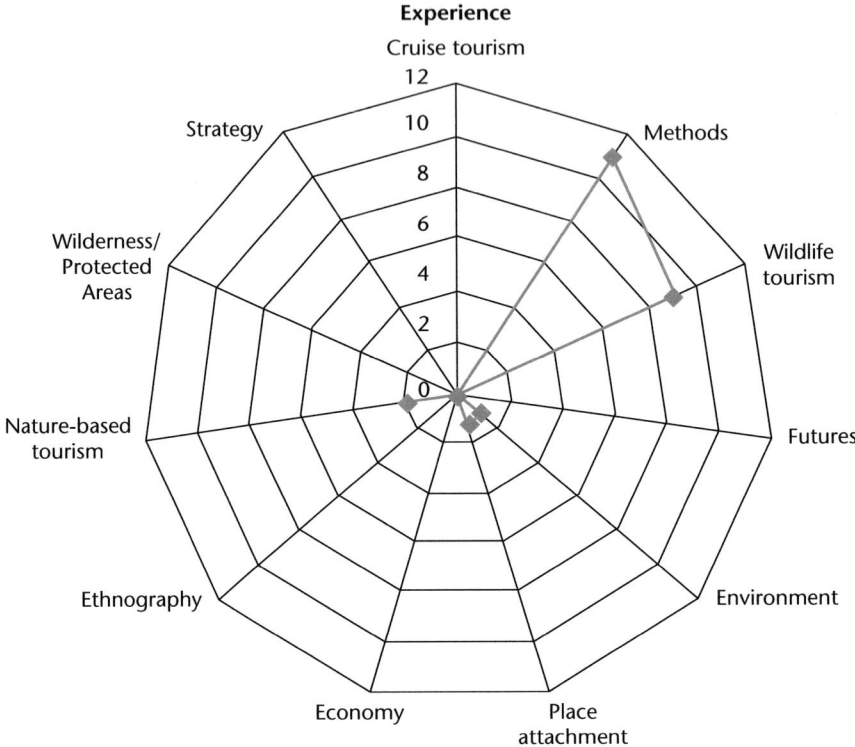

Figure 27.7 Experience theme and associated sub-themes

Concluding discussion

In 2005, it was claimed that tourism was starting to be viewed as a legitimate research activity in the Polar Regions whereby scholarship was characterized by empirical work that attempts to describe systematically and along themes, (often through case studies), and focused on largely practical issues and management. The shortfalls were a lack of empirically-based research and the hesitancy in applying existing tourism theory to the Polar Regions. While one can only specu-late about the reasons for such hesitancy, it may be explained by the fact that polar tourism is a relatively recent activity and that the phenomenon itself needed to be described and captured first before any further conceptual and analytic work could take place. Furthermore, the first descriptions of polar tourism activities were not necessarily undertaken by tourism researchers, who would be more likely to apply existing tourism theory to the new phenomenon. Instead, naturalists who worked on board cruise ships, or natural scientists employed by polar tourism research programmes, were among the first to describe aspects of polar tourism for those scholars who were unable to visit the Polar Regions and undertake in-situ work. As our updated review illustrates, research has continued to build on and develop traditional research clusters but has also extended its reach into new realms, exploring the implications of global change for tourism (and cruise ship tourism in particular) and the need to build robust governance structures, again particularly in relation to the growth in the polar cruise sector. We suggest that polar tourism research has now emerged from its infancy and is maturing to a point where scholarship is more likely to be underpinned by empirical research, to be situated theoretically, and to connect to a wider disciplinary base than in the past, and we witness, e.g., the integration of sea ice studies and ecological approaches taken with tourism studies as well as evidence of trans-disciplinary collaborations (Stewart et al., 2007). This maturing phase aligns with the growing momentum in the polar social sciences in general, but importantly, it has been nurtured and supported by the emergence of the IPTRN in 2007.

Our thematic analysis of polar tourism research (2005–2015) reveals that the calls from Stewart et al. (2005) have largely been addressed through the building of scholarship around established themes such as management, development and impacts in addition to more fully understanding visitor experience and, perhaps most impressively, the implications of global change for tourism in the Polar Regions. Not surprisingly, in light of the growth the polar travel sector experienced over the last couple of decades, which was accompanied by not only high-profile accidents and incidents but also the promise of economic returns, research related to polar tourism governance has emerged strongly over the past decade. There is also the sense, evidenced in part by the emergence of papers published in a wide variety of non-tourism journals, that researchers are working with colleagues from other disciplines to help address the multi-faceted nature of polar tourism. Similarly, there is a small but emerging literature that examines issues comparatively, mainly within regions (through case studies) but also with pan-regional studies that extend their focus across national boundaries in the Arctic, and occasional trans-polar comparative studies.

In an attempt to identify a research agenda for the next decade, we suggest that five key research themes require more scrutiny (see Table 27.1 for a summary). The first theme relates to tourism demand and, in particular, how polar tourism visitors are more diverse than they have been in the past (such as the increasing interest from Asia markets), and how tourism products have changed and evolved (such as new activities, citizen science, modes of transport). The second theme builds on the growing body of literature around polar nationalisms (see e.g. Elzinga 2013; Dodds 2014), and particularly banal nationalism which describes the easily overlooked and forgotten nationalism that influences the every-day lives of many citizens, the unnoticed 'flags' that adorn buildings and the norms that are assumed but build on or embellish national identity (Billig 1995), and its diffusion into a wide range of human activities in the Polar Regions, from scientific studies to commercial activities. So far, we do not have a solid understanding of the values brought in by new actors on the polar tourism arena and the role they, along with established stakeholders, can play from a geopolitical perspective. Examples from the Antarctic have shown us that tourism can be used as a political tool and considering the diffusion of state control in the process of globalization, we need to gain a better understanding of the role tourists and tour operators can play in the Arctic and Antarctic. The third theme around regulation, policy and governance expands our knowledge of effective and efficient governance regimes, with a special focus on the regulation of polar tourism. It takes into consideration that the majority of polar tourism operations are ship-based and integrates insights from political science, law, human geography and political anthropology with thorough empirical research to offer practical and solution-focussed assessments of regional and international governance of marine-based tourism. This includes assessments relating to matters around the UN Convention on the Law of the Sea (UNCLOS) and port-state vs. flag-state control. Traditionally, vessels are under the jurisdiction of the state whose flag they fly and whose port they are registered at (Farthing and Brownrigg 1997), which works well for areas beyond national jurisdiction but is problematic in areas under governance by international regimes, such as the Antarctic Treaty System which many flag states of Antarctic tourism vessels are not members of (see e.g. Swanson et al. 2015). Consequently, the issue of port-state control, i.e. "the competence of the port state to legislate and/or seek to enforce this jurisdiction over vessels visiting its port" (Pamborides 1999: 47), became a matter of interest to Antarctic policy makers as well as scholars. The third

Table 27.1 Overview of research themes requiring further attention.

1	New polar tourism markets and new interests
2	Values of tourism stakeholders and associated polar tourism politics
3	Regulation, policy and governance of polar tourism
4	Global change in relation to polar tourism activities
5	New technology and its implications on experience, behaviour and tourism management

theme on regulation, policy and governance links to the recent SCAR Horizon Scan, which directly focuses on tourism with the following question: "How will regulatory mechanisms evolve to keep pace with Antarctic tourism?" The fourth theme addresses global change in relation to tourism which, as environmental conditions in the Polar Regions continue changing, remains an important area of research. While some researchers have begun looking into the carbon footprint of polar tourism (e.g. Farreny et al. 2011), enormous gaps in our understanding of the contribution of polar tourism to climate change remain. Equally, we lack information about the value of the awareness raised by tour operators about climate change in the Polar Regions and about the nexus between awareness and action taken by the tourists after their visits to the Arctic or Antarctic. Finally, a fifth theme is represented by our need to examine the influence new technology may have on tourist experience and behaviour on the one hand, and on the management of polar tourism, which might be enhanced through novel, remote monitoring and observing solutions, on the other.

The research called for under these five themes is necessarily trans–disciplinary and integrative. Aside from in-depth regional case studies, trans-regional and trans-polar knowledge should be built on to develop meaningful comparisons and policy recommendations. The latter would further be aided by an even broader approach, which looks outside the Polar Regions for lessons that could be learnt from tourism research and management in other remote and environmentally significant destinations. Last but not least, the research themes identified above would benefit from contributions and support by stakeholders actively involved in tourism operations, management or regulation, and requires the sustained growth of the polar tourism research community, which requires funding to ensure that the required capacity is being built and maintained.

References

Baldacchino, G. (ed.) 2006. *Extreme Tourism: Lessons from the World's Cold Water Islands*. London: Routledge.
Bauer, T. 2001. *Tourism in the Antarctic: Opportunities, Constraints, and Future Prospects*. London: Routledge.
Billig, M. 1995. *Banal Nationalism*. London and Thousand Oaks, CA: Sage.
Crosbie, K. and Splettoesser, J. 2011. 'Antarctic tourism introduction' in Maher, P. T., Stewart, E.J. and Lück, M. (eds) *Polar Tourism: Human, Environmental and Governance Dimensions*. New York: Cognizant Communication Corporation. 105–120.
Dodds, K. 2014. 'Militant geography and frontier vigilantism: "Australian Antarctic territory" and the "southern flank"' in Powell, R. and Dodds, K. (eds) *Polar Geopolitics? Knowledges, Resources and Legal Regimes*. Cheltenham, UK: Edward Elgar.
Elzinga, A. 2013. 'Rallying around a flag? On the persistent gap in scientific internationalism between word and deed' in Brady, A-M. (ed.) *The Emerging Politics of Antarctica*. London: Routledge. 193–219.
Farreny, R., Oliver-Solà, J., Lamers, M., Amelung, B., Gabarrell, X., Rieradevall, J., Boada, M. and Benayas, J. 2011. 'Carbon dioxide emissions of Antarctic tourism' *Antarctic Science* 23(6): 556–566.
Farthing, B. and Brownrigg, M. 1997. *Farthing on International Shipping*. London: LLP.
Grano, O. 1981. 'External influence and internal change in the development of geography' in Stoddart, D.R. (ed.) *Geography, Ideology and Social Concern*. New York: Barnes and Noble Books.

Grenier, A.A. and Müller, D. 2011. *Polar Tourism: A Tool for Regional Development*. Québec City, QC: Presses de l'Université du Québec.

Hall, C.M. and Johnston, M.E. 1995. *Polar Tourism: Tourism in the Arctic and Antarctic Regions*. Chichester, UK: John Wiley & Sons.

Hall, M.C. and Page, S.J. 2002. *The Geography of Tourism and Recreation: Environment, Place and Space*. Abingdon, UK: Routledge.

Hall, M.C. and Saarinen, J. 2010. *Tourism and Change in Polar Regions: Climate, Environments and Experiences*. Abingdon, UK: Routledge.

Harman, J.R. 2003. 'Whither geography?' *The Professional Geographer* 55(4): 415–421.

Humphreys, B.H., Pedersen, Å.Ø., Prokosch, P.P. and Stonehouse, B. 1998. *Linking tourism and conservation in the Arctic*. Proceedings from Workshops. (Meddelelser No. 159.) Tromsø, Norway: Norsk Polarinstitutt.

Jabour, J. 2014. 'Strategic management and regulation of Antarctic tourism' in Tin, T., Maher, P.T., Liggett, D. and Lamers, M. (eds) *Antarctic Futures: Human Engagement with the Antarctic Environment*. Dordrecht, Netherlands: Springer. 273–286.

Jellum, C. 2012. 'Seeking Antarctica: an investigation into visitors' travel and recreation value-based motivations'. PhD thesis, University of Otago, Dunedin, New Zealand.

Johnston, M. E. 2011. 'Arctic tourism introduction' in Maher, P.T., Stewart, E.J. and Lück, M. (eds) *Polar Tourism: Human, Environmental and Governance Dimensions*. New York: Cognizant Communication Corporation. 17–32.

Krupnik, I., Allison, I., Bell, R., Cutler, P., Hik, D., López-Martínez, J., Rachold, V., Sarukhanian, E. and Summerhayes, C. 2011. *Understanding Earth's Polar Challenges: International Polar Year 2007–2008 – Summary by the IPY Joint Committee*. Rovaniemi, Finland and Edmonton, AB: University of the Arctic/ CCI Press and ICSU/WMO Joint Committee for International Polar Year 2007–2008.

Lamers, M. and Amelung, B. 2010. 'Climate change and its implications for cruise tourism in the Polar Regions' in Lück, M., Maher, P.T. and Stewart, E.J. (eds) *Cruise Tourism in the Polar Regions: Promoting Environmental and Social Sustainability?* London: Earthscan. 147–163.

Lamers, M., Eijgelaar, E. and Amelung, B. 2012. 'Last chance tourism in Antarctica: cruising for change?' in Lemelin, R.H., Dawson, J. and Stewart, E.J. (eds) *Last Chance Tourism: Adapting Tourism Opportunities in a Changing World*. Abingdon, UK: Routledge. 25–41.

Liggett, D. 2013 'Policy challenges of tourism as a commercial activity in Antarctica' in Egdell, D.L. and Swanson, J.R. (eds) *Tourism Policy and Planning: Yesterday, Today and Tomorrow*. London: Routledge. 107–118.

Liggett, D. and Stewart, E. J. 2015. 'Polar tourism (research) is not what it used to be: the maturing of a field of study alongside an activity' *The Polar Journal* 5(2): 247–256.

Liggett, D. and Stewart, E. J. 2017. 'The changing face of political engagement in Antarctic tourism' in Dodds, K., Hemmings, A. and Roberts, P. (eds) *Handbook on the Politics of Antarctica*. London: Edward Elgar. 368–391.

Lück, M., Maher, P.T. and Stewart, E.J. (eds) 2010. *Cruise Tourism in Polar Regions: Promoting Environmental and Social Sustainability?* London: Earthscan.

Maher, P.T. 2007. 'Arctic tourism: A complex system of visitors, communities, and environments' Editorial foreword to special issue of *Polar Geography* 30(1–2): 1–5.

Maher, P.T. and Stewart, E.J. 2007. 'Polar tourism: research directions for current realities and future possibilities' *Tourism in Marine Environments* 4(2–3): 65–68.

Maher, P.T., Stewart, E.J. and Lück, M. 2011. *Polar Tourism: Human, Environmental and Governance Dimensions*. New York: Cognizant Communication Corporation.

Marsh, J. 2000. 'Tourism and national parks in polar regions' in Butler, R.W. and Boyd, S.W. (eds) *Tourism and National Parks: Issues and Implications*. New York: John Wiley & Sons Ltd. 125–136.

Müller, D.K. and Jansson, B. (eds) 2006. *Tourism in Peripheries: Perspectives from the Far North and South*. Wallingford, UK: CABI.

Müller, D.K., Lundmark, L. and Lemelin, R. H. (eds) 2013. *New Issues in Polar Tourism: Communities, Environments, Politics*. Dordrecht, Netherlands: Springer Science & Business Media.

Pamborides, G.P. 1999. *International Shipping Law: Legislation and Enforcement*. Athens, The Hague, London: Ant. N. Sakkoulas Publishers & Kluwer Law International.

Roura, R.M. 2011. The footprint of polar tourism: tourist behaviour at cultural heritage sites in Antarctica and Svalbard. PhD thesis, University of Groningen, Netherlands.

Smith, V.L. 1994. 'A sustainable Antarctica: science and tourism' *Annals of Tourism Research* 21: 221–230.

Snyder, J. and Stonehouse, B. (eds) 2007. *Prospects for Polar Tourism*. Wallingford, UK: CABI.

Stewart, E.J., Draper, D. and Dawson, J. 2010. 'Monitoring patterns of cruise tourism across Arctic Canada' in Lück, M., Maher, P.T. and Stewart, E.J. (eds) *Cruise Tourism in the Polar Regions: Promoting Environmental and Social Sustainability?* London: Earthscan. 133–145.

Stewart, E.J., Draper, D. and Johnston, M.E. 2005. 'A review of tourism research in the Polar Regions' *Arctic* 58(4): 383–394.

Stewart, E.J., Howell, S.E.L., Draper, D., Yackel, J. and Tivy, A. 2007. 'Sea ice in Canada's Arctic: implications for cruise tourism' *Arctic* 60(4): 370–380.

Stonehouse, B. 2001. 'Polar environments' in Weaver, D.B. (ed.) *The Encyclopaedia of Ecotourism*. Wallingford, UK: CABI. 219–234.

Stonehouse, B. and Snyder, J. 2010. *Polar Tourism: An Environmental Perspective* (Vol. 43). Clevedon, UK: Channel View Publications.

Swanson, J. R., Liggett, D. and Roldan, G. 2015. 'Conceptualizing and enhancing the argument for port state control in the Antarctic gateway states' *The Polar Journal* 5(2), http://dx.doi.org/10.1080/2154896X.2015.1082785.

28

Consulting Arctic energy
From political hearings to roundtable events

Arthur Mason

In 2002, Ed Kelly was in his mid-40s and senior economist for the consulting firm Cambridge Energy Research Associates. He could spin tightly knit sentences from memory and was quick to remind me of the weight others placed on his predictions. "I know better than to say a specific number", he once cautioned, "where 'Cambridge Energy says this is the figure that energy consumers will save', and then it's in all the newspapers". When describing the energy future, Kelly also displayed the characteristics of a dreamer. During one conversation I had with him in a San Francisco hotel room, he practically never moved from the window. He just stood there, tall and lean, staring out across the San Francisco Bay with one hand holding aside the white-laced curtain. At the time, I was energy coordinator to Alaska governor Tony Knowles (working with Congress on energy issues) and reporting to Larry Persily, the governor's assistant on Alaska oil and gas development. The three of us met at the Palace Hotel, where earlier that day Kelly had delivered his latest forecasts to executives at a Cambridge Energy roundtable event. The Palace Hotel combines business-class convenience with an old-world extravagance established during the California Gold Rush. Persily had requested a private-client meeting to discuss promoting government support of a multibillion-dollar Alaska natural gas pipeline. The discovery in the 1960s of a large natural gas reservoir located at Prudhoe Bay, Alaska, has inspired frequent exchanges between state officials and consultants about delivering Arctic gas thousands of miles to mid-continental markets.

Interactions like this have raise several questions that I have wrestled with during my ongoing observation of Arctic hydrocarbon development: what role do consultants have in government and industry promotion of global oil and gas development? What is the status of consultant knowledge that it requires provisioning in elite spaces like the Palace Hotel? What is the broader context of this type of knowledge provisioning for visualizing the energy future?

In this chapter I outline a few of the forms responsible for stabilizing the reality of Arctic hydrocarbon development since the turn of the millennium: first, the rise of the consultant expert as a type of Mephistopheles, a child of the idea of development through which monetizing frontier Arctic oil and natural gas can be posed in a new logic. Here, I am concerned with the way consultants seize upon a degenerated system of plausibility, such as transporting Alaska gas to continental markets, only to use its old characters to erect new logics and new systems of meaning. In the case of Alaska natural gas, a thought model of a *growth imperative* conceived in the early 2000s gives rise to a post-millennial era of metaphysical speculation about Arctic frontier developments. As such, consultants employ a logic that underlies the Hegelian conception of the world and of history, the logic of evolution and of a historical dialectic that promises constructive destruction.

Second, is the staging of knowledge provisioning in non-traditional policy locations like executive roundtables, which take place in hotels, galleries, and other elite spaces. I refer to these settings, where luxurious lifestyle intersects with energy planning, as elite premium networked spaces or an *energy salon*. My use of the term salon gestures to a sense of the pleasurable, the complex, and the place of cross-talk in a kind of trading-zone where purpose, action, and affect mingle towards instrumental or unintended ends, fostering communities of interpretation around petro-industry information. Here, I draw on Norbert Elias's (1983) emphasis on etiquette, spatial relations, and modes of visualization that serve as objects of representation for structuring an image of court society. Similarly, the modern-day energy salon is a type of elite experience emphasizing luxury and security, producing a lifestyle whose aesthetics reinforce the veracity of strategic knowledge in energy planning. The origins of roundtable gatherings in the 1990s restructurings of the energy system, and their development since, reflect a growing reliance on future perspectives that are fundamental for social coordination. These meetings are notable for their rarified venues – five-star hotels, museums, pavilions – in major cities across the globe. Details are elaborately staged for each occasion and align elite experience with expertise with the aim of objectivizing views in an industry characterized by controversy and risk.

Together, these two forms represent a shift in energy planning from political hearings suggesting new spaces and spokespersons of knowledge provisioning that reside alongside older and more established forms such as the congressional hearing and elected official. While the hearing and the roundtable share an interest in the display of rhetoric before a community with decision-making authority, the aim of the political hearing seeks to identify appropriate governance, typically in the form of legislation (a text-based framework that builds on juridical precedent), whereas the roundtable event distributes forecasts and the scenario which are visual-based projections of the future and assembled through economic evaluation.

In the case of the former, legislation is a type of system-building instrument that manages the future through regulatory procedure. It entails an idyllic time function whose characteristic relies on precedent or legacy time (heritage or right of inheritance). Regulatory procedure does not constitute a time-sequence that is developmental or maturational. Inherent in its own function are rules that generate and define the measure of projects, such as descriptions of market structure, regulation, and custom associated with bureaucratic procedure. In the case of the latter, scenarios envision a desirable state of things and then develop strategies for achieving it, what might be called back-casting (as distinct from forecasting). It is a narrative of efficacy that accepts humans as part of the forces that influence market evolution and authorizes them to intervene on behalf of what is desired. Roundtables, unlike hearings, assume that energy futures, while not fully understandable, are open, malleable, and outweigh determining aspects such as legislation. In this way, roundtables foster elite consensus around energy futures, enabling long-term economic visions, separating time into government inactivity versus market volatility, while creating a new visibility for intermediary experts who distribute a new sensitivity for manifest destinies of development. In the context of government aims, such perspectives (consciousness, views) shift personal loyalties founded on political destinies to those associated with complex configurations that are perpetually constructed through competent agents engaged in valuation practices.

Consultant expertise

At the time of my meeting with Ed Kelly described earlier, a photograph identifying two men as Ed Small and Wilson Condon began appearing with some frequency in local Alaska newspapers, oil and gas trade press, and on Arctic development websites. The photo was taken by David Harbor, a retired ARCO lobbyist and founder of the prominent Northern Gas Pipelines website. Ed Small was Director of Canadian energy for Cambridge Energy and a respected authority on the North American natural gas industry. With a BSc from the University of Saskatchewan, he was at the time advising the State of Alaska on the interface of energy markets between Canada and the United States. Wilson Condon was Commissioner of Alaska Department of Revenue, and responsible for overseeing what he called the state's "fiscal health". Public displays of statements by Condon drew the attention of Alaskans because of an annual disbursement to residents from a Permanent Fund dividend that relies on the state's fiscal well being. Condon, with a law degree from Stanford University, served previously as Attorney General (1980–1982) during the Alaska Governor Jay Hammond administration.

On one page of the Northern Gas Pipelines website, the caption of the photograph states the following: "Ed Small (Photo-L, with Revenue Commissioner Wil Condon), of Cambridge Energy Research Associates (CERA), retained by the Department of Revenue". Viewed on the computer, the on-screen appearance of Small and Condon portrays an aesthetic alignment and shared intimacy of thought. Both men slightly hunch forward and face the photographer at a three-quarter angle with their shoulders touching. They carry a grin suggesting careful reflection and self-restraint. The eyes of both men lie on the same horizontal plane, emphasizing an intentional gaze. The image is cropped just below their shoulders, making visible their suit jackets, starched white shirts and ties with paisley design.

At the time when the image was captured, both Small and Condon had provided testimony at a hearing convened by the Alaska legislature's Joint Committee on Natural Gas Pipelines ("Committee"). The aim of the Committee, created by resolution during the 22nd legislature, was to review and further efforts to commercialize Alaska's North Slope natural gas reserves. At the hearing, Ed Small provided an energy market update, describing the fundamentals. He stated there:

> [i]s a window of opportunity for Alaska gas, but it is not a done deal by any stretch of the imagination. We think there is a reasonable probability or possibility to see Alaska gas within the course of the next decade. But that has been the case for [the] last three decades.

Directly after Small's testimony, Condon discussed the state's progress in preparing answers on questions relating to taking an equity position in pipeline construction, financing, and the potential of the state's ability to provide public services or negatively affect its credit worthiness.

The above example is one of several that highlights a growing shift in Arctic energy policy toward increased reliance by government and industry on practices of consultant expertise. Industry restructuring, climate change, and price volatility are exerting an organizationally destabilizing effect on the way government and industry leaders visualize perspectives on the future of energy markets. Through an integrated set of technologies – scenario planning, executive roundtables, and internet analyses – consulting firms translate the uncertainties of stakeholders into their own network. They combine technical predictions with new modes of visualization and are important for both the knowledge they generate and the ritual-like learning environments they create. By absorbing the fragmented understandings of clients, these firms provide an objectivized view of how the industry operates. I want to emphasize the important role played by conceptualizations of the future in consultant advisory services. By producing a knowable and concrete future, consulting firms allow for the envisioning of disparate individuals as related through the simultaneity of time. Much like the production of what Benedict Anderson (1991) called "calendrical coincidence" in the lives of people near and far, consulting firms illustrate (and in the process produce) a collective subjectivity on the energy future – a subjectivity that is justificatory of ideals of progress, economic growth, and increased energy consumption.

But the industry's competitive structure has raised problems for an older market segment of energy producers and pipeline companies who have sought to develop new sources of Arctic gas supply. By renouncing control over energy prices, government dismantled an environment in which financial instruments like long-term contracts could diminish the high-stakes uncertainty of investing in large energy systems. As such, market risk had become critically privatized. In the case of Arctic gas development, it is extremely difficult to synchronize the long-term horizon of energy production, which is measured in years, with the short-term volatility of markets. Inherent uncertainties also relate to climate policies, and a mismatch between the European Union (EU) and Russia over liberalization. Shale oil production in the United States and the drop in the oil price from December 2014 also influence the profitability and urgency of Arctic development. Indeed, the choice of market itself, whether the Asia-Pacific region, the EU, or North America, is in part determined by expectations of how these uncertainties will be resolved. Intermediaries provide organizational significance for the way industry and government leaders stabilize these uncertainties. They are promise builders who draw up signposts about the state of the industry and its future development while making available a community of interpretation on a commodified basis.

I consider energy consultants "intermediary actors" because of their success in mobilizing expectations (Buenza and Garud 2007). Their visibility reflects a growing reliance on advisory services that try to identify uncertainties and provide industry actors with the capacity to be ready for them. By mediating an entire ensemble of relations within the industry, intermediaries are transnational agents who wield control over economies once regulated through the national state. Energy development is one of several industries reliant upon intermediaries, and understanding this reliance sheds light on a more general phenomenon across society. Interest in intermediaries is growing, but there is little understanding of their specific type of expertise and how they create knowledge from specialized forms of study. Neither is there much information about how predictions exert complex forms of influence on local and global developments. Scenarios and forecasts aimed at theoretically anticipating events suggest an authority that may not reflect the aspirations of local communities. From this perspective, the Arctic no longer refers to a geographical space, but to an informational regime, whose forms of knowledge reduce the Arctic to definitions of either valuable or risky frontiers.

In Figure 28.1 I identify energy futures as a process of assembling and mobilizing through *internal practices* inside consulting firms and of mobilizing and performing through *external practices* outside consulting firms. This framework highlights how agendas are set by experts whose authority is based not on their structural position but on their theoretical knowledge and independent stance. In the past, the culture of power in energy planning reflected forms of collusion whose decision-making authority relied on structural positions of bureaucratic- and capitalist-led organization (sub-governments, managerial consensus, natural monopoly). Energy systems were (and still are) highly regulated by a national political community in which expertise is embedded in the political organizational form. Yet, the A-M-P draws attention to how agendas are set not by political institutions or the history of industry, but by experts, highlighting facets (accessible through ethnography) of what happens when consultants engage with political and industry leaders.

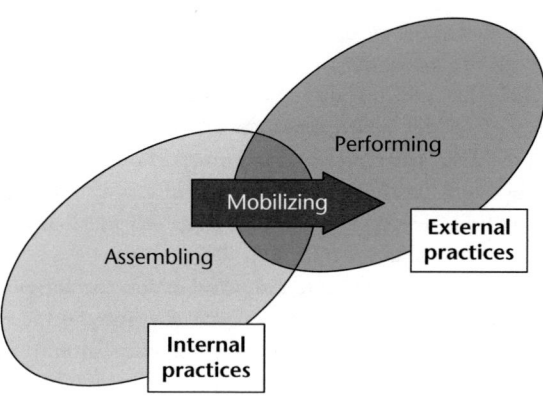

Figure 28.1 Assembling and mobilizing through *internal* practices inside of firms. Mobilizing and performing through *external* practices outside of firms.

Executive roundtables

Energy roundtable events are gatherings where industry participants take on a special ritual-like character by their actual placement within environments of luxury and security and from their proximity to the company of experts. Indeed, even entering into the four-star hotels, pavilions, and art galleries where energy events take place requires detachment by passing through several layers of security involving turnstiles, bodyguards, registration, and mandatory identification badges with barcodes. After security, there are the elaborately decorated spaces, hallways, and meeting rooms, often dressed with abstract paintings, sculptures, and various artistic installations, promotional banners, and brochures. Here, various sources of light intensify the rarity of these spaces. In addition to the ambient light emanating from the crystals of old-world chandeliers, there are accented lights from spot lamps that shine directly onto plenary speakers and wall paintings. Other light sources include computer images projected onto walls as backdrops and promotional banners. Such projections may be interrupted through live feed visuals of event speakers that are beamed into the room over the internet from different time zones.

While the aesthetics of these spaces may illicit an ambivalent response, the PowerPoint images flashing on projection screens, whose meanings are explained in detail by market analysts, do not. Each slide of an expert presentation appears as a hieroglyph of coded signs that requires deciphering (number, acronym, graph). The complexity of each visual offers a decisive contrast to the event logo, which is not only immediately recognizable but replicated on all manner of promotional items, from banners and brochures to tote bags, pens, and writing pads. In this way, such installations lay the ground for expert knowledge to deliver a sense of wonderment about the functioning of the global energy industry. The meaning of wonder is that instead of taking things for granted, spectators are surprised and taken aback by what they see. Maurice Merleau-Ponty (2009) writes that a reduction of perceptual impression through fabrication leads in thought and action to "a wonder in the face of the world". A sense of wonder certainly adds an ethical dimension to decision-making in the sense that participants likely consider their actions on the basis of new understandings. In short, these settings are not just hotel rooms but elaborate installations whose fabrications of luxury consolidate fragmented understandings by connecting expert judgment to the delicate and fleeting experiences of an elite lifestyle.

Energy futures are detailed expectations that represent strategic resources for attracting attention from (financial, political) sponsors to stimulate agenda-setting processes. They are also commodities whose value is uncertain and only indirectly based on their cost of purchase. Thus, in the context of energy planning, where foresight knowledge is characterized by uncertainty, incidental features may assist in the exercise of judgment. For this reason, I identify the ritual-like forms of roundtable events under the rubric of impression management or what Lucien Karpik (2010) calls "judgment devices" – suggesting that an ideal of trust or believability can mediate between objects of manufacture and their desire.

While the energy roundtable is an existing empirical event, the *energy salon* is an object that I created for producing academic knowledge in the form of ethnographic typologies of judgment devices in energy planning. In this, I follow standard practice in anthropology of requiring a stable subject to create purposeful data – *culture* being the most notable of these methodological fictions. By constructing the salon-space, I draw attention to a common structure surrounding expert performance in which incidental features (security, promotional images, etiquette) allow energy knowledge to form a center so as to appear agentive, singular, and authoritative. In this way, I argue that the executive roundtable is the accretion of a new material-epistemic investment; a historical development in diminishing the high-stakes uncertainty of post-mature energy systems. That is, where qualitative knowledge of investing in large-scale energy systems (pipelines, offshore installations) is highly unpredictable, the piling up of so-called judgment devices works to shore up the legitimacy of expert predictions.

From hearings to roundtables

The Arctic is a region dramatically altered through climate change, even as extractive industries and the nations that rely on them frame the Arctic as an alternatively valuable or risky frontier. As a result of the opening of formerly ice-bound waters, transcontinental shipping, resource capture, and scientific observation are accelerating in what many describe as a rush for the Arctic. The rush will likely be slower than the term implies due to lower prices and reduced global demand in the oil market. Still, the threat posed by commercial conquest warrants an examination of the role expert knowledge plays in provisioning a new common sense vision of Arctic hydrocarbon development that entrenches authoritative interpretations.

Development narratives that license the intervention of experts in debates about Arctic resource management suggest a role for social authority, intellectual technology, and practical strategy in reconfiguring the Arctic into hydrocarbon-rich and accessible landscapes. The logics of the prophesy dimension are raising concern over privatized knowledge systems in debates about community plans (Powell 2008), extractive industry, and infrastructure development and security, as well as the efficacy of integrating indigenous knowledge systems with Western institutional apparatuses (Dokis 2010). Since the middle of the twentieth century, publics underwriting research have increasingly acquired a stake in what science produces, making scientists aware of the necessity of public support (Jasanoff 2011). Patterns of democratization are apparent across the Arctic, where redefinitions of local knowledge create new partners in scientific enquiry (Clifford 2014).

The rise of consultant expertise, however, suggests greater control over access and production of Arctic knowledge that is privatized via commodification. Public accessibility exists, but without authority to determine limits of access. Such analyses, while not publicly available, may be shared within the consultant community, pointing to the collective nature of this research process whereby analysts critically scrutinize each other's work prior to publication. The unabashed economic motivation behind the rise of such knowledge reflects a postwar expansion of expert systems as part of a broader movement to a knowledge economy. The growth of this type of economy itself provides justification for an apparent contradiction, on the one hand, of increased democratization of expertise, and on the other, of its privatization.

Since the 1960s, policy discussions in Arctic arenas of oil and natural gas development in Alaska and Western Canada have taken place in government settings, in gatherings convened by federal and local governments to discuss regulatory, market, and environmental perspectives. Similar discussions in Norway and Russia concerning Barents Sea developments began in the 1980s. Hearings take place in government buildings whose rituals serve constitutional requirements. During testimony, presenters read aloud from carefully worded speeches whose contents are submitted to committee staff prior to delivery. Elected officials ask questions that do not invite open-ended discussion from presenters or members of an audience, thus dividing the congregation into a face-off between sovereign officials and those present at the discretion of the former. In my experience, all testimonies were scripted deliveries, and presenters offered only brief responses to questions.

In contrast, executive roundtables take place inside rooms that privilege fleeting experiences, such as hotels and art galleries as evidenced by the transformation these spaces make from uniformly arranged rows of chairs into banquet seating with tables laid out with cutlery and linens. Unlike hearings, where politicians serve as interrogators, the center stage of the roundtable is composed of a group of executives, politicians, and analysts seated together in a semi-circular configuration facing an audience interested in dialogue. These settings have a casual air as evidenced by the close proximity of speakers to audience members, the constant checking of communication devices (tablets, smart phones), and verbal exchange among panelists. This type of knowledge provisioning shares similar features to the university seminar, wherein a speaker delivers an unscripted text over fifteen minutes that is accompanied throughout by PowerPoint images and then engages contrasting reactions from the audience. In contrast to the university seminar, however, roundtable events involve theatrical lighting, elaborate promotional banners, and backdrops capable of advertising corporate sponsorships. In this way, a shift from hearings to roundtables involves a move from discursive to visual techniques, a shift that I consider in terms of a "post-discursive turn" (Mason 2016).

I attribute the rise of these events to changes beginning in the 1970s in arenas associated with energy restructuring, environmental perception, and communication technology that provide opportunities for advisory experts – mainly economists – to assemble, package, and perform knowledge about energy provisioning. Such changes include the reorganization of information related to energy systems development: for example, coal, no longer intimately associated with labour politics, is now seen instead as a fuel, alongside oil and natural gas, in a context of electric power generation. Such shifts coincide with the disembedding of a culture of expertise during the 1980s and with changes in preferences for representing knowledge visually from text to graphic design. What many refer to as the Reagan era of "defunding the left", for example, I see as a shift to the private sector of a first generation of government risk assessors, initially trained in response to a regulatory need for technical evaluation in the wake of the 1970s Congressional environmental legislations. As this relates to visualization, juridical representation of knowledge through text begins to appear backward-looking, especially when compared to the imagery employed by economists, whose graphic projections are forward-looking.

Also taking place at this time is the increasing relevance of economic knowledge in the service of government – as evidenced through the application of cost-benefit analysis in decision-making, particularly in environmental regulation, suggesting a shift from juridical evaluation to favoring economic efficiency through mathematical modeling. The introduction of this shift, what was to be called *reform*, was especially welcomed in the arenas of energy production where industry sought to leaven the "dead hand of regulation", as the perceived cause of the 1970s energy supply crisis was already being referred to by policy analysts of the time (Wilson 1971). Thus, disembedding, as I employ the term, refers to the (neoliberal) marketing of scientific evaluation among persons whose expertise remained indispensable to satisfying the requirements associated with regulation.

During this same period, environmental activist organizations came to rely on scientific representations of truth to generate an empathy for the planet. As such, the authenticity of their message begins to rely less on the drama of personal risk-taking. This is especially so among organizations which acquire media enterprises, and which therefore no longer have to depend upon an independent news estate to stage the verification of their truths. As evidence, consider the shift in activist organizations that during the 1990s sought to alter the dynamics of collective life by fostering an ecological sensibility by reference to media stunts (Wapner 1995). Today, converting mass audiences toward planetary concern requires employing images that stress an empathy for the graph, such as those prominently on display in the documentary *An Inconvenient Truth*. Other efforts aimed at assembling knowledge include the establishment of government houses directed toward collecting reliable data from ever wider aspects of energy arenas – in part by imposing mandatory accounting on firms and creating a rationalized reporting system, as happened in the wake of the OPEC embargo. The institutionalization of energy knowledge provisioning through the establishment of the US Energy Information Agency, for example, provides a new opportunity for independent experts to repackage otherwise hard-to-find data.

What I see here is differentiation between, on the one hand, economic experts who transcend proprietary limitations and, on the other, legal expertise which remains tethered to attorney-client privilege and which is thus incapable of advocating an industry-wide horizon of expectation. The use of visuals (graphs) to frame the economic future as accessible in one glance intensifies this differentiation. As such, within the energy system, legal knowledge is increasingly back-ended, represented by "partial intellectuals" (Bauman 1987: 114) whom industry and government rely on to deliver historical precedent. By contrast, economic knowledge emerges as vanguard, that is, capable of making claims in the form of statements intended not for clients, but on behalf of impersonal forces (the market, the future). From this, epistemic interests appear distanced from proprietary claims, lending to knowledge a prestige value whose independent stance appears as a natural attribute.

It is within this context that a newly established role for the executive roundtable event emerges. Identifying knowledge with a neutral value creates the possibility for inviting competing parties to gather around individual announcements in a shared setting. Where competing parties can stage their individual market condition without threat to proprietary position, the experience is visual, sequestered, and shared. Nevertheless, the quality of this new type of expression, what I call the energy salon, represents reliability where the conditions of veracity are abstracted from any actually existing assembly of data and, as such, are dependent upon the event itself.

Inside the energy salon, aesthetic exposure is a prerequisite for ensuring reliability. Within, the energy salon offers an expanded typology of delicate experiences that serve as a trademark for the veracity of expert knowledge, including attention to the spatial arrangement of security. At energy events, for example, the proximity to knowledge aligns with a movement from ostentatious forms of enforcement toward more delicate forms of identification. At the street entrance, the bodies of security guards are conspicuously robust, whereas the corporeality of attendants who deliver booklets to clients appears thin and feminine. This materiality of security entails its own redundancy in the distancing of delicacy, where metal turnstiles located at the entrance of buildings give way to barcode scanning inside conference rooms. The cost for event attendance is itself stratospheric ($15,000 for three days at the Oslo Energy Forum, for example), suggesting that price, in combination with security and luxury, contributes to the delicacy of semantic provisioning.

In this manner, energy roundtables may be seen as applying symbolic authority to the abstractness of expert judgment, providing a mode of obedience where relations can be made visible through (conspicuous) expenditures in performance. Pierre Bourdieu often contrasts these two forms of authority – the sensuousness of luxury and opacity of expertise – by reference to a threshold of modernity, identifying a shift from a pre-capitalist elite to a modern bureaucratic estate as a transformation in the exercise of power from "symbolic" to "overt" domination (Bourdieu 1990: 122–134). Likely, legitimacy at the roundtable relies on both types of authority, as much on post-capitalist knowledge assemblies as on pre-capitalist modes of enchantment – what Alfred Whitehead (1926: 11) recognized as the "staging of verification" in scientific experiments or what Bruno Latour (1987: 73) variously refers to as "inscription".

References

Anderson, B. 1991. *Imagined Communities*. London: Verso.

Bauman, Z. 1987. *Legislators and Interpreters*. Ithaca, NY: Cornell University Press.

Beunza, D. and R. Garud 2007. 'Intermediary role of securities analysts' *Sociological Review* 55(s2): 13–39.

Bourdieu, P. 1990. *The Logic of Practice*. Stanford, CA: Stanford University Press.

Clifford, J. 2014. *Returns: Becoming Indigenous in the Twenty-First Century*. Cambridge, MA: Harvard University Press.

Dokis, C. 2010. 'Modern day treaties' *Geography Research Forum* 10: 32–49.

Elias, N. 1983. *Court Society*. Amsterdam: Boeken.

Jasanoff, S. 2011. 'Cosmopolitan knowledge' in Dryzek et al. (eds) *Oxford Handbook of Climate Change and Society*. Oxford, UK: Oxford University Press. 128–143.

Karpik, L. 2010. *Valuing the Unique: The Economics of Singularities*. Princeton, NJ: Princeton University Press.

Latour, B. 1987. *Science in Action. How to Follow Scientists and Engineers through Society*. Cambridge, MA: Harvard University Press.

Mason, A. 2016 'Arctic energy image' *Polar Geography* 39(2): 130–143.

Merleau-Ponty, M. as cited in G. Olsson 2009. *The Visible and the Invisible*. Stockholm: KTH.

Powell, R. 2008. 'Configuring an Arctic commons?' *Political Geography* 27: 827–832.

Wapner P. 1995. 'Politics beyond the state' *World Politics* 47(3): 311–340.

Whitehead, A. 1926. *Science and the Modern World*. Dublin: Mentor Books.

Wilson, J. 1971. 'The dead hand of regulation' *The Public Interest* 25(Fall): 54–78.

29

Social and environmental impact assessments in the Arctic

Anne Merrild Hansen, Sanne Vammen Larsen and Bram Noble

Introduction

Exploration, and subsequent exploitation, of mineral and hydrocarbon resources is not new in the Arctic. Activities related to mineral and oil extraction have affected the lives, societies, and cultures of people living in close proximity to projects planned, on-going, or decommissioned in the region since the nineteenth century (Avango et al. 2014). Arctic countries with mineral extraction and oil and gas extraction histories and contemporary projects or planned activities include Finland, Sweden, Norway, Canada, Alaska, Russia, and Greenland (Haley et al. 2011), while other large-scale industries such as aluminium smelting have been developed in Iceland. The projects implemented have sometimes created wealth and brought economic benefit to northern communities and regions but have also brought considerable disruption to the environment and to people's livelihoods (Mortensen 2013; Vanclay et al. 2015). It is not only at the community level that extractive activities influence the Arctic. At broader national and international scales, the impacts of Arctic development are often managed through investment in national trust funds, facilitate the establishment and development of infrastructure, and contribute to national budgets through taxes and revenues (Duhaime and Caron 2006). Further, international extractive business networks have been found to be important strategic forums for policy-making. Extractive industries are expected to continue to influence the livelihoods of the people of the Arctic considerably in the future, both locally and through broader national and international policy and development contexts (Andrew 2014; Duhaime and Caron 2006; Nuttall 2012).

Projects, in the context of impact assessment, usually refer to physical undertakings, including, for example, the actual exploration and development of a mineral or energy resource and the associated infrastructure needed – such as roads, housing facilities, power supplies, and waste management facilities. The impacts resulting from individual projects depend on the nature of the activities undertaken, the effectiveness of any mitigation, and the characteristics of the potentially impacted communities and environment, especially in terms of their vulnerability and resilience (Hansen et al. 2016; Vanclay 2002). Arctic communities have in various cases experienced the impacts of project developments involving multinational corporations. The social and biophysical impacts of such extractive projects can be controversial and complex. The way these projects are managed can therefore enhance and/or retard social development options and trends (Esteves and Vanclay 2012). A key issue in this regard, as emphasised in Hansen and colleagues (2016: 25) is that companies can move on to other projects when reserves are exhausted or if mistakes are made, while a community generally only has one chance at development – this is why it is of utmost importance to get it right the first time. To manage the impacts of extractive projects before, during, and after extraction, Environmental Impact Assessment (EIA) and Social Impact Assessment (SIA) is employed across the Arctic to assess, mitigate, and monitor the impacts of resource development. In a review of Arctic guidelines for impact assessment, Koivurova (2008) suggested that impact assessment tools, including both EIA and SIA, will likely be some of the most critical tools for managing development in the Arctic.

Our aim in this chapter is to describe how impact assessments are implemented to manage extractive projects in the Arctic, and thereby mitigate social and environmental disruption, and contribute to desired development in affected communities. First, we provide a general introduction to the impact assessment tools. We then describe how SIA and EIA are applied in the Arctic countries with focus on four selected issues. Next a case study from Greenland is presented, reflecting on the challenges of securing local benefits in relation to projects, and to illustrate some of our points and arguments. Finally, in the concluding section, we contextualise our presentation and discussion.

Impact assessment: purpose, procedure, and effectiveness

The purpose of impact assessment tools, such as SIA and EIA, is to ensure informed decision-making, through mainstreaming environmental and social considerations, and to assess and mitigate the potential negative impacts of a development and enhance benefits in co-operation with local communities. The overall purpose of the impact assessment tools is to promote a form of sustainable development for a project, through influencing decisions and the design of projects, plans, programmes, and other development initiatives (Bond and Pope 2012; João et al. 2011; Senecal et al. 1999). The impacts involved may be remediable or irremediable; they can be short-term, long-term, or even permanent; they are often cumulative and interact with other environmental and social impacts; they can vary in many other ways; and when dealing with projects they are often site-specific (Vanclay 2002).

Impact assessment, such as EIA and SIA, follows a common procedure which has a general linear process. Based on the discussion of best practice by Senecal et al. (1999), this includes the following steps:

1 Screening: Determining whether or not a proposed project should be subject to an assessment.
2 Scoping: Establishing the scope of the assessment, including the alternatives to be assessed and the baseline against which the impacts are to be measured.
3 Impact analysis: Identifying and assessing possible impacts of the project.
4 Mitigation: Establishing mitigation measures to avoid, minimise, or offset predicted impacts.
5 Documentation: Documenting the results in an environmental report.
6 Review: Reviewing the process and reporting on whether it stands up to the demands for quality.
7 Decision-making: Using the impact assessment and the report when deciding on whether to approve or reject the project.
8 Follow-up: Monitoring the project development to ensure it is carried out as assumed and monitoring its impacts.

However, in reality, this generic process should be adapted to the particular project under assessment and review and the specific decision-making process concerned with it, and there are iterations between the various steps in the procedure. Also, public participation is an integrated part of impact assessment. It is integrated in different phases of the procedure. In some Arctic jurisdictions such as Norway, Greenland, Canada, and Alaska, there is public participation in relation to the scoping phase; and in most jurisdictions there is an opportunity for public review and comment on the published impact statement and in decision-making (Hansen and Larsen 2016). Public participation in the Arctic includes accessing local knowledge, understanding how local communities might respond or adapt to changes, and encouraging connections between companies and locals (Olsen and Hansen 2014).

There are many ways to discuss and assess whether an impact assessment based on the above provisions is effective – whether it is a good quality impact assessment. Chanchitpricha and Bond (2013), based on literature review and analysis, propose four different categories of effectiveness:

- Procedural effectiveness: Is the assessment process carried out as intended; does it for example have proper public participation, or deliver relevant information to decision-making?
- Substantive effectiveness: Does the assessment achieve its aims or purpose, for example achieving a more sustainable project?
- Transactive effectiveness: Does the assessment deliver the intended outcomes in a resource efficient manner?
- Normative effectiveness: Does the assessment reach the normative goals for impact assessment that different actors in society hold?

As described in this section, impact assessment is built on a common framework of purpose, procedures, and measures of effectiveness. However, the specific context and implementation of impact assessment varies in different jurisdictions, and in the following section we go into greater depth with some characteristics of impact assessment in the Arctic.

Key impact assessment principles, challenges, and opportunities

In this section we describe how SIA and EIA are applied in legislation and practice in the Arctic. We have chosen to focus on topics that are crucial and current to impact assessment in general and in the Arctic in particular.

Identifying alternatives

Assessment of alternatives allows for comparison of likely impacts and mitigation measures (Glasson et al. 2013). Alternatives can contribute to the assessment of impact significance and identification of mitigation measures. Alternatives are considered an essential part of an impact assessment process as some impacts can only be defined through understanding relative differences between alternatives; the lack of data and detailed knowledge of ecological and socio-cultural conditions and functions may challenge the evaluation of impact significance (Arctic Environment Protection Strategy 1997; Lawrence 2013). The recognition, discussion, and consideration of alternatives are therefore, according to international guidelines, expected to be an integrated part of any impact assessment process (de Jesus et al. 2015). Also, the Arctic EIA guidelines published by the Arctic Council specify that alternatives should be identified as a part of the scoping process: "The scope of an assessment should include all potential environmental, socio-cultural and economic impacts, especially impacts on the traditional uses of resources and livelihoods of indigenous people and also the consideration of alternatives" (Arctic Environment Protection Strategy 1997: 6).

EIA and SIA are impact assessments related to specific projects, and the potential alternatives to a proposed extractive project which companies may consider in their assessments are therefore not necessarily alternatives to the proposed activity in general, but rather alternative ways of undertaking the proposed activity (Glasson et al. 2013; Vanclay 2003). Alternatives can be defined as different approaches to the realisation of a project. For extractive activities, alternative sites are not usually available within a region as many of the activities related to extraction are very site-specific, particularly the location of the mineral/oil reserve, but also shipping, since port facilities often depend on deep-water access, and also related facilities such as oil refineries, chemical treatment of minerals, housing of workers, etc. An alternative that is always to be considered is the 'no action' alternative, also known as the 'zero alternative'. In addition to the zero alternative, assessments can include for example technical and technological alternatives such as choices on scale, appearance, timing, waste discharges, processing methods, and traffic management.

During the scoping phase alternatives are to be specified together with the project description and the terms of reference for the impact assessment process. The alternatives to the proposed project are to be described in sufficient detail to "identify potential direct and indirect impacts, including cumulative effects". International guidelines from the International Association for Impact Assessment recommends that impact assessment processes involve the affected communities and other stakeholders in the process of alternatives generation and selection (de Jesus et al. 2015). In Norway and Alaska, the EIA legislation for extractive activities specifically requires that, at a minimum, the zero alternative is to be described and incorporated in the assessments (BOEM 2008; Norwegian Ministry of the Environment 2003). In Alaska it is also emphasised that alternatives need to be identified and narrowed down to only include and address realistic and relevant alternatives early in the process, to ensure that resources are used in an appropriate manner (BOEM 2008). Similarly, in Canada, under the federal assessment process, proponents are required to consider, and assess the impacts of, alternative means of carrying out the project that are technically and economically feasible (Canadian Minister of Justice 2012).

In Norway, Canada, and Alaska, the responsibility to conduct impact assessments lies with the applying company or agency. There are no requirements for other actors to be involved in the identification generation or selection of alternatives prior to public hearings. There are no specific requirements either as to how the alternatives should be assessed and used in the processes of SIA or EIA (Hansen and Larsen 2016). This does not necessarily mean that the international guidelines are not followed in practice. Recent research points to a practice where the scope and content of impact assessment processes and documents are continuously negotiated between proponents and governments, explaining why alternative methods and technologies may be considered and incorporated even if they are not reflected in a transparent manner in the documents (Bidstrup and Hansen 2014). However, examples of inclusion of alternatives in social impact assessments are seen in relation to Arctic projects.

United Kingdom-based company Cairn Energy described in an SIA for an exploration drilling programme offshore West Greenland in 2011 that: "The selection of alternatives for drilling locations, drill units, mud selection and support operations are described below. One alternative to be considered is the No Development Option; the implications (positive and negative) of exploration not proceeding" (Cairn 2011: 3–16).

Addressing cumulative impacts

The cumulative impacts of mineral and energy exploration and development can add up quickly in the Arctic. Cumulative impacts are the additive and often synergistic impacts that result from the adverse environmental and socioeconomic changes attributed to multiple development projects (Duinker and Greig 2006). They can result from individually minor but collectively significant actions that take place over time, or within a particular geographic area. Cumulative impacts are often long-lasting legacy effects and adversely affect the resiliency of Arctic communities and ecosystems.

Consideration of the cumulative impacts of development, including mineral and energy exploration and extraction, and its associated infrastructure requirements and activities (e.g. road and port construction, shipping), is required to some extent under the impact assessment legislation and guidelines of most Arctic nations (Noble and Hanna 2015; Solodyankina and Koeppel 2009). Several Arctic mineral and energy development projects have been subject to cumulative impact assessment, from Canada's first diamond mine at Ekati in the Northwest Territories (CARC 1996), to the more recent liquefied natural gas operation in Yamal in northern Russia (ENVIRON 2014). Typically, when a project is proposed, the proponent is asked to consider whether and how the project's actions or impacts might interact or overlap, spatially or temporally, with the actions or impacts of other past, present, and reasonably foreseeable projects. Although simple in concept, cumulative impact assessment has proven somewhat intractable in practice, partly because of its focus on individual projects and on finding ways to reduce their incremental impacts, rather than focusing also on the condition of the receiving environment and on its capacity to absorb those incremental impacts (Gunn et al. 2014).

The limitations of cumulative impact assessment in the Arctic are well known (Ehrlich 2010; National Research Council 2003; Noble et al. 2013). It makes much more sense that cumulative impacts are addressed at the strategic and regional levels of policy and planning, versus solely at the project scale (Duinker et al. 2013). A number of regionally-focused, planning- and science-based initiatives have thus emerged across the Arctic, as a means to provide the monitoring, planning framework, and baseline data to understand and assess cumulative impacts more effectively. Some of these programmes include, for example, the National Research Council committee on the cumulative effects of oil and gas development on Alaska's North Slope, the Beaufort Sea regional environmental assessment initiative, Canada's Northwest Territories cumulative impact monitoring programme, and Norway's integrated management plan for the North Sea and Skagerrak (Fidler and Noble 2013; National Research Council 2003; Norwegian Ministry of the Environment 2013). An enduring challenge, however, is that although such programmes are valuable for understanding the state of ecosystems, development pressures, and Arctic communities, it remains unclear as to whether and how they actually inform decision-making or direct regulatory activities for the purpose of managing the cumulative impacts of development projects (Noble and Hanna 2015). As a result, decisions about where, when, and under what conditions and requirements industrial activities are permitted in the Arctic continue to unfold on a project-by-project basis. Cumulative impacts, by definition, cannot be addressed in isolation (Dales 2011).

Addressing climate change

Climate change is one of the key environmental and societal challenges today, and it is a significant challenge also in the Arctic region. According to the Intergovernmental Panel on Climate Change (IPCC), the Arctic region is warming more rapidly than the global mean, and the main impacts from this warming are shrinking ice sheets and glaciers, higher permafrost temperatures, changes in snow cover, and coastal erosion or other sea-level impacts (IPCC 2014). These changes lead to key risks for the Arctic being changes in marine and terrestrial ecosystems as well as impacts on health, well-being, and livelihoods. The Arctic is especially vulnerable because the rate of such changes will be very rapid, making timely adaptation of human and natural systems challenging (IPCC 2014).

In general, there are three ways in which impact assessment can contribute to addressing the challenge of climate change (European Commission 2013; Larsen and Kørnøv 2009):

- Mitigation: Predicting and assessing the expected impact from emissions of greenhouse gasses resulting from the project and mitigating them, i.e. through avoiding or reducing emissions.
- Adaptation: Predicting and assessing the impacts of climate change on the project and mitigating these impacts through adapting the project and surroundings to cope with coming climate changes.
- Baseline adaptation: Adjusting the baseline for the assessment to reflect the future changes in natural and human systems caused by climate change, thus providing the basis for prediction of impacts on these systems.

Looking at how climate change is integrated in the impact assessment regimes in the Arctic, in the majority of countries there is not an explicit demand for attention to climate change. This is the case in Canada, Alaska, Greenland, and Norway. Though in some jurisdictions, including Canada, requirements to address climate change are increasingly appearing in the terms of reference established for individual project impact assessments, including, for example, the EIA for the Mackenzie Gas Project (MGP) in the Northwest Territories (IGC et al. 2004). In contrast, in Sweden and Finland, who are members of the European Union, there is a demand for EIA to include assessment of inter alia "land, soil, water, air and climate" (Directive 2014/52/EU 2014, article 3). There are few recent studies of the actual practice of integrating climate change in EIA and SIA specifically in the Arctic. Through a range of case studies, Ohsawa and Duinker (2014) assess that climate change mitigation is a common part of EIA in Canada. This echoes the results of studies in other jurisdictions, where it has been established that mitigation is common in EIA, while adaptation is less common and baseline adaptation is rare (see e.g. Larsen 2014 and Agrawala et al. 2012). Thus climate change is a relevant environmental issue in the Arctic, and one that can be handled in impact assessments in multiple ways. However, there are few explicit demands for this integration, and practice is relatively unknown.

Engaging the public and inclusion of traditional knowledge

Meaningful community engagement and the incorporation of the traditional knowledge of indigenous communities, are cornerstones of impact assessment. Meaningful engagement implies, amongst other things, that those communities potentially affected by development are enlisted into the project planning and impact assessment process, and that project proponents are open to modifying their project and developing new plans or even discarding existing ones based on the knowledge and values of those affected (Noble and Udofia 2015). Meaningful engagement is essential for proponents who wish to earn a social license to operate, and it can contribute to improved project design, better impact mitigation, and increased legitimacy of a project (Prno and Slocombe 2012; Rozema et al. 2012).

When affected communities and their local values and traditional knowledge are *not* sufficiently integrated into impact assessment processes, the outcome can be detrimental for project proponents. Ehrlich (2010), for example, reports on a series of failed impact assessment applications for mineral exploration in the Upper Thelon River region in Canada's Northwest Territories. The proponents argued that their projects would have negligible to even positive socio-cultural effects on local indigenous communities; however, the communities disagreed, arguing that their cultural and spiritual values of the land base were not adequately considered by the proponents. The project applications were rejected.

Notwithstanding the importance of engaging communities, in a recent review of impact assessment across Arctic nations, Noble and Hanna (2015) identify the growing requirements for consultation with Arctic communities, combined with increasing numbers of project applications, as generating concerns about the capacity of Arctic communities to engage meaningfully in impact assessment. Kwiatkowski et al. (2009), for example, suggest that many indigenous communities lack the financial and human resource capacity to become engaged, and to remain engaged, due, in part, to the very size and complexity of the assessments carried out. *Road to Improvement*, a review of regulatory systems across the Canadian North which was submitted to Canada's federal Minister of Indian Affairs and Northern Development in 2008, identified similar concerns about the capacity of indigenous communities to document traditional knowledge to assist in impact assessment (McCrank 2008); and, in 2011, the Alaska Forum on the Environment identified concerns about communities not only needing to see the benefits of their engagement in impact assessment, but also needing the resources to participate (IWG 2011). Unfortunately, the problem is not new – reporting on experiences in northern Finland, Huttunen (1999) identified that community engagement in the Sierilä hydropower assessment contributed the realisation of local empowerment and self-management but cautioned that communities often lack the capacity to participate. In 2016, the Canadian Environmental Assessment Agency (CEAA) conducted a review, which included the input of indigenous stakeholders, of how best to integrate Aboriginal traditional knowledge in its principles and guidelines for environmental impact assessment, and there are hopes that the processes companies engage in with communities will be strengthened and result in more meaningful engagement. Meaningful engagement not only means providing opportunities for those affected by development to become engaged, it also means ensuring that they have the capacity to do so.

Securing local benefits: a case study from Greenland

The Government of Greenland, *Naalakkersuisut*, is encouraging an expansion of extractive industries in Greenland. The objective of the government's minerals strategy is to achieve a strengthened economic basis for Greenland's future social development. It is the *Mineral Resources Act* 2009 which regulates all extractive activities in Greenland. The operating companies are required to undertake an SIA and enter an impact benefits agreement as part of any application for mineral production, oil exploration drilling, and oil production. The emphasis is on securing local benefits through a high degree of local content in the projects and this is primarily understood as either direct or indirect employment opportunities for Greenlanders in the industry.

A recent case study from Northwest Greenland (presented in Hansen and Tejsner 2016) focused on the challenge of securing local benefits in relation to planned exploration drillings in the Upernavik district, where four operating companies including Maersk Oil, ConocoPhillips, Cairn Energy, and Shell held licenses to a total number of five blocks in 2014. The planned exploration activities were expected to lead potentially to production and possibly cause significant social impacts in the area. Oil exploration was later cancelled due to dropping oil prices;

however, the Upernavik district was at the time expected to become particularly exposed to offshore industrial initiatives due to the location and potential high level of activity in relation to the oil industry here. The aspirations of the residents, the Upernavimmiut, were investigated to understand the prerequisites for possible recruitment of locals to work in or for the industry while also securing local interests as part of any agreement.

The Upernavik district stretches 450 kilometres and covers the town of Upernavik and nine smaller settlements. About 2800 people reside in the district. The inhabitants here, as in the rest of Greenland, are mainly Inuit by ethnicity (approximately 98%). The main occupation in the area is hunting and fishing. In Upernavik, as elsewhere along the west coast, people's traditional or continued reliance upon and daily experiences with the sea and its resources is of great importance to social life and the mixed cash–subsistence economy of the local households. Hunting quotas regulate the harvest of selected species in Greenland, but while some species, such as seal or arctic cod remain abundant, other animals, such as narwhals and belugas, continue to remain subject to government regulations. All whale meat and *mattak* (whale skin with blubber, a local delicacy) is sold locally. Whale quotas are set by the *Naalakkersuisut* annually and subsequently distributed to local districts where the municipal authorities decide on the quotas allocated under commercial and leisure hunting licenses. Other species which local residents also hunt include birds, walrus, seals, and polar bears. Commercial fishing is primarily focused on Greenland halibut but both commercial fishing and recreational fishing also rely on species such as capelin, lumpsucker, redfish, and cod. Subsistence fishing takes place in the area and is considered an important supplement for many households.

The study found that fishing and hunting did not appear to be considered a livelihood by the locals, but rather a desired way of life. A necessity for changes to take place to secure a continued existence and improved development in the settlements was still articulated. The Upernavimmiut expressed a positive attitude towards the oil industry, which they saw as a potential facilitator of some of the locally needed changes. It was pointed out by many that it was not possible for the locals to combine work for industry with fisheries and hunting due to the existing licensing system, as the licenses can only be upheld if the main part of their income (more than 50%) is based on hunting and fishing – not taking into account the contribution to the subsistence economy. They also continued to highlight the fishing and hunting licensing system as a barrier for working in other sectors and hence to achieving the needed change in order to continue to uphold a partly traditional way of living. Despite feeling stressed by ongoing technological and demographic changes, it may be assumed that were the locals to choose between either working in industry or as fishermen and hunters today, they would still choose the latter. Securing a high degree of local content in oil projects in the Upernavik district and thereby securing local benefits would therefore not be achieved by the means emphasised by the Greenland government alone, but require more strategic initiatives which, together with an understanding of local aspirations, are essential for securing that the projects would lead to local benefits.

Conclusion

In this chapter we have introduced and provided an overview of key issues relating to social and environmental impact assessments, providing examples and a case study of how impact assessments work in parts of the Arctic. We have described how impact assessment internationally is developed as a common framework for managing projects so as to mitigate undesired impacts and enhance the possibility of achieving desired outcomes. Yet the tools are implemented in different contexts across the Arctic and therefore vary between jurisdictions. Still common issues emerge which can be considered general key challenges to impact assessment in the Arctic. We emphasise the importance and need for addressing alternatives in the impact assessment processes, which is in accordance with international and Arctic guidelines for impact assessment, and stress the need for transparency in the dialogue that occurs between proponents and regulators when adjustments are made during impact assessment processes. We also highlight the importance of addressing cumulative impacts and find that even if cumulative impacts are identified in extensive programmes, it is still not clear to what extent they are feeding into project or strategic decision-making processes. This is essential as cumulative impacts cannot be assessed or mitigated either on a case-by-case basis or in isolation. Climate change in impact assessment is a parameter which we also point to as of central importance in the Arctic. Mitigation, adaptation, and baseline-adaptation are fundamental concepts to consider in this regard. We finally discuss the necessity of meaningful engagement of potentially affected communities in impact assessment processes, stressing that meaningful engagement takes note of traditional and local knowledge and includes provision of opportunities for locals to become engaged as well as the capacity to do so. Following this discussion of key issues, we then presented a case study from Upernavik district in Northwest Greenland. The case study points to the need for understanding the affected communities to be able to promote local benefits as these are not achieved alone by means such as job creation. Residents in the communities are not necessarily interested or willing to work in industry if it compromises traditional activities.

In relation to the challenges of the Arctic impact assessment system as described, there are lessons to learn from other parts of the world, but there are similarly lessons to be learnt for other regions from the Arctic. This is the case, particularly when it comes to the involvement of communities with different and conflicting values and interests. In the Arctic methods have been developed over many years in relation to community and wider public involvement and in relation to coupling impact assessment and benefit sharing. Furthermore, Arctic impact assessments seem to be clearer and more experienced when it comes to addressing and mitigating social impacts and recognising the importance of these. For example, the European impact assessments have formerly been focused more on 'hardcore' environmental impacts but are increasingly recognising the need for also capturing the social impacts and taking these into consideration in decision-making processes. This is also the case in the Canadian Arctic, where in recent years projects have been rejected based primarily on concerns about potential cumulative impacts on indigenous societies and cultures. There are therefore also lessons to learn from the Arctic in relation to coupling distribution of impacts to the discussion of benefit sharing.

In conclusion, impact assessments are valuable tools to manage the impacts of industrial development in a rapidly changing Arctic environment; however, there are several enduring challenges that need to be overcome to ensure more effective and meaningful impact assessment processes for Arctic communities.

References

Agrawala, S., A.M. Kramer, G. Prudent-Richard, M. Sainsbury and V. Schreitter 2012. 'Incorporating climate change impacts and adaptation in environmental impact assessments: opportunities and challenges' *Climate and Development* 4(1): 26–39.

Andrew, R. 2014. *Socio-Economic Drivers of Change in the Arctic.* AMAP Technical Report No. 9 (2014), Arctic Monitoring and Assessment Programme (AMAP), Oslo, Norway.

Arctic Environment Protection Strategy 1997. Guidelines for Environmental Impact Assessment (EIA) in the Arctic. *Sustainable Development and Utilization.* Finnish Ministry of the Environment, Finland.

Avango, D., L. Hacquebord, and U. Wråkberg 2014. 'Industrial extraction of Arctic natural resources since the sixteenth century: technoscience and geo-economics in the history of northern whaling and mining' *Journal of Historical Geography* 44: 15–30.

Bidstrup, M. and A.M. Hansen 2014. 'The paradox of strategic environmental assessment' *Environmental Impact Assessment Review* 47: 29–35.

BOEM 2008 *National Environmental Policy Act, Handbook H-1790–1.* Washington, DC.

Bond, A. and J. Pope 2012. 'The state of the art of impact assessment in 2012' *Impact Assessment and Project Appraisal* 30(1): 1–4.

Cairn Energy 2011. Capricorn, Social Impact Assessment, Exploration Drilling Programme, Offshore West Greenland, 2011, Version 1. 1 March 2011. Available online: http://naalakkersuisut.gl/~/media/Nanoq/Files/Hearings/2012/Offentlig%20hoering%20af%20ansoegning%20om%20efterforsknings-boringer/SIA%20rapport%20engelsk.pdf.

Canadian Minister of Justice 2012. *Canadian Environmental Assessment Act.* Ottawa.

CARC (Canadian Arctic Resources Committee) 1996. 'Critique of the BHP environmental assessment: purpose, structure, and process' *Northern Perspectives* 24: 1–4.

Chanchitpricha C. and A. Bond 2013. 'Conceptualising the effectiveness of impact assessment processes' *Environmental Impact Assessment Review* 43: 65–72.

Dales J.T. 2011. 'Death by a thousand cuts: Incorporating cumulative effects in Australia's Environmental Protection and Biodiversity Conservation Act' *Pacific Rim Law and Policy Journal* 20(1): 149–178.

de Jesus, J., C. Bingham, P. Croal and R. Fuggle 2015. *Alternatives in Project EIA. International association for impact assessment,* FASTIPS No. 11 | November 2015. Available from www.iaia.org/uploads/pdf/FasTips_11_AlternativesinProjectEIA.pdf.

Directive 2014/52/EU. 2014. Directive 2014/52/EU of the European Parliament and of the council of 16 April 2014 amending Directive 2011/92/EU on the assessment of the effects of certain public and private projects on the environment. Brussels: The European Union.

Duhaime, G. and A. Caron 2006. 'The economy of the circumpolar Arctic' in *The Economy of the North.* Oslo: Statistics Norway. 17–26.

Duinker P., E.L. Burbidge, S.R. Boardley and L.A. Greig 2013. 'Scientific dimensions of cumulative effects assessment: toward improvement in guidance for practice' *Environmental Reviews* 21: 40–52.

Duinker P. and L. Greig 2006. 'The impotence of cumulative effects assessment in Canada: ailments and ideas for deployment' *Environmental Management* 37(2): 153–161.

Ehrlich A. 2010. 'Cumulative cultural effects and reasonably foreseeable future developments in the Upper Thelon Basin, Canada' *Impact Assessment and Project Appraisal* 28(4): 279–286.

ENVIRON 2014. *Yamal LNG: Environmental and Social Impact Assessment.* Available at http://yamallng.ru/403/docs/ESIA%20ENG%20.pdf.

Esteves A.M., D. Franks and F. Vanclay 2012. 'Social impact assessment: the state of the art' *Impact Assessment and Project Appraisal* 30(1): 35–44.

European Commission 2013. Guidance on Integrating Climate Change and Biodiversity into Environmental Impact Assessment. Brussels, European Union.

Fidler C. and B.F. Noble 2013. 'Stakeholder perceptions of current planning, assessment and science initiatives in Canada's Beaufort Sea' *Arctic* 66(2): 179–190.

Glasson, J., R. Therivel and A. Chadwick 2013. *Introduction to Environmental Impact Assessment.* London: Routledge.

Gunn A., D. Russell and L. Greig 2014. 'Insights into integrating cumulative effects and collaborative co-management for migratory tundra caribou herds in the Northwest Territories, Canada' *Ecology and Society* 9(4): 4. http://dx.doi.org/10.5751/ES-06856-190404.

Haley, S., M. Klick, N. Szymoniak and A. Crow 2011. 'Observing trends and assessing data for Arctic mining' *Polar Geography* 34(1–2): 37–61.

Hansen A.M. and S.V. Larsen 2016. Miljøvurdering af off-shore olie aktiviteter – En benchmarking af reguleringsregimer i Grønland, Norge, Canada, Danmark og Alaska. Aalborg, The Danish Center for Environmental Assessment.

Hansen A and P. Tejsner 2016. 'Identifying challenges and opportunities for residents in Upernavik as oil companies are making the first entrance into Baffin Bay' *Arctic Anthropology* 23(1): 84–94.

Hansen A.M., F. Vanclay, P. Croal and A.S. Skjervedal 2016. 'Managing the social impacts of the rapidly expanding extractive industries in Greenland' *Extractive Industries and Society* 3(1): 25–33.

Huttunen A. 1999. 'The effectiveness of public participation in the environmental impact assessment process: a case study of the projected Sierilä hydropower station at Oikarainen, northern Finland' *Acta Borealia: A Nordic Journal of Circumpolar Societies* 16(2): 27–41.

IGC, MVEIRB and Canadian Environmental Assessment Agency 2004. *Environmental Impact Assessment Terms of Reference for the Mackenzie Gas Project.* Available at reviewboard.ca.

IPCC 2014. Climate Change 2014: Synthesis Report. Contribution of Working Groups I, II and III to the Fifth Assessment Report of the Intergovernmental Panel on Climate Change [Core Writing Team, R.K. Pachauri and L.A. Meyer (eds.)]. IPCC, Geneva, Switzerland.

IWG (Interagency Working Group on Environmental Justice) 2011. *Agency Responses to Comments Received During the 2011 Alaska Forum on the Environment. EJ IWG Community Dialogue*, 7–11 February 2011, Anchorage, AK and Washington, DC: U.S. Environmental Protection Agency. www.doi.gov/pmb/oepc/upload/alaska-forum-2011-agency-responses.pdf.

João, E., F. Vanclay and L. den Broeder 2011. 'Emphasising enhancement in all forms of impact assessment' *Impact Assessment and Project Appraisal* 29(3): 170–180.

Koivurova, T. 2008. 'Transboundary environmental assessment in the Arctic' *Impact Assessment and Project Appraisal* 26(4): 265–275.

Kwiatkowski, R., T. Constantine, D.M. Peace and C. Bourassa 2009. 'Canadian indigenous engagement and capacity building in health impact assessment' *Impact Assessment and Project Appraisal* 27(1): 57–67.

Larsen, S.V. 2014. 'Is environmental impact assessment fulfilling its potential? The case of climate change in renewable energy projects' *Impact Assessment and Project Appraisal* 32(3): 234–240.

Larsen, S.V. and L. Kørnøv 2009. 'SEA of river basin management plans: incorporating climate change' *Impact Assessment and Project Appraisal* 27(4): 291–299.

Lawrence, D.P. 2013. *Impact Assessment*. New York: John Wiley.

McCrank, N. 2008. 'Road to improvement: the review of the regulatory systems across the North. *Report to the Honourable Chuck Strahl*' Ottawa: Indian Affairs and Northern Development Canada.

Mortensen, B.O.G. and Hansen, T.T. 2013. *The Yearbook of Polar Law*. Leiden, Netherlands: Martinus Nijhoff Publishers. 359–386.

National Research Council 2003. Cumulative Environmental Effects of Oil and Gas Activities on Alaska's North Slope. Washington, DC: National Academies Press.

Noble, B., S. Ketilson, A. Aitken and G. Poelzer 2013. 'Strategic environmental assessment opportunities and risks for Arctic offshore energy planning and development' *Marine Policy* 39: 296–302.

Noble B.F. and K. Hanna 2015. 'Environmental assessment in the Arctic: a gap analysis and research agenda' *Arctic* 68(3): 341–355.

Noble B.F. and A. Udofia 2015. *Protectors of the Land: Toward an EA Process That Works for Aboriginal Communities and Developers*. Ottawa: MacDonald-Laurier Institute.

Norwegian Ministry of the Environment 2003. *Environmental Impact Assessment*. Oslo.

Norwegian Ministry of the Environment 2013. 'Integrated management of the marine environment of the North Sea and Skagerrak, Management Plan' Meld.St.37 (2012–2013) Report to the Storting. Available at www.regjeringen.no/contentassets/f9eb7ce889be4f47b5a2df5863b1be3d/en-gb/pdfs/stm201220130037000engpdfs.pdf.

Nuttall M. 2012. 'Imagining and governing the Greenlandic resource frontier' *The Polar Journal* 2(1): 113–124.

Ohsawa, T. and P. Duinker 2014. 'Climate-change mitigation in Canadian environmental impact assessments' *Impact Assessment and Project Appraisal* 32: 222–233.

Olsen A.H. and A.M. Hansen 2014. 'Stakeholder perceptions of public participation in environmental impact assessment: a case study of offshore oil exploration industry in Northwest Greenland' *Impact Assessment and Project Appraisal* 32(1): 72–80.

Prno J. and D.S. Slocombe 2012. 'Exploring the origins of "social license to operate" in the mining sector: perspectives from governance and sustainability theories' *Resources Policy* 37(3): 346–357.

Rozema J.G., A. Bond, M. Cashmore and J. Chilvers 2012. 'An investigation of environmental and sustainability discourses associated with the substantive purposes of environmental assessment' *Environmental Impact Assessment Review* 33(1): 80–90.

Senecal P., B. Goldsmith, S. Conover, B. Sadler and K. Brown 1999. *Principles of Environmental Impact Assessment Best Practice*. Fargo, ND: International Association for Impact Assessment.

Solodyankina S. and J. Koeppel 2009. 'The environmental impact assessment process for oil and gas extraction projects in the Russian Federation: possibilities for improvement' *Impact Assessment and Project Appraisal* 27(1): 77–83

Vanclay F. 2002. 'Conceptualizing social impacts' *Environmental Impact Assessment Review* 22: 183–211.

Vanclay, F. 2003. 'International principles for social impact assessment' *Impact Assessment and Project Appraisal* 21(1): 5–12.

Vanclay, F., A.M. Esteves, I. Aucamp and D. Franks 2015. *Social Impact Assessment: Guidance for Assessing and Managing the Social Impacts of Projects*. Fargo, ND: International Association for Impact Assessment.

30

Northern fisheries

Alf Håkon Hoel

While there are no commercial fisheries in the central Arctic Ocean (Anon 2017) – and the signing of an international agreement at the end of November 2017 will likely see that remains the case until the 2030s – the subarctic seas surrounding it have some of the world's richest fishing grounds (Vilhjamsson et al. 2005). In the North Atlantic, ocean currents bring warm waters northwards into the Arctic, providing for a relatively warm climate compared to the North Pacific. Fisheries therefore occur much further north in the Atlantic, up to about 80° North, than in the Pacific where fisheries are limited to the area to the south of the Bering Strait at the Arctic Circle.

The coastal states adjacent to the central Arctic Ocean – Norway, the Russian Federation, USA, Canada, and Denmark/Greenland – are all major fishing nations. In a global perspective, these states have relatively well-developed fisheries management regimes, with high scores in global assessments of management performance (Pitcher et al. 2009), and many major fisheries are certified by international eco-labelling schemes.[1] Most Arctic fisheries occur in waters under the jurisdiction of the coastal states, although significant fishing activity also takes place in areas beyond national jurisdictions in the North Atlantic. The area beyond national jurisdictions in the central Arctic Ocean is ice-covered for most of the year and no fishing occurs there, although conservation groups express concern at the possibility of international interest in the area given the observed trend of thinning sea ice.

Changes in the physical environment such as water temperatures can have a profound effect on living marine resources. A warm period in the North Atlantic in the 1920s and 1930s brought a rapid increase and expansion of cod into the waters around Greenland, and a major fishery developed (ACIA 2005). Arctic marine ecosystems are subject to major, natural fluctuations on several timescales. Now, anthropogenic climate change comes on top of natural variability, and is generally expected to bring a northwards expansion of fish stocks (ACIA 2005; Cheung et al. 2009). A northwards expansion has been observed in the distribution of several fish species (Hollowed and Sundby 2014; Fossheim et al. 2015). A substantial reduction in ice cover is expected to continue in the years ahead, with the central Arctic Ocean becoming virtually ice free in summer by mid-century (IPCC 2013). This chapter provides an overview of major commercial fisheries in the high north, the global and regional management frameworks that apply to these fisheries, as well as an account of the process regarding potential future fisheries in the central Arctic Ocean.

The oceans and the fisheries

Some of the Arctic countries are major fishing nations on a global scale. According to the FAO 2016, "State of the World's Fisheries and Aquaculture", the USA is number 3, Russia number 4, and Norway number 10 on the ranking list of global marine catches, with 5, 4 and 2.3 million tonnes respectively in 2014. On the same list, Iceland is number 19 (1 million tonnes), Canada number 21 (835,000 tonnes) and Denmark, including Greenland and the Faroes which are part of the Kingdom of Denmark, number 24 (745,000 tonnes). Altogether, the Arctic countries catch some 13–14 million tonnes in their marine fisheries, amounting to more than 15% of the global marine catch of 81.5 million tonnes in 2014. In addition, aquaculture is increasingly important in the north in Norway, Iceland, the Faroes, Northwest Russia and Eastern Canada. For all countries mentioned above a significant part of their total marine catch stems from Northern waters. The fisheries in Alaska, for example, are among the most important in the USA, as are the Barents Sea fisheries in Norway.

"Northern fisheries" are difficult to delineate. First, no commonly agreed definition of "Northern" or of the Arctic exists (Osherenko and Young 1989). One approach is to use latitude, for example the Arctic Circle at 66°33′39″ N,[2] and apply "Northern" to those fisheries occurring to the north of this latitude. That would exclude significant fisheries around Iceland, Greenland, and in the Bering Sea, however, so a wider understanding of "Northern" is used here. Second, fisheries' statistics are not built around geographic definitions that provide for a neat classification of fisheries into "Northern" or "Arctic" categories, and there are differences from country to country in how they are organized. It is therefore difficult to determine how much of the catch from different countries can be attributed to the Arctic or the North.

For the purpose of discussing the management and harvest of living marine resources in the north ("Northern fisheries"), and in order to include significant fisheries, an understanding of "Northern" as including the sub-Arctic seas surrounding the central Arctic Ocean is necessary. This means including the central Arctic Ocean as well as the Bering Sea,[3] the East Siberian Sea, the Chukchi Sea, the Beaufort Sea, the Davis Strait, Baffin Bay and Labrador Sea, the Greenland Sea, the waters around Iceland and Northern parts of the Norwegian Sea, the Barents Sea, the Kara Sea, and the Laptev Sea. The oceans and seas included in this definition comprise an area of some 20 million km^2 (AOR 2011: 13). Statistics from the countries fishing as well as from various other sources such as the International Council for the Exploration of the Sea (ICES) are used. Figure 30.1 shows the broad outline of the central Arctic Ocean and the surrounding seas.

The central Arctic Ocean, to the north of the continents, is characterized by broad shelves and a deep ocean basin that is ice-covered for most of the year (Bluhm et al. 2015). Ice maximum in spring is some 15 million km^2. The lowest extent is in September at some 4 million km^2, about half of the extent in the mid-1980s when satellite-based measurements started.[5] The two main gateways into the central Arctic Ocean are the Bering Strait on the Pacific side and the Fram Strait and the Barents Sea on the Atlantic side. The water masses in the central Arctic Ocean are highly stratified with little vertical mixing, due to river inputs of freshwater and ice melt. Ice cover along with stratification limits biological primary production, which is low compared with that in the surrounding seas (AOR 2011: 12).

While the central Arctic Ocean has low biological productivity and therefore limited potential for sustaining large populations of fish or marine mammals, the surrounding shelves and sub-Arctic seas are often biologically rich and sustain major fisheries. Due to the Atlantic current which brings warm waters from the south into the northeast Atlantic, this region has a much warmer climate than seas at the same latitude on the Pacific side of the Arctic Ocean. Also,

Figure 30.1 The Arctic marine environment. Source: Arctic Marine Strategic Plan 2015.[4]

commercial fisheries take place at higher latitudes in the northeast Atlantic than in the North Pacific or in the northwest Atlantic. In the northeast Atlantic, commercial fisheries now regularly occur up to 80° N, while in the North Pacific their northernmost extension is to the south of the Arctic Circle at 66°33′ in the Bering Strait.

As for the fisheries taking place in these ecosystems, the coastal areas of the central Arctic Ocean have a number of subsistence fisheries for mostly anadromous species, amounting to a few thousand tonnes annually (Zeller et al. 2011). Also, on the shelf to the north of Svalbard, shrimp are fished commercially – this is probably currently the world's northernmost commercial fishery. In terms of volume, the most important fisheries in the sub-Arctic seas as defined above are pollock in the Bering Sea and Atlantic cod in the Barents Sea as well as herring and mackerel in the northern Norwegian Sea.

In the Barents Sea, Atlantic cod, haddock, capelin, Greenland halibut and redfish are considered shared fish stocks between Norway and Russia. The Norway-Russia Joint Fisheries Commission (see later) sets annual Total Allowable Catches (TACs) for these stocks. The 2016 TAC of Atlantic cod in the Barents Sea was 894,000 tonnes, while the haddock 2016 TAC was 244,000 tonnes. The capelin TAC was set to zero, redfish to 30,000 tons and Greenland halibut to 22,000 tons. For these fisheries the Joint Commission also allocates quotas to third countries (the EU,[6] Iceland, the Faroes and Greenland).[7] The total TACs for jointly managed fish stocks in the Barents Sea is some 1.2 million tonnes. In addition, a number of fisheries are managed by the coastal state in its own waters, such as saithe and various shellfish fisheries, including red king crab and snow crab. The latter are invasive species of substantial and growing commercial importance (Sundet and Hoel 2016).

To the southwest, in the northern Norwegian Sea, major pelagic fisheries for mackerel, herring and blue whiting take place. These fish stocks have expanded considerably in recent decades, and also have a more northwestern distribution than previously. A significant share of the catches of these species can therefore now be termed "Northern", although it is difficult to determine exact amounts. The average annual total catch of mackerel was 719,000 tonnes in the 1980–2014 period (Meld.St.20: 66). The corresponding figure for herring for the 1950–2014 period was 720,000 tonnes (Meld.St.20: 64), and for blue whiting 952,000 tonnes in the 1978–2014 period (Meld.St.20: 68). The fishing nations are Norway, Russia, the Faroes, Iceland and Greenland, as well as fishing activity by EU vessels. An average annual figure for these three fisheries, then, would be approximately 2 million tonnes annually. Only parts of this would be in the northern part of the Norwegian Sea, but that part is increasing over time.

In the North Pacific, there are no commercial fisheries to the north of the Bering Strait. In the Bering Sea, the ocean to the north of the Aleutian chain, which is divided between Russia and the USA, the 2016 TAC for pollock in US waters in the Bering Sea was 1,340 million tonnes.[8] Other important fisheries include Pacific cod, sablefish and Greenland turbot, as well as salmon in coastal waters. The total groundfish TACs for 2016 amount to 2 million tonnes in US waters. The Bering Sea also has significant crab fisheries, with snow crab as the most important species (34,000 tons in 2014–2015) in US waters.[9] On the Russian side of the Bering Sea boundary (Western Bering Sea and Eastern Kamchatka), pollock TAC was 750,000 tonnes in 2016. Other significant fisheries here included cod (77,000 tonnes) and herring (51,000 tonnes).[10] Salmon and crab fisheries are also significant in this region. Total catches from the Bering Sea are in the order of magnitude of 2.8–3 million tonnes, with pollock as the dominating species.

Iceland is situated just south of the Arctic Circle, and all fisheries taking place in its waters can therefore be considered "Northern" according to the understanding we apply here. For the fishing year 2015–2016, the most important fishery was for cod (TAC 190,000 tonnes), followed by redfish (46,000 tonnes), saithe (44,000 tonnes) and haddock (29,000 tonnes).[11] The allocated quotas for 2015–2016 amounted to some 550,000 tonnes.

Greenland and the Faroe Islands are parts of the Kingdom of Denmark, but manage their own fisheries and are not subject to the Danish membership of the European Union.[12] In Greenland, the most important fishery is that of shrimp, with a total quota of 90,000 tonnes in 2016.[13] Other important fisheries include cod (41,000 tonnes TAC in 2016), and Greenland halibut (53,000 tonnes TAC in 2016), as well as redfish, capelin, shellfish, halibut, crabs and others – altogether some 35,000 tonnes. The total Greenland catch for all species in 2016 was some 220,000 tonnes.

In Canada, there are no commercial fisheries in the Arctic. An ban on the opening of commercial fisheries has been in place since 2011[14] and was extended in 2014.[15] Canadian fisheries statistics for Eastern Canada has Newfoundland as the northernmost region for which statistics are registered, and in 2015, commercial landings here totalled 255,000 tonnes.[16] The northern fisheries in Eastern Canada used to be dominated by cod, but since cod stocks collapsed in the early 1990s due to a regime shift to colder waters and overfishing, the stocks have been slow to rebuild and quotas are still down to a few thousand tonnes (DFO 2016). Significant northern fisheries in this region (Newfoundland) include cod, Greenland turbot, herring, capelin, shrimp and crabs.[17]

While insignificant in economic terms, marine mammals are important to indigenous people and local communities in the high North. Several whale species as well as seals are harvested in Iceland, Greenland, Canada, the USA, Russia and Norway (NAMMCO 2006). In Norway and Iceland minke whales are harvested in small-scale commercial fisheries. In the other Arctic countries whaling is of a subsistence nature. Commercial sealing takes place in Norway and Canada.

It is difficult to arrive at an exact figure of "Northern" or "Arctic" fish catches or landings, for reasons outlined earlier. However, the order of magnitude can be assessed to be in the ballpark of 6–7 million tonnes, based on the above discussion:

Barents Sea (Norway, Russia, third countries): 1.2 million tonnes

Norwegian Sea: 1–2 million tonnes

Bering Sea: 2.8–3 million tonnes

Iceland waters: 0.5 million tonnes

Greenland waters: 0.2 million tonnes

Eastern Canada (Newfoundland) waters: 0.25 million tonnes

This is a significant part of the annual global fish catch of some 80 million tonnes, and an even larger share of the total global whitefish catch. In addition, there is a growing aquaculture production of Atlantic salmon in particular. Since populations in these northern regions are small, meaning, in turn, a small market, most of the fish landed and produced here has to be exported to markets further south. The fisheries and aquaculture production of northern waters are therefore of global significance, and increasingly so.

Legal framework

There is a comprehensive, global legal framework for the management of living marine resources (Hoel and VanderZwaag 2014). The 1982 Law of the Sea Convention, which entered into force in 1994 and now has 168 parties,[18] is the "constitution for the oceans", providing the ground rules for the ownership, management and use of the oceans (Churchill and Lowe 1989). All Arctic countries except the USA are parties to the Convention.

The Convention provides for the establishment of 200 nautical mile Exclusive Economic Zones (EEZs) where coastal states have sovereign rights over the natural resources in those zones. All Arctic states have established EEZs or similar arrangements for their maritime areas. Beyond the EEZs are the High Seas, and in the north there are four such areas: in the central Arctic Ocean there is a high seas area of 2.8 million km^2. This is currently ice covered for most of the year and no fishing takes place there (Anon 2017). In the Norwegian Sea, substantial fisheries take place in the "Banana Hole" beyond national jurisdiction (320,000 km^2). Similarly, in the Barents Sea, the "Loop Hole" high seas area of some 60,000 km^2 has fishing grounds, as do the high seas "Donut Hole" in the Bering Sea (some 140,000 km^2) and the High Seas area in the Northwest Atlantic.

The 1982 Law of the Sea Convention provides an obligation for coastal states to manage resources in the waters under their jurisdiction sustainably and to utilize them in an optimal manner. Also, in the case of transboundary resources, there is an obligation to co-operate with other countries (Hoel and VanderZwaag 2014). Such transboundary resources can be shared (in the EEZs of two or more countries) or straddling (in the waters of at least one coastal state and in the high seas) fish stocks. In addition, there are specific provisions for highly migratory fish stocks (listed in an annex to the Convention), marine mammals, anadromous fish and sedentary species, among others.

During the 1980s, rapidly increasing fisheries on the high seas on straddling fish stocks undermined the conservation efforts of coastal states. In response to this, and in order to make the provisions of the convention regarding straddling fish stocks and highly migratory fish stocks more effective, an implementing agreement – the 1995 UN Fish Stocks Agreement (UNFSA) – was negotiated between 1993 and 1995 and entered into force in 2001. Importantly, the UNFSA establishes that a precautionary approach is to be applied in the management of living marine resources, and it also addresses an ecosystem approach to management.[19] A second feature of the UNFSA is that international co-operation in the management of high seas fisheries is to be organized in Regional Fisheries Management Organisations or Arrangements (RFMO/As), of which there are now about 20 globally. The UNFSA also provides for strengthening of enforcement of fisheries regulations (Hoel and VanderZwaag 2014). Building on this, and specifically targeting IUU fisheries, a Port State Measures Agreement to Prevent, Deter, and Eliminate IUU Fishing was negotiated in 2009 and entered into force in 2016.[20]

In addition to these legally binding instruments, a significant number of other, not legally-binding instruments have been developed since the 1990s at the global level. The role of the FAO has been particularly important in this regard, as it has developed four International Plans of Action[21] based on its 1995 Code of Conduct for Responsible Fisheries.[22] The FAO has also developed a number of guidelines, among other things for implementing the precautionary approach to fisheries (2003), for deep sea fisheries (2009) and for flag state performance (adopted in 2014).[23] Also, the UN General Assembly every year adopts a comprehensive fisheries resolution (as well as an oceans' resolution), setting out the aspirations of the global community for the management of living marine resources.[24]

Following up on the regional approach of the UNFSA, existing RFMO/As have been modernized since the 1990s, bringing their statutory provisions up to the standards provided by UNFSA in this respect. Also, most RFMO/As have been subject to one or more performance reviews addressing their implementation of the global standards laid down in the Law of the Sea Convention and in the UNFSA (Hoel 2010). In addition, a number of new RFMO/As have been negotiated in regions where such bodies were missing.

Regional organizations are also important in marine science. In the North Atlantic, the International Council for the Exploration of the Sea (ICES), which was established in 1902, draws on the collective scientific capacity of their member countries to provide scientific advice for the management of marine ecosystems and the living marine resources there (Rozwadowski 2002).[25] Scientific advice on management measures is given to countries as well as international co-operative bodies. All Arctic states are members of ICES. In the North Pacific, the Pacific International Science Organization (PICES) provides for co-operation in marine science in that region.[26] PICES does not however currently have the advisory capacity that ICES has.

International co-operation in the management of large cetaceans is vested in the International Whaling Commission (IWC) at the global level.[27] At a regional level the North Atlantic Marine Mammals Commission (NAMMCO) is a platform for co-operation on management

(NAMMCO 2006). Member nations are Norway, Greenland, the Faroe Islands and Iceland.[28] In addition to these instruments concerned with the management of living marine resources, there are also a significant number of international instruments, global as well as regional, dealing with environmental issues, science and economic aspects that also apply in the Arctic (AOR 2011).

Management

In the Arctic, the global framework described here has been extensively applied by the coastal states. The Arctic countries are prominent actors in the development of this global framework as well as in its implementation. All Arctic countries have established 200-mile zones or similar arrangements. The USA, while not a party to the Convention, regards the provisions of the Convention as customary international law. All Arctic states are party to the 1995 UN Fish Stocks Agreement and most also to the 2009 Port State Measures Agreement which entered into force in 2016.[29]

A recent study (Melnychuk et al. 2016) surveying fisheries management regimes in 28 countries, demonstrated that successful management regimes are associated with the presence of three critical properties of management regimes: the establishment and maintenance of knowledge about the resource being exploited, regulations limiting access to and harvest of the resource, and arrangements for enforcing the regulations. This finding is in keeping with previous studies (e.g. Christie 1973), as well as more recent ones (e.g. Selig et al. 2016). The Arctic coastal states all seem to have domestic management regimes with these attributes or similar features, enabling them to implement measures adopted in international bodies. A particularly important aspect is the adoption and implementation of management plans with harvest control rules that commit management authorities to a precautionary approach to resource management (Kvamsdal et al. 2016).

Many populations of living marine resources in the Arctic are transboundary, and international co-operation on their management is therefore important. A significant number of bilateral arrangements for the management of transboundary stocks exist, such as the Norway-Russia Joint Fisheries Commission, which manages the fisheries in the Barents Sea.[30] The Commission was established in 1975 and is tasked with the management of fish stocks and seals that are shared between Norway and Russia. It sets total quotas for several shared fish stocks based on management plans with harvest control rules and allocates catch shares to the two coastal states as well as to third countries. The Commission also provides for a number of other management measures as well as co-operation between the two countries on enforcement of regulations. The management measures are based on scientific advice from the International Council for the Exploration of the Sea, and the co-operation is also underpinned by collaboration between marine research institutions in the two countries (Hammer and Hoel 2012). The Commission is generally considered as an example of successful management, with most stocks in management at healthy levels (Hønneland 2012).

In the high seas areas, the Northeast Atlantic Fisheries Commission (NEAFC) is mandated to manage fisheries in that region, including in the Atlantic sector of the high seas area in the central Arctic Ocean.[31] The most important fisheries in the NEAFC area are for herring, mackerel and blue whiting. Similarly in the Northwest Atlantic, the Northwest Atlantic Fisheries Organization (NAFO) manages high seas fisheries there.[32] In the Bering Sea, the 1994 Convention on Conservation and Management of the Pollock Resources of the Central Bering Sea[33] provides for an RFMO/A there but does not actively manage fisheries, as a de facto moratorium on fisheries has been established. Additionally, in 1988, the USA and Russia signed the Agreement Between the Government of the United States of America and the Government of the Union of Soviet Socialist Republics on Mutual Fisheries Relations,[34] establishing a US-Russia Intergovernmental Consultative Committee on fisheries.[35] The main objective of the Agreement is to maintain a fisheries relationship that benefits both countries. The USA and Russia co-operate on scientific research, consult on fisheries matters beyond their EEZs and beyond the EEZ of any third party to ensure proper conservation and management, and co-operate to address Illegal, Unreported, and Unregulated (IUU) fishing activities.[36] There are also a number of other bilateral and regional fisheries management arrangements or bodies with a mandate for Arctic and Northern areas, including the North Atlantic Salmon Organization (NASCO). Norway alone is party to more than 15 fisheries agreements with annual negotiations over management measures.

The central Arctic Ocean process

In the 2008 Ilulissat Declaration the five coastal states to the central Arctic Ocean emphasized their responsibilities following from the UN Law of the Sea.[37] On the basis of this and in response to concerns regarding potential, future fisheries in the central Arctic Ocean, the five coastal states had a number of meetings from 2010 onwards to discuss the potential for fisheries to emerge in the 2.8 million km² high seas in the central Arctic Ocean and possible approaches to manage an eventual, future fishery and prevent unregulated fishing (Hoel 2015). An important aspect is that the Northeast Atlantic Fisheries Commission has a mandate in the Atlantic sector of the high seas area.

A parallel scientific process was initiated to address questions raised by the coastal states in this regard. Four scientific meetings have been held between 2011 and 2016, concluding that commercial fisheries were unlikely to emerge in the short term in the high seas area in the central Arctic Ocean. A key issue in this respect is the number of factors limiting the possibility for expansion of fish stocks northwards, including availability of food, water temperatures and topographic conditions (Hollowed et al. 2013). The scientific meetings have also emphasized the need for strengthening research and monitoring. A scientific meeting in 2016 (the fourth scientific meeting) developed a science plan laying out the finer details of how potential resources can be mapped, something that will require survey activities in the central Arctic Ocean.

In 2015 the five coastal states signed the Declaration Concerning the Prevention of Unregulated High Seas Fishing in the Central Arctic Ocean.[38] The declaration provides that the coastal states will only allow their vessels to fish in the high seas of the central Arctic Ocean after an arrangement to manage fisheries, such as an RFMO/A, has been established. It also establishes a Joint Program of Scientific Research to address questions relating to the potential for commercial fisheries in the high seas in the central Arctic Ocean. Realizing that states other than the coastal states have rights in the high seas, the Declaration also signalled an intent to work with those with interests in fisheries in high seas areas. Following this, late in 2015 a new set of talks was initiated, which also included Japan, the Republic of Korea, China, Iceland and the EU. By April 2017 four meetings had been held, addressing a range of issues pertaining to future management for the high seas area in the central Arctic Ocean, including the establishment of a joint programme of scientific research and monitoring. All delegations have committed to prevent unregulated fishing in the high seas in the central Arctic Ocean and promote the conservation and sustainable use of living marine resources and to safeguard healthy marine ecosystems in the area. A fifth scientific meeting was held in Ottawa, Canada in October 2017, addressing future activities in mapping and monitoring ecosystems in the central Arctic Ocean.

On 30 November 2017, the ten parties arrived at an agreement to prevent unregulated fishing in the high seas area in the central Arctic Ocean. The agreement commits the parties to abstain from fishing for 16 years following its entry into force. After the elapse of the 16 years, the agreement is automatically renewed every fifth year in the absence of protestations or the negotiation of an RFMO/A for whole or parts of the area that is currently not covered by an RFMO/A. The agreement also commits the parties to substantial efforts in fisheries and ecosystem research and monitoring. The agreement builds on the Law of the Sea Convention and the UN Fish Stocks Agreement and represents an additional building block in the evolution of the global framework for fisheries management. It is unique in the sense of tackling a potential problem many years ahead of its possible emergence.

Future developments

The Arctic states are all major fishing nations, and their fisheries taking place in the sub-Arctic seas are globally important. Also, most major fisheries are well managed, with fish stocks in sustainable conditions. This is most of all due to the development of management regimes that operate on the basis of sound scientific knowledge about fish stocks and marine ecosystems, provide for effective regulation of fishing activity, as well as enforcement of those regulations. These management regimes – domestic, bilateral and regional – are grounded in the global legal framework for fisheries, the 1982 UN Law of the Sea Convention and the 1995 UN Fish Stocks Agreement.

An important question concerning the future development of northern fisheries is how to adapt to the effects of climate change (ACIA 2005). Changes in the geographic distribution of fish stocks and species assemblages, driven by changes in water temperatures, are well documented (Fossheim et al. 2015). While there are a number of factors limiting geographic expansion of fish stocks (Hollowed et al. 2013), a development where some fish stocks expand into the central Arctic Ocean could be foreseen. This is likely to occur mostly in waters under the jurisdiction of the coastal states.

Another important issue for future developments in Northern fisheries, then, is the potential for new species to appear in Northern and Arctic waters. This could happen by natural expansion, as in the case of several species in the Northeast Atlantic, or it could be by introduction as in the case of king crab in the Barents Sea (Sundet and Hoel 2016). This, as well as expansion of commercial fisheries to previously unfished areas, could impact biodiversity and vulnerable habitats and pose new challenges for management (Christiansen et al. 2013). Scientific research and monitoring will be critical in providing new insights into living marine resources in the High Arctic.

Also pointing to the future is the growing production of fish and other seafood resources from aquaculture, in northern Norway in particular but also in Eastern Canada, Iceland and northwest Russia. North Norway already produces more than 500,000 tonnes of Atlantic salmon annually.[39] While production is nowhere near this in the other countries, growing demand for seafood is likely to bring further growth in aquaculture production in the north, making the Arctic even more important from a global, marine food security perspective.

Notes

1 See webpage of Marine Stewardship Council, list of certified fisheries: www.msc.org/track-a-fishery/fisheries-in-the-program/certified.
2 The region north of which one experiences at least one day 24-hour sunlight or at least one day with the sun below the horizon.
3 The USA defines the Bering Sea to the north of the Aleutian chain as "Arctic" in domestic legislation.
4 www.pame.is/index.php/projects/arctic-marine-strategic-plan.

5 https://nsidc.org/arcticseaicenews/.
6 The EU has a "Common Fisheries Policy" where the member states have ceded authority to manage fisheries to the European Commission. In international fisheries negotiations, it is therefore the EU that is the party, not the individual member countries.
7 Figures from the protocols of the meetings of the Joint Norway–Russia Fisheries Commission, available at the Commission Web site at: www.jointfish.no/nno/OM-FISKERIKOMMISJONEN/PROTOKOLLER.
8 TAC figures for the Bering Sea groundfish are available at the website of the North Pacific Fisheries Management Council: www.npfmc.org/bering-seaaleutian-islands-groundfish/.
9 Information from the web site of the North Pacific Fisheries Management Council: www.npfmc.org/safe-stock-assessment-and-fishery-evaluation-reports/.
10 https://gain.fas.usda.gov/Recent%20GAIN%20Publications/Russian%20Fisheries%20Sector%20Production%20and%20Trade%20Update_Moscow_Russian%20Federation_10-27-2016.pdf.
11 www.fiskistofa.is/english/quotas-and-catches/total-catch-and-quota-status/.
12 This gives rise to the peculiar situation where Greenland and the Faroes are represented by Denmark in international fisheries talks, while the EU represents Denmark.
13 All figures from Greenland's fisheries authorities: http://naalakkersuisut.gl/~/media/Nanoq/Files/Attached%20Files/Fiskeri_Fangst_Landbrug/DK/2016/TAC%20på%20rejer%20for%202016_DK_2.pdf and http://naalakkersuisut.gl/~/media/Nanoq/Files/Attached%20Files/Fiskeri_Fangst_Landbrug/DK/2016/.
14 www.cbc.ca/news/canada/north/beaufort-sea-commercial-fishing-banned-1.1028286.
15 www.wsj.com/articles/canada-restricts-large-scale-fishing-in-large-area-of-beaufort-sea-1413574925.
16 www.dfo-mpo.gc.ca/stats/commercial/land-debarq/sea-maritimes/s2015aq-eng.htm.
17 www.dfo-mpo.gc.ca/stats/commercial/land-debarq/sea-maritimes/s2015aq-eng.htm.
18 As of February 2017.
19 www.un.org/Depts/los/convention_agreements/convention_overview_fish_stocks.htm.
20 www.fao.org/fishery/psm/agreement/en.
21 www.fao.org/fishery/code/ipoa/en.
22 www.fao.org/fishery/code/en.
23 www.fao.org/fishery/code/guidelines/en.
24 www.un.org/Depts/los/general_assembly/general_assembly_resolutions.htm.
25 www.ices.dk/Pages/default.aspx.
26 http://meetings.pices.int.
27 https://iwc.int/home.
28 www.nammco.no.
29 As per May 2017 Canada has not ratified but plans to do so in the near future.
30 www.jointfish.com/eng.
31 www.neafc.org.
32 www.nafo.int.
33 www.afsc.noaa.gov/REFM/CBS/Docs/Convention%20on%20Conservation%20of%20Pollock%20in%20Central%20Bering%20Sea.pdf.
34 www.nmfs.noaa.gov/ia/slider_stories/2013/04/agreement.pdf.
35 www.nmfs.noaa.gov/ia/agreements/bilateral_arrangements/russia/russiabilat.pdf.
36 www.nmfs.noaa.gov/ia/agreements/bilateral_arrangements/russia/us_russia.html.
37 www.oceanlaw.org/downloads/arctic/Ilulissat_Declaration.pdf.
38 www.regjeringen.no/globalassets/departementene/ud/vedlegg/folkerett/declaration-on-arctic-fisheries-16-july-2015.pdf.
39 www.ssb.no/jord-skog-jakt-og-fiskeri/statistikker/fiskeoppdrett/aar/2016-10-28?fane=tabell&sort=nummer&tabell=281378.

References

ACIA 2005. *Arctic Climate Impact Assessment*. Cambridge, UK: Cambridge University Press.
Anon 2017. *Final Report of the Fourth Meeting of Scientific Experts on the Fish Stocks in the Central Arctic Ocean*. www.afsc.noaa.gov/Arctic_fish_stocks_fourth_meeting/.
AOR 2011. Arctic Ocean Review_I_Report_to_Ministers_2011.pdf.

Bluhm, B., K.N. Kosobokova and E.C. Carmack 2015. 'A tale of two basins: an integrated physical and biological perspective of the deep Arctic Ocean' *Progress in Oceanography* 139: 89–121.

Cheung, W., V. Lam, J.L. Sarmiento, K. Kearney, R. Watson, D. Zeller and D. Pauly 2009. 'Large-scale redistribution of maximum fisheries catch potential in the global ocean under climate change' *Global Change Biology* doi: 10.1111/j.1365-2486.2009.01995.x.

Churchill, R. and V. Lowe 1989. *The Law of the Sea*. Manchester, UK: Manchester University Press.

Christiansen, J.S., K.W. Mecklenburg and O.V. Karamushko 2013. 'Arctic marine fishes and their fisheries in light of global change' *Global Change Biology*, doi: 10.1111/gcb.12395.

Christie, F. 1973. *Alternative Arrangements for Marine Fisheries: An Overview*. Washington, DC: Resources for the Future.

DFO 2016. *2016: Stock Assessments of Northern Cod*. Canadian Science Advisory Secretariat: Science Advisory Report 2016/026. https://fisheryimprovementprojects.org/wp-content/uploads/CSAC-SAR-2016026-COD.pdf.

FAO 2016. *State of the World's Fisheries and Aquaculture*. Rome: FAO.

Fossheim, M., R. Primicerio, E. Johannesen, R.B. Ingvaldsen, M.M Aschan and A.V. Dolgov 2015. 'Recent warming leads to a rapid borealization of fish communities in the Arctic' *Nature Climate Change* 5(7): 673–677.

Hammer, M. and A.H. Hoel 2012. 'The development of scientific cooperation under the Norway–Russia fisheries regime in the Barents Sea' *Arctic Review of Law and Politics* 3(2): 244–274.

Hoel, A.H. 2010. 'Performance reviews of regional fisheries management organizations' in D.A. Russel and D. VanderZwaag (eds) *Recasting Transboundary Fisheries Management Arrangements in Light of Sustainability Principles*. Leiden, Netherlands: Martinus Nijhoff Publisheries. 449–472.

Hoel, A.H. 2015. 'Fish and fisheries management in the Arctic Basin' in L.M. Helgesen, K. Holmén and O.A. Misund (eds) *The Ice is Melting*. Bergen, Norway: Fagbokforlaget.

Hoel, A.H. and D. VanderZwaag 2014. 'Global legal dimensions of fisheries and conservation governance: navigating the currents of rights and responsibilities.' in S. Garcia, A. Charles and J. Rice (eds) *Governance in Fisheries and Marine Biodiversity Conservation*. Chichester, UK: Wiley-Blackwell. 96–109.

Hollowed, A.B., B. Planque and H. Loeng 2013. 'Potential movement of fish and shellfish stocks from the sub-Arctic to the Arctic Ocean' *Fisheries Oceanography* 22(5): 355–370.

Hollowed, A.B. and S. Sundby 2014. 'Change is coming to the northern oceans' *Science* 344(6188): 1084–1085.

Hønneland, G. 2012. *Making Fisheries Agreements Work. Post-Agreement Bargaining in the Barents Sea*. Cheltenham, UK: Edward Elgar.

IPCC 2013. *Climate Change 2013: The Physical Science Basis*. Contribution of Working Group I to the Fifth Assessment Report of the Intergovernmental Panel on Climate Change [T.F. Stocker, D. Qin, G.-K. Plattner, M. Tignor, S.K. Allen, J. Boschung, A. Nauels, Y. Xia, V. Bex and P.M. Midgley (eds.)]. Cambridge, UK: Cambridge University Press, doi:10.1017/CBO9781107415324.

Kvamsdal, S.F., A. Eide, N.-A. Ekerhovd, K. Enberg, A. Gudmundsdottir, A.H. Hoel, K.E. Mills, F.J. Mueter, L. Ravn-Jonsen, L.K. Sandal, J.E. Stiansen and N. Vestergaard 2016. 'Harvest control rules in modern fisheries management' *Elementa: Science of the Anthropocene* 4: 000114, doi: 10.12952/journal.elementa.000114.

Meld.St.20. Noregs fiskeriavtalar for 2016 og fisket etter avtalane i 2014 og 2015. (Norway's fishery agreements for 2016 and the fisheries under the agreements in 2014 and 2015). www.regjeringen.no/conten tassets/6a052d53d85145999f09b3fa8e2d22eb/nn-no/pdfs/stm201520160020000dddpdfs.pdf.

Melnychuk, M.C., E. Peterson, M. Elliot and R. Hilborn 2016. 'Fisheries management impact on target species status' *PNAS*, www.pnas.org/cgi/doi/10.1073/pnas.1609915114.

NAMMCO 2006. *User Knowledge and Scientific Knowledge in Management and Decision-Making*. Conference Proceedings, Reykjavik 4–7 January 2003. www.nammco.no/assets/Publications/Conference-Proceedings.pdf.

Osherenko, G. and O. Young 1989. *The Age of the Arctic. Hot Conflicts and Cold Realities*. Cambridge, UK: Cambridge University Press.

Pitcher, T., D. Kalikoski, G. Pramod and K. Short 2009. 'Not honouring the code' *Nature* 475(29): 658–659.

Rozwadowski, H. 2002. *The Sea Knows No Boundaries: A Century of Marine Science Under ICES*. Seattle, WA: University of Washington Press.

Selig, E.R., K.M. Kleisner, O. Ahoobim, F. Arocha, A. Cruz-Trinidad, R. Fujita, M. Hara, L. Katz, P. McConney, B.D. Ratner, L.M. Saavedra-Díaz, A.M. Schwarz, D. Thiao, E. Torell, E. Troëng and S. Villasante 2016. 'A typology of fisheries management tools: using experience to catalyse greater success' *Fish and Fisheries* 18(3): 543–570.

Sundet, J.H. and A.H. Hoel 2016. 'The Norwegian management of an introduced species: the Arctic red king crab fishery' *Marine Policy*, http://dx.doi.org/10.1016/j.marpol.2016.04.041.

Vilhjalmsson, H., Hoel, A.H., (lead authors) et al. 2005. *Fisheries and Aquaculture. I: Artic Climate Impact Assessment*. Cambridge, UK: Cambridge University Press 2005. 691–780.

Zeller, D., S. Booth, E. Pakhomov, W. Swartz and D. Pauly 2011. 'Arctic fisheries catches in Russia, USA, and Canada: Baselines for neglected ecosystems' *Polar Biology* 34(7): 945–954.

31

The future of Antarctica
Minerals, bioprospecting, and fisheries

Sanjay Chaturvedi

Introduction

The term 'future'[1] invokes both hope and fear with regard to a time that is 'yet to be'. Correspondingly, strategies to prepare for a future could vary from precautionary, 'risk-based' techno-managerial approaches, through 'expertise' driven proactive governance strategies, to aggressive geopolitical and strategic preemptive doctrines. Suspicions and fears with regard to the 'real' motives of some actors and stakeholders can sometimes invite an illusory, anticipatory geopolitics. This derives its legitimacy and authority from, and in turn provides a fillip to, emotional-imaginative geographies of various kinds of nationalism[2] – including resource nationalism – that constructs a desirable or undesirable future in one's own imagination, much before that anticipated 'future' actually arrives in the form of a 'present'.

In the case of the Antarctic any serious reflection on the question of the region's future involves *looking back* as much as *looking forward*. This is particularly obvious and compelling in the case of 'resources futures'[3] *in* and *of* the Antarctic. Whereas the former relates to the resource endowment of the Antarctic region and resource diplomacy of the Antarctic regional regime, including the role played by the movers and shakers of the Antarctic Treaty System (ATS),[4] the latter is shaped by unfolding global trends in both 'Political Economy'[5] and 'Moral Economy',[6] tempered with trends in global demography.[7] Needless to say the two kinds of future interact and overlap in a dynamic and intricate pattern, making predictions difficult.

Whereas the key focus of this chapter is on resources futures (minerals, fisheries and bioprospecting) *in* and *of* the Antarctic – and the dynamic interplay between the two – the key underlying assumption is that the 'Future of Antarctica' is going to be shaped by these resources futures in a large measure but not subsumed by them. One is talking of geographical 'possibilism' here and not some kind of determinism.[8] One also needs to be conscious of not only epistemological pluralism while approaching the question of 'future' (i.e. different ways of knowing) but also ontological pluralism (i.e. different things to be known). What is at stake therefore is not only the future of the physical realms of Antarctica – continental and maritime – and their 'exceptional' polar attributes but also futures of polar ecology, ecosystem services, the Antarctic regime, values,[9] power-knowledge equations, science-geopolitics interface, diverse and diversifying 'peaceful'[10] commercial enterprises such as tourism and so on. How these futures evolve, interact, converge or diverge, compete and/or collide with one another remains to be seen.

What is beyond doubt is that at the heart of polar moral economy is the new geological epoch called the Anthropocene and ethical considerations attached to it.[11] With climate change making the Antarctic more *accessible* (Chaturvedi 2012), geopolitically driven perceptions of resource-hungry Asian economies, intricately combined with the ethical imperatives of providing human security (Foster 2012), especially to billions in the 'Majority World' or the Global South (Chaturvedi and Doyle 2015; Verbitsky 2014), are on the rise and demand scrutiny.

Resources futures of the Antarctic: international geopolitical economy of production, consumption and flows

Much of the contemporary geopolitical discourse around 'Antarctic futures' is predicated on some kind of resource use inevitability, bordering determinism. How will Antarctica and its resource base, especially minerals, be perceived, approached and 'developed' around the year 2050, on a planet inhabited by almost 10 billion people (Jones and Anderson 2015)? As of now, alarming narratives of 'global' resource scarcity have contributed to a paradoxical situation in both the Arctic and the Antarctic. Fossil fuels, the key catalysts behind the reinvigorated interest in the Polar Regions, are also the major culprits responsible for global warming (or at least, the use of them by humans is the root cause of rising temperatures). The conventional moral framing of the Polar Regions as truly exceptional social–natural science laboratories, where alternatives to highly unsustainable economic growth-oriented models of 'development' in the era of anthropogenic climate change could be sought and sustained, seems to be being eclipsed by a geo-economic framing of inevitable hydrocarbon exploitation. The 'collective behavioural change' towards a sustainable future in the Antarctic does not seem to be happening despite a growing sense of urgency related to climate change:

> Whether either Antarctic or global ecosystems could sustain a massive Antarctic hydrocarbon 'bonanza' is not a matter for open debate, especially within the ATS. Meanwhile, marine living resource harvesting continues to expand and ATS parties argue over the scale and extent of marine protected areas around the Southern Ocean. In the face of interest in both future hydrocarbon extraction and current and near-term fishing activity, the increasingly serious and compelling evidence of the effects of anthropogenic climate change on the Antarctic ice sheets and marine environments appears to be having no impact in policy terms. Viewed objectively, this presents a first-order failure of the imagination of the current Antarctic policy system.
>
> *(Hemmings et al. 2017: 9–10)*

Debunking the myth of the 'end of geography' due to globalization, the physical geographies of the Antarctic will continue to matter. However, the moment one re-locates Antarctica on the map of dynamic functional geographies of globalizing international political economy, "interlinked resource systems" and the "coming resource crunch" (Lee et al. 2012), a different picture of resources futures *in* the Antarctic emerges. What the Chatham House Report had to say on 'Resources Futures' in 2012 resonates reasonably well even today. Underlining that "resource politics matters" and "the spectre of resource insecurity has come back with a vengeance", the report pointed out: "The world is undergoing a period of intensified resource stress, driven in part by the scale and speed of demand growth from emerging economies and a decade of tight commodity markets" (Lee et al. 2012: x). It concluded on a rather pessimistic note:

Resource politics, not environmental preservation or sound economics, are set to dominate the global agenda and are already playing themselves out through trade disputes, climate negotiations, market manipulation strategies, aggressive industrial policies and the scramble to control frontier areas. The quest for resources will put ecologically sensitive areas under continuous pressure unless a cooperative approach is taken, not least in the Polar Regions, major forests and international fisheries.

(Lee et al. 2012: xiii)

Marine resources futures in the Antarctic, once approached within the larger context of the global ocean and marine fish stocks, force us again to think in broader systemic terms. According to the Food and Agriculture Organization of the United Nations (FAO) the "current state of the world's marine fish stocks" has not shown a marked improvement overall, "despite notable progress in some areas" (FAO 2016; see also the chapter by Hoel on Northern fisheries in this volume).[12]

Whereas a geopolitical economy of resources revolves around the question of who gets what, when, *where* and how from a given resource endowment, a moral economy perspective adds to the equation 'why' (why some and not others) while problematizing 'who'. The moral and ethical considerations involved in the questions raised above do cause contestations within the spaces of 'national jurisdiction' but become far more complex and compelling in what are called the 'Global Commons' or spaces beyond sovereign territorial boundaries of jurisdiction (Chaturvedi and Painter 2007). After all there is a profound difference between self-proclaimed geopolitical-legal access to a resource base and a 'legitimate' entitlement to certain resources. Götz (2015: 158) points out that the objective behind 'moral economy' is to:

[e]mploy economic options in moral ways, that is, to the economic supply side. The "moral" would thus no longer be something per se uneconomic or laudable, but an alternative way of "utility maximization" through the construction of altruistic meaning for economic transactions.

In the case of the Antarctic, Armstrong (2017: 8) argues that we need "to sustain a sense of possibility" and explore the ways and means of resource use (e.g. 'rational use' in the case of Antarctic marine living resources under the CCAMLR regime or bioprospecting) in which "injustice might be ameliorated and equality promoted".

The moral economy in the Antarctic, despite huge potential and enormous possibilities, remains entangled with highly convoluted political geographies of disputed ownership and contested territoriality. A 'sense of possibility' arises due to values enshrined in various Antarctic instruments such as 'open access' for science (more notional than real perhaps given the considerable financial, human and intellectual investment involved in securing access to the Antarctic space to demonstrate substantial scientific interest), sharing of knowledge and information, and 'peaceful' uses of the space–place in the best interests of humankind. As pointed out by Foster (2012: 155):

[t]he concept of common concern of humanity is already immanent in Antarctic law and policy. This concept could potentially be harnessed to help develop a new thread of international discussion in pursuit of Antarctic policies that will embrace both human and environmental concerns.

The pivotal notion around which various understandings and framings of 'common' (e.g. 'common concern', 'common interest', 'common stake' and 'common heritage') continue to evolve – and revolve – is "the interest of all mankind" under the Preamble to the Antarctic Treaty.

As Wolfrum (2017: 149) points out, besides keeping the Antarctic regime "open to all those states who have shown a substantial interest in Antarctic matters", it is for the Antarctic Treaty Consultative Parties (ATCPs) to ensure that as they further develop the Antarctic legal regime:

> [t]hey must continue to seek to accommodate the interests of the world community and to continue to receive general acceptance for any new development . . . [and be] . . . aware of the dangers which may arise from unilateral actions with regard to maritime areas south of a latitude of 60°S.

It is critically important, at the same time, to be aware that:

> [t]he notion of an Anthropos, or "humanity", as global, unified "geological force" threatens to mask the diversity and differences in the actual conditions and impacts of humankind, and does not do justice to the diversity of local and regional contexts.
>
> *(Biermann et al. 2016: 341)*

In a world marked by growing inequalities "the notion of the Anthropocene, once seen in light of social inequalities and regional differences, allows for novel analysis of issue-based problems in the context of a global understanding, in both academic and political terms" (ibid.).

Antarctic minerals, resource diplomacy and the future of mining

In the 'collective memory' of Antarctic powers, the resource diplomacy of the 1970s and 1980s, especially the episode of the rise and fall of the minerals issue, carries a number of lessons that have an important bearing on negotiated resources futures. During the 1970s, Antarctica, then perceived largely as a 'Continent of Science', went through a major discursive transformation as a result of an extensive geological and biological research process and audit. In the wake of the global oil crisis, a somewhat incomplete, but nonetheless galvanizing Antarctic mineral map – iron ore, chromium, copper, gold, nickel and platinum besides sizable hydrocarbon deposits, primarily oil and coal – was soon filled in by neo-Malthusian perceptions of an increasingly populated world resulting in large-scale consumption and causing alarming depletion of resources. Diplomatic initiatives aimed at a minerals regime during the early 1970s, were driven more by the anticipatory geopolitics of fear, especially among the claimant states, of a resource future marked by unregulated mining in Antarctica. This went straight to the heart of contested territoriality. Few had a clue then about the sites and structures for the purpose, the costs involved and the environmental-ecological consequences that might follow (Chaturvedi 1996).

Least surprising, the moment ATCPs turned to the question of the Antarctic resource future there was a sudden surge of interest among the 'outsiders' in joining the Antarctic regime, which until then looked and functioned more like a privileged club with a premium membership enjoyed by the original twelve, including seven territorial claimants and two 'semi-claimants'.[13] India, Brazil, China and Uruguay, once admitted as consultative members, ushered in the hope among some that as representatives of the Global South they would work towards making the Antarctic regime more representative, less hierarchical and more accountable through burden sharing, joint knowledge production and south-south co-operation.

The Convention on the Regulation of Antarctic Mineral Resource Activities (CRAMRA) prescribed tough procedures to be adopted on environmental protection before any patch of land or shore area could be identified for exploration and development. But, it also established the property-rights regime that was the sine qua non for investment security in an area of otherwise unresolved jurisdiction. In January 1988, when CRAMRA was opened for signatures, dispatch of an expedition to service Greenpeace's World Park base at Cape Evans signified the arrival of a new non-state contributor and claimant to knowledge production in the ATS; the NGO umbrella organization, the Antarctic and Southern Ocean Coalition (ASOC).

The prospects of CRAMRA collapsed in 1989, when a number of governments, perhaps most critically original signatories and claimants France and Australia, refused to ratify it on the grounds of environmental conservation, and sought instead international support for the establishment of an 'Antarctic Wilderness Park'. The crisis of consensus was a forceful reminder that the Antarctic Treaty, despite legal-geopolitical innovation achieved under Article IV, had made its first order values a 'permanent' hostage to the colonial legacy of territorial claims, counter-claims and 'rights'. The announcement on 4 July 1991 of the US decision to sign the Protocol on Environmental Protection to the Antarctic Treaty (cited hereafter as the Protocol) restored the dialogic politics and consensual diplomacy to the ATS. This crisis of consensus over the minerals issue had been much more threatening than the campaigns of the critical lobby on the "Question of Antarctica", led by Malaysia[14] in the UN, for the obvious reason that this time, the divide and dispute was internal to the ATS and not between the ATS and those opposed to it (Chaturvedi 1996).

Concluded in the wake of consensus 'lost and found', the 1991 Madrid Protocol designates Antarctica as a "natural reserve devoted to peace and science" and commits its parties to "comprehensive" protection of the Antarctic environment and its (so far undefined) "dependent and associated ecosystems" – its intrinsic and extrinsic worth, including its wilderness, aesthetic value and its value as an area for scientific research, especially that which is essential to understanding the global environment. Termed as "an instrument of deferral" (Dodds 2017: 207), the Protocol categorically prohibits any "activity relating to mineral resources, other than scientific research" under Article 7. The prohibition remains in effect unless and until it is overturned via amendment to the Madrid Protocol. This could occur at any time or following a Review Conference (which may only be convened in 2048 or later).

Dodds and Nuttall (2016), using critical geopolitical perspectives, have interrogated the imaginative geographies of a 'scramble' for resources in the Polar Regions. In their view "if there is acceleration in some parts of the Polar Regions, there is evidence of deferral in other areas, namely Antarctica and the Southern Ocean" (ibid.: 137). However, Antarctic resource geopolitics might change, both materially and perceptually – sooner rather than later – as happened in the case of marine and mineral resources in the not too distant past, due to trends in international geopolitical economy.

Strategic recall, fast-forward and the futures of fisheries

Any future prospects of ecologically-sustainable and socially-just development of Antarctic resources cannot be divorced from long-standing histories of imperialism (Chaturvedi 1996; Scott 2017), ecological harm and overexploitation (Chaturvedi 1996; Hofman 2017; Miller 1991). These histories are much older than the 'limits to growth' argument, the discovery of the ozone hole and the formidable challenge of climate change. Ably assisted by the technological advances of the day, Antarctic fur seals had been hunted to near extinction by the late nineteenth century for the profits derived from fetching blubber, skins and furs from the far south. As noted by Ainley and Pauly (2013), exploitation of the Southern Ocean began in the late 1700s, especially around the coastal waters of the Scotia and Bellingshausen Seas, and 'fishing down the food web' continued throughout the 1800s and 1900s. Marine mammals were the first to be targeted and overexploited, followed by many species of finfish and finally krill. By the late 1970s, the Southern Ocean was subjected to significant modern commercial interest.

"The finfish fishery experienced resurgence in the late 1980s when Patagonian tooth fish from around South Georgia and other islands within the CCAMLR area were marketed and targeted by both legal and pirate fishing vessels" (Brooks and Ainley 2017: 909; see also, Kock 1992).

A good deal of literature exists with regard to Antarctic marine resource diplomacy during the 1970s and 1980s (Chaturvedi 1996; Dodds 1997; Joyner 1992). As noted by Österblom and Olsson (2017: 409), one of the key motivating factors behind the management of krill was the "growing recognition of the vulnerabilities of Southern Ocean species and ecosystems". It had taken just a few fishing seasons to cause the serious depletion of a number of finfish species in the Southern Ocean. Moreover, it was becoming obvious that "managing krill sustainably was a 'last chance' to conserve one of the few remaining wilderness spaces on the planet" (ibid.). Equally galvanizing was the fact that in the mid-1970s, northern hemisphere countries such as the USSR, Poland and Japan – and also South Korea and Taiwan – were getting pushed out of their traditional fishing grounds due to the proclamation of 200-mile fishing zones around coastal states, and began heading south.

The 1982 Convention on the Conservation of Antarctic Marine Living Resources (CCAMLR), maintains the special legal and political status, rather status quo, of Antarctica, enshrined in Article IV of the Antarctic Treaty and applies to all Antarctic marine living resources South of the Antarctic Convergence, approximately 45–60° South, including finfish, molluscs and crustaceans. The Commission for the Conservation of Antarctic Marine Living Resources (CCAMLR) that was established as the governing body of the Convention functions on the basis of consensus.[15] This handicap has severely restricted its effective responses to IUU fishing, especially for Patagonian toothfish (Stephens 2017: 445). The adoption by CCAMLR of innovative conservation measures such as, catch documentation and vessel monitoring has been widely noted and appreciated.

Krill, the current focus of Antarctic commercial fishing (*Euphausia superba*), is one of the richest sources of protein in a highly unequal world facing serious food insecurity.[16] The commercial fishery for Antarctic krill was initiated soon after the Antarctic Treaty came into effect. During 1961–62 nearly "47 tonnes were taken by two research vessels from the USSR". In the following decade, "small catches of krill were reported by the USSR as part of the research phase of the fishery development". And by "early to mid-1970s a multivessel multination fishery for krill was active" (CCAMLR 2016).[17] Since the early 1990s, "the location of fishing has moved from the Indian Ocean to the Atlantic sector" (ibid.).

The 'White Gold', a name given to toothfish, including both Patagonian and Antarctic toothfish, has received a great deal of attention due to high prices of $50/kg, and considerable demand in the USA, Europe and Asia. The toothfish were extensively exploited until the late 1980s and "by the early 1990s, Patagonian waters were opened to international fleets and toothfish swiftly became fully exploited" (Brooks and Ainley 2017: 427).

The future of fisheries in the Southern Ocean is likely to be decided not only by the CCAMLAR but also by partnership or lack of it between state parties and the fishing industry. Whether self-regulation by fishing industry through agencies like the Coalition of Legal Toothfish Operators (COLTO) will be sufficient needs to be seen. The COLTO has taken some credit for bringing the levels of illegal, unreported and unregulated (IUU) fishing down over the years.

It looks like Antarctic Treaty Consultative Meetings (ATCMs) in comparison to the meetings of CCAMLR Commission, over the past decade or so, have become more sedate (Hemmings 2017a). The contours of interests and alignments among diverse interest groups (e.g. between fishing [old and new] and non-fishing countries) are more visible in the case of latter. The tension inherent in the ecosystem approach between logic of 'rational use' and 'conservation' (and in CCAMLR the term "conservation" is defined as including "rational use") may not be easily visible but it asserts itself from time to time. Norway is the largest fishing nation in the Southern Ocean both in terms of volume and trade. There has been a growing concern that steadily expanding fishing activities of Russia, China and Ukraine would soon increasingly acquire a strong foothold in the Southern Ocean dimension. But it is China that is often singled out for its huge resource appetite and its potentially unsettling consequences for the ATS (Brady 2017).

Antarctic bioprospecting and Antarctic moral economy: access, profit and equity

In the case of Antarctic bioprospecting,[18] science, resource geopolitics and conservation in the continent are entangled in an intricate pattern. The role of science and the scientists here is going to oscillate between the value of fundamental science and the lure of commercial-corporate interests. As Hemmings (2010: 11) notes, "Here, for the first time, science wears two hats, its traditional Antarctic bonnet, and the hard-hat of commercial self-interest" and in order to preempt "conflict of interests" some "formal mechanism" would be required along with "some deliberate mechanism to ensure that the interests of science as exploiter are not laundered through its standing as privileged participant in the ATS".

Bioprospecting poses potential but significant challenges to the authority, legitimacy and effectiveness of the Antarctic regime (Chaturvedi 2009; Jabour-Green and Nicol 2003; Herber 2006; Joyner 2012). Who owns the rich biodiversity of the Antarctic and on what legal, geopolitical and ethical (Guyomard 2010) grounds? This enormously complex question defies straightforward answers, especially due to the contested territoriality of Antarctica. A major future challenge facing the ATCPs is to ensure that the public good principle remains paramount in any consensus-based negotiation of an Antarctic bioprospecting regime. Imperatives of the long-established Antarctic scientific tenets of public funding, sharing of information and knowledge among both the state and non-state holders, and international transparency would demand collective awareness and proactive action. In the process, they might find it both practical and prudent to rethink, even modify, some of the provisions of the Madrid Protocol and its annexes.

It is instructive to note that ever since the Antarctic Treaty came into force in 1961, the term 'peaceful uses' of Antarctica has been interpreted to include commercial activities such as fishing in the Southern Ocean and tourism. Should private companies be allowed to make profit from species unique to the Antarctic as yet another 'peaceful' use of Antarctica and the Southern Ocean? What benefit sharing arrangements should be in place? Are private interests at liberty to appropriate without recompensing anybody? The analogy with fishing freedom that is sometimes made does not stand up to scrutiny, since, in the case of fish harvesting, there is a more detailed process of authorization, catch allocation and conditionality, reporting and a duty to conduct scientific research (all under CCAMLR), which currently is not the case with bioprospecting activity.

The task of putting into place a consensus-based legal-political arrangement (one that resists commercialization of polar biodiversity in support of Antarctic moral economy) for the Antarctic is quite complex. Since bioprospecting is an activity with potentially both environmental and resource implications, the Antarctic Treaty parties need to work out, sooner rather than later, a more comprehensive policy position, if not a regulatory framework, that can operate across the terrestrial and marine divide currently subject to quite separate legal instruments even within the ATS. Coordination with other international legal forums is seemingly an inevitable aspect of the formation of a comprehensive Antarctic Bioprospecting Policy regime.

The discussion on bioprospecting at the ATCMs so far (Joyner 2012) is marked by a rather intriguing juxtaposition of hope and fear, anticipation and avoidance, geopolitical compulsions and ethical imperatives. The first serious discussion on bioprospecting at the 28th ATCM in

2005, prompted by a working paper submitted by New Zealand and Sweden, led to Resolution 7 (2005) *Biological Prospecting in Antarctica*. It calls upon the Parties to draw the attention of their national Antarctic programmes and other research institutes engaged in Antarctic biological prospecting activities to obligations enshrined under Article III(1) of the Antarctic Treaty. Whereas Resolution 9 (ATCM 2009) sends a loud and clear message that the ATS is *the appropriate* framework for "managing the collection of biological material in the Antarctic Treaty Area and for considering its use". In other words, the interest and involvement by other frameworks such as the CBD, SACR and industry are not perceived as appropriate.

At the 40th ATCM (Beijing, 22 May–1 June 2017), the Netherlands pointed out that "there continues to be considerable and growing activity in patenting of uses and applications based on Antarctic genetic resources" (ATCM 2017a). The Final Report (ATCM 2017b: 39–40) notes that:

> The Netherlands also brought to the attention of Parties the status of the process of the development of an international legally binding instrument under the United Nations Convention on the Law of the Sea (UNCLOS) on the conservation and sustainable use of marine biological diversity of areas beyond national jurisdiction.

Furthermore, "The Meeting re-affirmed that the ATS is the competent framework within which to address the conservation and sustainable use of biodiversity in the Antarctic region". It is to state the obvious perhaps that UNCLOS's discussions have proved to be a key catalyst and the topic of bioprospecting has been included in the *Multi-Year Strategic Work Plan*, and the Parties encouraged to submit relevant Working Papers at the next ATCM.

There is a growing body of evidence to suggest that it is becoming increasingly difficult to deny or dismiss the critical link between resources futures *in* and *of* the Antarctic. In a seminal contribution, Steven L. Chown and his co-authors (Chown et al. 2017) note the exclusion of Antarctica and the Southern Ocean, which encompass 10% of the planet's surface, from various assessments of progress achieved against the 'Strategic Plan for Biodiversity', adopted under the auspices of the Convention on Biological Diversity. The strategic plan provides "the basis for taking effective action to curb biodiversity loss across the planet by 2020 – an urgent imperative". A key finding of the study is that "for a region so remote and apparently pristine as the Antarctic, the biodiversity outlook is similar to that for the rest of the planet". At the same time the search for agreed mechanisms to manage bioprospecting has remained elusive with only two hortatory resolutions by the ATCPs on the subject so far. Whereas "existing conservation arrangements preclude population-level impacts, but benefit sharing remains problematic and is unlikely to be addressed by 2020".

Resource geopolitics and 'Polar Orientalism':[19] post-colonial engagement and assertions of Antarctic nationalism

As pointed out by Salazar (2017: 137), "the future and futures of the Antarctic region are continually being prefigured, anticipated and contested – a reflection of Antarctica's emergence as a global matter of concern". However, as discussion so far has shown, there is both a complex geography (physical, legal and moral) and history to the question of Antarctic resources futures. Contested issues, cutting across the issues of minerals, fisheries and bioprospecting relate to access, ownership entitlement and equity. One of the most enduring geopolitical legacies of the Antarctic resource exploitation serving – and in return served by – various waves of imperialism (Scott 2011, 2017) is the disputed territoriality of the continent and consequential jurisdiction over the surrounding Southern Ocean. The seven territorial claims and counter claims, legally frozen, under Article IV of the Antarctic Treaty, remain geopolitically alive for all practical policy purposes.

Yet another layer of complexity has been added by noticeable power-political shifts in international geopolitical economy, compressed in casually deployed terms such as the 'Rise of Asia' and the 'Asian Century'. The broadening and deepening of 'Asian' interest in the Antarctic, despite the persistent gap between the physical/scientific presence in the region and geopolitical/diplomatic influence at the ATCMs (Chaturvedi 2013), appears to have aroused a good deal of concern, bordering on anxiety, in some quarters over the 'actual' and 'long-term' intentions of Asian states like China, South Korea and India in the Antarctic. Partly real and partly fabricated, these imaginative geographies, articulated in a highly alarmist tone in the mainstream media of some of the Antarctic claimant states, refer to demographic trends (billions in Asia and Africa) and predictions of resource scarcity in the era of climate change to construct a particular narrative of and about Antarctica that is devoid of any future.

Both a critical geopolitical analysis of the Polar Regions (Dodds and Nuttall 2016), and a post-colonial interrogation of the politics behind the production of geographical knowledge in the current Antarctic regime (Chaturvedi 2015; Dodds and Collis 2017) underline the need to "challenge those forms of polar geopolitics that reproduce uncritical 'scramble' discourses, while resisting the temptation to exaggerate, to simplify and to marginalize" (Dodds and Nuttall 2016: 188). Failure to question such imaginative geographies and political reductionism inherent in them runs the risk of further reinforcing 'Polar Orientalism' (Dodds and Hemmings 2013), that remains "fundamentally suspicious of East Asian states, Ukraine and Russia and their motivations for being involved in Antarctic and Southern Ocean activities" (Dodds and Nuttall, 2016: 167).

If left unquestioned, this form of unexamined speculative geopolitics could result in some kind of functional paralysis in the ATS, especially due to the largely unfounded fear of Asian, especially Chinese, presence in the Antarctic. Little surprise perhaps, at the 40th Beijing ATCM, the host country, conscious perhaps of anxieties surrounding its plans for the Antarctic despite long-standing presence and participation in the ATS, emphasized "China's status as a firm advocate of Antarctic environmental protection", advocated the "adoption of the "Green Expedition" concept, which aimed to reduce the environmental impacts of expedition-related activities in Antarctica, underlined "that the peaceful use of Antarctica was the fundamental prerequisite for all human activities in the region" and "encouraged Parties to *further enhance mutual trust and assume a stronger sense of shared responsibility, stepping up dialogue and consultation, and promoting joint plans and solutions to tackle challenges in the region*" (emphasis added).

Conclusions

The sustainable future of the Antarctic, the future of sustainability in the Antarctic and the sustainability of the Antarctic political regime remain entangled. The Antarctic geopolitical exceptionalism – applied to both the region and the regime – that has so far zealously maintained and monitored the boundary between resources futures *in* and *of* the Antarctic, is withering away slowly but surely.

The 'Resources Futures in the Antarctic' around the year '2048' – when the current ban on mining in the Antarctic might be subjected to a review – could be approached by the Antarctic regime along two significantly different pathways: one dictated by a politics of hope that is accommodative of diversity, and the other driven by a geopolitics of fear that insists on homogenization. Both, however, have certain ground realities to reckon with, including the growing global focus on the southern Polar Regions in the era of climate change and resource scarcities – both real and/or imagined.

The first and foremost basic reality is that the key movers and shakers of resources futures *of* the Antarctic – the key players of international geopolitical economy including the G-7 countries with the heaviest carbon footprints – include those who matter the most in shaping resources futures *in* the Antarctic. This paradox needs more reflection and research. The club-mentality of the Antarctic regime and the twelve originals has been questioned in several ways including the fact that as of today the number of 'latecomers' – the new consultative parties with proven credentials of 'substantial scientific interest' – stands surpassed by seventeen, with a significant number from the Global South including India, China and Brazil. Malaysia is getting ready to apply for consultative status and the countries that have recently acceded to the Antarctic Treaty include Pakistan, Turkey and Kazakhstan. Iran is reported to have plans for setting up a permanent scientific base in the Antarctic.

What would happen to a consensus-based governance system confronted with an increasingly complex agenda including climate change, bioprospecting and tourism is difficult to predict. Will the ATS, already showing the symptoms of 'hollowing' out (Hemmings 2017b), remain robust enough to face new ethical and geopolitical challenges to its authority, legitimacy and effectiveness in an increasingly warming world, inhabited by 7 billion people (and even more by 2050), with a vast majority living on the continents of Asia and Africa? Resources futures *in* the Antarctic continue to be shaped by contested territoriality, an intriguing interplay between increasingly assertive nationalisms, and a relatively subdued but significant internationalism, and tension between logics of development and conservation. How effectively will the Antarctic regime cope with the challenges in view of the fact that "nationalism is increasing globally" and "in the Antarctic, we face the continuing sensitivities of unresolved state-to-state relations around territory, resource management, environmental protection – all within the context of [the] profound effect of climate change in the region" (Hemmings et al. 2015: 554).

To conclude the inconclusive, the prospects of a sustainable future for the Antarctic, and thereby a future of sustainability – futures where moral economy continues to assert its autonomy despite the overwhelming presence of and assertions by geopolitical economy – seem to be served better by a politics of hope and not by a geopolitics of unfounded fear whipped up by uncritical, unrestrained nationalisms. In the case of the Antarctic, it is comforting to recall that the ATS has earlier managed an extremely intense and divisive geopolitics of the Cold War better than anywhere else on the globe. As Randers (2012: 351) points out in his 'Global Forecast' for a world around 2015:

> In an increasingly crowded [warmer] world collective well-being will be more important than individual rights . . . the main challenge is mental . . . don't let the possibility of impending disaster crush your spirits. Don't let the prospect of a suboptimal long-term future kill your hope. Hope for the unlikely! Work for the unlikely! Remember too, that even if we do not succeed in our fight for a better world, there will still be a future world. And there will still be world with a future – just less beautiful and less harmonious that it could have been.

Acknowledgement

I am enormously grateful to Dr. Alan D. Hemmings for meticulously reading through the draft paper and providing invaluable comments and insights. The responsibility for flaws and errors rests solely with the author. I am thankful to Poorva Chaturvedi for her excellent editorial input. The views expressed in this chapter are those of the author and do not represent the views of any Ministry or Department of the Government of India.

Notes

1 Invoked in a plural sense throughout the text.
2 See Hemmings et al. 2015.
3 The term 'Resources Futures' has been used in *Chatham House 2012 Report*. Available at: http://resourcesfutures.org/#!/introduction.
4 See Haward 2017.
5 Lee et al. 2012.
6 See Armstrong 2017.
7 United Nations 2015. https://esa.un.org/unpd/wpp/publications/files/key_findings_wpp_2015.pdf.
8 See Fekadu 2014.
9 See Liggett and Hemmings 2013.
10 Given the lack of willingness on the part of the architects of the Antarctic Treaty to define "peaceful purposes" it could possibly imply anything short of another ambiguous, undefined term "measures of a military nature".
11 X. Bai et al. 2016.
12 FAO notes that the share of fish stocks within biologically sustainable levels decreased from 90 percent in 1974 to 68.6 percent in 2013.
13 The term 'semi-claimants', coined by Alan D. Hemmings, describes the position adopted by Russia and the USA to reserve their right to make future claims while refusing to acknowledge the existing seven territorial claims.
14 See Hamzah 2011.
15 As rightly pointed out, a holistic mandate notwithstanding, "the CAMLR convention does not regulate whales or seals, which are regulated by separate conventions already in effect prior to the establishment of CAMLR, including the international whaling commission (IWC)",
16 "In 2013, the year of the latest comprehensive data on global poverty, 767 million people are estimated to have been living below the international poverty line of US$1.90 per person per day" (World Bank Group 2016: 35). https://openknowledge.worldbank.org/bitstream/handle/10986/25078/9781464809583.pdf.
17 "The CCAMLR database holds data on krill catches starting in 1973. Just over half of this catch was reported by the USSR (51%), with Japan (21%), Norway (9.5%), Republic of Korea (5.6%), Poland (3.4%) and Ukraine (3.4%) the other major fishing nations. The only CCAMLR Members that have fished for more than ten years are Japan (40 years), Poland (33 years), Korea (27 years), USSR (18 years), Chile (18 years) and Ukraine (14 years) ... Within the past decade (including seasons 2005–2014), 41% of the total catch has been taken by Norway, 21% by Korea and 11% by Japan" (CCAMLR 2016: 2).
18 Bioprospecting has been defined as "the search for valuable chemical compounds and genetic materials from plants, animals and microorganisms; the extraction and testing of those compounds and materials for biological activity; and the search for, recovery, testing and commercial development of, biological materials" (Hemmings and Rogan-Finnemore 2005).
19 Dodds and Hemmings 2013.

References

Ainley, D.G. and D. Pauly 2013. 'Fishing down the food web of the Antarctic continental shelf and slope' *Polar Record* 50: 92–107.
Armstrong, C. 2017. *Justice & Natural Resources: An Egalitarian Theory.* Oxford, UK: Oxford University Press.
ATCM 2005. *Biological Prospecting in Antarctica, XXVIII ATCM/WP13* (Submitted by New Zealand and Sweden).
ATCM 2009. *Final Report of the 32nd Antarctic Treaty Consultative Meeting.* Baltimore, MD, USA 6 April–17 April 2009, p. 288.
ATCM 2017a. *An Update on the Status and Trends, Biological Prospecting in Antarctica and Recent Policy Developments at International Level.* Information Paper 168 presented by the Netherlands, XXXX ATCM, Beijing, China.
ATCM 2017b. *Final Report of the Fortieth, Antarctic Treaty Consultative Meeting, Volume I.* Beijing, China, 22 May–1 June 2017.
Bai, X., S. van der Leeuw, K.F. O'Brien, F. Berkhout, F. Biermann, E. S. Brondizio, C. Cudennec, J. Dearing, A. Duraiappah, M. Glaser, A. Revkin, W. Steffen, and J. Syvitski 2016. 'Plausible and desirable futures in the Anthropocene: a new research agenda' *Global Environmental Change* 39: 351–362.

Biermann, F., B. Xuemei, N. Bondre, W. Broadgate, C.T.A. Chen, O.P. Dube, J.W. Erisman, M. Glaser, S. van der Hel, M.C. Lemos, S. Seitzinger and K.C. Seto 2016. 'Down to earth: contextualizing the Anthropocene' *Global Environmental Challenge* 39: 341–350.

Brady, A.M. 2017. 'The past in the present: Antarctica in China's national narrative' in K. Dodds, A. Hemmings and P. Roberts (eds) *Handbook on the Politics of Antarctica*. Cheltenham: Edward Elgar. 284–300.

Brooks, C.M. and D.G. Ainley 2017. 'Fishing the bottom of the earth: the political challenges of ecosystem-based management' in K. Dodds, A. Hemmings and P. Roberts (eds) *Handbook on the Politics of Antarctica*. Cheltenham, UK: Edward Elgar. 422–438.

CCAMLR 2016. *Krill Fishery Report 2016. Convention on the Conservation of Antarctic Marine Living Resources*, Hobart, Tasmania.

Chaturvedi, S. 1996. *The Polar Regions: A Political Geography*. Chichester, UK: John Wiley & Sons.

Chaturvedi. S. 2009. 'Biological prospecting in the southern Polar Region: science-geopolitics interface' in J.M. Shadian and M. Tennberg (eds) *Legacies and Change in Polar Sciences: Historical, Legal and Political Reflections on the International Polar Year*. Aldershot, UK: Ashgate. 171–188.

Chaturvedi, S. 2012. 'The Antarctic climate security dilemma and the future of Antarctic governance' in A.D. Hemmings, D.R. Rothwell and K.N. Scott (eds) *Antarctic Security in the Twenty First Century*. London: Routledge. 257–283.

Chaturvedi, S. 2013. 'Antarctic as global knowledge commons: geopolitics, science and trusteeship' in R. Ramesh, M. Sudhakar and S. Chattopadhyay (eds) *Scientific and Geopolitical Interests in Arctic and Antarctic*. Delhi: Lights Research Foundation. 11–32.

Chaturvedi, S. and T. Doyle 2015. *Climate Terror: A Critical Geopolitics of Climate Change*. London: Palgrave Macmillan.

Chaturvedi, S. and J. Painter 2007. 'Whose world, whose order? Spatiality, geopolitics and the limits of world order concept' *Cooperation and Conflict* 42(4): 375–395.

Chown S.L. et al. 2017. 'Antarctica and the strategic plan for biodiversity' *PLoS Biol* 15(3): e2001656.

Dodds, K. 1997. *Geopolitics of Antarctica: Views from the Southern Oceanic Rim*, Chichester, UK: John Wiley & Sons.

Dodds, K. 2017. 'Antarctic geopolitics' in K. Dodds, A. Hemmings and P. Roberts (eds) *Handbook on the Politics of Antarctica*. Cheltenham, UK: Edward Elgar. 199–214.

Dodds, K. and C. Collis 2017. 'Post-colonial Antarctica' in K. Dodds, A. Hemmings and P. Roberts (eds) *Handbook on the Politics of Antarctica*. Cheltenham, UK: Edward Elgar. 50–68.

Dodds, K. and A.D. Hemmings 2013. 'Britain and the British Antarctic territory in the wider geopolitics of the Antarctic and the Southern Ocean' *International Affairs* 89(6): 1429–1444.

Dodds, K. and M. Nuttall 2016. *The Scramble for the Poles: The Geopolitics of the Arctic and Antarctic*. Cambridge, UK: Polity.

FAO 2016. *The State of World Fisheries and Aquaculture: Contributing to Food Security and Nutrition for All*, available at: www.fao.org/3/a-i5555e.pdf.

Fekadu, K. 2014. 'The paradox in environmental determinism and possibilism: a literature review' *Journal of Geography and Regional Planning* 7(7): 132–139.

Foster, C.E. 2012. 'Antarctic resources and human security' in A.D. Hemmings, D.R. Rothwell and K.N. Scott (eds) *Antarctic Security in the Twenty First Century*. London: Routledge. 154–171.

Götz, N. 2015. 'Moral economy: its conceptual history and analytical prospects' *Journal of Global Ethics* 11(2): 147–162.

Guyomard, A-I. 2010. 'Ethics and bioprospecting in Antarctica' *Ethics in Science and Environmental Politics* 10(31): 31–44.

Hamzah, B. 2011. 'Malaysia and the 1959 Antarctic Treaty: a geopolitical interpretation' *The Polar Journal* 1(2): 287–300.

Haward, M. 2017. 'The originals: the role and influence of the original signatories to the Antarctic Treaty' in K. Dodds, A. Hemmings and P. Roberts (eds) *Handbook on the Politics of Antarctica*. Cheltenham, UK: Edward Elgar. 232–240.

Hemmings, A.D. 2010. 'Does bioprospecting risk moral hazard for science in the Antarctic Treaty system?' *Ethics in Science and Environmental Politics* 10: 5–12.

Hemmings, A.D. 2017a. 'The hollowing of Antarctic governance' in P.S. Goel, R. Ravindra and S. Chattopadhyay (eds) *Science and Geopolitics of The White World: Arctic-Antarctic-Himalaya*. New York: Springer.

Hemmings, A.D. 2017b. 'The Antarctic Treaty system [The Year in Review 2015]' *New Zealand Yearbook of International Law* 13: 272–281.

Hemmings, A.D., S. Chaturvedi, E. Leane, D. Liggett and J.F. Salazar 2015. 'Nationalism in today's Antarctic' in G. Alfredson and T. Koivurova (eds) *The Yearbook of Polar Law*. Leiden, the Netherlands and Boston, MA: Brill Nijhoff. 531–555.

Hemmings, A.D., K. Dodds and P. Roberts 2017. 'Introduction: the politics of Antarctica' in K. Dodds, A.D. Hemmings and P. Roberts (eds) *Handbook on the Politics of Antarctica*. Cheltenham, UK: Edward Elgar. 1–20.

Hemmings, A.D. and M. Rogan-Finnemore (eds) 2005. *Antarctic Bioprospecting. Gateway Antarctica Special Publication Series 0501*, Christchurch, New Zealand: University of Canterbury.

Herber, B.P. 2006. 'Bioprospecting in Antarctica: the search for a policy regime' *Polar Record* 42: 221, 139–146.

Hofman, R.J. 2017. 'Sealing, whaling and krill fishing in the Southern Ocean: past and possible future effects on catch regulations' *Polar Record* 53(268): 88–99.

Jabour-Green, J. and D. Nicol 2003. 'Bioprospecting in areas outside national jurisdiction: Antarctica and the Southern Ocean' *Melbourne Journal of International Law* 4(1): 76–111.

Jones, S. and M. Anderson 2015. 'Global population set to hit 9.7 billion by 2050 despite fall in fertility' *The Guardian*, 29 July, available at www.theguardian.com/global-development/2015/jul/29/un-world-population-prospects-the-2015-revision-9-7-billion-2050-fertility.

Joyner, C. 1992. *Antarctica and the Law of the Sea*. Dordrecht, the Netherlands: Martinus Nijhoff Publishers.

Joyner, C. 2012. 'Bioprospecting as a challenge to the Antarctic Treaty' in A.D. Hemmings, D.R. Rothwell and K.N. Scott (eds) *Antarctic Security in the Twenty First Century*. London: Routledge. 197–214.

Kock, K-H. 1992. *Antarctic Fish and Fisheries*. Cambridge, UK: Cambridge University Press.

Lee, B., F. Preston, J. Kooroshy, R. Bailey and G. Lahn 2012. *Resources Futures: A Chatham House Report*, available at: http://resourcesfutures.org/downloads/CHJ204_Resources_Futures_WEB_28.01.13.pdf.

Liggett, D. and A.D. Hemmings, eds. 2013. *Exploring Antarctic Values*. Christchurch, New Zealand, Gateway Antarctic Special Publication, available at: http://antarctica-hasseg.com/wp-content/uploads/2013/05/SSAG-proceedings-2013.pdf.

Miller, D.G.M. 1991. 'Exploitation of Antarctic marine living resources: a brief history and a possible approach to managing the krill fishery' *South African Journal of Marine Science* 10(1): 321–339.

Österblom, H. and O. Olsson 2017. 'CCAMLR: an ecosystem approach to the Southern Ocean in the Anthropocene' in K. Dodds, A.D. Hemmings and P. Roberts (eds) *Handbook on the Politics of Antarctica*. Cheltenham, UK: Edward Elgar. 408–421.

Randers, J. 2012. *A Global Forecast for the Next Forty Years: 2052*. White River Junction, VT: Chelsea Green Publishing.

Salazar, J.F. 2017. 'Mediating Antarctica in digital culture: politics of representation and visualisation in arts and science' in K. Dodds, A.D. Hemmings and P. Roberts (eds) *Handbook on the Politics of Antarctica*. Cheltenham, UK: Edward Elgar. 125–141.

Scott, S.V. 2011. 'Ingenious or innocuous? Article IV of the Antarctic Treaty as imperialism' *The Polar Journal* 4(2): 51–62.

Scott, S.V. 2017. 'Three waves of Antarctic imperialism' in K. Dodds, A.D. Hemmings and P. Roberts (eds) *Handbook on the Politics of Antarctica*. Cheltenham, UK: Edward Elgar. 37–49.

Stephens, T. 2017. 'An icy reception or a warm embrace' in K. Dodds, A.D. Hemmings and P. Roberts (eds) *Handbook on the Politics of Antarctica*. Cheltenham, UK: Edward Elgar. 439–452.

United Nations Department of Economic and Social Affairs, Population Division 2015. *World Population Prospects: The 2015 Revision, Key Findings and Advance Tables*. Working Paper No. ESA/P/WP.241. https://esa.un.org/unpd/wpp/publications/files/key_findings_wpp_2015.pdf.

Verbitsky, J. 2014. 'Just transitions and a contested space: Antarctica and the Global South' *The Polar Journal* 4(2): 319–334.

Wolfrum, R. 2017. 'Common interest and common heritage in Antarctica' in K. Dodds, A.D. Hemmings and P. Roberts (eds) *Handbook on the Politics of Antarctica*. Cheltenham, UK: Edward Elgar. 142–151.

World Bank Group 2016. *Talking on Inequality: Poverty and Shared Prosperity*. Washington International Bank for Reconstruction and Development/World Bank.

32

Conservation and environmental governance in the Polar Regions

Mark Nuttall

Global environmental conservation is characterized by a large number of increasingly complex arrangements and assemblages of regional, national and international organizations, institutions, agreements and regimes. These are concerned variously with the governance of the environment or have a mandate for dealing with specific issues related to environmental protection, conservation, biodiversity and wildlife management. Some of these agreements, organizations and institutions have responsibility for – and jurisdiction over – broader aspects of ecosystem management, biodiversity conservation, the protection and conservation of wildlife, the sustainable use of resources, long-range air pollutants or adaptation strategies in response to climate change. Others have a particular concern with forests, wetlands, oceans, coastal ecosystems, migratory birds or fisheries (in a general sense), or particular species of animal such as polar bears, whales or tigers, or fish stocks considered to be endangered as a result of overfishing, such as cod, or various species of salmon. Environmental assessment and ongoing monitoring of threats to ecosystems and habitats and the status of wildlife are key components of national and international environmental governance regimes and conservation management (e.g., Warner 2013). For example, the 1973 Endangered Species Act (ESA) of the United States provides for the conservation of species that are defined as endangered or threatened throughout all or a significant portion of their range, as well as the conservation of the ecosystems on which they depend (the ESA lists around 2,300 species as endangered or threatened). In the case of marine species listed by the ESA, the US National Oceanic and Atmospheric Administration (NOAA) has responsibility for monitoring the status of over 159 species considered endangered or threatened; these include a number of Arctic species, such as the bowhead whale and the Beringia and Okhotsk bearded seal populations.

Some of the world's most environmentally-sensitive areas, as well as a number of ecosystems and species that have been identified by national or international agencies as threatened, are found in the Polar Regions. As several chapters in this book make clear, current climate change trends, threats to biodiversity and the cryosphere, the rapid loss of sea ice, diminishing glacial mass, warming oceans, pollution in polar marine and terrestrial ecosystems, and the negative effects of human activities add increasing pressure to animal populations and species habitat in both the Arctic and Antarctic. They also place pressure on existing environmental initiatives in addressing some of the most pressing and urgent environmental challenges we face on the planet. Polar geopolitics and specific regimes, institutions and governance mechanisms (such as the Law of the Sea, fisheries management and the management of tourism in Polar Regions), are dealt with at length in several chapters in this handbook. In this chapter, I provide a brief survey of some aspects of environmental governance and conservation in the Polar Regions and of its changing nature in light of increased pressures on Arctic ecosystems and wildlife which arise from climate change, pollution, habitat disturbance and loss, and human activities, including resource extraction and the over-exploitation of resources, and consider ways in which indigenous and local perspectives are being incorporated into environmental monitoring and community-based research.

Environmental institutions and the Polar Regions

Many international agreements and institutions concerned with environmental management and governance have relevance for the Polar Regions, and cover aspects of polar ecosystems (such as wetlands in countries of the circumpolar North under the Ramsar Convention[1] to which all eight Arctic states are parties), or wildlife management (for example, whales in both the Arctic and Antarctic under the regimes and procedures of the International Whaling Commission [IWC]), or fisheries (see the chapter on northern fisheries by Hoel in this volume) and conservation, or with protected sites subject to classification by organizations such as the International Union for Conservation of Nature (IUCN) and adhering to categories of an international standard (such as nature reserves restricted mainly for science, wilderness areas, national parks, natural monuments, habitat/species management areas, protected landscapes and seascapes). The IUCN, for instance, also produces reports and assessments and sets out guidelines and recommendations for biodiversity and ecosystem protection that focus on areas it defines as wilderness, World Heritage Sites, large landscapes and seascapes, or the design of large-scale protected areas. These often identify areas of environmental concern in the Polar Regions or adjacent lands and seas (e.g., Kormos et al. 2017; Speer et al. 2017). In the Antarctic – or more specifically the Southern Ocean – some activities are regulated by international conventions in addition to the instruments of the Antarctic Treaty System, such as the Law of the Sea Convention, the International Convention for the Regulation of Whaling, or the International Convention for the Regulation of Pollution from Ships (MARPOL). However, many other international agreements, such as the Ramsar Convention, the World Heritage Convention, or the Convention on the Conservation of Migratory Species of Wild Animals (also known as the CMS or the Bonn Convention) do not apply. This is because sovereignty in Antarctica is not recognized under the Antarctic Treaty. When international conventions do apply in the world's southern circumpolar regions, this is in relation to sub-Antarctic islands and surrounding waters where sovereignty has been established and where sovereign states have become signatories to and have ratified the relevant convention or agreements (Hansom and Gordon 1998: 269).

Countries that ratify and become signatories to international conventions, such as those agreed upon by the United Nations or more polar-specific arrangements such as the Antarctic Treaty System (ATS), are bound legally to observe their principles and act within the rules and regulations which have been set out, often following considerable political negotiation. In the case of the International Convention on the Regulation of Whaling (ICW), for example, an integral part of the convention is its legally-binding Schedule. This sets out specific and very prescribed measures that the IWC has decided upon collectively and which are considered necessary for the regulation of whaling and the conservation of whale stocks.

One prominent example of a large multilateral agreement, and one which has current relevance for both the Arctic and Antarctic, is the 1982 United Nations Convention on the Law of the Sea (UNCLOS). This deals with a range of environmental governance issues relating to the world's oceans and their use, including legal issues concerning sovereignty over continental shelves and subsea resources (see the chapter by Donald Rothwell in this volume). UNCLOS formalized Exclusive Economic Zones (EEZs), for instance. In the case of wildlife, even when states establish their own environmental institutions charged with regulation and governance for animals within their borders, and which are not part of a transnational or international agreement, international bodies and organizations nonetheless often have a major role in making sure they function accordingly. The Convention on International Trade in Endangered Species of Wild Fauna and Flora (CITES), whose secretariat is administered by the United Nations Environment Programme (UNEP), is a good example of an international network of environmental governance. A country which is a signatory to CITES has to work to ensure both national and international efforts for conservation of particular species, including CITES regulations concerning trade (such as narwhal tusks in Greenland and Canada, for example), are adhered to.

In the case of migratory species that cross national borders into other jurisdictions and also into international waters, such as many species of fish, specific forms of environmental governance are required. The United Nations Fish Stocks Agreement (UNFA) is one example of how states are required to manage the activities of fishing vessels that pursue highly migratory fish stocks, such as cod, halibut and flounder. Highly migratory fish stocks as also known as straddling stocks because they "straddle" the outer limits of the waters of coastal states and range beyond into the adjacent high seas (see Hoel's chapter in this volume and the recent discussions to manage and prohibit fisheries in Arctic waters, including the central Arctic Ocean). The UNFA obliges states to take a precautionary approach to fisheries management and to ensure they control and regulate the activities of their fishing vessels when they operate in international waters. UNFA came into force in December 2001 and fifty-nine states have currently ratified it.

Other bodies, such as IUCN, which was created in 1948 and has UN Observer status, are member organizations that bring together governments and civil society groups. In essence, IUCN is a neutral forum in which governments, scientists, non-governmental organizations (NGOs), local communities, indigenous peoples' organizations, and other groups concerned with the environment, conservation and sustainability can meet and work together to identify issues and work towards management and conservation efforts. One distinctive agreement which links together nature conservation and the preservation of cultural properties is the Convention concerning the Protection of the World Cultural and Natural Heritage (1972) – or the World Heritage Convention – which was established under the UNESCO regime. The dominant activity under the convention is the World Heritage List concerning natural areas and landmarks as well as cultural sites that have been defined as unique and in need of protection or conservation. In seeking and achieving World Heritage List status for a site, parties to the convention are encouraged to integrate the protection of the cultural and natural heritage of the area into regional planning programmes, set up staff and services at their sites, undertake scientific and technical conservation research and work to ensure heritage sites have a presence or relevance in the local community. The International Council of Monuments and Sites (ICOMOS) evaluates and designates cultural sites and IUCN evaluates natural sites. There are currently five World Heritage Sites which lie above the Arctic Circle, including Russia's Wrangel Island Reserve, which was inscribed in 2004. Another is Greenland's Ilulissat Icefjord in Disko Bay. Greenland became part of the convention in 1979 and in 2003 submitted a request for the ice fjord to be included in the list. The ice fjord system became a protected area in May 2003, wildlife management and environmental monitoring programmes were put in place, and the area was designated a World Heritage Site by IUCN in July 2004.

A recent IUCN report has intended to advance recognition for the conservation of globally significant natural marine sites in the Arctic, a region which the authors point out is underrepresented on the World Heritage List. The report summarized a scientific assessment of globally significant Arctic ecosystems with respect to the natural criteria for World Heritage status. It described seven areas in the Arctic Ocean of such global significance that they may be priorities for inscription on the World Heritage List (Speer et al. 2017). These sites are the Bering Sea ecoregion, Greenland's Northeast Water polyna ecoregion, the Northern Baffin Bay ecoregion, the Disko Bay and Store Hellefiskebanke ecoregion, the Scoresbysund polynya ecoregion, High Arctic archipelagos, and the Great Siberian Polyna.

Antarctica

While Antarctica is subject to a complex system of environmental governance as a result of inter-national co-operation under the Antarctic Treaty and its associated protocols and instruments (see the chapters by Anita Dey Nuttall, Klaus Dodds, Donald Rothwell, and Sanjay Chaturvedi), since 1961 those instruments, agreements and initiatives dealing with conservation and environmental protection have been added to the Antarctic governance system. When the treaty was first nego-tiated, conservation measures were not built into it; rather, the early signatories recognized that while environmental protection was necessary in parts of the Antarctic continent, any agreements would have to be negotiated and agreed upon within the Antarctic Treaty System as a way of furthering its principles and objectives (Hansom and Gordon 1998: 267). Although a number of agreements and conservation measures have been drawn up and implemented in Antarctica and comprise an "environmental sub-regime within the Antarctic Treaty System" (Elliott 1994: 4), as Elliott discusses, "both policy and practice on environmental protection were constrained by the political interests which dominated the regime" (ibid.). By the 1980s, global concern over the future of Antarctica was growing; increased international campaigning activities by environmen-tal NGOs highlighting threats to Antarctica, the discovery in 1984 by British Antarctic Survey scientists of a recurring ozone hole above the Antarctic during spring, international opposition to whaling operations in the Southern Ocean, and anxieties over future exploration for miner-als, all played a significant part in the emergence of Antarctica as a unique space influencing (and affected by) global processes and in need of protection. The Antarctic environment became politicized and issues of conservation and environmental protection became ways through which developing nations and environmental NGOs could also challenge the ATS.

The environmental sub-regime instituted within the framework of the ATS now includes the Agreed Measures on the Conservation of Antarctic Flora and Fauna (and which first set aside protected areas of special interest in 1964), the Convention for the Conservation of Antarctic Seals (CCAS) and the Convention on the Conservation of Antarctic Marine Living Resources (CCAMLR). The Protocol on Environmental Protection to the Antarctic Protocol (also known as the Environmental Protocol) was adopted in 1991 and came into force in 1998. This designates Antarctica, according to Article 2 of the protocol, as a "natural reserve, devoted to peace and science" (see also the chapters in this volume by Dey Nuttall, Dodds, and Chaturvedi). Article 3 sets out principles applicable to human activities in Antarctica, while Article 7 prohibits all activities relating to Antarctic mineral resources, except those carried out for scientific research. Until 2048, the protocol can only be modified by unanimous agreement of all Consultative Parties to the Antarctic Treaty. Furthermore, the prohibition on mineral resource activities can-not be removed unless a binding legal regime on Antarctic mineral resource activities is in force. In 1991, Annex V to the Environment Protocol replaced earlier categories of protected areas under the ATS and provided for the designation of Antarctic Specially Protected Areas (ASPA) and Antarctic Specially Managed Areas (ASMA). Special areas may also be designated under the provisions of CCAS and CCAMLR – the Ross Sea region's Marine Protected Area (MPA) created by CCAMLR in 2016 (and which came into force in December 2017) is one notable, recent example. Providing protection for the habitat and foraging ranges of large populations of Adélie and emperor penguins, Weddell and crabeater seals, killer whales, Antarctic petrels and Antarctic toothfish, this is the world's largest marine protected area and includes the 500,000 square kilometer (193,000 square mile) Ross Ice Shelf. A new proposal to create a wildlife sanctuary and prohibit fishing in the Weddell Sea is also currently under discussion.

The environmental sub-regime of the ATS does not necessarily protect all of Antarctica entirely from resource use, which is why organizations such as Greenpeace run global campaigns calling for greater protection of the continent's marine ecosystems. CCAMLR was established by international convention in 1982 with the objective of conserving Antarctic marine life. This was in response to increasing commercial interest in Antarctic krill resources and the history of over-exploitation of several other marine resources in the Southern Ocean, such as whales, fur seals and some species of fish. CCAMLR does not prohibit or exclude fishing or the harvest of Antarctic krill; rather, it practices an ecosystem-based management approach and expects harvesting of marine resources to be done in a sustainable manner and take account of the effects of fishing on other components of the ecosystem. CCAMLR's mandate has broadened as it has developed from an organization originally concerned largely with krill to one concerned with a wide range of Antarctic species, habitats and ecosystems. The CCAMLR Ecosystem Monitoring Programme (CEMP) was set up in the late 1980s to monitor, detect and record significant changes in critical components of the Antarctic marine ecosystem, and to serve as a basis for the conservation of Antarctic marine living resources; and it seeks to determine whether changes are due to the harvesting of commercial species, or whether changes in the marine ecosystem can be attributed to environmental variability (Agnew 1997).

While this more diverse approach to ecosystem management has contributed to the perception of CCAMLR being an effective model for international fisheries management (Österblom and Olsson 2017), and while CCAMLR has appeared to do well with regulating the take of Antarctic krill and is making progress in efforts to establish marine protected areas, Brooks and Ainley (2017) argue that it was not that well prepared for the rapid development of the lucrative fishery for Antarctic toothfish. Brooks and Ainley discuss how the CCAMLR Commission regulates the high seas catch of krill quite effectively and in accordance with ecosystem-based management; however, they describe how the regulation of most fisheries for Patagonian toothfish is influenced by their location within Exclusive Economic Zones (EEZs) around sub-Antarctic islands, which are all outside the jurisdiction of the Antarctic Treaty System, thus hindering CCAMLR's attempts to establish marine protected areas and implementing extensive ecosystem-based management in many parts of the southern circumpolar regions. Yet Österblom and Olsson (2017: 416) point out that CCAMLR has never really been tested over its effectiveness to manage Antarctic krill fishing and question how far its authority would stretch if there was increased global interest in krill and a corresponding pressure on krill stocks and the ecosystem (it is difficult to catch krill without affecting other parts of the marine ecosystem) – as they put it, "Krill fishing has yet to become the next 'gold rush' activity in marine resource use and it has been argued that it never will". Yet, Antarctic krill fishing is increasing, and the catch has grown by 40% since 2010[2] with Norway, China, Russia and South Korea being the main krill fishing nations – krill are valuable for omega-3 related products and for the manufacture of feed for farmed seafood, poultry and pets. Krill stocks also fluctuate. As Ryabov et al. (2017) argue, intraspecific competition for food is the main driver of the krill cycle, while external climatological factors possibly affect and modulate its phase and synchronization over large scales. Their concern is that, given that the cycle amplitude increases with reduction of krill loss rates, a decline of apex predators (krill are the main diet for fish, whales, seals and penguins in the Southern Ocean) is likely to affect the krill cycle and destabilize the marine food web, with drastic consequences for the entire Antarctic ecosystem. Krill are also considered the keystone species in the Southern Ocean ecosystem, a major grazer of primary production, feeding on micro-size phytoplankton and, in turn, providing food for predators. Krill – and phytoplankton – need sea ice and cold seas, but the ice is thinning, melting and disappearing and Antarctic waters are warming. How intensive fishing of krill – or indeed fishing for other species – would also affect the ecosystem in light of such dramatic environmental changes, and how CCAMLR would be able to respond by regulating those fisheries activities further to allow for protection of the ecosystem, sustainable fisheries and global food security, would be a significant test for the institutions and practices of Antarctic conservation. However, there is optimism for CCAMLR – Österblom and Olsson, for instance, highlight CCAMLR's significance not just for conservation in Antarctica, but for global marine conservation efforts: "International marine governance institutions are in great need of role models, and CCAMLR has the track record and ability to continue developing a leadership position by acting as a source of inspiration for others" (2007: 417).

The Arctic

Conservation and environmental management in the Arctic, on the other hand, is largely more a matter for individual states and their implementation of national legislation. This includes national parks in Canada's Arctic, the ESA in the case of Alaska, or specific nationwide conservation regimes such as, again in the case of Alaska by way of example, the US national wildlife refuge system, of which the Arctic National Wildlife Refuge (ANWR) is part. Other aspects of conservation and environmental governance adhere to international agreements, such as those related to the International Whaling Commission (Arctic states and indigenous organizations participate in discussions over annual whale catch quotas), Ramsar, or CITES, for instance, or specific circumpolar regimes that involve a few Arctic states and are concerned with migratory wildlife management. Individual Arctic states have also established protected areas or specific wildlife sanctuaries or wildlife reserves. Russia, for example, has a system of *zapovedniki* (nature reserves[3]) and *zakazniks* (state nature reserves), while in Greenland the Northeast Greenland National Park was established in 1974. This is the world's largest national park and has the status of a biosphere area under UNESCO's Man and the Biosphere (MAB) programme; it was also the first national park to be established in the Kingdom of Denmark and remains the only one in Greenland.

In some Arctic regions, such as in the case of Canada and Alaska, federal agencies work in co-operation with indigenous communities and the regional wildlife boards or environmental departments established following land claims negotiations and settlements, while in other parts of the Arctic such as Greenland this participatory approach involving hunters and fishers is in its infancy and wildlife management seldom takes note of local knowledge (e.g., Nuttall 2016). In 1941, for example, the US banned the commercial hunting of walrus, which were later protected further under the MMPA. The Walrus Islands State Game Sanctuary was established in 1960 and protects a group of seven islands and their surrounding waters – key walrus habitat – in northern Bristol Bay. Walrus hunting is restricted to subsistence purposes by Alaska Natives and the annual harvest is monitored by the US Fish and Wildlife Service (USFWS) which focuses on management in co-operation with the Alaska Eskimo Walrus Commission (AEWC). A co-management agreement for the conservation and management of walrus was signed between the USFWS and the AEWC in 1997.

Examples of specific circumpolar-wide agreement between Arctic states to manage wildlife include initiatives concerned with narwhals, belugas and polar bears. In the case of narwhals and belugas, the Canada-Greenland Joint Commission on Conservation and Management of

Narwhal and Beluga (JCNB) provides management advice to Canada and Greenland, with scientific advice provided by a JCNB working group and the North Atlantic Marine Mammal Commission's (NAMMCO) scientific working group on aspects such as stock delineation, possible threats to beluga and narwhal populations, and total allowable catches. In Greenland, for example, government quotas are then based on the JCNB recommendations and Greenland's hunting council and the municipal authorities distribute the quotas. The conservation and management regime that has responsibility for polar bears makes them one of the most carefully monitored, studied and managed marine mammals in the Arctic. In the 1960s, the five nations with polar bear populations (Denmark/Greenland, the US, Canada, Norway and Russia) grew concerned that the animals were declining in number due to overharvesting, mainly from commercial hunts. In 1967, IUCN formed a specialist group to co-ordinate research, conservation and management of polar bears at an international level. Norway and Russia eventually banned commercial hunting within their Arctic territories and the US, Canada and Denmark (for Greenland) limited polar bear hunting for subsistence purposes by Inuit hunters, with Canada implementing an annual quota system (followed later by the US and Greenland). These five nations signed the Agreement on the Conservation of Polar Bears in 1973. The Agreement was one of the first international regimes to include ecological principles and calls for the protection of the ecosystems upon which polar bears depend and, specifically, to protect special habitat components. It allows for the hunting, killing and capturing of polar bears for scientific and conservation purposes, to protect other resources, and for subsistence harvest by local people using traditional methods or where people have a tradition of hunting polar bears. Many polar bear scientists claim that the Agreement has been effective because resource users, and those involved in research and management, were committed to finding a solution to improve polar bear conservation. In addition to the Agreement, polar bears are also protected in Alaska under the US Marine Mammal Protection Act (MMPA) of 1972. Today, however, polar bears are increasingly under threat from climate change, pollutants and disappearing sea ice, and these environmental pressures challenge the effectiveness of the polar bear management regime, which was negotiated and enacted when hunting was the primary threat to polar bears throughout their range. As Derocher et al. (2013) point out, the rapid loss of sea ice as primary polar bear habitat has changed the nature of conservation and management dramatically, and there is an urgency for new conservation measures and management responses. They conclude that:

> Considering the global attention paid to polar bears, managers will be forced to respond to sudden changes in environmental conditions that negatively affect polar bears. We believe that managers and policy makers, who have anticipated the effects, consulted with stakeholders, defined conservation objectives, created enabling legislation, and considered possible management actions will be most able to effectively respond to large-scale negative changes.

> *(2013: 373)*

International procedures for environmental governance that apply to Arctic states, especially those that cut across national boundaries and jurisdictions, have arisen from agreements signed between a number of countries. For example, the International Whaling Commission (IWC) is an intergovernmental institution established by the International Convention for the Regulation of Whaling, which was signed in Washington DC on 2 December 1946. While this oversees the management of whaling globally and ensures the conservation of some species of whale, the category of Aboriginal Subsistence Whaling permits the hunting of certain species of whales by indigenous peoples for customary and cultural needs under strict circumstances and limited quota arrangements.

Although some Arctic states are often bound together within conventions and management regimes that deal with migratory Arctic species (such as polar bears, caribou and fisheries), and while Arctic marine legal issues may be dealt with under the United Nations Convention on the Law of the Sea (UNCLOS), there is no regional Arctic-wide system of environmental governance or conservation similar to the Antarctic, and this is one way the Arctic is different to the southern circumpolar regions. The Arctic Council, for example, is a forum for co-operation on matters related to the environment and sustainable development, but it has no legally-binding powers, neither does it have a mandate to initiate legislation related to environmental governance (see Koivurova's chapter in this volume). The Arctic Council does, though, regularly compile scientific and indigenous knowledge of the state of the Arctic environment and its wildlife. For example, the Arctic Biodiversity Assessment (ABA) was published in 2013 under the auspices of the Arctic Council's Conservation of Arctic Flora and Fauna (CAFF) working group – a comprehensive assessment of the status and trends of Arctic biodiversity, it includes a set of policy recommendations for Arctic biodiversity for Arctic states to act upon (see also the chapter in this handbook on biodiversity in the Polar Regions by Hans Meltofte, who was the ABA chief scientist).

Arctic Council assessments also emphasize the global interrelationships between Arctic wildlife and Arctic systems and the rest of the planet (such as the Arctic Climate Impact Assessment).[4] CAFF, for example, has initiated a project called the Arctic Migratory Birds Initiative (AMBI) to address urgent conservation trends and needs in light of declining migratory bird populations in the circumpolar Arctic and sub-Arctic regions (Johnston et al. 2015). The emphasis of AMBI is not just on the circumpolar North, however, and it calls attention to global migration routes, such as the importance of conservation measures for areas in coastal areas in Africa or tidal flats in Asia which are important sites for migrating shorebirds on their way to and from the Arctic. Through AMBI, CAFF aims to provide an international forum for Arctic and non-Arctic states to discuss urgent action on conservation of habitat for migratory birds. CAFF has identified three flyways between the Arctic regions and more southerly places – an Africa Eurasian Flyway, an Americas Flyway and an East Asian-Australasian Flyway – and a Circumpolar Flyway. It aims to identify priority conservation issues, obtain greater understanding of species habitat and different bird populations, and is working to ensure that habitat for birds such as lesser white-fronted geese, spoon-billed sandpipers and bar-tailed godwits are not just protected and managed in places such as Iceland, Arctic Russia, Alaska and northern Canada, but that coastal ecosystems in Africa, Central and South America, China and Australia are protected or management plans for bid habitat in places such as Iceland, Singapore and Guinea-Bissau are implemented. For example, CAFF is co-operating with Iceland to ensure policies are put in place to avoid or limit risk to breeding water birds from changes in land use in the Icelandic lowlands, especially those arising from the Icelandic government's afforestation policy. Similarly, CAFF is aiming to work with South Korea and China to manage intertidal areas and coastal ecosystems that are critical places for Arctic shorebirds, with Russia to ensure habitat protection and the elimination of the illegal hunting of birds, and with other Arctic states with fishing industries to assess fisheries by-catch of seabirds and to work out by-catch mitigation measures.

This illustrates the significance of the Arctic Council as a forum in which dialogue between Arctic states and non-Arctic states (and involving the Arctic's indigenous peoples) can proceed on matters of international conservation, environmental management and species protection. Such dialogue and co-operation between Arctic and non-Arctic states reminds us of the importance of broadening discussion of Arctic international issues beyond how non-Arctic states are becoming increasingly influential in Arctic affairs and asserting purely economic and strategic interests. By admitting non-Arctic states such as China, South Korea, India and Singapore into the Arctic Council as Observers, the Arctic Council has stronger possibilities to engage with them on circumpolar issues that are not just specific to the Arctic regions, but which are global concerns and should be global priorities. And while it is in the interests of Arctic states that, for example, the Luannan Coast or the tidal flats of the Jiangsu coast ecosystem are protected, it is probably just as much in China's interest that intertidal areas of the West Kamchatka and north Sakhalin coasts or the Yukon-Kuskokwim Delta are also protected as habitat for shorebirds and for their importance to the wider ecosystem. Conservation and environmental management processes and initiatives are never entirely free of a political dimension, and they are often muddied by political interests. While conservation in the Polar Regions must also be examined within the context of geopolitical and strategic interests, it requires us to approach, frame and understand those interests in a broader context of polar geopolitics.

Indigenous peoples and Arctic environmental governance

A number of significant and far-reaching political changes since the 1970s have led to the settlement of land claims between indigenous peoples and Arctic states and the formation of regional governments in Alaska, Canada and Greenland (see Chapter 5 in Part I of this handbook). By and large, these arrangements recognize indigenous rights to use traditional lands and waters, to practice traditional subsistence activities such as hunting and fishing, and often include legislation concerning the ways that living and non-living resources are managed. A greater degree of local involvement and the incorporation of indigenous knowledge in resource use assessments and management has been introduced, including in some cases the devolution of decision-making and management authority to the local community or regional level. Innovative co-management regimes have been implemented that allow for the sharing of responsibility for resource management between indigenous and other resources users and state institutions. Examples include work done by the Alaska Eskimo Whaling Commission (which is concerned with management and conservation of bowhead whales), the Alaska Eskimo Walrus Commission, the Porcupine Caribou Management Board in Canada (federal, territorial and First Nations governments work with local communities to conserve caribou habitat and manage harvesting practices), the Inuvialuit Game Council (in Canada's Northwest Territories), and the Nunavut Wildlife Management Board (an institution of public government established under the Nunavut Agreement and which acts as the main instrument of wildlife management in the Nunavut Settlement Area).

In principle, co-management arrangements for wildlife and environmental management involve greater recognition of indigenous rights to resource use and allow for collaborative efforts in decision-making. This presents opportunities for co-operation between indigenous peoples, scientists and policy-makers concerned with the sustainable use and management of living resources and the monitoring of Arctic ecosystem processes. This is not to say things always run smoothly, as the example of the management of narwhal hunting in Greenland and Canada illustrates, where it is subject to both national and international regulations. The IUCN has red-listed the narwhal population in West Greenland as Critically Endangered and the global population is listed as Data Deficient. While a catch quota system has been in place for the Canadian Eastern Arctic since the 1970s, quotas for narwhal hunting in Greenland were introduced by Greenland's government in 2005. Although indigenous knowledge is taken into consideration in narwhal management deliberations in Canada, in Greenland local knowledge does not figure in decision-making. Quotas and regulations are based on scientific research and advice and narwhal hunters are calling for their involvement – and for the recognition of the importance of their knowledge – in research, monitoring and decision-making processes (Nuttall 2016). In Nunavut, though, Dale and Armitage (2011) discuss the challenge of knowledge co-production and the implications for learning and adapting in the context of narwhal co-management. Knowledge co-production is the collaborative process of bringing a plurality of knowledge sources and types together to address a defined problem and build an integrated or systems-oriented understanding of that problem, but Dale and Armitage show how compartmentalized views of knowledge continue to constrain adaptive and collaborative management.

In Greenland, Canada and Alaska, Inuit groups have outlined and put into practice many of their own environmental strategies and policies to safeguard the future of Inuit resource use, and to ensure a workable participatory approach between indigenous peoples, scientists and policy-makers to sustainable resource management and development. From an Inuit viewpoint, threats to wildlife and the environment do not necessarily come from hunting, but from industrial activities, airborne and seaborne pollutants entering the Arctic from industrial areas far to the south, as well as from the impacts of global climate change. Extraction of non-renewable resources, such as oil and gas and minerals, poses other challenges. In recent years Inuit organizations have sought ways to counteract such threats and devise strategies for environmental protection and sustainable development. They argue that adequate and appropriate systems of environmental management and sustainability are only possible if based on local knowledge and Inuit cultural values. By so doing, the Inuit claim their right to be recognized internationally as resource conservationists. This approach has been made more effective through the work of the Inuit Circumpolar Council (ICC). One current ICC project is concerned with the North Water polyna between Northwest Greenland and Arctic Canada. The Greenlandic name for the North Water is Pikialasorsuaq, which means "the great upwelling". This refers to how the mixing of water currents results in the upwelling of nutrients and so producing the attractive conditions and feeding opportunities favourable to marine mammals, fish and birds in the region, and which make it a key hunting and fishing area for Inuit in northern Nunavut and Northwest Greenland. Concerned about the effects of climate change and the prospects of further oil and mineral exploration and increased shipping in the area, ICC's Pikialasorsuaq Commission is working to put forward an Inuit strategy for the management and future of the North Water, while in northern Nunavut the Canadian government and the Qikiqtani Inuit Association have agreed on the boundaries for a future marine conservation area in Tallurutiup Imanga/Lancaster Sound. Inuit from Canada's eastern Arctic expect to have a major role to play in the defining of conservation principles and in the overall management of the area.

Community-based monitoring in the Arctic

The Arctic is not only experiencing rapid climate change, there is increasing interest in non-renewable resource development in the region. As environmental changes become apparent in terrestrial, marine and freshwater ecosystems, and as resource exploration and development activities intensify, there is urgent need to map, monitor and assess the environmental and social and economic effects and understand the socio-economic and cultural impacts on communities and people's livelihoods. Danielsen et al. (2007) argue that while there is a need for a greater understanding of the status of environments and wildlife under threat, concerns have also been raised regarding the exclusion of local people from environmental decision-making processes (and the issue of narwhal hunting and environmental monitoring discussed above is a good example of this). In recent years, community-based monitoring programmes and participatory research initiatives have been seen as effective ways of addressing this exclusion and for setting in place long-term monitoring systems.

As Conrad and Hilchey (2011) point out, community-based approaches enhance initiatives to monitor and manage natural resources, track species at risk and conserve protected areas. The advantage of a community-based approach is that observation and monitoring practices, to varying degrees, include local people most affected by climate change and other environmental and socio-economic transformations – and as community-based monitoring is motivated by concern for places and people experiencing environmental threats, it can also contribute to efforts to overcome longstanding conflict between diverse stakeholder groups (Berkes 2007; Bliss et al. 2008). Whitelaw et al. (2003) show how community-based ecosystem monitoring activities in Canada increased in response to a number of factors including the needs of decision-makers for timely information on local environmental changes; limited use of government monitoring data and information by decision-makers; government cuts to monitoring programmes; the increasingly recognized need to include stakeholders in planning and management processes; and the desire of local community members to contribute to environmental protection and long-term environmental monitoring. And as Danielsen et al. (2007) show, participatory biodiversity monitoring represents a cost-effective alternative when conventional scientific monitoring is impossible (e.g., when scientists are restrained by cost and logistics from being in Arctic field sites for long periods), but it is also a powerful complementary approach to conventional monitoring and is capable of generating a much higher level of conservation management intervention.

The extent of community involvement in and control over monitoring programmes differs along a scale from limited participation (when projects are implemented by scientists and agencies external to a community, and when no consultation has taken place) to initiatives developed, operated and led by communities with little or no participation by external scientists (Johnson et al. 2016). At the heart of community-based observation and monitoring, though, is a recognition that programmes for wildlife conservation and environmental management should acknowledge locally-situated engagement with the environment and the resources people depend upon, and that they should be informed by indigenous and local knowledge and acknowledge community priorities. In collaboration with scientific research projects, communities can be placed within wider regional, national and international networks, allowing local voices to be heard and local concerns to be expressed. Community-based ecosystem monitoring, for example, refers to a range of observation and measurement activities and techniques involving participation by community members and is designed to understand and assess the ecological and social factors affecting a community and influencing resource use. The success of community-based monitoring, however, depends on a number of factors, ranging from understanding social, cultural and economic diversity within communities to the availability and use of appropriate technology, through to capacity-building, the nature of community involvement and long-term funding.

In their discussion of observations from a range of community-based ecosystem monitoring activities throughout the US, Bliss et al. (2008) discuss factors leading to the emergence of community-based ecosystem monitoring, multi-party monitoring and its role in building social capital, the monitoring process itself, the integration of social and ecological factors, and ongoing challenges in community-based monitoring. Such challenges include achieving effective, diverse community participation, integrating social indicators into ecosystem monitoring and analysis, identifying an appropriate level of procedural rigour for the implementation and maintenance of specific monitoring objectives, and exactly how monitoring can be integrated effectively into decision-making processes in practice. In an examination of integrated management in the Canadian North, Berkes et al. (2007) assess its contribution to the advancement of knowledge and practice regarding the role of indigenous knowledge and community-based monitoring. They point to how work in managing the Beaufort Sea, designated a Large Ocean Management Area under Canada's Oceans Action Plan, is a particularly good example of a consultative planning process, especially with regard to how special attention has been given to the involvement of indigenous peoples in the process. Drawing attention to the role of indigenous knowledge in management and conservation, they use the problem of Arctic marine food web contamination to illustrate the strengths and limitations of traditional ecological knowledge and its relationship to science. Their discussion of community-based monitoring draws to some extent from the seminal Voices from the Bay initiative in the 1990s, which involved Inuit and Cree of Hudson Bay and James Bay in discussions of ecosystem management, and Inuit observations of climate change studies in the Canadian western Arctic. The examples Berkes and colleagues discuss address integrated coastal management and the health of ocean ecosystems, showing how stakeholder participation and knowledge help widen the range of knowledge and deepen the nature of the research process to understand and assist in monitoring environmental change.

Effective community-based monitoring demands an understanding of the nature of community and indeed what constitutes communities, as well as the need for training and capacity-building. In their discussion of community-based monitoring activities in Canada, Pollock and Whitelaw (2011) point to the need to give attention to community diversity and argue that community-based monitoring requires an approach that is context-specific, iterative and adaptive. Given these emergent characteristics, they argue for an enhanced conceptual framework based on four themes: community mapping, participation assessment, capacity-building and information delivery. Providing an especially good example of community-based monitoring that has relevance for the Arctic, McKenzie et al. (2000) describe a programme called Seagrass-Watch in Queensland, Australia in which community groups and volunteers are trained to assist fisheries scientists to establish a reliable early warning system on the status of seagrass resources. Intertidal seagrass habitats are mapped and monitored, and a body of information is built up to assess and evaluate change. The success of the programme relies on community volunteers collecting information for coastal management on changes in seagrass meadow characteristics, such as the extent of coverage, position and depth of habitat, species composition, estimates of abundance, presence of dugong feeding trails and possible human impacts.

Across the Arctic today a number of community-based monitoring programmes are in place and seek to observe and gather indigenous and local knowledge of environmental change (Johnson et al. 2016). These programmes operate not only in order to understand changes to ecosystems, but to provide information that communities can use to assist them in responding to those changes as well as to provide such information for feeding into decision-making and natural resource management. In the Arctic Council, CAFF has initiated and supported a number of activities through the Circumpolar Biodiversity Monitoring Program (CBMP) that promotes community-based monitoring. Working with indigenous organizations, the CBMP collects indigenous knowledge and information from community-based programmes and integrates this into CAFF's monitoring and assessment activities (ibid.). An online atlas of Arctic community-based monitoring is available at www.arcticcbm.org. Providing information about a wide variety of programmes, and serving as a resource for communities, conservationists, environmental and wildlife managers, and decision-makers, the atlas is intended to help raise awareness about the nature and vital role of community-based and community-led monitoring and its importance for the development of Arctic observing systems. The project is led by ICC-Canada, together with the Exchange for Local Observations and Knowledge of the Arctic (ELOKA) and Inuit Tapiriit Kantami's (ITK) Inuit Qaujisarvingat: Inuit Knowledge Centre (ITK is Canada's national Inuit organization).

Not all community-based monitoring has a concern with environmental change alone, however, or with wildlife, and efforts to track the effects of social, political and economic processes on people's livelihoods are crucial alongside observations of ecosystem shifts and changes in wildlife behaviour and habitat. Significantly, no Greenlandic initiative was considered in the comprehensive review conducted by Johnson et al. (ibid.), as part of a Sustaining Arctic Observing Networks (SAON) task, of community-based monitoring programmes in the Arctic. Indeed, the situation on environmental monitoring in Greenland today is that there is an absence of robust community-based approaches in the country, and Greenland is certainly a considerable way from having community-led programmes in place. However, discussion is beginning to take place about the development of appropriate community-based methodologies for Greenland that can inform policy- and decision-making processes for the implementation of community-based observing and monitoring programmes that can feed into wider Greenlandic monitoring and assessment of environmental change and wildlife stocks and habitats, and to the development of interdisciplinary research (Nuttall 2016).

Conclusions

Environmental institutions and governance processes set out to protect or conserve environments, ecosystems and species, and while there are some examples of how this can work well in the Polar Regions (in the case [for the most part] of polar bear management in the Arctic, for example, the co-management of bowhead whales on the north coast of Alaska, nature-based tourism in the Russian Arctic[5] or conservation instruments under the Antarctic Treaty System), they can also provoke disagreement and cause conflicts of interest (such as over narwhal management in Greenland, for instance, or over who identifies and defines what is threatened or endangered and needs to be managed or protected). They can also fail to recognize that attempts to ensure the sustainable management of one resource or one element of an ecosystem, may lead to unsustainable practice or environmental degradation in another, while the effectiveness of local resource regimes can be hindered by national and international institutions (e.g., Young 2003). In the case of fisheries or marine mammal hunting in the Arctic, or other activities based on the harvesting of living resources, environmental management regimes and conservation policies can often restrict or inhibit the rights of people who fish and hunt for a living. Institutions may often appear to be innovative but can fail to understand and take into account the underlying causes of a problem or the social, cultural and environmental effects of implementing measures of environmental management and governance. The challenge for global governance, as Jasanoff and Martello and their contributors (2004) see it, is to seek new ways of being attentive to the local while developing institutions which transcend it.

Climate change, ocean acidification, pollution and resource extraction also bring new pressures and challenges to the Polar Regions. Environmental change and broader, far-reaching social, economic and global processes are testing those institutional arrangements and conservation initiatives that make up polar governance systems (Dodds and Nuttall 2016; Chown 2017), requiring them to broaden their mandates and evolve into entities that are quite different from how they were when founded originally. This is illustrated by CCMALR, which was first concerned mainly with the harvest of Antarctic krill, or the international agreement concerned with polar bear management, which had its origins in anxieties over commercial hunting. Neither agreement was worked out and implemented in a context of rapid climate change, for instance, or other pressures on the environment, but both now need to be attentive to broader aspects of ecosystem-based management, complex ecosystem inter-linkages and human-environment relations. And as oil and mineral exploration and production continue in the Arctic, robust domestic and international legal instruments will need to be maintained to monitor and deal with the regulation of the industry and to ensure the conservation of circumpolar biodiversity (while also recognizing that indigenous livelihoods and community and regional economies in many regions of the circumpolar North are increasingly dependent on extractive industries).

The long-range transport of pollutants and pesticides into the Arctic marine environment has led to their bioaccumulation in certain marine species, posing significant health risks to people and communities who depend upon these species as a food source, while land-based sources of pollution from within and without the Arctic (from industrial activities, sewage, port facilities, etc.) also present a major concern to the health of the marine environment. Such threats from pollution have an impact on ecosystems, animal habitats and movement, and also have consequences for food security and human health, thus seriously constraining the abilities of indigenous peoples to achieve sustainable livelihoods and maintain well-being.

One further issue relates to scale, of how far it is possible to talk of a system of environmental governance for the Arctic or for the Antarctic within a context of global environmental management. Here scale becomes critical to a discussion of environmental governance – institutions are also involved in a political process of defining and scaling the things and objects that are governed, as well as the places and spaces they live within, compose and configure. Furthermore, recent debates in the social sciences and humanities concerning the nature and place of animals in ecosystems and human-animal relations (e.g., Philo and Wilbert 2000; Hurn 2012; Corbey and Lanjouw 2013), and anthropological concern with multispecies ethnography and interest in other non-human entities such as ice, as well as indigenous knowledge and ontologies, disturb assumptions of how the world looks and how everything that fills it is supposed to function and behave. This challenges scientific understanding of animal behaviour not just in the Arctic, but in the Antarctic too, suggesting that environmental governance systems and conservation biology need to consider how animals live out their lives as beings in the world, and how they are entangled in complex relations with other non-humans as well as people, rather than as species in abstract spaces or habitats. It also requires us to look at the environment in different ways than conventional management institutions tend to do. In the Arctic, for example, animals are not viewed merely as economic resources by northern peoples, and "wildlife" is not an indigenous categorization informing local understandings of being, living and acting in the world that best describes the intricate human-animal relations that infuse circumpolar places (Anderson and Nuttall 2004; Kalland and Sejersen 2005), and so "wildlife management" is something that is often contested from a local community viewpoint. Neither, from an indigenous perspective, is it necessarily appropriate to merely describe and reduce polar landscapes and marine environments to ecosystems or cryospheric environments. Organizations such as the World Wide Fund for Nature (WWF), for example, have been pushing for a set of circumpolar-wide agreements to protect and conserve some species of Arctic wildlife by setting aside specific protected areas (such as ecosystems that are crucial habitat for polar bears and narwhals) or prohibit oil exploration and development in some northern waters and lands, or have launched initiatives to designate parts of the Arctic such as glacial and sea icescapes as protected areas (such as WWF's Last Ice Area project[6]).

Such views, however, are often at odds with indigenous people's perceptions and knowledge of Arctic worlds and how they think about and relate to animals, sea ice, glaciers, water and land. In Northwest Greenland, for example, which is a world of extraordinary encounters between the human and non-human, people, animals, ice, rocks, the wind and everything that makes up the world are participants that share and shape it (e.g., Nuttall 2017). In the Arctic, and as community-based perspectives on environmental management show, conservation policies and action need to be based on community consultation and dialogue and should be informed by an understanding of how people not only experience a complexity of change and how they reflect upon the global processes affecting their lives, but how their understandings of human-environment relations inform the richness of local knowledge. And how well we act on conservation matters in the Antarctic too will also depend on how we reflect on the nature and complexity of what we classify and define as animals, ice, land and marine environments.

Notes

1 The Ramsar Convention, or the Convention on Wetlands, is an intergovernmental treaty providing for the protection and wise use of wetlands and their resources. It was adopted in Ramsar, Iran in 1971 and came into force in 1975.
2 See the figures for the catch of Antarctic krill available on the CCAMLR website: www.ccamlr.org/en/fisheries/krill-%E2%80%93-biology-ecology-and-fishing.
3 The *zapovednik* is a protected area in which the utilization of natural resources is prohibited, although nature-based tourism is permitted. In a *zakaznik*, on the other hand, temporary or permanent limitations can be imposed upon economic activities, such as hunting, fishing, grazing reindeer, logging or extractive industries.
4 See ACIA 2005 *Arctic Climate Impact Assessment: Scientific Report*. Cambridge, UK: Cambridge University Press.
5 See, for example, A. Pashkevich, O. Stjernström and L. Lundmark 2016 'Nature-based tourism, conservation and institutional governance: a case study from the Russian Arctic' *The Polar Journal* 6(1): 112–130.
6 For information about WWF 'Last Ice Area' initiative, see www.wwf.ca/conservation/arctic/lia/.

References

Agnew, D.J. 1997. 'The CCAMLR Ecosystem Monitoring Programme' *Antarctic Science* 9(3): 235–242.

Anderson, D.G. and M. Nuttall, eds. 2004. *Cultivating Arctic Landscapes: Knowing and Managing Animals in the Circumpolar North*. Oxford, UK: Berghahn.

Berkes, F. 2007. 'Community-based conservation in a globalized world' *PNAS* 104(39): 15188–15193.

Berkes, F., M. Kislalioglu Berkes and H. Fast 2007. 'Collaborative integrated management in Canada's North: the role of local and traditional knowledge and community-based monitoring' *Coastal Management* 35(1): 143–162.

Bliss, J., G. Appelt, C. Hartzell, P. Harwood, P. Jahnige, D. Kittredge, S. Lewandowski and M.L. Soscia 2008. 'Community-based ecosystem monitoring' *Journal of Sustainable Forestry* 12(3–4): 143–167.

Brooks, C.M. and D.G. Ainley 2017. 'Fishing in the bottom of the earth: the political challenges of ecosystem-based management' in K. Dodds, A.D. Hemmings and P. Roberts (eds) *Handbook on the Politics of Antarctica*. Cheltenham, UK: Edward Elgar.

Chown, S.L. 2017. 'Antarctic environmental challenges in a global context' in K. Dodds, A.D. Hemmings and P. Roberts (eds) *Handbook on the Politics of Antarctica*. Cheltenham, UK: Edward Elgar.

Conrad, C. and K. Hilchey 2011. 'A review of citizen science and community-based environmental monitoring: issues and opportunities' *Environmental Monitoring and Assessment* 176(1): 273–291

Corbey, R. and A. Lanjouw, eds. 2013. *The Politics of Species: Reshaping our Relationships with Other Animals*. Cambridge, UK: Cambridge University Press.

Dale, A. and D. Armitage 2011. 'Marine mammal co-management in Canada's Arctic: knowledge co-production for learning and adaptive capacity' *Marine Policy* 35(4): 440–449.

Danielsen, F., M.M. Mendoza, A. Tagtag, P.A. Alviola, D.S. Balete, A.E. Jensen, M. Enghoff and M.K. Poulsen 2007. 'Increasing conservation management action by involving local people in natural resource monitoring' *Ambio* 36(7): 566–570.

Derocher, A., J. Aars, S.C. Amstrup, A. Cutting, N.J. Lunn, P.K. Molnár, M.E. Obbard, I. Stirling, G.W. Thiemann, D. Vongraven, Ø. Wiig and G. York 2013. 'Rapid ecosystem change and polar bear conservation' *Conservation Letters* 6(5): 368–375.

Dodds, K. and M. Nuttall 2016. *The Scramble for the Poles: The Geopolitics of the Arctic and Antarctic*. Cambridge, UK: Polity.

Elliott, L.M. 1994. *International Environmental Politics: Protecting the Antarctic*. London: St. Martin's Press.

Hansom, J.D. and J.E. Gordon 1998. *Antarctic Environments and Resources: A Geographical Perspective*. Harlow, UK: Longman.

Hurn, S. 2012. *Humans and Other Animals: Human-Animal Interactions in Cross-Cultural Perspective*. London: Pluto Press.

Jasanoff, S. and M.L. Martello, eds. 2004. *Earthly Politics: Local and Global Environmental Governance*. Cambridge, MA: The MIT Press.

Johnson, N., C. Beha, F. Danielsen, E-M. Krümmel, S. Nickels and P.L. Pulsifer 2016. *Community-Based Monitoring and Indigenous Knowledge in a Changing Arctic: A Review for the Sustaining Arctic Observing Networks*. Ottawa: Inuit Circumpolar Council.

Johnston, V., E. Syroechkovskiy, N. Crockford, R.B. Lanctot, S. Millington, R. Clay, G. Donaldson, M. Ekker, G. Gilchrist, A. Black and R. Crawford. 2015. *Arctic Migratory Birds Initiative (AMBI): Workplan 2015–2019*. CAFF Strategies Series No. 6. Akureyri, Iceland: Conservation of Arctic Flora and Fauna.

Kalland, A. and F. Sejersen 2005. *Marine Mammals and Northern Cultures*. Edmonton, AB: CCI Press.

Kormos, C.F., T. Badman, T. Jaeger, B. Bertzky, R. van Merm, E. Osipova, Y. Shi, P. Bille Larsen 2017. *World Heritage, Wilderness and Large Landscapes and Seascapes*. Gland, Switzerland: IUCN.

McKenzie, L.J., W.J. Lee Long, R.G. Coles and C.A. Roder 2000. 'Seagrass-watch: community based monitoring of seagrass resources' *Biol. Mar. Medit.* 7(2): 393–396.

Nuttall, M. 2016. 'Narwhal hunters, seismic surveys and the Middle Ice: monitoring environmental change in Greenland's Melville Bay' in S.A. Crate and M. Nuttall (eds) *Anthropology and Climate Change: From Encounters to Actions*. London and New York: Routledge.

Nuttall, M. 2017. *Climate, Society and Subsurface Politics in Greenland: Under the Great Ice*. London and New York: Routledge.

Österblom, H. and O. Olsson 2017. 'CCAMLR: an ecosystem approach to the Southern Ocean in the Anthropocene' in K. Dodds, A.D. Hemmings and P. Roberts (eds) *Handbook on the Politics of Antarctica*. Cheltenham, UK: Edward Elgar.

Philo, C. and C. Wilbert (eds.) 2000. *Animal Spaces, Beastly Places: New Geographies of Human-Animal Relations*. London: Routledge.

Pollock, R.M. and G.S. Whitelaw 2011. 'Community-based monitoring in support of local sustainability' *Local Environment: The International Journal of Justice and Sustainability* 10(3): 211–228.

Ryabov, A.B., A.M. de Roos, B. Meyer, S. Kawaguchi and B. Blasius 2017. 'Competition-induced starvation drives large-scale population cycles in Antarctic krill' *Nature Ecology & Evolution* 1, doi:10.1038/s41558-017-0177.

Speer, L., R. Nelson, R. Casier, M. Gavrilo, C. von Quillfeldt, J. Cleary, P. Halpin, P. and P. Hooper 2017. *Natural Marine World Heritage in the Arctic Ocean*. Gland, Switzerland: IUCN.

Warner, R. 2013. 'Environmental assessments in the marine areas of the polar regions' in E. Molenaar, A.G.O. Elferink and D.R. Rothwell (eds) *The Law of the Sea and Polar Regions: Interactions Between Global and Regional Regimes*. Leiden, Netherlands: Martinus Nijhoff Publishers.

Whitelaw, G., H. Vaughan, B. Craig and D. Atkinson 2003. 'Establishing the Canadian Community Monitoring Network' *Environmental Monitoring and Assessment* 88(1): 409–418.

Young, O.R. 2003. 'Environmental governance: the role of institutions in causing and confronting environmental problems' *International Environmental Agreements: Politics, Law and Economics* 3: 377–393.

Part IV
Polar scientific frontiers

33

Technology and the discovery of Antarctic subglacial landscapes

Martin J. Siegert

Introduction

Around 98% of the land in Antarctica is covered by thick ice. To measure it, geophysical surveys that 'see through' and allow scientists to peer deep down into the ice are needed. Over the last century, a history of landscape and glacial discovery in Antarctica has formed, which is linked closely with technological developments. In this chapter, I summarize how advances in field instrumentation, especially over the past 50 years, have led to a much greater understanding of the continent beneath the ice. I also look briefly at technologies being developed today and how these may be used in an Antarctic context in coming decades.

Early history of discovery

The discovery of the Antarctic continent in 1820/21, by US sealer Nathaniel Palmer and English sailors Edward Bransfield and William Smith, was possible as a consequence of seafaring technology that allowed ships to voyage south across the icy Southern Ocean in search of marine mammal 'resources'. Indeed, much of our early attention to Antarctica was tightly coupled to this lucrative business in sealing and whaling in the waters around Antarctica and some sub-Antarctic islands, driving the development of ship technology, building logistics for substantial remote operations (from ship-based and coastal operations), and furthering our knowledge of this hostile environment and how to cope with it. Despite increasing presence across the sub-Antarctic Islands in the nineteenth century, it wasn't until 1895 that the first people set foot on the Antarctic continent at Cape Adare (a party led by Norwegian Carsten Borchgrevink, including the New Zealander Alexander von Tunzelmann who probably was the first to step ashore), ushering in the heroic age of exploration (this was the first documented or substantiated landing – while the actual first landing was likely by the crew of *Cecilia* at Hughes Bay in 1821). A mere seven years later, the first serious attempt to reach the South Pole took place; the 1902 Discovery Expedition, involving Robert Falcon Scott, Edward Wilson and Ernest Shackleton, which penetrated deep into the Ross Ice Shelf and reached 82°S, 450 miles from the Pole. This was followed by Shackleton's Nimrod Expedition, which established a new route from the Ross Ice Shelf through the Transantarctic Mountains to the ice-sheet plateau, and which got to within 97 miles of the South Pole in 1907–09; Roald Amundsen's successful expedition, which pioneered a new path to the plateau and reached the South Pole on 14 December 1911; and Scott's ill-fated Terra Nova campaign, which followed Shackleton's route and got to the South Pole 34 days after Amundsen on 18 January 1912 only to perish on the return journey.

We learned a great deal from these expeditions about how to undertake deep-field reconnaissance in Antarctica. In particular, from the early British explorers, and also from Australian geologist Douglas Mawson, we also began to build our knowledge of the continent's natural environment (geology, glaciology, oceanography, meteorology, botany etc.). While these pioneering expeditions can be characterized by many things (e.g., heroics, discovery, duty, science etc.) none of them were associated with substantial successful technological development. This all changed after World War I, however, with the introduction of aircraft in Antarctica.

Introduction of aircraft in Antarctica

The first flight in Antarctica was by Australian Sir Hubert Wilkins and American Carl Ben Eielson in November 1928 using a Lockheed Vega and a crude runway on the beach of Deception Island. Although the flight only lasted 20 minutes, allowing a couple of circuits of the island, the demonstration of the utility of planes opened an age of technology in Antarctic exploration. Remarkably, a year later in November 1929, Admiral Richard E. Byrd and a crew of three flew a Ford Trimotor for 19 hours from the Little America base camp on the Ross Ice Shelf over the South Pole and back again. Seven years afterwards, in November 1935, Lincoln Ellsworth and co-pilot Herbert Hollick-Kenyon made the first trans-continental flight in Antarctica. Over a two-week mission they flew from Dundee Island to Little America, discovering in the process the Heritage Range of the Ellsworth Mountains (which Ellsworth named after his father).

Within a handful of years, aircraft had transformed our appreciation of Antarctica from being largely inaccessible to one in which access was highly feasible. Simultaneous with advances in aircraft and their use in Antarctica, great strides were made in the development of clothing and overland traversing, making the (relatively) safe exploration of the deep continental interior possible. By the end of the 1930s, however, although we had a demonstrable ability to work in many parts of Antarctica, we still knew very little about the shape, size, depth and thickness of the ice sheet, or indeed whether Antarctica was actually even a single continent.

Geophysical surveying of the ice sheet

While the intervention of World War II led to a cessation of Antarctic exploratory missions, it also saw a period of significant developments in field geophysical equipment, data processing and analysis. Shortly after the end of the war, Australian glaciologist Gordon Robin, through his PhD investigations as part of the Norwegian-British-Swedish expedition to Dronning Maud Land in East Antarctica, perfected the use of seismic sounding to measure ice thickness (Robin 1958). Seismic waves (sound waves) travel well through dense ice but are attenuated by soft firn and snow. Consequently, to increase the signal to noise ratio, two boreholes (~50 m deep) need to be drilled; one for the charge, one for the receiver(s). The experiment is simple; a small explosion sets off a sound wave, which travels down to the ice base where it is reflected and subsequently recorded by the receiver. The two-way travel time is noted, and converted into distance as the speed of sound in ice is known reasonably well. Thus, a measure of ice thickness is possible using a simple seismic reflection test, adapted for the harsh polar field conditions by Robin. While the process of data acquisition is time-consuming (two boreholes for each data point, meaning that a single datum would need at least a day to record) by aligning measurements along a survey line a profile of ice-sheet thickness, and therefore bed topography, could be derived. In this way, the first cross-sections of the Antarctic subglacial landscape were obtained.

A decade after the end of World War II, a major international collaboration was developed, called the International Geophysical Year (IGY, 1957–8) which, although it was a global initiative, targeted Antarctica as a location where basic discovery was needed and possible (it was coincident with the fourth International Polar Year, IPY) (Naylor et al. 2008). The IPY's mission statement was "to observe geophysical phenomena and to secure data from all parts of the world; to conduct this effort on a coordinated basis by fields, and in space and time, so that results could be collated in a meaningful manner". This inclusive approach led to several exploratory scientific missions across Antarctica, using the seismic techniques described by Robin a few years earlier. Two overland traverses were most notable. A US expedition crossing West Antarctica, involving a young glaciologist named Charles Bentley, and a Russian survey from the coast to the centre-point of East Antarctica (the Pole of Relative Inaccessibility), which had among its party Andrei Kapitsa. The data collected by these surveys transformed our knowledge of the continent, proving it to be a single landmass, showing the ice to be several kilometres thick (at Vostok Station, for example, it was measured as ~3.7 km) and, in large parts of West Antarctica, revealing the bed to be over two kilometres below the level of the sea. The appearance and set-up of these expeditions had far more in common with today's scientific missions than that of the early explorers 50 years previously. Using snow-cat tractor-trains, guided by crevasse-searching radars, the scientific parties were able to cross hundreds of kilometres in relative comfort and with ample personal and scientific supplies. We use overland traversing in much the same way today. Then, as now, the time-limiting factor to the ice-sheet information gained from seismic studies was the need for borehole drilling.

In the early 1960s, UK physicist Stan Evans and Robin, by now Director of the Scott Polar Research Institute in Cambridge, began experiments to understand the electrical properties of cold ice, and how VHF radio waves could be used to measure ice thickness. VHF radio waves (50–150 MHz) travel very well through cold ice (<−10°C) but reflect off boundaries of dielectric contrast (such as at the ice-bed interface). Radio-echo sounding (RES), as it was known (also called ice-penetrating radar), is able to chart ice thickness, therefore, in an analogous way to seismic sounding. The major advantages of RES over seismic sounding was that it did not require the drilling of boreholes and could be deployed on a moving platform to obtain cross-section information during transit. The most significant innovation by Evans and Robin was to consider how RES could be mounted and used effectively on aircraft. In the late 1960s their Cambridge team, supported by funding and logistics from the US Antarctic Research Programme, demonstrated the use of airborne RES with instant and revolutionary success.

Radio-echo sounding and Antarctic exploration

Using RES on an aircraft, the rate and quality of data acquisition improved enormously. What took at least a day to get a data point now took less than a second and with equal accuracy (an improvement by five orders of magnitude). What took a season to build a transect now took a single sortie; and where a profile of the ice sheet may have been constructed with a few dozen seismic data-points, now it could be put together with many thousands of RES reflections. Early RES trial flights, using a Super Constellation L-1049 aircraft, were targeted at the very centre of the East Antarctic ice sheet, that Russian traversing had covered a decade before.

In so doing, Robin and his team proved continental-wide coverage by aircraft mounted with RES was feasible, and that the data were remarkable. So followed one of the world's key decades in Antarctic glaciological and continental discovery.

Systematic profiling of the ice sheet took place in four field seasons: 1971–72, 1974–75, 1977–78 and 1978–79. Over the decade, further advances in RES equipment were made, primarily through the work of Danish physicist Preben Gudmundsen, and navigation was improved (the early flights used 'dead reckoning', which was replaced by an Inertial Navigation System or 'INS'). The aircraft of choice by now was a long-range Hercules C130 transporter, supplied by the US Navy. Thus, a US-UK-Danish collaboration surveyed about 40% of East Antarctica and 80% of West Antarctica, defining the subglacial landscape for the first time and making profound discoveries about the way in which ice flowed. For example, lakes beneath the ice were discovered in 1969 (the bright flat reflections over water being easily distinguishable from the ice rock interface), 17 of them were documented from East Antarctica in 1971/72 (Oswald and Robin 1973) and Lake Vostok (the gigantic subglacial lake in East Antarctica) was detected in 1974 (Robin et al. 1977); buried crevasses were discovered, testifying to the 'switching off' of a major fast-flowing ice stream, evidencing hitherto unappreciated ice-dynamic change (Rose 1979); and a compilation of data led to the first reliable sub-continental map of ice thickness, subglacial topography, surface elevation and internal layering (i.e., the Antarctic Geophysical Folio; Drewry 1983).

The RES data collected by Robin and his international team were so good that they appear similar visually to modern RES data. Indeed, RES data collected today use much of the same basic equipment as that devised in the 1970s. Over the past 40 years there have been two major upgrades to the technology, however. First, the 1970s data were stored using an analogue recording method; the radar oscilloscope information was cast onto rolling 35 mm negatives, from which cross-section prints could be developed (special equipment was built in Cambridge, allowing horizontal compression of the lengthy film records). In the 1980s, 'analogue to digital' conversion was possible, making analogue film records obsolete. Although initially limited by computer storage capacity, causing 1980s RES data to be highly pixilated (and often less informative than sharply focused analogue records), recent digital RES data emulate the fidelity of the analogue records. Second, navigation systems used in the 1970s meant that, after flying for several hours, the aircraft's position was only known with an accuracy of less than 5 km. This issue was solved with the availability of satellite-derived GPS, resulting in continuously accurate location data (to within a few metres). In addition to these changes, RES has also benefited from the development of coherent radar systems, allowing the phase of the radio wave reflections to be determined (thus allowing bed conditions to be evaluated), polarimetric systems, which can measure ice-crystal fabrics, and numerous processing techniques, resulting in high-definition topographic information.

Given the success of the 1970s survey, it is surprising to observe that large-scale RES studies in Antarctica were largely halted in the 1980s (Turchetti et al. 2008). There are several reasons for this: a focusing of spatially restricted hypothesis-driven research; the termination of funding (US funding of essentially a UK programme was and is unusual); the completion of the most logistically feasible work (Dean et al. 2008); and the retirement of Gordon Robin in 1982. While glaciology advanced as a subject considerably in the 1980s and 1990s, our appreciation of the Antarctic subglacial topography and basal ice-sheet conditions did not change significantly. This is best illustrated by comparing the subglacial bed topography published by Drewry (1983) with the next update (BEDMAP; Lythe and Vaughan 2001), which was 18 years later and not hugely different (and indeed retained large regions where data were either extremely sparse or absent).

Combining satellite data with geophysical knowledge

In the early 1990s, the European Space Agency launched the European Remote Sensing satellites ERS-1 (in July 1991) and ERS-2 (in April 1995), which provided an extremely accurate measurement of the surface elevations of land, sea and ice. This yielded unprecedented knowledge about the surface elevation of the polar ice sheets (noting a data gap at the most extreme latitudes due to the satellite orbit paths). One notable discovery was an extremely large, flat region with a clearly defined perimeter at, and for 240 km north of, Vostok Station. This flat surface feature is the ice-surface expression of Lake Vostok and it allowed us to be certain that data collected over the lake decades before were indeed from a single body of water; RES data collected in the 1970s defining the lake's edges in several places were shown to match the change in ice surface slopes. This also meant that seismic data collected at Vostok Station by Russian scientists in the early 1960s were also from the lake. While radio waves do not travel well in water, sound waves do, making seismic data useful for determining water depths. Kapitsa et al. (1996) collated all known information about Lake Vostok (satellite altimetry, RES and seismic data) to form a seminal paper on shape and size of the lake, noting it to be over 500 m deep at Vostok Station, which makes it one of the world's top ten largest lakes.

Later satellite missions, including the Canadian Radarsat, NASA's IceSat, and ESA's Cryosat have provided further knowledge of the surface elevation of polar ice masses, and the ~30-year time series of information that has resulted has been used to quantify ice loss in Antarctica, in particular over the Amundsen Sea region of West Antarctica (Pritchard et al. 2012; McMillan, Chapter 14 of this volume). Such loss of ice has consequences for global sea level, which demands that our ability to model the flow of ice, and understand the processes governing such loss, is improved.

However, while our ability to model the ice sheet numerically improved with high performance computing and software developments, model results were affected adversely by the fundamental bed input data, which remained absent over large regions. This posed a serious problem for ice-sheet models and, in particular, their ability to predict ice-sheet changes. Hence, our ability to determine the processes by which ice sheets modulate global sea level, as quantified by satellite data, was hampered by a lack of bed elevation information. Obtaining a fuller and more accurate measurement of the whole of the Antarctic bed became a priority, and large-scale surveys of the ice sheet became a scientific necessity.

The fourth International Polar Year

Fifty years following the IGY/IPY, a new international programme, the 4th International Polar Year (2007–08) took place to integrate the international community further toward large research problems that national programmes found challenging to resolve. While the IPY was a multi-disciplinary effort, involving all aspects of Antarctic, sub-Antarctic and Southern Ocean research, one of the research priorities concerned large-scale surveying of the 'unknown' regions of subglacial Antarctica. Two projects emerged: AGAP (US, UK, Australia, China and Germany), which undertook airborne geophysical surveys across the Gamburtsev Mountains (Rose et al. 2013) and ICECAP (US, UK, France, Australia, and latterly China and India), which focused on the low lying topography where the East Antarctic Ice Sheet may be prone to change at its margin (Young et al. 2011). In addition, a series of other, smaller-scale, surveys of both East and West Antarctica were also undertaken, adding detail to otherwise poorly measured regions of the ice base. As a consequence of these new data, by 2013 a new collated bed–elevation dataset was formed (BEDMAP II; Fretwell et al. 2013). While BEDMAP II represents a major upgrade in our appreciation of subglacial Antarctica, two distinct regions remained about which little or no data existed; the so-called 'Poles of Ignorance' at Princess Elizabeth Land and the Recovery Ice Stream sectors. Plans to fill these knowledge gaps are presently underway, and it seems highly likely the first complete picture of subglacial topography in Antarctica will be ready by 2020.

Technology has obviously influenced the discovery of subglacial Antarctica, with a pivotal moment being the use of airborne RES over seismic investigations during the 1960s. Despite the likely complete coverage of data across Antarctica in the coming years, this would only yield a 'first-order' view of topography, as over the bulk of the continent geophysical flight-lines are separated by tens of kilometres, meaning substantial data free blocks within surveyed regions. For example, the US-UK-Danish survey was undertaken with a transect spacing of 50 km, meaning the survey regions are covered with 2500 km² blocks without data. Maps of topography get around this issue by interpolating between data, giving the appearance of continual information (Fretwell et al. 2013). Recent surveys within these data holes have demonstrated substantial errors in BEDMAP II, thus continued targeted survey of the Antarctic ice sheet remains important.

Future plans

The fact that our knowledge of subglacial Antarctica remains incomplete 50 years after our ability to measure it was developed is testament to the facts that such research is extremely challenging logistically as well as being very costly. Several new technologies offer a way forward to resolve both issues, however. First, plans exist to upgrade RES equipment from those offering one-dimensional point measurements (providing cross sections along transects) to a two-dimensional side-scan system, allowing data acquisition either side of transects as well as along them. Second, automated drones are starting to be deployed in the Polar Regions, leading to increased continuous flight times and a (potential) reduction in costs. Third, while satellites already regularly measure the ice surface, future innovation may lead to RES equipment installed within satellites, so that equally high-resolution information from the ice base can be obtained. Fourth, rates of data transfer into and out of Antarctica, which today are very low (needing direct satellite communication, such as Iridium), are likely to be increased both across the continent and to other locations, improving scientific decisions on data gathering during fieldwork. What is clear is that the next 50 years are likely to see the resolution of bed knowledge in Antarctica increased substantially, benefiting our ability to model the flow of ice and predict change to its size and configuration. While this chapter has focused exclusively on Antarctica, it should be noted that the history of RES surveys of the Greenland ice sheet is similar, that knowledge of the Greenland bed is good yet incomplete, and that the reasons to drive further data collection there are as relevant as in Antarctica.

References

Dean, K., Naylor, S. and Siegert, M.J. 2008. 'Data in Antarctic science and politics' *Social Studies of Science* 38(4): 571–604.

Drewry, D.J. 1983. *Antarctica: Glaciological and Geophysical Folio*. Cambridge, UK: Scott Polar Research Institute, University of Cambridge.

Fretwell P., Pritchard, H.D., Vaughan, D.G., Bamber, J.L., Barrand, N.E., Bell, R., Bianchi, C., Bingham, R.G., Blankenship, D.D., Casassa, G., Catania, G., Callens, D., Conway, H., Cook, A.J., Corr, H.F.J., Damaske, D., Damm, V., Ferraccioli, F., Forsberg, R., Fujita, S., Gim, Y., Gogineni, P., Griggs, J.A., Hindmarsh, R.C.A., Holmlund, P., Holt, J.W., Jacobel, R.W., Jenkins, A., Jokat, W., Jordan, T., King, E.C., Kohler, J., Krabill, W., Riger-Kusk, M., Langley, K.A., Leitchenkov, G., Leuschen, C.,

Luyendyk, B.P., Matsuoka, K., Mouginot, J., Nitsche, F.O., Nogi, Y., Nost, O.A., Popov, S.V., Rignot, E., Rippin, D.M., Rivera, A., Roberts, J., Ross, N., Siegert, M.J., Smith, A.M., Steinhage, D., Studinger, M., Sun, B., Tinto, B.K., Welch, B.C., Wilson, D., Young, D.A., Xiangbin, C. and Zirizzotti, A. 2013. 'Bedmap2: improved ice bed, surface and thickness datasets for Antarctica' *The Cryosphere* 7: 375–393.

Kapitsa, A., Ridley, J.K., Robin, G. de Q., Siegert, M.J. and Zotikov, I. 1996. 'Large deep freshwater lake beneath the ice of central East Antarctica' *Nature* 381: 684–686.

Lythe, M.B. and Vaughan, D.G. 2001. 'BEDMAP: a new ice thickness and subglacial topographic model of Antarctica' *Journal of Geophysical Research* 106: 11335–11351.

Naylor, S., Dean, K. and Siegert, M.J. 2008. 'The IGY and the ice sheet: surveying Antarctica' *Journal of Historical Geography* 34: 574–595.

Oswald, G.K.A, and Robin, G. de Q. 1973. 'Lakes beneath the Antarctic Ice Sheet' *Nature* 245: 251–254.

Pritchard, H.D., Ligtenberg, S.R.M., Fricker, H.A., Vaughan, D.G., van den Broeke, M.R. and Padman, L. 2012. 'Antarctic Ice-Sheet loss driven by basal melting of ice shelves' *Nature* 484: 502–505.

Robin, G. de Q. 1958. 'Seismic shooting and related investigations' in *Glaciology III*. Oslo: Norsk Polarinstitutt.

Robin, G. de Q., Drewry, D.J. and Meldrum, D.T. 1977. 'International studies of ice sheet and bedrock' *Philosophical Transactions of the Royal Society of London*. 279, 185–196.

Rose, K.C., Ferraccioli, F., Jamieson, S.S.R., Bell, R.E., Corr, H., Creyts, T.T., Braaten, D., Jordan, T.A., Fretwell, P.T. and Damaske, D. 2013. 'Early East Antarctic Ice Sheet growth recorded in the landscape of the Gamburtsev Subglacial Mountains' *Earth and Planetary Science Letters* 375: 1–12.

Rose, K.E. 1979. 'Characteristics of ice flow in Marie Byrd Land, Antarctica' *Journal of Glaciology* 24(90): 63–75.

Turchetti, S., Dean, K., Naylor, S. and Siegert, M.J. 2008. 'Accidents and opportunities: a history of the Radio Echo Sounding (RES) of Antarctica, 1958–1979' *British Journal of the History of Science* 41: 417–444.

Young, D.A., Wright, A.P., Roberts, J.L., Warner, R.C., Young, N., Greenbaum, J.S., Schroeder, D.M., Holt, J.W., Sugden, D.E., Blankenship, D.D., van Ommen, T. and Siegert, M.J. 2011. 'A dynamic early East Antarctic Ice Sheet suggested by ice covered fjord landscapes' *Nature* 474: 72–75.

34

Sediment and ice cores (past polar climates)

Robert McKay

Introduction

Forecasting the contribution of melting polar ice sheets to future sea-level rise remains one of the greatest uncertainties regarding future anthropogenic climate change (Church et al. 2013). The Antarctic and Greenland ice sheets cover 9.5% of the global land surface, and if melted would raise sea levels by 66 m. Satellite observations indicate that both the Greenland and Antarctic ice sheets have lost mass over the past two decades, and this loss is now accelerating (Vaughan et al. 2013; McMillan, Chapter 14 of this volume). While instrumental records show global sea-level is rising at 3.4 mm/yr on the back of rapidly rising greenhouse gases, the current rates are relatively modest compared to those that have occurred in the geological past, which suggest continued acceleration of ice loss into the future is likely.

Studies of ice and sediment core archives demonstrate that Earth's Polar Regions are capable of large climatic shifts triggered by relatively small changes in external forcings. Over the past 50 years, there have been over two dozen deep ice cores and almost 100 deep sediment drill cores collected from the Arctic and Antarctic regions (Figure 34.1), with many hundreds more short sediment and ice cores. Despite these efforts, there are large gaps in our understanding of how the Polar Regions have responded to past climate change. Geological records have extended our understanding beyond the 800,000 year-long ice core records back to 65 million years. These Cenozoic (last 65 Ma) geological archives provide examples of natural climate states globally warmer than the present, which are associated with atmospheric CO_2 concentrations above pre-industrial levels. They have been critical to understanding Earth's evolution from the "greenhouse" climates of 65–34 million years ago when CO_2 levels were in excess of 750 ppm, to the development and subsequent evolution of the polar ice sheets after 34 million years ago, as CO_2 has declined to pre-industrial levels ~280ppm.

This chapter focuses on the last 5 million years of Earth history. This time period is considered the most accessible and relevant for providing insights into future climate change because:

1   It contains examples of warmer-than-present climates when atmospheric CO_2 was ~400ppm (today's level), and global temperature reached an equilibrium of +2–3°C above pre-industrial levels.

2   The configuration of continents and oceans was broadly similar to today.

3   Geological and ice core records for the last 1 million years are particularly well-resolved, and allow the response of different elements of the Earth system (e.g. carbon cycle, surface temperature, polar ice sheets and sea-level) to past natural changes in radiative forcing (e.g. solar/orbital, greenhouse gases) to be evaluated during past glacial to interglacial transitions and abrupt climate cycles.

During the Last Glacial Maximum (26,500 to 19,000 years ago) the ice sheets were greatly expanded relative to today, resulting in sea levels ~130 m below present-day levels (Lambeck et al. 2014). Most of this ice advance occurred in the high-latitudes of North America and Eurasia, which were covered with continental-scale ice sheets. Prior to the Last Glacial Maximum, ice sheets fluctuated greatly in size, with hundreds of ice ages (glacials) and warm interglacials occurring over the past 34 million years (Zachos et al. 2001). These ice age cycles occur at periods of tens of thousands of years, paced by changes in Earth orbital parameters that control variations in the timing, duration and intensity of seasons, which act to cause the growth or decay of polar ice sheets (Hays et al. 1976). However, the energy variations associated with these orbital cycles are not large enough to drive ice sheet cycles on their own, and instead they drive atmosphere and ocean feedbacks that act to amplify the response of Earth's cryosphere, particularly due to their influence on the global carbon cycle (Denton et al. 2010). Understanding the chain of events that led to past ice sheet retreat events, and the rates at which these occurred, remains a central focus of study for paleoclimate researchers (see Whitehouse, Chapter 16 of this volume).

Ice core records provide high-resolution (at times seasonal to annual) archives of changing atmospheric conditions through time that overlap with the instrumental record but are currently limited to recording climatic variations over the past 800,000 years. Sediment cores are of lower resolution and have larger inherent uncertainties, but while they overlap with the ice cores they also extend our view further back in time to when the Earth system responded to larger climate perturbations similar in magnitude to what we can expect in the future. In particular, emphasis has been placed on understanding fast (e.g. sea ice) and slow (e.g. ocean heat uptake, ice sheets) polar feedbacks (see Richard Hodgkins, Chapter 19 of this volume) and the extent of polar temperature amplification during times when CO_2 last exceeded 400 ppm and the planet was >2°C warmer – a climate state that Earth has not experienced for ~3 million years.

The geographies of the two Polar Regions are dramatically different, and this has a fundamental influence on how the ice sheets respond to external forcings, and how they have evolved through geological time. The shallow Arctic Ocean is surrounded by continents, and its surface is currently warming at twice the rate of the global average with sea ice extent declining at rates approaching 3.5–4.1% per decade (Vaughan et al. 2013; Bingham, Chapter 12 of this volume). Conversely, Antarctica is isolated by the deep well-mixed Southern Ocean, which is taking up more anthropogenic heat and carbon dioxide than oceans in other latitudes, which acts to slow the pace of regional Antarctic surface warming (see Marshall, Chapter 15 of this volume). Consequently, sea ice extent has increased by ~1.2–1.8% per decade since 1979 (Vaughan et al. 2013), a trend that reversed for the first time in 2016, when Antarctic sea ice extent sharply declined. Overall, changes in Antarctic sea ice extent are regionally variable, and some regions in West Antarctica have experienced the significant decreases for decades (Turner et al. 2016). Critically, these decreases are occurring (e.g. in the Amundsen Sea) where warm ocean waters are upwelling onto the continental shelf and causing the retreating and thinning of the marine margins of the West Antarctic Ice Sheet (Rignot et al. 2014). While the Greenland Ice Sheet is melting from the top down, as well as at its marine margin where it meets the warm Arctic oceans and seas, the Antarctic surface remains cold, and melting is occurring from the bottom up at its oceanic margin (McMillan, Chapter 14 of this volume). This highlights the importance of understanding the influence of the Southern Ocean circulation and sea ice on Antarctica's vast marine-based ice sheets, which could ultimately contribute up to +22 m global sea-level rise. Models and paleoclimate reconstructions show that Antarctica will eventually experience amplified warming two to three times the global average, as the Arctic is experiencing now (Masson-Delmotte et al. 2013). Understanding the time of emergence of that amplified surface warming signal is critical to predicting how the Antarctic ice sheets will contribute to future sea-level rise.

Ice core proxies for past climate

Greenland contains the only modern ice sheet in the Northern Hemisphere, which has a volume of 2.96×10^6 km^3; it reaches thicknesses exceeding 3 km, equivalent to 7.3 m of global sea level (Bamber et al. 2013). Two-thirds of the ice sheet is situated on bedrock elevated above present-day sea level, and ablation of the ice sheet is a combination of surface melt and iceberg calving at its marine margin. The Antarctic ice sheets are an order of magnitude larger, with a volume of 26.92×10^6 km, exceeding 4 km in thickness, and contain 58.3 m of total sea level equivalent, 22.7 m of which is currently grounded below sea level (Fretwell et al. 2013).

Greenland receives warmer air masses than Antarctica, and therefore experiences higher snow precipitation rates, allowing for high-resolution ice cores to be obtained. However, most of this ice is relatively young with maximum ice age estimates of 150,000 to 200,000 years old. In contrast, East Antarctic Ice Sheet (EAIS) cores provide climate archives extending back ~800,000 years (Jouzel et al. 2007). However, there is a trade-off with most East Antarctic cores having much lower accumulation rates and therefore lower temporal resolution. Recent efforts in West Antarctica have attempted to address this issue, with high-resolution records covering the past 70,000 years now allowing direct comparison to the Greenland records to understand oceanic and atmospheric teleconnections between the hemispheres (WAIS Divide Project Members 2015).

Ice cores capture seasonal variations in snowfall, which is eventually buried by further snowfall and becomes compressed and forms ice. This ice has clearly defined annual layers that can be counted back in time and analyzed for a range of past environmental conditions (Figure 34.2). By measuring the isotopic and geochemical composition of melted ice core samples, researchers can assess changes in past temperatures and atmospheric circulation. Isotopes in a water molecule will consist of atoms of hydrogen or oxygen containing the same number of protons, but different numbers of neutrons. The consequence of this is each atom has a different mass that can be measured by mass spectrometry. Using the principles of Rayleigh fractionation (Cuffey and Paterson 2010), a simple interpretation is that in warmer climates there is more energy available to evaporate heavier isotopes of hydrogen (^{2}H, known as deuterium) and oxygen (^{18}O) from the ocean surface into the atmosphere. These heavier isotopes are then transported in the atmosphere where they are precipitated into the ice sheet. Conversely, during cooler climates, less of the heavier isotopes are evaporated from the ocean, and they precipitate out at lower latitudes. Therefore the isotopic compositions of the ice sheets contains less of the heavier isotopes, and more ^{1}H and ^{16}O. By measuring the ratio of these water isotopes in the layered ice core record, this simple relationship can provide insights into changing global temperature through time.

Another unique aspect of ice cores is that air bubbles trapped in the ice record preserve a direct record of the chemical composition of the atmosphere, in particular that of CO_2 and methane, which is used to examine relationships between these greenhouse gases and temperature change through time. Numerous other proxies, relating mostly to atmospheric processes, can be derived from ice cores, including past windiness from dust content and trace elements.

Sediment core proxies for past climate

As with ice cores, stable isotopes form one of the key datasets used to derive climate and ice sheet histories from sediment cores. As noted above, ice sheets preferentially incorporate isotopically light oxygen (^{16}O) from the ocean during periods of glacial expansion, and subsequently the oceans become enriched in ^{18}O relative to ^{16}O (Emiliani 1966). This signal of change (termed $\delta^{18}O$) is recorded in the shells of marine carbonate microfossils, but also contains a signal of the water temperature in which the shell was precipitated. Thus, enriched values of $\delta^{18}O$ represent either a cooling in water temperatures and/or increased ice volume. Carbon isotopes are also important but are more complex to interpret. The ratio of ^{13}C to ^{12}C, expressed $\delta^{13}C$, effectively measures the residence time a water mass has been isolated from the atmosphere. Although biological processes complicate this signal, it is commonly used to provide an indication of oceanic current strength and overturning rates, whereby surface waters sink into the deep ocean in one region (carrying heat and CO_2), and deep water (which contains dissolved CO_2 that is subsequently vented into the atmosphere) is brought to the surface in another region.

Closer to the Polar Regions, direct indications of cryospheric change can be identified by shifts in terrigenous sediment supply to the ocean. The abundance of gravel in marine sediment cores adjacent to the continental margin of a polar ice sheet generally reflects the proximity of a glacier calving into the ocean. This environmental proxy, along with other physical characteristics of the sediment, has provided compelling evidence of past variations in ice sheet size and extent (Jansen et al. 2000; Patterson et al. 2014; Weber et al. 2014). Where the ice sheet has advanced across the continental shelf, sediment types recovered in drill cores provide absolute constraints of past expansion and retreat of marine-based ice sheets through time (e.g. Naish et al. 2009; The RAISED Consortium 2014). Microfossil assemblage data from marine and terrestrial sediments can give indication of past temperature and salinity changes, as well as the presence or absence of sea ice in a region (e.g. Gersonde and Zielinski 2000; Dowsett et al. 2012). Novel geochemical techniques derived from biomarkers, which are "molecular fossils" of organic remains, have provided some of most compelling evidence for past changes in atmospheric CO_2 levels, high latitude temperatures and sea ice changes (Muller et al. 2009; Pagani et al. 2009; Seki et al. 2010; Shevenell et al. 2011; McKay et al. 2012a). However, developments of these proxies are in their infancy, and calibrations to past environmental variables have large uncertainties.

Evidence for rapid millennial-scale climate change events

The first long-term isotope records derived from a deep ice core in Greenland were collected at Camp Century by the US Army Corps of Engineering as part of Project Iceworm, an experiment to test the feasibility of developing a network of ballistic nuclear missile silos within the Greenland Ice Sheet. A 1390 m long ice core was collected and investigated by Danish paleoclimatologist Dr. Willi Dansgaard at the University of Copenhagen, who used oxygen isotopes to identify events during the Holocene (past 11,700 years) that were previously documented in historical archives and tree ring records, such as the Little Ice Age (1450 to 1850 CE) and the Medieval Climate Anomaly (950 to 1250 CE). However, towards the base of the core a far larger isotopic depletion (cooling approximately 11,000 years ago) occurred, marking the transition from the modern day interglacial to past glacial conditions, but also showed rapid, cyclic (~1500 year) isotopic excursions indicating warm events during the last glacial and deglaciation (Dansgaard 1964).

Following this successful demonstration of the utility of ice cores to reconstruct past temperatures, the Greenland Ice Sheet Project (GISP) was developed, which was a decade long collaboration between Denmark, Switzerland and the United States. The 2040 m long DYE-3 ice core was drilled between 1979 and 1981 and recovered a record back to ~90,000 years ago. DYE-3 also contained the same millennial-scale (~1500 year) isotopic excursions representing cold "stadials" punctuated by warm "interstadials" roughly every 1500 years, as noted in the Camp Century record, indicating these climatic features were of regional extent rather than a local depositional signal (Dansgaard et al. 1982). These excursions, termed Dansgaard–Oeschger (DO) cycles were approximately half the amplitude of the glacial to interglacial change, and provided the first insights into the rates of rapid climate change events that have occurred in the past. Subsequent ice cores in Greenland (GISP2, NGRIP, GRIP) have improved the resolution of these earlier records, and indicate that 25 DO cycles occurred during the last glacial period, some of which were associated with rapid regional atmospheric warming of ~6°C over Greenland within a few years to decades (Steffensen et al. 2008). After the abrupt onset of each of the DO events, the warming slowed but continued for several hundred years before reversing promptly back into glacial conditions, albeit at a slower rate than the onset of these events (Figure 34.3).

Similar duration, but lower amplitude events are also observed in Antarctic ice cores, and to determine the phasing between the Greenland and Antarctic millennial scale events, methane has proved a useful tool for the dating and synchronization of these bipolar records. As methane has a short residence time in the atmosphere, variations in this greenhouse gas can be assumed to be a globally synchronous signal. Correlation of the methane records in Antarctic and Greenland ice cores demonstrated that the warming events in Antarctica could be matched one-to-one to the DO events, but they were out of phase, with Greenland leading (Jouzel et al. 2007). Termed Antarctic Isotopic Maximum (AIM), the pattern of these events differs from DO events. AIM events have a more gradual warming and cooling patterns (Figure 34.3). As a warm DO event rapidly initiated in Greenland, Antarctica gradually cooled, and conversely as the DO event ended, Antarctica began to warm again. This led to the development of the "bipolar seesaw" hypothesis, whereby it was proposed a cooling in the Northern Hemisphere leads to a gradual heat build-up in the Southern Hemisphere (Broecker 1998). The mechanism for this is proposed to lie largely in the ocean, via the Atlantic Meridional Overturning Circulation (AMOC). This oceanic circulation pattern exchanges heat between the hemispheres via a north to south exchange of water in the Atlantic Ocean. During periods of enhanced North Atlantic circulation (i.e. stronger AMOC), heat is removed from the Southern Ocean and transported to the Northern Hemisphere – and the opposite is proposed to occur during periods of a weaker AMOC. However, the large thermal inertia of the Southern Ocean, and therefore slower response time relative to the North Atlantic, helps to explain the difference in the pattern of these events in each hemisphere. Most Antarctic ice cores have much lower accumulation rates than those in Greenland, and thus determining the exact nature of this phasing has proved difficult.

The recent WAIS Divide ice core was the first Antarctic record to have a high enough temporal resolution to precisely constrain the phasing of temperature change in Antarctic versus the Greenland records (WAIS Divide Project Members 2015). This core showed that during the AIM events, the onset of gradual Antarctic cooling occurred approximately two centuries after the rapid DO warm event initiated. The same duration lag also occurs after the rapid cooling in the DO events, with onset of Antarctic warming beginning ~200 years later. This Southern Hemisphere lag points to a Northern Hemisphere driver for these events via ocean circulation (e.g. AMOC), with the delay in the Southern Hemisphere being the response time of the AMOC and a consequence of thermal inertia in the deep well-mixed South Ocean.

To understand the oceanic drivers, marine sediment cores have provided additional insights. Although of lower temporal resolution than ice cores, they can provide reconstructions of ice sheet dynamics and shifting oceanic circulation patterns through time. During the last glacial period, sediment cores from the deep North Atlantic indicate some of the DO events are associated with discrete layers of sand and gravel in otherwise muddy deposits (Heinrich 1988). These coarser grains are interpreted as being deposited by armadas of icebergs discharging from the Laurentide Ice Sheet that covered much of North America in the last ice age (Whitehouse, Chapter 16 of this volume). Termed Heinrich events, a total of ten events occurred during the last glacial period (Figure 34.3). These layers appear to be associated with the cold phases of the DO events, termed stadials, due to the coeval occurrence of cold polar marine plankton that appear to migrate southwards, pointing toward a large fresh glacial meltwater flux into the Atlantic (Bond et al. 1992). The exact cause of these events has long been debated, with one hypothesis suggesting the ice sheet underwent cycles of "binging" and "purging" as a consequence of internal ice sheet dynamics rather than a direct response to climate (MacAyeal 1993). This hypothesis suggests that when an ice sheet slowly builds up it reaches a critical thickness and starts to melt at its base. This increased basal melting results in enhanced ice flow and the ice sheet becomes highly unstable and collapses. However, recent studies have indicated these events may have a more direct climatic or ocean trigger. Geochemical analysis of plankton in sediment cores shows that the Heinrich events were preceded by a slow down of the AMOC (Zahn et al. 1997) which, although this cooled surface waters, resulted in a build up of subsurface heat in the ocean (Marcott et al. 2011). It was this heat build up that may have led to a rapid collapse of fringing ice shelves and the margin of marine-based ice sheets in the North Atlantic (Bassis et al. 2017). There is some debate about whether similar iceberg discharge events occurred in the Southern Ocean, and if they were associated with millennial-scale shifts in the dynamics of the Antarctic ice sheets (Kanfoush et al. 2000). Notwithstanding, an open debate about the exact origin and nature of DO cycles and their link to Antarctica, their far-field recognition in marine records of the Cariaco Basin, Santa Barbara Basin, the Mediterranean and cave records in China suggests abrupt synchronous and widespread reorganization of the coupled ocean-atmosphere system in the northern hemisphere.

Last glacial termination

The most recent, and intensely studied of the large climate shifts in Earth's geological history was the transition out of the last ice age 20,000 years ago, when the Northern Hemisphere ice sheets and parts of the marine-based ice sheet margins in the Antarctic melted (Whitehouse, Chapter 16 of this volume). Global average temperature increased by ~4.4°C during this event, but the warming was at least double that in Polar Regions (Waelbroeck et al. 2009; Masson-Delmotte et al. 2013; Cuffey et al. 2016). The trigger for this event was a periodic change in Earth's orbit. Although the energy changes associated with these periodic orbital shifts are small, they appear to have initiated a chain of events and powerful feedbacks that acted to greatly amplify the initial orbital change (Richard Hodgkins, Chapter 19 of this volume).

These amplifying processes are called positive feedbacks, and although the exact sequence of feedbacks that occurred during the last deglacial are not fully understood, the following simplified sequence of events is currently proposed to have occurred. A warm Southern Hemisphere summer initiated by orbital forcing resulted in reduced sea ice in the Southern Ocean, allowing for increased ventilation CO_2 from the deep ocean (WAIS Divide Project Members 2013). This amplified the response of the climate system to weak orbital changes, and initiated the melting of the Northern Hemisphere ice sheets at 20,000 years (Shakun et al. 2012). This melting initiated a series of cold freshwater perturbations to the ocean and altered wind fields around the planet. (McManus et al. 2004; Denton et al. 2010). A decrease in wind strength may have also led to decreased dust deposition over the Southern Ocean, the largest biosphere on the planet

(Anderson et al. 2009). Reduced iron delivery by dust thus starved the photosynthesizing plankton from an essential nutrient, with the consequence of less primary productivity and reduced CO_2 uptake from the atmosphere to the ocean (Jaccard et al. 2016). This CO_2 feedback further amplified global warming. (Denton et al. 2010).

Continued warming would also reduce sea ice extent, which creates a physical barrier restricting ventilation of stored carbon in the ocean, and thus an additional input of CO_2 is placed into the atmosphere (Sigman and Boyle 2000). The loss of this sea ice would also replace a white ocean with a dark ocean, reducing the reflectivity of the Earth's surface and causing the planet to absorb more incoming solar radiation.

The millennial-scale isotopic excursions observed during the glacial period also appear to have been an important feature of the termination of Last Glacial Maximum. This gives insights into the complexities of the oceanic change and the inter-hemispheric linkages between the Northern and Southern Hemisphere ice sheets. During the early deglacial, an ice-rafting Henrich event (H1) was noted, before the onset of a rapid warming of the Bølling-Allerød interval between 14,700–12,800 years ago (Figure 34.3). This warming was then rapidly terminated by the Younger Dryas cold period (12,800–11,500 years ago) which was characterized in Greenland by a ~10°C cooling within 50 years, although some of this cooling was the result of a northward shift in wind fields bringing warmer air masses to Greenland (Steffensen et al. 2008). In the Southern Hemisphere, the Antarctic Cold Reversal (14,700 to 13,000 years ago) coincides with the Bølling-Allerød warmth in the Northern Hemisphere, but as with earlier DO events, the onset and end of this event was much more gradual than in the north (Blunier et al. 1997) (Figure 34.3). This chain of events is inferred to be the consequence of the "bipolar seesaw". When Northern Hemisphere ice sheet melt accelerated during the Heinrich iceberg event (H1) it added a freshwater cap to the Atlantic Ocean, slowing the AMOC (Denton et al. 2010). This resulted in less efficient heat transport between the North Atlantic and Southern Ocean and led to a gradual Antarctic warming and reduced heat transport to the high northern latitudes (Johnson and McClure 1976; Carlson et al. 2008). Once this melt event slowed, sediment cores indicate the AMOC began to accelerate, and the onset of the Antarctic Cold Reversal began (McManus et al. 2004).

The global sea level changes during the last glacial termination are well constrained by sea level records derived from coral reefs in tropical regions. Ice sheet melt since the last ice age raised sea level by ~130 m over a period of ~13,000 years, at an average rate of ~1 m of sea level rise per century (Lambeck et al. 2014). An interval known as Meltwater Pulse 1a (MWP-1a) between 14,650 and 14,310 years ago saw rates of sea level rise exceeding 4 m per century over a period of ~340 years (Deschamps et al. 2012). While the majority of post ice age sea level rise was sourced from melting Northern Hemisphere ice sheets, there remains a debate regarding the origin of MWP-1a (Clark et al. 2002; Mackintosh et al. 2011; Weber et al. 2014). Rates of this magnitude were likely the consequence of marine ice sheet instability processes, whereby ice sheets that lie in an overdeepened marine continental shelf are inherently unstable and once retreat is triggered it becomes a runaway event (Thomas and Bentley 1978). Determining the origin of the ice that contributed to MWP-1a is relevant for the future, as there remain significant amounts of ice in Antarctica that are potentially still susceptible to this marine ice sheet instability, whereas only one third of Greenland is susceptible owing to its more terrestrial nature. Thus, if MWP-1a was derived from the Antarctic Ice Sheet, the geological record provides an insight into what the natural system may still be capable of in a rapidly warming future world.

Sediment cores on the Antarctic continental shelves suggest the majority of Antarctic ice sheet retreat occurred after the MWP-1a, pointing to a northern source for this event (The RAISED Consortium 2014). However, these records do suffer from large dating uncertainties, and there is sparse data from key regions where the majority of post glacial marine-based ice sheet retreat is thought to have occurred. Sediment cores collected from the Scotia Sea indicate that armadas of icebergs did discharge from Antarctica at a millennial timescale during the last glacial termination, with the largest of these peaks occurring during the MWP-1a event (Weber et al. 2014). While such a proxy cannot give an estimate to the amount of sea level equivalent the Antarctic Ice Sheet contributed to this event, it does indicate that Antarctica likely contributed, at least in part, to this event, with a recent ice sheet model simulation suggesting up to 1 m/century (Golledge et al. 2014).

Last interglacial climates

Although records of the last ice age and glacial termination give an indication of magnitude and rate of large, rapid climate events in Earth's recent history, they are unable to give an indication of the Polar Regions' response to significantly warmer-than-present climates. If the present-day ice sheets were susceptible to collapse in such past climates, what were the thresholds or mechanisms that led to that collapse?

A key focus of the Greenland ice core research community has been to obtain a pristine record of the last interglacial period ~120,000 years ago, a time interval known as the Eemian. The Eemian had a marginally warmer-than-present climate, when atmospheric CO_2 was similar to the preindustrial Holocene levels of ~280 ppm, and sea levels were thought to be 6–9 m higher than present (Kopp et al. 2009; Dutton and Lambeck 2012). Globally, average atmospheric warming was ~1–2°C (Masson-Delmotte et al. 2013; Otto-Bliesner et al. 2013), but ice core evidence from Greenland suggests significant polar amplification with temperatures up to 8±4°C warmer than present (NEEM Community Members, 2013), and Antarctic 3–4°C warmer (Sime et al. 2009). Such elevated sea levels require either loss of most of the Greenland Ice Sheet, or parts of the marine-based ice sheets in Antarctica – or a combination of both. Sediment cores collected from southern Greenland indicate glacial erosion was active in Greenland during the Eemian, and therefore a significantly sized ice sheet must have persisted through this time (Colville et al. 2011). This interpretation was supported by the recovery of heavily-deformed Eemian-aged ice in the NGRIP and NEEM ice cores obtained in Northern Greenland that imply elevations approximately 400±250 m lower than today (NEEM Community Members 2013). These ice core data only allows a ~2 m contribution from the Greenland Ice Sheet (relative to the present day) to global sea level rise during the Last Interglacial and implies at least a 4 m contribution from the Antarctic Ice Sheets. Direct evidence of Antarctic Ice Sheet loss at this time is lacking, but the presence of diatom-rich sediments from short sediment cores collected from beneath the parts of West Antarctic Ice Sheet (WAIS) now grounded indicates some loss of this marine-based ice sheet has occurred within the past 1 million years (Scherer et al. 1998). However, it has proven difficult to pinpoint the exact timing of this. A similar situation exists for ice core data, and as yet no ice of Eemian age has been recovered from areas of the West Antarctic that are thought to have been most susceptible to collapse. Consequently, it remains a primary focus of the ice and sediment core community to obtain a record of this critical time interval and assess if the WAIS was reduced in volume relative to today.

Extending beyond the ice core records

At present, only Antarctic ice cores extend back to 800,000 years ago, and over this time period there have been eight ice age cycles, and atmospheric CO_2 has varied between 170 and 300 ppm in an almost linear relationship with temperature (Jouzel et al. 2007; Lüthi et al.

2008). For times older than the ice core records, we rely on sedimentary drill cores to reconstruct atmospheric greenhouse gas composition, surface temperature, and ice sheet dynamics. Ice proximal sediments in ANDRILL-1B collected from under the Ross Ice Shelf (Figure 34.2), overlapped with and extended the ice core record further back. For the last 800,000 years, approximately eight cycles of glacial till and ice shelf muds were observed (McKay et al. 2012b) indicating that the WAIS oscillated between a grounded ice sheet and a floating ice shelf during the climate cycles identified in the ice core. However, due to the erosive nature of this record, it is equivocal if there was wholesale loss of the Ross Ice Shelf and retreat of the WAIS during this time. However, the presence of diatom oozes deposited ~1 million years ago beneath the present-day Ross Ice Shelf provides the most compelling evidence that the WAIS did collapse during a past warm "super interglacial" (Naish et al. 2009; McKay et al. 2012b). Such inferences are supported by model experiments that indicate that the loss of the "buttressing" Ross Ice Shelf at the ANDRILL-1B site would result in a marine ice sheet instability and a runaway retreat of WAIS (Pollard and DeConto 2009). To obtain a high-resolution record of the Antarctic climate through this event, the ice core communities are actively seeking to recover ice cores that go back to 1.5 million years (Fischer et al. 2013). This will also further our understanding of atmospheric temperatures and greenhouse gas levels prior to a well-described global cooling and intensification of glaciation ~800,000 years ago, known as the Mid-Pleistocene Transition.

Past greenhouse worlds

Beyond the timeframe covered by ice records, the geological record becomes more difficult to interpret due to the lower resolution nature of sediment core studies, and because the proxies used to reconstruct climate have larger uncertainties. To obtain past CO_2 values, indirect estimates are made from either boron isotopes in foraminifera, which quantifies the past pH of the ocean, or carbon isotopes from organic molecular remains (alkenones) of marine phytoplankton. While these methods have large errors, they have reliably been used to show that it has been 3 million years since atmospheric CO_2 last exceeded 400 ppm (Seki et al. 2010). At this time, microfossil assemblages and geochemical temperature proxies in sediment cores indicate the global climate was 2–3°C warmer than today, but also that warming was approximately double that in polar regions (Dowsett et al. 2012; Masson-Delmotte et al. 2013).

Global sea level records also have greater uncertainty at this time, as they are based on sequence stratigraphy methods or oxygen isotopes from marine fossils, which can be influenced by numerous factors other than sea level, including tectonic processes, mantle convection, glacio-isostatic adjustments and oceanic temperature changes. However, global compilations of these multiple lines of evidence point towards peak sea levels 3.5–2.5 million years ago that were 20±10 m above present day, requiring loss of Greenland ice (7 m sea level equivalent), the marine-based sectors of WAIS (3–4 m sea level equivalent), and possibly sectors of the marine-based margin EAIS (Miller et al. 2012).

The ANDRILL record in the Ross Sea indicates there were indeed multiple retreat events of the WAIS at this time, while analysis of diatom assemblages and geochemical data from that core revealed that surface waters in the Ross Sea were ~5°C warmer than present during the interglacials of peak Pliocene warmth (McKay et al. 2012a). Although the deglaciation of Greenland and WAIS on their own can account for the lower error limits of Pliocene sea levels, sediment cores collected offshore of East Antarctica revealed that the deeply incised basins currently beneath the EAIS may have also deglaciated significantly. Geochemical fingerprinting of Pliocene-aged sediments indicates that the composition of sediment in the drill cores could only have originated from rocks currently under the ice sheet hundreds of kilometres inland from the present calving line, and were eroded by an ice sheet that had retreated significantly (Cook et al. 2013). Thus, if all the marine-based sectors of EAIS melted (Fretwell et al. 2013), together with loss of WAIS and Greenland, this would be consistent with the upper limit of uncertainty (30 m) in Pliocene sea level rise estimates. However, resolving the exact extent of this retreat, and the contribution of EAIS subglacial basins requires direct sampling of geological strata from strategically selected regions beneath the marine-based sectors of the EAIS (McKay et al. 2016).

In the Northern Hemisphere, a sediment core collected in Lake El'gygytgyn in Northeast Russia provides a glimpse into the extent of polar amplification during the Pliocene warmth, with fossil pollen and geochemical data indicating summer atmospheric temperatures were 8°C warmer than present, and were likely too warm to sustain large continental ice sheets in the Northern Hemisphere (Brigham-Grette et al. 2013). Indeed, sediment cores in the North Atlantic and North Pacific indicated that the first onset of ice-rafted debris being delivered to the Northern Hemisphere oceans only occurred ~2.7 million years ago (Jansen et al. 2000; Haug et al. 2005). It is proposed that the trigger for the initiation of North Hemisphere Ice Sheet was atmospheric CO_2 dropping below a threshold of ~280 ppm, although the exact mechanism to initiate this change in the carbon cycle remains debated (DeConto et al. 2008; McKay et al. 2012a).

Summary of future challenges

Ice and sediment coring in the Polar Regions has been a key tool of paleoclimatic researchers for the past 50 years, but many uncertainties remain in our knowledge of past polar response to environmental perturbations. To constrain unequivocally the details of past ice sheet collapse events in warmer than present climates (e.g. the last interglacial, Pliocene warmth), further records are required from beneath the ice sheet in regions where it is currently grounded. However, this requires numerous logistical and technical challenges to be overcome, and a careful consideration of the environmental impact such activities would have in the pristine subglacial environment (McKay et al. 2016). Improvement of sediment-based proxies for sea ice extent and past

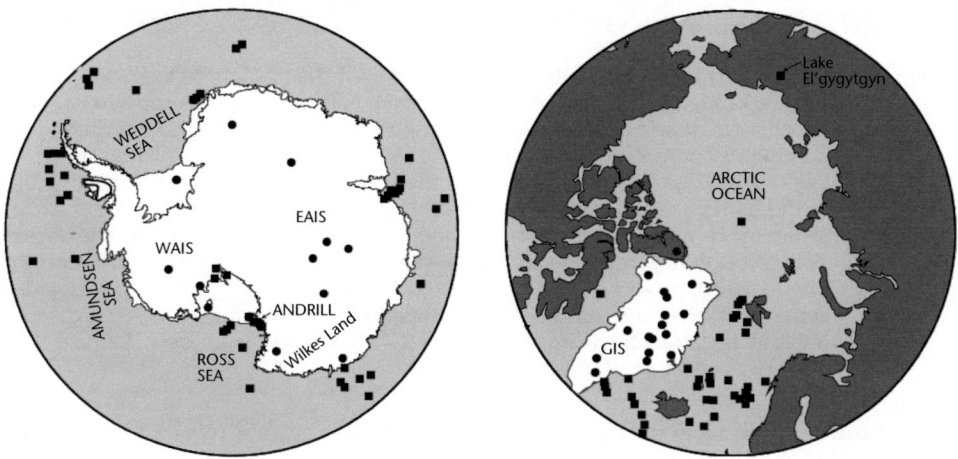

Figure 34.1 Map of the Antarctic (left) and Arctic (right), with location of deep drill sediment drill cores (black squares) and deep ice cores (black circles). Logistic and technical challenges need to be overcome in order to address the lack of sediment cores from beneath the modern-day ice sheets and the sea ice-covered regions of the central Arctic Ocean (McKay et al. 2016).

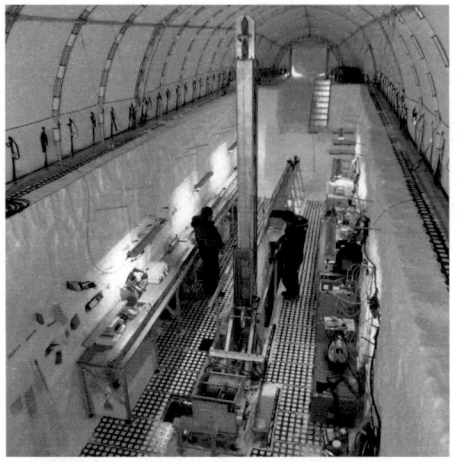 

Figure 34.2 (Upper left) Ice coring system for the Roosevelt Island Climate Evolution (RICE) project in West Antarctica. A covered trench was dug into the upper snow layers in order to maintain a cold environment, which is essential for good drilling conditions and preserving the scientific properties of the core. (Upper right) The layered RICE ice core emerging from the coring barrel, with core cutters visible at the barrel head. (Lower left) A weighted gravity core system to collect a short sediment core (< 1 m) from the seafloor beneath the Ross Ice Shelf. The access hole through the ice shelf was made with a hot water drill system. (Lower right) The ANDRILL project used a hot water drill and an industry standard rotary drill system to collect a 1285 m long drill sediment core from beneath the Ross Ice Shelf, that provided a history of the West Antarctic Ice Sheet over the past 13 million years. Photographs by author.

temperatures has been rapid over the last decade, but further work is required to continue to reduce uncertainties in these proxies. Such proxies will likely be key to determining the magnitude of the environmental forcings and processes that triggered a reduction in past ice volumes relative to the present day. For the ice core community, obtaining ice cores back to 1.5 million years will enable direct comparisons of atmospheric conditions that occurred during known periods of past West Antarctic Ice Sheet collapse, and the rates at which such change occurred.

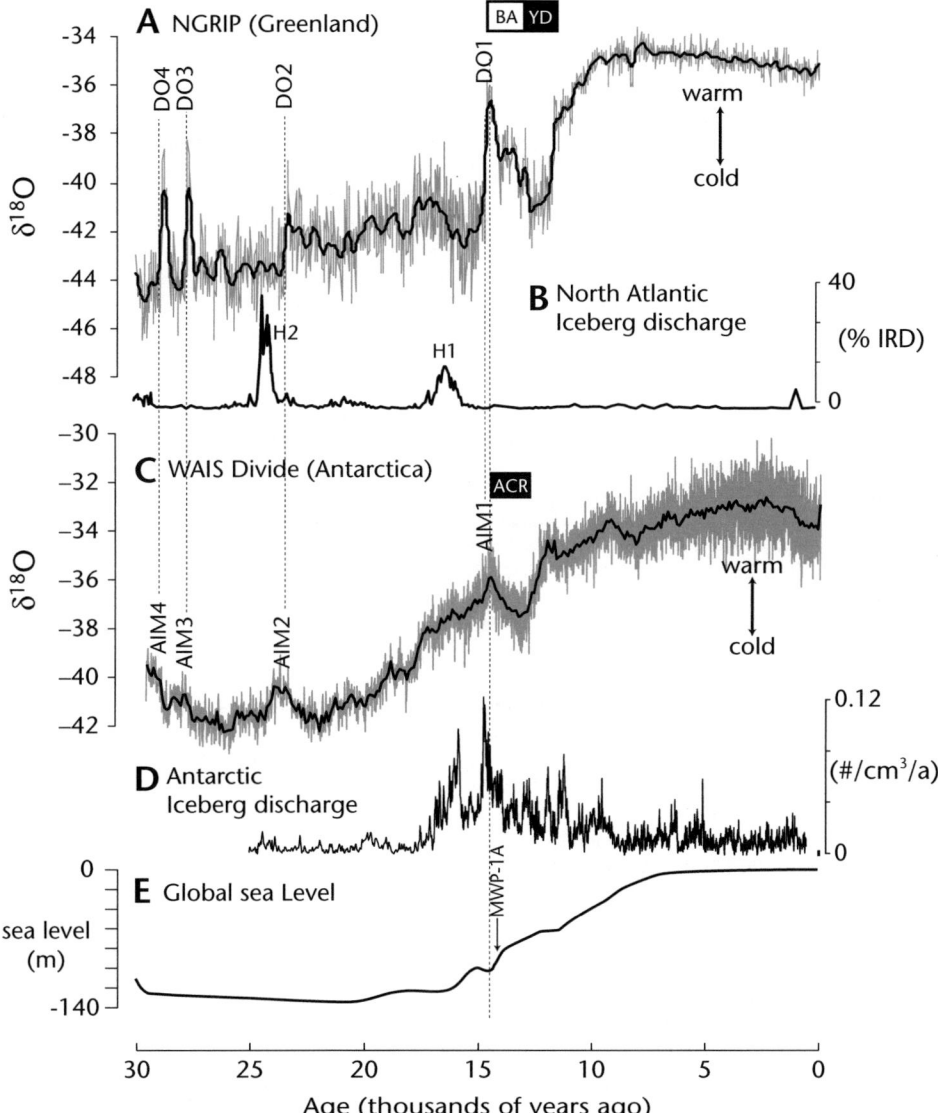

Figure 34.3 Proxies used to determine past shifts in temperature, ice sheet dynamics, and sea level. A) δ^{18}O (a temperature proxy) from the NGRIP ice core in Greenland, showing Dansgaard Oeschger (DO) events during the last glacial, and the Bølling-Allerød (BA) and Younger Dryas (YD) events during the deglaciation (North_Greenland_Ice-Core_Project_members 2004). B) Ice rafted debris (IRD) layers in the North Atlantic with Heinrich events H2 and H1 (de Abreu et al. 2003). C) δ^{18}O from the WAIS Divide ice core (West Antarctica), showing the out-of-phase nature of millennial scale climate shifts, including Antarctic Isotopic Maxima (AIM) and the Antarctic Cold Reversal (ACR) (WAIS Divide Project Members 2015). D) Ice rafted debris layers in the South Atlantic showing peaks during the deglaciation (Weber et al. 2014). E) A global composite of sea level rise records, highlighting the timing of Meltwater Pulse 1a (MWP-1a) (Lambeck et al. 2014). Figure produced by author.

References

Anderson, R.F., Ali, S., Bradtmiller, L.I., Nielsen, S.H.H., Fleisher, M.Q., Anderson, B.E. and Burckle, L.H. 2009. 'Wind-driven upwelling in the Southern Ocean and the deglacial Rise in atmospheric CO_2' *Science* 323: 1443–1448.

Bamber, J.L., Griggs, J.A., Hurkmans, R., Dowdeswell, J.A., Gogineni, P., Howat, I., Mouginot, J., Paden, J., Palmer, S., Rignot, E. and Steinhage, D. 2013. 'A new bed elevation dataset for Greenland' *The Cryosphere* 7: 499–510.

Bassis, J.N., Petersen, S.V. and Mac Cathles, L. 2017. 'Heinrich events triggered by ocean forcing and modulated by isostatic adjustment' *Nature* 542: 332–334.

Blunier, T., Schwander, J., Stauffer, B., Stocker, T., Dällenbach, A., Indermühle, A., Tschumi, J., Chappellaz, J., Raynaud, D., Barnola, J-M. 1997. 'Timing of the Antarctic cold reversal and the atmospheric CO_2 increase with respect to the Younger Dryas Event' *Geophysical Research Letters* 24: 2683–2686.

Bond, G., Heinrich, H., Broecker, W., Labeyrie, L., McManus, J., Andrews, J., Huon, S., Jantschik, R., Clasen, S., Simet, C., Tedesco, K., Klas, M., Bonani, G. and Ivy, S. 1992. 'Evidence for massive discharges of icebergs into the North Atlantic Ocean during the last glacial period' *Nature* 360: 245–249.

Brigham-Grette, J., Melles, M., Minyuk, P., Andreev, A., Tarasov, P., DeConto, R., Koenig, S., Nowaczyk, N., Wennrich, V., Rosén, P., Haltia, E., Cook, T., Gebhardt, C., Meyer-Jacob, C., Snyder, J. and Herzschuh, U. 2013. 'Pliocene warmth, polar amplification, and stepped Pleistocene cooling recorded in NE Arctic Russia' *Science* 340: 1421–1427.

Broecker, W.S. 1998. 'Paleocean circulation during the Last Deglaciation: a bipolar seesaw?' *Paleoceanography* 13, 119–121.

Carlson, A.E., Oppo, D.W., Came, R.E., LeGrande, A.N., Keigwin, L.D., Curry, W.B. 2008. 'Subtropical Atlantic salinity variability and Atlantic meridional circulation during the last deglaciation' *Geology* 36: 991–994.

Church, J.A., Clark, P.U., Cazenave, A., Gregory, J.M., Jevrejeva, S., Levermann, A., Merrifield, M.A., Milne, G. A., Nerem, R.S., Nunn, P.D., Payne, A.J., Pfeffer, W.T., Stammer, D., Unnikrishnan, A.S. 2013. 'Sea level change' in Stocker, T.F., Qin, D., Plattner, G.-K., Tignor, M., Allen, S.K., Boschung, J., Nauels, A., Xia, Y., Bex, V. and Midgley, P.M. (eds) *Climate Change 2013: The Physical Science Basis*. Contribution of Working Group I to the Fifth Assessment Report of the Intergovernmental Panel on Climate Change. Cambridge, UK and New York: Cambridge University Press. 1137–1216.

Clark, P.U., Mitrovica, J.X., Milne, G.A. and Tamisiea, M.E. 2002. 'Sea-level fingerprinting as a direct test for the source of global meltwater pulse IA' *Science* 295: 2438–2441.

Colville, E.J., Carlson, A.E., Beard, B.L., Hatfield, R.G., Stoner, J.S., Reyes, A.V. and Ullman, D.J. 2011. 'Sr-Nd-Pb isotope evidence for ice-sheet presence on southern Greenland during the last interglacial' *Science* 333: 620–623.

Cook, C.P., van de Flierdt, T., Williams, T., Hemming, S.R., Iwai, M., Kobayashi, M., Jimenez-Espejo, F.J., Escutia, C., González, J.J., Khim, B.-K., McKay, R.M., Passchier, S., Bohaty, S.M., Riesselman, C.R., Tauxe, L., Sugisaki, S., Galindo, A.L., Patterson, M.O., Sangiorgi, F., Pierce, E.L., Brinkhuis, H., Klaus, A., Fehr, A., Bendle, J.A.P., Bijl, P.K., Carr, S.A., Dunbar, R.B., Flores, J.A., Hayden, T.G., Katsuki, K., Kong, G.S., Nakai, M., Olney, M.P., Pekar, S.F., Pross, J., Röhl, U., Sakai, T., Shrivastava, P.K., Stickley, C.E., Tuo, S., Welsh, K. and Yamane, M. 2013. 'Dynamic behaviour of the East Antarctic ice sheet during Pliocene warmth' *Nature Geoscience*, advance online publication. doi:10.1038/ngeo1889.

Cuffey, K.M., Clow, G.D., Steig, E.J., Buizert, C., Fudge, T.J., Koutnik, M., Waddington, E.D., Alley, R.B. and Severinghaus, J.P. 2016. 'Deglacial temperature history of West Antarctica' *Proceedings of the National Academy of Sciences of the United States of America* 201609132. doi:10.1073/pnas.1609132113.

Cuffey, K.M. and Paterson, W.S.B. 2010. *The Physics of Glaciers*. Cambridge, MA: Academic Press.

Dansgaard, W. 1964. 'Stable isotopes in precipitation' *Tellus* 16: 436–468.

Dansgaard, W., Clausen, H.B., Gundestrup, N., Hammer, C.U., Johnsen, S.F., Kristinsdottir, P.M. and Reeh, N. 1982. 'A new Greenland deep ice core' *Science* 218: 1273–1277.

de Abreu, L., Shackleton, N.J., Schönfeld, J., Hall, M. and Chapman, M. 2003. 'Millennial-scale oceanic climate variability off the Western Iberian margin during the last two glacial periods' *Marine Geology* 196: 1–20.

DeConto, R.M., Pollard, D., Wilson, P.A., Palike, H., Lear, C.H. and Pagani, M. 2008. 'Thresholds for Cenozoic bipolar glaciation' *Nature* 455: 652–656.

Denton, G.H., Anderson, R.F., Toggweiler, J.R., Edwards, R.L., Schaefer, J.M. and Putnam, A.E. 2010. 'The last glacial termination' *Science* 328: 1652–1656.

Deschamps, P., Durand, N., Bard, E., Hamelin, B., Camoin, G., Thomas, A.L., Henderson, G.M., Okuno, J. and Yokoyama, Y. 2012. 'Ice-sheet collapse and sea-level rise at the Bolling warming 14,600 years ago' *Nature* 483: 559–564.

Dowsett, H.J., Robinson, M.M., Haywood, A.M., Hill, D.J., Dolan, A.M., Stoll, D.K., Chan, W-L., Abe-Ouchi, A., Chandler, M.A., Rosenbloom, N.A., Otto-Bliesner, B.L., Bragg, F.J., Lunt, D.J., Foley, K.M. and Riesselman, C.R. 2012. 'Assessing confidence in Pliocene sea surface temperatures to evaluate predictive models' *Nature Climate Change* 2: 365–371.

Dutton, A. and Lambeck, K. 2012. 'Ice volume and sea level during the last interglacial' *Science* 337: 216–219.

Emiliani, C. 1966. 'Paleotemperature analysis of Caribbean cores P6304-8 and P6304-9 and a generalized temperature curve for the past 425,000 years' *The Journal of Geology* 74: 109–124.

Fischer, H., Severinghaus, J., Brook, E., Wolff, E., Albert, M., Alemany, O., Arthern, R., Bentley, C., Blankenship, D., Chappellaz, J., Creyts, T., Dahl-Jensen, D., Dinn, M., Frezzotti, M., Fujita, S., Gallee, H., Hindmarsh, R., Hudspeth, D., Jugie, G., Kawamura, K., Lipenkov, V., Miller, H., Mulvaney, R., Parrenin, F., Pattyn, F., Ritz, C., Schwander, J., Steinhage, D., van Ommen, T. and Wilhelms, F. 2013. 'Where to find 1.5 million yr old ice for the IPICS "Oldest-Ice" ice core' *Climate of the Past* 9: 2489–2505.

Fretwell, P., Pritchard, H.D., Vaughan, D.G., Bamber, J.L., Barrand, N.E., Bell, R., Bianchi, C., Bingham, R.G., Blankenship, D.D., Casassa, G., Catania, G., Callens, D., Conway, H., Cook, A.J., Corr, H.F.J., Damaske, D., Damm, V., Ferraccioli, F., Forsberg, R., Fujita, S., Gim, Y., Gogineni, P., Griggs, J.A., Hindmarsh, R.C.A., Holmlund, P., Holt, J.W., Jacobel, R.W., Jenkins, A., Jokat, W., Jordan, T., King, E.C., Kohler, J., Krabill, W., Riger-Kusk, M., Langley, K.A., Leitchenkov, G., Leuschen, C., Luyendyk, B.P., Matsuoka, K., Mouginot, J., Nitsche, F.O., Nogi, Y., Nost, O.A., Popov, S.V., Rignot, E., Rippin, D.M., Rivera, A., Roberts, J., Ross, N., Siegert, M.J., Smith, A.M., Steinhage, D., Studinger, M., Sun, B., Tinto, B.K., Welch, B.C., Wilson, D., Young, D.A., Xiangbin, C. and Zirizzotti, A. 2013. 'Bedmap2: improved ice bed, surface and thickness datasets for Antarctica' *The Cryosphere* 7: 375–393.

Gersonde, R. and Zielinski, U. 2000. 'The reconstruction of late Quaternary Antarctic sea-ice distribution: the use of diatoms as a proxy for sea-ice' *Palaeogeography, Palaeoclimatology, Palaeoecology* 162: 263–286.

Golledge, N.R., Menviel, L., Carter, L., Fogwill, C.J., England, M.H., Cortese, G. and Levy, R.H. 2014. 'Antarctic contribution to meltwater pulse 1A from reduced Southern Ocean overturning' *Nature Communications* 5. doi:10.1038/ncomms6107

Haug, G.H., Ganopolski, A., Sigman, D.M., Rosell-Mele, A., Swann, G.E.A., Tiedemann, R., Jaccard, S.L., Bollmann, J., Maslin, M.A., Leng, M.J. and Eglinton, G. 2005. 'North Pacific seasonality and the glaciation of North America 2.7 million years ago' *Nature* 433: 821–825.

Hays, J.D., Imbrie, J. and Shackleton, N.J. 1976. 'Variations in the Earth's orbit: pacemaker of the Ice Ages' *Science* 194: 1121–1132.

Heinrich, H. 1988. 'Origin and consequences of cyclic ice rafting in the Northeast Atlantic Ocean during the past 130,000 years' *Quaternary Research* 29: 142–152.

Jaccard, S.L., Galbraith, E.D., Martínez-García, A. and Anderson, R.F. 2016. 'Covariation of deep Southern Ocean oxygenation and atmospheric CO2 through the last ice age' *Nature* 530: 207–210.

Jansen, E., Fronval, T., Rack, F. and Channell, J.E.T. 2000. 'Pliocene-Pleistocene ice rafting history and cyclicity in the Nordic Seas during the last 3.5 Myr' *Paleoceanography* 15: 709–721.

Johnson, R.G. and McClure, B.T. 1976. 'A model for Northern Hemisphere continental ice sheet variation' *Quaternary Research* 6: 325–353.

Jouzel, J., Masson-Delmotte, V., Cattani, O., Dreyfus, G., Falourd, S., Hoffmann, G., Minster, B., Nouet, J., Barnola, J.M., Chappellaz, J., Fischer, H., Gallet, J.C., Johnsen, S., Leuenberger, M., Loulergue, L., Luethi, D., Oerter, H., Parrenin, F., Raisbeck, G., Raynaud, D., Schilt, A., Schwander, J., Selmo, E., Souchez, R., Spahni, R., Stauffer, B., Steffensen, J.P., Stenni, B., Stocker, T.F., Tison, J.L., Werner, M. and Wolff, E.W. 2007. 'Orbital and millennial Antarctic climate variability over the past 800,000 years' *Science* 317: 793–796.

Kanfoush, S.L., Hodell, D.A., Charles, C.D., Guilderson, T.P., Mortyn, P.G. and Ninnemann, U.S. 2000. 'Millennial-scale instability of the Antarctic Ice Sheet during the last glaciation' *Science* 288: 1815–1819.

Kopp, R.E., Simons, F.J., Mitrovica, J.X., Maloof, A.C. and Oppenheimer, M. 2009. 'Probabilistic assessment of sea level during the last interglacial stage' *Nature* 462: 863–867.

Lambeck, K., Rouby, H., Purcell, A., Sun, Y. and Sambridge, M. 2014. 'Sea level and global ice volumes from the Last Glacial Maximum to the Holocene' *Proceedings of the National Academy of Sciences of the United States of America* 111: 15296–15303.

Lüthi, D., Le Floch, M., Bereiter, B., Blunier, T., Barnola, J.-M., Siegenthaler, U., Raynaud, D., Jouzel, J., Fischer, H., Kawamura, K. and Stocker, T.F. 2008. 'High-resolution carbon dioxide concentration record 650,000–800,000 years before present' *Nature* 453: 379–382.

MacAyeal, D.R. 1993. 'Binge/purge oscillations of the Laurentide Ice Sheet as a cause of the North Atlantic's Heinrich events' *Paleoceanography* 8: 775–784.

Mackintosh, A., Golledge, N., Domack, E., Dunbar, R., Leventer, A., White, D., Pollard, D., DeConto, R., Fink, D., Zwartz, D., Gore, D. and Lavoie, C. 2011. 'Retreat of the East Antarctic ice sheet during the last glacial termination' *Nature Geoscience* 4: 195–202.

Marcott, S.A., Clark, P.U., Padman, L., Klinkhammer, G.P., Springer, S.R., Liu, Z., Otto-Bliesner, B.L., Carlson, A.E., Ungerer, A., Padman, J., He, F., Cheng, J. and Schmittner, A. 2011. 'Ice-shelf collapse from subsurface warming as a trigger for Heinrich events' *Proceedings of the National Academy of Sciences of the United States of America* 108: 13415–13419.

Masson-Delmotte, V., Schulz, M., Abe-Ouchi, A., Beer, J., Ganopolski, A., González Rouco, J., Jansen, E., Lambeck, K., Luterbacher, J. and Naish, T. 2013. 'Information from paleoclimate archives' in Stocker, T.F., Qin, D., Plattner, G-K., Tignor, M., Allen, S.K., Boschung, J., Nauels, A., Xia, Y., Bex, V. and Midgley, P.M. (eds) *Climate Change 2013: The Physical Science Basis*. Contribution of Working Group I to the Fifth Assessment Report of the Intergovernmental Panel on Climate Change. Cambridge, UK and New York: Cambridge University Press. 383–464.

McKay, R., Naish, T., Carter, L., Riesselman, C., Dunbar, R., Sjunneskog, C., Winter, D., Sangiorgi, F., Warren, C., Pagani, M., Schouten, S., Willmott, V., Levy, R., DeConto, R. and Powell, R.D. 2012a. 'Antarctic and Southern Ocean influences on Late Pliocene global cooling' *Proceedings of the National Academy of Sciences of the United States of America* 109: 6423–6428.

McKay, R., Naish, T., Powell, R., Barrett, P., Scherer, R., Talarico, F., Kyle, P., Monien, D., Kuhn, G., Jackolski, C. and Williams, T. 2012b. 'Pleistocene variability of Antarctic Ice Sheet extent in the Ross Embayment' *Quaternary Science Reviews* 34: 93–112.

McKay, R.M., Barrett, P.J., Levy, R.S., Naish, T.R., Golledge, N.R. and Pyne, A. 2016. 'Antarctic Cenozoic climate history from sedimentary records: ANDRILL and beyond' *Philosophical Transactions of the Royal Society A* 374, 20140301. doi:10.1098/rsta.2014.0301.

McManus, J.F., Francois, R., Gherardi, J-M., Keigwin, L.D. and Brown-Leger, S. 2004. 'Collapse and rapid resumption of Atlantic meridional circulation linked to deglacial climate changes' *Nature* 428: 834–837.

Miller, K.G., Wright, J.D., Browning, J.V., Kulpecz, A., Kominz, M., Naish, T.R., Cramer, B.S., Rosenthal, Y., Peltier, W.R. and Sosdian, S. 2012. 'High tide of the warm Pliocene: implications of global sea level for Antarctic deglaciation' *Geology*. doi:10.1130/G32869.1.

Muller, J., Masse, G., Stein, R. and Belt, S.T. 2009. 'Variability of sea-ice conditions in the Fram Strait over the past 30,000 years' *Nature Geoscience* 2: 772–776.

Naish, T., Powell, R., Levy, R., Wilson, G., Scherer, R., Talarico, F., Krissek, L., Niessen, F., Pompilio, M., Wilson, T., Carter, L., DeConto, R., Huybers, P., McKay, R., Pollard, D., Ross, J., Winter, D., Barrett, P., Browne, G., Cody, R., Cowan, E., Crampton, J., Dunbar, G., Dunbar, N., Florindo, F., Gebhardt, C., Graham, I., Hannah, M., Hansaraj, D., Harwood, D., Helling, D., Henrys, S., Hinnov, L., Kuhn, G., Kyle, P., Laufer, A., Maffioli, P., Magens, D., Mandernack, K., McIntosh, W., Millan, C., Morin, R., Ohneiser, C., Paulsen, T., Persico, D., Raine, I., Reed, J., Riesselman, C., Sagnotti, L., Schmitt, D., Sjunneskog, C., Strong, P., Taviani, M., Vogel, S., Wilch, T. and Williams, T. 2009. 'Obliquity-paced Pliocene West Antarctic ice sheet oscillations' *Nature* 458: 322–328.

NEEM Community Members, N. community, 2013. 'Eemian interglacial reconstructed from a Greenland folded ice core' *Nature* 493: 489–494.

North Greenland Ice-Core Project members, Andersen, K.K., Azuma, N., Barnola, J.-M., Bigler, M., Biscaye, P., Caillon, N., Chappellaz, J., Clausen, H.B., Dahl-Jensen, D., Fischer, H., Flückiger, J., Fritzsche, D., Fujii, Y., Goto-Azuma, K., Grønvold, K., Gundestrup, N.S., Hansson, M., Huber, C., Hvidberg, C.S., Johnsen, S.J., Jonsell, U., Jouzel, J., Kipfstuhl, S., Landais, A., Leuenberger, M., Lorrain, R., Masson-Delmotte, V., Miller, H., Motoyama, H., Narita, H., Popp, T., Rasmussen, S.O., Raynaud, D., Röthlisberger, R., Ruth, U., Samyn, D., Schwander, J., Shoji, H., Siggard-Andersen, M.-L., Steffensen, J.P., Stocker, T., Sveinbjörnsdottir, A.E., Svensson, A., Takata, M., Tison, J.-L.,

Thorsteinsson, T., Watanabe, O., Wilhelms, F. and White, J. 2004. 'High-resolution record of the Northern Hemisphere climate extending into the last interglacial period' *Nature* 431: 147–151.

Otto-Bliesner, B.L., Rosenbloom, N., Stone, E.J., McKay, N.P., Lunt, D.J., Brady, E.C. and Overpeck, J.T. 2013. 'How warm was the last interglacial? New model–data comparisons' *Philosophical Transactions of the Royal Society A* 371, 20130097. doi:10.1098/rsta.2013.0097.

Pagani, M., Caldeira, K., Berner, R. and Beerling, D.J. 2009. 'The role of terrestrial plants in limiting atmospheric CO2 decline over the past 24 million years' *Nature* 460: 85–88.

Patterson, M.O., McKay, R., Naish, T., Escutia, C., Jimenez-Espejo, F.J., Raymo, M.E., Meyers, S.R., Tauxe, L., Brinkhuis, H. and IODP Expedition 318 Scientists, 2014. 'Orbital forcing of the East Antarctic ice sheet during the Pliocene and Early Pleistocene' *Nature Geoscience* 7: 841–847.

Pollard, D. and DeConto, R.M. 2009. 'Modelling West Antarctic ice sheet growth and collapse through the past five million years' *Nature* 458: 329–332.

Rignot, E., Mouginot, J., Morlighem, M., Seroussi, H. and Scheuchl, B. 2014. 'Widespread, rapid grounding line retreat of Pine Island, Thwaites, Smith, and Kohler glaciers, West Antarctica, from 1992 to 2011' *Geophysical Research Letters* 41: 3502–3509.

Scherer, R.P., Aldahan, A., Tulaczyk, S., Possnert, G., Engelhardt, H. and Kamb, B. 1998. 'Pleistocene collapse of the West Antarctic Ice Sheet' *Science* 281: 82–85.

Seki, O., Foster, G.L., Schmidt, D.N., Mackensen, A., Kawamura, K. and Pancost, R.D. 2010. 'Alkenone and boron-based Pliocene pCO2 records' *Earth and Planetary Science Letters* 292: 201–211.

Shakun, J.D., Clark, P.U., He, F., Marcott, S.A., Mix, A.C., Liu, Z., Otto-Bliesner, B., Schmittner, A. and Bard, E. 2012. 'Global warming preceded by increasing carbon dioxide concentrations during the last deglaciation' *Nature* 484: 49–54.

Shevenell, A.E., Ingalls, A.E., Domack, E.W. and Kelly, C. 2011. 'Holocene Southern Ocean surface temperature variability west of the Antarctic Peninsula' *Nature* 470: 250–254.

Sigman, D.M. and Boyle, E.A. 2000. 'Glacial/interglacial variations in atmospheric carbon dioxide' *Nature* 407: 859–869.

Sime, L.C., Wolff, E.W., Oliver, K.I.C. and Tindall, J.C. 2009. 'Evidence for warmer interglacials in East Antarctic ice cores' *Nature* 462: 342–345.

Steffensen, J.P., Andersen, K.K., Bigler, M., Clausen, H.B., Dahl-Jensen, D., Fischer, H., Goto-Azuma, K., Hansson, M., Johnsen, S.J., Jouzel, J., Masson-Delmotte, V., Popp, T., Rasmussen, S.O., Röthlisberger, R., Ruth, U., Stauffer, B., Siggaard-Andersen, M.-L., Sveinbjörnsdóttir, Á.E., Svensson, A. and White, J.W.C. 2008. 'High-resolution Greenland ice core data show abrupt climate change happens in few years' *Science* 321: 680–684.

The RAISED Consortium, 2014. 'Reconstruction of Antarctic Ice Sheet Deglaciation (RAISED): a community-based geological reconstruction of Antarctic ice sheet deglaciation since the Last Glacial Maximum' *Quaternary Science Reviews* 100: 1–9.

Thomas, R.H. and Bentley, C.R. 1978. 'A model for Holocene retreat of the West Antarctic Ice Sheet' *Quaternary Research* 10: 150–170.

Turner, J., Hosking, J.S., Marshall, G.J., Phillips, T. and Bracegirdle, T.J. 2016. 'Antarctic sea ice increase consistent with intrinsic variability of the Amundsen Sea Low' *Climate Dynamics* 46: 2391–2402.

Vaughan, D.G., Comiso, J. C., Allison, I., Carrasco, J., Kaser, G., Kwok, R., Mote, P., Murray, T., Paul, F., Ren, J., Rignot, E., Solomina, O., Steffen, K. and Zhang, T. 2013. 'Observations: cryosphere', in Stocker, T.F., Qin, D., Plattner, G.-K., Tignor, M., Allen, S.K., Boschung, J., Nauels, A., Xia, Y., Bex, V. and Midgley, P.M. (eds) *Climate Change 2013: The Physical Science Basis*. Contribution of Working Group I to the Fifth Assessment Report of the Intergovernmental Panel on Climate Change. Cambridge, UK and New York: Cambridge University Press. 317–382.

Waelbroeck, C., Paul, A., Kucera, M., Rosell-Melé, A., Weinelt, M., Schneider, R., Mix, A.C., Abelmann, A., Armand, L., Bard, E., Barker, S., Barrows, T.T., Benway, H., Cacho, I., Chen, M.-T., Cortijo, E., Crosta, X., Vernal, A. de, Dokken, T., Duprat, J., Elderfield, H., Eynaud, F., Gersonde, R., Hayes, A., Henry, M., Hillaire-Marcel, C., Huang, C.-C., Jansen, E., Juggins, S., Kallel, N., Kiefer, T., Kienast, M., Labeyrie, L., Leclaire, H., Londeix, L., Mangin, S., Matthiessen, J., Marret, F., Meland, M., Morey, A.E., Mulitza, S., Pflaumann, U., Pisias, N.G., Radi, T., Rochon, A., Rohling, E.J., Sbaffi, L., Schäfer-Neth, C., Solignac, S., Spero, H., Tachikawa, K. and Turon, J.-L. 2009. 'Constraints on the magnitude and patterns of ocean cooling at the Last Glacial Maximum' *Nature Geoscience* 2: 127–132.

WAIS Divide Project Members, 2013. 'Onset of deglacial warming in West Antarctica driven by local orbital forcing' *Nature* 500: 440–444.

WAIS Divide Project Members, 2015. 'Precise interpolar phasing of abrupt climate change during the last ice age' *Nature* 520: 661–665.

Weber, M.E., Clark, P.U., Kuhn, G., Timmermann, A., Sprenk, D., Gladstone, R., Zhang, X., Lohmann, G., Menviel, L., Chikamoto, M.O., Friedrich, T. and Ohlwein, C. 2014. 'Millennial-scale variability in Antarctic ice-sheet discharge during the last deglaciation' *Nature* 510: 134–138.

Zachos, J., Pagani, M., Sloan, L., Thomas, E. and Billups, K. 2001. 'Trends, rhythms, and aberrations in global climate 65 Ma to Present' *Science* 292: 686–693.

Zahn, R., Schönfeld, J., Kudrass, H-R., Park, M-H., Erlenkeuser, H. and Grootes, P. 1997. 'Thermohaline instability in the North Atlantic during meltwater events: stable isotope and ice-rafted detritus records from Core SO75-26KL, Portuguese Margin' *Paleoceanography* 12: 696–710.

35

Subglacial access and investigation

Keith Makinson

Introduction

Isolated from the outside world by hundreds or even thousands of metres of ice, subglacial environments remain some of the most inaccessible, unknown environments on Earth. Remote sensing provides only basic and limited geophysical information about the vast ice sheets (Siegert, Chapter 33 of this volume), the sediments and bedrock upon which they rest (Jordan, Chapter 11 of this volume), or the ocean upon which they float (Bingham, Chapter 12 of this volume). At the ice sheet base, geothermal heat warms the ice, causing melting over large areas where the thick ice sheet provides sufficient insulation from the severely cold surface conditions. In the presence of liquid water these extreme environments, potentially isolated from the outside world for millions of years, may have evolved new microbial ecosystems with unique characteristics. In deep ice-covered valleys, pooling water can form subglacial lakes that may hold sedimentary records of climate history and ice sheet extent over many hundreds of thousands of years. Beneath the ice shelves, the floating extensions of the ice sheet, melting and freezing exchanges vast quantities of heat and fresh water, producing globally important ocean water masses. Furthermore, ice shelves exert an important restraint on the discharge of grounded ice into the ocean through a buttressing effect, and therefore on changes in global sea level. Only by gaining direct access to measure and sample these subglacial environments can hypotheses be tested and our understanding of subglacial systems and processes be advanced. This chapter provides a historical overview of access drilling technologies, some of the key science discoveries, the present status of subglacial access work on ice shelves, ice sheets, and clean access technologies, as well as future aspirations for subglacial access.

History of subglacial access drilling

The scientific desire to access and investigate subglacial environments directly has given rise to numerous ice drilling techniques and technological advances. The first documented ice drilling system was used in 1841 by Louis Agassiz who attempted, using a simple string of iron rods, to drill to the bed of Unteraargletscher, in the Swiss Alps, to measure the ice thickness. Despite drilling to 60 m, his attempts to reach the glacier bed were unsuccessful. Not until 1895 did Adolf Blümcke and Hans Hess, using a hand powered rotary drill on Hintereisferner, become the first to drill through a glacier successfully, with holes eventually reaching depths of 224 m. In 1948 the first electro-thermal drill reached a glacier bed at a depth of 137 m near the Jungfrujoch Research Station, Switzerland. This drill used an electrically heated drill head, or hot point, to penetrate snow and ice by melting. An alternative hot point drill pumped hot water through a closed loop insulated hose to heat the drill head. The return water was then reheated at the surface (Clarke 1987). With the advent of seismic techniques in the late 1920s and pulsed radars in the 1960s, which were able to provide ice thickness measurements (Siegert, Chapter 33 of this volume), the primary use of holes became the installation of instrument cables and sensors within or at the base of glaciers.

During the late 1940s and 1950s it became clear that ice sheets preserved annual climate information over many thousands of years. This discovery led to the development of drills that recovered ice cores, with early systems reaching depths of 100–400 m. During the International Geophysical Year in 1958, the first drilling to reach the base of either of the two great polar ice sheets took place at Little America V, near the front of Ross Ice Shelf, Antarctica, where ice core drilling stopped at approximately 255 m, after sea water rose 188 m into the hole (Bentley and Koci 2007). As drilling depths increased beyond 300 m, increasing overburden pressure resulted in significant borehole closure. To maintain borehole stability a drilling fluid, normally based around a mixture of kerosene and other hydrocarbons, was used to balance the pressure. The first deep coring to penetrate the full thickness of a grounded ice sheet took place in the summer of 1966. After several seasons of drilling at Camp Century in northwest Greenland, an electro-mechanical ice core drill, using a fluid filled hole, reached 1391 m and recovered frozen bed material. The same ice core drill reached bed rock in February 1968 at Byrd Station in central West Antarctica. At a depth of 2164 m, the final section of ice core contained rock debris, while further attempts to recover a sub-ice core failed. Subsequent multiyear ice coring programmes have reached depths of over 3000 m and contain 800,000 year climate records at some locations such as Dome C (in central East Antarctica; see McKay, Chapter 34 of this volume). At the Russian Vostok Station in central East Antarctica the deepest ice coring reached 3623 m in 1998 after 8 years of drilling. The site was located above the largest known subglacial lake (Kapitsa et al. 1996), and drilling was suspended to determine how and if to drill into Vostok Subglacial Lake.

As interest in subglacial access and sampling grew, a new drilling method capable of making deep access holes in hours or days rather than in months or years was required. Rapid subglacial access eventually became a reality with the advent of hot water drilling. This technique, initially developed for use on Alpine glaciers during the early 1970s, used hot water to deliver huge amounts of energy, over long distances, to melt the ice at the base of the hole. The essential components of a hot water drill (HWD) are: water derived from local snow and ice, water pump, water heater, and a hose with a weighted drill nozzle that delivers a high speed jet of hot water to melt ice ahead of the drill (Figure 35.1). Lowering the drill at a predefined, slow enough rate ensures the drill hangs freely within the hole forming a straight and vertical hole, with gravity providing the steering mechanism. The cold admixture of drill and melt water then flows slowly up the melted hole toward the surface whilst also supporting the hole's walls.

In 1972, a simple, light weight HWD was successfully used on Glacier de St-Sorlin to reach depths of 120 m quickly and easily even through ice containing sand or silt (Gillet 1975). Other drills soon achieved drilling speeds of 100 m hr^{-1} and depths of 400 m, using summer surface melt water to supply the drill. When this supply was unavailable and water recovery from the borehole was unreliable, continuous snow melting was needed, necessitating a doubling of both system heating capacity and fuel (Taylor 1984).

During the Ross Ice Shelf Project, the Browning flame-jet drill was used in December 1977 at Site J9, 430 km from the open sea, to make the first non-coring penetration of an Antarctic ice shelf. The drill, similar to a rocket engine, burned fuel oil with compressed air at the base

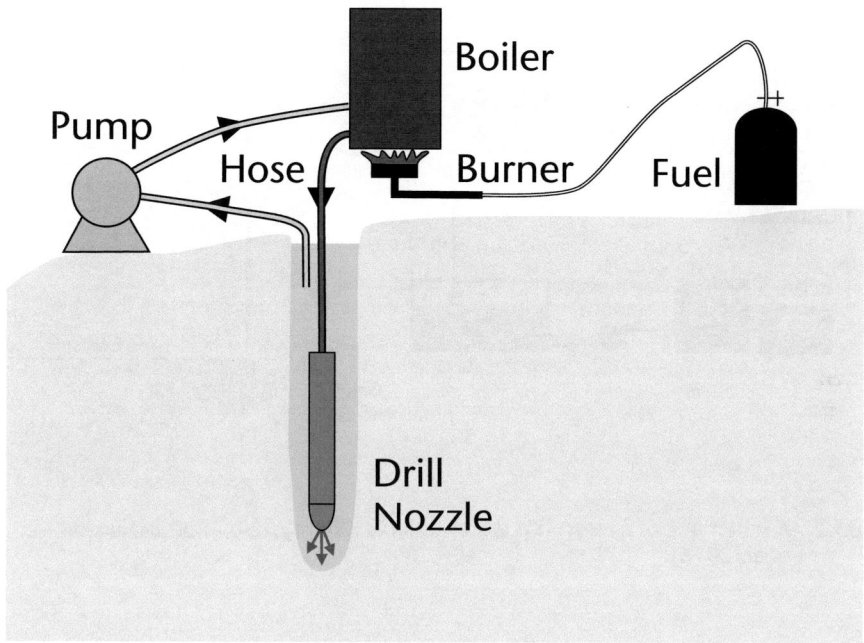

Figure 35.1 A schematic of the first documented hot water drill system (after Gillet 1975).

of the melt–water filled hole, making an access hole within 9 hours, through the 420 m thick ice shelf. A second 30 cm diameter hole was drilled and maintained against refreezing by occasional flame reaming to re-widen the access hole over a three week period (Browning and Somerville 1978). The hole, however, was heavily contaminated by unburned fuel, soot and other combustion products from the flame drill and the diameter was highly irregular. In the following 1977–78 season, three much cleaner 90 cm diameter access holes were made using the Browning 2 MW hot water drill, each in around 15 hours. To form a regular shaped hole in the porous firn region in the upper part of the ice shelf, the first 30 m of the hole was drilled using a 38 cm diameter shower head nozzle, after which a single jet nozzle was used. To ensure a continuous water supply to the boiler, drill water was continuously recirculated by pumping it from a depth of 50 m to the surface via a secondary hole interconnected by a subsurface cavity or 'Rodrigues Well' (Browning et al. 1979). This is a method used in almost all hot water drilling (Figure 35.2).

Observations at J9 provided the first confirmation of life deep beneath the large Antarctic ice shelves. Baited traps collected many amphipods and an isopod, with cameras capturing images of fish, shrimp and crustaceans (Lipps et al. 1979). The acquisition of sediment cores, water samples and oceanographic observations of the coldest and densest water masses in the world marked the beginning of direct sub-ice shelf observations.

Following the J9 successes, several research groups developed and successfully used hot water drills to provide access through many Antarctic ice shelves (Bentley and Koci 2007), which range in thickness from as little as a few tens of metres to over 2000 m at some deep grounding lines, occupying about 40% of the continental shelf and 44% of the coastline (Figure 35.3). These sub-ice shelf observations are key to understanding circulation patterns, seasonal and interannual variability, and melting and freezing processes. With over 50% of Antarctic ice

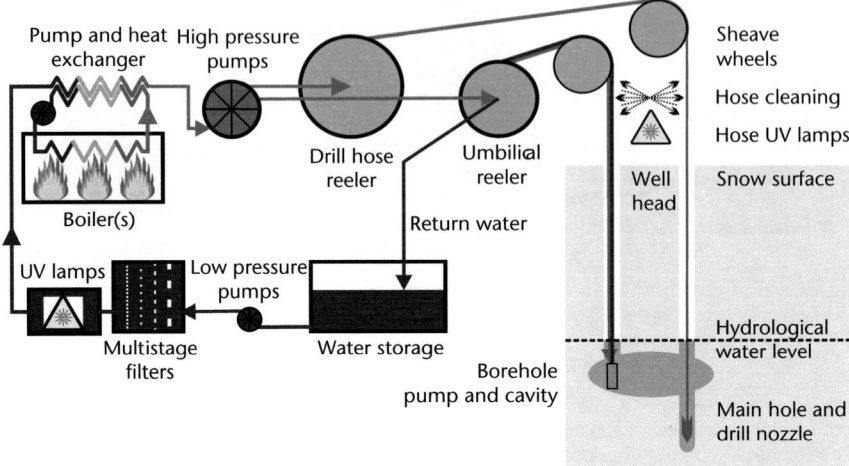

Figure 35.2   A schematic of a clean hot water drill (CHWD) system (after Makinson et al. 2016).

mass loss occurring as basal melting (McMillan, Chapter 14 of this volume), it is essential to understand the pivotal role ice-ocean interactions play in ice shelf and ice sheet evolution, and the production of globally significant water masses.

The advancement of hot water drilling beyond 1000 m was pioneered in the late 1980s on grounded ice. One group working along the Siple Coast ice streams in west Antarctica drilled 119 boreholes during the field seasons 1988–2001, reaching depths of 1200 m, and a second group on Jakobshavn Isbrae in western Greenland reached depths of 1630 m (Bentley and Koci 2007).

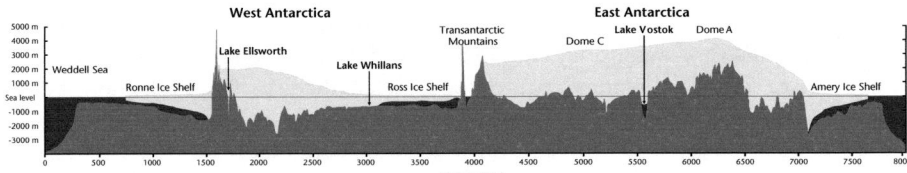

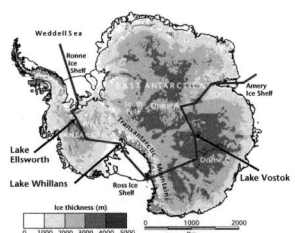

Figure 35.3   A cross section of the Antarctic Ice Sheet and ice thickness map based on the Bedmap2 (Fretwell et al. 2013). Both figures show the location of subglacial lakes Ellsworth, Whillans and Vostok

Both groups used relatively light weight drilling systems transported by snow mobile and sleds or small helicopters. Instruments were installed to measure ice temperatures, shear within the ice, and basal water pressure. Basal sediment cores were collected and various instruments deployed into the ice stream bed to investigate till properties and subglacial hydrology.

With increasing depth, thermal power and fuel requirements quickly increased along with drill system size and logistical support. To minimise these demands and optimise drilling operations, calculations using the thermal properties of the drill equipment, drill water and surrounding ice were initially applied to systems used in the Alps and western Greenland to predict the hole size and freezeback rates accurately. Most recently, at the South Pole, a drilling model (Greenler et al. 2014) coupled with increased drill reliability, achieved maximum fuel savings of 44% on the project's base line figure of 27000 litres per hole. In total, 89 holes, 60 cm in diameter, were drilled to 2500 m for the installation of IceCube, a cubic-kilometre neutrino detector.

Recent subglacial lake access programmes

Over recent decades there has been a growing awareness of hydrological systems beneath large parts of the Antarctic and Greenland ice sheets. Active drainage systems have long been the target of drilling campaigns on valley glaciers but the first indications of water beneath the Antarctic Ice Sheet came from airborne radar surveys over 40 years ago that showed the presence of subglacial lakes (Oswald and Robin 1973). Now, over 400 lakes are known to exist (Siegert et al. 2014). Modelling studies also suggest that about 55% of the Antarctic Ice Sheet is likely to be underlain by water, where the background geothermal heat flow and frictional heating through ice flow is sufficient to raise basal ice temperatures to the pressure melting point (Pattyn 2010). Some locations, such as Lake Vostok, are thought to be isolated, but widespread subglacial melting can ultimately produce a dynamic hydrological system with subglacial lakes and drainage networks radiating out to the ice sheet margin.

Regarded as extreme habitats for microbial life and repositories of important paleoclimate records, interest in accessing subglacial environments has grown over the last 20 years. This led the U.S. National Academy of Sciences and the Scientific Committee on Antarctic Research to issue guidelines on cleanliness requirements for the exploration and direct sampling of subglacial aquatic environments, which presented significant challenges to existing drilling technologies.

Firmly established as a mature access drilling technique, hot water drills were considered the most viable option for clean, rapid, access through thick ice sheets. The locally sourced drill water is already pasteurised to around 90°C, but adding both submicron filtration to remove dust, bacteria and some viruses, and exposure to strong ultra violet (UV) light, ensures microorganisms are either removed or killed. Modifications to the drill equipment such as pumps and boilers, together with new operational protocols, including cleaning drill hoses and UV exposure around the well head, provides compliance with the international guidelines for subglacial access (Figure 35.2). The first clean hot water drill (CHWD) was used and tested on the Langjökull Ice Cap in Iceland. After comprehensive drill cleaning protocols were adopted, test results confirmed that microbes were reduced to undetectable levels (Thorsteinsson et al. 2007).

Recently, three Antarctic projects were devised to directly access very different Antarctic subglacial lakes (Figure 35.3). The first attempt was by a Russian team at Vostok Station, located over Lake Vostok, which is over 200 km long and up to 1000 m deep. The ice core drilling of borehole 5G began in 1990, several years before the presence of the lake was officially acknowledged, with a hiatus between 1998 and 2005 while plans were drawn up on how to enter the lake cleanly. To prevent contamination of the lake with the hydrocarbon based drilling fluid, its level in the borehole was lowered prior to penetration into the lake (Lukin and Vasiliev 2014). On 6 February 2012, the first entry into Lake Vostok was achieved at the depth of 3769.3 m (Bulat 2016). On reaching the lake, water rushed 363 m up the borehole, a level considerably higher than initially anticipated but later determined to be the result of the rapid withdrawal of the drill acting as a piston within the hole. Upon recovery to the surface the drill was found to have lake water frozen to the drill bit. The turbulent inrush created an emulsion of lake water and drill fluid that was allowed to freeze over the following austral winter. During the 2012–2013 season, 32 m of frozen lake water was cored until the drill deviated away from the original borehole. The lake entry experiment was repeated again on 25 January 2015 with water rising 70 m up the borehole and given a few days to freeze. A further 12 m of lake water ice core was recovered until water again came up into the borehole. In both cases, the cores contained large bubbles of kerosene with numerous micro droplets of drilling fluid present within the lake water ice. Microbiological studies found previously unidentified and unclassified bacteria that may be indigenous to Lake Vostok. However, this ice coring technology using kerosene and other chemicals as a drilling fluid, "proved to be inappropriate to collect liquid water in general and clean samples in particular" (Bulat 2016). It was also recognised that sampling water only from the lake surface is very limiting, as samples should be collected throughout the water column and especially close to the lake bed sediments where biological activity is likely to be greatest. This technique is clearly not a desirable approach for subglacial lake access, providing neither rapid access nor clean samples, and severely limiting the experiments that can be undertaken.

In December 2012, a British team attempted the second subglacial access at Lake Ellsworth (SLE) in West Antarctica. Much smaller than Lake Vostok, Lake Ellsworth is 14.7 km long and up to 3.1 km wide, with a maximum water depth of 156 m (Siegert et al. 2012). At the drill site, almost 2000 m above sea level, the ice sheet was 3155 m thick and the underlying lake was 146 m deep. To access the lake, a specially designed hot-water drill with clean deployment methodologies was developed to enter the lake cleanly and safely, and in compliance with international subglacial exploration guidelines. The drill was capable of delivering 210 l/min at 90°C or about 1.35 MW of heating capacity to melt a 36 cm diameter access hole through ice with minimum temperatures of −32°C to the lake below over a three to four day period.

The drill site was selected in the central part of the lake to maximise the probability of recovering a long undisturbed sedimentary record from the lake floor, whilst reducing the risk from possible basal freezing and the build-up of buoyant air hydrates or clathrates. Formed as trapped air bubbles are pressurised with depth, clathrates eventually form, becoming a crystalline solid below depths of 1500 m. In subglacial lakes, melting ice releases the air trapped in clathrates and the dissolved gas concentration increases to a maximum of approximately 2.5 L of gas per kilogram of lake water, almost regardless of depth (fizzy drinks contain about 3 L kg^{-1}). Beyond this concentration, and at depths greater than 1.5 km, further basal melting results in the accumulation of clathrates at the lake surface. When accessing a lake by hot water drilling however, heat from the drilling process will destabilise these clathrates, generating large volumes of gas, potentially leading to the risk of a surface blowout where depressurisation could lead to high-speed ejection of water and gas from the borehole. At SLE for example, the risk of clathrates would have been considered high if the lake had been hydrologically closed for 400 kyr, with melt/freeze rates of 15 cm yr^{-1}, low if closed for 100 kyr, and negligible if closed only for the Holocene (10 kyr). With melt/freeze rates of 4 cm yr^{-1} at SLE, the risks reduce, to low, very low, and negligible (Siegert et al. 2012). If uncertainly remained, then cold drilling below the clathrate disassociation temperature of 5°C during lake entry would mitigate the potential blowout danger.

The aim of the Lake Ellsworth project was to deploy purpose built sterile probes and samplers that would collect samples throughout the lake's water column and a sub-lake sediment core up to 3 m long. On 25 December 2012, however, drilling was halted and then abandoned after a succession of equipment issues culminated in a failure to link with a subsurface cavity at 300 m depth. Consequently, the full deep drilling capabilities of this 3.4 km CHWD and sampling equipment remain untested.

The Whillans Ice Stream Subglacial Access Research Drilling (WISSARD) programme from the United States was the third project attempting subglacial lake access in Antarctica. On 27 January 2013 Subglacial Lake Whillans (SLW), a small shallow ephemeral lake on Whillans Ice Stream, close to Ross Ice Shelf in West Antarctica, was accessed using a CHWD. Lying beneath 801 m of ice on the lower portion of the Whillans Ice Stream, SLW is part of an extensive and rapidly evolving subglacial drainage network with periodic drainage events that lower the lake level by about 5 m. The 180 T, 1.4 MW hot-water drilling system was transported 1000 km by a large over snow traverse from McMurdo Station to the drill site (Rack 2016) and used filtration, UV, pasteurisation, and hose cleaning to meet the required cleanliness and environmental protection standards. A 60 cm diameter borehole was melted through the ice sheet, enabling in situ measurements of the 2.2 m water column properties and collection of water and sediment samples, making it the first subglacial Antarctic lake to be accessed and sampled cleanly. The unequivocal results reveal the presence of a viable microbial ecosystem in SLW, indicating that subglacial ecosystems contain possibly globally-relevant pools of carbon and microbes that can mobilise elements from sediments and contribute a significant nutrient flux to the Southern Ocean (Christner et al. 2014).

Future plans and new technologies

From these three recent subglacial lake access research programmes, it is clear that subglacial access and sampling remains technically and logistically challenging. Currently, CHWDs remain the most viable option for clean, rapid, access through thick ice sheets but are untested beyond 800 m and have no experience of the potential hazards associated with clathrates in lakes at over 1500 m depth. Building on the legacy of these programmes, the ultimate goal is to develop CHWDs and other access technologies that routinely and rapidly drill deep and clean subglacial access holes in a safe, efficient, and predictable way (Makinson et al. 2016).

Miniaturisation of down-hole instruments and samplers is fundamental to greater efficiencies by reducing borehole diameters, drill equipment volumes, logistical support, and fuel usage. The development and adoption of advances in logistical solutions such as new sled designs to reduce towing forces, will further improve logistical efficiency and mobility.

In regions where basal ice temperatures are well below the melting point and therefore devoid of basal drainage systems and shallow groundwater aquifers, other drilling methods can be used to access and sample the bed material. One such technology, new to ice drilling, began Antarctic testing in the 2016–2017 austral summer. The US Rapid Access Ice Drill (RAID), based on a modified industry-standard diamond rock-coring system as used for mineral exploration, is capable of making several deep boreholes in a single field season. Threaded metal drill pipe sections are added at the surface as the drill cuts its way down through ice while a drill fluid maintains the correct operating temperatures at the drill head and transports ice cuttings to the surface where they are separated from the recirculating fluid. To occasionally take short ice cores for paleoclimate study and bedrock cores for subglacial geology, age dating and crustal history, drilling tools are lowered through the drill string to the base of the hole. Cores are then recovered to the surface using a wireline retrieval system. RAID is designed to be mobile, providing fast borehole access through thick ice up to 3,300 m in about one week. The 8.9 cm diameter holes will then remain fluid filled for future down-hole measurements such as geothermal heat flow, accumulation history, and ice deformation processes. Although firmly restricted to frozen bed locations, RAID offers an alternative method for rapid subglacial access.

Conventional drilling technologies (e.g. CHWD, RAID) still require large volumes of equipment and fuel or drilling fluid, and significant supporting logistics. Light weight technologies in the form of instrumented thermal probes that melt ice on contact with minimal deployment infrastructure offer a realistic way of dramatically reducing logistic demands. For environmental protection, probes can be sterilised prior to deployment, and forward contamination is minimised as the hole refreezes behind the probe.

New and novel probe technologies are already being investigated, developed, and tested with the long-term goal of deployment on planetary science missions to the Martian ice cap and the subsurface oceans of outer planetary moons such as Europa and Enceladus. In the medium term, deployments to Antarctica and its subglacial lakes will provide a useful analogue for such missions. These systems are generally based on a Philberth thermal probe (Philberth 1976), where a heated hot point tip melts ice on contact. This type of probe, which also contains instrumentation and a reel of cable, effectively sinks under its own weight, freely paying out the cable used for both the transmission of the electric power to the probe and data from the onboard instrumentation during penetration of the ice. In 1968, such a probe reached a depth of 1005 m and provided ice temperatures from within the Greenland Ice Sheet. In the 1990s, Germany's Alfred-Wegener-Institute developed a similar probe, SUSI (Sonde Under Shelf Ice), that successfully penetrated the 225 m thick ice shelf at Neuymayer Station with sensors to measure oxygen and carbon dioxide content of the ice/water mixture, the amount of inorganic nutrients, pH-values, and instruments for the investigation of microorganisms. In 2001, the Jet Propulsion Laboratory field tested Cryobot, an improved Philberth probe design that used a self-contained method of hot water jetting to drill into the ice that also held dust and particulates in suspension, therefore drilling deeper than passive systems alone. Building directly on these ideas and currently under development is VALKYRIE (Very-deep Autonomous Laser-powered Kilowatt-class Yo-yoing Robotic Ice Explorer), a NASA-funded project to develop key technologies for an autonomous ice penetrator (Stone et al. 2014). Using a 5 kW laser as its primary power source, an optical fibre is spooled from the descending probe. Traveling down the optical fibre, the laser light energy is used to heat multiple hot-water jets that allow the probe to steer a melt route through the ice, guided by a radar that senses obstacles in its path. To run the on-board electronics and water jet pumps, photovoltaic cells convert some laser light to electricity. Ultimately, this 1.6 m tall, 0.45 m diameter probe will carry instrumentation packages, samplers or even deploy smaller highly instrumented robots. Another manoeuvrable probe currently undergoing field trials is IceMole, which combines melting and mechanical propulsion in the form of an ice screw at the head of the probe. Differential heating at the melting head and side-wall heaters allow changes in direction (Dachwald et al. 2014), with radar obstacle avoidance, target detection, and navigation also under development.

It is highly likely that versions of these instrumented probes will eventually be used in Antarctica to deliver multi-parameter instrumentation and analysis packages cleanly through the ice sheet and into subglacial lakes, and possibly recover samples, as part of their ongoing development for future planetary science missions. The experience gained from recent subglacial access programmes now provides a much better understanding of how to access and sample these unique subglacial environments. The results from these initial lake entries continue to raise yet more fundamental scientific questions, thus providing the impetus to develop new drills and probes that can deliver clean, reliable deep-ice access, in-situ data acquisition, and sample recovery, which are all essential for future subglacial access, exploration, and scientific discovery.

References

Bentley, C.R. and B.R. Koci 2007. 'Drilling to the beds of the Greenland and Antarctic ice sheets: a review' *Annals of Glaciology* 47: 1–9.

Browning, J.A., R.A. Bigl and D.A. Somerville 1979. 'Hot water drilling and coring at site J-9 Ross Ice Shelf' *Antarctic Journal of the United States* 14(5): 60–61.

Browning, J.A. and D.A. Somerville 1978. 'Access hole drilling through Ross Ice Shelf' *Antarctic Journal of the United States* 13(4): 55.

Bulat, S.A. 2016. 'Microbiology of the subglacial Lake Vostok: first results of borehole-frozen lake water analysis and prospects for searching for lake inhabitants' *Philosophical Transactions of the Royal Society of London A: Mathematical, Physical and Engineering Sciences* 374(2059).

Christner, B.C., J.C. Priscu, A.M. Achberger, C. Barbante, S.P. Carter, K. Christianson, A.B. Michaud, J.A. Mikucki, A.C. Mitchell, M.L. Skidmore, T.J. Vick-Majors and W.S. Team 2014. 'A microbial ecosystem beneath the West Antarctic ice sheet' *Nature* 514(7522): 310–313.

Clarke, G.K.C. 1987. 'A short history of scientific investigations on glaciers' *Journal of Glaciology: Special issue commemorating the Fiftieth Anniversary of the International Glaciological Society (1987)*, 33(51): 4–24.

Dachwald, B., J. Mikucki, S. Tulaczyk, I. Digel, C. Espe, M. Feldmann, G. Francke, J. Kowalski and C. Xu 2014. 'IceMole: a maneuverable probe for clean in situ analysis and sampling of subsurface ice and subglacial aquatic ecosystems' *Annals of Glaciology* 55(65): 14–22.

Fretwell, P. et al. 2013. 'Bedmap2: improved ice bed, surface and thickness datasets for Antarctica' *Cryosphere* 7: 375–393.

Gillet, F. 1975. 'Steam, hot-water and electrical thermal drills for temperate glaciers' *Journal of Glaciology* 14(70): 171–179.

Greenler, L., T. Benson, J. Cherwinka, A. Elcheikh, F. Feyzi, A. Karle and R. Paulos 2014. 'Modeling hole size, lifetime and fuel consumption in hot-water ice drilling' *Annals of Glaciology* 55(68): 115–123.

Kapitsa, A.P., J.K. Ridley, G. de Q. Robin, M.J. Siegert and I.A. Zotikov 1996. 'A large deep freshwater lake beneath the ice of central East Antarctica' *Nature* 381(6584): 684–686.

Lipps, J.H., T.E. Ronan and T.E. Delaca 1979. 'Life below the Ross Ice Shelf, Antarctica' *Science* 203(4379): 447–449.

Lukin, V.V. and N.I. Vasiliev 2014. 'Technological aspects of the final phase of drilling borehole 5G and unsealing Vostok Subglacial Lake, East Antarctica' *Annals of Glaciology* 55(65): 83–89.

Makinson, K., D. Pearce, D.A. Hodgson, M.J. Bentley, A.M. Smith, M. Tranter, M. Rose, N. Ross, M. Mowlem, J. Parnell and M.J. Siegert 2016. 'Clean subglacial access: prospects for future deep hot-water drilling' *Philosophical Transactions of the Royal Society of London A: Mathematical, Physical and Engineering Sciences* 374(2059).

Oswald, G.K.A. and G. de Q. Robin 1973. 'Lakes beneath the Antarctic Ice Sheet' *Nature* 245(5423): 251–254.

Pattyn, F. 2010. 'Antarctic subglacial conditions inferred from a hybrid ice sheet/ice stream model' *Earth and Planetary Science Letters* 295(3–4): 451–461.

Philberth, K. 1976. 'The thermal probe deep-drilling method by EGIG in 1968 at Station Jarl-Joset, central Greenland' in Splettstoesser, J.F. (ed.) *Ice-Core Drilling*. Lincoln, NB: University of Nebraska Press. 117–131.

Rack, F.R. 2016. 'Enabling clean access into Subglacial Lake Whillans: development and use of the WISSARD hot water drill system' *Philosophical Transactions of the Royal Society of London A: Mathematical, Physical and Engineering Sciences* 374(2059).

Siegert, M.J., R.J. Clarke, M. Mowlem, N. Ross, C.S. Hill, A. Tait, D. Hodgson, J. Parnell, M. Tranter, D. Pearce, M.J. Bentley, C. Cockell, M.-N. Tsaloglou, A. Smith, J. Woodward, M.P. Brito and E. Waugh 2012. 'Clean access, measurement, and sampling of Ellsworth Subglacial Lake: a method for exploring deep Antarctic subglacial lake environments' *Reviews of Geophysics* 50: RG1003.

Siegert, M.J., K. Makinson, D. Blake, M. Mowlem and N. Ross. 2014. 'An assessment of deep-hot-water drilling as a means to undertake direct measurement and sampling of Antarctic glacial lakes: experiences and lessons learned from the Lake Ellsworth field season 2012-13.' *Annals of Glaciology* 55 (65): 59–73.

Stone, W.C., B. Hogan, V. Siegel, S. Lelievre and C. Flesher 2014. 'Progress towards an optically powered cryobot' *Annals of Glaciology* 55(65): 1–13.

Taylor, P.L. 1984. 'A hot water drill for temperate ice' in Holdsworth, G., K.C. Kuivinen and J.H. Rand (eds) *Proceedings of the 2nd International workshop on Ice Drilling Technology*. Calgary, AB: US Army Cold Regions Research and Engineering Laboratory. 105–117.

Thorsteinsson, T., S.O. Elefsen, E. Gaidos, B. Lanoil, T. Johannesson, V. Kjartansson, V.P. Marteinsson, A. Stefansson and T. Thorsteinsson 2007. 'A hot water drill with built-in sterilization: design, testing and performance' *Jokull* 57: 71–82.

36

Upper atmosphere physics and chemistry

Sheila Kirkwood

Introduction

The upper atmosphere of the Earth is invisible to the vast majority of the planet's inhabitants. Almost all of the visible signs of the atmosphere's presence – clouds, rain, snow, rainbows, lightning – are located in the lowermost part of the atmosphere, below about 8–16 km height. This 'weather layer' is known as the troposphere and contains about 80% of the atmosphere. The rest lies at higher altitudes in what are known as the stratosphere (up to about 50 km height), mesosphere (up to about 90 km), thermosphere (up to about 600 km) and exosphere (above 600 km). At polar latitudes, residents and visitors are made well aware of the presence of the upper atmosphere by a number of very visible, and at times dramatically colourful features – polar stratospheric clouds, noctilucent clouds at the upper edge of the mesosphere and the aurora, extending from the upper mesosphere into the thermosphere. Meteors burn up in the mesosphere and thermosphere and their trails are visible over all parts of the globe but the results – clouds of meteor smoke – make their way to the Earth's surface primarily through the upper atmosphere in the Polar Regions. This has surprising consequences which we will return to later.

The aurora, in particular, was the driving force for the early establishment of permanent scientific research institutes in the Arctic, starting in the early twentieth century. Understanding the relation between activity on the Sun and disturbances of the Earth's magnetic field, which are closely related to the aurora, was a major part of the International Geophysical Year in 1957–1958, which was the first truly global scientific collaborative effort involving all aspects of Earth science. In Antarctica, ten new permanent stations were established in the run-up to this effort, more than doubling the number on the continent. The twelve countries which had been active in Antarctica in this scientific effort became the original signatories to the Antarctic Treaty in 1959 (Dodds, Chapter 20 this volume; Dey Nuttall, Chapter 23 this volume).

Upper atmosphere research in the Polar Regions has always been dependent on the development of new technologies – from the first photographic techniques for recording the aurora from the ground, through the first balloon and rocket-borne instruments reaching the upper atmosphere, to the development of modern radar, lidar and passive spectroscopic techniques for remote sensing. All of these techniques were used in both Polar Regions and are still used there today.

Although huge progress in understanding the upper atmosphere was made (and is still made) using the many ground-based techniques mentioned above, most of them provide measurements

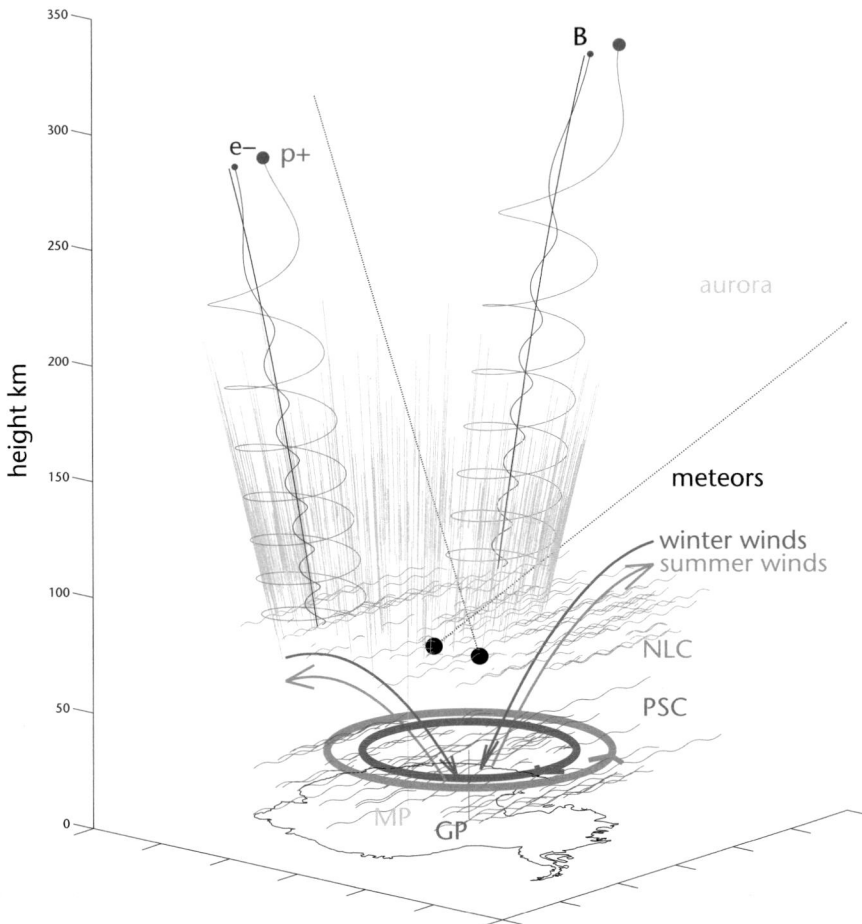

Figure 36.1 Diagram of the main features controlling the physics of the upper atmosphere. Lines (B) indicate magnetic field lines with electrons (e–) and protons (p+) from space spiraling down into the atmosphere. The visible signature of this particle precipitation, the aurora, has its highest occurrence rate in a ring centered on the geomagnetic pole (MP), the 'auroral zone'. Prevailing wind systems in the stratosphere and mesosphere form the 'polar vortex', centered on the geographic pole (GP). The visible signatures of the very cold temperatures are at about 85 km height in summer – noctilucent clouds, NLC – and at 20–30 km height in winter – polar stratospheric clouds. Black dots and dotted black lines indicate meteors entering the atmosphere and vaporizing in the upper atmosphere. The illustration refers to Antarctica – in the Arctic, MP and GP are closer together and the wind systems are more variable and less symmetric about the pole. (Figure produced by the author).

only at a few locations with particularly problematic gaps in the high Arctic and inland Antarctica, where the possibilities for making measurements are restricted. Probably the most dramatic step forward in our ability to study the polar upper atmosphere, especially regarding the possibilities to compare and contrast the two poles, comes from the advent of satellite-based remote-sensing. Remote-sensing of the upper atmosphere from space has progressed from the simple column-ozone sensors on weather satellites (since 1978), through dedicated missions for height-resolved passive sensing of winds, temperatures and chemical composition (since the 1990s) to active missions using radar and lidar (since the early 2000s). Most of the dedicated atmospheric satellite missions have operated in close-to-polar orbits and provide comparable measurements every day over both Polar Regions.

The main features of the polar upper atmosphere, as it is currently understood, are illustrated in Figure 36.1. The figure is drawn for Antarctica, but the only difference for the Arctic would be less distance between the geomagnetic and geographic poles, and less symmetry in the circumpolar wind systems. The configuration of the Earth's magnetic field, with field lines becoming steep-to-vertical near the magnetic poles, channels high-energy electrons and protons from space into the polar atmosphere, leading to ionization and energizing of the ambient gases. Light emission from these gases constitutes the aurora. The rotation of the Earth about its spin axis leads to wind systems which form vortices around the geographic poles. The inclination of the spin axis leads to several months of darkness in the winter and several months of perpetual sunlight in the summer. In winter the air in the stratosphere within the polar vortex becomes extremely cold and subsides, allowing polar stratospheric clouds to form. In summer, the global circulation of the mesosphere and stratosphere induces upwelling over the polar regions, and extremely cold temperatures are found at the boundary between the polar mesosphere and thermosphere. This leads to the formation of the highest clouds in the atmosphere, polar mesospheric clouds (also known as noctilucent clouds).

It is not the intention here to give an exhaustive review of all of the research topics which have interested polar-atmospheric scientists over the years. For example, we will not discuss research on the aurora, which has continued up to the present day but with the emphasis moving out into space, to the source regions of the energetic particles (see e.g. Keiling et al. 2012 for a recent overview). Rather, the focus will be on a selection of new frontiers – the surprises and the new research directions which have emerged in recent years, particularly those where understanding what happens in the polar upper atmosphere matters for the rest of the globe. These are grouped into three themes: stratospheric clouds, the ozone hole and storm tracks; noctilucent clouds, meteor smoke and geo-engineering; solar variability, relativistic electrons and climate.

Stratospheric clouds, the ozone hole and storm tracks

The main source of ozone in the Earth's atmosphere is the action of solar ultraviolet light on oxygen in the stratosphere and mesosphere. (A small amount of ozone is produced by pollution near the Earth's surface, but this is not discussed here). In the natural (pre-industrial) state, ozone is produced mainly over the tropics (where solar radiation is most intense) and is transported to the Polar Regions by global wind systems in the upper atmosphere. Ozone accumulates in the polar stratosphere and the natural condition is for the polar stratosphere to have higher concentrations of ozone than other latitudes. In the 1980s a dramatic departure from this pattern started to be noticed in Antarctica, with ozone amounts in spring drastically reduced compared to earlier years. This came to be known as the 'ozone hole'. In the following decades, enormous research efforts went into understanding the causes of the ozone disappearance – it turned out to be due to very small amounts of chlorine released in the stratosphere from freons, industrially produced gases used as coolants and as inert carriers, for example for spraying other chemicals. The freons, and the ultraviolet light to release the chlorine, are spread over the whole globe, and indeed contributed to ozone loss world-wide, but the dramatic, almost complete loss of ozone over Antarctica is due to the unfortunate coincidence of chlorine from freons and stratospheric clouds, which accelerate the rate of ozone destruction many times over. Ozone destruction has not been as extreme over the Arctic because the stratospheric wind-system, the polar vortex, is not as stable there as over Antarctica. In the northern hemisphere, the asymmetries in land, sea and topography are much larger which leads to wave disturbances in the wind fields. As a result, the Arctic winter stratosphere generally does not get as cold, and there are far fewer stratospheric clouds. Freons were banned for most uses in 1989 and the levels of chlorine in the stratosphere are starting to decline although it is expected to be many decades before they are back at pre-1980s levels. Monitoring of stratospheric ozone continues and, as yet, no clear improvement has been demonstrated, with the ozone hole continuing to form over Antarctica each spring (Figure 36.2). In fact, in March 2011, ozone depletion over the Arctic reached levels previously associated only with the Antarctic spring. This was due to an unusually stable polar vortex during the 2010/2011 winter and illustrates that the ozone layer is still vulnerable. The current state of understanding is summarized in regular reports by the World Meteorological Institute (WMO 2014).

The focus on understanding the stratospheric ozone layer, necessitated by the discovery of the ozone hole, and the 'active experiment' on our atmosphere that ozone depletion has provided, has

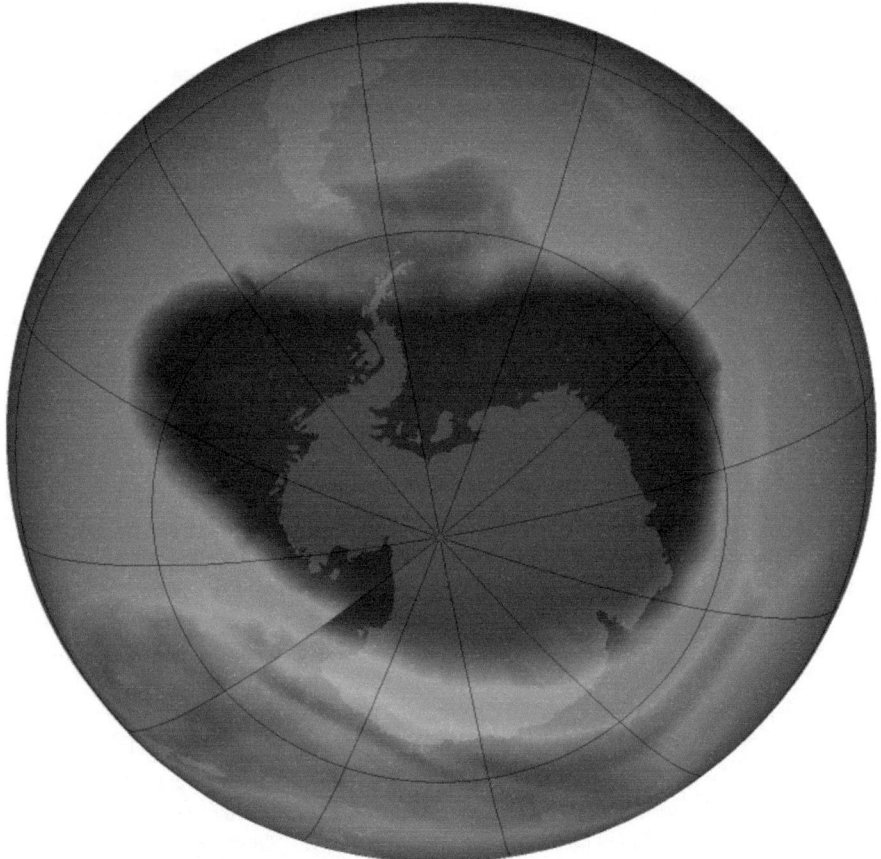

Figure 36.2   The ozone hole over Antarctica on 4 October 2014. The shading over Antarctica indicates where there is the least ozone, while the shading at the bottom of the image is where there is more ozone. The data are from the OMI instrument (KNMI / NASA) onboard the Aura satellite, provided by NASA Ozone Watch: http://ozonewatch.gsfc.nasa.gov/.

also led to new understanding of ways in which conditions in the stratosphere can influence weather systems in the troposphere below. In both hemispheres, patterns of sea-level pressure difference between mid- and polar latitudes, known as the Northern and Southern Annular modes (NAM and SAM) have been found to dominate variations in both weather and climate (e.g. Baldwin 2001; Bingham, Chapter 12 of this volume). Systematic changes have been observed in SAM over the latter part of the twentieth century (e.g. Marshall 2003), corresponding to a strengthening of the westerly winds which encircle Antarctica, and a poleward shift of the high-latitude storm tracks.

The changes in SAM have been convincingly attributed primarily to the cooling of the polar stratosphere caused by the Antarctic ozone hole (with a lesser contribution from increased greenhouse gases, Polvani et al. 2011). The effects on regional climate in the southern hemisphere are thought to be wide reaching and not confined to Antarctica – for example summer warming over southern Africa has been linked to the changes in SAM (Manatsa et al. 2013). In the northern hemisphere, although changes in the NAM have been found, these are more variable than in the southern hemisphere. At the same time, ozone depletion is much less severe in the Arctic and no clear link between NAM variability and ozone depletion has been made. However, if the amplitude of the Arctic ozone depletion were to increase, such as happened in 2011, modeling studies show that an effect on the NAM is expected (Karpechko et al. 2014; Smith and Polvani 2014), in particular with a poleward shift of the storm tracks over the North Atlantic, bringing more Atlantic storms to northern Europe and fewer to southern Europe.

Noctilucent clouds, meteor smoke and geo-engineering

The first scientifically recorded observations of noctilucent clouds, recognized as being at very high altitudes (around 85 km) were made in 1885 and it was from the beginning conjectured that they were formed of water ice (for a review on the early research on noctilucent clouds, see Gadsden and Schröder 1989). As awareness of greenhouse gas increases in the atmosphere grew in the 1980s, it was suggested that increased methane would be broken down by ultraviolet light in the upper atmosphere and lead to increased water vapour, which, together with increased cooling from carbon dioxide, would lead to noctilucent clouds becoming more common over time (Thomas et al. 1989). Although initial analysis of visual reports suggested that an increase was already happening, later analysis of the visual record argued against this (Kirkwood et al. 2008). However, visual records cover only the mid-latitude edges of the noctilucent cloud region and only the northern hemisphere – they generally form only at latitudes higher than 50°; the midnight Sun at polar latitudes during the summer makes the sky too bright for visual observations from latitudes higher than about 60°, and there are very few possibilities for observation from the ground between 50° and 60° latitudes in the southern hemisphere. At polar latitudes, the clouds can only be observed by instruments – radars, lidars or satellites. So far, the radar and lidar records are too short to discern any increasing trend. Satellite records, based on observations made for other purposes by a series of spacecraft with overlapping but not entirely compatible observing windows, have suggested a strong upward trend at high latitudes (Shettle et al. 2009). However, recent reanalysis of the satellite record with the help of new, dedicated satellite measurements suggests that, once the sensitivity to variations in solar radiation are accounted for, any residual trend in occurrence rate is small, less than 1% per decade, and not statistically significant (Hervig and Stevens 2014).

The possibility of an anthropogenic effect on noctilucent clouds (although as yet unproven!) provided a strong impetus for research on the topic and provided a huge increase in the numbers, variety and quality of associated observations. One particular aspect of this research is of particular interest – meteoric smoke. It was long suspected that some kind of very small particles must be present in the upper mesosphere to provide nuclei for the condensation of cloud particles. 'Meteor smoke' formed by the ablation and re-condensation of meteors was suggested (Hunten et al. 1981), but for many years there was no way to observe whether this was present at all. Sounding rocket and radar measurements have provided a number of increasingly convincing clues during recent years but the most convincing evidence has been provided by the AIM (Aeronomy of Ice in the Mesosphere) satellite, a dedicated US mission for noctilucent cloud research. This has shown the presence of small particles throughout the depth of the polar mesosphere, distributed in height and season exactly as would be expected for meteor smoke (Hervig et al. 2009), and embedded in noctilucent cloud ice particles (Hervig et al. 2012).

Many tonnes of meteoric material are dumped in the atmosphere every day. Exactly how much is not well known with estimates varying from 5 to 300 tonnes, depending on the technique used to measure the input. In addition to nucleating noctilucent clouds, meteoric smoke may be involved in the stratospheric aerosols (including clouds) and polar ozone depletion, and in fertilizing the Southern Ocean (for a review, see Plane 2012). The large discrepancy between the highest and lowest estimates of the amount of meteoric material entering the atmosphere hampers our ability to understand whether meteor dust is significant or not in relation to other sources affecting the stratosphere (e.g. sulphate aerosol) or the ocean (e.g. aeolian dust). In both of these areas, there have been suggestions of geo-engineering solutions to the ongoing increase in carbon dioxide in the atmosphere. Crutzen (2006) has suggested injecting sulphur into the stratosphere to increase the stratospheric aerosol content and reflect more of the Sun's radiation back to space, while Lampitt et al. (2008) have discussed ocean fertilization to increase carbon dioxide uptake. In both cases, present knowledge of the natural processes, and possible future changes due to shifts in atmospheric circulation, is far from adequate to allow confidence in the likely consequences of implementing such techniques.

In the case of the natural source provided by meteoric dust, global transport effects are still inadequately understood. Figure 36.1 illustrates the role of the polar upper atmosphere in this transport with a global wind system which brings air and meteoric smoke from lower latitudes towards the pole and down towards the lower atmosphere in the winter – where meteoric aerosol has indeed been observed, concentrated in the polar vortex in the lower stratosphere (Curtius et al. 2005). However, if the rate of vertical transport through the polar mesosphere and stratosphere and the location of the main transport to the surface are as understood in current models, there are order of magnitude discrepancies between the amounts of meteoric dust observed in the mesosphere, and the amounts observed in polar ice-cores and deep see sediments. Ongoing research addresses possible candidates for faster vertical transport such as mixing due to the action of gravity waves (e.g. Grygalashvyly et al. 2012), and details of the transport from the stratosphere to the troposphere. Stratospheric air (and the dust it carries) enters the troposphere primarily in 'tropopause fold' structures, which are associated with the storm tracks around the polar regions. Although global meteorological assimilations of atmospheric measurements have been used to track this transport (e.g. Škerlak et al. 2014), these assimilations do not correctly capture the tropopause folds at polar latitudes (see e.g. Mihalikova et al. 2012). Nor do they give an accurate representation of rain or snowfall, which is needed to transport dust to the surface. Much more detailed models are still needed to map the final destination of meteoric dust accurately, which is a prerequisite for understanding its role in the atmosphere-ocean system.

Solar variability, relativistic electrons and climate

Energy from the Sun is the primary energy source for the Earth's climate system and, ultimately, for all life on Earth. It might therefore seem obvious that variations in the Sun's energy output should cause variations in climate. While this is almost certainly true on geological time scales, during historical times the main part of the solar radiation feeding into the climate system, the total solar irradiance, seems to have varied very little (by less than 0.2% during the last 400 years, see e.g. Solanki et al. 2013). Variability in the short-wavelength (ultraviolet/X-ray) part of solar radiation, which is absorbed by the Earth's upper atmosphere, has been much higher, up to 100% in some wavelength intervals. The effects of the variability of the short-wavelength radiation on the global stratosphere and ozone layer, and links to some tropospheric climate patterns, have been successfully modeled and are reasonably well understood (Gray et al. 2010; Solanki et al. 2013).

However, solar radiation is not the only aspect of solar energy output which affects the Earth. The Sun also emits particles – mainly electrons and protons – which stream out from the Sun forming the 'solar wind'. Huge variations in the numbers and energies of these particles occur. Typical solar wind speeds are around 400 km/s, with densities of a few particles per cubic centimeter, but high-speed streams up to about 800 km/s are frequently ejected from 'coronal holes', and occasional 'coronal mass ejections' can send out proton streams with speeds up to 3000 km/s (solar proton events). Both kinds of disturbance affect the upper atmosphere in the Polar Regions. The very fast protons follow the Earth's magnetic field lines directly reaching low in the mesosphere and sometimes even the stratosphere. The more common high-speed streams carry with them magnetic field lines from the Sun. They interact with the Earth's magnetic field and cause increases in energy in electrons trapped there (in regions known as the radiation belts), reaching energies up to Mev ('relativistic energies'), and these electrons are precipitated into the polar mesosphere. Both the protons and electrons expend their energy by ionizing or dissociating the atoms and molecules of the upper atmosphere gases, which leads to changes in atmospheric chemistry which can lead to ozone destruction.

There has been variability over recent decades of the energy available to the Earth from solar radiation, and from the solar wind. Although the total amount of energy available is much greater for solar radiation than for the solar wind, the variability (in PW) is similar between the two. While the variability in total irradiance has most effect in the tropics, the variability in the solar wind primarily affects the upper atmosphere in the polar regions. It is by now well established that solar proton events reduce stratospheric ozone concentration and the largest event yet observed (in 1859) could have had significant effects on temperatures at the Earth's surface on a global scale (a few °C reduction, Calisto et al. 2012). However, most solar proton events are much weaker than this and they are a relatively rare occurrence, affecting the Earth between zero and about 30 days a year, depending on the phase of the eleven-year solar cycle. Precipitation of energized electrons from the radiation belts, due to the effects of high-speed streams, on the other hand, is generally less intense but occurs much more often. Recent research suggests that the relativistic electron precipitation caused by variability in the solar wind can reduce stratospheric ozone, similarly to solar proton events, and also have an effect on regional climate (e.g. Rozanov et al. 2012), which is of the same order as the effects of variability in solar radiation (Seppälä and Clilverd 2014). There appear to be overall downward trends in both solar radiation and solar wind energy – it remains to be seen whether this will continue over the coming decades and further complicate the process of understanding the relative contributions of natural and anthropogenic climate change.

Concluding remarks

As in many other branches of polar science, the need to understand the causes and consequences of climate change has heavily influenced research interests in the polar atmosphere in recent decades. Initial research directions based on pure curiosity – understanding of the visible phenomena of the aurora, stratospheric and noctilucent clouds and meteors – have found applications in unexpected directions. In particular, polar stratospheric ozone, influenced by anthropogenic pollution, by energetic particles from space, and perhaps even by meteor smoke, has been found to have an unexpected role in modifying regional weather patterns and climate far outside the polar regions. One of the most important tasks for the future is to maintain a close watch and an open mind over processes in the polar upper atmosphere – there are likely many new surprises yet to come.

References

Baldwin, M.P. 2001. 'Annular modes in global daily surface pressure' *Geophysical Research Letters* 28: 4115–4118.

Calisto, M., Verronen, P.T., Rozanov, E. and Peter, T. 2012. 'Influence of a Carrington-like event on the atmospheric chemistry, temperature and dynamics' *Atmospheric Chemistry and Physics* 12: 8679–8686.

Crutzen, P.J. 2006. 'Albedo enhancement by stratospheric sulphur injections: a contribution to resolve a policy dilemma?' *Climatic Change* 77: 211–219.

Curtius, J., Weigel, R., Vössing, H-J., Wernli, H., Werner, A., Volk, C-M., Konopka, P., Krebsbach, M., Schiller, C., Roiger, A., Schlager, H., Dreiling, V. and Borrmann, S. 2015. 'Observations of meteoric material and implications for aerosol nucleation in the winter Arctic lower stratosphere derived from in situ particle measurements' *Atmospheric Chemistry and Physics*, 5: 3053–3069.

Gadsden, M., and Schröder, W. 1989. *Noctilucent Clouds*. Berlin: Springer Verlag.

Gray, L.J. et al. 2010. 'Solar influences on climate' *Reviews of Geophysics*, 48, RG4001, doi:10.1029/2009RG000282.

Grygalashvyly, M., Becker, E. and Sonnemann, G.R. 2012. 'Gravity wave mixing and effective diffusivity for minor chemical constituents in the mesosphere/lower thermosphere' *Space Science Reviews* 168(1–4): 333–362.

Hervig, M.E., Deaver, L.E., Bardeen, C.G., Russell, J.M., Bailey, S.M. and Gordley, L.L. 2012. 'The content and composition of meteoric smoke in mesospheric ice particles from SOFIE observations' *Journal of Atmospheric and Solar-Terrestrial Physics* 84–85: 1–6.

Hervig, M.E., Gordley, L.L., Deaver, L.E., Siskind, D.E., Stevens, M.H., Russell, J.M., Bailey, S.M., Megner, L. and Bardeen, C.G. 2009. 'First satellite observations of meteoric smoke in the middle atmosphere' *Geophysical Research Letters* 36: 18805–18810.

Hervig, M.E. and Stevens, M.H. 2014. 'Interpreting the 35 year SBUV PMC record with SOFIE observations' *Journal of Geophysical Research: Atmospheres* 119: 12689–12705.

Hunten, D.M., Turco, R.P. and Toon, O.B. 1981. 'Smoke and dust particles of meteoric origin in the mesosphere and thermosphere' *Journal of the Atmospheric Sciences* 37: 1342–1357.

Karpechko, A.Y., Perlwitz, J. and Manzini, E. 2014. 'A model study of tropospheric impacts of the Arctic ozone depletion 2011' *J. Geophys. Res. Atmos.* 119: 7999–8014.

Keiling. A., Donovan, E., Bagenal, F. and Karlsson, T. eds. 2012. *Auroral Phenomenology and Magnetospheric Processes: Earth and Other Planets, Geophysical Monograph Series*, Vol. 197. Washington, DC: American Geophysical Union.

Kirkwood, S., Dalin, P. and Rechou, A. 2008. 'Noctilucent clouds observed from the UK and Denmark – trends and variations over 43 years' *Annales Geophysicae* 26: 1243–1254.

Lampitt, R.S., Achterberg, E.P., Anderson, T.R., Hughes, J.A., Iglesias-Rodriguez, M.D., Kelly-Gerreyn, B.A., Lucas, M., Popove, E.E., Sanders, R., Shepherd, J.G., Smythe-Wright, D. and Yool, A. 2008. 'Ocean fertilization: a potential means of geoengineering' *Philosophical Transactions of the Royal Society A* 366(1882): 3919–3945.

Manatsa, D., Morioka, Y., Behera, S. K., Yamagata, T. and Matarira, C.H. 2013. 'Link between Antarctic ozone depletion and summer warming over southern Africa' *Nature Geoscience* 6: 934–939.

Marshall, G.J. 2003. 'Trends in the Southern Annular Mode from observations and reanalyses' *Journal of Climate* 16: 4134–4143.

Mihalikova, M., Kirkwood, S., Arnault, J. and Mikhaylova, D. 2012. 'Observation of a tropopause fold by MARA VHF wind-profiler radar and ozonesonde at Wasa, Antarctica: comparison with ECMWF analysis and a WRF model simulation' *Annales Geophysicae* 30: 1411–1421.

Plane, J.M.C. 2012. 'Cosmic dust in the earth's atmosphere' *Chemical Society Reviews* 41: 6507–6518.

Polvani, L.M., Waugh, D.W., Correa, G.J.P. and Son, S-W. 2011. 'Stratospheric ozone depletion: the main driver of twentieth-century atmospheric circulation changes in the Southern Hemisphere' *Journal of Climate* 24: 795–812.

Rozanov, E., Calisto, M., Egorova, T., Peter, T. and Schmutz, W. 2012. 'Influence of the precipitating energetic particles on atmospheric chemistry and climate' *Surveys in Geophysics* 33: 483–501.

Seppälä, A. and Clilverd, M.A. 2014. 'Energetic particle forcing of the Northern Hemisphere winter stratosphere: comparison to solar irradiance forcing' *Frontiers of Physics* 2: 25. doi: 10.3389/fphy.2014.00025.

Shettle, E.P., DeLand, M.T., Thomas, G.E. and Olivero, J.J. 2009. 'Long term variations in the frequency of polar mesospheric clouds in the Northern Hemisphere from SBUV' *Geophysical Research Letters* 36, LO2803, doi:10.1029/2008GL036048.

Škerlak, B., Sprenger, M. and Wernli, H. 2014. 'A global climatology of stratosphere-troposphere exchange using the ERA-Interim data set from 1979 to 2011' *Atmospheric Chemistry and Physics* 14: 913–937.

Smith, K. and Polvani, L. 2014. 'The surface impacts of Arctic stratospheric ozone anomalies' *Environmental Research Letters* 9, doi:10.1088/1748-9326/9/7/074015.

Solanki, S.K., Kriviva, N.A. and Haigh, J.D. 2013. 'Solar irradiance variability and climate' *Annual Review of Astronomy and Astrophysics* 51: 311–351.

Thomas, G.E., Olivero, J.J., Jensen, E.J., Schroeder, W. and Toon, O.B. 1989. 'Relation between increasing methane and the presence of ice clouds at the mesopause' *Nature* 338: 490–492.

WMO 2014. *Scientific Assessment of Ozone Depletion: 2014, Global Ozone Research and Monitoring Project, Report No. 55.* Geneva: World Meteorological Organization, www.wmo.int/pages/prog/arep/gaw/ozone_2014/full_report_TOC.html.

37

Ocean-land interactions and the Arctic carbon cycle

Frans-Jan W. Parmentier

Introduction

The future direction of the Arctic carbon cycle cannot be fully understood without an integrated view of the marine and terrestrial environment. Feedbacks across the ocean-land boundary may amplify the response to climate change in diverse and unexpected ways. Will a loss of sea ice lead to more or less favourable conditions for plant growth, and will higher temperatures exacerbate permafrost thaw and raise methane emissions? Do changes in rainfall lead to more or less organic matter flowing into the Arctic Ocean, where receding coastlines already add vast amounts of carbon to coastal shelf regions? While the potential for changes is high, there are currently no straightforward answers to these questions. This chapter explores the complexity of connections between the ocean and land of the North Pole region, and possible impacts on greenhouse gas exchange and lateral carbon flows thereof.

While this book covers both Polar Regions, this chapter will focus on the Arctic since ocean-land interactions are more important for the Arctic than the Antarctic carbon cycle. The South Pole region has not experienced a large reduction in sea ice (Parkinson and Cavalieri 2012), and warming has been modest in comparison to the Arctic (Serreze and Barry 2011). Besides, the impact of sea ice decline on the carbon cycle would be very dissimilar between the two regions due to diametric differences. While Antarctica is a frozen continent with little vegetation surrounded by ocean, the Arctic Ocean is a dynamic environment surrounded by land with vast expanses of vegetation, and an enormous amount of carbon locked away in the permafrost below (Hugelius et al. 2014). Rivers connect immense landmasses to the Arctic coastal region where one-third of the world's coastlines are located. The high degree of entwinement between the terrestrial and marine Arctic underlines the strong potential for ocean-land interactions to affect the carbon cycle.

Nowhere is this clearer than for sea ice decline, which has caused a wide diversity of changes in the North Pole region. With a loss of ~14% per decade (Stroeve et al. 2014), sea ice retreats rapidly, and the high latitudes warm at twice the global rate as a result (Deser et al. 2010; Screen and Simmonds 2010). Indications are that this warming has already affected the greenhouse gas exchange of terrestrial ecosystems (Bhatt et al. 2014; Parmentier et al. 2013). More directly, sea ice decline worsens coastal erosion, since receding ice exposes permafrost shores to violent

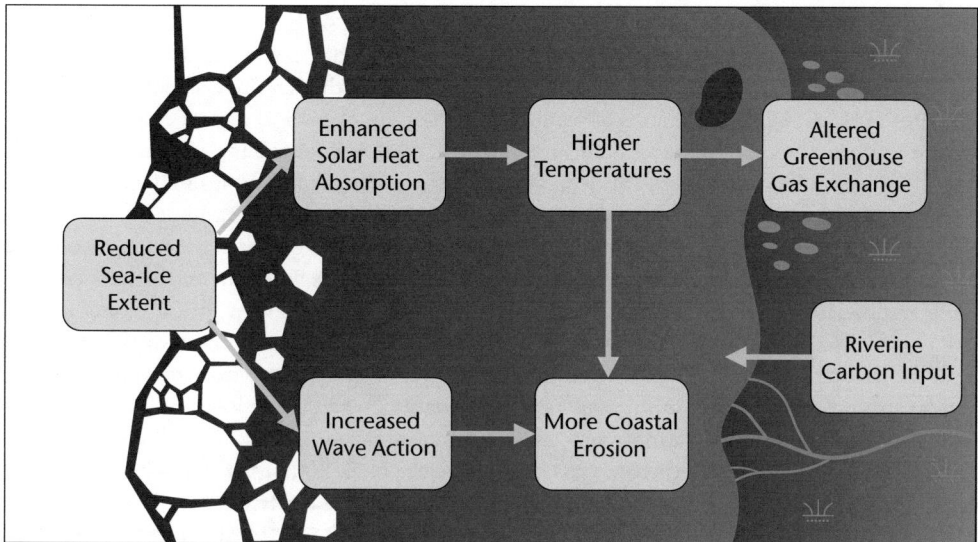

Figure 37.1   Schematic drawing of ocean-land interactions that affect the Arctic carbon cycle. A reduction in sea ice leads to more absorption of solar heat and higher temperatures, which ultimately affects plant productivity, respiration and methane emissions on land. Higher temperatures and increased wave action due to sea ice retreat cause more coastal erosion, while changes in terrestrial hydrology may alter the riverine input of carbon into the Arctic Ocean. Figure produced by the author.

storms (Barnhart et al. 2014). Changes on land affect the marine domain as well: Arctic rivers transport huge amounts of carbon from land to ocean, which may be remobilized in coastal shelf regions (Vonk and Gustafsson 2013). Albedo changes on land contribute to Arctic warming and even more sea ice melt (Jeong et al. 2014). Extending far beyond the North Pole region, the retreat of sea ice has also been suggested to affect mid-latitude weather (Francis et al. 2009: Screen 2013), which underlines the wide consequences of sea ice decline for the Arctic and beyond. These examples show that the marine and terrestrial environments of the Arctic are intricately connected to each other as depicted in Figure 37.1.

Polar amplification and the terrestrial environment

Snow and ice feedbacks in the Arctic

Reductions in sea ice extent and snow cover have drastically altered the energy balance of the Arctic, which strongly contributes to an amplified warming of the region compared to the rest of the world (Déry and Brown 2007; Screen et al. 2012). Sea ice decline contributes to polar amplification since the albedo (surface reflectance) of sea ice and open water are vastly different. While white sea ice reflects most sunlight, dark open water absorbs most sunlight. Thus, the immense reduction in sea ice extent has caused a massive increase of the heat input into the Arctic Ocean (Perovich and Polashenski 2012; Pistone et al. 2014), raising local temperatures. Likewise, when snowmelt advances, it exposes a dark surface with lower albedo earlier in the year, which leads to an increase in the absorption of solar energy, and higher temperatures (Déry and Brown 2007).

Other explanations for polar amplification have been proposed, such as increased atmospheric and oceanic heat transport from lower latitudes (Spielhagen et al. 2011; Yang et al. 2010), increasing humidity and cloud cover (Graversen and Wang 2009), and thermal inversions in winter time (Bintanja et al. 2011). Although it is beyond the scope of this chapter to discuss each of these processes and their relation to polar amplification in detail, there is a convincing argument to be made that the albedo effect − from sea ice decline in particular − has had a strong influence on Arctic warming, affecting greenhouse gas exchange as a result. This has to do with location and timing.

First of all, location: Numerous studies on polar amplification consider warming throughout the entire atmospheric column (see e.g. Graversen and Wang 2009; Pithan and Mauritsen 2014; Yang et al. 2010). This makes sense from a meteorological point of view, but the only relevant temperature change for photosynthesizing plants and methane-producing microorganisms would be at the surface − not high up in the atmosphere. A warming influence from outside the Arctic, such as atmospheric heat transport, may be an important contributor to long-term Arctic warming (Yang et al. 2010), but it cannot explain the strong surface warming. The albedo effect does account for surface warming (Screen et al. 2012; Serreze et al. 2009), but climate model studies have obtained highly varying estimates of the contribution of the retreat of snow and sea ice to polar amplification (Bintanja et al. 2011; Graversen and Wang 2009; Pithan and Mauritsen 2014; Taylor et al. 2013; Winton 2006). However, these studies employed future simulations with CO_2 concentrations much higher than present day, when sea ice extent and snow cover will be greatly reduced. An investigation of the past few decades is more appropriate to assess the contribution of albedo feedbacks to the large changes already observed in Arctic ecosystems. In that case, a clear surface-based warming is identified both in models and atmospheric reanalysis products − caused largely by sea ice decline (Deser et al. 2010; Screen et al. 2012; Serreze et al. 2009).

Second, timing: Ocean-atmosphere feedbacks vary through the year, since the manner in which the weather influences sea ice, and sea ice influences the weather, changes with the seasons. For example, the amount of sea ice loss during summer is strongly related to springtime weather conditions. When the melt season has an early and strong onset, more melt ponds will form on top of the ice. Melt ponds have a lower albedo than ice. More ponds lead to more absorption of solar heat, and more melt of sea ice throughout the summer (Schröder et al. 2014). Springtime weather conditions are, therefore, a strong determinant of the sea ice extent in autumn (Kapsch et al. 2013). This means that autumn sea ice conditions are not governed by the weather at that time, but rather the reverse is true − where the amount of open water influences the weather. When summer ends and the amount of available sunlight diminishes, a large and relatively warm body of open water remains. The additional heat absorbed during summer is reradiated back into the atmosphere, raising air temperatures above the ocean, and ultimately the adjacent land. The highest impact of sea ice decline on Arctic warming is therefore in autumn and into winter, although a possible influence in spring remains (Screen et al. 2012).

Apart from sea ice, snow-related feedbacks also have a major influence on the terrestrial Arctic. Arctic soils are frozen for most of the year while the top layer briefly thaws during the summer months (see Johansson, Chapter 18 of this volume). Extension of this brief period due to earlier snowmelt may have a strong impact on the carbon cycle. The extra energy absorbed by a snow-free surface leads to higher soil temperatures and a deeper active layer − ultimately affecting carbon cycle processes. Counter-intuitively, a deeper active layer can also occur due to a thicker snow pack in the preceding winter. Snow acts as an insulator, shielding the soil from the lowest winter temperatures, which raises the annual average soil temperature, leads to permafrost thaw, and wetter vegetation that emits more methane (Johansson et al. 2013; Nauta et al. 2014). Less snow in spring, and more in winter, can therefore strongly impact permafrost stability and carbon cycle processes.

Reductions in snow cover and sea ice extent may be intricately connected to each other. Sea ice retreat has been associated with increases in humidity (Screen and Simmonds 2010), possibly affecting the amount of snow in the following winter (Liu et al. 2012). In turn, the amount of snow on top of the ice affects annual dynamics in sea ice (Webster et al. 2014). These interactions between sea ice extent and snow cover make it difficult to disentangle them from each other. However, the relative importance of sea ice decline may rise since the long-term thinning trend in sea ice thickness implies that less warming is needed to melt larger areas (Laxon et al. 2013). Despite these complexities, all indications are that warming associated with sea ice decline and changing snow cover is likely to affect the carbon cycle.

Sea ice decline and mid-latitude weather

Recent research has indicated that sea ice decline may have an impact beyond the North Pole region, affecting mid-latitude weather extremes such as stronger cold spells in winter, more heat waves in summer, and increased rainfall (Cohen et al. 2014; Francis et al. 2009; Screen 2013). Climate extremes can have a strong impact on ecosystems (Reichstein et al. 2013), which suggests the possibility for an influence of sea ice decline on the carbon cycle at mid-latitudes. Three different mechanisms for sea ice-related climate extremes at the mid-latitudes have emerged: changes in storm tracks, an altered jet stream, and variations in planetary waves. When storm tracks shift northwards, this leads to mild winters in the mid-latitudes but harsh ones in the Arctic, and the reverse is true when a southward shift occurs. These shifts are strongly controlled by the North Atlantic Oscillation/Arctic Oscillation (NAO/AO), an atmospheric mode defined by the difference in sea level pressure between Iceland and the Azores. If sea ice decline affects barometric pressure, this may in turn affect the NAO/AO, and thus storm tracks (Vihma 2014). A more popular hypothesis is that polar amplification may have caused a weakening of the jet stream. The jet stream is driven by the temperature gradient between the tropics and the poles, and polar amplification reduces that difference. This may lead to a weaker and wavier jet stream, more persistent weather patterns, followed by an increase in weather extremes (Francis and Vavrus 2015). Finally, reduced sea ice extent and associated increases in snow cover (Liu et al. 2012) may have led to larger planetary waves, an altering of the stratospheric polar vortex, and subsequent atmospheric circulation anomalies (Cohen et al. 2014).

The plausibility of the above hypotheses, however, remains somewhat ambiguous due to the strong natural variability of the atmosphere, and the relatively short time since the emergence of polar amplification. At the moment, the most likely candidates for linkages between sea ice decline and climate extremes are changes in the jet stream and planetary waves (Cohen et al. 2014), especially in wintertime. Improved model simulations and longer time-series will help to further test these hypotheses and determine the extent of the influence of sea ice decline beyond the North Pole region. Since the impact of sea ice decline on high-latitude climate is much more certain, it is currently more feasible to identify interactions between the Arctic Ocean and the terrestrial carbon cycle within the North Pole region rather than beyond.

Sea ice decline and terrestrial CO$_2$ exchange

Trends in plant productivity and respiration

While the transition of the Arctic Ocean towards a seasonally sea ice-free state continues unabated, consequences for the uptake and release of CO$_2$ by terrestrial ecosystems remain uncertain. A response to sea ice-induced warming is expected, since both photosynthesis and respiration are temperature sensitive processes. For example, when a climate shifts towards a more optimal range for photosynthesis, and plants grow longer, the net uptake of CO_2 from the atmosphere may increase. Indeed, observations at the plot scale have shown that warmer summers have led to taller and more abundant vegetation in the Arctic (Elmendorf et al. 2012). Shrubs across the tundra biome, however, show a varied response to amplified summer warming (Myers-Smith et al. 2015). The mixed response to warming from this important vegetation type complicates extrapolation of plot-scale results to the entire Arctic, and verification of a link between sea ice decline and circumpolar plant growth.

At the pan-Arctic scale, ocean-land interactions are better investigated through careful analysis of satellite data. Plants display a strong contrast between the amount of light that is reflected in the visible red and near-infrared spectral regions, and this is expressed in satellite products as the Normalized Difference Vegetation Index (NDVI). A higher NDVI corresponds to higher photosynthetic activity and is a good indicator of plant productivity across tundra biomes (Mbufong et al. 2014). Satellites have provided long-term datasets of NDVI, and trends therein can be compared with other satellite observations of sea ice decline. Investigations have shown significant correlations between NDVI and sea ice concentrations throughout the Arctic over the 1982–2008 period (Bhatt et al. 2010; Post et al. 2013). This indicates that plant productivity increased while sea ice declined in response to more summer warmth.

Recent trends in NDVI, however, revealed unexpected changes in the relationship between plant productivity and sea ice decline. The strong link between sea ice and NDVI that appeared in trends from 1982 to 2008 lessened when additional data up to 2011 were included (Bhatt et al. 2010, 2013). In these three additional years, spring sea ice continued its decline throughout most of the Arctic, but long-term trends in NDVI weakened rather than strengthened, most strikingly in the Canadian high Arctic (Bhatt et al. 2013). These reductions in NDVI, commonly known as arctic browning, may have occurred as a result of lower summer warmth around the Arctic, possibly related to increased cloud cover that reduced incoming sunlight and photosynthesis. While it is possible that cloud cover changes were related to sea ice decline – if more open water led to more evaporation and clouds – another explanation may be that changes in large scale atmospheric circulation altered weather patterns (Bhatt et al. 2013). Alternatively, it has become increasingly clear that arctic browning may be event-driven, caused by extreme winter events that damage vegetation, pest outbreaks and fires (Phoenix and Bjerke 2016), and this has been confirmed regionally (Bjerke et al. 2014). Outside of the tundra biome, regional sea ice decline has been linked to other downward trends in NDVI. NDVI trends adjacent to Hudson Bay and Hudson Strait indicate a decline in forest growth due to higher temperatures and moisture stress, amplified by regional sea ice loss (Girardin et al. 2014). These recent developments show that a connection between sea ice decline and plant productivity may not be that straightforward.

One of the complicating factors is the fact that polar amplification during the growing season has been rather limited (Cohen et al. 2014; Screen et al. 2012). Changes during the shoulder seasons and lagged responses to those may be much more important: On the one hand, earlier sea ice retreat in spring and simultaneous advancing of snowmelt can extend the growing season, allowing for longer plant growth (Bhatt et al. 2010). On the other hand, a delay in snowfall at the end of summer, e.g. due to sea ice-induced warming, is not as likely to increase photosynthesis. By September, the amount of sunlight in the Arctic declines quickly, plants start to senesce, and higher temperatures cannot promote photosynthesis. However, higher temperatures do stimulate respiration – a release of CO_2. Unfortunately, respiration cannot be observed from space in the same way as plant growth is monitored with NDVI. Indications from observations and models are, however, that soil respiration may contribute significant amounts of carbon to the atmosphere following permafrost thaw (Schuur et al. 2015), with a clear trend in early winter (Commane et al. 2017), possibly enhanced by sea ice decline (Bhatt et al. 2014; Parmentier et al. 2013).

Net carbon sequestration and sea ice decline

The terrestrial carbon sink of the Arctic may shift towards more or less uptake of CO_2, whether respiration or photosynthesis is stimulated the most by warming. Unfortunately, observation programmes do not provide consistent results to resolve this problem. In Northeast Greenland, for example, a monitoring project showed that the net uptake of CO_2 over a 5-year period increased when the growing season lengthened (Groendahl et al. 2007). Yet, after a few additional years of flux measurements, the link between length of growing season and net carbon uptake disappeared (Lund et al. 2012). While respiration remained stimulated by higher temperatures, the same was not true for photosynthesis, and the strength of the carbon sink weakened. Around the same time, a study comparing eight years of CO_2 measurements in Northeastern Siberia concluded that while photosynthesis and respiration clearly differed with varying summer warmth, the two fluxes kept each other in balance. Consequently, longer growing seasons did not increase net carbon uptake (Parmentier et al. 2011). One of the possible explanations for the varied response of Arctic vegetation to changes in growing season length may be due to interannual variations in leaf production and photosynthetic capacity (Humphreys and Lafleur 2011). This leads to quite different year-to-year responses to environmental conditions and complicates a simple comparison between growing season length and net carbon uptake over the short term.

Studies aimed at the long-term perspective may provide some additional insight. A 20-year warming experiment in Alaskan tundra, for example, showed that plant biomass had increased over time, combined with a higher litter input to the soil (Sistla et al. 2013). At the same time, however, respiration increased throughout the soil, including a 'biotic awakening' of the deeper mineral layer related to a thicker active layer at the end of summer. Surprisingly, despite increases in both carbon input and decomposition throughout the active layer, the total amount of soil carbon remained the same – suggesting an increase in net carbon storage when combined with the additional plant growth. It remains possible, however, that this particular experiment has not yet attained a steady-state situation and decomposition may still outpace the additional carbon inputs in the future (Sistla et al. 2013).

What may be concluded from the above studies about the kind of impact that warming from sea ice retreat will have on the the exchange of CO_2 from the terrestrial Arctic? Many of these studies were not designed to investigate the particular warming that may follow from sea ice decline, but include the combined result from a variety of factors influencing the polar climate. Since sea ice decline is most likely to affect temperature trends in springtime and autumn (Screen et al. 2012), its direct impact will be most apparent in those seasons. Warmer autumns may lead to deeper thaw depths and increased respiration, but earlier snowmelt leads to prolonged periods of plant growth. Due to lagged responses to sea ice decline, e.g. through its influence on snow cover, the effect on the net carbon uptake of the terrestrial Arctic is not easily determined. Monitoring programmes seem to suggest that net carbon uptake has not yet changed much, but long-term trends in soil and biomass carbon pools may be negatively affected by permafrost thaw (Schuur et al. 2015), extreme winter events (Phoenix and Bjerke 2016) and tundra fires (Hu et al. 2010). Continued monitoring and model improvement remain necessary to pinpoint the direction in which the Arctic carbon sink will change following sea ice decline.

Sea ice decline and terrestrial methane emissions

The potential for sea ice-induced warming to affect methane emissions

Other than for CO_2 exchange, the impact of sea ice decline on Arctic methane emissions appears more straightforward. The Arctic is a source of methane (McGuire et al. 2012) and higher temperatures stimulate methane-producing microbes in the ground. Models indicate, therefore, that emissions have increased while sea ice declined (Parmentier et al. 2013). Then again, methane emissions are very sensitive to changes in hydrology since production occurs in the anaerobic, waterlogged, part of the soil. When methane diffuses upwards it may be oxidized while it passes through the top oxic layer of the soil. This oxidation layer deepens when the soil dries out and the water table lowers, while higher temperatures stimulate the methane-oxidizing microorganisms within. Combined, these processes could lead to a reduction in emissions from Arctic permafrost environments (Whalen and Reeburgh 1990). However, the reverse is just as probable: Permafrost thaw has been associated with a lowering of the surface, increased wetness, and higher methane emissions (Christensen et al. 2004). Permafrost environments exhibit highly heterogeneous geomorphology and vegetation, and the effect of a warming will vary accordingly (see Johansson, Chapter 18 of this volume). Indeed, a satellite-driven model study showed that methane emissions in the Arctic over the period 2003–2011 increased due to a wetting of the surface and higher summer temperatures (Watts et al. 2014). In the sub-Arctic, however, surface drying and cooling reduced emissions, largely compensating for the increases further up north. Due to these varying trends, methane emissions are not likely to respond uniformly to sea ice decline.

Correlations between methane emissions and sea ice decline

Overall, both observations and models indicate that high-latitude methane emissions have increased since the 1990s (McGuire et al. 2012), but monitoring programmes are spread too thin throughout the Arctic, and time periods covered are too short, to verify from measurements alone whether a connection to sea ice decline may have caused this. Possibly, both sea ice decline and rising methane emissions responded to higher temperatures, rather than one amplifying the other. A practical way to solve this conundrum is to compare sea ice concentrations to methane emissions simulated by biogeochemical process models. Models can represent the large heterogeneity of the Arctic environment, and account for the fragmented distribution of methane-emitting wetlands. Moreover, models can be forced with climate data spanning decades, thus providing a long-term dataset that can be correlated with the satellite record of sea ice concentrations. Figure 37.2 shows the results from such a study, with clear covariances between simulated methane emissions and sea ice throughout the northern high latitudes (Parmentier et al. 2015). Many areas show strong negative correlations, indicating that methane emissions went up while sea ice levels went down. Surprisingly, and contrary to expectations, some areas show positive correlations, which suggests a decrease in methane emissions with sea ice decline. However, these areas represent dry parts of the Arctic with few wetlands and exhibit net oxidation of methane in dry aerated parts of the soil. These positive correlations depict an increase in methane oxidation in response to higher temperatures, a logical consequence of sea ice decline. This uptake of methane is rather low and does not compensate for the increased methane emissions in other parts of the Arctic (Parmentier et al. 2015). In general, methane emissions in vast areas of the Arctic appear to covary strongly with sea ice decline.

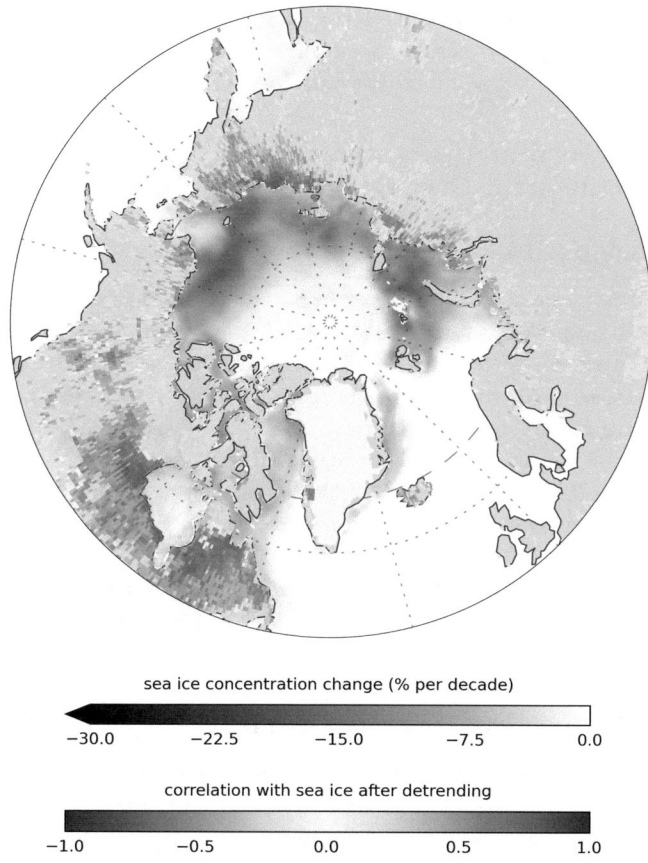

sea ice concentration change (% per decade)

−30.0   −22.5   −15.0   −7.5   0.0

correlation with sea ice after detrending

−1.0   −0.5   0.0   0.5   1.0

Figure 37.2 Correlations between May and October terrestrial methane emissions and sea ice concentration from 1981 to 2010. The linear trend in sea ice concentration is shown to indicate areas of high retreat. Note that high correlations do not necessarily equal high emissions. Original from Parmentier et al. (2015)

Simulated methane emissions for the years 2005–2010 are 1.7 Tg CH_4 yr^{-1} higher than the 1980s due to the retreat of sea ice (Parmentier et al. 2015) – equivalent to ~10% of all tundra emissions (McGuire et al. 2012). This increase was not spread evenly throughout the year, however. Temperature increases in sea ice-affected areas were predominantly higher towards the autumn, and the relative increase in modelled methane emissions was highest in that season. Indeed, observations support the possibility for significant amounts of methane to be emitted during the autumn (Mastepanov et al. 2013; Sturtevant et al. 2012). Although correlations with sea ice were also found in springtime, relative increases in methane emissions were not strongly amplified in sea ice-affected areas during that time. The weather in springtime has a strong influence on sea ice melt but causes earlier snowmelt on land as well. This provides an additional feedback, which makes it difficult to disentangle the combined effect of snow and ice interactions in that time of year. While Arctic methane emissions do not necessarily respond uniformly to higher temperatures, due to the interplay with hydrology, models suggest that – in general – methane emissions have risen due to recent sea ice decline. Since further sea ice retreat may lead to a warmer and wetter Arctic (Bintanja and Selten 2014), a reasonable assumption is that high latitude methane emissions will continue to rise in the near future.

Lateral carbon flows from the terrestrial environment to the Arctic Ocean

Coastal erosion and sea ice decline

So far, this chapter has dealt with the indirect impact of sea ice decline on the carbon cycle of the terrestrial Arctic through a warming feedback. However, sea ice decline has a much more direct effect on the release of terrestrial carbon through the erosion of ice-rich permafrost shores. Sea ice close to these shores protects against strong wave impacts, especially during the stormy Arctic autumn and winter. Retreat of sea ice, however, exposes shores to longer periods of wave action, while storm activity may be rising (Barnhart et al. 2014). The current coastal erosion rate of 0.5 m yr^{-1}, already much higher than for temperate coasts and exacerbated by permafrost thaw, is therefore expected to increase. This releases more carbon into the Arctic Ocean (Lantuit et al. 2013), and directly affects the marine carbon cycle. The impact on climate from these erosional processes depends on whether this carbon is preferentially mineralized and emitted into the atmosphere as CO_2 rather than buried in ocean sediments (Vonk et al. 2012, 2014).

Riverine input of carbon into the Arctic Ocean

Besides coastal erosion, the terrestrial environment further manifests its influence on the marine carbon cycle through the vast amount of organic matter and nutrients transported by Arctic rivers. Rivers have a disproportionately large influence on the Arctic Ocean: despite containing 1% of the global ocean volume, the basin receives about 10% of global river discharge (Holmes et al. 2011). Changes in the amount of carbon carried by rivers may therefore have a strong impact. In the East Siberian Arctic Shelf, the vast majority of organic matter is dominated by terrestrial sources (Vonk et al. 2012). Terrestrial organic matter originating from the permafrost region may be mobilized through erosion, fire or permafrost thaw. Once released into Arctic streams and rivers, this carbon starts off on a long journey towards the Arctic Ocean during which it may be released as a greenhouse gas, temporarily or permanently buried in sediments, or transferred to the ocean. In this way, lateral flows alter the location and timing of greenhouse gas release from thawing permafrost to the atmosphere, and connect the land and ocean (Vonk and Gustafsson 2013).

An intensification of the hydrological cycle may amplify the amount of organic carbon flowing into the Arctic Ocean. Indications are that the amount of water flowing into the Arctic Ocean has increased in recent decades (Déry et al. 2009; Peterson et al. 2002), which has been attributed to an increased northward transport of moisture as a result of climate change (McClelland et al. 2004). The impact of additional precipitation on lateral carbon flows may depend on the degree of permafrost thaw. For example, the amount of dissolved carbon in permafrost-free watersheds in West Siberia was found to be higher than in areas still influenced by the presence of permafrost (Frey et al. 2007). However, it is unclear whether this carbon is emitted back into the atmosphere once it reaches the ocean. On geological timescales, for instance, Arctic rivers may have been an important sink for CO_2 (Hilton et al. 2015). Also, carbon released from thawing permafrost dominates the East Siberian Arctic Shelf, but it settles towards the bottom rapidly – despite its high lability (Vonk et al. 2012, 2014). This limits the potential for a strong atmospheric feedback. Although it is likely that a warmer, wetter Arctic will lead to changes in the lateral transport of organic carbon, its fate will be strongly determined by local factors (Tank et al. 2012). The inclusion of lateral ocean-land processes in Earth system models is necessary to further our insight on how the interplay of the ocean and land will affect the Arctic carbon cycle.

Conclusions

Ocean-land interactions in the Arctic integrate the terrestrial and marine environments. The immense loss of sea ice in the Arctic Ocean has led to a warming feedback that has amplified high-latitude temperatures. Higher temperatures affect the terrestrial carbon cycle through altered plant productivity, increased respiration, and higher methane emissions. Although connections to the ocean appear likely, the sum of these parts is uncertain due to the high variability in each. Plant productivity and respiration are two large opposing fluxes and compensate for changes in each other. Still, methane emissions appear to have increased as a result of sea ice decline. In turn, the terrestrial environment affects the marine carbon cycle through the large amount of organic matter moving into the Arctic Ocean by coastal erosion and riverine input. These flows are expected to increase with further climate change, but the possibility for an atmospheric feedback is strongly controlled by the tendency of this carbon for either burial or mineralization. As shown in this chapter, the intricate ways in which the Arctic marine and terrestrial environment are connected shows that the Arctic carbon cycle, and future trends therein, cannot be understood without a thorough understanding of ocean-land interactions.

References

Barnhart, K.R., Overeem, I. and Anderson, R.S. 2014. 'The effect of changing sea ice on the physical vulnerability of Arctic coasts' *The Cryosphere* 8(5): 1777–1799.

Bhatt, U.S., Walker, D.A., Raynolds, M.K., Bieniek, P., Epstein, H., Comiso, J., Pinzon, J., Tucker, C. and Polyakov, I. 2013. 'Recent declines in warming and vegetation greening trends over pan-Arctic tundra' *Remote Sensing* 5(9): 4229–4254.

Bhatt, U.S., Walker, D.A., Raynolds, M.K., Comiso, J.C., Epstein, H.E., Jia, G., Gens, R., Pinzon, J.E., Tucker, C.J., Tweedie, C.E. and Webber, P.J. 2010. 'Circumpolar Arctic tundra vegetation change is linked to sea ice decline' *Earth Interactions* 14(8): 1–20.

Bhatt, U.S., Walker, D.A., Walsh, J.E., Carmack, E.C., Frey, K.E., Meier, W.N., Moore, S.E., Parmentier, F.J.W., Post, E., Romanovsky, V.E. and Simpson, W.R. 2014. 'Implications of Arctic sea ice decline for the Earth system' *Annual Review of Environment and Resources* 39(1): 57–89.

Bintanja, R., Graversen, R.G. and Hazeleger, W. 2011. 'Arctic winter warming amplified by the thermal inversion and consequent low infrared cooling to space' *Nature Geoscience* 4(11): 758–761.

Bintanja, R. and Selten, F.M. 2014. 'Future increases in Arctic precipitation linked to local evaporation and sea-ice retreat' *Nature* 509(7501): 479–482.

Bjerke, J.W., Karlsen, S.R., Høgda, K.A., Malnes, E., Jepsen, J.U., Lovibond, S., Vikhamar-Schuler, D. and Tømmervik, H. 2014. 'Record-low primary productivity and high plant damage in the Nordic Arctic Region in 2012 caused by multiple weather events and pest outbreaks' *Environmental Research Letters* 9(8), 084006, doi:10.1088/1748–9326/9/8/084006.

Christensen, T.R., Johansson, T.R., Akerman, H.J., Mastepanov, M., Malmer, N., Friborg, T., Crill, P.M. and Svensson, B.H. 2004. 'Thawing sub-Arctic permafrost: effects on vegetation and methane emissions' *Geophysical Research Letters* 31, L04501, doi:10.1029/2003GL018680.

Cohen, J., Screen, J.A., Furtado, J.C., Barlow, M., Whittleston, D., Coumou, D., Francis, J., Dethloff, K., Entekhabi, D., Overland, J. and Jones, J. 2014. 'Recent Arctic amplification and extreme mid-latitude weather' *Nature Geoscience* 7(9): 627–637.

Commane, R., Lindaas, J., Benmergui, J., Luus, K.A., Chang, R.Y.W., Daube, B.C., Euskirchen, E.S., Henderson, J.M., Karion, A., Miller, J.B., Miller, S.M., Parazoo, N.C., Randerson, J.T., Sweeney, C., Tans, P., Thoning, K., Veraverbeke, S., Miller, C.E. and Wofsy, S.C. 2017. 'Carbon dioxide sources from Alaska driven by increasing early winter respiration from Arctic tundra' *PNAS* 114(21): 5361–5366.

Déry, S.J. and Brown, R.D. 2007. 'Recent Northern Hemisphere snow cover extent trends and implications for the snow-albedo feedback' *Geophysical Research Letters* 34(22), L22504, doi:10.1029/2007GL031474.

Déry, S.J., Hernández Henríquez, M.A., Burford, J.E. and Wood, E.F. 2009. 'Observational evidence of an intensifying hydrological cycle in northern Canada' *Geophysical Research Letters* 36(13), L13402, doi:10.1029/2009GL038852.

Deser, C., Tomas, R., Alexander, M. and Lawrence, D. 2010. 'The seasonal atmospheric response to projected Arctic sea ice loss in the late twenty-first century' *Journal of Climate* 23(2): 333–351.

Elmendorf, S.C., Henry, G.H.R., Hollister, R.D., Björk, R.G., Boulanger-Lapointe, N., Cooper, E.J., Cornelissen, J.H.C., Day, T.A., Dorrepaal, E., Elumeeva, T.G., Gill, M., Gould, W.A., Harte, J., Hik, D.S., Hofgaard, A., Johnson, D.R., Johnstone, J.F., Jónsdóttir, I.S., Jorgenson, J.C., Klanderud, K., Klein, J.A., Koh, S., Kudo, G., Lara, M., Lévesque, E., Magnússon, B., May, J.L., Mercado-Díaz, J.A., Michelsen, A., Molau, U., Myers-Smith, I.H., Oberbauer, S.F., Onipchenko, V.G., Rixen, C., Martin Schmidt, N., Shaver, G.R., Spasojevic, M.J., Þórhallsdóttir, Þ.E., Tolvanen, A., Troxler, T., Tweedie, C.E., Villareal, S., Wahren, C-H., Walker, X., Webber, P.J., Welker, J.M. and Wipf, S. 2012. 'Plot-scale evidence of tundra vegetation change and links to recent summer warming' *Nature Climate Change* 2(6): 453–457.

Francis, J.A., Chan, W., Leathers, D.J., Miller, J.R. and Veron, D.E. 2009. 'Winter Northern Hemisphere weather patterns remember summer Arctic sea-ice extent' *Geophysical Research Letters* 36(7), L07503, doi:10.1029/2009GL037274.

Francis, J.A. and Vavrus, S.J. 2015. 'Evidence for a wavier jet stream in response to rapid Arctic warming' *Environmental Research Letters* 10(1), 014005, doi:10.1088/1748–9326/10/1/014005.

Frey, K.E., Siegel, D.I. and Smith, L.C. 2007. 'Geochemistry of west Siberian streams and their potential response to permafrost degradation' *Water Resources Research* 43, W03406, doi:10.1029/2006WR004902.

Girardin, M.P., Guo, X.J., De Jong, R., Kinnard, C., Bernier, P. and Raulier, F. 2014. 'Unusual forest growth decline in boreal North America covaries with the retreat of Arctic sea ice' *Global Change Biology* 20(3): 851–866.

Graversen, R.G. and Wang, M. 2009. 'Polar amplification in a coupled climate model with locked albedo' *Climate Dynamics* 33(5): 629–643.

Groendahl, L., Friborg, T. and Soegaard, H. 2007. 'Temperature and snow-melt controls on interannual variability in carbon exchange in the high Arctic' *Theoretical and Applied Climatology* 88(1–2): 111–125.

Hilton, R.G., Galy, V., Gaillardet, J., Dellinger, M., Bryant, C., O'Regan, M., Gröcke, D.R., Coxall, H., Bouchez, J. and Calmels, D. 2015. 'Erosion of organic carbon in the Arctic as a geological carbon dioxide sink' *Nature* 524(7563): 84–87.

Holmes, R.M., McClelland, J.W., Peterson, B.J., Tank, S.E., Bulygina, E., Eglinton, T.I., Gordeev, V.V., Gurtovaya, T.Y., Raymond, P.A., Repeta, D.J., Staples, R., Striegl, R.G., Zhulidov, A.V. and Zimov, S.A. 2011. 'Seasonal and annual fluxes of nutrients and organic matter from large rivers to the Arctic Ocean and surrounding seas' *Estuaries and Coasts* 35(2): 369–382.

Hu, F.S., Higuera, P.E., Walsh, J.E., Chapman, W.L., Duffy, P.A., Brubaker, L.B. and Chipman, M.L. 2010. 'Tundra burning in Alaska: Linkages to climatic change and sea ice retreat' *Journal of Geophysical Research: Biogeosciences* 115, G04002, doi:10.1029/2009JG001270.

Hugelius, G., Strauss, J., Zubrzycki, S., Harden, J. W., Schuur, E.A.G., Ping, C.L., Schirrmeister, L., Grosse, G., Michaelson, G.J., Koven, C.D., O'Donnell, J.A., Elberling, B., Mishra, U., Camill, P., Yu, Z., Palmtag, J. and Kuhry, P. 2014. 'Estimated stocks of circumpolar permafrost carbon with quantified uncertainty ranges and identified data gaps' *Biogeosciences* 11(23): 6573–6593.

Humphreys, E.R. and Lafleur, P.M. 2011. 'Does earlier snowmelt lead to greater CO2 sequestration in two low Arctic tundra ecosystems?' *Geophysical Research Letters* 38(9), L09703, doi:10.1029/2011GL047339.

Jeong, J-H., Kug, J-S., Linderholm, H.W., Chen, D., Kim, B-M. and Jun, S-Y. 2014. 'Intensified Arctic warming under greenhouse warming by vegetation-atmosphere-sea ice interaction' *Environmental Research Letters* 9(9), 094007, doi:10.1088/1748–9326/9/9/094007.

Johansson, M., Callaghan, T.V., Bosiö, J., Åkerman, H.J., Jackowicz-Korczyński, M. and Christensen, T.R. 2013. 'Rapid responses of permafrost and vegetation to experimentally increased snow cover in sub-arctic Sweden' *Environmental Research Letters* 8(3), 035025, doi:10.1088/1748–9326/8/3/035025.

Kapsch, M-L., Graversen, R.G. and Tjernström, M. 2013 'Springtime atmospheric energy transport and the control of Arctic summer sea-ice extent' *Nature Climate Change* 3(8): 744–748.

Lantuit, H., Overduin, P.P. and Wetterich, S. 2013. 'Recent progress regarding permafrost coasts' *Permafrost and Periglacial Processes* 24(2): 120–130.

Laxon, S.W., Giles, K.A., Ridout, A.L., Wingham, D.J., Willatt, R., Cullen, R., Kwok, R., Schweiger, A., Zhang, J., Haas, C., Hendricks, S., Krishfield, R., Kurtz, N., Farrell, S. and Davidson, M. 2013. 'CryoSat-2 estimates of Arctic sea ice thickness and volume' *Geophysical Research Letters* 40(4): 732–737.

Liu, J., Curry, J.A., Wang, H., Song, M. and Horton, R.M. 2012 'Impact of declining Arctic sea ice on winter snowfall' *PNAS* 109(11): 4074–4079.

Lund, M., Falk, J.M., Friborg, T., Mbufong, H.N., Sigsgaard, C., Soegaard, H. and Tamstorf, M.P. 2012. 'Trends in CO2 exchange in a high Arctic tundra heath, 2000–2010' *Journal of Geophysical Research: Biogeosciences* 117, G02001, doi:10.1029/2011JG001901.

Mastepanov, M., Sigsgaard, C., Tagesson, T., Strom, L., Tamstorf, M.P., Lund, M. and Christensen, T.R. 2013. 'Revisiting factors controlling methane emissions from high-Arctic tundra' *Biogeosciences* 10(7): 5139–5158.

Mbufong, H.N., Lund, M., Aurela, M., Christensen, T.R., Eugster, W., Friborg, T., Hansen, B.U., Humphreys, E.R., Jackowicz-Korczyński, M., Kutzbach, L., Lafleur, P.M., Oechel, W.C., Parmentier, F.J.W., Rasse, D.P., Rocha, A.V., Sachs, T., van der Molen, M.K. and Tamstorf, M.P. 2014. 'Assessing the spatial variability in peak season CO2 exchange characteristics across the Arctic tundra using a light response curve parameterization' *Biogeosciences* 11(17): 4897–4912.

McClelland, J.W., Holmes, R.M., Peterson, B.J. and Stieglitz, M. 2004. 'Increasing river discharge in the Eurasian Arctic: consideration of dams, permafrost thaw, and fires as potential agents of change' *Journal of Geophysical Research: Atmospheres* 109, D18102, doi:10.1029/2004JD004583.

McGuire, A.D., Christensen, T.R., Hayes, D., Heroult, A., Euskirchen, E., Kimball, J.S., Koven, C., LaFleur, P., Miller, P.A., Oechel, W., Peylin, P., Williams, M. and Yi, Y. 2012. 'An assessment of the carbon balance of arctic tundra: comparisons among observations, process models, and atmospheric inversions', *Biogeosciences* 9(8): 3185–3204.

Myers-Smith, I.H., Elmendorf, S.C., Beck, P.S.A., Wilmking, M., Hallinger, M., Blok, D., Tape, K.D., Rayback, S.A., Macias-Fauria, M., Forbes, B.C., Speed, J.D.M., Boulanger-Lapointe, N., Rixen, C., Lévesque, E., Schmidt, N.M., Baittinger, C., Trant, A.J., Hermanutz, L., Collier, L.S., Dawes, M.A., Lantz, T.C., Weijers, S., Jørgensen, R.H., Buchwal, A., Buras, A., Naito, A.T., Ravolainen, V., Schaepman-Strub, G., Wheeler, J.A., Wipf, S., Guay, K.C., Hik, D.S. and Vellend, M. 2015. 'Climate sensitivity of shrub growth across the tundra biome', *Nature Climate Change* 5(9): 887–891.

Nauta, A.L., Heijmans, M.M.P.D., Blok, D., Limpens, J., Elberling, B., Gallagher, A., Li, B., Petrov, R.E., Maximov, T.C., van Huissteden, J. and Berendse, F. 2014. 'Permafrost collapse after shrub removal shifts tundra ecosystem to a methane source' *Nature Climate Change* 5(1): 67–70.

Parkinson, C.L. and Cavalieri, D.J. 2012. 'Antarctic sea ice variability and trends, 1979–2010' *The Cryosphere* 6(4): 871–880.

Parmentier, F.J.W., Christensen, T.R., Sørensen, L.L., Rysgaard, S., McGuire, A.D., Miller, P.A. and Walker, D.A. 2013. 'The impact of lower sea-ice extent on Arctic greenhouse-gas exchange' *Nature Climate Change* 3(3): 195–202.

Parmentier, F.J.W., van der Molen, M.K., van Huissteden, J., Karsanaev, S.A., Kononov, A.V., Suzdalov, D.A., Maximov, T.C. and Dolman, A.J. 2011. 'Longer growing seasons do not increase net carbon uptake in the northeastern Siberian tundra' *Journal of Geophysical Research: Biogeosciences* 116, G04013, doi:10.1029/2011JG001653.

Parmentier, F.J.W., Zhang, W., Mi, Y., Zhu, X., Huissteden, J., Hayes, D.J., Zhuang, Q., Christensen, T.R. and McGuire, A.D. 2015. 'Rising methane emissions from northern wetlands associated with sea ice decline' *Geophysical Research Letters* 42(17): 7214–7222.

Perovich, D.K. and Polashenski, C. 2012. 'Albedo evolution of seasonal Arctic sea ice' *Geophysical Research Letters* 39(8), L08501, doi:10.1029/2012GL051432.

Peterson, B.J., Holmes, R.M., McClelland, J.W., Vörösmarty, C.J., Lammers, R.B., Shiklomanov, A.I., Shiklomanov, I.A. and Rahmstorf, S. 2002. 'Increasing river discharge to the Arctic Ocean' *Science* 298(5601): 2171–2173.

Phoenix, G.K. and Bjerke, J.W. 2016. 'Arctic browning: extreme events and trends reversing arctic greening' *Global Change Biology* 22(9): 2960–2962.

Pistone, K., Eisenman, I. and Ramanathan, V. 2014. 'Observational determination of albedo decrease caused by vanishing Arctic sea ice' *PNAS* 111(9): 3322–3326.

Pithan, F. and Mauritsen, T. 2014. 'Arctic amplification dominated by temperature feedbacks in contemporary climate models' *Nature Geoscience* 7(3): 181–184.

Post, E., Bhatt, U.S., Bitz, C.M., Brodie, J.F., Fulton, T.L., Hebblewhite, M., Kerby, J., Kutz, S.J., Stirling, I. and Walker, D.A. 2013. 'Ecological consequences of sea-ice decline' *Science* 341(6145): 519–524.

Reichstein, M., Bahn, M., Ciais, P., Frank, D., Mahecha, M.D., Seneviratne, S.I., Zscheischler, J., Beer, C., Buchmann, N., Frank, D.C., Papale, D., Rammig, A., Smith, P., Thonicke, K., van der Velde, M., Vicca, S., Walz, A. and Wattenbach, M. 2013. 'Climate extremes and the carbon cycle' *Nature* 500(7462): 287–295.

Schröder, D., Feltham, D.L., Flocco, D. and Tsamados, M. 2014. 'September Arctic sea-ice minimum predicted by spring melt-pond fraction' *Nature Climate Change* 4(5): 353–357.

Schuur, E.A.G., McGuire, A.D., Schädel, C., Grosse, G., Harden, J.W., Hayes, D.J., Hugelius, G., Koven, C.D., Kuhry, P., Lawrence, D.M., Natali, S.M., Olefeldt, D., Romanovsky, V.E., Schaefer, K., Turetsky, M.R., Treat, C.C. and Vonk, J.E. 2015. 'Climate change and the permafrost carbon feedback' *Nature* 520(7546): 171–179.

Screen, J.A. 2013. 'Influence of Arctic sea ice on European summer precipitation' *Environmental Research Letters* 8(4), 044015, doi:10.1088/1748–9326/8/4/044015.

Screen, J.A., Deser, C. and Simmonds, I. 2012. 'Local and remote controls on observed Arctic warming' *Geophysical Research Letters* 39(10), L10709, doi:10.1029/2012GL051598.

Screen, J.A. and Simmonds, I. 2010. 'The central role of diminishing sea ice in recent Arctic temperature amplification' *Nature* 464(7293): 1334–1337.

Serreze, M.C., Barrett, A.P., Stroeve, J.C., Kindig, D.N. and Holland, M.M. 2009. 'The emergence of surface-based Arctic amplification' *The Cryosphere* 3(1): 11–19.

Serreze, M.C. and Barry, R.G. 2011. 'Processes and impacts of Arctic amplification: a research synthesis' *Global and Planetary Change* 77: 85–96.

Sistla, S.A., Moore, J.C., Simpson, R.T., Gough, L., Shaver, G.R. and Schimel, J.P. 2013. 'Long-term warming restructures Arctic tundra without changing net soil carbon storage' *Nature* 497(7451): 615–618.

Spielhagen, R.F., Werner, K., Sorensen, S.A., Zamelczyk, K., Kandiano, E., Budeus, G., Husum, K., Marchitto, T.M. and Hald, M. 2011. 'Enhanced modern heat transfer to the Arctic by warm Atlantic water' *Science* 331(6016): 450–453.

Stroeve, J.C., Markus, T., Boisvert, L., Miller, J. and Barrett, A. 2014. 'Changes in Arctic melt season and implications for sea ice loss' *Geophysical Research Letters* 41(4): 1216–1225.

Sturtevant, C.S., Oechel, W.C., Zona, D., Kim, Y. and Emerson, C.E. 2012. 'Soil moisture control over autumn season methane flux, Arctic Coastal Plain of Alaska' *Biogeosciences* 9(4): 1423–1440.

Tank, S.E., Frey, K.E., Striegl, R.G., Raymond, P.A., Holmes, R.M., McClelland, J.W. and Peterson, B.J. 2012. 'Landscape-level controls on dissolved carbon flux from diverse catchments of the circumboreal' *Global Biogeochemical Cycles* 26(4), GB0E02, doi:10.1029/2012GB004299.

Taylor, P.C., Cai, M., Hu, A., Meehl, J., Washington, W. and Zhang, G.J. 2013. 'A decomposition of feedback contributions to polar warming amplification' *Journal of Climate* 26(18): 7023–7043.

Vihma, T. 2014. 'Effects of Arctic sea ice decline on weather and climate: a review' *Surveys in Geophysics* 35(5): 1175–1214.

Vonk, J.E. and Gustafsson, Ö. 2013. 'Permafrost-carbon complexities' *Nature Geoscience* 6(9): 675–676.

Vonk, J.E., Sánchez-García, L., van Dongen, B.E., Alling, V., Kosmach, D., Charkin, A., Semiletov, I.P., Dudarev, O.V., Shakhova, N., Roos, P., Eglinton, T. ., Andersson, A. and Gustafsson, O. 2012. 'Activation of old carbon by erosion of coastal and subsea permafrost in Arctic Siberia' *Nature* 489(7414): 137–140.

Vonk, J.E., Semiletov, I.P., Dudarev, O.V., Eglinton, T.I., Andersson, A., Shakhova, N., Charkin, A., Heim, B. and Gustafsson, Ö. 2014. 'Preferential burial of permafrost-derived organic carbon in Siberian-Arctic shelf waters' *Journal of Geophysical Research: Oceans* 119(12): 8410–8421.

Watts, J.D., Kimball, J.S., Bartsch, A. and McDonald, K.C. 2014. 'Surface water inundation in the boreal-Arctic: potential impacts on regional methane emissions' *Environmental Research Letters* 9(7), 075001, doi:10.1088/1748-9326/9/7/075001.

Webster, M.A., Rigor, I.G., Nghiem, S.V., Kurtz, N.T., Farrell, S.L., Perovich, D.K. and Sturm, M. 2014. 'Interdecadal changes in snow depth on Arctic sea ice' *Journal of Geophysical Research: Oceans* 119(8): 5395–5406.

Whalen, S.C. and Reeburgh, W.S. 1990. 'Consumption of atmospheric methane by tundra soils' *Nature* 346(6280): 160–162.

Winton, M. 2006. 'Amplified Arctic climate change: what does surface albedo feedback have to do with it?' *Geophysical Research. Letters* 33(3), L03701, doi:10.1029/2005GL025244.

Yang, X-Y., Fyfe, J.C. and Flato, G.M. 2010. 'The role of poleward energy transport in Arctic temperature evolution' *Geophysical Research Letters* 37, L14803, doi:10.1029/2010GL043934.

38

Back to the future

Detecting past Arctic environmental change and investing in future observations

*Terry V. Callaghan, Margareta Johansson
and Nadya Matveyeva*

Introduction

World attention is focused on changes in the Arctic's environment because the change in air temperature is twice as fast here compared to the global average (Overland et al. 2016; Marshall, Chapter 15 in this volume) and the consequences of the changes provide challenges and opportunities for Arctic residents and the global community (e.g. Callaghan et al. 2014). However, most of our information on past changes in climate and their environmental consequences for the cryosphere and ecosystems come from dedicated land-based observations generally stretching back about 60 years and even more recent satellite observations that started only about 30 years ago (e.g. Xu et al. 2013). Exceptionally, environmental monitoring activities at early-established research stations were initiated over 100 years ago, for example at the Abisko Station in Northern Sweden (Callaghan et al. 2010) and at the Arctic Station in West Greenland (http://arktiskstation.ku.dk/english/). Such long-term observations show two periods of climate warming in the Arctic during the twentieth century and place current warming in a longer-term context. For even longer periods of observation, proxies of environmental changes can be derived from tree-rings (about 7,400 years, Grudd et al. 2002), shrub rings (more than 400 years, Buras and Wilmking 2015) and other plant proxies (about 500 years, e.g. *Cassiope tetragona* growth (Havström et al. 1995) and precipitation environment (Welker et al. 1995). Peat profiles and lake sediment cores give proxies stretching back thousands of years (Briner et al. 2014) and ice cores now yield records from 800,000 years ago (Luthi et al. 2008; Whitehouse, Chapter 16 this volume; McKay, Chapter 34 this volume). Archaeology also provides evidence of past environments such as past biodiversity, particularly of now locally extirpated animals hunted and observed by prehistoric communities (Hadingham 1979).

In addition to these "conventional" methods of detecting past environmental changes, there are many other methods that use direct measures from various records not intended for the purpose. These methods include paintings, photographs and expedition reports. They also include observations made by researchers over the past half-century, before the paradigm of climate change appeared, that can be repeated – sometimes by the same individuals. To capture this type of information on environmental changes in the Arctic, and to stimulate re-visits to old sites, an International Polar Year (IPY) project "Back to the Future" (BTF) was formulated and funded. The results of the initial project were published in a special issue of the journal *Ambio* by Callaghan and Tweedie (2011). However, the project had legacy and further studies developed (e.g. Matveyeva and Zanokha 2013a, 2013b, 2017; Matveyeva et al. 2014; Matveyeva 2017). In addition, many long-term data sets existed that were not analysed before the BTF Project, e.g. snow depth and ice layers (Johansson, C. et al. 2011) and permafrost temperature (Johansson, M. et al. 2011). One of the aims of the BTF Project was to stimulate an additional legacy through inter-generational interactions. Researchers that had initiated studies many decades earlier, worked with early career researchers to pass on stewardship of old sites and data sets. In fact, this chapter exemplifies such inter-generational interactions as co-author Terry Callaghan was the founder of INTERACT, a network of 84 northern research stations that monitor environmental changes, and co-author Margareta Johansson is now the coordinator of this network (www.eu-interact.org).

An important aspect of the BTF approach is that the evidence of changes – or no changes – is determined independently of the climate change issue as the original sites, paintings, photographs, and data sets were established before the climate change paradigm predominated. In contrast, studies carried out since climate change became a dominant issue focus on areas that have changed in the way expected and do not necessarily form a random sample. Thus, about seven times more publications focus on greening of the Arctic than no changes or browning (Gatti and Callaghan in prep.), whereas satellite images show that only 37% of the Arctic has greened significantly between 1982 and 2012 (Xu et al. 2013), and more recent studies show a net browning in recent years (Epstein et al. 2015; Phoenix and Bjerke 2016).

The aims of this chapter are: 1) to describe the Back to the Future approach with illustrations of different data sets and their conclusions; and 2) to stimulate the growth of such studies. We do not provide a comprehensive review of BTF but give example stories from the first-hand experience of the authors. The numerous and important works based on palaeo-proxies of past environmental changes are outside the scope of this study.

Art

Artwork gives clues to past environments but includes the biases of the artist's eye. Sometimes, fantasy interplays with reality and the credibility is diminished. Nonetheless, outside the Arctic, early Saharan cave paintings made 12,000–6,000 thousand years ago (Coulson and Campbell 2013) indicate ecosystems with grasslands and herbivores in areas that are now desert areas and this dramatic environmental change can be validated by proxies (Brooks et al. 2005). Sadly, such examples of artwork appear to be lacking in the Arctic. Instead, drawings and paintings by expeditions from outside the Arctic can be found. Unfortunately, most dramatize the Arctic's harsh environments and the dangers of ice and polar bears to impress people in the South. A rare exception is a painting of Kongsfjord, Svalbard by Chydenius from 12 August 1861 (Chydenius 1861, cited in Liljequist 1993; Figure 38.1a). The mountains can be validated by recent photographs (Figure 38.1b) and therefore glacier retreat can be inferred with some certainty. Comparatively recent (since 1967) retreat of the glacier can be validated by direct measurements (Hagen and Lefauconnier 1995). Knowledge of changes in biodiversity (Myers-Smith et al. 2011, 2015; Elmendorf et al. 2012a) with climate warming enable an artistic impression to be made of a possible future environment (Figure 38.1c), thereby encompassing a period of actual and projected environmental changes over about 200 hundred years. Although the quantitative aspects of changes are difficult to determine, the visual impact of past changes and expected future changes are important communication aids for policy makers and educators.

Terry V. Callaghan et al.

Photographs

Many photographs exist of environments throughout the Arctic. Even though the intentions of the photographs were not to record the environments for future reference, backgrounds and foregrounds in many photos provide good evidence of past environments. Usually, the spots from where photographs were made can be identified and used for repeat photography. A successful example of this approach was provided in tree-line studies based in northern Sweden (Van Bogaert et al. 2011). An archive of photographs taken by Borg Mesch, an early tour guide and photographer, is held by Kiruna Commune, northern Sweden. The photographs were taken at the end of the nineteenth century and in the beginning of the twentieth century and many show important environmental details in an area of Swedish Lapland where development has not obscured the field of view of the photographs. Comparisons of tree-line dynamics on three neighbouring mountains over the past 100 years (Van Bogaert et al. 2011), for example, show an upward advance of tree-line as expected during climate warming, a downward movement of tree-line which was not expected but probably resulted from winter warming and increased insect damage to trees and a completely stable tree-line where scree slopes prevent upward movement.

Although it is difficult to determine causes of changes in ecosystems and environments from Back to the Future studies (correlations with climate warming can be misleading), some early photographs give indications of possible causes. Again using Borg Mesch photographs, Emanuelsson (1987) used repeat photography to demonstrate dramatic forest thickening and growth in northern Sweden during the twentieth century. However, close inspection of the Borg Mesch photographs show a Sámi life style that has now passed with goats in the mountains and Sámi kotas (summer sleeping tents made of birch logs and bark). Consequently, the changes that were demonstrated by the repeat photography were probably due to the synergistic effect of climate warming and declining intensity of land use.

Site re-visits

During the International Biological Programme Tundra Biome (IBP) Project (1967 to 1974), the concept of climate warming did not exist broadly and population dynamics, at least in plants, was in its infancy with major works appearing shortly afterwards (Harper 1977; Grime 1978; but see Callaghan 1976). The emphasis of the botanical studies was therefore on recording plant community structure (cover, abundance and frequency of species: Andreev and Aleksandrova 1981; Bliss 1984) and productivity – both production *per se* (Tikhomirov et al. 1981; Wielgolaski et al. 1981) and production processes (Tieszen et al. 1981). Many research plots were established throughout the Arctic by the then young researchers and they were marked with simple wooden or metal pegs – GPS did not exist then. Occasionally, photographs were taken. A challenge during the BTF Project, was for the original researchers to go back to their sites together with members of the next generation of researchers (often physically demanding!) and to relocate their old sites and plots. Once found, the plots could be recorded again using the "same eyes" and same methods as earlier. However, legacy could be achieved by using "new" GPS technology to mark the plots for the future and handing stewardship of sites and data to the next generation.

West Greenland

In 1967, sites were established near to the Arctic Station on Disko Island, West Greenland during the IBP "Bipolar Botanical Project" (Callaghan et al. 1976). In 1970, neighbouring

494

Figure 38.1a

Figure 38.1b

Figure 38.1c

Figure 38.1 Historical and hypothetical environmental changes at the head of Kongsfjord, Svalbard, looking east from the south shore near Ny Ålesund; a. Painting from Chydenius from 12 August 1861 cited in Liljequist 1993; b. Photo by T.V. Callaghan from 2001; c. Hypothetical future changes (art work by C.P. Callaghan). The time slices show actual and hypothetical glacier retreat, and hypothetical tall shrub growth and red fox invasion.

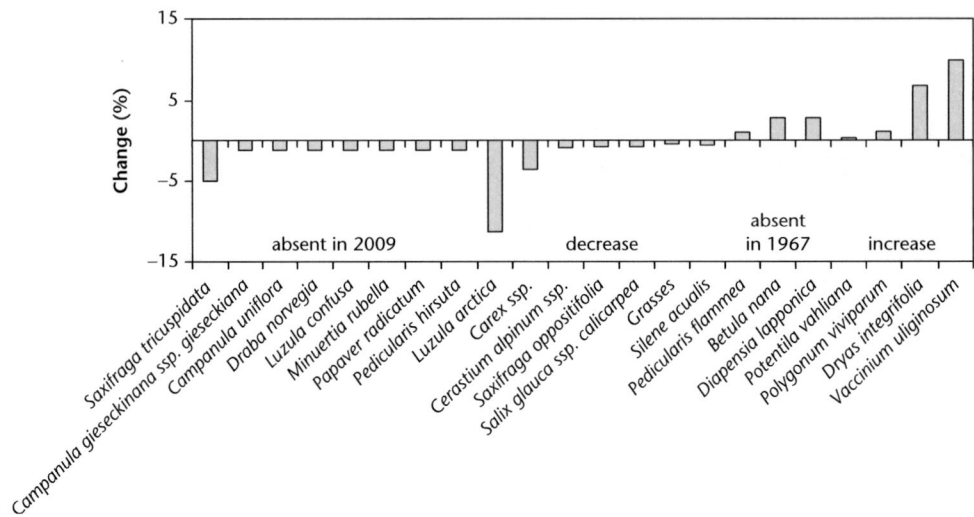

Figure 38.2   Example data set from the site re-visit to Disko Island, West Greenland showing changes in frequency of species at the fell-field site. The data set shows an overall loss of and decline in species.

sites were established, and species frequency and cover were recorded together with the phenology and biometrics of the grass *Phleum alpinum* (Callaghan 1972). In 2009, the original researcher returned with two younger researchers to re-record two sites, a rich herb meadow site and a fell-field site (Callaghan et al. 2011a). Photographic evidence and species recording (Figure 38.2) showed surprising stability of habitat and vegetation over the interim 39 years despite recent rapid climatic warming. However, detailed quantification of plant community changes showed small differences in species frequency and cover-abundance. At the fell-field site, total cover was still about 50% after 39 years with the dominance of low-growing plant species but there was a small shift from a fell-field towards a heath community. Biodiversity had decreased over time with a loss of 6 species, although 3 species not present in 1970 were recorded in 2009. At the herb-rich meadow site, changes in the plant community were also minor but with some indication of a drying of the habitat. The stability/small changes contrast with the "greening of the Arctic" concept that previously prevailed (Jia et al. 2003). As opposed to the relative stability at this site, re-recording of the phenology of *Phleum alpinum* showed dramatic shifts in timing with early flowering development responding to an earlier onset of the growing season. However, measurements of biometry showed no significant difference in the morphology of the species. Also, no viable seeds have (to the authors' knowledge) been recorded from this, the world's northernmost population of the grass. Indeed, detailed maps made in 1970 showed that measurements were made on the same clones as those investigated in 1970. This raises the question of the importance of the multitude of studies on plant phenology. The greatest recorded age of a clonal Arctic plant is probably that for two *Carex* species which survived for an estimated 3,000 years (Jónsdóttir et al. 2000) so, with a limited need for seed production, phenology is probably more important for trophic interactions, for example those involving pollinators, than for the local population dynamics of the plant species.

The relative stability of vegetation structure recorded on Disko Island is not an isolated case, although such cases are more rarely recorded than "greening". Similar examples exist for two sites (Tareya and Dickson) on the Taimyr Peninsula (Matveyeva and Zanokha 2013a,

2013b; Matveyeva et al. 2014; see later) and Svalbard (Prach et al. 2010). In the latter example, vegetation had changed little over a ca. 70-year period.

Southeast Greenland

An important current debate on Arctic environmental change concerns conflicting data from multi-decadal remote sensing that show on one hand an overall drying of the tundra (although with variation across the different categories of permafrost distribution (Hinzman et al. 2005; Smith et al. 2005) and, on the other hand, a net increase in the formation of small water bodies (Polishchuk et al. 2015). Whether the tundra is drying or becoming wetter is critically important for water sources for people, biodiversity, productivity, biogeochemical cycling and surface-air energy exchange (Parmentier, Chaper 37 in this volume). Using multi-decadal repeat photography and sometimes vegetation re-surveys and sediment analyses, the few BTF studies show a net drying (Smol and Douglas 2007; Callaghan et al. 2011a; Daniëls and de Molienaar 2011). In Tasiilaq, Southeast Greenland, Daniëls and de Molienaar used old literature sources from 1900 and repeat visits from 1968 to show a rather stable vegetation during the past 40 years, while vegetation mapping of a pond showed conspicuous changes denoting drying due to increased evaporation and decreased snow accumulation. Callaghan et al. (2011a) found a similar trend on Disko Island while Lougheed et al. (2011) found increased pond water temperature, increased water column nutrient concentration, the presence of at least one new chironomid species and increased macrophyte cover in Alaska after 40 years. However, perhaps the most dramatic finding was made by Smol and Douglas (2007) who investigated a pond (Camp Pond) on Ellesmere Island and found that it had vanished between 1979 and 2006: sediment analysis showed that it had existed for millennia. In contrast, it is well-known that thermokarst ponds are forming as permafrost thaws but repeat photography and renewed vegetation recording over multiple decades appear to be lacking. The projected changes for the future in high Arctic areas include increased active layer thickness and additional degradation of permafrost that will provide great reservoirs for subsurface storage of ground water and near surface soil moisture. This is likely to result in increased soil moisture on plateaus, vegetation changes that will increase evapotranspiration and the succession of wetting and drying of landscapes (Bring et al. 2016).

Taimyr Peninsula

Challenges with the BTF approach sometimes existed in accessing research areas studied earlier. The "Tareya" biogeocenological field research station in the mid-course of the Pyasina River (Western Taimyr) was active for a period of 13 years (1965–1977) and contributed to two international networks (IBP (International Biological Programme) and MAB (Man and the Biosphere)). Researchers in different branches of science (botanists, zoologists, microbiologists, soil scientists, climatologists, hydrologists and geomorphologists) performed a large-scale comprehensive study under the leadership of B. A. Tikhomirov. Many data on biodiversity were obtained there and the results were presented in four thematic proceedings (Biogeocenoses of Taimyr tundra and their productivity, 1971, 1973; Structure and functions of biogeocenoses of Taimyr, 1978; Biogeocenoses of Taimyr tundra, 1980) and in more than 200 papers. Nadya V. Matveyeva was there last in 1970 before a re-visit in 2010 when she was accompanied by four younger colleagues during the BTF Project. Reaching the site was much more difficult than 40 years ago when it was possible by regular flight from Norilsk to a small airport situated on an island 35 km downstream in the Pyasina River.

Over the decades, significant changes in landscape had occurred. The island has been partly eroded, and in 1992 both the airport and polar weather station (operating 1955–1992) stopped functioning. Also, and surprisingly, the interfluves between stream valleys and wet depressions had changed over a wide area of about 20 km². Repeat photography from the same spots show that the previously level, smooth upland (~50 m above sea level) interfluvial surface (Figure 38.3a–b) had become polygonized (Figure 38.3c–d). In neighbouring areas, marked polygonization had occurred and large areas had been transformed into a network of regularly-shaped flat-topped mounds of 7–10 m diameter separated by depressions of 2–3 m in width and 0.3–0.9 m in depth (Matveyeva et al. 2013). The general view of hundreds of mounds is similar to Siberian baidzharakh massifs along the slopes of stream valleys. Satellite images (Quick Bird of Google Earth on 8.11.2003) clearly showed the phenomenon existed in 2003. However, evidence from a few colleagues (personal communications) who visited Tareya successively in 1976, 1983 and 1994 confirm the absence of any signs of such changes. Consequently, this process was rapid, occurring within less than nine years. Similar extensive landscape transformations were also found near the sea port of Dickson that was revisited in 2012 after the last former visit in 1980 (Matveyeva and Zanokha 2013b). In 2012, huge baidzharakh massifs were common on the stream interfluves that were flat surfaces in 1980. There were also some other transformations in the micro- and nanorelief. However, the rapid and extensive polygonization of the interfluves was the most spectacular modification.

Vegetation (Matveyeva and Zanokha 2013a, 2017; Matveyeva et al. 2014) was re-recorded in the vicinity of both Tareya and Dickson using relevés that were sampled 40–32 years previously. Despite the dramatic transformations seen in the landscape topography, flora and vegetation (plant cover structure – the pattern of communities within a landscape, and composition – the set of community types) remained stable with only slight changes in species distributions within the landscape. A general greening in newly-formed depressions was observed, but due to a lower accumulation of dead matter rather than to any changes in composition or species abundance. In Tareya, 162 species of the former 213 had the same local distribution and abundance as earlier. Five new species were recorded (all in the floodplain of the Pyasina River) and a few formerly rare species were not relocated, most likely because the 2010 visit was considerably shorter than the original visits (two weeks compared with 3 months within 5 field seasons). Some changes in a few *Carex* species' abundance were recorded in polygonal mires on flooded low polygon centres that had coalesced (*C. stans*) due to the rim going down and fragmentation, and on the rims of frost-boils in dwarf shrub-sedge-moss communities presumably due to succession by *Carex ensifolia* subsp. *arctisibirica*.

In Dickson also, no significant changes were recorded in the composition of the vascular plant flora (Matveyeva and Zanokha 2017). No new additions were recorded, but ten species rare in the past 127 species were not found. Of the 117 species recorded in 2012, 24 species had slight differences in pattern within the landscape while 93 were distributed as earlier. Also, no changes have so far been recorded in the composition and structure of plant communities, including on the transformed interfluves.

The invertebrate fauna was also studied during the revisit to Tareya in 2010. So far, only data on Collembola have been published (Babenko 2013). Collembolan assemblages of all plant associations studied in 2010 clearly differ from the complexes that were registered in the same territory at the end of the 1960s. The number of species recorded in 2010 was about double that in 1968–1969 (Ananjeva 1973) – 117 against 62, and only two previously recorded species were not found in 2010. However, the approaches to the taxonomy of this group have changed drastically in recent decades: more than half the species recorded in 2010 were described after Ananjeva's publication, including about 15 based on the material collected during her study. In addition, many species which were found only in some highly specific communities (e.g. sandy beaches) do not seem to have been studied at all in 1968–1969. Also, changes in the density of previously abundant species may be the result of incomplete extraction of collembolans from soil samples or by the limited use of net-sweeping for studying the inhabitants of grass stands. According to Babenko (2013), even some considerable changes of relative abundance observed might still fall within the range of natural interannual fluctuations, the extent of which in the high latitudes is almost completely unknown. Consequently, in addition to any causes of change, the general trend of changes is not exactly clear, and these cannot be regarded as a response to a unidirectional influence of any single factor, for example, as a direct result of "global warming", which in most interpretations implies expansion of the "southern" forms into the high latitudes (Callaghan and Johansson 2009).

No reliable cases of expansion of southern collembolan and plant species from the South can be presently demonstrated in Tareya. Both for plant cover and invertebrate populations, the relatively few changes in species composition and abundance are due to topographic transformations, which in turn are the consequence of surface permafrost thaw and dynamics evident as ice-wedge degradation (Liljedahl et al. 2016). In general, however, the observations made within the Taimyr and Dickson study areas demonstrate a stable flora and vegetation, a mobile landscape and uncertainty in "before and after" comparisons of Collembola.

The important vegetation study by Matveyeva and colleagues (Matveyeva and Zanokha 2013a, 2013b, 2017; Matveyeva et al. 2014; Matveyeva 2017) confirms a prediction made already during the first IPCC assessment, that species would not be able to relocate at the same rate as climate change (Melillo et al. 1990) but importantly adds that landscape change can be faster than species relocation. The study by Matveyeva and colleagues reinforces the studies in West Greenland and Svalbard that showed surprising stability of vegetation in contrast to the "Greening of the Arctic" concept. In addition, they demonstrate that without re-visits to sites that were recorded before satellite images existed, there would be no understanding of how quickly tundra landscapes can be transformed.

Swedish Lapland

In contrast to the Back to the Future examples that showed stability of flora and vegetation, a project at a tree-line near Abisko in Swedish Lapland showed consistent dramatic changes. The study was based on data collected in 1977 and a re-visit by a PhD student in 2009 with help from the original researcher. The study using detailed maps showed that in the 34-year period, some shrubs and trees had expanded their cover by six times while aspen, *Populus tremula*, not formerly present at the site, had become established (Rundqvist et al. 2011). An early study near this site by Sandberg (1963) using repeat photography had shown increases in the altitudinal cover of tree-line shrub growth (*Salix* species) during a period of climatic *cooling*. Although the Rundqvist et al. (2011) study was conducted during a period of warming, other factors contributing to greening were explored. Local indigenous knowledge and aerial photography showed the existence of a former (pre-World War II) reindeer-herding corral. This indicated that there was intense activity by Sámi people and their reindeer herds more than 70 years ago and that the rapid greening now seen is probably a result of the synergy between vegetation recovery from former intense land use and climate warming.

Other Back to the Future studies showed increases in plant cover, productivity and range extensions, particularly by woody species (Danby et al. 2011; Hedenås et al. 2011; Myers-Smith et al. 2011) that agreed in general with the results of warming experiments that predicted vegetation responses (Elmendorf et al. 2012b) and control plots of these experiments (Elmendorf et al. 2012a). However, the snap-shot approach of BTF did not identify the specific effects of extreme events on vegetation that are now being reported (e.g. Bokhorst et al. 2015) or the recent "browning of the Arctic" which was highlighted by Xu et al. (2013) and detailed by Epstein et al. (2015).

(c)

(d)

Figure 38.3a–d The landscape surface at Tareya in 1967 (a and b) and 2010 (c and d).

Data-mining

The BTF Project included several studies that "discovered" old data sets, digitized them, carried out analyses and made data and analyses available in publications. The examples include important records of snow stratigraphy over a period of 50 years (Johansson, C. et al. 2011). The analysis of long records of snow stratigraphy showed an increase in the thickness and frequency of hard snow/ice layers in the snow profile over time. These layers denote partial – and even complete – snow melt during the Arctic winter and the changes were documented at a time when extreme winter warming events and their impacts on vegetation and animals were being observed by researchers as well as reindeer herders (e.g. Bokhorst et al. 2009). Since these studies, the importance of such extreme events has been recognized and in 2013–2014 the deaths of 65,000 reindeer in the Yamal-Nenets Autonomous Okrug were recorded due to one extreme icing event (that starved reindeer: Sokolov et al. 2016). A multitude of data-mining possibilities remain unexplored although some initiatives are being implemented. These include analysing ships' log books for weather conditions and sea ice observations (NOAA 2017), and analysing harbour masters' records for coastal ice conditions. Also, numerous records of the environment remain at research stations and have not yet been digitized or analysed.

Data mining was combined with site re-visits in Swedish Lapland, where permafrost dynamics were recorded in a BTF Project that operated with another International Polar Year Project, "TSP", the Thermal State of Permafrost (Johansson, M. et al. 2011). In 1980, three boreholes were drilled in three mires and temperatures were recorded. Data were combined with those from five new boreholes that allowed year-round temperatures to be recorded. Mean annual temperatures in the mires were close to 0°C at 5 m depth and warming was found in both the upper and lower parts of the boreholes (down to 15 m). When the new boreholes were drilled, it was impossible to include one former site (Katterjokk) because permafrost had disappeared since 1980 (Åkerman and Johansson 2008)! Also, at another site, permafrost thickness between 1980 and 2009 had decreased from 15 m to 9 m with an accelerating thaw during the past decade. All the measurements indicated permafrost thaw and vulnerability to continued warming (Johansson, M. et al. 2011).

Conclusions

About 15 studies were initiated during the BTF Project and published in a special issue of the journal *Ambio* (Callaghan and Tweedie 2011; Callaghan et al. 2011b). Other studies inspired by the BTF Project were also published (e.g. Van Bogaert et al. 2011; Babenko 2013; Matveyeva and Zanokha 2013a, 2013b, 2017; Matveyeva et al. 2014; Matveyeva 2017). The studies include major contributions to the existing knowledge base.

BTF studies present a view of changes that span over a century, from the painting of Kongsfjord, Svalbard in 1861, through the photos of Borg Mesch between 1899 and 1919, to records of vegetation, invertebrates, snow, permafrost, wetlands and landscape surface over the past ca. 60 years. The landscapes, sites and data sources available were selected before research on climate warming became dominant. The data sets are therefore independent of any preconceived bias in choosing sites where greening is evident (Gatti et al. in prep.). For the past ca. three decades, there has been a concept of "ground-truthing" proxies of the state of, and changes in the ground surface recorded remotely from satellites. However, the bias in ground-based publications on the greening of the Arctic (Gatti et al. in prep.) exemplified by the BTF Project and satellite recordings (Xu et al. 2013; Epstein et al. 2015) suggest that a concept of "space-truthing" of ground observations is now necessary. Indeed, recent analyses of satellite images have shown that the browning of the Arctic noted by Xu et al. (2013) has, since 2012, dominated the greening (Epstein et al. 2015). Furthermore, Phoenix and Bjerke (2016) suggest that the browning could, at least in part, result from the extreme winter warming events exemplified by the BTF snow records (Johansson, C. et al. 2011).

The long-term view provided by the BTF Project integrates long-term trends with extreme events and drivers of changes across all seasons while presenting evidence of no change. The long-term view also shows that simple assumptions and correlations between climate warming and vegetation responses cannot substitute for detailed determinations of the "pathology" of vegetation changes where these exist. These determinations can come from experiments such as experimental warming of ecosystems (e.g. continuous, mild treatment – Elmendorf et al. 2012b; episodic, extreme treatment – Bokhorst et al. 2009), and indigenous knowledge. Examples of the failure of simple correlations between warming and vegetation dynamics include the repeat photography over 100 years by Van Bogaert et al. (2011) which shows tree-line upward advance, downwards retreat, and stability on neighbouring mountains during the same climate change regime. Also, during the 1950s–1960s, repeat photography showed an altitudinal advance of shrubs at the tree-line (Sandberg 1963) during a period of *cooling*, whereas Rundqvist et al. (2011) and Hallinger et al. (2010) showed a similar multi-decadal attitudinal advance of shrubs at the same tree-line but during a period of climate *warming*! These specific examples show how there could be areas of uncertainty at the pan-Arctic scale. For example, are the areas of greening responding physiologically to increased summer temperatures or loss of millions of herbivores (reindeer: Sokolov et al. 2016; muskoxen: Rennert et al. 2009 and lemmings: Kausrud et al. 2008) due to winter warming events (Hansen et al. 2013)?

The BTF Project showed some surprises. For example, without BTF studies in the Russian Far North, it would be difficult to determine the rapidity of transformations in the landscape surface due to permafrost thaw as the first records were made before satellite images were available. Also, these studies and others showed that in many areas of the Arctic, vegetation has not changed as would be expected from climate warming data. Also, although (given in only one case) plant phenology advanced during climate warming, the effect was not relevant to the performance of the species. This points strongly to the necessity for a more rigorous assessment of what to monitor to detect climate impacts in terms of ecological consequences. Perhaps also surprising is that two of the Greenlandic sites (Disko Island and Tasiilaq) had experienced recent damage by direct human intervention – water bottling from a spring on Disko Island and trampling at Tasiilaq (T.V. Callaghan, pers. obs.; F. Daniëls, pers. comm.). However, this represents another bias in sampling change as many sites are relatively close to human habitations that provide necessary logistics.

Finally, perhaps the greatest impact of the BTF Project is that it is built on the legacy of networking originating in the IBP Programme. It now explicitly seeks to maximize its future legacy by investing in new networks such as INTERACT (www. eu-interact.org) and by facilitating inter-generational interactions to build capacity for detecting and understanding Arctic environmental changes.

Acknowledgements

The IPY project 512 "BTF" was kindly funded by the Swedish Science Research Council, grant number 327–2007–833, and many of the individual projects were independently funded by numerous sources. We thank Craig Tweedie and Pat Webber who were co-founders of BTF and many participating BTF researchers.

References

Åkerman, H.J. and Johansson, M. 2008. 'Thawing permafrost and thicker active layers in sub-arctic Sweden' *Permafrost and Periglacial Processes* 19: 279–292.

Ananjeva, S.I. 1973. 'Collembola of the Western Taimyr' *Biogeocenoses of Taimyr Tundra and Their Productivity* 2: 152–165. [In Russian.]

Andreev, V.N. and Aleksandrova, V.D. 1981. 'Geobotanical division of the Soviet Arctic' in L.C. Bliss, O.W. Heal and J.J. Moore (eds) *Tundra Ecosystems: A Comparative Analysis*. Cambridge, UK: Cambridge University Press. 25–34.

Arctic Station web site 2017. 'History of the Arctic Station' http://arktiskstation.ku.dk/om/historie/.

Babenko, A.B. 2013. '"Collembola of the Western Taimyr": forty years later' *Entomological Review* 93(6): 737–754.

Biogeocenoses of Taimyr Tundra and Their Productivity. 1971. Tikhomirov, B.A. (ed.). Leningrad. 239 pp. [In Russian].

Biogeocenoses of Taimyr Tundra and Their Productivity, vol. 2. 1973. Tikhomirov, B.A. (ed.). Leningrad. 207 pp. [In Russian].

Biogeocenoses of Taimyr Tundra. 1980. Leningrad. 254 pp. [In Russian].

Bliss, L.C. 1984. 'North American and Scandinavian tundras and polar deserts' in L.C. Bliss, O.W. Heal and J.J. Moore (eds) *Tundra Ecosystems: A Comparative Analysis*. Cambridge, UK: Cambridge University Press. 8–24.

Bokhorst, S., Bjerke, J.W., Tömmervik, H., Callaghan T.V. and Phoenix, G.K. 2009. 'Winter warming events damage sub-Arctic vegetation: consistent evidence from an experimental manipulation and a natural event' *Journal of Ecology* 97: 1408–1415.

Bokhorst, S., Phoenix, G.K., Berg, M., Callaghan, T.V., Kirby-Lambert, C. and Bjerke, J. 2015. 'Climatic and biotic extreme events moderate long-term responses of above- and belowground sub-Arctic heathland communities to climate change' *Global Change Biology* 21: 4063–4075.

Briner, J.P., McKay, N.P., Axford, Y., Bennike, O., Bradley, R.S., de Vernal, A., Fisher, D., Francus, P., Frechette, B., Gajewski, K., Jennings, A., Kaufman, D.S., Miller, G., Rouston, C. and Wagner, B. 2014. 'Holocene climate change in Arctic Canada and Greenland' *Quaternary Science Reviews* 147: 340–364.

Bring, A., Fedorova, I., Dibike, Y., Hinzman, L., Mard, J., Mernild, S.H., Prowse, T., Semenova, O., Stuefer, S.L., Woo, M.K. 2016. 'Arctic terrestrial hydrology: a synthesis of processes, regional effects, and research challenges' *Journal of Geophysical Research: Biogeosciences* 121(3): 621–649.

Brooks, N., Chiapello, I., di Lernia, S., Drake, N., Legrand, M., Moulin, C., and Pospero, J. 2005. 'The climate-environment-society nexus in the Sahara from prehistoric times to the present day' *The Journal of North African Studies* 10(3–4): 253–292.

Buras, A. and Wilmking, M. 2015. 'Recent influence of climate on shrub growth around the North-Atlantic Region' in T.V. Callaghan and H. Savela (eds) *INTERACT Stories of Arctic Science*. Aarhus University, Denmark: DCE – Danish Centre for Environment and Energy. 112–113.

Callaghan, T.V. 1972. 'Ecophysiological and taxonomic studies on bipolar Phleum alpinum L'. PhD thesis. University of Birmingham, UK.

Callaghan, T.V. 1976. 'Strategies of growth and population dynamics of tundra plants. 3. Growth and population dynamics of *Carex bigelowii* in an alpine environment' *Oikos* 27: 402–413.

Callaghan, T.V., Bergholm, F., Christensen, T.R., Jonasson, C., Kokfelt, U. and Johansson, M. 2010. 'A new climate era in the sub-Arctic: accelerating climate changes and multiple impacts' *Geophysical Research Letters* 37, L14705. DOI:10.1029/2009GL042064.

Callaghan, T.V., Christensen, T.R. and Jantze, E.J. 2011a. 'Plant and vegetation dynamics on Disko Island, West Greenland: snapshots separated by over 40 years' in T.V. Callaghan and C.E. Tweedie (eds) 'Multi-decadal changes in tundra environments and ecosystems: the International Polar Year Back to the Future Project'. *Ambio* 40(6): 624–637.

Callaghan, T.V. and Johansson, M. 2009. 'The changing living tundra: a tribute to Yuri Chernov' in *Species and Communities in Extreme Environments*, 14–48. Moscow–Sofia.

Callaghan, T.V., Myneni, R., Xu, A. and Johansson, M. 2014. 'The age of the Arctic: challenges and opportunities in Arctic and global communities' in A. Karlqvist and E. Kessler (eds) *Redrawing the Map. Climate, Human Migration, Food Security – the 11th Royal Colloquium*, May 2013. Stockholm: Kessler and Karlqvist. 79–86.

Callaghan, T.V., Smith, R.I.L. and Walton, D.W.H. 1976. 'The IBP Bipolar Botanical Project' *Philosophical Transactions of the Royal Society, Series B* 274: 15–319.

Callaghan, T.V. and Tweedie, C.E. (eds) 2011. 'Multi-decadal changes in tundra environments and eco-systems: the International Polar Year Back to the Future Project' *Ambio* 40(6): 555–716.

Callaghan, T.V., Tweedie, C.E., Åkerman, J., Andrews, C., Bergstedt, J., Butler, M.G., Christensen, T.R. et al. 2011b. 'Multi-decadal changes in tundra environments and ecosystems: synthesis of the International Polar Year Back to the Future Project' in T.V. Callaghan and C.E. Tweedie (eds) 'Multi-decadal changes in tundra environments and ecosystems: the International Polar Year Back to the Future Project' *Ambio* 40(6): 705–716.

Chydenius 1861. 'Svenska expeditionen till Spetsbergen 1861' cited in Liljequist, G.H. 1993. *High Latitudes: A History of Swedish Polar Travels and Research*. Stockholm: Swedish Polar Research Secretariat.

Coulson, D. and Campbell, A. 2013. *Rock Art of the Tassili n Ajjer, Algeria*. http://africanrockart.org/wp-content/uploads/2013/11/Coulson-article-A10-proof.pdf.

Danby, R.K., Koh, S., Hik, D.S. and Price, L.W. 2011. 'Four decades of plant community change in the Alpine Tundra of Southwest Yukon, Canada' *Ambio*, DOI:10.1007/s13280-011-0172-2.

Daniëls, F.J.A. and de Molienaar, G. 2011. 'Flora and vegetation of Tasillaq, formerly Angmagssalik, Southeast Greenland: a comparison of data between around 1900 and 2007' *Ambio* 40: 650–659.

Elmendorf, S.C. et al. 2012a. 'Plot-scale evidence of tundra vegetation change and links to recent summer warming' *Nature Climate Change* 2: 453–457.

Elmendorf, S.C. et al. 2012b. 'Global assessment of experimental climate warming on tundra vegetation: heterogeneity over space and time' *Ecology Letters* 15: 164–175.

Emanuelsson, U. 1987. 'Human influence on vegetation in the Torneträsk area during the last three centuries' *Ecological Bulletins* 38: 95–111.

Epstein, H.E., Bhatt, U.S., Raynolds, M.K. et al. 2015. 'Tundra greenness' in M.O. Jeffries, J. Richter-Menge and J.E. Overland (eds) *Arctic Report Card: Update for 2015*. Silver Spring, MD: NOAA. Available at: www.arctic.noaa.gov/reportcard/.

Gatti, R., Callaghan, T.V. and Phoenix, G.K. in prep. 'Hidden challenges in Arctic ecology: exploring greening and browning of the Arctic'.

Grime, J.P. 1978. *Plant Strategies and Vegetation Processes*. Chichester, UK: Wiley and Sons.

Grudd, H., Briffa, K.R., Karlén, W., Bartholin, T.S., Jones, P.D. and Kromer, B. 2002. 'A 7400-year tree-ring chronology in northern Swedish Lapland: natural climatic variability expressed on annual to millennial timescales' *Holocene* 12(6): 657–665.

Hadingham, E. 1979. *Secrets of the Ice Age: The World of the Cave Artists*. London: William Heinemann Ltd.

Hagen, J.O. and Lefauconnier, B. 1995. 'Reconstructed runoff from the high Arctic Basin Bayelva based on mass-balance measurements' *Nordic Hydrology* 26(4–5): 285–296.

Hallinger, M., Manthey, M. and Wilmking, M. 2010. 'Establishing a missing link: warm summers and winter snow cover promote shrub expansion into alpine tundra in Scandinavia' *New Phytologist* 186: 890–899.

Hansen, B.B., Grøtan, V., Aanes, R., Sæther, B.-E., Stien, A., Fuglei, E., Ims, R.A., Yoccoz, N.G. and Pedersen, Å.Ø. 2013. 'Climate events synchronize the dynamics of a resident vertebrate community in the High Arctic' *Science* 339 (6117): 313–315.

Harper, J.L. 1977. *Population Biology of Plants*. London: Academic Press.

Havström, M., Callaghan, T.V., Svoboda, J. and Jonasson, S. 1995. 'Little Ice Age temperature estimated by growth and flowering differences between subfossil and extant shoots of *Cassiope tetragona*, an arctic heather' *Functional Ecology* 9: 650–654.

Hedenås, H., Olsson, H., Jonasson, C., Bergstedt, J., Dahlberg, U. and Callaghan, T.V. 2011. 'Changes in tree growth, biomass and vegetation over a thirteen-year period in sub-Arctic Sweden' *Ambio* 40(6): 566–574.

Hinzman, L.D., Bettez, N.D., Bolton, W.R., Chapin, F.S., Dyurgerov, N.B., Fastie, C.L., Griffith, B., Hollister, R.D. et al. 2005. 'Evidence and implications of recent climate change in northern Alaska and other arctic regions' *Climate Change* 72: 251–298.

Jia, G.S.J., Epstein, H.E. and Walker, D.A. 2003. 'Greening of arctic Alaska, 1981–2001' *Geophysical Research Letters* 30(20): 2067. DOI: 10.1029/2003GL018268.

Johansson, C., Pohjola, V.A., Jonasson, C. and Callaghan, T.V. 2011. 'Multi-decadal changes in snow characteristics in Sub-Arctic Sweden' *Ambio* 40(6): 566–574.

Johansson, M., Åkerman, J., Keuper, F., Christensen, T.R., Lantuit, H. and Callaghan, T.V. 2011. 'Past and present permafrost temperatures in the Abisko area: redrilling of boreholes' *Ambio* 40(6): 558–565.

Jónsdóttir, I.S., Augner, M., Fagerström, T., Persson, H. and Stenström, A. 2000. 'Genet age in marginal populations of two clonal *Carex* species in the Siberian Arctic' *Ecography* 23: 402–412.

Kausrud, K.L., Mysterud, A., Steen, H., Vik, J.O., Østbye, E., Cazelles, B., Framstad, E., Eikeset, A.M., Mysterud, I., Solhøy, T. and Stenseth, N.C. 2008. 'Linking climate change to lemming cycles' *Nature* 456: 93–97.

Liljedahl, A.K., Boike J., Daanen, R.P., Fedorov, A.N., Frost, G.V., Grosse, G., Hinzman, L.D., Iijma, Y., Jorgenson, J.C., Matveyeva, N., Necsoiu, M., Raynolds, M.K., Romanovsky, V., Schulla, J., Tape, K., Walker, D.A., Wilson, C.J., Yabuki, H. and Zona, D. 2016. 'Pan-Arctic ice-wedge degradation in warming permafrost and its influence on tundra hydrology' *Nature Geoscience.* 9(4): 1–7.

Liljequist, G.H. 1993. *High Latitudes: A History of Swedish Polar Travels and Research.* Stockholm: Swedish Polar Research Secretariat.

Lougheed, V.L., Butler, M.G., McEwen, D.C. and Hobbie, J.E. 2011. 'Changes in tundra pond limnology: re-sampling Alaskan ponds after 40 years' *Ambio* 40(6): DOI:10.1007/s13280-011-0165-1.

Luthi, D., Le Floch, M., Bereiter, B., Blunier, T., Barnola, J.M., Siegenthaler, U., Raynaud, D., Jouzel, J., Fischer, H., Kawamura, K. and Stocker, T.F. 2008. 'High-resolution carbon dioxide concentration record 650,000–800,000 years before present' *Nature* 453(7193): 379–382.

Matveyeva, N.V. 2017. 'Plant cover response to the ice wedge degradation within the Arctic'. *West Siberian Peatlands and Carbon Circle: Past and Present. Proceedings of the Fifth International Field Symposium* (Khanty-Mansiysk, June 19–29, 2017), Tomsk: 34–36. [In Russian].

Matveyeva N.V., Cherosov, M.M. and Telyatnikov, M.Y. 2013. 'The Russian input to the Arctic Vegetation Archive and an example of the value of plot data for assessing climate change on the Taymyr Peninsula'. Arctic vegetation archive (AVA) workshop, Krakov, Poland, April 14–16, *CAFF Proceedings Report.* 10: 76–80.

Matveyeva, N.V. and Zanokha, L.L. 2013a. 'Plant cover stability under significant landscape transformation in Western Taymyr tundras'. Trudy Vserossiiskoi nauchnoi konferentsii, *Biodiversity of ecosystems of the Far North: inventarization, monitoring, protection* (Syktyvkar, Komi Republik), 3–7 July 2013. Syktyvkar. 96–106. [In Russian].

Matveyeva, N.V. and Zanokha, L.L. 2013b. 'Changes in vascular plant flora in the vicinity of Dickson (western taymyr) within the period of 32 years'. Trudy Vserossiiskoi nauchnoi konferentsii, *Biodiversity of ecosystems of the Far North: inventarization, monitoring, protection* (Syktyvkar, Komi Republik), 3–7 July 2013. Syktyvkar. 201–208. [In Russian].

Matveyeva, N.V. and Zanokha, L.L. 2017. 'Changes in vascular plant flora in Dickson vicinity (Western Taymyr) in between 1980 and 2012 years' *Bot. Journ.* 102(6): 812–846.

Matveyeva, N.V., Zanokha, L.L. and Yanchenko, Z.A. 2014. 'Changes in vascular plant flora in the region of the Taymyr biogeocenological field station (mid course of the Pyasina River, western Taymyr) in between 1970 and 2010 years' *Bot. Journ.* 99(8): 841–867. [In Russian].

Melillo, J., Callaghan, T.V., Woodward, F.I., Salati, E. and Sinha, S.K. 1990. 'The effects on ecosystems' in J. Houghton, G.J. Jenkins and J.J. Ephraums (eds) *Climate Change, the IPCC Scientific Assessment.* Cambridge, UK: Cambridge University Press. 282–310.

Myers-Smith, I.H. et al. 2011. 'Shrub expansion in tundra ecosystems: dynamics, impacts and research priorities' *Environmental Research Letters* 6(4): 045509. DOI: 10.1088/1748-9326/6/4/045509.

Myers-Smith, I.H., Elmendorf, S.C., Beck, P.S.A. et al. 2015. 'Climate sensitivity of shrub growth across the tundra biome' *Nature Climate Change* 5: 887–891.

NOAA 2017. www.noaanews.noaa.gov/stories2012/20121022_oldweatherprojectlaunch.html.

Overland, J., Hanna, E., Hanssen-Bauer, I., Kim, S.-J., Walsh, J.E., Wang, M., Bhatt, U.S. and Thoman R.L. 2016. 'Surface air temperature' *Arctic Report Cards* http://arctic.noaa.gov/Report-Card/Report-Card-2016/ArtMID/5022/ArticleID/271/Surface-Air-Temperature.

Phoenix, G.K and Bjerke, J.W. 2016. 'Arctic browning: extreme events and trends reversing arctic greening' *Global Change Biology* 22: 2960–2962.

Polishchuk, Y.M., Bryksina, N.A. and Polishchuk, V.Y. 2015. 'Remote analysis of changes in the number of small thermokarst lakes and their distribution with respect to their sizes in the cryolithozone of Western Siberia' *Atmospheric and Oceanic Physics* 51(9): 999–1006.

Prach, K., Kosnar, J., Klimesova, J. and Hais, M. 2010. 'High Arctic vegetation after 70 years: a repeated analysis from Svalbard' *Polar Biology* 33: 635–639.

Rennert, K.J., Roe, G., Putkonen, J., and Bitz, C.M. 2009. 'Soil thermal and ecological impacts of rain on snow events in the circumpolar Arctic' *Journal of Climate* 22(9): 2302–2315.

Rundqvist, S., Hedenås, H., Sandström, A., Emanuelsson, U., Eriksson, H., Jonasson, C. and Callaghan, T.V. 2011. 'Tree and shrub layer expansion over the past 34 years at the tree-line in Abisko, sub-Arctic' in T.V. Callaghan and C.E. Tweedie (eds) 'Multi-decadal changes in tundra environments and ecosystems: the International Polar Year Back to the Future Project', *Ambio* 40(6): 683–692.

Sandberg, G. 1963. 'Växtvärlden I Abisko nationalpark' in K. Curry-Lindahl (ed.) *Natur I Lappland*, II. Uppsala, Sweden: Bokförlaget Svensk Natur. [In Swedish]. 885–909.

Smith, L.C., Sheng, Y., MacDonald, G.M., and Hinzman, L.D. 2005. 'Disappearing Arctic lakes' *Science* 308(5727): 1429. DOI: 10.1126/science.1108142.

Smol, J.P. and Douglas, M.S.V. 2007. 'Crossing the final ecological threshold in high Arctic ponds' *Proceedings of the National Academy of Sciences of the United States of America* 104: 12395–12397.

Sokolov, A.A., Sokolova, N.A., Ims, R.A., Brucker, L. and Ehrich, D. 2016. 'Emergent rainy winter warm spells may promote boreal predator expansion into the Arctic' *Arctic* 69(2): 121–129.

Structure and functions of biogeocenoses of Taimyr tundra. 1978. Tikhomirov, B.A. and Tomilin, B.A. (eds). Leningrad. 303 pp. [In Russian].

Tieszen, L.L., Lewis, M.C., Miller, P.C., Mayo, J., Chapin, F.S. III and Oechel, W. 1981. 'An analysis of processes of primary production in tundra growth forms' in L.C. Bliss, O.W. Heal and J.J. Moore (eds) *Tundra Ecosystems: A Comparative Analysis*. Cambridge, UK: Cambridge University Press. 285–356.

Tikhomirov, B.A., Shamurin, V.F. and Aleksandrova, V.D. 1981. 'Phytomass and primary production of tundra communities, USSR, 1981' in L.C. Bliss, O.W. Heal and J.J. Moore (eds) *Tundra Ecosystems: A Comparative Analysis*. Cambridge, UK: Cambridge University Press. 227–238.

Van Bogaert, R., Haneca, K., Hoogesteger, J., Jonasson, C., De Dapper, M. and Callaghan T.V. 2011. 'A century of tree line changes in sub-Arctic Sweden show local and regional variability and only a minor role of 20th century climate warming' *Journal of Biogeography* 38: 907–921.

Welker, J.M., Heaton, T.H.E., Spiro, B. and Callaghan, T.V. 1995. 'Indirect effects of winter climate on the d13C and dD characteristics of annual growth segments in the long-lived, arctic plant *Cassiope tetragona*: a preliminary analysis' in B. Frenzel (ed.) 'Problems of stable isotopes in tree rings, lake sediments and peat bogs as climatic evidence for the Holocene' *Proceedings of the European Science Foundation Workshop, Bern, Switzerland, April 1993*. European Science Foundation, European Paleoclimate Programme, Gustav Fischer Verlag, Stuttgart. 105–119.

Wielgolaski, F.E., Bliss, L.C., Svoboda, J. and Doyle, G. 1981. 'Primary production of tundra' in L.C. Bliss, O.W. Heal and J.J. Moore (eds) *Tundra Ecosystems: A Comparative Analysis*. Cambridge, UK: Cambridge University Press. 187–226.

Xu, L., Myneni, R.B., Chapin III, F.S., Callaghan, T.V., Pinzon, J.E., Tucker, C.J., Zhu, Z., Bi, J., Ciais, P., Tømmervik, H. 2013. 'Temperature and vegetation seasonality diminishment over northern lands nature' *Climate Change*. DOI: 10.1038/NCLIMATE1836.

Index

ICRW *see* International Convention for the
Regulation of Whaling
ICSU *see* International Council for Science
identity: indigenous peoples 76–7, 81, 96; maps
83; Sámi 94, 98
igneous rocks 152, 154
IGY *see* International Geophysical Year
Ilulissat Declaration (2008) 277, 286, 350, 398
ILUOP *see* Inuit Land Use and Occupancy
Project
IMO *see* International Maritime Organization
impact assessment 88, 380–90, 425
imperialism 37, 48, 49, 267–8, 270, 407; *see also*
colonialism
India: Antarctic Treaty 273, 297; Arctic
Council 10, 286, 423–4; as ATCP 406, 412;
continental separation 155; ICECAP 439;
international co-operation 306; law of the sea
278; national Antarctic programmes 299, 303;
suspicion of 411
Indian Ocean 159–60, 306
indigeneity 331, 332, 342, 345
indigenous peoples 2, 4, 6, 8–9, 13–14, 67–80,
116, 128; Arctic Council 10, 322; community-
based monitoring 426–7; cultural heritage
125–6; early settlers 19–20; education 97–8,
108, 109, 110–11, 112–13; environmental
governance 421, 424–5; global warming
impact on 3, 169; health 90, 91–102; impact
assessment 385–6; indigenous cartographies
81–9; indigenous diplomacies 331–47; IUCN
418; land claims 4–5; participation in decision-
making 5, 13, 71, 79, 345, 424; Persistent
Organic Pollutants 289; SLiCA 109; species
extinctions 140; sustainable development 311,
318; sustainable livelihoods 429; whaling 422;
wildlife conservation 429; *see also* Inuit; Sámi
inequalities 90, 95, 406
infrastructure 30, 32; archaeological sites 123;
cultural heritage sites 125; impact assessment
380; impact of permafrost on 243–4; mining
120; national Antarctic programmes 297, 298,
299, 300
Inglefield, Edward 27
Ingold, T. 82
Innes, Hammond 57, 61
Innuksuk, Rhoda 334
insolation 215–16; *see also* solar radiation
INTERACT 11, 493, 503
interdisciplinary research 12, 220, 366–7, 368, 428
interfluves 497–8
Intergovernmental Panel on Climate Change
(IPCC) 187, 198, 199; anthropogenic
influence 206, 252, 253; Arctic snow cover
203; impact assessment 384–5; permafrost
258; Southern Annular Mode 205; species
relocation 499

International Arctic Science Committee (IASC) 11
International Biological Programme (IBP) 494–5,
497, 503
international co-operation 8, 9–11, 43, 288;
Antarctic Treaty 265–6, 271–2; Arctic
Council 285–7, 289–90; early Antarctic
exploration 39; fisheries 396, 397–8;
International Geophysical Year 42, 296;
national Antarctic programmes 306;
sustainable development 318–19, 321
International Code for Ships Operating in Polar
Waters (Polar Code) 4, 5, 281, 316
International Convention for the Prevention of
Pollution from Ships (MARPOL) 281–2, 417
International Convention for the Regulation of
Whaling (ICRW) 279–80, 417
International Council for Science (ICSU) 11
International Council for the Exploration of the
Sea (ICES) 10, 392, 396, 397
International Council of Monuments and Sites
(ICOMOS) 418
International Geophysical Year (IGY) 32, 35, 271,
437, 469; Antarctic Treaty 7–8, 42, 266, 272,
295; drilling 460; international co-operation
296; national Antarctic programmes 294, 297;
science stations 123
international law 5; Arctic Council 287–8;
indigenous diplomacies 331–2, 335–7, 345; law
of the sea 275–83; territorial claims 269
International Maritime Organization (IMO) 4,
281, 290, 316, 318
International Polar Tourism Research Network
(IPTRN) 365, 367
International Polar Year (IPY) 11, 28, 31,
37, 437; Antarctica 42; archaeological sites
121; "Back to the Future" Project 492–3;
education 108; technology 439–40; Thermal
State of Permafrost project 240, 502; tourism
359, 364, 365
International Relations (IR) 331, 332, 333,
335, 345
International Union for Conservation of Nature
(IUCN) 417, 418–19, 422, 424
International Whaling Commission (IWC) 396,
413n15, 417, 421, 422
internationalism: Antarctic resource futures 412;
indigenous 331, 333
interwar period 30–2, 41–2
Inughuit 82
Inuit 20, 24, 32, 68, 71, 337–8; Boas' work
29, 83; community-based monitoring 427;
cultural and linguistic continuity 98–9;
cultural survival 77; diplomacy 334; education
98, 111; employment 95; environmental
governance 425; health issues 91–2; housing
95; impact assessment 387; loss of sea ice 1;
mapping 86–7; marine conservation areas 5;